Controversies and Recent Advances in Senescence and Aging

Controversies and Recent Advances in Senescence and Aging

Editors

Nicole Wagner
Kay-Dietrich Wagner

MDPI • Basel • Beijing • Wuhan • Barcelona • Belgrade • Manchester • Tokyo • Cluj • Tianjin

Editors

Nicole Wagner
Institute of Biology Valrose
University Cote d'Azur
Nice
France

Kay-Dietrich Wagner
Institute of Biology Valrose
University Cote d'Azur
Nice
France

Editorial Office
MDPI
St. Alban-Anlage 66
4052 Basel, Switzerland

This is a reprint of articles from the Special Issue published online in the open access journal *Cells* (ISSN 2073-4409) (available at: www.mdpi.com/journal/cells/special_issues/Senescence_Aging).

For citation purposes, cite each article independently as indicated on the article page online and as indicated below:

LastName, A.A.; LastName, B.B.; LastName, C.C. Article Title. *Journal Name* **Year**, *Volume Number*, Page Range.

ISBN 978-3-0365-7169-0 (Hbk)
ISBN 978-3-0365-7168-3 (PDF)

Contents

Nicole Wagner and Kay-Dietrich Wagner
Controversies and Recent Advances in Senescence and Aging
Reprinted from: *Cells* **2023**, *12*, 902, doi:10.3390/cells12060902 1

Francisco J. Carrillo-Salinas, Siddharth Parthasarathy, Laura Moreno de Lara, Anna Borchers, Christina Ochsenbauer and Alexander Panda et al.
Short-Chain Fatty Acids Impair Neutrophil Antiviral Function in an Age-Dependent Manner
Reprinted from: *Cells* **2022**, *11*, 2515, doi:10.3390/cells11162515 7

Minwoo Baek, Wijeong Jang and Changsoo Kim
Dual Oxidase, a Hydrogen-Peroxide-Producing Enzyme, Regulates Neuronal Oxidative Damage and Animal Lifespan in *Drosophila melanogaster*
Reprinted from: *Cells* **2022**, *11*, 2059, doi:10.3390/cells11132059 27

Qin Ke, Lili Gong, Xingfei Zhu, Ruili Qi, Ming Zou and Baoxin Chen et al.
Multinucleated Retinal Pigment Epithelial Cells Adapt to Vision and Exhibit Increased DNA Damage Response
Reprinted from: *Cells* **2022**, *11*, 1552, doi:10.3390/cells11091552 37

Maria Donatella Semeraro, Gunter Almer, Wilfried Renner, Hans-Jürgen Gruber and Markus Herrmann
Influences of Long-Term Exercise and High-Fat Diet on Age-Related Telomere Shortening in Rats
Reprinted from: *Cells* **2022**, *11*, 1605, doi:10.3390/cells11101605 51

Siyu Yan, Song Lin, Kexin Chen, Shanshan Yin, Haoyue Peng and Nanshuo Cai et al.
Natural Product Library Screens Identify Sanguinarine Chloride as a Potent Inhibitor of Telomerase Expression and Activity
Reprinted from: *Cells* **2022**, *11*, 1485, doi:10.3390/cells11091485 67

Tom Zimmermann, Michaela Pommer, Viola Kluge, Chafia Chiheb, Susanne Muehlich and Anja-Katrin Bosserhoff
Detection of Cellular Senescence in Human Primary Melanocytes and Malignant Melanoma Cells In Vitro
Reprinted from: *Cells* **2022**, *11*, 1489, doi:10.3390/cells11091489 85

Hasan Safwan-Zaiter, Nicole Wagner, Jean-François Michiels and Kay-Dietrich Wagner
Dynamic Spatiotemporal Expression Pattern of the Senescence-Associated Factor p16Ink4a in Development and Aging
Reprinted from: *Cells* **2022**, *11*, 541, doi:10.3390/cells11030541 101

Kay-Dietrich Wagner and Nicole Wagner
The Senescence Markers p16INK4A, p14ARF/p19ARF, and p21 in Organ Development and Homeostasis
Reprinted from: *Cells* **2022**, *11*, 1966, doi:10.3390/cells11121966 121

Rui Chen and Thomas Skutella
Synergistic Anti-Ageing through Senescent Cells Specific Reprogramming
Reprinted from: *Cells* **2022**, *11*, 830, doi:10.3390/cells11050830 147

Xiaojing Hong, Lihui Wang, Kexiong Zhang, Jun Liu and Jun-Ping Liu
Molecular Mechanisms of Alveolar Epithelial Stem Cell Senescence and Senescence-Associated Differentiation Disorders in Pulmonary Fibrosis
Reprinted from: *Cells* **2022**, *11*, 877, doi:10.3390/cells11050877 . . . . **167**

Arbi Aghali, Maunick Lefin Koloko Ngassie, Christina M. Pabelick and Y. S. Prakash
Cellular Senescence in Aging Lungs and Diseases
Reprinted from: *Cells* **2022**, *11*, 1781, doi:10.3390/cells11111781 . . . . **191**

Ines Martic, Pidder Jansen-Dürr and Maria Cavinato
Effects of Air Pollution on Cellular Senescence and Skin Aging
Reprinted from: *Cells* **2022**, *11*, 2220, doi:10.3390/cells11142220 . . . . **207**

Lena Guerrero-Navarro, Pidder Jansen-Dürr and Maria Cavinato
Age-Related Lysosomal Dysfunctions
Reprinted from: *Cells* **2022**, *11*, 1977, doi:10.3390/cells11121977 . . . . **227**

Hongfu Jin, Wenqing Xie, Miao He, Hengzhen Li, Wenfeng Xiao and Yusheng Li
Pyroptosis and Sarcopenia: Frontier Perspective of Disease Mechanism
Reprinted from: *Cells* **2022**, *11*, 1078, doi:10.3390/cells11071078 . . . . **247**

Ashley Brauning, Michael Rae, Gina Zhu, Elena Fulton, Tesfahun Dessale Admasu and Alexandra Stolzing et al.
Aging of the Immune System: Focus on Natural Killer Cells Phenotype and Functions
Reprinted from: *Cells* **2022**, *11*, 1017, doi:10.3390/cells11061017 . . . . **261**

Nanshuo Cai, Yifan Wu and Yan Huang
Induction of Accelerated Aging in a Mouse Model
Reprinted from: *Cells* **2022**, *11*, 1418, doi:10.3390/cells11091418 . . . . **287**

cells

Editorial

Controversies and Recent Advances in Senescence and Aging

Nicole Wagner * and Kay-Dietrich Wagner *

CNRS, INSERM, iBV, Université Côte d'Azur, 06107 Nice, France
* Correspondence: nwagner@unice.fr (N.W.); kwagner@unice.fr (K.-D.W.)

Aging is the leading predictive factor of many chronic diseases that account for most of the morbidity and mortality worldwide, i.e., neurodegeneration, cardiovascular, pulmonary, renal, and bone diseases, as well as cancers. Oxidative stress and reactive oxygen species generation, over-production of inflammatory cytokines, the activation of oncogenes, DNA damage, telomere shortening, and the accumulation of senescent cells are all widely accepted mechanisms contributing to aging. Senescence is mainly thought to be provoked by negative cellular stress but might also be induced by physiological developmental stimuli. Senescence is characterized by irreversible cell cycle arrest independent of quiescence and terminal differentiation. However, more recent observations suggest that the status of developmental and cancer senescent cells might not be irreversible. Aside from cell cycle arrest, senescent cells are characterized by morphological changes and molecular damage, metabolic alterations, and a specific secretory phenotype (SASP). Senescent cells contribute to embryonic development and participate in tissue repair and tumor sup-pression, but they are also involved in detrimental tissue decline during aging. Thus, the application of senolytic or senostatic drugs to halt or reverse age-related pathologies could represent an interesting therapeutic option. This Special Issue of *Cells* compiles novel and exciting insights into the mechanisms of aging and senescence.

Carrillo-Salinas and colleagues describe the effects of short-chain fatty acids on the function of neutrophils in young and older women with relevance for HIV infection risk [1]. HIV infection risk is high in younger women, but new HIV infections in older women are rising worldwide. Vaginal microbiota represent a defense against infections including HIV. Alterations in the physiological vaginal bacterial populations occur alongside other stimuli in older women as well. High concentrations of short-chain fatty acids are the result of vaginal dysbiosis. The authors compared the response of neutrophils from younger and older woman to short-chain fatty acids. In response to HIV stimulation, short-chain fatty acids reduced the chemokine secretion of neutrophils of young and older women. In addition, incubation with pathological concentrations of short-chain fatty acids diminished the activation and migration of neutrophils from older women and reduced the secretion of alpha defensins as molecules with antiviral activity. These interesting results do not only show that vaginal dysbiosis via short-chain fatty acids reduces neutrophil function but that these perturbations become more prominent with increasing age. The data also suggest that the re-establishment of physiological vaginal microbial flora with a resulting decrease in short-chain fatty acids might be a relatively simple way to reduce to some extend the risk of infections, especially in older populations.

Oxidative stress is generally assumed to increase with aging and induce senescence which might have consequences for one's lifespan. Reducing neuronal oxidative stress is known to extend the lifespan in Drosophila [2]. However, the exact source for reactive oxygen species in this model is not fully understood. Baek and colleagues published in this Special Issue of *Cells* the identification of dual oxidase (Duox) as a source for reactive oxygen species (ROS) in Drosophila melanogaster [3]. Duox is activated by intracellular calcium to produce H_2O_2. Duox heterozygous male flies showed an expected reduced expression of Duox by 50%, as well as decreased ROS and H_2O_2 production. They survived longer under

Citation: Wagner, N.; Wagner, K.-D. Controversies and Recent Advances in Senescence and Aging. *Cells* **2023**, *12*, 902. https://doi.org/10.3390/cells12060902

Received: 13 March 2023
Accepted: 14 March 2023
Published: 15 March 2023

standard conditions as well when they were exposed to ROS-producing food. Whether neuroinflammation and senescence in this model are reduced remains to be determined. Nevertheless, the data nicely support the critical involvement of reactive oxygen species in lifespan determination in this model. Unfortunately, the relations between antioxidants, ROS production, and human health and longevity seem to be more complex [4,5].

Ke et al. determined ROS production in retinal pigment epithelial (RPE) cells [6]. They compared mononucleated and multinucleated RPE cells and determined that under baseline conditions, ROS production was similar in both cell types, while multinucleated cells had higher ROS production and DNA damage after irradiation. Surprisingly, in mice, the number of multinucleated cells was not age-dependent and the comparison between different species revealed that multinucleation seems to be a characteristic of nocturnal animals, while in humans and other diurnal species the fraction of multinucleated cells is low. Differences in ROS production and DNA damage were only detectable after the irradiation of mononucleated and multinucleated cells. As multinucleation was independent of the age of the mice, this represents another example of the dissociation of reactive oxygen species production, DNA damage, and age-related phenotypes.

Besides the deleterious effects of ROS, telomere shortening is considered to be a hallmark of aging [7]. Exercise is believed to have positive effects on telomere length and the associated shelterin complex proteins, while the opposite is the case for obesity. Nevertheless, shelterin genes show a very dynamic spaciotemporal expression pattern throughout the lifespan [8], and the effects of exercise on telomere length differ largely across multiple studies and have mostly been measured in peripheral blood cells. The group of researchers working alongside Markus Herrmann reported a careful study using exercised and sedentary rats fed with either a standard or a high-fat diet [9]. The rats were exercised for quite a long period of 10 months and telomere length and mRNA expression of telomerase, as well as the shelterin genes Terf-1 and Terf-2, were measured in multiple organs. A high-fat diet in the non-exercised control group induced telomere shortening and reduced mRNA expression for telomerase, Terf-1, and Terf-2 only in visceral fat, while in most organs no conclusive effects were observed in telomere length, telomerase, Terf-1, and Terf-2 expression in response to exercise or a high-fat diet. Nevertheless, it seems possible that such a difference might occur in response to training in very old age. A challenge for the future will be to establish training protocols and dietary interventions which might increase telomere length and delay aging.

Unfortunately, long telomeres and high telomerase activity might not only protect against aging but are also characteristic of cancer cells. Therefore, the inhibition of telomerase activity could represent an attractive therapeutic target for anti-tumor applications. Yan et al. screened a library of 800 natural compounds for potential inhibitors of telomerase activity [10]. They identified sanguinarine chloride as being an inhibitor of telomerase expression and activity. This compound inhibits the growth of several cancer cell lines in vitro and of xenograft tumors in vivo. The safety and efficacity as a potential drug candidate for anti-tumor therapy in humans remains to be determined in future studies.

Another important factor driving aging is cellular senescence. Senescence is characterized by the growth arrest of cells, which was first described in fibroblasts in long-term culture [11,12] and the expression of characteristic markers and secretion of a variety of diverse molecules, the so-called senescence-associated secretory phenotype (SASP) [13]. A problem with the characterization of senescent cells is that not a single highly specific marker exists to identify these cells. Thus, the International Cell Senescence Association released a consensus statement remarking that a combination of more than two typical markers is required to identify a cell as being senescent [14]. Zimmermann and colleagues carefully investigated markers of senescence in melanocytes and melanoma cells. They induced senescence in human melanocytes via the overexpression of mutant BRAFV600E and in melanoma cells via the chemotherapeutic agent etoposide [15]. Both cell types showed increased beta-galactosidase (β-Gal) activity. As this is very common but poorly understood in the field, although all cells were exposed to the senescence-inducing stimuli, only a fraction became β-Gal positive. The cell cycle inhibitor p16INK4A was induced in both models, and as a third marker for the SASP, the authors suggested CXCL2. The use of the two independent cell systems allows important conclusions to be drawn on the choice of senescence biomarkers when working with melanocytic systems, and thus the paper serves as important guidance in the field. In the future, an enormous amount of work regarding the detection of senescence in response to different stimuli and in different cell and organ systems will have to be conducted.

Although the authors describe p16INK4A as a marker for senescence and p16Ink4a-targeting models are frequently used to eliminate senescent cells (reviewed in [13]), we reported a careful analysis of p16Ink4a expression in several organs starting from embryonic development (Embryonic Day 10) until old age in mice [16]. The expression of p16 was highly dynamic in all organs in the embryonic and postnatal stages and increased dramatically in old mice, which at this time point agrees with senescence and SASP factor expression. The expression of p19 and p21 was less variable and increased to a moderate extent in old age. Interestingly, high p16Ink4a protein expression during embryonic development coincided with organ differentiation. In old mice, we observed a predominant expression of p16 mRNA and protein in liver endothelial cells versus non-endothelial cells. This is in agreement with a recent p16 ablator mouse model, which affects liver sinusoidal endothelial cells the most prominently [17]. The expression of p16Ink4a in early life was confirmed recently in a highly sensitive reporter system in fibroblasts of the lung. These p16Ink4a-positive cells surprisingly had an enhanced capacity to sense tissue inflammation and respond through their increased secretory capacity to promote epithelial regeneration [18].

In addition, we reviewed the roles of p16INK4A, p14ARF/p19ARF, and p21 in organ development and homeostasis in this Special Issue of *Cells* [19]. We analyzed the knowledge surrounding p16INK4A, p14ARF/p19ARF, and p21 in embryonic and organ development and described in detail the data reported in the literature and the different animal models targeting these senescence-associated proteins. We highlight the most recent advancements and controversial findings, which have largely contributed to a broader understanding of the senescence mechanism and the roles of p16, p19Arf, and p21 therein. Interestingly, senescent cells do not only have detrimental effects but are also required for physiological functions and are involved in tissue repair. The SASP is not a uniform set of secret factors but differs depending on p16 or p21. The beneficial effects of senescent cell removal are most likely due to a normalization of the SASP and not merely attributed to the removal of these "non-functional" cells; finally, p21-dependent senescence is not an irreversible mechanism which leads to the clearance of the p21-expressing cells via macrophages, but can be reversible [20].

Chen and Skutella propose in their review partial senescent cell reprogramming as a strategy for anti-aging therapies [21]. They suggest that partial reprogramming can produce a secretory phenotype that facilitates cellular rejuvenation. They carefully point out that only partial reprogramming is desired to avoid tumor risk and organ failure and describe approaches for achieving this goal. The authors review the strategies for reversing senescence and the potential underlying mechanisms, identify candidates for this approach, and develop clinical translational strategies to achieve partial reprogramming of senescent cells with the aim of increasing people's healthy lifespan and reducing frailty. This review has already attracted a very broad audience.

Hong and colleagues focus in their review on the molecular mechanisms of alveolar epithelial stem cell senescence and the senescence-associated differentiation disorders in pulmonary fibrosis [22]. The topic is of high actual interest as SARS-CoV-2 viral infections induce acute pulmonary epithelial cell senescence, which is followed by fibrosis and largely determines the disease outcome. The authors focus on the TGF-β signaling pathway inducing the suppression of telomerase activity and thereby inducing senescence of the alveolar epithelial stem cell and pulmonary fibrosis. Alternatively, dysregulation of the shelterin complex protein TPP1 mediating the DNA damage response, pulmonary senescence, and fibrosis is discussed. They highlight studies indicating that the development of senescence-associated differentiation disorders is reprogrammable and reversible by inhibiting stem cell replicative senescence in pulmonary fibrosis, and provide a framework for the targeted intervention of the molecular mechanisms of alveolar stem cell senescence and pulmonary fibrosis.

Cellular senescence in aging lungs and lung diseases is reviewed in this Special Issue of *Cells* by Aghali and colleagues [23]. They provide a successful overview of cellular senescence, as well as the known signaling pathways and biomarkers of senescence. The role of cellular senescence in chronic obstructive pulmonary disease (COPD) and idiopathic pulmonary fibrosis (IPF) is reviewed in detail. Furthermore, the implications of mitochondrial alterations and mitochondrial DNA mutations in senescence and aging in the lung are discussed. Finally, the authors provide a clinically important outlook regarding senescence as a potential therapeutic target in lung diseases.

The group working with Maria Cavinato published two review articles in this Special Issue of *Cells* entitled "Controversies and Recent Advances in Senescence and Aging". The first article deals with the topic of senescence. The authors introduce the relation between cellular senescence and skin aging and analyze in great detail the major components of air pollution on lungs and mainly skin aging. Air pollution and the consequences for senescence and aging is a highly relevant topic for countries with increased industrialization and intensified transport. Fortunately, the authors also provide guidance for tackling the consequences of air pollution on the skin by reviewing the available information on therapeutics and cosmetics in this specific field [24].

In the second article, the group summarizes the current knowledge surrounding age-related lysosomal dysfunctions [25]. Deregulated nutrient sensing, mitochondrial dysfunction, and altered intercellular communication are additional characteristics of senescent cells, which can be attributed to lysosomal dysfunction. The authors introduce lysosomal components, their structure, and lysosomal biogenetic and metabolic pathways. They describe the function of lysosomes in endocytosis, autophagy, mitophagy, and mitochondrial dysfunction, and explain in detail the lysosomal dysfunctions related to aging and senescence. As a major pathway for senescence, mTORC signaling is discussed. In terms of potential therapeutic interventions, it is interesting to note that the treatment of senescent cells with mTORC1 inhibitors ameliorates senescence phenotypes and extends the lifespan in mice [26]. In addition, increased β-galactosidase activity in the lysosomes of senescent cells might represent an opportunity to activate highly specific pro-drugs as senolytic compounds [27].

Jin and colleagues review the relation between pyroptosis, autophagy, and sarcopenia in aging [28]. Pyroptosis—cellular inflammatory necrosis—represents a form of regulated cell death, which plays a role in the ageing progress. It is closely related to age-related diseases such as cardiovascular diseases, Alzheimer's disease, osteoarthritis, and sarcopenia. Sarcopenia refers to an aging-related loss of muscle mass. Autophagy of skeletal muscle cells can inhibit the activation of the pyroptosis pathway. The authors discuss the mechanisms of aggravated oxidative stress and poor skeletal muscle perfusion in ageing muscle, which activate the nod-like receptor (NLRP) family to trigger pyroptosis, and the role of chronic low-grade inflammation in this process.

Brauning et al. discuss natural killer cells' phenotypes and functions in aging [29]. The age-related impairment of the immune function (immunosenescence) is one important cause of age-related morbidity and mortality. Despite an increased number of natural killer (NK) cells in aged individuals, their function is impaired with reduced cytokine secretion and decreased target cell cytotoxicity. NK cells are the central actors in the immunosurveillance of senescent cells, thus also linking the mechanisms of senescence and aging together. This excellent review describes the recent advances and open questions in understanding the interplay between systemic inflammation, senescence burden, and NK cell dysfunction in the context of aging. A profound understanding of the factors driving NK cell aging is a pre-requisite for developing potential therapies countering age-related diseases.

Last but not least, Cai and colleagues review mouse models of accelerated aging [30]. Mice are frequently used in aging and senescence research due to their similarities to humans, their short lifespan, and the ease of reproduction. Nevertheless, models of accelerated aging are highly valuable in order to decrease time and costs in aging research. This review provides excellent guidance and a description of the available models for researchers working in the field.

Taken together, the Special Issue "Controversies and Recent Advances in Senescence and Aging" comprises an excellent collection of original articles and reviews highlighting different novel aspects in the fields of senescence and aging research. They will hopefully stimulate discussions and further research in these fields which are extremely important for a constantly aging human population.

Author Contributions: Conceptualization, K.-D.W. and N.W.; formal analysis, K.-D.W. and N.W.; investigation, K.-D.W. and N.W.; writing—original draft preparation, K.-D.W. and N.W.; writing—review and editing, K.-D.W. and N.W.; project administration, K.-D.W. and N.W.; funding acquisition, K.-D.W. and N.W. All authors have read and agreed to the published version of the manuscript.

Funding: This research was funded by Fondation pour la Recherche Medicale, grant number FRM DPC20170139474 (K.-D.W.), Fondation ARC pour la recherche sur le cancer, grant number n°PJA 20161204650 (N.W.), Gemluc (N.W.), Plan Cancer. INSERM (K.-D.W.), Agence Nationale de la Recherche, grant R19125AA "Senage" (K.-D.W.), and Fondation ARC pour la recherche sur le cancer, grant number n°PJA 20161204650 (K.-D.W.).

Conflicts of Interest: The authors declare no conflict of interest.

References

1. Carrillo-Salinas, F.J.; Parthasarathy, S.; Moreno de Lara, L.; Borchers, A.; Ochsenbauer, C.; Panda, A.; Rodriguez-Garcia, M. Short-Chain Fatty Acids Impair Neutrophil Antiviral Function in an Age-Dependent Manner. *Cells* **2022**, *11*, 2515. [CrossRef]
2. Liu, Z.; Zhou, T.; Ziegler, A.C.; Dimitrion, P.; Zuo, L. Oxidative Stress in Neurodegenerative Diseases: From Molecular Mechanisms to Clinical Applications. *Oxid. Med. Cell. Longev.* **2017**, *2017*, 2525967. [CrossRef]
3. Baek, M.; Jang, W.; Kim, C. Dual Oxidase, a Hydrogen-Peroxide-Producing Enzyme, Regulates Neuronal Oxidative Damage and Animal Lifespan in. *Cells* **2022**, *11*, 2059. [CrossRef]
4. Klein, E.A.; Thompson, I.M.; Tangen, C.M.; Crowley, J.J.; Lucia, M.S.; Goodman, P.J.; Minasian, L.M.; Ford, L.G.; Parnes, H.L.; Gaziano, J.M.; et al. Vitamin E and the risk of prostate cancer: The Selenium and Vitamin E Cancer Prevention Trial (SELECT). *JAMA* **2011**, *306*, 1549–1556. [CrossRef]
5. Cook, N.R.; Albert, C.M.; Gaziano, J.M.; Zaharris, E.; MacFadyen, J.; Danielson, E.; Buring, J.E.; Manson, J.E. A randomized factorial trial of vitamins C and E and beta carotene in the secondary prevention of cardiovascular events in women: Results from the Women's Antioxidant Cardiovascular Study. *Arch. Intern. Med.* **2007**, *167*, 1610–1618. [CrossRef]
6. Ke, Q.; Gong, L.; Zhu, X.; Qi, R.; Zou, M.; Chen, B.; Liu, W.; Huang, S.; Liu, Y.; Li, D.W. Multinucleated Retinal Pigment Epithelial Cells Adapt to Vision and Exhibit Increased DNA Damage Response. *Cells* **2022**, *11*, 1552. [CrossRef]
7. Wang, Q.; Zhan, Y.; Pedersen, N.L.; Fang, F.; Hägg, S. Telomere Length and All-Cause Mortality: A Meta-analysis. *Ageing Res. Rev.* **2018**, *48*, 11–20. [CrossRef]
8. Wagner, K.D.; Ying, Y.; Leong, W.; Jiang, J.; Hu, X.; Chen, Y.; Michiels, J.F.; Lu, Y.; Gilson, E.; Wagner, N.; et al. The differential spatiotemporal expression pattern of shelterin genes throughout lifespan. *Aging* **2017**, *9*, 1219–1232. [CrossRef]
9. Semeraro, M.D.; Almer, G.; Renner, W.; Gruber, H.J.; Herrmann, M. Influences of Long-Term Exercise and High-Fat Diet on Age-Related Telomere Shortening in Rats. *Cells* **2022**, *11*, 1605. [CrossRef]
10. Yan, S.; Lin, S.; Chen, K.; Yin, S.; Peng, H.; Cai, N.; Ma, W.; Songyang, Z.; Huang, Y. Natural Product Library Screens Identify Sanguinarine Chloride as a Potent Inhibitor of Telomerase Expression and Activity. *Cells* **2022**, *11*, 1485. [CrossRef]
11. Hayflick, L. The limited in vitro lifetime of human diploid cell strains. *Exp. Cell Res.* **1965**, *37*, 614–636. [CrossRef]
12. Hayflick, L.; Moorhead, P.S. The serial cultivation of human diploid cell strains. *Exp. Cell Res.* **1961**, *25*, 585–621. [CrossRef]
13. Safwan-Zaiter, H.; Wagner, N.; Wagner, K.D. P16INK4A-More Than a Senescence Marker. *Life* **2022**, *12*, 1332. [CrossRef]
14. Gorgoulis, V.; Adams, P.D.; Alimonti, A.; Bennett, D.C.; Bischof, O.; Bishop, C.; Campisi, J.; Collado, M.; Evangelou, K.; Ferbeyre, G.; et al. Cellular Senescence: Defining a Path Forward. *Cell* **2019**, *179*, 813–827. [CrossRef]
15. Zimmermann, T.; Pommer, M.; Kluge, V.; Chiheb, C.; Muehlich, S.; Bosserhoff, A.K. Detection of Cellular Senescence in Human Primary Melanocytes and Malignant Melanoma Cells In Vitro. *Cells* **2022**, *11*, 1489. [CrossRef]
16. Safwan-Zaiter, H.; Wagner, N.; Michiels, J.F.; Wagner, K.D. Dynamic Spatiotemporal Expression Pattern of the Senescence-Associated Factor p16Ink4a in Development and Aging. *Cells* **2022**, *11*, 541. [CrossRef]

17. Grosse, L.; Wagner, N.; Emelyanov, A.; Molina, C.; Lacas-Gervais, S.; Wagner, K.-D.; Bulavin, D.V. Defined p16High Senescent Cell Types Are Indispensable for Mouse Healthspan. *Cell Metab.* **2020**, *32*, 87–99.e6. [CrossRef]
18. Reyes, N.S.; Krasilnikov, M.; Allen, N.C.; Lee, J.Y.; Hyams, B.; Zhou, M.; Ravishankar, S.; Cassandras, M.; Wang, C.; Khan, I.; et al. Sentinel. *Science* **2022**, *378*, 192–201. [CrossRef]
19. Wagner, K.D.; Wagner, N. The Senescence Markers p16INK4A, p14ARF/p19ARF, and p21 in Organ Development and Homeostasis. *Cells* **2022**, *11*, 1966. [CrossRef]
20. Sturmlechner, I.; Zhang, C.; Sine, C.C.; van Deursen, E.J.; Jeganathan, K.B.; Hamada, N.; Grasic, J.; Friedman, D.; Stutchman, J.T.; Can, I.; et al. p21 produces a bioactive secretome that places stressed cells under immunosurveillance. *Science* **2021**, *374*, eabb3420. [CrossRef]
21. Chen, R.; Skutella, T. Synergistic Anti-Ageing through Senescent Cells Specific Reprogramming. *Cells* **2022**, *11*, 830. [CrossRef] [PubMed]
22. Hong, X.; Wang, L.; Zhang, K.; Liu, J.; Liu, J.P. Molecular Mechanisms of Alveolar Epithelial Stem Cell Senescence and Senescence-Associated Differentiation Disorders in Pulmonary Fibrosis. *Cells* **2022**, *11*, 877. [CrossRef] [PubMed]
23. Aghali, A.; Koloko Ngassie, M.L.; Pabelick, C.M.; Prakash, Y.S. Cellular Senescence in Aging Lungs and Diseases. *Cells* **2022**, *11*, 1781. [CrossRef] [PubMed]
24. Martic, I.; Jansen-Dürr, P.; Cavinato, M. Effects of Air Pollution on Cellular Senescence and Skin Aging. *Cells* **2022**, *11*, 2220. [CrossRef]
25. Guerrero-Navarro, L.; Jansen-Dürr, P.; Cavinato, M. Age-Related Lysosomal Dysfunctions. *Cells* **2022**, *11*, 1977. [CrossRef]
26. Harrison, D.E.; Strong, R.; Sharp, Z.D.; Nelson, J.F.; Astle, C.M.; Flurkey, K.; Nadon, N.L.; Wilkinson, J.E.; Frenkel, K.; Carter, C.S.; et al. Rapamycin fed late in life extends lifespan in genetically heterogeneous mice. *Nature* **2009**, *460*, 392–395. [CrossRef] [PubMed]
27. Cai, Y.; Zhou, H.; Zhu, Y.; Sun, Q.; Ji, Y.; Xue, A.; Wang, Y.; Chen, W.; Yu, X.; Wang, L.; et al. Elimination of senescent cells by β-galactosidase-targeted prodrug attenuates inflammation and restores physical function in aged mice. *Cell Res.* **2020**, *30*, 574–589. [CrossRef]
28. Jin, H.; Xie, W.; He, M.; Li, H.; Xiao, W.; Li, Y. Pyroptosis and Sarcopenia: Frontier Perspective of Disease Mechanism. *Cells* **2022**, *11*, 1078. [CrossRef]
29. Brauning, A.; Rae, M.; Zhu, G.; Fulton, E.; Admasu, T.D.; Stolzing, A.; Sharma, A. Aging of the Immune System: Focus on Natural Killer Cells Phenotype and Functions. *Cells* **2022**, *11*, 1017. [CrossRef]
30. Cai, N.; Wu, Y.; Huang, Y. Induction of Accelerated Aging in a Mouse Model. *Cells* **2022**, *11*, 1418. [CrossRef]

cells

MDPI

Article

Short-Chain Fatty Acids Impair Neutrophil Antiviral Function in an Age-Dependent Manner

Francisco J. Carrillo-Salinas [1], Siddharth Parthasarathy [1,2], Laura Moreno de Lara [1,3], Anna Borchers [1], Christina Ochsenbauer [4], Alexander Panda [5,6] and Marta Rodriguez-Garcia [1,2,*]

1 Department of Immunology, Tufts University School of Medicine, Boston, MA 02111, USA
2 Immunology Program, Tufts Graduate School of Biomedical Sciences, Boston, MA 02111, USA
3 Immunology Unit, Biomedical Research Centre (CIBM), University of Granada, 18071 Granada, Spain
4 Department of Medicine, Hem/Onc & CFAR, University of Alabama at Birmingham, Birmingham, AL 35233, USA
5 Tufts Medical Center/Division of Pulmonary and Critical Care (PCCM), Boston, MA 02111, USA
6 Tufts Clinical and Translational Science Institute (CTSI), Boston, MA 02111, USA
* Correspondence: marta.rodriguez_garcia@tufts.edu

Abstract: Half of the people living with HIV are women. Younger women remain disproportionally affected in endemic areas, but infection rates in older women are rising worldwide. The vaginal microbiome influences genital inflammation and HIV infection risk. Multiple factors, including age, induce vaginal microbial alterations, characterized by high microbial diversity that generate high concentrations of short-chain fatty acids (SCFAs), known to modulate neutrophil function. However, how SCFAs may modulate innate anti-HIV protection by neutrophils is unknown. To investigate SCFA-mediated alterations of neutrophil function, blood neutrophils from younger and older women were treated with SCFAs (acetate, butyrate and propionate) at concentrations within the range reported during bacterial vaginosis, and phenotype, migration and anti-HIV responses were evaluated. SCFA induced phenotypical changes preferentially in neutrophils from older women. Butyrate decreased CD66b and increased CD16 and CD62L expression, indicating low activation and prolonged survival, while propionate increased CD54 and CXCR4 expression, indicating a mature aged phenotype. Furthermore, acetate and butyrate significantly inhibited neutrophil migration in vitro and specifically reduced α-defensin release in older women, molecules with anti-HIV activity. Following HIV stimulation, SCFA treatment delayed NET release and dampened chemokine secretion compared to untreated neutrophils in younger and older women. Our results demonstrate that SCFAs can impair neutrophil-mediated anti-HIV responses.

Keywords: short-chain fatty acids; neutrophil; HIV; aging; women

Citation: Carrillo-Salinas, F.J.; Parthasarathy, S.; Moreno de Lara, L.; Borchers, A.; Ochsenbauer, C.; Panda, A.; Rodriguez-Garcia, M. Short-Chain Fatty Acids Impair Neutrophil Antiviral Function in an Age-Dependent Manner. *Cells* **2022**, *11*, 2515. https://doi.org/10.3390/cells11162515

Academic Editors: Nicole Wagner and Kay-Dietrich Wagner

Received: 19 May 2022
Accepted: 11 August 2022
Published: 13 August 2022

1. Introduction

New human immunodeficiency virus (HIV) infections have been reduced by 40% since 1998, but around 1.5 million people were still newly infected with HIV in 2020 [1]. Although younger women are at higher risk in endemic areas, new HIV infections in older women are rising worldwide [2–5], a fact to take into account given the increase in the size of the elderly population expected in the upcoming decades [6,7].

The vaginal microbiota is a dynamic community of bacteria that works as a first-line defense against invading pathogens, along with the epithelial mucosal barrier and the immune mucosal response [8–10]. The vaginal microbiota, dominated by the *Lactobacilli* species that maintain high concentrations of lactic acid [11] and a low pH in the lower tract, is considered to be beneficial and reduce the risk of HIV acquisition [12–15]. However, multiple stimuli, including antibiotics, sexual activity, vaginal hygiene, menstrual cycle, and oral contraceptives, can alter these bacterial populations, resulting in vaginal dysbiosis and increased risk of HIV acquisition [12,16–21]. Data also indicate that the vaginal microbiome changes after menopause, with a decreased presence of *Lactobacillus* species compared to premenopausal women [22–24]. This fact, together with changes in immune cell responses, the pro-inflammatory environment, and a reduction of protective mediators in the female reproductive tract (FRT), may increase the risk of HIV acquisition in postmenopausal women [7,23,25–28].

An important consequence associated with vaginal dysbiosis is the increased presence of short-chain fatty acids (SCFAs), metabolic products of anaerobic bacterial fermentation. SCFAs are physiologically present in the genital mucosa at low concentrations (0–4 mM), with acetic, butyric and propionic acids being the most abundant [11,29–32]. However, under conditions that induce vaginal dysbiosis, the reduction in Lactobacilli is associated with a drop in lactate concentration and an increased vaginal pH that facilitates the growth of facultative and anaerobic bacteria, resulting in abnormally elevated concentrations of acetate, butyrate and propionate (20–140 mM) [11,29–32].

SCFAs present at mucosal surfaces act on epithelial cells to modify barrier function [9] and also diffuse into the subepithelial compartment to act directly on immune cells by interaction with the G-protein coupled receptors GPR43, GPR41 and GPR109A [33–35]. SCFAs have been described to play a critical role as immunomodulators to prevent mucosal inflammation in the gut [36]. However, high concentrations of SCFAs in the FRT seem to promote inflammation [9,31,37]. Genital inflammation is known to increase susceptibility to HIV infection [15,38,39], but very little is known about the effect of high concentrations of SCFAs on innate immune responses and the susceptibility to HIV infection in women [31].

We recently demonstrated that neutrophils from the FRT of healthy women release neutrophil extracellular traps (NETs) in response to HIV stimulation [40]. NET release is a process characterized by the extracellular ejection of DNA associated with granular proteins with antimicrobial activity [40–42], which has been shown to inactivate HIV in vitro [40,41]. In contrast, studies of inflammation in the context of sexually transmitted infections (STIs) describe associations between neutrophil-derived molecules in cervicovaginal secretions with an increased risk of HIV acquisition [43–45]. These apparently opposite findings highlight the gap in our knowledge about how alterations in the microbiome may affect neutrophil function and anti-HIV responses. Importantly, neutrophils highly express GPR43, the main SCFA receptor [35], representing a likely candidate to be modulated by changes in the microbial metabolome. However, it is unknown whether the anti-HIV potential of neutrophils is modified by SCFAs.

In this context, we hypothesize that pathological concentrations of SCFAs due to changes in genital microbiota can modulate neutrophil responses to HIV and directly impact the risk of HIV acquisition. To begin to answer this question, we optimized an in vitro model to evaluate the effects of pathological concentrations of SCFA on neutrophil function and anti-HIV activity in younger and older women.

We found that pathological concentrations of SCFAs reduced neutrophil activation in an age-dependent manner, inhibited neutrophil migration and reduced the release of NETs and innate antiviral molecules. Our findings provide proof of concept that genital microbial alterations that induce an increase in SCFA concentrations may impair neutrophils' physiological functions and reduce their antiviral potential.

2. Materials and Methods

2.1. Study Subjects

All investigations involving human subjects were conducted according to the principles expressed in the Declaration of Helsinki and carried out with the approval of the Institutional Review Board of Tufts University (protocol code: MODCR-01-11201, approved on 20 October 2014). Volunteer healthy and HIV-seronegative women were included in the study, and informed consent was obtained from all subjects. Information regarding age was provided, but no other information was disclosed. Women included in the study were classified as younger (n = 17; 18–28 years-old; median = 24) or older (n = 18; 65–72 years-old; median = 68).

2.2. Neutrophil Isolation from Human Peripheral Blood and Treatment with SCFAs

Venous blood of healthy women was collected into 10 mL EDTA tubes (BD Vacutainer; Franklin Lakes, NJ, USA). Polymorphonuclear cell (PMN) isolation was performed by positive selection using CD15 MicroBeads (Miltenyi Biotec; Auburn, CA, USA) and a whole blood column kit following the manufacturer's instructions (Miltenyi Biotec). This isolation method was effective with 92.84% viability (Figures 1a and S1a) and >94% neutrophil enrichment (as $CD45^+$ $CD15^+$ $CD66b^+$ cells), determined by flow cytometry. Purified neutrophils were resuspended in HBSS culture medium (Hanks' Balanced Salt Solution, Gibco; Waltham, MA, USA) for imaging analysis or in X-VIVO 15 media for cell culture (Lonza; Bend, OR, USA) and stimulated with 25 mM of sodium acetate, sodium butyrate or sodium propionate for 1 or 3 h as indicated for further analysis.

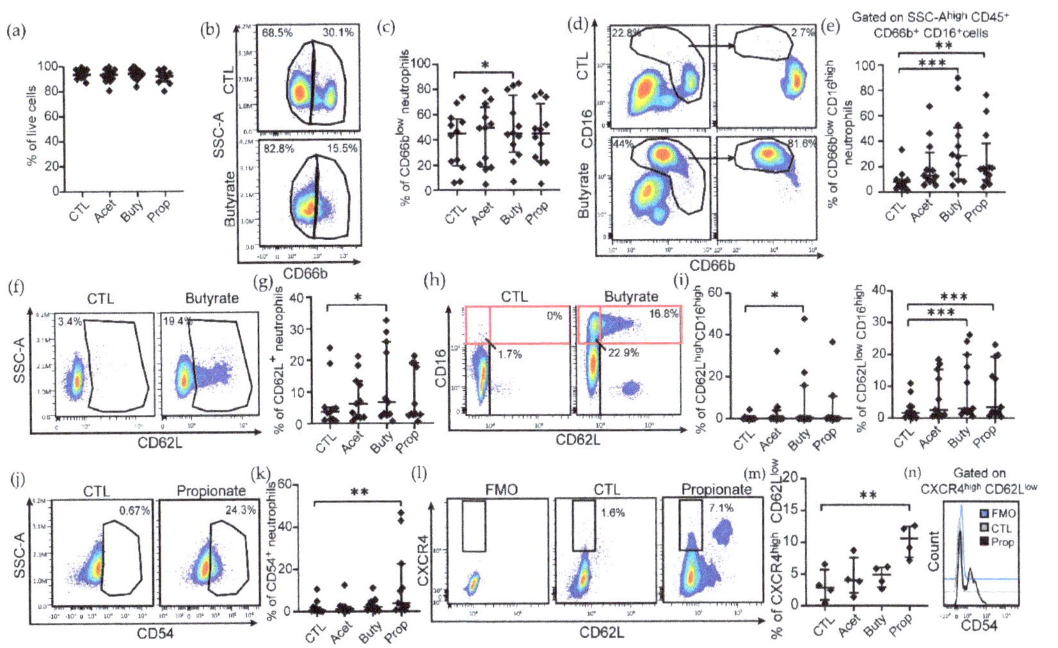

Figure 1. High concentrations of butyrate and propionate induce phenotypical changes in human neutrophils. (**a**) Quantification of cell death induced by high concentration of SCFA (n = 12). (**b**) Representative example of flow cytometry plots showing CD66b expression on neutrophils in control condition (CTL) and after butyrate treatment (25 mM) for 3 h. (**c**) Percentage of $CD66b^{low}$ neutrophils after treatment with SCFAs (acetate, butyrate, and propionate) 25 mM for 3 h (n = 12). (**d**) Representative example of flow cytometry plots and (**e**) percentage of $CD66b^{low}$ $CD16^{high}$ neutrophils after SCFA treatment (n = 12). (**f**) Flow cytometry plots and (**g**) changes in percentage of $CD62L^+$ neutrophils induced by SCFAs. Effect of pathological concentration of SCFAs on ((**h**,**i**); n = 11) $CD16^{high}$ $CD62L^{high}$ or $CD62L^{low}$ neutrophil population and (**j**,**k**) $CD54^+$ neutrophils (n = 11). (**l**) Representative flow cytometry plots and (**m**) quantification of the percentage of $CXCR4^{high}$ $CD62L^{low}$ neutrophils after treatment with pathological concentration of SCFAs (n = 4). (**n**) Changes induced by propionate treatment in CD54 expression of neutrophils gated on $CXCR4^{high}$ $CD62L^{low}$ population. Each dot represents a different patient (age of patients: 18–72 years old). Non-parametric paired Friedmann test was used, * $p \leq 0.05$, ** $p \leq 0.01$, *** $p \leq 0.001$. CTL: control; Acet: acetate 25 mM; Buty: butyrate 25 mM; Prop: propionate 25 mM.

2.3. Determination of Neutrophil Phenotype by Flow Cytometry

Neutrophils were fixed with 4% PFA for 30 min at 4 °C, washed and stained for 20 min in the dark with the following anti-human antibodies: CD45-APC-Cy7 (clone 2D1; Biolegend, San Diego, CA, USA), CD54-BV421 (clone HA58; Biolegend), CD62L-BV711 (clone SK11; Biolegend), CXCR4-BV-785 (clone 12G5; Biolegend), CD66b-APC (clone REA306; Miltenyi Biotec), CD15-FITC (Miltenyi Biotec) and CD16-PE (clone 3G8; BD Biosciences, Franklin Lakes, NJ, USA). A live/dead fixable blue dead cell stain kit (Thermo Scientific; Waltham, MA, USA) was used to assess cell death in cultures before fixation. Fluorescence Minus One (FMO) controls were used to identify and gate positive populations (Figures 1 and Figure S1b). Analysis was performed on an LSRII flow cytometer (BD; Ashland, Wilmington, DE, USA) or Aurora cytometer (Cytek Biosciences; Fremont, CA, USA) and assessed using FlowJo software (BD) or OMIQ (www.omiq.ai (accessed on 13 July 2021)). The expression of surface markers was measured by the percentage of positive cells.

2.4. Migration Assay

Neutrophil migration was evaluated using a Transwell assay inserted into an ultra-low attachment 24-well plate (Corning Inc., Corning, NY, USA). Cells were seeded at a density of 4×10^5/well in X-VIVO 15 media (Lonza) into the upper chamber of a Transwell insert (5 μm pore size; Corning, Inc.), and XVIVO-15 with sodium acetate, sodium butyrate or sodium propionate at a final concentration of 25 mM was added to the lower chamber to study if SCFAs at this concentration could act as a chemoattractant. After 3 h at 37 °C, the cells from both chambers were collected and stained for immune phenotyping by flow cytometry. The migration ratio was calculated by dividing the number of cells in the bottom chamber by the sum of cells in the top + bottom chamber and normalizing to the control group.

2.5. Generation of GFP-Labeled VLPs

Modified pNL43 provirus-based plasmid for expression of GFP-labeled viral-like particles (VLPs) and encoding NL43 Env in cis (referred to as pNL4GagGSGFP/K795) was described previously [46]. Briefly, the enhanced GFP (EGFP) coding sequence is expressed in the frame at the 3′end of the gag, replacing the protease and most of the reverse transcriptase coding region. The Ψ-signal on the RNA and the complete gag open reading frame (ORF) remain intact. Furthermore, a plasmid with an inactivated Env ORF, resulting in no expression of functional Env protein (referred to as pNL4GagGSGFPDelta-env/K806), was derived from K795 for pseudotyping and complemented with pBaL.26 Env expression plasmid (NIH AIDS Reagent program, catalog number 11,446, contributed by Dr. John Mascola) [47]. Non-infectious, EGFP-labelled VLPs were produced by transfection, concentrated by ultracentrifugation, and enumerated essentially as described [46].

2.6. Time-Lapse Microscopy of NETs

Human purified neutrophils from blood were plated in a 96-well plate (Corning Inc.; Corning, NY, USA) and stimulated with 25 mM of sodium acetate, sodium butyrate or sodium propionate (Sigma-Aldrich; St. Louis, MO, USA), in the presence or absence of GFP-labeled HIV-viral like particles (HIV-VLPs). Cytotox red reagent (Essen Bioscience; Ann Arbor, MI, USA) was used to label DNA. Images were collected every 3–5 min for at least 3 h at 37 °C using a 10x objective with the IncuCyte S3 (Sartorius; Bohemia, NY, USA). Extracellular DNA-labeled red signal and GFP-VLP signal were quantified to determine the NET-HIV area with the Incucyte software as described [40].

2.7. HIV Stimulation

HIV-1-BaL (R5) isolates were obtained from the AIDS Research and Reference Reagent Program, Division of AIDS, NIAID, NIH, from Dr. Suzanne Gartner, Dr. Mikulas Popovic and Dr. Robert Gallo [48] and propagated in PBMCs as described [49]. Purified blood neutrophils were stimulated with HIV-1 BaL for 1h at an MOI of 0.5, after which the culture supernatants were collected and stored at −80 °C until used for cytokine and chemokine analysis by Luminex. Uninfected controls were incubated for the same length of time in media without the virus.

2.8. *Quantification of Cytokines and Chemokines by Luminex*

Supernatants from SCFA- and HIV-stimulated neutrophils were centrifuged at 18,000× g for 10 min to remove any cell debris and NETs. Then, supernatants were transferred to a new plate for HIV inactivation with 0.5% Triton X-100 (Sigma) for 30 min at 4 °C. Two different panels of cytokines and chemokines were measured using Millipore human cytokine multiplex kits (EMD Millipore Corporation; Billerica, MA, USA) following the manufacturer's instructions. Panel 1: sCD40L, MIP-1α, MIP-1β, GM-CSF, IFNγ, TNFα, IL-1β, IL-5, IL-6, IL-8, IL-10, IL-22, GROα, IFNα2, IL-13, IL-27 and PDFG-AB/BB. Panel 2: IL-8, MCP-1, MIP-1α, RANTES, MDC and MIG. Signal was measured using the MAGPIX system by Luminex (Luminex Corporation; Austin, TX, USA) and quantified with Luminex xPONENT software.

2.9. *ELISA*

The concentration of α-defensins 1–3 was quantified in cell-free culture supernatants after 3 h of treatment with acetate, butyrate or propionate 25 mM using the commercial Human alpha-defensin 1 DuoSet ELISA (R&D Systems; Minneapolis, MN, USA) following the manufacturer's instructions. The control group was incubated with X-VIVO 15 media. The concentration of sCD62L was also measured using the Human L-Selectin/CD62L DuoSet ELISA (R&D Systems) from the same supernatants used for α-defensin and Luminex.

2.10. *Quantification of GPR43 by Western Blot*

Human neutrophil pellets were resuspended in extraction buffer containing RIPA buffer, 1% NP-40, 1mM PMSF, 1x phosphatase inhibitor (PhosSTOP, Millipore Sigma; Burlington, MA, USA), and 1× protease inhibitor (cOmplete, Millipore Sigma) and were lysed for 45 min on ice. The cell lysate was centrifuged at 12,000× g at 4 °C for 5 min. Protein concentration was determined with a protein assay reagent (Pierce 660 nm, Thermo Scientific), and 10–20 µg were mixed with 1× Laemmli SDS sample buffer and heated at 95 °C for 5 min. Samples were run in 4–20% acrylamide gels (Mini-PROTEAN TGX Precast Protein Gels, Bio-Rad; Hercules, CA, USA) at 200 V for 1 h. After electrophoresis, proteins were transferred to a PVDF membrane using a rapid transfer system (Trans-Blot Turbo Transfer System, Bio-Rad). Membranes were blocked using a blocking solution (Pierce Fast Blocking Buffer, Thermo Scientific) for 5 min at RT, washed with 0.1% Tween20 in TBS (pH 7.6), and then incubated overnight with 1:1000 polyclonal rabbit anti-human GPR43 antibody (Thermo Scientific) in a 5% BSA, 0.05% NaN3, 0.1% Tween-20 TBS solution. After washing, the membrane was incubated with 1:7500 anti-Rabbit IgG (H + L) (IRDye 800CW, Li-Cor; Lincoln, Dearborn, MI, USA) in a 1% dry milk, 0.1% Tween 20, TBS solution for 45 min at RT. Protein quantification was performed using the Li-Cor Odyssey system. Protein levels were relativized to unstimulated control.

2.11. *Statistical Analysis*

Data analysis was performed using the GraphPad Prism 9 software. Data are represented as median ± interquartile range (IQR). A two-sided p-value ≤ 0.05 was considered statistically significant. Non-parametric Mann–Whitney U test or Wilcoxon's matched pair test was used for comparison of two groups, and non-parametric Kruskal-Wallis or Friedman tests followed by Dunn's post-test were used for comparison of three or more groups. * $p \leq 0.05$; ** $p \leq 0.01$; *** $p \leq 0.001$. Grubb's analysis (alpha = 0.05) was used to identify potential outliers.

3. Results

3.1. Pathological Concentrations of SCFAs Induce Phenotypical Changes in Human Blood Neutrophils

In order to identify phenotypical changes in neutrophils under conditions of high concentration of SCFAs, we incubated blood neutrophils from healthy women (range of age: 18–72 years old) with three different SCFAs (butyrate, propionate or acetate) at a pathological concentration (25 mM) and compared the expression of several activation markers by flow cytometry (Figure S1a; gating strategy). This concentration was selected based on previous literature indicating a range concentration of SCFA of 20–140 mM during vaginal dysbiosis [11,29–32]. First, we observed that pathological concentrations of SCFAs did not induce cell death in neutrophils after 3 h of treatment compared to the control group (Figure 1a). In untreated neutrophils, two subpopulations were identified based on CD66b expression, CD66b^{low} and CD66b^{high} (Figure 1b, top panel). Treatment with butyrate specifically induced a significant decrease in the expression of CD66b (Figure 1b; bottom panel), increasing the proportion of CD66b^{low} neutrophils (Figure 1c), while no significant changes were observed for CD66b expression when neutrophils were treated with acetate or propionate (Figure 1c). In addition, butyrate and propionate treatment modified CD16 expression, inducing a shift from the CD66b^{high}CD16^{low} population found in untreated neutrophils (Figure 1d; top panels) to CD66b^{low}CD16^{high} in butyrate and propionate-treated neutrophils (Figure 1d,e; bottom panels). Butyrate treatment also increased CD62L expression on neutrophils (Figure 1f,g), while no significant changes were detected for the other SCFAs (Figure 1g).

Because CD16 and CD62L can define different neutrophil subsets with distinct effector functions [50], we next determined changes in the co-expression of these two markers (Figure 1h). Consistent with our observations with each individual marker, we detected a significant increase in the CD62L^{high}CD16^{high} neutrophil population after butyrate treatment (Figure 1h,i; left graph), indicating a mature and partially activated phenotype. Although we identified an outlier in Figure 1i (left graph), the difference remained significant after excluding from the analysis this outlier data point in the data set ($p = 0.016$). In addition, butyrate and propionate treatment also increased the proportion of CD62L^{low} CD16^{high} neutrophils (Figure 1i; right graph).

Finally, we analyzed the expression of CD54 (intracellular adhesion molecule 1, ICAM-1) in neutrophils, a marker of activation and migration. Only propionate induced a significant increase in the proportion of CD54^{+} neutrophils (Figure 1j,k), while no effect was observed with the other SCFAs. Since upregulation of CD54 and downregulation of CD62L in combination with CXCR4 [51] are markers that indicate "aged" neutrophils, an overly active population of circulating neutrophils with an expanded lifespan, we further explored the expression of CXCR4 following SCFA treatment to confirm the induction of an aged phenotype. Only pathological concentrations of propionate significantly increased the proportion of CXCR4^{high}CD62L^{low} neutrophils (Figure 1l,m), which also showed higher expression of CD54 (Figure 1n), characteristic of neutrophils with an aged phenotype [51].

Taken together, our results suggest that butyrate at pathological concentrations reduces activation of neutrophils and increases maturation, while propionate induces phenotypical alterations characteristic of "aged" neutrophils.

3.2. Effects of SCFA Treatment Are Enhanced in Neutrophils from Older Women

Recognizing that as women age, immune functions and the composition of the vaginal microbiome are modified [7,24], we stratified the women in our study into younger (average of 24.78 years old) and older groups (average of 66.22 years old) and evaluated phenotypical changes to determine if age had any potential effects on susceptibility to a pathological concentration of SCFAs. Interestingly, the CD66^{high} (Figure 2a) and CD66b^{low} (Figure 2b) neutrophil populations were very conserved in younger women and did not change after treatment with SCFAs. In contrast, older women showed high variability in the levels of CD66b expression, and these levels were significantly reduced after butyrate treatment, with a decrease in the proportion of CD66b^{high} neutrophils (Figure 2a) and an increase in CD66b^{low} neutrophils (Figure 2b). Furthermore, the observed effect of butyrate and propionate increasing CD16 expression on CD66b^{low} neutrophils (Figure 1c,d) was enhanced in neutrophils from older women (butyrate: 12.2-fold change, propionate: 8-fold change) compared to younger women (butyrate: 3-fold change, propionate: 2-fold change) (Figure 2c). Similarly, butyrate and propionate treatment only increased CD62L expression on neutrophils in older women, with no significant changes in younger women (Figure 2d). However, when we analyzed changes in the co-expression of CD62L and CD16, we observed high variability of a CD62L^{high} CD16^{high} neutrophil population in older women after SCFA treatment, while this population was almost absent in younger women (Figure 2e).

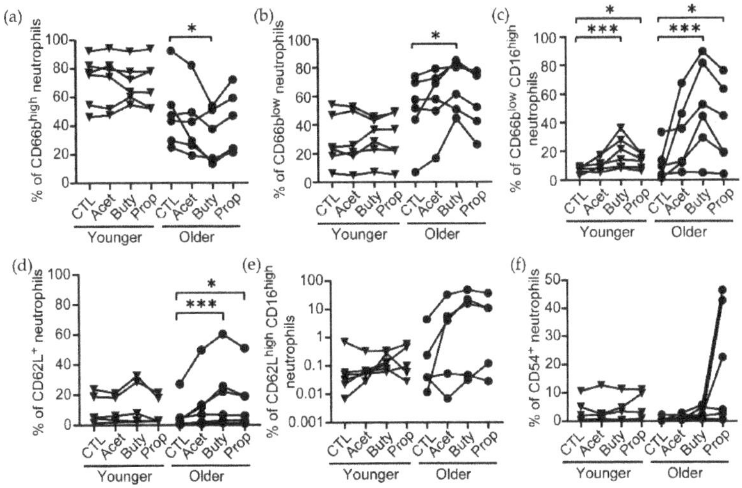

Figure 2. Pathological concentrations of butyrate and propionate preferentially modifies the phenotype of neutrophils from older women. Effect of SCFAs on the percentage of (**a**) CD66b^{high}, (**b**) CD66b^{low}, (**c**) CD66b^{low}CD16^{high}, (**d**) CD62L^{+}, (**e**) CD62L^{high}CD16^{high}, and (**f**) CD54^{+} neutrophils after 3 h of treatment (25 mM) in neutrophils from younger (triangles) and older women (circles), quantified by flow cytometry. Each dot represents a different patient (younger = 6; older = 5–6). Non-parametric paired Friedmann test was used, followed by Dunn's post-test for multiple comparisons; * $p \leq 0.05$, *** $p \leq 0.001$. CTL: control; Acet: acetate 25 mM; Buty: butyrate 25 mM; Prop: propionate 25 mM.

Finally, although propionate induced an increase in the proportion of CD54^{+} neutrophils in the older population, this change did not reach statistical significance (Figure 2f).

Collectively, our data indicate that neutrophils from older women are more responsive to high concentrations of SCFAs than younger women, specifically to butyrate and propionate, resulting in a non-activated mature phenotype.

3.3. *Pathological Concentrations of SCFAs Inhibit Neutrophil Migration In Vitro*

CD62L (L-selectin) and CD54 (ICAM-1) are adhesion molecules involved in neutrophil transmigration [52]. Since we observed changes in CD62L and CD54 expression following SCFA treatment, we next investigated neutrophil migration in the context of pathological concentrations of SCFAs. To determine if pathological concentrations of SCFAs would act as a chemoattractant for neutrophils, we used a transwell system, plating neutrophils on the top chamber and a high concentration of SCFAs (25 mM) or control media in the bottom chamber, and the phenotype of neutrophils was determined by flow cytometry following the same gating strategy used for Figures 1, 2 and S1a. After 3 h, high concentrations of acetate and butyrate significantly inhibited neutrophil migration compared to the control condition, while propionate did not affect neutrophil migration compared to control media (Figure 3a).

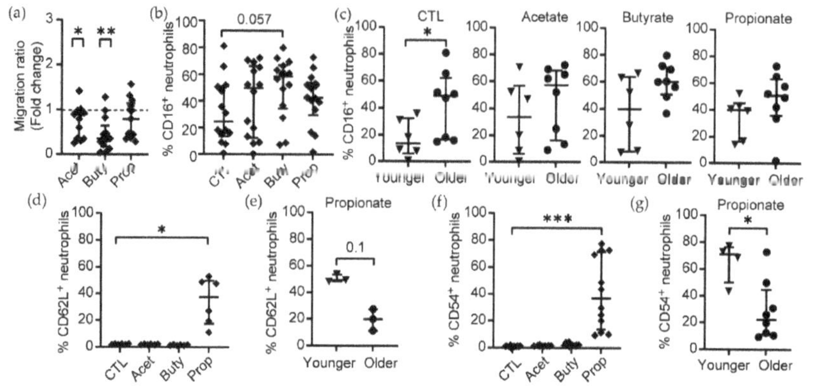

Figure 3. High concentrations of acetate and butyrate inhibit neutrophil migration. (**a**) Migration ratio of neutrophils in the presence of SCFAs 25 mM for 3 h (n = 14). (**b**) Expression of CD16 on neutrophils that migrated to the bottom chamber after 3 h in the absence (control: CTL) or presence of different SCFAs (25 mM) measured by flow cytometry (n = 14). (**c**) Effect of age on CD16$^+$ neutrophils treated with SCFAs 25 mM (younger = 6; older = 8). Migrated neutrophils in the bottom chamber of the transwell were collected and the percentage of (**d**) CD62L$^+$ neutrophils in the absence or presence of SCFAs was quantified (n = 6). (**e**) Effect of age on CD62L$^+$ neutrophils that migrated towards propionate (younger = 3; older = 3). The same analysis was conducted for CD54$^+$ neutrophils (**f**, n = 12), and effect of age on CD54$^+$ neutrophils that were chemoattracted by propionate (**g**) was examined (younger = 4; older = 8). Each dot represents a different patient. (**a**,**c**) Wilcoxon t-test and (**b**,**d**–**g**) non-parametric paired Friedmann test were used; * $p \leq 0.05$, ** $p \leq 0.01$, *** $p \leq 0.001$. CTL: control; Acet: acetate 25 mM; Buty: butyrate 25 mM; Prop: propionate 25 mM.

Next, we analyzed changes in CD16, CD62L and CD54 expression in the neutrophils that migrated. We observed a trend towards an increased proportion of CD16$^+$ neutrophils in the butyrate condition, although it did not reach statistical significance compared to untreated controls (Figure 3b). When we evaluated the effects of age, untreated neutrophils from older women showed significantly higher expression of CD16 after migration compared to neutrophils from younger women (Figure 3c). However, this difference was abrogated in the presence of SCFAs (Figure 3c).

In contrast, we found a significant increase in CD62L$^+$ neutrophils after migration in the propionate condition (Figure 3d), which was independent of age (Figure 3e), while these populations were absent in the acetate and butyrate conditions. In addition, we observed a higher proportion of CD54$^+$ neutrophils after migration to propionate (Figure 3f), and this CD54$^+$ neutrophil population was significantly more abundant in younger compared to older women (Figure 3g).

Taken together, these results demonstrate that high concentrations of acetate and butyrate inhibit neutrophil migration, while propionate does not affect migration capacity but modifies the phenotype of migrated neutrophils with increased expression of CD62L and CD54 in an age-dependent manner.

3.4. High Concentrations of SCFAs Modify Innate Secretion Profile by Neutrophils from Older Women

Following cellular activation and migration, CD62L is rapidly cleaved off and released, displaying immunomodulatory properties [53], mainly inhibiting leukocyte recruitment. Because we detected changes in activation phenotype and surface expression of CD62L after SCFA treatment of neutrophils, we next measured levels of soluble CD62L (sCD62L) in supernatants. The concentration of sCD62L was significantly decreased when neutrophils were treated with butyrate or propionate for 3 h, but not acetate (Figure 4a). Interestingly, when samples from women were stratified by age, SCFAs did not alter sCD62L secretion in neutrophils from younger women (Figure 4b), but sCD62L was significantly reduced after acetate, butyrate and propionate treatment in older women compared to untreated controls (Figure 4c).

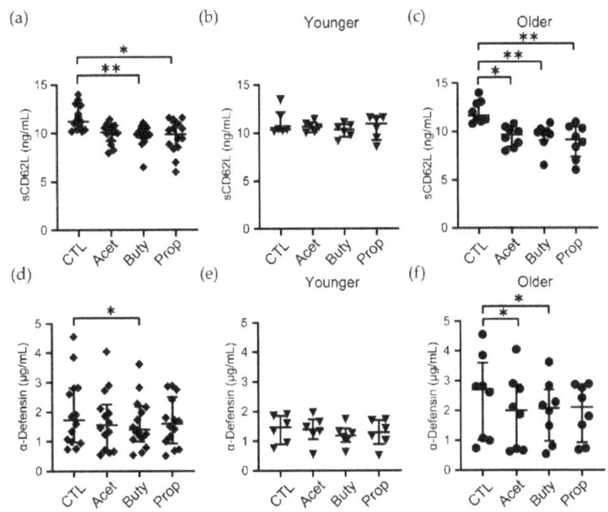

Figure 4. Pathological concentrations of SCFAs significantly reduced the release of sCD62L and α-defensin 1–3 by neutrophils from older women. (**a**) Quantification of sCD62L released by SCFA-treated neutrophils and stratified by younger (**b**) and older women (**c**), measured by ELISA. (**d**) Quantification of α-defensin 1–3 released by SCFA-treated neutrophils and stratified by younger (**e**) and older women (**f**) after 3 h, measured by ELISA. Each dot represents a different patient (n = 14; younger = 6, older = 8). Non-parametric paired Friedmann test was used; * $p \leq 0.05$, ** $p \leq 0.01$. CTL: control; Acet: acetate 25 mM; Buty: butyrate 25 mM; Prop: propionate 25 mM.

To evaluate if other innate secreted molecules could also be affected by SCFA treatment, we measured the secretion of α-defensin 1–3, an antimicrobial peptide very abundant in neutrophils that displays broad spectrum microbicidal activity, including anti-HIV activity [54]. When all women were analyzed together, only butyrate significantly reduced α-defensin 1–3 release by neutrophils after 3 h (Figure 4d). After separating women based on age, no changes were observed in the younger group (Figure 4e), while acetate and butyrate treatments significantly reduced α-defensin 1–3 release by neutrophils from older women (Figure 4f).

Next, we determined if classical pro-inflammatory molecules were modified by SCFA treatment. A panel of cytokines and chemokines was measured in supernatants from SCFAs-treated neutrophils by Luminex. However, none of the secreted cytokines and chemokines analyzed, which included the classical pro-inflammatory cytokines GM-CSF, IFN-γ, IL-1β, IL-6, IL-8 and TNFα, changed after 3 h of treatment with SCFAs (Figure S2). No significant differences were found between younger and older women, although neutrophils from older women showed a trend to secrete lower levels of most of the molecules analyzed (Figure S2).

Overall, these data demonstrate that pathological concentrations of SCFAs modify secretion profiles of innate molecules selectively in neutrophils from older women but do not alter the secretion of pro-inflammatory cytokines and chemokines.

3.5. High Concentrations of SCFAs Delay NET Release and Chemokine Secretion in Response to HIV

Since vaginal dysbiosis is known to increase the vaginal concentrations of SCFAs [55,56] and the risk of HIV acquisition [13,17,19,21], we next investigated if the high SCFA environment could affect neutrophil antiviral response to HIV. SCFA-treated neutrophils from younger and older women were challenged with HIV-VLPs, and the release of NETs was measured with time-lapse microscopy as described [40]. Consistent with our previous results, we observed that following HIV stimulation, neutrophils actively released NETs, which entrapped HIV-VLPs (Figure 5a). NET release started within minutes after HIV exposure and was sustained for at least 3 h after HIV stimulation (Figure 5b). Quantification of NET-HIV complexes demonstrated a significant upregulation after HIV exposure compared to unstimulated controls (Figure 5c).

Next, we asked if SCFA could modify NET release by themselves or in response to HIV. In the absence of HIV stimulation, high concentrations of acetate, butyrate or propionate did not increase NET release in comparison to control untreated neutrophils (Figure S3). In addition, pathological concentrations of SCFAs did not modify the overall magnitude of HIV-induced NET release during the first 3 h after stimulation, whether women were analyzed together (Figure 5d) or separated into younger and older groups (Figure 5e).

Next, we evaluated if SCFA treatment could modify the timing of NET release. HIV-induced NETs were significantly upregulated as early as 5 min following HIV stimulation in untreated and butyrate-treated neutrophils (Figure 5f, white and black symbols). In contrast, acetate- and propionate-treated neutrophils showed a 15 min delay in NET release after HIV stimulation (Figure 5f). Furthermore, propionate-treated neutrophils not only delayed but significantly decreased initial NET release following HIV stimulation compared to the HIV control condition (Figure 5f). Then, we evaluated whether SCFA treatment affected neutrophils from younger and older women differently. In the absence of SCFAs, HIV-induced NET release was detected 5 min after stimulation in neutrophils from younger women (Figure 5g) and 15 min after stimulation in neutrophils from older women (Figure 5h). Interestingly, acetate, butyrate and propionate treatments delayed 1 h HIV-induced NET release in neutrophils from younger women (Figure 5g), while butyrate induced a 30 min delay and propionate induced a 1h delay in the anti-HIV response of old neutrophils (Figure 5h).

As a control, we determined whether HIV could modify the protein level of the main receptor for SCFAs, GPR43, which is highly expressed by neutrophils [35]. HIV did not change the expression of GPR43 in neutrophils after 1h of stimulation, quantified by Western blot (Figure 5i,j).

In order to study if additional mechanisms involved in neutrophil-mediated anti-HIV responses were altered by a high concentration of SCFAs, neutrophils were stimulated with replication-competent HIV-BaL for 1 h and a selected panel of chemokines relevant for chemoattraction of HIV-target cells, and direct anti-HIV activity was measured in supernatants by Luminex [39]. We observed a significant increase in the release of IL-8 (CXCL8), MCP-1 (CCL2), MIP1α (CCL3), RANTES (CCL5), MDC (CCL22) and MIG (CXCL9) by neutrophils in response to HIV (Figure 6), while other cytokines, such as TNFα or IL-10, were undetectable (data not shown). Interestingly, HIV stimulation in the presence of a high concentration of SCFAs dampened the release of these molecules, with a significant reduction for most molecules detected in the presence of butyrate and a significant reduction in MDC secretion in the presence of butyrate and propionate (Figure 6).

Taken together, our findings indicate that a pathological concentration of SCFAs significantly delays HIV-induced NET release by neutrophils from both younger and older women and reduces the release of chemokines in response to HIV stimulation.

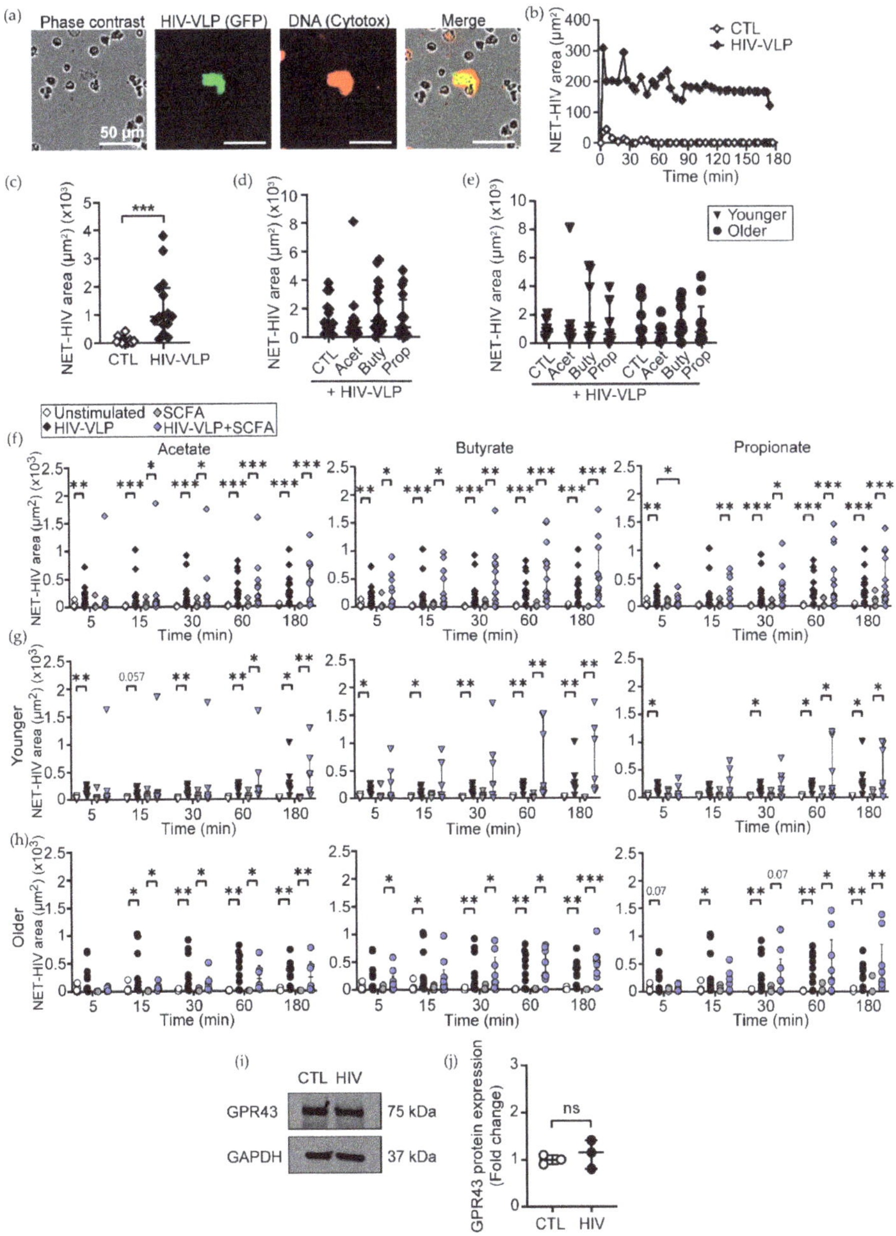

Figure 5. Pathological concentrations of SCFA induce a significant delay in HIV-induced NET release by neutrophils. (**a**) Representative image of NET formation and co-localization with HIV-VLPs, quantified with IncuCyte S3 system. (**b**) Total NET-HIV area of unstimulated controls (white dots) and HIV-VLP (black dots) is represented over time. (**c**) Quantification of HIV-NET area in the first 3 h and (**d**) after treatment with 25 mM of acetate, butyrate or propionate after stimulation with HIV-VLPs (n = 16). (**e**) Effect of age on NET release by SCFA-treated neutrophils. (**f**) Comparison of HIV-induced NET release by time intervals in all women (n = 16) and separated in (**g**) younger (n = 7)

and (**h**) older (n = 9) groups. (**i**) Representative Western blot and (**j**) neutrophil GPR43 protein quantification after 1 h of stimulation with HIV-VLPs, relative to unstimulated CTL (n = 3). Scale bar: 50 μm. Each dot represents a different patient (n = 16; younger = 7; older = 9). Wilcoxon's matched-pairs signed-rank test was used for two-group comparisons, and Kruskall–Wallis with Dunn's post-test was used to compare three or more groups. * $p \leq 0.05$, ** $p \leq 0.01$, *** $p \leq 0.001$. CTL: control; Acet: acetate 25 mM; Buty: butyrate 25 mM; Prop: propionate 25 mM.

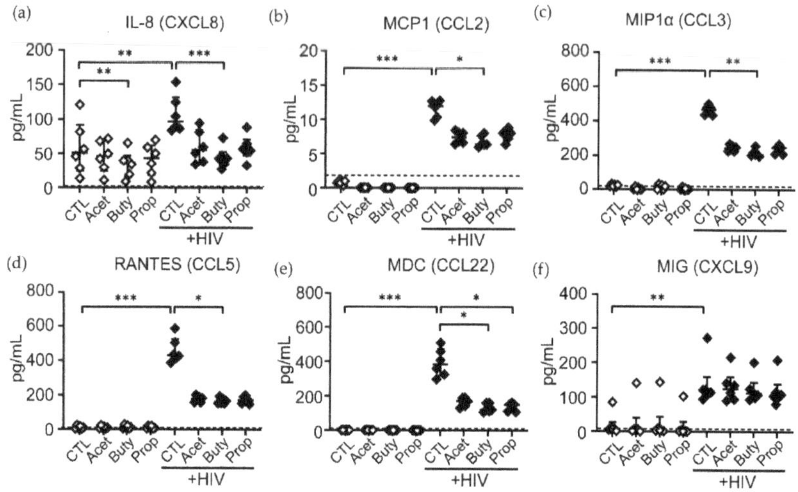

Figure 6. High concentrations of butyrate significantly reduces the release of chemokines in response to HIV stimulation. Neutrophils were stimulated with HIV-BaL in the presence or absence of SFCAs for 1 h, and cell-free supernatants were used to quantify the concentration of the following molecules by Luminex: (**a**) IL-8 (CXCL8), (**b**) MCP-1 (CCL2), (**c**) MIP1α (CCL3), (**d**) RANTES (CCL5), (**e**) MDC (CCL22) and (**f**) MIG (CXCL9). Each dot represents a different patient (n = 6). Dotted line: limit of detection. Non-parametric paired Friedmann test was used, * $p \leq 0.05$, ** $p \leq 0.01$, *** $p \leq 0.001$. CTL: control; Acet: acetate 25 mM; Buty: butyrate 25 mM; Prop: propionate 25 mM.

4. Discussion

Our study demonstrates that pathological concentrations of SCFAs modify neutrophil activation, secretion profile and anti-HIV responses in an age-dependent manner. We found that different SCFAs exert specific effects and that neutrophils from older women are more susceptible to modulation by SCFA treatment. Our findings are relevant to understanding how changes in the composition of the genital microbiome that alter the metabolome may affect neutrophil-mediated innate protection in the genital tract of younger and older women.

It is well known that women with vaginal dysbiosis and bacterial vaginosis (BV) have a shift in their vaginal microbiota, with an increased number and diversity of facultative and anaerobic bacteria [8]. Furthermore, high-diversity bacterial communities in the FRT are strongly associated with pro-inflammatory genital cytokines that activate immune cells in vivo [15,57]. Several studies have shown that there is a shift in metabolites from lactate toward mixed SCFAs during vaginal dysbiosis [30,31,55,56,58], which include a wide range in concentrations (20–140 mM) of acetate, butyrate and propionate [11,29–31].

To understand how changes in the microbiome and microbial metabolites may affect neutrophil-mediated innate protection against infections, in this study, we optimized an in vitro model to evaluate the potential effects of a high-SCFA-concentration environment on neutrophils by treating blood neutrophils with three different SCFAs (acetate, butyrate and propionate), known to be increased in conditions with vaginal dysbiosis and at concentrations described as pathological in the lower tract of the FRT [9,11,29,30]. Under these conditions, we observed modifications specific to each SCFA. Butyrate induced phenotypical changes in neutrophils, with decreased CD66b and increased CD16 and CD62L expression. CD66b is an adhesion molecule and a marker of neutrophil activation [59], while CD16 is involved in neutrophil survival [60], and CD62L is a marker for transmigration and maturation that is downregulated as neutrophils age [61,62]. Therefore, our results suggest that a high concentration of butyrate induces low activation and longer survival of non-aged mature neutrophils. Consistently, previous studies have shown upregulation of CD16 expression on vaginal neutrophils in women with BV [63] and inhibition of neutrophil apoptosis after treatment of peripheral neutrophils with high concentrations of butyrate in vitro [64]. Although Aoyama et al. [65] reported a significantly higher percentage of neutrophil apoptosis after treatment with a high concentration of butyrate, this only happened after 20 h in culture, and no differences were observed only after 3 h, which is in line with our observations. Furthermore, propionate treatment very specifically upregulated CD54 (ICAM-1) expression, which is typically upregulated in transmigrating neutrophils but not in circulating or tissue-resident neutrophils [66].

Recently, aged neutrophils have been considered important in inflammatory responses [51]. This pro-inflammatory subset, defined as CXCR4^{high} CD62L^{low} CD54^{high} neutrophils, displays increased capacity to phagocytize and migrate [51], but can mediate tissue damage by producing NETs and reactive oxygen species (ROS) under conditions of sterile inflammation [67]. In this line, preventing the recruitment of aged neutrophils has been demonstrated to be protective against tissue damage [62]. Interestingly, we observed that pathological concentrations of propionate induced a phenotype of aged neutrophils. A previous unpublished study has reported a higher number of aged neutrophils in cervicovaginal fluid and cervical cytobrush in women with vaginal dysbiosis [68], which could be deleterious for the genital mucosa under vaginal dysbiosis.

Interestingly, phenotypical changes were mainly observed in older women in the presence of SCFAs, while neutrophils from younger women only showed a minimal increase in a CD66^{low} CD16^{high} population after butyrate and propionate treatment, suggesting enhanced sensitivity to the effects of SCFA with aging. The reason for this remains unsolved, but it could be related to changes in the expression of SCFA receptors. Future studies are needed to address this question.

Upon neutrophil activation, surface expression of CD62L is quickly reduced mainly through proteolytic cleavage [53,69], which results in the release of a functionally active soluble extracellular fragment, known as sCD62L. sCD62L is detected in the plasma of healthy humans [70,71] and plays two major roles: preventing lymphocyte recirculation [72] and inhibiting transendothelial migration of other leukocytes [71]. Consistent with the increase in expression of membrane-bound CD62L in neutrophils from older women after SCFA treatment, we detected a significant reduction in the release of sCD62L. Given that sCD62L has been involved in the regulation of leukocyte adhesion and migration, the functional consequences of reduced sCD62L in the FRT of older women remain to be investigated.

In addition to changes in phenotype, we observed reduced migration of neutrophils towards SCFA-rich environments in both younger and older women. Particularly, high concentrations of acetate and butyrate, but not propionate, significantly inhibited neutrophil migration. These findings are in agreement with previous studies with human peripheral neutrophils [35] and animal models [73–75], demonstrating that pathological concentrations of SCFAs inhibit neutrophil migration. Interestingly, while propionate did not affect neutrophil migration, we describe for the first time that unstimulated neutrophils which migrated towards propionate experienced phenotypical changes with upregulation of CD54 in an age-dependent manner, suggesting that these neutrophils are prepared for transendothelial migration.

Taken together, our study and previous studies from others suggest that physiological levels of SCFAs can act as chemoattractants for immune cells and specifically for neutrophils [35,76], while pathological concentrations prevent neutrophil migration independently of age. This anti-chemotactic action could be a contributing factor to the described low presence of neutrophils in vaginal secretions from women with BV [11,37,77], although reduced neutrophil presence during BV has not been confirmed in all studies [10,63]. Importantly, while our in vitro model evaluated the individual effects of each SCFA, neutrophils in the genital tract would be exposed to a combination of SCFAs, and therefore, the overall result remains to be elucidated. However, given that both butyrate and acetate inhibited migration, we speculate that SCFAs in combination at pathological concentrations would likely inhibit neutrophil recruitment as a final result. Nevertheless, it will be important to determine which combinations and specific concentrations of SCFAs attract or inhibit neutrophil chemotaxis in vivo since some microbial alterations associated with STIs are accompanied by an increased presence of neutrophils in genital secretions, while other alterations are not.

SCFAs are known to play an important role in the host as immunomodulators [36]. Despite their well-described anti-inflammatory properties at physiological levels, SCFAs are less inhibitory when found at high concentrations [9,31,74,78]. Previous studies have reported that peripheral blood mononuclear cells (PBMCs) increased their production of pro-inflammatory cytokines, including IL-1β, IL-6 and IL-8, after 6 to 18 h of treatment with pathological concentrations of SCFAs alone [31]. In contrast to these findings, we did not observe an increase in classical pro-inflammatory cytokines following SCFA treatment for 3 h, suggesting that pathological levels of SCFAs do not enhance the pro-inflammatory secreted profile of neutrophils. We did observe, however, that butyrate and acetate treatment reduced the secretion of α-defensin 1–3, particularly in neutrophils from older women. α-Defensins 1–3 are antimicrobial peptides with broad antimicrobial activity, including anti-HIV activity [45,54,79,80], and are important for the anti-HIV activity of neutrophils [41]. Our finding of reduced α-defensin 1–3 secretion by neutrophils after SCFA treatment may indicate reduced HIV inactivation potential.

Furthermore, we detected a delay in NET release after HIV stimulation when neutrophils were challenged in the presence of SCFAs. In contrast to a previous report [81] showing increased NET release after neutrophil treatment with SCFAs in the mM concentration range, we did not observe any significant differences after SCFA treatment in the absence of HIV stimulation. The reason behind these disparate results remains to be determined but may be due to the different methodologies used to quantify NETs. We have recently demonstrated that neutrophils from the FRT and blood display direct anti-HIV activity through the release of NETs [40]. Therefore, our results suggest that tissue environments with pathological concentrations of SCFAs can reduce NET release and impair anti-HIV defense by neutrophils. Interestingly, while propionate delayed NET release in both younger and older women, butyrate and acetate preferentially affected NET release in neutrophils from younger women. The mechanisms responsible for these differences remain unknown but recognizing that the composition of the vaginal metabolome changes as women age, this finding may be relevant to understanding how anti-HIV protection changes with age.

Another novel finding in our study is that we demonstrate that HIV stimulation of neutrophils in vitro induces the secretion of the chemokines IL-8, MCP-1, MIP1α, RANTES, MDC and MIG, and this effect is dampened in the presence of a high concentration of SCFAs, particularly butyrate. This finding adds to the evidence suggesting that SCFAs reduce the ability of neutrophils to respond to HIV stimulation.

Lastly, our study has several limitations. While our results with blood neutrophils offer insight into how environments with high SCFA concentrations may modify neutrophil phenotype and function, future studies with neutrophils obtained from the genital tract of women with vaginal dysbiosis are needed to confirm our results. Further, we address the effects of individual SCFAs, but in vivo, we would expect a combination of SCFAs and different combinations depending on the microbiome alterations [32]. Studies that analyze the effects of vaginal secretions (with a known metabolome) on neutrophils are needed. We describe age-dependent effects; however, the sample size included in this study is limited, and the ages of the women recruited for our study were very polarized between younger women (18–35 years old) and older women (54–72 years old). Further studies with larger sample sizes are needed to confirm our results and to define changes in middle-aged and perimenopausal groups, also at high risk for infection in the US [82]. Finally, we have described a number of factors that are modified in a high-SCFA environment that could be relevant for inflammation and antiviral defense; however, additional factors and underlying mechanisms for the described changes remain to be investigated, such as, for example, the potential role of galectin-9 modifications in inflammatory cytokine/chemokine release and neutrophil recruitment [83–86].

5. Conclusions

Overall, our study demonstrates that pathological concentrations of SCFAs can induce phenotypic changes in neutrophils in an age-dependent manner, specifically butyrate and propionate, which differentially upregulate CD16, CD62L and CD54, classical markers of activation, maturation and aging. Furthermore, high concentrations of SCFAs reduce anti-HIV responses in neutrophils, including reduced secretion of the antimicrobial peptide α-defensin 1–3, reduced secretion of chemokines and delayed NET release following HIV stimulation. Our findings provide proof of concept that genital microbial alterations which induce an increase in SCFA concentrations may impair neutrophil physiological functions and reduce their antiviral potential.

Supplementary Materials: The following supporting information can be downloaded at: https://www.mdpi.com/article/10.3390/cells11162515/s1, Figure S1: Gating strategy and survival of neutrophils after treatment with SCFAs; Figure S2: Effect of age on cytokine and chemokine release by neutrophils after treatment with pathological concentrations of SCFAs; Figure S3: Pathological concentrations of SCFA did not change basal NET release by unstimulated neutrophils.

Author Contributions: Conceptualization, F.J.C.-S. and M.R.-G.; methodology, F.J.C.-S., S.P., L.M.d.L. and A.B.; software, F.J.C.-S., L.M.d.L. and S.P.; validation, F.J.C.-S. and M.R.-G.; formal analysis, F.J.C.-S. and M.R.-G.; investigation, F.J.C.-S. and M.R.-G.; resources, C.O., A.P. and M.R.-G.; data curation, F.J.C.-S. and M.R.-G.; writing—original draft preparation, F.J.C.-S. and M.R.-G.; writing—review and editing, F.J.C.-S. and M.R.-G.; funding acquisition, M.R.-G. All authors have read and agreed to the published version of the manuscript.

Funding: This research was funded by the National Institutes of Health (NIH), grant number R01AG060801 (M.R.-G.), and P30 AI27767 Birmingham Center for AIDS Research—Basic Science Core-Virology Facility (C.O.).

Institutional Review Board Statement: All investigations involving human subjects were conducted according to the principles expressed in the Declaration of Helsinki and carried out with the approval of the Institutional Review Board of Tufts University (protocol code: MODCR-01-11201, approved on 20 October 2014).

Informed Consent Statement: Informed consent was obtained from all subjects involved in the study.

Data Availability Statement: Not applicable.

Acknowledgments: We are grateful for the excellent technical support in the preparation of GFP-VLP by Jie Zeng, UAB Department of Medicine and CFAR. We thank the Flow Cytometry Core at Tufts University for the support.

Conflicts of Interest: The authors declare no conflict of interest.

References

1. World Health Organization. HIV/AIDS. 2021. Available online: https://www.who.int/news-room/fact-sheets/detail/hiv-aids (accessed on 1 January 2022).
2. UNAIDS. *Women and Girls and HIV*; UNAIDS: Geneva, Switzerland, 2018.
3. Govender, K.; Beckett, S.; Reddy, T.; Cowden, R.G.; Cawood, C.; Khanyile, D.; Kharsany, A.B.M.; George, G.; Puren, A. Association of HIV Intervention Uptake with HIV Prevalence in Adolescent Girls and Young Women in South Africa. *JAMA Netw. Open* **2022**, 5, e228640. [CrossRef]

4. CDC. Diagnoses of HIV infection among adults aged 50 years and older in the United States and dependent areas, 2010–2014. In *Surveillance Supplemental Report*; CDC: Atlanta, GA, USA, 2016; Volume 21.
5. Tavoschi, L.; Gomes Dias, J.; Pharris, A. New HIV diagnoses among adults aged 50 years or older in 31 European countries, 2004–2015: An analysis of surveillance data. *Lancet HIV* **2017**, *4*, e514–e521. [CrossRef]
6. He, W.; Goodkind, D.; Kowal, P. *An Aging World: 2015*; US Census Bureau: Suitland, MD, USA, 2016.
7. Rodriguez-Garcia, M.; Patel, M.V.; Shen, Z.; Wira, C.R. The impact of aging on innate and adaptive immunity in the human female genital tract. *Aging Cell* **2021**, *20*, e13361. [CrossRef] [PubMed]
8. Amabebe, E.; Anumba, D.O.C. The Vaginal Microenvironment: The Physiologic Role of Lactobacilli. *Front. Med.* **2018**, *5*, 181. [CrossRef] [PubMed]
9. Delgado-Diaz, D.J.; Tyssen, D.; Hayward, J.A.; Gugasyan, R.; Hearps, A.C.; Tachedjian, G. Distinct Immune Responses Elicited From Cervicovaginal Epithelial Cells by Lactic Acid and Short Chain Fatty Acids Associated With Optimal and Non-optimal Vaginal Microbiota. *Front. Cell. Infect. Microbiol.* **2019**, *9*, 446. [CrossRef]
10. Gustin, A.; Cromarty, R.; Schifanella, L.; Klatt, N.R. Microbial mismanagement: How inadequate treatments for vaginal dysbiosis drive the HIV epidemic in women. *Semin. Immunol.* **2021**, *51*, 101482. [CrossRef]
11. Al-Mushrif, S.; Eley, A.; Jones, B.M. Inhibition of chemotaxis by organic acids from anaerobes may prevent a purulent response in bacterial vaginosis. *J. Med. Microbiol.* **2000**, *49*, 1023–1030. [CrossRef]
12. Chen, C.; Song, X.; Wei, W.; Zhong, H.; Dai, J.; Lan, Z.; Li, F.; Yu, X.; Feng, Q.; Wang, Z.; et al. The microbiota continuum along the female reproductive tract and its relation to uterine-related diseases. *Nat. Commun.* **2017**, *8*, 875. [CrossRef]
13. Gosmann, C.; Anahtar, M.N.; Handley, S.A.; Farcasanu, M.; Abu-Ali, G.; Bowman, B.A.; Padavattan, N.; Desai, C.; Droit, L.; Moodley, A.; et al. Lactobacillus-Deficient Cervicovaginal Bacterial Communities Are Associated with Increased HIV Acquisition in Young South African Women. *Immunity* **2017**, *46*, 29–37. [CrossRef]
14. Tyssen, D.; Wang, Y.Y.; Hayward, J.A.; Agius, P.A.; DeLong, K.; Aldunate, M.; Ravel, J.; Moench, T.R.; Cone, R.A.; Tachedjian, G. Anti-HIV-1 Activity of Lactic Acid in Human Cervicovaginal Fluid. *mSphere* **2018**, *3*, e00055-18. [CrossRef]
15. Anahtar, M.N.; Byrne, E.H.; Doherty, K.E.; Bowman, B.A.; Yamamoto, H.S.; Soumillon, M.; Padavattan, N.; Ismail, N.; Moodley, A.; Sabatini, M.E.; et al. Cervicovaginal bacteria are a major modulator of host inflammatory responses in the female genital tract. *Immunity* **2015**, *42*, 965–976. [CrossRef] [PubMed]
16. Zhou, X.; Brown, C.J.; Abdo, Z.; Davis, C.C.; Hansmann, M.A.; Joyce, P.; Foster, J.A.; Forney, L.J. Differences in the composition of vaginal microbial communities found in healthy Caucasian and black women. *ISME J.* **2007**, *1*, 121–133. [CrossRef] [PubMed]
17. McKinnon, L.R.; Achilles, S.L.; Bradshaw, C.S.; Burgener, A.; Crucitti, T.; Fredricks, D.N.; Jaspan, H.B.; Kaul, R.; Kaushic, C.; Klatt, N.; et al. The Evolving Facets of Bacterial Vaginosis: Implications for HIV Transmission. *AIDS Res. Hum. Retrovir.* **2019**, *35*, 219–228. [CrossRef]
18. Eastment, M.C.; McClelland, R.S. Vaginal microbiota and susceptibility to HIV. *AIDS* **2018**, *32*, 687–698. [CrossRef] [PubMed]
19. Sewankambo, N.; Gray, R.H.; Wawer, M.J.; Paxton, L.; McNaim, D.; Wabwire-Mangen, F.; Serwadda, D.; Li, C.; Kiwanuka, N.; Hillier, S.L.; et al. HIV-1 infection associated with abnormal vaginal flora morphology and bacterial vaginosis. *Lancet* **1997**, *350*, 546–550. [CrossRef]
20. Salazar, A.S.; Nogueira, N.F.; Rodriguez, V.J.; Mantero, A.; Cherenack, E.M.; Raccamarich, P.; Maddalon, M.; Brophy, T.; Montgomerie, E.; Klatt, N.R.; et al. A Syndemic Approach to Explore Factors Associated with Bacterial Vaginosis. *AIDS Behav.* **2022**, *26*, 3110–3118. [CrossRef]
21. Low, N.; Chersich, M.F.; Schmidlin, K.; Egger, M.; Francis, S.C.; van de Wijgert, J.H.; Hayes, R.J.; Baeten, J.M.; Brown, J.; Delany-Moretlwe, S.; et al. Intravaginal practices, bacterial vaginosis, and HIV infection in women: Individual participant data meta-analysis. *PLoS Med.* **2011**, *8*, e1000416. [CrossRef]
22. Gupta, S.; Kumar, N.; Singhal, N.; Kaur, R.; Manektala, U. Vaginal microflora in postmenopausal women on hormone replacement therapy. *Indian J. Pathol. Microbiol.* **2006**, *49*, 457–461.
23. Thurman, A.R.; Yousefieh, N.; Chandra, N.; Kimble, T.; Asin, S.; Rollenhagen, C.; Anderson, S.M.; Herold, B.C.; Freiermuth, J.L.; Starkman, B.S.; et al. Comparison of Mucosal Markers of Human Immunodeficiency Virus Susceptibility in Healthy Premenopausal versus Postmenopausal Women. *AIDS Res. Hum. Retrovir.* **2017**, *33*, 807–819. [CrossRef]
24. Hudson, P.L.; Ling, W.; Wu, M.C.; Hayward, M.R.; Mitchell, A.J.; Larson, J.; Guthrie, K.A.; Reed, S.D.; Kwon, D.S.; Mitchell, C.M. Comparison of the Vaginal Microbiota in Postmenopausal Black and White Women. *J. Infect. Dis.* **2021**, *224*, 1945–1949. [CrossRef]
25. Jais, M.; Younes, N.; Chapman, S.; Cu-Uvin, S.; Ghosh, M. Reduced levels of genital tract immune biomarkers in postmenopausal women: Implications for HIV acquisition. *Am. J. Obs. Gynecol.* **2016**, *215*, 324.e1–324.e10. [CrossRef] [PubMed]
26. Rodriguez-Garcia, M.; Connors, K.; Ghosh, M. HIV Pathogenesis in the Human Female Reproductive Tract. *Curr. HIV/AIDS Rep.* **2021**, *18*, 139–156. [CrossRef] [PubMed]
27. Rodriguez-Garcia, M.; Barr, F.D.; Crist, S.G.; Fahey, J.V.; Wira, C.R. Phenotype and susceptibility to HIV infection of CD4+ Th17 cells in the human female reproductive tract. *Mucosal Immunol.* **2014**, *7*, 1375–1385. [CrossRef] [PubMed]
28. De Lara, L.M.; Parthasarathy, R.S.; Rodriguez-Garcia, M. Mucosal Immunity and HIV Acquisition in Women. *Curr. Opin. Physiol.* **2021**, *19*, 32–38. [CrossRef] [PubMed]
29. Chaudry, A.N.; Travers, P.J.; Yuenger, J.; Colletta, L.; Evans, P.; Zenilman, J.M.; Tummon, A. Analysis of vaginal acetic acid in patients undergoing treatment for bacterial vaginosis. *J. Clin. Microbiol.* **2004**, *42*, 5170–5175. [CrossRef]

30. Mirmonsef, P.; Gilbert, D.; Zariffard, M.R.; Hamaker, B.R.; Kaur, A.; Landay, A.L.; Spear, G.T. The effects of commensal bacteria on innate immune responses in the female genital tract. *Am. J. Reprod. Immunol.* **2011**, *65*, 190–195. [CrossRef]
31. Mirmonsef, P.; Zariffard, M.R.; Gilbert, D.; Makinde, H.; Landay, A.L.; Spear, G.T. Short-chain fatty acids induce pro-inflammatory cytokine production alone and in combination with toll-like receptor ligands. *Am. J. Reprod. Immunol.* **2012**, *67*, 391–400. [CrossRef]
32. Baldewijns, S.; Sillen, M.; Palmans, I.; Vandecruys, P.; Van Dijck, P.; Demuyser, L. The Role of Fatty Acid Metabolites in Vaginal Health and Disease: Application to Candidiasis. *Front. Microbiol.* **2021**, *12*, 705779. [CrossRef]
33. Brown, A.J.; Goldsworthy, S.M.; Barnes, A.A.; Eilert, M.M.; Tcheang, L.; Daniels, D.; Muir, A.I.; Wigglesworth, M.J.; Kinghorn, I.; Fraser, N.J.; et al. The Orphan G protein-coupled receptors GPR41 and GPR43 are activated by propionate and other short chain carboxylic acids. *J. Biol. Chem.* **2003**, *278*, 11312–11319. [CrossRef]
34. Ahmed, K.; Tunaru, S.; Offermanns, S. GPR109A, GPR109B and GPR81, a family of hydroxy-carboxylic acid receptors. *Trends Pharmacol. Sci.* **2009**, *30*, 557–562. [CrossRef]
35. Le Poul, E.; Loison, C.; Struyf, S.; Springael, J.Y.; Lannoy, V.; Decobecq, M.E.; Brezillon, S.; Dupriez, V.; Vassart, G.; Van Damme, J.; et al. Functional characterization of human receptors for short chain fatty acids and their role in polymorphonuclear cell activation. *J. Biol. Chem.* **2003**, *278*, 25481–25489. [CrossRef] [PubMed]
36. Koh, A.; De Vadder, F.; Kovatcheva-Datchary, P.; Backhed, F. From Dietary Fiber to Host Physiology: Short-Chain Fatty Acids as Key Bacterial Metabolites. *Cell* **2016**, *165*, 1332–1345. [CrossRef] [PubMed]
37. Amabebe, E.; Anumba, D.O.C. Female Gut and Genital Tract Microbiota-Induced Crosstalk and Differential Effects of Short-Chain Fatty Acids on Immune Sequelae. *Front. Immunol.* **2020**, *11*, 2184. [CrossRef] [PubMed]
38. Dabee, S.; Passmore, J.S.; Heffron, R.; Jaspan, H.B. The Complex Link between the Female Genital Microbiota, Genital Infections, and Inflammation. *Infect. Immun.* **2021**, *89*, e00487-20. [CrossRef]
39. Selhorst, P.; Masson, L.; Ismail, S.D.; Samsunder, N.; Garrett, N.; Mansoor, L.E.; Abdool Karim, Q.; Abdool Karim, S.S.; Passmore, J.S.; Williamson, C. Cervicovaginal Inflammation Facilitates Acquisition of Less Infectious HIV Variants. *Clin. Infect. Dis.* **2017**, *64*, 79–82. [CrossRef]
40. Barr, F.D.; Ochsenbauer, C.; Wira, C.R.; Rodriguez-Garcia, M. Neutrophil extracellular traps prevent HIV infection in the female genital tract. *Mucosal Immunol.* **2018**, *11*, 1420–1428. [CrossRef]
41. Saitoh, T.; Komano, J.; Saitoh, Y.; Misawa, T.; Takahama, M.; Kozaki, T.; Uehata, T.; Iwasaki, H.; Omori, H.; Yamaoka, S.; et al. Neutrophil extracellular traps mediate a host defense response to human immunodeficiency virus-1. *Cell Host Microbe* **2012**, *12*, 109–116. [CrossRef]
42. Brinkmann, V.; Reichard, U.; Goosmann, C.; Fauler, B.; Uhlemann, Y.; Weiss, D.S.; Weinrauch, Y.; Zychlinsky, A. Neutrophil extracellular traps kill bacteria. *Science* **2004**, *303*, 1532–1535. [CrossRef]
43. Arnold, K.B.; Burgener, A.; Birse, K.; Romas, L.; Dunphy, L.J.; Shahabi, K.; Abou, M.; Westmacott, G.R.; McCorrister, S.; Kwatampora, J.; et al. Increased levels of inflammatory cytokines in the female reproductive tract are associated with altered expression of proteases, mucosal barrier proteins, and an influx of HIV-susceptible target cells. *Mucosal Immunol.* **2016**, *9*, 194–205. [CrossRef]
44. Fan, S.R.; Liu, X.P.; Liao, Q.P. Human defensins and cytokines in vaginal lavage fluid of women with bacterial vaginosis. *Int. J. Gynaecol. Obs.* **2008**, *103*, 50–54. [CrossRef]
45. Levinson, P.; Kaul, R.; Kimani, J.; Ngugi, E.; Moses, S.; MacDonald, K.S.; Broliden, K.; Hirbod, T.; Kibera, H.I.V.S.G. Levels of innate immune factors in genital fluids: Association of alpha defensins and LL-37 with genital infections and increased HIV acquisition. *AIDS* **2009**, *23*, 309–317. [CrossRef] [PubMed]
46. Forthal, D.N.; Landucci, G.; Ding, H.; Kappes, J.C.; Wang, A.; Thung, I.; Phan, T. IgG2 inhibits HIV-1 internalization by monocytes, and IgG subclass binding is affected by gp120 glycosylation. *AIDS* **2011**, *25*, 2099–2104. [CrossRef] [PubMed]
47. Li, Y.; Svehla, K.; Mathy, N.L.; Voss, G.; Mascola, J.R.; Wyatt, R. Characterization of antibody responses elicited by human immunodeficiency virus type 1 primary isolate trimeric and monomeric envelope glycoproteins in selected adjuvants. *J. Virol.* **2006**, *80*, 1414–1426. [CrossRef] [PubMed]
48. Gartner, S.; Markovits, P.; Markovitz, D.M.; Kaplan, M.H.; Gallo, R.C.; Popovic, M. The role of mononuclear phagocytes in HTLV-III/LAV infection. *Science* **1986**, *233*, 215–219. [CrossRef]
49. Rodriguez-Garcia, M.; Biswas, N.; Patel, M.V.; Barr, F.D.; Crist, S.G.; Ochsenbauer, C.; Fahey, J.V.; Wira, C.R. Estradiol reduces susceptibility of CD4+ T cells and macrophages to HIV-infection. *PLoS ONE* **2013**, *8*, e62069. [CrossRef]
50. Mortaz, E.; Alipoor, S.D.; Adcock, I.M.; Mumby, S.; Koenderman, L. Update on Neutrophil Function in Severe Inflammation. *Front. Immunol.* **2018**, *9*, 2171. [CrossRef]
51. Uhl, B.; Vadlau, Y.; Zuchtriegel, G.; Nekolla, K.; Sharaf, K.; Gaertner, F.; Massberg, S.; Krombach, F.; Reichel, C.A. Aged neutrophils contribute to the first line of defense in the acute inflammatory response. *Blood* **2016**, *128*, 2327–2337. [CrossRef]
52. Schnoor, M.; Alcaide, P.; Voisin, M.B.; van Buul, J.D. Crossing the Vascular Wall: Common and Unique Mechanisms Exploited by Different Leukocyte Subsets during Extravasation. *Mediat. Inflamm.* **2015**, *2015*, 946509. [CrossRef]
53. Lee, D.; Schultz, J.B.; Knauf, P.A.; King, M.R. Mechanical shedding of L-selectin from the neutrophil surface during rolling on sialyl Lewis x under flow. *J. Biol. Chem.* **2007**, *282*, 4812–4820. [CrossRef]
54. Chang, T.L.; Vargas, J., Jr.; DelPortillo, A.; Klotman, M.E. Dual role of alpha-defensin-1 in anti-HIV-1 innate immunity. *J. Clin. Investig.* **2005**, *115*, 765–773. [CrossRef]

55. Yeoman, C.J.; Thomas, S.M.; Miller, M.E.; Ulanov, A.V.; Torralba, M.; Lucas, S.; Gillis, M.; Cregger, M.; Gomez, A.; Ho, M.; et al. A multi-omic systems-based approach reveals metabolic markers of bacterial vaginosis and insight into the disease. *PLoS ONE* **2013**, *8*, e56111. [CrossRef] [PubMed]
56. Vitali, B.; Cruciani, F.; Picone, G.; Parolin, C.; Donders, G.; Laghi, L. Vaginal microbiome and metabolome highlight specific signatures of bacterial vaginosis. *Eur. J. Clin. Microbiol. Infect. Dis.* **2015**, *34*, 2367–2376. [CrossRef] [PubMed]
57. Masson, L.; Mlisana, K.; Little, F.; Werner, L.; Mkhize, N.N.; Ronacher, K.; Gamieldien, H.; Williamson, C.; McKinnon, L.R.; Walzl, G.; et al. Defining genital tract cytokine signatures of sexually transmitted infections and bacterial vaginosis in women at high risk of HIV infection: A cross-sectional study. *Sex. Transm. Infect.* **2014**, *90*, 580–587. [CrossRef] [PubMed]
58. Srinivasan, S.; Morgan, M.T.; Fiedler, T.L.; Djukovic, D.; Hoffman, N.G.; Raftery, D.; Marrazzo, J.M.; Fredricks, D.N. Metabolic signatures of bacterial vaginosis. *MBio* **2015**, *6*, e00204-15. [CrossRef] [PubMed]
59. Yoon, J.; Terada, A.; Kita, H. CD66b regulates adhesion and activation of human eosinophils. *J. Immunol.* **2007**, *179*, 8454–8462. [CrossRef]
60. Dransfield, I.; Buckle, A.M.; Savill, J.S.; McDowall, A.; Haslett, C.; Hogg, N. Neutrophil apoptosis is associated with a reduction in CD16 (Fc gamma RIII) expression. *J. Immunol.* **1994**, *153*, 1254–1263.
61. Casanova-Acebes, M.; Pitaval, C.; Weiss, L.A.; Nombela-Arrieta, C.; Chevre, R.; A-González, N.; Kunisaki, Y.; Zhang, D.; van Rooijen, N.; Silberstein, L.E.; et al. Rhythmic modulation of the hematopoietic niche through neutrophil clearance. *Cell* **2013**, *153*, 1025–1035. [CrossRef]
62. Zhang, D.; Chen, G.; Manwani, D.; Mortha, A.; Xu, C.; Faith, J.J.; Burk, R.D.; Kunisaki, Y.; Jang, J.E.; Scheiermann, C.; et al. Neutrophil ageing is regulated by the microbiome. *Nature* **2015**, *525*, 528–532. [CrossRef]
63. Beghini, J.; Giraldo, P.C.; Riboldi, R.; Amaral, R.L.; Eleuterio, J., Jr.; Witkin, S.S.; Guimaraes, F. Altered CD16 expression on vaginal neutrophils from women with vaginitis. *Eur. J. Obs. Gynecol. Reprod. Biol.* **2013**, *167*, 96–99. [CrossRef]
64. Moulding, D.; Quayle, J.A.; Stringer, R.E.; Hart, C.A.; Edwards, S.W. Regulation of neutrophil apoptosis by sodium butyrate. *Biologicals* **1996**, *24*, 301–306. [CrossRef]
65. Aoyama, M.; Kotani, J.; Usami, M. Butyrate and propionate induced activated or non-activated neutrophil apoptosis via HDAC inhibitor activity but without activating GPR-41/GPR-43 pathways. *Nutrition* **2010**, *26*, 653–661. [CrossRef] [PubMed]
66. Silvestre-Roig, C.; Hidalgo, A.; Soehnlein, O. Neutrophil heterogeneity: Implications for homeostasis and pathogenesis. *Blood* **2016**, *127*, 2173–2181. [CrossRef] [PubMed]
67. Kolaczkowska, E.; Jenne, C.N.; Surewaard, B.G.; Thanabalasuriar, A.; Lee, W.Y.; Sanz, M.J.; Mowen, K.; Opdenakker, G.; Kubes, P. Molecular mechanisms of NET formation and degradation revealed by intravital imaging in the liver vasculature. *Nat. Commun.* **2015**, *6*, 6673. [CrossRef] [PubMed]
68. Farr, C.; Noel-Romas, L.; Birse, K.; De Leon, M.; Kratzer, K.; Perner, M.; Ayele, H.; Wiwchar, C.; Pymar, H.; Berard, A.R.; et al. Increased cervical neutrophil survival during bacterial vaginosis in Canadian women from the THRIVE study. *J. Immunol.* **2020**, *204*, 157.13.
69. Kahn, J.; Ingraham, R.H.; Shirley, F.; Migaki, G.I.; Kishimoto, T.K. Membrane proximal cleavage of L-selectin: Identification of the cleavage site and a 6-kD transmembrane peptide fragment of L-selectin. *J. Cell. Biol.* **1994**, *125*, 461–470. [CrossRef]
70. Ivetic, A. A head-to-tail view of L-selectin and its impact on neutrophil behaviour. *Cell. Tissue Res.* **2018**, *371*, 437–453. [CrossRef]
71. Schleiffenbaum, B.; Spertini, O.; Tedder, T.F. Soluble L-selectin is present in human plasma at high levels and retains functional activity. *J. Cell. Biol.* **1992**, *119*, 229–238. [CrossRef]
72. Ivetic, A.; Hoskins Green, H.L.; Hart, S.J. L-selectin: A Major Regulator of Leukocyte Adhesion, Migration and Signaling. *Front. Immunol.* **2019**, *10*, 1068. [CrossRef]
73. Correa, R.O.; Vieira, A.; Sernaglia, E.M.; Lancellotti, M.; Vieira, A.T.; Avila-Campos, M.J.; Rodrigues, H.G.; Vinolo, M.A.R. Bacterial short-chain fatty acid metabolites modulate the inflammatory response against infectious bacteria. *Cell. Microbiol.* **2017**, *19*, e12720. [CrossRef]
74. Vinolo, M.A.; Rodrigues, H.G.; Hatanaka, E.; Hebeda, C.B.; Farsky, S.H.; Curi, R. Short-chain fatty acids stimulate the migration of neutrophils to inflammatory sites. *Clin. Sci.* **2009**, *117*, 331–338. [CrossRef]
75. Vinolo, M.A.; Ferguson, G.J.; Kulkarni, S.; Damoulakis, G.; Anderson, K.; Bohlooly, Y.M.; Stephens, L.; Hawkins, P.T.; Curi, R. SCFAs induce mouse neutrophil chemotaxis through the GPR43 receptor. *PLoS ONE* **2011**, *6*, e21205. [CrossRef] [PubMed]
76. Sina, C.; Gavrilova, O.; Forster, M.; Till, A.; Derer, S.; Hildebrand, F.; Raabe, B.; Chalaris, A.; Scheller, J.; Rehmann, A.; et al. G protein-coupled receptor 43 is essential for neutrophil recruitment during intestinal inflammation. *J. Immunol.* **2009**, *183*, 7514–7522. [CrossRef] [PubMed]
77. Amabebe, E.; Anumba, D.O.C. Mechanistic Insights into Immune Suppression and Evasion in Bacterial Vaginosis. *Curr. Microbiol.* **2022**, *79*, 84. [CrossRef]
78. Vinolo, M.A.; Hatanaka, E.; Lambertucci, R.H.; Newsholme, P.; Curi, R. Effects of short chain fatty acids on effector mechanisms of neutrophils. *Cell. Biochem. Funct.* **2009**, *27*, 48–55. [CrossRef]
79. Klotman, M.E.; Chang, T.L. Defensins in innate antiviral immunity. *Nat. Rev. Immunol.* **2006**, *6*, 447–456. [CrossRef] [PubMed]
80. Rodriguez-Garcia, M.; Climent, N.; Oliva, H.; Casanova, V.; Franco, R.; Leon, A.; Gatell, J.M.; Garcia, F.; Gallart, T. Increased alpha-defensins 1-3 production by dendritic cells in HIV-infected individuals is associated with slower disease progression. *PLoS ONE* **2010**, *5*, e9436. [CrossRef] [PubMed]

81. Iniguez-Gutierrez, L.; Godinez-Mendez, L.A.; Fafutis-Morris, M.; Padilla-Arellano, J.R.; Corona-Rivera, A.; Bueno-Topete, M.R.; Rojas-Rejon, O.A.; Delgado-Rizo, V. Physiological concentrations of short-chain fatty acids induce the formation of neutrophil extracellular traps in vitro. *Int. J. Immunopathol. Pharm.* **2020**, *34*, 2058738420958949. [CrossRef]
82. Andany, N.; Kennedy, V.L.; Aden, M.; Loutfy, M. Perspectives on menopause and women with HIV. *Int. J. Womens Health* **2016**, *8*, 1–22. [CrossRef]
83. Iqbal, A.J.; Krautter, F.; Blacksell, I.A.; Wright, R.D.; Austin-Williams, S.N.; Voisin, M.B.; Hussain, M.T.; Law, H.L.; Niki, T.; Hirashima, M.; et al. Galectin-9 mediates neutrophil capture and adhesion in a CD44 and beta2 integrin-dependent manner. *FASEB J.* **2022**, *36*, e22065. [CrossRef] [PubMed]
84. Fichorova, R.N.; DeLong, A.K.; Cu-Uvin, S.; King, C.C.; Jamieson, D.J.; Klein, R.S.; Sobel, J.D.; Vlahov, D.; Yamamoto, H.S.; Mayer, K.H. Protozoan-Viral-Bacterial Co-Infections Alter Galectin Levels and Associated Immunity Mediators in the Female Genital Tract. *Front. Cell. Infect. Microbiol.* **2021**, *11*, 649940. [CrossRef]
85. Dunsmore, G.; Rosero, E.P.; Shahbaz, S.; Santer, D.M.; Jovel, J.; Lacy, P.; Houston, S.; Elahi, S. Neutrophils promote T-cell activation through the regulated release of CD44-bound Galectin-9 from the cell surface during HIV infection. *PLoS Biol.* **2021**, *19*, e3001387. [CrossRef] [PubMed]
86. Bozorgmehr, N.; Mashhouri, S.; Perez Rosero, E.; Xu, L.; Shahbaz, S.; Sligl, W.; Osman, M.; Kutsogiannis, D.J.; MacIntyre, E.; O'Neil, C.R.; et al. Galectin-9, a Player in Cytokine Release Syndrome and a Surrogate Diagnostic Biomarker in SARS-CoV-2 Infection. *MBio* **2021**, *12*, e00384-21. [CrossRef] [PubMed]

cells

Article

Dual Oxidase, a Hydrogen-Peroxide-Producing Enzyme, Regulates Neuronal Oxidative Damage and Animal Lifespan in *Drosophila melanogaster*

Minwoo Baek, Wijeong Jang and Changsoo Kim *

School of Biological Sciences and Technology, Chonnam National University, Gwangju 61186, Korea; minwoo.back@gmail.com (M.B.); kimoe@naver.com (W.J.)
* Correspondence: changgk2001@hanmail.net

Abstract: Reducing the oxidative stress in neurons extends lifespan in *Drosophila melanogaster*, highlighting the crucial role of neuronal oxidative damage in lifespan determination. However, the source of the reactive oxygen species (ROS) that provoke oxidative stress in neurons is not clearly defined. Here, we identify *dual oxidase* (*duox*), a calcium-activated ROS-producing enzyme, as a lifespan determinant. Due to the lethality of *duox* homozygous mutants, we employed a *duox* heterozygote that exhibited normal appearance and movement. We found that *duox* heterozygous male flies, which were isogenized with control flies, demonstrated extended lifespan. Neuronal knockdown experiments further suggested that *duox* is crucial to oxidative stress in neurons. Our findings suggest *duox* to be a source of neuronal oxidative stress associated with animal lifespan.

Keywords: lifespan; duox; ROS; *Drosophila melanogaster*

Citation: Baek, M.; Jang, W.; Kim, C. Dual Oxidase, a Hydrogen-Peroxide-Producing Enzyme, Regulates Neuronal Oxidative Damage and Animal Lifespan in *Drosophila melanogaster*. *Cells* **2022**, *11*, 2059. https://doi.org/10.3390/cells11132059

Academic Editors: Nicole Wagner and Kay-Dietrich Wagner

Received: 28 May 2022
Accepted: 27 June 2022
Published: 29 June 2022

1. Introduction

Hydrogen peroxide (H_2O_2) is a reactive oxygen species (ROS) that functions in signaling pathways by specifically oxidizing redox-sensitive proteins in diverse biological processes [1–4]. However, the upregulated or dysregulated production of H_2O_2 in combination with diminished anti-ROS activity confers oxidative stress that abrogates redox-dependent signaling or nonspecifically and irreversibly damages macromolecules, in turn accelerating the aging process and decreasing lifespan [5–9]. Neurons are particularly susceptible to oxidative stress, and the elevation of the antioxidant power in neurons has been shown to extend lifespan [10–14]. Various anti-ROS enzymes and anti-ROS transcriptional networks are well known, yet the source of ROS production that provokes oxidative stress in neurons is not yet completely understood.

ROS production is inducible by members of the NADPH oxidase (NOX) family; such induced production is distinct from the steady-state ROS production that occurs through oxidative respiration in mitochondria [15]. Dual oxidase (Duox) is a member of the Nox family, with an EF-hand motif that is activated by intracellular calcium to produce H_2O_2 [1,16,17]. It is widely expressed in the nervous system in rats, zebrafish, and *Drosophila* [18–21]. Inflammation, wounds, and various signals that increase intracellular calcium ion concentration all activate *duox* [22–25] and, in particular, UV irradiation, ROS, and p38 MAPK signaling increase its expression [18,26–30]. *Duox* is known to mediate pain hypersensitivity due to UV-induced damage [31] and, in the CNS, axon regeneration after wounds [32,33]. However, the full role of *duox* in the nervous system still remains largely unknown. In *Drosophila*, there is only a single *duox* gene, in contrast to the two genes present in mammals [21]. Here, we show that through its generation of H_2O_2, *duox* contributes to oxidative stress in neurons and is a determinant of *Drosophila melanogaster* lifespan.

2. Materials and Methods

2.1. Fly Strains and Maintenance

Tub-Gal4, elav-Gal4, Mef-Gal4, UAS-duox^{RNAi} (#38907 and #32903), and *duox^{KG07745}* were purchased from the Bloomington Drosophila Stock Center. *UAS-duox^{RNAi}* (#38907 and #32903) and *duox^{KG07745}* were outcrossed with w^{1118} for ten generations to eliminate background effects. Flies were raised on standard cornmeal fly food at 25 °C and 50% humidity in a 12-hour light/dark-cycle incubator.

2.2. Lifespan and Oxidative Stress Assays

For lifespan assays, 20 flies per vial were transferred to fresh food vials every two days. Dead flies were counted every day. For survival under oxidative stress, 5-day-old flies were starved for 6 h at 25 °C in vials containing water-soaked tissues and then transferred to vials containing either normal food with 5 mM methyl viologen hydrate (paraquat) or 5% sucrose-agar medium with 5% hydrogen peroxide [34]. Dead flies were counted every 12 h.

2.3. Locomotive Activity

Negative geotactic movement assays were performed as described previously [35]. Briefly, a vial (9.5 cm × 2 cm) containing 10 flies was tapped to send the flies to the bottom. Flies that crossed the 8-centimeter line from the bottom within 10 s were scored. In total, 100 male flies were assessed in 10 vials (10 flies per vial).

2.4. RT-PCR and Real-Time PCR

Total RNA was extracted from 20 flies using Trizol reagent (Invitrogen, Grand Island, NY, USA) and cDNA was obtained by reverse transcription (TOPscript RT Mix; Enzynomics, Seoul, Korea). Real-time PCR was conducted using SYBR green PCR Master Mix on the ABI PRISM 7500 system (Applied Biosystem, Foster City, CA, USA). The primer pairs used were as follows. *Duox*: CAGACCGAGAAACAGCGCTAC, AAACAGCCGGCTGAGCCTGCG; *rp49*: CACCAGGAACTTCTTGAATCCGG, AGATCGTGAAGAAGCGCACC; *catalase*: CGGCTTCCAATCAGTTGAT, GATGTGAACTTCCTGGATGAG; *dSOD1*:CAAGGGCACGGT TTTCTTC, TCCGGACCGCACTTCAATC; *dSOD2*: AATTTCGCAAACTGCAAGC, ACCACCAAGCTGATTCAGC.

2.5. Determination of Total ROS Levels

Total ROS was measured as previously described [36]. Briefly, 10 flies or the heads of 50 flies were homogenized in 200 μL of 50 mM Tris-HCl, pH 7.4, on ice. Extracts were centrifuged at 13,200 rpm for 10 min at 4 °C. Homogenates were incubated with H_2DCFDA (Life Technologies, Carlsbad, CA, USA) at 5 μM in a 200-microliter reaction volume at 37 °C for 15 min in darkness, carried out on 96-well plates. Florescence intensity was monitored on a Gemini XPS fluorescence microplate reader (Molecular Devices, Sunnyvale, CA, USA) set for 488 nm excitation and 520 nm emission. Florescence intensity was normalized by protein amount measured by Bradford method [37]. All experiments were carried out with three biological replicates.

2.6. Determination of H_2O_2 Levels

H_2O_2 was measured as previously described [36]. Briefly, 10 flies were homogenized in phosphate saline buffer (50 mM, pH 7.4) containing aminotriazol (2 mg/mL) and incubated at 4 °C for 15 min. The extracts were centrifuged at 13,200 rpm for 10 min at 4 °C to obtain supernatants. H_2O_2 levels were measured using a hydrogen peroxide assay kit (Cayman Chemical, Ann Arbor, MI, USA) on a microplate reader (VersaMax; Molecular Devices) set to 590 nm. H_2O_2 levels were normalized according to protein amount measured by Bradford method [37]. All experiments were carried out with three biological replicates.

2.7. Determination of Protein Oxidation

Carbonyl groups in proteins were detected using an oxyblot protein oxidation detection kit (Milipore, Billerica, MA, USA) according to the manufacturer's instructions. Briefly, proteins were extracted from 10 flies or ~50 fly heads using RIPA buffer, reacted with 2,4-dinitophenylhydrazine (5 mM) at 25 °C for 100 min, and then subjected to SDS/PAGE and Western blotting using an antibody to the dinitrophenyl moiety. The blots were developed with ECL (Amersham, Buckinghamshire, UK) and the images were captured with a Vilber Lourmat Fusion FX (Vilber Lourmat, Eberhardzell, Baden-Württemberg, Germany). Band intensities were determined with the Fusion program (Vilber Lourmat). Finally, blots were stripped with a stripping buffer (Thermo Fisher scientific, Waltham, MA, USA), and then by a separate Western blot carried out using β-actin antibody (Santa Cruz, Dallas, TX, USA), which was used as loading control. All experiments were carried out with three biological replicates.

2.8. Determination of DNA Oxidation

The level of 8-hydroxydeomyguanosine (8-OHdG) was measured using the OxiSelect™ Oxidative DNA Damage ELISA kit (Cell Biolabs, San Diego, CA, USA). Briefly, genomic DNA from 10 flies or ~50 fly heads was extracted using PureLink™ genomic DNA kits (Invitrogen, Carlsbad, CA, USA), treated at 95 °C for 5 min, and digested with nuclease P1 (20 mM sodium acetate, pH 5.2) and alkaline phosphatase (100 mM Tris, pH 7.5). Samples were centrifuged for 5 min at 6000× *g* and the supernatants were used for ELISA assays following the manufacturer's instructions. The amount of 8-OHdG in each sample was calculated based on an 8-OHdG standard curve.

2.9. TUNEL Labeling

Adult fly brains were dissected in M3 medium (Merck, Boston, MA, USA), fixed in 4% paraformaldehyde in phosphate buffered saline (PBS, 50 mM, pH 7.4) for 20 min at room temperature, washed with PBS, and permeabilized by 2 min incubation in PBS containing Triton X-100 (0.1%) on ice. After washing with PBS three times, the samples were first incubated in Na citrate (0.1 M, pH 6.0) for 10 min at 65 °C, followed by incubation in Tris-HCl buffer (0.1 M, pH 7.5) containing 3% BSA and 20% bovine serum albumin for 30 min. After washing three times with PBS, the samples were incubated in TUNEL assay solution (In Situ Cell Death Detection Kit; Roche, Basel, Switzerland) for 1 h at 37 °C, in accordance with the manufacturer's recommendations. After washing three times with PBS, the samples were mounted in mounting solution (Prolong Gold Antifade reagent; Life Technologies) and examined by confocal microscopy (Leica TCS SP5; Leica Microsystems, Morrisville, NC, USA).

2.10. Statistical Analysis

The statistical significance of differences in standard error of the mean (SEM) values was obtained using Student's *t*-test (two-tailed), analysis of variance (ANOVA) with post hoc analysis (Dunnett's tests, two-tailed). Kaplan-Meier survival curves were used for lifespan and survival analysis with log-rank tests. Prism software (GraphPad version 6.0, San Diego, CA, USA) was used to calculate *p*-values. Results were considered statistically significant at levels * $p < 0.05$, ** $p < 0.01$, *** $p < 0.001$, and **** $p < 0.0001$.

3. Results

To find genes involved in lifespan extension, we carried out a small-scale screening of mutants with extended lifespan, which identified *duox* heterozygous ($duox^{KG07745/+}$) male flies of interest. In the transgenic allele, the P-element KG07745 was inserted into the second intron of the *duox* locus, which reduced the *duox* transcripts to ~half the level in control flies (Figure 1A–C). As the insertion was downstream of the initiation codon, the Duox protein structure was disrupted, suggesting that the $duox^{KG07745}$ allele was probably null or severely hypomorphic with regards to protein function The. $Duox^{KG07745}$ homozygotes were lethal, while the *duox*^{KG07745} heterozygotes were normal in both appearance and movement (Figure S1).

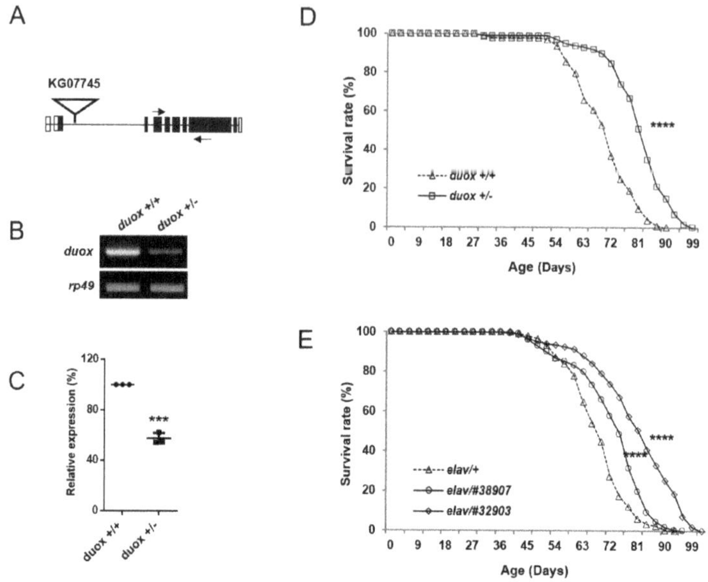

Figure 1. Extension of male lifespan with *duox* heterozygosity and neuronal *duox* knockdown. (**A**) Diagram showing the location of the P-element (KG07745), which is inserted in the second intron of *duox*. Empty and filled squares denote noncoding and coding regions of exons, respectively. Arrows denote primers used for RT-PCR. (**B**) Agarose-gel image of RT-PCR products following amplification of the *duox* gene from control (w^{1118}, denoted $duox^{+/+}$) and heterozygous ($duox^{KG07745/+}$, denoted $duox^{+/-}$) flies. *rp49* was used as a loading control. Three independent experiments carried out with similar results. (**C**) qRT-PCR of *duox* gene, with *rp-49* used as a control. Relative expression denotes *duox* transcript level. Averages and standard errors were derived from three independent experiments. Student's *t*-test, *** $p < 0.001$. (**D**) Lifespan of wild-type and *duox* heterozygous male flies. Twenty flies/vial, total twenty vials. Log-rank test, **** $p < 0.0001$. $duox^{+/+}$ denotes w^{1118} flies. $duox^{+/-}$ denotes $duox^{+/KG07745}$ flies, which were isogenized ten times with control flies (w^{1118}). (**E**) Lifespans of *elav-Gal4 > UAS-duox*^{RNAi} flies. *elav/#38907* and *elav/#32903* denote *elav-Gal4 > UAS-duox*^{RNAi} #38907, *elav-Gal4/+; UAS-duox*^{RNAi} #32903/+. Twenty flies/vial, total ten vials, male flies. Log-rank test, **** $p < 0.0001$. *UAS-duox*^{RNAi} (*#38907, 32903*) flies were isogenized with control w^{1118} flies ten times to reduce genetic background differences.

Initial screening involved a small number ($n = 40$) of unisogenized *duox* heterozygous males. To ensure rigor, we isogenized the *duox* heterozygous flies with the control flies (w^{1118}) ten times to reduce genetic background differences, and increased the population to 400 flies, with 20 flies per vial given fresh food every two days. We focused on the *duox* heterozygous male flies due to the small effect of the allele on the lifespans of the heterozygous female flies (Figure S2A). The control male flies (w^{1118}) lived for a median of 70 days; meanwhile, the isogenized *duox* heterozygous males had an extended median lifespan of 81 days, which was an increase of 15% (Figure 1D, Table S1). Next, we employed the bipartite *Gal4/UAS* expression system to knock down the *duox* function across the whole body, including in both muscle and neurons. Rigorously, we isogenized the *UAS-duox*^{RNAi} (*#38907, 32903*) flies with w^{1118} flies ten times to reduce the genetic background differences between the *Gal4 > UAS-duox*^{RNAi} flies and the control *Gal4 >* w^{1118} flies. Driving the whole-body *duox*^{RNAi} (#38907 and #32903) expression with the whole-body *Gal4* (*tub-Gal4*) was lethal. A second experiment using muscle-only *Gal4* (*mef-Gal4*) did not demonstrate pronounced lifespan extension in the males or females (Figure S2C,D). However, driving the *duox* RNAi with the pan-neuronal *Gal4* (*elav-Gal4*) extended the male lifespan by 10% and 20%, respectively, for the $duox^{RNAi}$ #38907 and #32903 relative to the *elav-Gal4/+* control (Figure 1E, Table S1); the lifespan extension in the females was not pronounced (Figure S2B).

As Duox is a ROS-producing enzyme, the *duox* heterozygotes ($duox^{+/-}$) would be expected to produce less ROS than the control flies ($duox^{+/+}$). The global ROS was measured using the non-fluorescent compound 2′,7′-dichlorodihydrofluorescein diacetate (DCFH-DA), which reacted with the ROS to produce fluorescent dichlorofluorescein (DCF). The ROS-dependent DCF fluorescence intensity was normalized to the protein concentration in each extract. The whole-fly extracts from the *duox* heterozygotes produced less ROS than those from the controls (Figure 2A). The head extracts from the flies with the neuronal knockdown of the *duox* also exhibited less ROS than the controls (Figure 2B). Similarly, the assays of the H_2O_2 production revealed that the *duox* heterozygote whole-fly extracts produced less H_2O_2 than those from the controls (Figure 2C). The same was true of the head extracts from the flies with the neuronal knockdown of the *duox* (Figure 2D). ROS mediates the oxidation of amino acid residues in proteins and of the bases in DNA [38–40], and this oxidation was decreased in the whole-body extracts of the *duox* heterozygous male flies (Figure 3A,C). Similarly, the neuronal knockdown of the *duox* via RNAi resulted in reduced protein and DNA oxidation in the head extracts (Figure 3B,D).

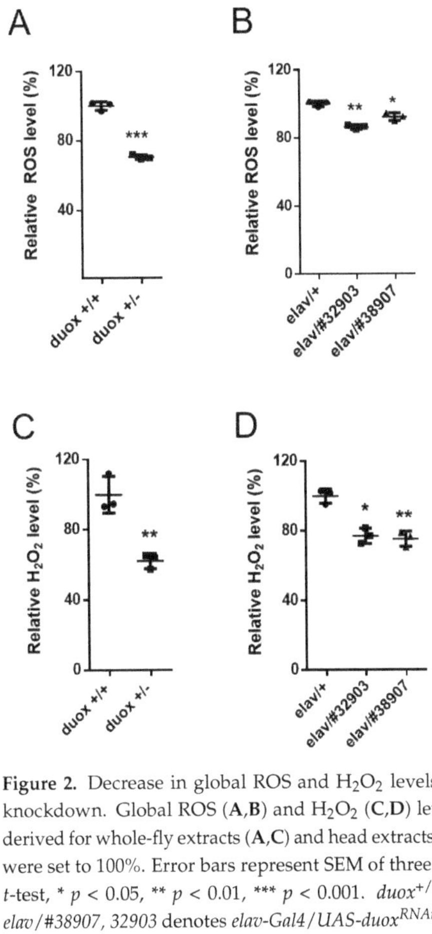

Figure 2. Decrease in global ROS and H_2O_2 levels with *duox* heterozygosity and neuronal *duox* knockdown. Global ROS (**A**,**B**) and H_2O_2 (**C**,**D**) levels normalized to protein concentration were derived for whole-fly extracts (**A**,**C**) and head extracts (**B**,**D**). Values for $duox^{+/+}$ (**A**,**C**) and *elav*/+ (**B**,**D**) were set to 100%. Error bars represent SEM of three replicates using 5-day-old male flies. Student's *t*-test, * $p < 0.05$, ** $p < 0.01$, *** $p < 0.001$. $duox^{+/+}$ and $duox^{+/-}$ denote w^{1118} and $duox^{+/kG07745}$. *elav*/#38907, 32903 denotes *elav-Gal4*/*UAS-duox*^{RNAi} #38907, *elav-Gal4*/+; *UAS-duox*^{RNAi} #32903/+.

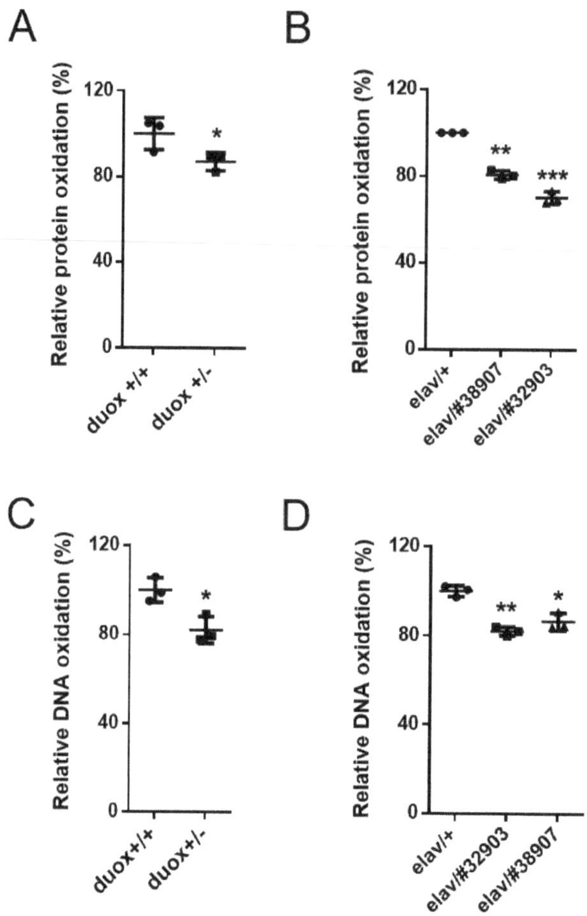

Figure 3. Decrease in oxidative damage with *duox* heterozygosity and neuronal *duox* knockdown. Relative protein and DNA oxidation levels in whole-fly extracts (**A**,**C**) and in head extracts (**B**,**D**). Data were set to $duox^{+/+}$ (**A**,**C**) and elav/+ (**B**,**D**) at 100%. Error bars represent SEM of three replicates using 60-day-old (**A**) and 45-day-old male flies (**B**–**D**). Three replicates. Error bars represent SEM. Student's *t*-test, * $p < 0.05$, ** $p < 0.01$, and *** $p < 0.001$. $duox^{+/+}$ and $duox^{+/-}$ denote w^{1118} and $duox^{+/KG07745}$, respectively. *elav/#38907, 32903* denotes *elav-Gal4/ UAS-*$duox^{RNAi}$ *#38907, elav-Gal4/+; UAS-*$duox^{RNAi}$ #32903/+.

Given that the *duox* heterozygous males experienced less oxidative damage, next, we examined whether these males exhibited increased survival when provided with ROS-producing food. When given food-containing paraquat, an agent that produces cellular ROS, the $duox^{KG07745}$ heterozygous males exhibited increased survival; they also showed increased survival on H_2O_2-containing food (Figure 4A,B). Oxidative stress is relevant to apoptotic cell death and neurodegeneration, suggesting that apoptotic cells might be reduced by *duox* reduction. Indeed, the apoptotic cells in the brain were reduced by the *duox* heterozygosity and by its neuronal knockdown (Figure S3). The observed ROS decrease could have been due to increased levels of anti-ROS enzymes. However, the levels of the anti-ROS enzyme catalase, dSOD1, and dSOD2 were constant with age regardless of the *duox* status (Figure S4), suggesting that the lower ROS level in the *duox* heterozygous flies was not due to anti-ROS enzymes, but to *duox* heterozygosity.

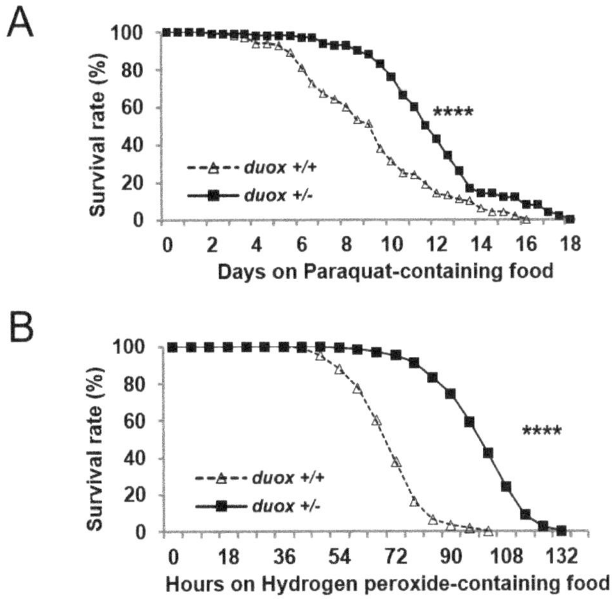

Figure 4. ROS resistance associated with *duox* heterozygosity. Survival of 5-day-old male flies on food containing 10 mM paraquat (**A**) and 10% hydrogen peroxide (H_2O_2) (**B**). $n = 200$ (20 flies/vial, total 10 vials). Log-rank test, **** $p < 0.0001$. $duox^{+/+}$ denotes w^{1118} flies. $duox^{+/-}$ denotes $duox^{+/kG07745}$ flies, which were isogenized ten times with control flies (w^{1118}).

4. Discussion

Neurons are vulnerable to oxidative stress, which is related to neuronal aging, neuropathies, and lifespan. Considerable attention has been paid to antioxidant defense systems, but the ROS sources that oxidize macromolecules in neurons are not yet well defined. Here, we identify Duox, a calcium-activated NADPH oxidase, as a determinant of neuronal oxidative stress and lifespan. We found that neuronal oxidative damage was reduced and lifespan was extended by the neuronal knockdown of *duox* in *Drosophila melanogaster*. Our findings suggest that *duox* is a source of ROS production in neurons, which affects lifespan.

It is intriguing that oxidative stress, particularly stress in neurons, should impact lifespan. It was previously shown that lifespan can be extended by enhancing anti-oxidative power in neurons, specifically through the neuronal overexpression of Peroxiredoxin 4 (*prx4*), an anti-ROS protein, and of glutamate-cysteine ligase, an enzyme that catalyzes the biosynthesis of the reducing agent, glutathione [13,41]. Similarly, the neuronal knockdown of Kelch-like ECH-associated protein 1 (*keap1*) increases longevity [8,12]; this gene encodes an inhibitor of the cap 'n' collar (CncC) transcription factor, the *Drosophila* homolog of the mammalian nuclear factor E2-related factor 2 (NRF2), which provides a major anti-ROS cellular system by directly inducing expression of anti-ROS genes. Thus, our findings are in line with previous reports that reducing oxidative stress in neurons is beneficial to lifespan.

We further observed that lifespan extension is pronounced in *duox* heterozygous males but less so in heterozygous females, and a similar trend was seen with the neuronal knockdown of the *duox*. Such sexual dimorphism in longevity has previously been observed in other contexts. For example, increased reducing power in neurons via the increased activation of CncC due to *keap1* mutation results in extended lifespan for males, but not females [42]. A recent report similarly found that oltipraz, an NRF2-activating drug, extends lifespan more in males than in females [43]. It appears that sexual dimorphism exists in regard to neuronal oxidative-stress-dependent lifespan.

It is worth mentioning the *duox*-dependent lifespan extension in *C. elegans*. In this model nematode, lifespan is expectably shortened upon the substantive increase of *duox* activity [44]. However, and much more unexpectedly, a modest increase in *duox* results in extended lifespan [44,45]. Why is a low-level increase in *duox* activity beneficial to lifespan? The underlying mechanism involves the activation of the expression of SKN-1, the *C. elegans* homolog of NRF2 [44]. NRF2 is a transcription factor that responds to ROS and is known to extend lifespan [9,45–47]. Therefore, extension of lifespan with a low-level increase in *duox* is achieved by strengthening anti-ROS power via NRF2 activation. Thus, our findings concerning *Drosophila duox* in combination with other reports regarding *C. elegans duox* support the notion that oxidative stress regulates lifespan.

Oxidative stress can occur due to an imbalance between ROS production and the cellular antioxidant defense network. In *Drosophila*, NRF2-dependent anti-ROS power declines with age, suggesting that oxidative stress is more likely to occur in older animals [8,43,48,49]. *Duox* activity and expression are increased in gut and embryo epithelial cells undergoing apoptotic cell death, in mammalian neurons through the stabilization of *duox* transcripts by ROS, and in gut epithelial cells by PLCβ-calcium and p38 MAPK signaling [18,24,27,29,30]. In addition, *duox* is elevated in aged brains and in *Drosophila* models of Alzheimer's disease [50]. Thus, chronic inflammation, protein-aggregate-induced cell death, and various signals involving PLCβ-calcium or p38 MAK could aberrantly activate *duox* activity and expression in neurons, which could confer oxidative stress and lifespan determination. It would be informative to examine whether neuronal *duox* knockdown reduces neuronal inflammation, a hallmark of which is the activation of microglial cells or astrocytes. However, the pursuit of this idea is hindered by the lack of antibodies or markers for activated microglial and astrocytic cells, which is a limitation of these studies.

Supplementary Materials: The following supporting information can be downloaded at: https://www.mdpi.com/article/10.3390/cells11132059/s1, Figure S1. Negative geotactic movement. Wild-type ($duox^{+/+}$) and *duox* heterozygous ($duox^{+/-}$) male flies were used for negative geotactic movement. n = 100 (10 flies/vial, total 10 vials for each point). $duox^{+/+}$ denotes w^{1118} flies. $duox^{+/-}$ denotes *duox* $^{kG07745/+}$ flies, which were isogenized with control (w^{1118}) flies; Figure S2. Lifespan of female (A–C) and male flies (D). (A) Wild-type and *duox* heterozygous flies. n = 400, female flies. $duox^{+/+}$ denotes w^{1118} flies. $duox^{+/-}$ denotes *duox* $^{kG07745/+}$ flies, which were isogenized with control (w^{1118}) flies. (B,C) Lifespan of *elav-Gal4>UAS-duox*^{RNAi} (*elav/#38907, 32903*) and *mef-Gal4> UAS-duox*^{RNAi} (*mef/#38907, 32903*). n = 200, female flies. (D) Lifespan of *mef-Gal4> UAS-duox*^{RNAi} (*mef/#38907, 32903*). n = 200, male flies. *UAS-duox*^{RNAi} (*#38907, 32903*) flies were isogenized with control w^{1118} flies ten times to reduce genetic background differences; Figure S3. Decrease of TUNEL-positive cells in the brain with *duox* heterozygosity and neuronal *duox* reduction. (A,B) Confocal images of whole-mount staining of *Drosophila* (40-day-old) brains with the DNA dye DAPI (blue) and TUNEL (red). (C,D) Quantification of TUNEL-positive cells in panels A & B. Error bars represent SEM of three replicates. Student's *t*-test, *** $p < 0.001$. $duox^{+/+}$ and $duox^{+/-}$ denote w^{1118} and $duox^{+/kG07745}$, respectively. *elav/#38907, 32903* denotes *elav-Gal4/ UAS-duox*^{RNAi} *#38907, elav-Gal4/+; UAS-duox*^{RNAi} *#32903/+*; Figure S4. Anti-ROS enzyme levels are not affected by *duox* heterozygosity. Agarose gel images of RT-PCR products of anti-ROS genes obtained from male flies. Similar results were obtained among the three replicates. *dSOD* denotes *Drosophila superoxide dismutase*. *rp49* was used as a loading control. $^{+/+}$ and $^{+/-}$ denote w^{1118} and $duox^{+/kG07745}$, respectively. Table S1. Quantification of Lifespan.

Author Contributions: Conceptualization, M.B. and C.K. Experiments, M.B. and W.J. Writing, M.B. and C.K. All authors have read and agreed to the published version of the manuscript.

Funding: This work was supported by grants from the National Research Foundation of Korea (NRF-2021R1A2C1010334 to CK, NRF-2020R1I1A1A01074292 to WJ).

Acknowledgments: We thank the Bloomington *Drosophila* Stock Center and Vienna *Drosophila* RNAi Center for flies.

Conflicts of Interest: The authors declare no conflict of interest.

References

1. Bedard, K.; Krause, K.H. The NOX Family of ROS-Generating NADPH Oxidases: Physiology and Pathophysiology. *Physiol. Rev.* **2007**, *87*, 245–313. [CrossRef] [PubMed]
2. Zhang, J.; Wang, X.; Vikash, V.; Ye, Q.; Wu, D.; Liu, Y.; Dong, W. ROS and ROS-mediated cellular signaling. *Oxid. Med. Cell. Longev.* **2016**, *2016*, 4350965. [CrossRef] [PubMed]
3. Sies, H. Hydrogen peroxide as a central redox signaling molecule in physiological oxidative stress: Oxidative eustress. *Redox Biol.* **2017**, *11*, 613–619. [CrossRef]
4. Holmström, K.M.; Finkel, T. Cellular mechanisms and physiological consequences of redox-dependent signalling. *Nat. Rev. Mol. Cell Biol.* **2014**, *15*, 411–421. [CrossRef] [PubMed]
5. Van der Vliet, A.; Janssen-Heininger, Y.M. Hydrogen peroxide as a damage signal in tissue injury and inflammation: Murderer, mediator, or messenger? *J. Cell. Biochem.* **2014**, *115*, 427–435. [CrossRef] [PubMed]
6. Colavitti, R.; Finkel, T. Reactive Oxygen Species as Mediators of Cellular Senescence. *IUBMB Life* **2005**, *57*, 277–281. [CrossRef]
7. Sohal, R.S.; Orr, W.C. The redox stress hypothesis of aging. *Free Radic. Biol. Med.* **2012**, *52*, 539–555. [CrossRef] [PubMed]
8. Rahman, M.M.; Sykiotis, G.P.; Nishimura, M.; Bodmer, R.; Bohmann, D. Declining signal dependence ofNrf2-MafS-regulated gene expression correlates with aging phenotypes. *Aging Cell* **2013**, *12*, 554–562. [CrossRef]
9. Kubben, N.; Zhang, W.; Wang, L.; Voss, T.C.; Yang, J.; Qu, J.; Liu, G.-H.; Misteli, T. Repression of the Antioxidant NRF2 Pathway in Premature Aging. *Cell* **2016**, *165*, 1361–1374. [CrossRef]
10. Rego, A.C.; Oliveira, C.R. Mitochondrial dysfunction and reactive oxygen species in excitotoxicity and apoptosis: Implications for the pathogenesis of neurodegenerative diseases. *Neurochem. Res.* **2003**, *28*, 1563–1574. [CrossRef]
11. Liu, Z.; Zhou, T.; Ziegler, A.C.; Dimitrion, P.; Zuo, L. Oxidative Stress in Neurodegenerative Diseases: From Molecular Mechanisms to Clinical Applications. *Oxid. Med. Cell. Longev.* **2017**, *2017*, 2525967. [CrossRef]
12. Spiers, J.G.; Breda, C.; Robinson, S.; Giorgini, F.; Steinert, J.R. Drosophila Nrf2/Keap1 Mediated Redox Signaling Supports Synaptic Function and Longevity and Impacts on Circadian Activity. *Front. Mol. Neurosci.* **2019**, *12*, 86. [CrossRef] [PubMed]
13. Orr, W.C.; Radyuk, S.; Prabhudesai, L.; Toroser, D.; Benes, J.J.; Luchak, J.M.; Mockett, R.J.; Rebrin, I.; Hubbard, J.G.; Sohal, R.S. Overexpression of Glutamate-Cysteine Ligase Extends Life Span in *Drosophila melanogaster*. *J. Biol. Chem.* **2005**, *280*, 37331–37338. [CrossRef] [PubMed]
14. Wang, M.C.; Bohmann, D.; Jasper, H. JNK Signaling Confers Tolerance to Oxidative Stress and Extends Lifespan in Drosophila. *Dev. Cell* **2003**, *5*, 811–816. [CrossRef]
15. Buvelot, H.; Jaquet, V.; Krause, K.H. Mammalian NADPH Oxidases. *Methods Mol. Biol.* **2019**, *1982*, 17–36. [CrossRef]
16. De Deken, X.; Corvilain, B.; Dumont, J.E.; Miot, F. Roles of DUOX-Mediated Hydrogen Peroxide in Metabolism, Host Defense, and Signaling. *Antioxid. Redox Signal.* **2014**, *20*, 2776–2793. [CrossRef]
17. Geiszt, M.; Witta, J.; Baffi, J.; Lekstrom, K.; Leto, T.L. Dual oxidases represent novel hydrogen peroxide sources supporting mucosal surface host defense. *FASEB J.* **2003**, *17*, 1–14. [CrossRef]
18. Damiano, S.; Fusco, R.; Morano, A.; De Mizio, M.; Paternò, R.; De Rosa, A.; Spinelli, R.; Amente, S.; Frunzio, R.; Mondola, P.; et al. Reactive Oxygen Species Regulate the Levels of Dual Oxidase (Duox1-2) in Human Neuroblastoma Cells. *PLoS ONE* **2012**, *7*, e34405. [CrossRef]
19. Weaver, C.J.; Leung, Y.F.; Suter, D.M. Expression dynamics of NADPH oxidases during early zebrafish development. *J. Comp. Neurol.* **2015**, *524*, 2130–2141. [CrossRef]
20. Conner, G.E. Regulation of dual oxidase hydrogen peroxide synthesis results in an epithelial respiratory burst. *Redox Biol.* **2021**, *41*, 101931. [CrossRef]
21. Kim, S.H.; Lee, W.J. Role of DUOX in gut inflammation: Lessons from Drosophila model of gut-microbiota interactions. *Front. Cell. Infect. Microbiol.* **2014**, *3*, 116. [CrossRef] [PubMed]
22. Razzell, W.; Evans, I.R.; Martin, P.; Wood, W. Calcium Flashes Orchestrate the Wound Inflammatory Response through DUOX Activation and Hydrogen Peroxide Release. *Curr. Biol.* **2013**, *23*, 424–429. [CrossRef] [PubMed]
23. Sirokmány, G.; Pató, A.; Zana, M.; Donkó, Á.; Bíró, A.; Nagy, P.; Geiszt, M. Epidermal growth factor-induced hydrogen peroxide production is mediated by dual oxidase 1. *Free Radic. Biol. Med.* **2016**, *97*, 204–211. [CrossRef]
24. Ha, E.M.; Lee, K.A.; Park, S.H.; Kim, S.H.; Nam, H.J.; Lee, H.Y.; Kang, D.; Lee, W.J. Regulation of DUOX by the Galphaq-phospholipase Cbeta-Ca^{2+} pathway in Drosophila gut immunity. *Dev. Cell.* **2009**, *16*, 386–397. [CrossRef] [PubMed]
25. Juarez, M.T.; Patterson, R.A.; Sandoval-Guillen, E.; McGinnis, W. Duox, Flotillin-2, and Src42A Are Required to Activate or Delimit the Spread of the Transcriptional Response to Epidermal Wounds in Drosophila. *PLoS Genet.* **2011**, *7*, e1002424. [CrossRef]
26. Xiao, X.; Huang, M.; Fan, C.; Zuo, F. DUOX2 participates in skin aging induced by UVB in HSF2 cells by activating NF-kappaB signaling. *Exp. Ther. Med.* **2021**, *21*, 157. [CrossRef]
27. Ha, E.M.; Lee, K.A.; Seo, Y.Y.; Kim, S.H.; Lim, J.H.; Oh, B.H.; Kim, J.; Lee, W.J. Coordination of multiple dual oxidase-regulatory pathways in responses to commensal and infectious mi-crobes in drosophila gut. *Nat. Immunol.* **2009**, *10*, 949–957. [CrossRef]
28. Kizhedathu, A.; Chhajed, P.; Yeramala, L.; Sain Basu, D.; Mukherjee, T.; Vinothkumar, K.R.; Guha, A. Duox-generated reactive oxygen species activate ATR/Chk1 to induce G2 arrest in Drosophila tra-cheoblasts. *eLife* **2021**, *10*, e68636. [CrossRef]
29. Fogarty, C.E.; Diwanji, N.; Lindblad, J.L.; Tare, M.; Amcheslavsky, A.; Makhijani, K.; Brückner, K.; Fan, Y.; Bergmann, A. Extracellular Reactive Oxygen Species Drive Apoptosis-Induced Proliferation via Drosophila Macro-phages. *Curr. Biol.* **2016**, *26*, 575–584. [CrossRef]

30. Amcheslavsky, A.; Lindblad, J.L.; Bergmann, A. Transiently "Undead" Enterocytes Mediate Homeostatic Tissue Turnover in the Adult Drosophila Midgut. *Cell Rep.* **2020**, *33*, 108408. [CrossRef]
31. Jang, W.; Baek, M.; Han, Y.S.; Kim, C. Duox mediates ultraviolet injury-induced nociceptive sensitization in Drosophila larvae. *Mol. Brain* **2018**, *11*, 16. [CrossRef]
32. Rieger, S.; Sagasti, A. Hydrogen Peroxide Promotes Injury-Induced Peripheral Sensory Axon Regeneration in the Zebrafish Skin. *PLoS Biol.* **2011**, *9*, e1000621. [CrossRef] [PubMed]
33. Yang, L.Q.; Chen, M.; Ren, D.L.; Hu, B. Dual Oxidase Mutant Retards Mauthner-Cell Axon Regeneration at an Early Stage via Modulating Mito-chondrial Dynamics in Zebrafish. *Neurosci. Bull.* **2020**, *36*, 1500–1512. [CrossRef] [PubMed]
34. Slack, C.; Werz, C.; Wieser, D.; Alic, N.; Foley, A.; Stocker, H.; Withers, D.J.; Thornton, J.M.; Hafen, E.; Partridge, L. Regulation of Lifespan, Metabolism, and Stress Responses by the Drosophila SH2B Protein, Lnk. *PLoS Genet.* **2010**, *6*, e1000881. [CrossRef] [PubMed]
35. Ali, Y.O.; Escala, W.; Ruan, K.; Zhai, R.G. Assaying Locomotor, Learning, and Memory Deficits in Drosophila Models of Neurodegeneration. *J. Vis. Exp.* **2011**, *49*, e2504. [CrossRef] [PubMed]
36. Casani, S.; Gómez-Pastor, R.; Matallana, E.; Paricio, N. Antioxidant compound supplementation prevents oxidative damage in a Drosophila model of Parkinson's disease. *Free Radic. Biol. Med.* **2013**, *61*, 151–160. [CrossRef]
37. Bradford, M.M. A rapid and sensitive method for the quantitation of microgram quantities of protein utilizing the principle of protein-dye binding. *Anal. Biochem.* **1976**, *72*, 248–254. [CrossRef]
38. Korolainen, M.A.; Nyman, T.; Nyyssonen, P.; Hartikainen, E.S.; Pirttila, T. Multiplexed Proteomic Analysis of Oxidation and Concentrations of Cerebrospinal Fluid Proteins in Alzheimer Disease. *Clin. Chem.* **2007**, *53*, 657–665. [CrossRef]
39. Smith, M.A.; Sayre, L.M.; Anderson, V.E.; Harris, P.L.; Beal, M.F.; Kowall, N.; Perry, G. Cytochemical Demonstration of Oxidative Damage in Alzheimer Disease by Immunochemical Enhancement of the Carbonyl Reaction with 2,4-Dinitrophenylhydrazine. *J. Histochem. Cytochem.* **1998**, *46*, 731–735. [CrossRef]
40. Furukawa, A.; Hiraku, Y.; Oikawa, S.; Luxford, C.; Davies, M.J.; Kawanishi, S. Guanine-specific DNA damage induced by gamma-irradiated histone. *Biochem. J.* **2005**, *388 Pt 3*, 813–818. [CrossRef]
41. Klichko, V.I.; Orr, W.C.; Radyuk, S.N. The role of peroxiredoxin 4 in inflammatory response and aging. *Biochim. Biophys. Acta (BBA)-Mol. Basis Dis.* **2015**, *1862*, 265–273. [CrossRef] [PubMed]
42. Sykiotis, G.P.; Bohmann, D. Keap1/Nrf2 signaling regulates oxidative stress tolerance and lifespan in Drosophila. *Dev. Cell* **2008**, *14*, 76–85. [CrossRef] [PubMed]
43. Cheng, Y.; Pitoniak, A.; Wang, J.; Bohmann, D. Preserving transcriptional stress responses as an antiaging strategy. *Aging Cell* **2021**, *20*, e13297. [CrossRef] [PubMed]
44. Sasakura, H.; Moribe, H.; Nakano, M.; Ikemoto, K.; Takeuchi, K.; Mori, I. Lifespan extension by peroxidase/dual oxidase-mediated ROS signaling through pyrroloquinoline quinone in *C. elegans*. *J. Cell Sci.* **2017**, *130*, 2631–2643. [CrossRef] [PubMed]
45. Ewald, C.Y.; Hourihan, J.M.; Bland, M.S.; Obieglo, C.; Katic, I.; Mazzeo, L.E.M.; Alcedo, J.; Blackwell, T.K.; Hynes, N.E. NADPH oxidase-mediated redox signaling promotes oxidative stress resistance and longevity through memo-1 in *C. elegans*. *eLife* **2017**, *6*, e19493. [CrossRef] [PubMed]
46. Blackwell, T.K.; Steinbaugh, M.J.; Hourihan, J.M.; Ewald, C.Y.; Isik, M. SKN-1/Nrf, stress responses, and aging in Caenorhabditis elegans. *Free Radic. Biol. Med.* **2015**, *88*, 290–301. [CrossRef]
47. Castillo-Quan, J.I.; Li, L.; Kinghorn, K.J.; Ivanov, D.K.; Tain, L.S.; Slack, C.; Kerr, F.; Nespital, T.; Thornton, J.; Hardy, J.; et al. Lithium Promotes Longevity through GSK3/NRF2-Dependent Hormesis. *Cell. Rep.* **2016**, *15*, 638–650. [CrossRef]
48. Przybysz, A.J.; Choe, K.P.; Roberts, L.J.; Strange, K. Increased age reduces DAF-16 and SKN-1 signaling and the hormetic response of Caenorhabditis elegans to the xenobiotic juglone. *Mech. Ageing Dev.* **2009**, *130*, 357–369. [CrossRef]
49. Ungvari, Z.; Bailey-Downs, L.; Gautam, T.; Sosnowska, D.; Wang, M.; Monticone, R.E.; Telljohann, R.; Pinto, J.T.; de Cabo, R.; Sonntag, W.E.; et al. Age-associated vascular oxidative stress, Nrf2 dysfunction, and NF-κB activation in the nonhuman primate *Macaca mulatta*. *J. Gerontol. A Biol. Sci. Med. Sci.* **2011**, *66*, 866–875. [CrossRef]
50. Barati, A.; Masoudi, R.; Yousefi, R.; Monsefi, M.; Mirshafiey, A. Tau and amyloid beta differentially affect the innate immune genes expression in Drosophila models of Alzheimer's disease and β-D Mannuronic acid (M2000) modulates the dysregulation. *Gene* **2021**, *808*, 145972. [CrossRef]

cells

MDPI

Article

Multinucleated Retinal Pigment Epithelial Cells Adapt to Vision and Exhibit Increased DNA Damage Response

Qin Ke †, Lili Gong *,†, Xingfei Zhu, Ruili Qi, Ming Zou, Baoxin Chen, Wei Liu, Shan Huang, Yizhi Liu and David Wan-Cheng Li *

State Key Laboratory of Ophthalmology, Zhongshan Ophthalmic Center, Sun Yat-Sen University, Guangzhou 510060, China; keqingzzoc@126.com (Q.K.); zhuxingfeigzzoc@163.com (X.Z.); qiruili10@163.com (R.Q.); zouming6991@163.com (M.Z.); chenbaoxin@gzzoc.com (B.C.); lvv961201@163.com (W.L.); huangshan@gzzoc.com (S.H.); liuyizhi@gzzoc.com (Y.L.)

* Correspondence: gonglili@gzzoc.com (L.G.); liwancheng@gzzoc.com (D.W.-C.L.)

† These authors contributed equally to this work.

Abstract: Multinucleated retinal pigment epithelium (RPE) cells have been reported in humans and other mammals. Rodents have an extremely high percentage of multinucleated cells (more than 80%). Both mouse and human multinucleated RPE cells exhibit specific regional distributions that are potentially correlated with photoreceptor density. However, detailed investigations of multinucleated RPE in different species and their behavior after DNA damage are missing. Here, we compared the composition of multinucleated RPE cells in nocturnal and diurnal animals that possess distinct rod and cone proportions. We further investigated the reactive oxygen species (ROS) production and DNA damage response in mouse mononucleated and multinucleated RPE cells and determined the effect of p53 dosage on the DNA damage response in these cells. Our results revealed an unrealized association between multinucleated RPE cells and nocturnal vision. In addition, we found multinucleated RPE cells exhibited increased ROS production and DNA damage after X-ray irradiation. Furthermore, haploinsufficiency of p53 led to increased DNA damage frequency after irradiation, and mononucleated RPE cells were more sensitive to a change in p53 dosage. In conclusion, this study provides novel information on in vivo PRE topography and the DNA damage response, which may reflect specific requirements for vision adaption and macular function.

Keywords: multinucleation; retinal pigment epithelium; reactive oxygen species; photoreceptor; DNA damage; p53

Citation: Ke, Q.; Gong, L.; Zhu, X.; Qi, R.; Zou, M.; Chen, B.; Liu, W.; Huang, S.; Liu, Y.; Li, D.W.-C. Multinucleated Retinal Pigment Epithelial Cells Adapt to Vision and Exhibit Increased DNA Damage Response. *Cells* **2022**, *11*, 1552. https://doi.org/10.3390/cells11091552

Academic Editors: Nicole Wagner and Kay-Dietrich Wagner

Received: 21 March 2022
Accepted: 28 April 2022
Published: 5 May 2022

1. Introduction

The retina pigment epithelium (RPE) is a pigmented cell monolayer between photoreceptors and the choroid of the retina. The RPE plays a key role in normal retina function due to its phagocytosis of photoreceptor outer segments (POS), cycling of retinoids for phototransduction, the constitution of the blood–retina barrier, and the maintenance of the immune-privileged status of the eye [1–3]. Therefore, dysfunction of the RPE is closely linked to multiple degenerative diseases of the retina, such as age-related macular degeneration (AMD) [4,5].

Most mammalian cells are mononucleated, while polyploidy is detected in megakaryocytes, hepatocytes, trophoblast giant cells, and cardiomyocytes [6,7]. In humans, multinucleated RPE was reported, where 3–5.3% of human RPE cells are bi-nucleated [8,9]. Interestingly, the percentage of multinucleated RPE is especially high in rodents, compromising more than 80% of total RPE cells, and the amount increases in an age-dependent manner [8,10]. The existence of the multinucleated RPE may be important for the phagocytosis of the RPE since oxidized POS increases RPE multinucleation in vitro [10]. A recent investigation of the human RPE implied that the existence of multinucleated RPE is in accordance with rod and cone photoreceptor density [9]. Therefore, we conducted a comparative study of the multinucleated RPE of nocturnal and diurnal animals with distinct rod and cone percentages in their retinas.

Increased mitochondrial and nuclear DNA damage has been detected in the RPE of degenerated retinas [11,12]. High oxygen tension in the macule and the unique phagocytosis function cause constant oxidative stress in the RPE, which is considered the major insult in DNA damage [12,13]. The efficient repair of a double-strand or single-strand DNA break is critical for preventing the genomic instability that can cause cell death, gene mutation, and cellular senescence. Whether mononucleated and multinucleated RPE cells exhibit altered DNA repair efficiency is largely unknown.

In the present study, we confirmed the existence of multinucleated RPE cells in humans of different ages. The composition of multinucleated RPE cells was further studied in mice from postnatal day 11 (before eye-opening) to 22 months and compared in nocturnal and diurnal animals. Our observation revealed that the multinucleation of RPE cells might be an adaptation to night vision. Finally, we found that multinucleated cells exhibited reduced DNA repair efficiency and were more sensitive to p53 dosage change upon DNA damage exposure in vivo. Haploinsufficiency of p53 leads to delayed DNA damage repair in multinucleated RPE cells compared to mononucleated RPE cells in the same eye.

2. Materials and Methods

2.1. Animals

C57BL/6J mice were used in this study. Mice were housed in standard cages in a specific pathogen-free facility on a 12-h light/dark cycle with ad libitum access to food and water. Mice were given 1 Gy of X-radiation (Rs2000 160 kV, 25 mA, and 1.22 Gy/min). At the end of the repair time, mice were euthanized, and their eyes were removed. All experimental procedures involving animals were approved by the Animal Use and Care Committee of Zhongshan Ophthalmic Center at the Sun Yat-Sen University, Guangzhou, China. Chickens, pigeons, pigs, and rabbits were bought from the market, zebrafish and rats were bought from Sun Yat-Sen University. For pigeons, chickens, rabbits and pigs, the eyes were harvested within three hours after slaughtering the animals. For mice, rats and zebrafish, the eyes were collected immediately after the death of the animals. The animal ages are: pigeon: 1 year, chicken: 1 year, pig: 10 months, rabbit: 6 months, zebrafish: 1 year, mouse: 2 months, rat: 2 months.

2.2. Genotyping

Mouse tail lysis buffer was added to mouse tail, incubating at 65 °C for more than 2 h and heating at 95 °C for 5 min. Briefly, 2 μL of genomic DNA were mixed with 20 μL of the Green Taq Mix (Vazyme, Nanjing, China, #P131-03), For detecting p53 gene, and primers and probes specific for p53-geno (primer-F GTGCCCTGTGCAGTTGTG and primer-R CTCGGGTGGCTCATAAGGTA), p53-neo (primer-F TGAATGAACTGCAGGACGAG and primer-R AATATCACGGGTAGCCAACG). For detecting *Pde6* gene, primer *Pde6brd1* F1: TACCCACCCTTCCTAATTTTTCTCAGC, *Pde6brd1* F2: GTAAACAGCAAGAGGCTTTATTGGGAAC, and *Pde6brd1* R: TGACAATTACTCCTTTTCCCTCAGTCTG were used. For detecting *Crb1* gene, primer *Crb1rd8* F1: GTGAAGACAGCTACAGTTCTGATC, *Crb1rd8* F2: GCCCCTGTTTGCATGGAGGAAACTTGGAAGACAGCTACAGTTCTTCTG, and *Crb1rd8* R: GCCCCATTTGCACACTGATGAC were used.

2.3. Fundus Photography and Fluorescein Angiography

Fundus images and fluorescein angiography were performed before X-ray treatment using the Micron IV retinal imaging microscope (Phoenix Research Laboratories, Pleasanton, CA, USA) [14]. After anesthesia with 1% sodium pentobarbital (70 μL/10 g), dilation of the pupils and lubrication of the cornea, the mice were taken for fundus photography first, and then I.P. was injected with 2% fluorescein sodium solution (Al-con laboratories, Fort Worth, TX, USA) (5 μL/g), and fluorescein angiographic images were recorded in 5 min.

2.4. Mouse Retina Protein Extraction and Western Blot Analysis

The retinas were dissected in PBS and suspended in 120 μL of RIPA buffer (per retina) containing proteinase inhibitor cocktail (Bimake, Shanghai, China, #B14002), protein phosphatase inhibitor A (Beyotime, Shanghai, China, #P1082) and protein phosphatase inhibitor C (Beyotime, Shanghai, China, #P1092). The total proteins were extracted sonication using an EpiSonic 2000 Sonication System (EPIGENTEK, Farmingdale, NY, USA) (Amplitude: 40%, 10 s on and 10 s off for 7 min in total). For Western blot (WB) analysis, it was performed as described previously with some modifications [14]. For each WB, 30–50 μg of total protein was used. The protein was separated by 12% SDS-PAGE and transferred to the PVDF membrane. The membrane was blocked by 5% milk in TBST for 1 h. After washing with TBST, the membrane was incubated with primary antibodies γH2Ax (Santa Cruz, Dallas, Texas, USA, sc-517348, 1:1000 dilution) and GAPDH (Proteintech, Rosemont, IL, USA, #60004-1-Ig, 1:2000 dilution). The secondary antibody was diluted in TBST (1:3000 dilution). After washing with TBST, enhanced chemiluminescence (ECL) detection was performed by using the Ultra sensitive ECL Chemiluminescence Kit (NCM Biotech, Suzhou, China, #P10300) according to the manufacturer's specifications. The exposure and development of PVDF membrane were performed using Tanon 5200 (Tanon, Shanghai, China).

2.5. Histology, Immunohistochemistry and Immunofluorescence

For immunohistochemistry (IHC) and immunofluorescence (IF), the eyes were fixed in the FAS eye fixation solution (Servicebio, Wuhan, China, #G1109), dehydrated using an increasing ethanol gradient and embedded in paraffin as previously described [14]. Three sections (thickness: 10 μm) through the optic disk of each eye were prepared. The antigen was retrieved by incubation at 95 °C in 10 mM sodium citrate buffer for 15 min, after which the slides were immunoassayed with primary antibodies Rhodopsin (Cell Signaling Technology, Danvers, MA, USA, #27182 1:200 dilution) at 4 °C overnight. The following IHC was conducted according to the manufacturer's protocol (GTVision TMIII, #GK500705) (Gene Tech, Shanghai, China). After development, the slides were counterstained with hematoxylin and observed under a Tissue-FAXS Q confocal microscope (TissueGnostics, Vienna, Austria). For the immunofluorescence, the slides were immunoassayed with primary antibodies γH2Ax (sc-517348, 1:50 dilution) at 4 °C overnight, followed by a 2-h incubation with the secondary antibody. The cell nucleus was labeled with DAPI (SIGMA, Saint Louis, MO, USA, #D9542). F-actin was labeled with fluorescein isothiocyanate phalloidin (YEASEN, Shanghai, China, # 40735ES75).

2.6. Animals' RPE Flat Mount Immunofluorescence

For mouse RPE IF, the procedure was performed as described previously [15]. The RPE flat mounts were incubated with primary antibodies γH2Ax (sc-517348 1:50 dilution) or 53bp1 (Bethyl, Montgomery, TX, USA, A300-272A-M, 1:200 dilution) overnight at 4 °C, followed by a 2-h incubation with the secondary antibody (Cell signaling # 4412S #8890S) and DAPI (SIGMA #D9542). Images were captured with a Tissue Fax confocal microscope. For chickens, pigeons, zebrafish, pigs, rats, rabbits RPE IF, the cell nucleus was labeled with DAPI, the epithelial cell was labeled with ZO1, or F-actin was labeled with fluorescein isothiocyanate phalloidin (YEASEN #40735ES75). Images were captured with TissueFAXS Q confocal microscope (TissueGnostics, Vienna, Austria). Image J (National Institutes of Health, Bethesda, MD, USA) was used to delineate cell profiles and measure the area of the cell for each age group of mice, more than 50 mono-nucleate RPE cells and multi-nucleate RPE cells were detected. For chickens, pigeons, zebrafish, pigs, rats, or rabbits RPE IF, the cell nucleus was labeled with DAPI, epithelial cell was labeled with ZO1. The slides were captured with a Leica DM4000 B LED (Leica, Wetzlar, Germany) or TissueFAXS Q confocal microscope (TissueGnostics, Vienna, Austria). For images were analyzed by TissueFAXS Viewer (TissueGnostics, Vienna, Austria) and ImageJ (National Institutes of Health, Bethesda, MD, USA).

2.7. Comet Assay

The RPE were dissected in PBS and suspended in 1 mL of 0.25% Trypsin for 1 h, 37 °C. After centrifuging and removing Trypsin, the RPE cells were diluted with cold PBS at 1×10^5/mL. Pay attention to avoiding light during the experiment. Using CometAssay® Kit (R&D, Minneapolis, MN, USA, #4250-050-K), the following procedure was conducted according to the manufacturer's protocol. The cell was labeled with SYBR® GREEN I (biosharp, Hefei, China, #BS358A). Images were captured with a TissueFAXS Q confocal microscope (TissueGnostics, Vienna, Austria). Images analyzed by TriTek Comet Score Freeware 1.6.1.13 (TriTek, Corp. Sumerduck, VA, USA).

2.8. Primary Cell Culture

The eyeballs were quickly dipped in 70% ethanol and then rinsed in sterilized PBS. The cornea, lens, iris, and neuron retina were removed and the remaining posterior eyecups in a 1.5 mL EP tube containing 1 mL of pre-warmed Trypsin were added. After incubation at 37 °C for 1 h, we resuspended the RPE cells in the tube by flipping, then gently aspirated the RPE/Trypsin solution to a new tube with a blue tip. Leave the choroid in the original tube. Collected RPE cells by centrifugation. 1500–2000 rpm, RT, 5 min. Washed the RPE pellet with 1 mL pre-warmed complete DMEM 2 times. (1500–2000 rpm, RT, 5 min.) Gently resuspend the washed RPE cells and seed them in the coated 12-well dish with a coverslip. After 5 days, we washed the unattached cells and debris with PBS.

2.9. Cell Proliferation Assay

The EdU cell proliferation staining was performed using an EdU kit (BeyoClick™ EdU Cell Proliferation Kit with Alexa Fluor 488, Beyotime Biotechnology, Shanghai, China, C0071S). Briefly, primary mouse RPE cells were seeded in 12-well plates for 5 days. Subsequently, cells were incubated with 10 μM EdU for 4.5 h, fixed with 4% paraformaldehyde for 15 min, and permeated with 0.3% Triton X-100 for another 15 min. The cells were incubated with α-tubulin overnight at 4 °C and followed by a 1h incubation with the secondary antibody in a dark place, later the Click Reaction Mixture for 30 min at room temperature and then incubated with Hoechst 33342 for 30 min. The slides were observed under a 40× oil objective lens with a ZEISS LSM 980 confocal microscope (ZEISS Microscopy, Jena, Germany). Image J was used to count Edu positive or negative cells.

2.10. Measurement of Intracellular ROS Levels

The intracellular ROS levels were measured using a Reactive Oxygen Species Assay Kit (Beyotime, Shanghai, China, S0033S). Briefly, the cells were seeded in 12-well plates as described in primary cell culture and exposed to 1 Gy X radiation and continued culture for 4 h. Following the treatment, the cells were incubated with 10 μM DCFH-DA for 30 min at 37 °C and then incubated with Hoechst 33342 for 30 min. The slides were captured with a Leica DM4000 B LED (Leica, Wetzlar, Germany). ImageJ was used to analyze the fluorescence integrity for ROS level.

2.11. Mitochondrial Membrane Potential

The mice were divided into three groups, PBS, sodium iodate (SI), and X-ray. Briefly, as for PBS and SI groups, we first intraperitoneally injected mice with PBS or 20 mg/kg SI, and then anesthetized mice with 1% sodium pentobarbital, and then dilated the pupils and lubricated the cornea. Later, 1 μL of 200 μM Mito-Tracker Red CMXRos (Beyotime, Shanghai, China, C1049B) was intravitreally injected and the RPE whole mounts were prepared 1 day-post injection. As for the X-ray group, intravitreal injection of Mito-Tracker Red CMXRos was performed 1 day before exposure to 1 Gy of X-ray irradiation. After dissecting of RPE whole mount, DAPI and FITC were counterstaining as described above. The slides were captured with a Leica DM4000 B LED (Leica, Wetzlar, Germany). Image J was used to analyze mitochondria number and average area.

2.12. Statistical Analysis

Results are expressed as mean ± SEM and mean ± SD unless otherwise indicated. GraphPad Prism 9.0software (GraphPad software, Inc., La Jolla, CA, USA) was used for statistical analysis as described in Results. All tests are two-tailed, unpaired *t*-tests unless otherwise indicated. *, $p < 0.05$; **, $p < 0.01$; ***, $p < 0.0001$.

3. Results

3.1. Distribution of Mononucleated and Multinucleated Cells in Mouse RPE

Firstly, we determined the distribution of mono- and multinucleated RPE cells in mice of different ages. The mice were confirmed by sequencing or PCR to exclude $Pde6b^{rd1}$ or $Crb1^{rd8}$ strains, which are naturally occurring retinal degeneration mouse mutants (Supplementary Figure S1). The RPE whole mount was obtained from postnatal day 11 (P11), 2-month (2M), and 22-month (22M)-old mice, and image regions were selected according to the distance to the optic nerve head (Figure 1a). Mice have a significantly higher percentage of multinucleated RPE cells compared to humans, and the highest amount was detected around the optic nerve, where 80% of RPE cells were multinucleated in all ages examined (Figure 1b). Furthermore, the number of multinucleated cells decreased in the peripheral regions, which is consistent with previous reports (Figure 1c) [10]. Interestingly, when observed at similar locations, no significant differences in the number of multinucleated cells were found at different ages (Figure 1d), suggesting that region rather than age impacts the existence of the multinucleated RPE in mice. When the RPE cell size was analyzed, multinucleated RPE cells exhibited a two-fold increased area than mononucleated cells in all ages examined (Figure 1e). In addition, significantly increased cell size was observed in old mice (22 M), for both mononucleated and multinucleated RPE cells. Finally, we compared cell proliferation in mononucleated and multinucleated RPE cells by 5-ethynyl-2 deoxyuridine (EdU) analysis. As shown in Figure 1f, mononucleated and multinucleated RPE cells show similar Edu-positive cell percentages when cultured in vitro, suggesting multinucleation does not affect DNA incorporation in RPE cells.

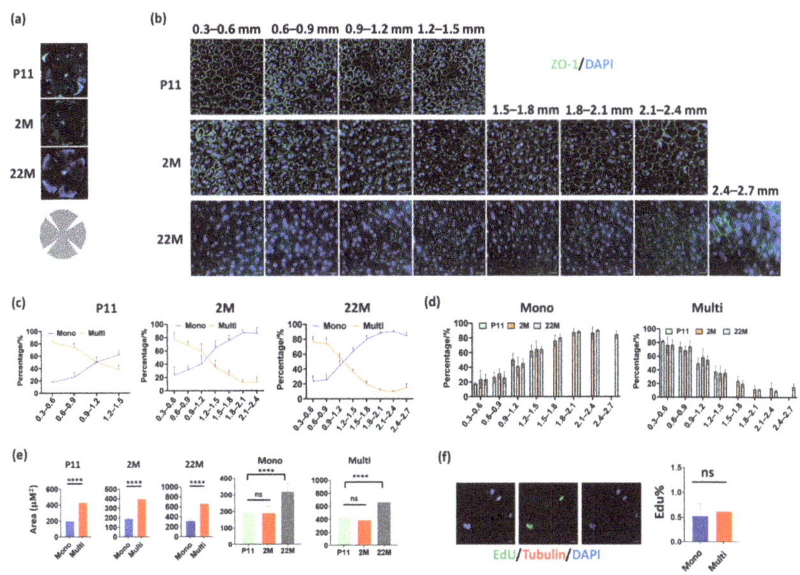

Figure 1. Distribution of mononucleated and multinucleated cells in mouse RPE of different ages. (a) IF analysis was performed on whole-mount RPE of postnatal day 11 (mice that had just opened their

eyes, P11) 2-month (2 M), and 22-month-old (22 M) mice. The RPE morphology was demonstrated by ZO1 staining, and the nuclei were stained with DAPI. Scale bar: 200 μM for P11 and 500 μM for 2 M and 22 M. The bottom image depicts a schematic graph showing different geographic locations of RPE flat mounts used in image analysis. (**b**) The mononucleated and multinucleated RPE cells are shown in different regions. The length indicates the relative distance from the optic nerve head. Scale bar: 20 μM. (**c**,**d**) Quantification of mononucleated and multinucleated RPE cells at different regions. (**e**) The cellular area of mononucleated and multinucleated RPE cells is indicated. ****: $p < 0.0001$, ns: not significant. For each group, more than 50 cells were quantified. (**f**) EdU staining shows DNA synthesis in primary mouse RPE cell cultures. The cell skeleton structure was labeled by α-tubulin staining and the nuclei were counterstained by DAPI. Right panels: quantification results of EdU-positive RPE cells. For each group, more than 20 cells were quantified. All Data are shown as mean ± SD.

3.2. A High Percentage of Multinucleated RPE Correlates with Nocturnal Vision

Although the mouse retina does not have a macula, the central region resembles the human macula in some aspects [15]. Interestingly, the highest frequencies of multinucleated cells in human RPE were found in the macula [9]. Further, multinucleated RPE cells are enriched in macula perifovea, where the highest amount of rods are located but are absent in the macular fovea, which only contains cone photoreceptors [9]. These results prompted us to examine the multinucleated RPE in diurnal and nocturnal vision animals, in which rod and cone photoreceptors show distinct compositions. Generally, nocturnal animals had a higher percentage of rods than diurnal animals [16]. Interestingly, our results show that nocturnal animals (mice, rats, and rabbits) had a significantly increased multinucleated RPE than diurnal animals (chickens, pigeons, pigs and zebrafish) (Figure 2a,b). Even in nocturnal animals, rodents (mice and rats), which have a lower number of cones than rabbits, showed a higher percentage of multinucleated RPE cells than rabbits (Figure 2c). Finally, we found a significant positive correlation between the percentage of multinucleated RPE and rod in six animals with known rod amounts (Figure 2d) [17–26]. Therefore, we concluded that the percentage of multinucleated RPE and rod photoreceptors is positively correlated in the retina.

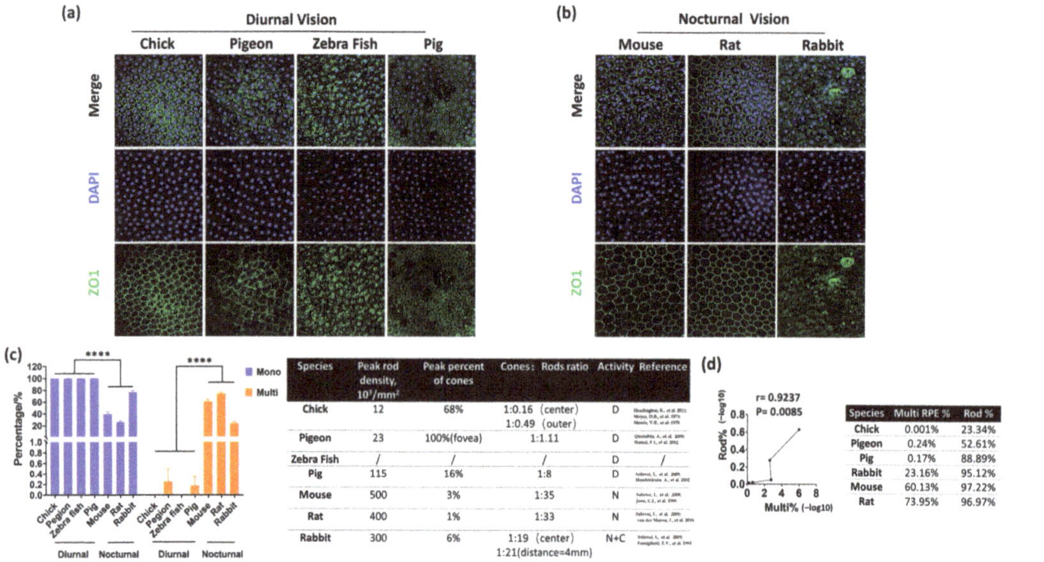

Figure 2. High percentage of multinucleated RPE correlates with nocturnal vision. RPE cells in diurnal

vision (**a**) and nocturnal vision (**b**) animals. The PPE morphology was indicated by ZO1 staining, and the nuclei were stained by DAPI. n = 3 for each animal. (**c**) comparison of mononucleated and multinucleated RPE with the key parameters characterizing the adaptation of the retina to nocturnal or diurnal vision. (**d**) Correlation of multinucleated RPE percentage with rods percentage in animals mentioned in (**c**). The data were presented as –log10 and two-tailed Pearson correlation coefficients were used to calculate the r and *p* values.

3.3. Multinucleated RPE Exhibits Increased DNA Damage Compared to Mononucleated RPE

The RPE of mice represents an ideal model system to study DNA damage response in polyploidy cells due to its high percentage of multinucleated cells. Therefore, we induced DNA damage by exposing mice to 1 Gy of X-ray ionizing radiation (IR), as this dosage has been reported to cause DNA double-strand break in mice retinas [27]. We first confirmed that IR led to DNA damage in mice retinas through immunofluorescence (IF) analysis using the DNA damage marker γH2Ax. We found the damaged DNA signal culminated at 1 h post-IR, then dramatically decreased 1 day later and was barely detected 3 days after IR (Figure 3a). This tendency was further confirmed by WB analysis (Figure 3b). Next, we determined RPE DNA damage through γH2Ax staining. Similar to the retina, the RPE displays distinct DNA damage as early as 1 h after irradiation (Figure 3c). The damaged DNA was gradually repaired as γH2Ax-positive cells decreased in number 1 day after IR and further decreased at 3 days post-IR (Figure 3c). Notably, the multinucleated RPE displayed a significantly higher level of γH2Ax-positive cells at all three time points, indicating that the multinucleated RPE may have reduced DNA repair efficiency compared to mononucleated cells (Figure 3c). 53BP1 is a key regulator for DNA damage repair, the 53BP1-decorated nuclear bodies mediate the formation of the DNA damage repair platform. Therefore, we investigated 53BP1 foci in mouse RPE whole mount. However, although multinucleated RPE cells exhibited increased γH2Ax foci, the 53BP1 foci were not significantly altered in multinucleated and mononucleated cells (Figure 3c). Finally, we performed an analysis of DNA double-strand breaks using a neutral comet assay which revealed an increased tail moment 1 h after IR for multinucleated RPE (Figure 3d). Moreover, at 1 and 3 days post-IR, the multinucleated RPE had a 1.32-fold and 1.49-fold greater tail moment than mononucleated cells, respectively (Figure 3e). Together, these results showed that DNA double-strand breaks are repaired less efficiently in multinucleated RPE cells than in mononucleated RPE cells although similar 53BP1 foci formations were observed.

3.4. Multinucleated RPE Cells Show Increased ROS Production after IR Exposure

Since increased DNA damage was observed in multinucleated RPE cells, we thus determined reactive oxygen species (ROS), a potent DNA damage inducer for DNA damage in RPE cells. Fluorescent ROS analysis demonstrated that primarily cultured mononucleated and multinucleated RPE cells show similar low ROS levels in normal conditions (Figure 4a,b). After IR exposure, dramatic upregulation of ROS was detected in RPE cells, where multinucleated cells show significantly higher levels of ROS (Figure 4a,b). Since ROS production contributes to mitochondrial damage, we further investigated the mitochondrial morphology in mouse RPE in vivo. As shown in Figure 4c, oxidative stress directly induced by oxidant sodium iodate, or by IR, leads to evident enlarged mitochondria area, possibly due to swelling of mitochondria upon damage insults. However, no significant alterations were found in mononucleated and multinucleated RPE cells (Figure 4d). Taken together, these results indicate that multinucleated RPE cells generated more ROS than mononucleated cells upon DNA damage insult.

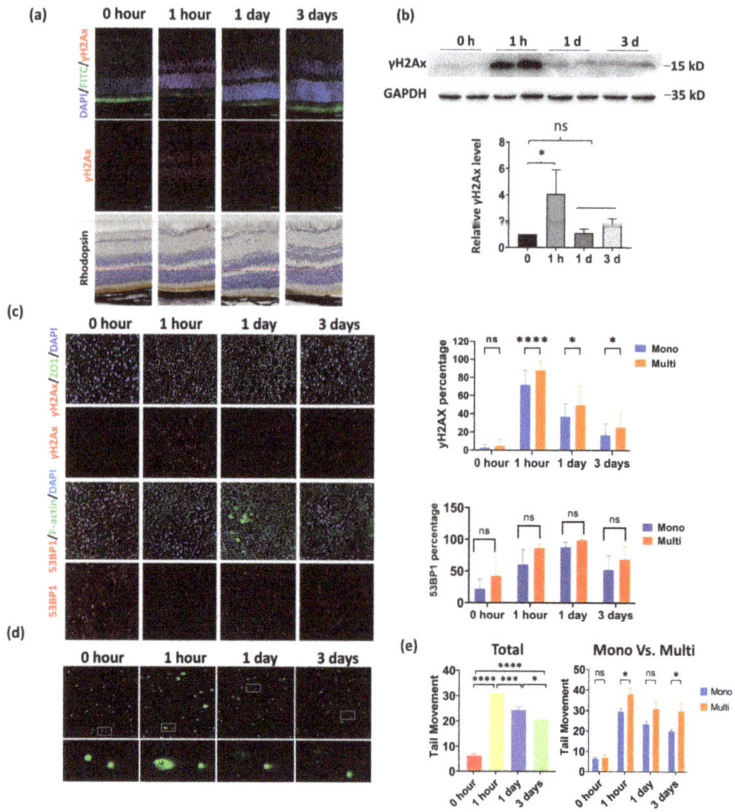

Figure 3. Multinucleated RPE cells exhibit reduced DNA repair efficiency compared to mononucleated RPE. DNA damage was induced by exposing the mice to 1 Gy of X-ray irradiation (IR). The retina and RPE were collected at 1 h, 1 day, and 3 days after IR. (**a**). IF and immunohistochemistry (IHC) analysis of retina cryosections. The DNA damage was indicated by γ-H2AX staining, and the cytoskeleton F-actin was labeled by FITC− phalloidin. The retina structure was further demonstrated by IHC staining of rhodopsin, the photoreceptor marker. Scale bar: 20 μm. (**b**). WB analysis of the retina treated as described above. The quantification results of three independent experiments were shown in the bottom panel. *: $p < 0.05$. ns: not significant (**c**). IF images of γH2AX and 53BP1 staining in the RPE at 1 h, 1 day, and 3 days after irradiation. Scale bar: 20 μm. Right panels: quantitative analysis of the DNA damage comparing mononucleated and multinucleated RPE was completed by counting the γH2AX-positive or 53BP1-positive cells. Forty regions in whole-mount RPE from 4 mice were randomly selected and quantified. * $p < 0.05$, **** $p < 0.0001$ and ns: not significant. (**d**). Comet assay showed DNA damage in RPE cells after IR. Neutral comet assay was performed using digested mouse RPE cells, which were collected at the indicated time point post-IR. The enlarged figure demonstrates a typical multinucleated and mononucleated cell. Scale bar: 100 μm. (**e**). Quantitative analysis of tail movement at the indicated time post-IR. At each time point, more than 80 mononucleated and 15 multinucleated RPE cells were counted, respectively. * $p < 0.05$, *** $p < 0.001$, **** $p < 0.0001$, ns: not significant. All data are shown as mean ± SD.

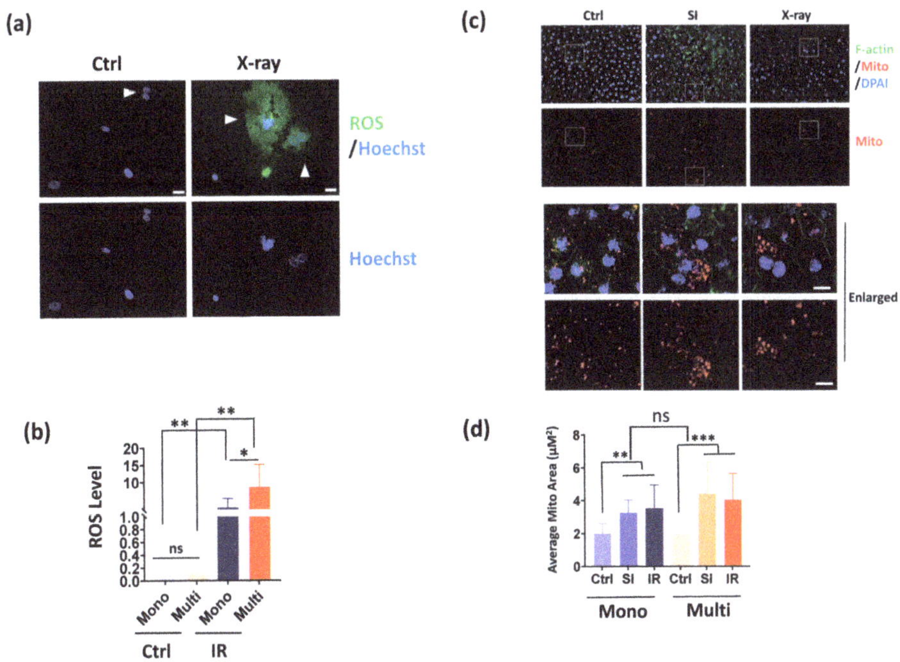

Figure 4. Multinucleated RPE cells generate more ROS than mononucleated RPE cells. Multinucleate mouse RPE cells exhibited increased ROS production than mononucleate cells after IR exposure. The primary mouse RPE cells were untreated (Ctrl) or exposed to 1 Gy of IR. After recovery in growth medium for 4 h, the ROS levels were detected and the nuclei were counterstained by Hoechst. Arrowheads indicate multinucleated RPE cells. (**b**) Quantitative results of ROS as indicated in (**a**). The green fluorescence intensity was quantified by Image J. For each group, about 15 cells were quantified. *: $p < 0.05$, ns: not significant. (**c**) Mitochondria morphology in mouse RPE whole mount with or without sodium iodate (SI) or IR exposure. The F-actin was labeled by FITC-phalloidin and the mitochondria were labeled by Mitotracker red. The nuclei were stained by DAPI. Scale bar: upper panels: 25 μM, lower upper panels: 10 μM. (**d**) Quantification results of mitochondria average area in mononucleated and multinucleated RPE cells. n = 15 cells per group. **: $p < 0.01$, ***: $p < 0.001$, ns: not significant. All data are shown as mean ± SD.

3.5. p53 Haploinsufficiency Leads to Increased DNA Damage in the RPE

p53 is a key gene controlling the DNA damage response. It is unknown whether or not p53 has a different effect on mononucleated and multinucleated RPE DNA damage. Because homozygous depletion of p53 in C57BL/6J mice led to severe eye abnormalities [28], we used p53 heterozygotes (p53+/−) in our investigation. However, fundus photography revealed that more than 60% of p53+/− mice (14 out of 21 mice) also have an ocular abnormality, including retinal pigment epithelial depigmentation, retina folds, colobomas, and abnormal vasculature (Figure 5a). HE staining further confirmed a retinal fold in those mice (Figure 5b). To exclude the effect of pre-existed ocular abnormalities, we selected p53+/− mice with normal fundus characteristics for IR exposure. IF analysis of RPE flat mounts showed a normal RPE structure in these p53+/− mice (Figure 5c). In control mice, γH2AX signals were barely detected in wild-type (WT) and p53+/− RPE, suggesting that haploinsufficiency of p53 does not spontaneously cause DNA damage (Figure 5c). After IR exposure, p53+/− RPE exhibited increased DNA damage in both mononucleated and multinucleated cells compared to WT RPE, and this higher level of DNA damage was observed at 1 h, 1 day, and 3 days post-IR (Figure 5d,e). These results highlight the requirement of p53 in efficient DNA damage repair in the RPE. Next, we compared DNA damage between mononucleated and multinucleated cells in p53+/− RPE. Similar to WT RPE, p53+/− multinucleated RPE showed a higher level of DNA damage than mononucleated RPE at 1 day and 3 days post-IR. However, when investigated 1 h after IR, mononucleated and multinucleated cells in p53+/− RPE exhibited comparable γH2AX signals (Figure 5f); this contrasts with WT RPE, in which multinucleated cells displayed a higher frequency of DNA damage than mononucleated cells (Figure 4c). These results suggest that mononucleated cells may be more sensitive to p53 reduction than multinucleated cells after DNA damage exposure.

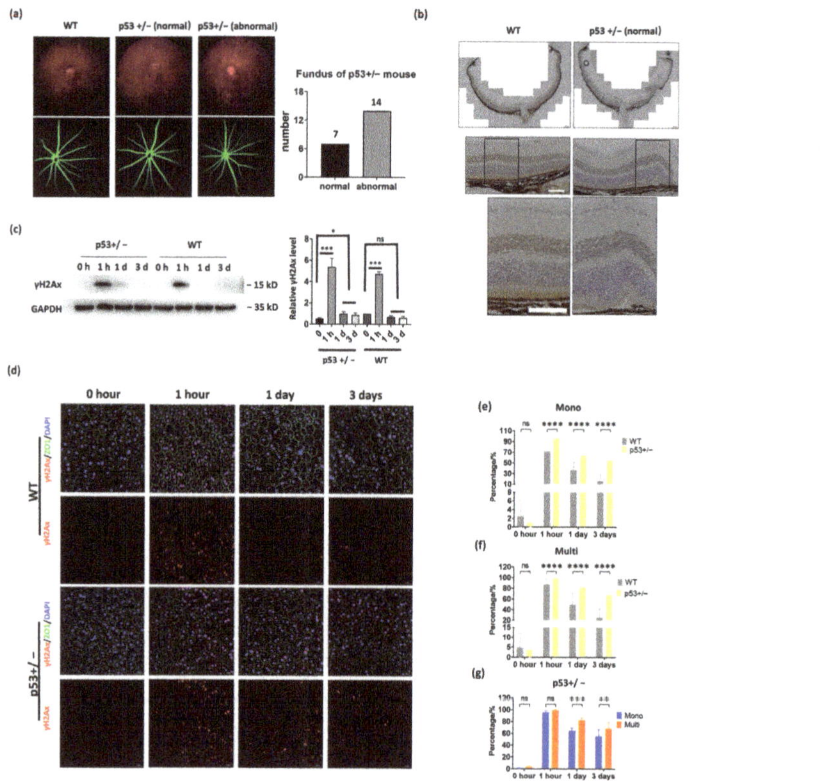

Figure 5. Induction of DNA damage in WT and p53+/− mouse retina. (**a**). Left: fundus photography (upper panels) and fluorescein angiography (lower panels) were performed to analyze WT and p53+/− mice eye morphology. Right: quantification results of the normal and abnormal eye fundus photography from p53+/− mice. (**b**). IHC of rhodopsin shows RPE and retina morphology in WT and p53+/− mice. Scale bar: upper: 200 μM, middle and bottom: 100 μM. (**c**–**g**). The WT and p53+/− mice were subject to 1 Gy of IR, and retinas or RPE cells were collected at the indicated time point post-IR. (**c**) WB analysis shows relative the indicated protein levels in mouse retinas. Right panels: the relative γH2Ax level was obtained by normalizing with GAPDH. ***: $p < 0.001$, ns: not significant. (**d**). Comparison of DNA damage response in WT and p53+/− RPE. IF analysis of γH2AX staining at the indicated time point post-IR. The cell cytoskeleton F-actin was labeled by FITC-phalloidin staining. Scale bar: 20 μm. (**e**–**g**). Quantitative analysis of the γH2AX-positive RPE cells in WT and p53+/− mice. WT: n = 4, p53+/−: n = 3. Total of 40 regions of WT group and 30 regions of p53+/− group were randomly selected, and the γH2AX-positive cells were counted and quantified. * $p < 0.05$, ** $p < 0.005$, *** $p < 0.001$, **** $p < 0.0001$, and ns: not significant. All data are shown as mean ± SD.

4. Discussion

In this study, we determined the composition of multinucleated RPE in several animal species. We revealed region but not age as the determining factor for the multinucleated cell amount in the RPE of mice. The multinucleated RPE is associated with the rods' percentage, which may be an adaption to nocturnal vision. Moreover, our results demonstrate that a multinucleated RPE has reduced efficiency in DNA damage repair, and p53 dosage change has a stronger impact on mononucleated cells than multinucleated cells after IR-induced DNA damage.

The existence of polyploidy cells might be a consequence of cellular stress or metabolic requirement. For example, polyploidy is important for cardiac muscle function under stressed conditions, and multinucleated mammary epithelial cells are essential for effective lactation [29,30]. Polyploidization may increase tissue metabolic capacity by enhancing transcriptional and translational output [31]. In this regard, the central region of the mouse retina possesses a higher photoreceptor cell density than the peripheral regions, underlining an increased phagocytic and metabolic burden for the central RPE [15]. Our study further correlated the distinct spatial distribution of multinucleated RPE with the rods' proportion and nocturnal vision adaption. Nocturnal animals have developed several unique ocular structures to maximize light collection; one example is the inverted heterochromatin structure in the rods' nuclei [22]. Currently, we do not know if a high proportion of multinucleated RPE is required for specific rod photoreceptor metabolism or for dim-light sensing in the dark. Nevertheless, to our knowledge, this is the first evidence linking multinucleated RPE with the rods' proportion and nocturnal vision. Further studies using neural retina leucine zipper (Nrl) knockout mice, in which rod photoreceptors are converted to cone photoreceptors [32], should directly address this point.

Due to a large number of multinucleated cells, the RPE of mice presents an ideal model to study whether multinucleated and mononucleated cells have different responses to DNA damage. Our results show, for the first time, that the multinucleated RPE cells exhibited increased DNA damage after IR. We speculate that multinucleated RPE cells are more prone to DNA damage in comparison with mononucleated RPE cells, due to the enhanced production of ROS, which is a potent DNA damage inducer. On the other side, we found the key DNA damage repair protein, 53BP1, showed similar foci formation efficiency in mononucleated and multinucleated RPE cells after IR, suggesting the impaired DNA damage repair is not due to the 53BP1 defect. A recent in vitro study showed that multinucleated human RPE1 cells exhibited more γH2AX-marked DNA damage and delayed formation and resolution of 53BP1 foci [33]. In that study, the in vitro cultured cell line was used and multinucleation was induced by disruption of normal cell mitosis [33]. We speculate that different cellular conditions, i.e., naturally occurred versus induced multinucleation, and in vitro versus in vivo environment, may result in different DNA damage responses in diverse multinucleated cells. The exact mechanism responsible for increased DNA damage in multinucleated RPE is still unknown, and systemic analysis of the DNA damage sensors, transducers, and effectors should provide valuable information.

Supplementary Materials: The following supporting information can be downloaded at: https://www.mdpi.com/article/10.3390/cells11091552/s1, Figure S1: Exclusion for Pde6brd1 and Crb1rd8 mouse strains.

Author Contributions: Q.K. and L.G. conducted the major experiments. X.Z., R.Q., M.Z. and W.L. performed WB analysis and IF analysis. B.C. and S.H. provided assisted with microscopy analysis. Y.L. discussed the manuscript. L.G and D.W.-C.L. designed the project and wrote the paper. All authors have read and agreed to the published version of the manuscript.

Funding: This work was supported by the National Natural Science Foundation of China (Grants #82070969, #81970787, #82000876), Guangdong Natural Science Foundation (2021A1515011793), the Joint Key Project of Natural Science Foundation of Guangdong; Province and Guangzhou City (2019B1515120014), and the Fundamental Research Fund of the State Key Laboratory of Ophthalmology (3030901010111), Zhongshan Ophthalmic Center, Sun Yat-sen University in China.

Institutional Review Board Statement: The Animal Use and Care Committee of Zhongshan Ophthalmic Center at the Sun Yat-Sen University Z2021088.

Data Availability Statement: All data from the study are given in the manuscript.

Acknowledgments: We thank the staff of Laboratory Animal Center and the staff of Core Facilities at State Key Laboratory of Ophthalmology, Zhongshan Ophthalmic Center for technical support. We thank Junjie Liang and Bixia Liang for providing the chicken and pigeon eyes.

Conflicts of Interest: We declare that we have not received any support from any companies for the submitted work and have no financial or nonfinancial interests that might be relevant to the submitted work.

References

1. Strauss, O. The role of retinal pigment epithelium in visual functions. *Ophthalmologe* **2009**, *106*, 299–304. [CrossRef] [PubMed]
2. Marneros, A.G.; Fan, J.; Yokoyama, Y.; Gerber, H.P.; Ferrara, N.; Crouch, R.K.; Olsen, B.R. Vascular Endothelial Growth Factor Expression in the Retinal Pigment Epithelium Is Essential for Choriocapillaris Development and Visual Function. *Am. J. Pathol.* **2005**, *167*, 1451–1459. [CrossRef]
3. Holtkamp, G.M.; Kijlstra, A.; Peek, R.; de Vos, A.F. Retinal pigment epithelium-immune system interactions: Cytokine production and cytokine-induced changes. *Prog. Retin. Eye Res.* **2001**, *20*, 29–48. [CrossRef]
4. Fuhrmann, S.; Zou, C.J.; Levine, E.M. Retinal pigment epithelium development, plasticity, and tissue homeostasis. *Exp. Eye Res.* **2014**, *123*, 141–150. [CrossRef] [PubMed]
5. Sharma, R.; Bose, D.; Maminishkis, A.; Bharti, K. Retinal Pigment Epithelium Replacement Therapy for Age-Related Macular Degeneration: Are We There Yet? *Annu. Rev. Pharmacol. Toxicol.* **2020**, *60*, 553–572. [CrossRef] [PubMed]
6. Orr-Weaver, T.L. When bigger is better: The role of polyploidy in organogenesis. *Trends Genet.* **2015**, *31*, 307–315. [CrossRef] [PubMed]
7. Schoenfelder, K.P.; Fox, D.T. The expanding implications of polyploidy. *J. Cell Biol.* **2015**, *209*, 485–491. [CrossRef]
8. Ts'o, M.O.; Friedman, E. The retinal pigment epithelium. I. Comparative histology. *Arch. Ophthalmol.* **1967**, *78*, 641–649. [CrossRef]
9. Starnes, A.C.; Huisingh, C.; McGwin, G., Jr.; Sloan, K.R.; Ablonczy, Z.; Smith, R.T.; Curcio, C.A.; Ach, T. Multi-nucleate retinal pigment epithelium cells of the human macula exhibit a characteristic and highly specific distribution. *Vis. Neurosci.* **2016**, *33*, e001. [CrossRef]
10. Chen, M.; Rajapakse, D.; Fraczek, M.; Luo, C.; Forrester, J.V.; Xu, H. Retinal pigment epithelial cell multinucleation in the aging eye—A mechanism to repair damage and maintain homoeostasis. *Aging Cell* **2016**, *15*, 436–445. [CrossRef]
11. Lin, H.; Xu, H.; Liang, F.Q.; Liang, H.; Gupta, P.; Havey, A.N.; Boulton, M.E.; Godley, B.F. Mitochondrial DNA damage and repair in RPE associated with aging and age-related macular degeneration. *Investig. Ophthalmol. Vis. Sci.* **2011**, *52*, 3521–3529. [CrossRef] [PubMed]
12. Hyttinen, J.M.; Blasiak, J.; Niittykoski, M.; Kinnunen, K.; Kauppinen, A.; Salminen, A.; Kaarniranta, K. DNA damage response and autophagy in the degeneration of retinal pigment epithelial cells—Implications for age-related macular degeneration (AMD). *Ageing Res. Rev.* **2017**, *36*, 64–77. [CrossRef] [PubMed]
13. Blasiak, J.; Piechota, M.; Pawlowska, E.; Szatkowska, M.; Sikora, E.; Kaarniranta, K. Cellular Senescence in Age-Related Macular Degeneration: Can Autophagy and DNA Damage Response Play a Role? *Oxidative Med. Cell. Longev.* **2017**, *2017*, 5293258. [CrossRef] [PubMed]
14. Sun, Q.; Gong, L.; Qi, R.; Qing, W.; Zou, M.; Ke, Q.; Zhang, L.; Tang, X.; Nie, Q.; Yang, Y.; et al. Oxidative stress-induced KLF4 activates inflammatory response through IL17RA and its downstream targets in retinal pigment epithelial cells. *Free Radic. Biol. Med.* **2020**, *147*, 271–281. [CrossRef]
15. Volland, S.; Esteve-Rudd, J.; Hoo, J.; Yee, C.; Williams, D.S. A Comparison of Some Organizational Characteristics of the Mouse Central Retina and the Human Macula. *PLoS ONE* **2015**, *10*, e0125631. [CrossRef]
16. Kaskan, P.M.; Franco, E.C.; Yamada, E.S.; Silveira, L.C.; Darlington, R.B.; Finlay, B.L. Peripheral variability and central constancy in mammalian visual system evolution. *Proc. R. Soc. B Boil. Sci.* **2005**, *272*, 91–100. [CrossRef]
17. Headington, K.; Choi, S.S.; Nickla, D.; Doble, N. Single cell imaging of the chick retina with adaptive optics. *Curr. Eye Res.* **2011**, *36*, 947–957. [CrossRef]
18. Meyer, D.B.; May, H.C., Jr. The topographical distribution of rods and cones in the adult chicken retina. *Exp. Eye Res.* **1973**, *17*, 347–355. [CrossRef]
19. Morris, V.B. Symmetry in a receptor mosaic demonstrated in the chick from the frequencies, spacing and arrangement of the types of retinal receptor. *J. Comp. Neurol.* **1970**, *140*, 359–398. [CrossRef]
20. Querubin, A.; Lee, H.R.; Provis, J.M.; O'Brien, K.M. Photoreceptor and ganglion cell topographies correlate with information convergence and high acuity regions in the adult pigeon (Columba livia) retina. *J. Comp. Neurol.* **2009**, *517*, 711–722. [CrossRef]
21. Harosi, F.I.; Novales Flamarique, I. Functional significance of the taper of vertebrate cone photoreceptors. *J. Gen. Physiol.* **2012**, *139*, 159–187. [CrossRef] [PubMed]

22. Solovei, I.; Kreysing, M.; Lanctot, C.; Kosem, S.; Peichl, L.; Cremer, T.; Guck, J.; Joffe, B. Nuclear architecture of rod photoreceptor cells adapts to vision in mammalian evolution. *Cell* **2009**, *137*, 356–368. [CrossRef] [PubMed]
23. Hendrickson, A.; Hicks, D. Distribution and density of medium- and short-wavelength selective cones in the domestic pig retina. *Exp. Eye Res.* **2002**, *74*, 435–444. [CrossRef]
24. Jeon, C.J.; Strettoi, E.; Masland, R.H. The major cell populations of the mouse retina. *J. Neurosci.* **1998**, *18*, 8936–8946. [CrossRef] [PubMed]
25. van der Merwe, I.; Lukats, A.; Blahova, V.; Oosthuizen, M.K.; Bennett, N.C.; Nemec, P. The topography of rods, cones and intrinsically photosensitive retinal ganglion cells in the retinas of a nocturnal (Micaelamys namaquensis) and a diurnal (Rhabdomys pumilio) rodent. *PLoS ONE* **2018**, *13*, e0202106. [CrossRef]
26. Famiglietti, E.V.; Sharpe, S.J. Regional topography of rod and immunocytochemically characterized "blue" and "green" cone photoreceptors in rabbit retina. *Vis. Neurosci.* **1995**, *12*, 1151–1175. [CrossRef]
27. Frohns, A.; Frohns, F.; Naumann, S.C.; Layer, P.G.; Lobrich, M. Inefficient double-strand break repair in murine rod photoreceptors with inverted heterochromatin organization. *Curr. Biol.* **2014**, *24*, 1080–1090. [CrossRef]
28. Ikeda, S.; Hawes, N.L.; Chang, B.; Avery, C.S.; Smith, R.S.; Nishina, P.M. Severe ocular abnormalities in C57BL/6 but not in 129/Sv p53-deficient mice. *Investig. Ophthalmol. Vis. Sci.* **1999**, *40*, 1874–1878.
29. Millay, D.P.; O'Rourke, J.R.; Sutherland, L.B.; Bezprozvannaya, S.; Shelton, J.M.; Bassel-Duby, R.; Olson, E.N. Myomaker is a membrane activator of myoblast fusion and muscle formation. *Nature* **2013**, *499*, 301–305. [CrossRef]
30. Rios, A.C.; Fu, N.Y.; Jamieson, P.R.; Pal, B.; Whitehead, L.; Nicholas, K.R.; Lindeman, G.J.; Visvader, J.E. Essential role for a novel population of binucleated mammary epithelial cells in lactation. *Nat. Commun.* **2016**, *7*, 11400. [CrossRef]
31. Pandit, S.K.; Westendorp, B.; de Bruin, A. Physiological significance of polyploidization in mammalian cells. *Trends Cell Biol.* **2013**, *23*, 556–566. [CrossRef] [PubMed]
32. Daniele, L.L.; Lillo, C.; Lyubarsky, A.L.; Nikonov, S.S.; Philp, N.; Mears, A.J.; Swaroop, A.; Williams, D.S.; Pugh, E.N., Jr. Cone-like morphological, molecular, and electrophysiological features of the photoreceptors of the Nrl knockout mouse. *Investig. Ophthalmol. Vis. Sci.* **2005**, *46*, 2156–2167. [CrossRef] [PubMed]
33. Hart, M.; Adams, S.D.; Draviam, V.M. Multinucleation associated DNA damage blocks proliferation in p53-compromised cells. *Commun. Biol.* **2021**, *4*, 451. [CrossRef] [PubMed]

cells

Article

Influences of Long-Term Exercise and High-Fat Diet on Age-Related Telomere Shortening in Rats

Maria Donatella Semeraro, Gunter Almer, Wilfried Renner, Hans-Jürgen Gruber * and Markus Herrmann

Clinical Institute of Medical and Chemical Laboratory Diagnostics (CIMCL), Medical University of Graz, 8036 Graz, Austria; maria.semeraro@medunigraz.at (M.D.S.); gunter.almer@medunigraz.at (G.A.); wilfried.renner@medunigraz.at (W.R.); markus.herrmann@medunigraz.at (M.H.)
* Correspondence: hans.gruber@medunigraz.at

Abstract: (1) Obesity and exercise are believed to modify age-related telomere shortening by regulating telomerase and shelterins. Existing studies are inconsistent and limited to peripheral blood mononuclear cells (PBMCs) and selected solid tissues. (2) Female Sprague Dawley (SD) rats received either standard diet (ND) or high-fat diet (HFD). For 10 months, half of the animals from both diet groups performed 30 min running at 30 cm/s on five consecutive days followed by two days of rest (exeND, exeHFD). The remaining animals served as sedentary controls (coND, coHFD). Relative telomere length (RTL) and mRNA expression of telomerase (TERT) and the shelterins TERF-1 and TERF-2 were mapped in PBMCs and nine solid tissues. (3) At study end, coND and coHFD animals showed comparable RTL in most tissues with no systematic differences in TERT, TERF-1 and TERF-2 expression. Only visceral fat of coHFD animals showed reduced RTL and lower expression of TERT, TERF-1 and TERF-2. Exercise had heterogeneous effects on RTL in exeND and exeHFD animals with longer telomeres in aorta and large intestine, but shorter telomeres in PBMCs and liver. Telomere-regulating genes showed inconsistent expression patterns. (4) In conclusion, regular exercise or HFD cannot systematically modify RTL by regulating the expression of telomerase and shelterins.

Keywords: telomeres; telomerase; shelterin; moderate exercise; high-fat diet; Sprague Dawley rats

Citation: Semeraro, M.D.; Almer, G.; Renner, W.; Gruber, H.-J.; Herrmann, M. Influences of Long-Term Exercise and High-Fat Diet on Age-Related Telomere Shortening in Rats. *Cells* **2022**, *11*, 1605. https://doi.org/10.3390/cells11101605

Academic Editors: Nicole Wagner and Kay-Dietrich Wagner

Received: 15 April 2022
Accepted: 9 May 2022
Published: 10 May 2022

1. Introduction

The shortening of telomeres, protective nucleoprotein structures at the end of all chromosomes, is a hallmark of aging that compromises genomic integrity and alters the expression of many genes [1]. Due to the inability of the DNA polymerase to fully replicate the 3′ end of chromosomes, telomeres progressively shorten with every cell division until they reach a critical threshold below which they lose their DNA-protecting properties and send cells into senescence or apoptosis [2]. Numerous studies have shown that short and dysfunctional telomeres are linked to premature atherosclerosis, diabetes, and hypertension [3–7]. Furthermore, telomere length is inversely related to mortality risk [8–11].

Aging is an individual process that can be influenced by modifiable lifestyle factors, such as physical activity, nutrition, stress, sleep, smoking and others [12–21]. Physical inactivity and obesity are established triggers of metabolic dysfunction, chronic inflammation, and oxidative stress, which increase the risk of atherosclerosis, diabetes, hypertension, dementia, and other age-related diseases [22,23]. Based on previous studies, it has been speculated that physical inactivity and obesity also accelerate telomere attrition and promote telomeric dysfunction [24]. Conversely, it has been proposed that regular exercise and a balanced diet promote healthy aging not only through beneficial effects on body composition, metabolic function, vascular function, blood pressure, inflammatory processes, and mental stress [25], but also through the preservation of telomere length and function [13,26–30]. It has further been hypothesized that lifestyle-induced effects on telomeres are mediated through telomere-regulating proteins, such as telomerase and shelterins [16,18,19,21,27,31]. Telomerase can counteract telomere shortening by adding new hexanucleotides to the telomeric ends. With the help of shelterins, a complex of six individual proteins, telomeres assume a unique three-dimensional structure that is essential for their function. Upon binding of the shelterin complex to the TTAGGG motif, telomeric DNA folds backward forming a structure known as t-loop [32,33]. A breakdown of the t-loop structure, called telomere uncapping, represents a critical mechanism that promotes age-related vascular dysfunctions, cellular senescence and inflammation beyond telomere shorting [34].

Obesity and physical inactivity are highly prevalent in modern societies [35]. According to the World Health Organization (WHO), approximately 30% of the global population is obese [36] and the Centre of Disease Control in the US has estimated an overall prevalence of physical inactivity in the US of approximately 25% [37]. Despite intensive research activities, the mechanisms that mediate the increased risk of chronic degenerative diseases in obese and inactive individuals are incompletely understood. Previous studies have nurtured the idea that both of these lifestyle factors could increase disease risk and mortality through an enhancement of telomere shortening that compromises genomic integrity [38]. However, the results of observational studies are controversial, and experimental evidence that establishes a mechanistic link between obesity, physical inactivity and accelerated telomere shortening is largely lacking. Several observational studies showed an inverse relationship between telomere length (TL) in leucocytes (LTL) and BMI [39,40], whereas others found the opposite [41] or no significant association [42,43]. Inverse correlations have also been reported for LTL and different indices of body composition, such as body fat content, waist circumference, waist-to-hip ratio, and nuchal fat thickness [43–50]. In contrast, two mouse models of obesity and metabolic syndrome failed to show accelerated telomere shortening despite an upregulation of telomerase and senescence-associated genes, such as checkpoint kinase 2 (*Chk2*), *p53*, and *p21* [22,23].

Considering that exercise is a highly cost-effective way to improve health and to prolong life [51–55], obese individuals are often prescribed a physical activity program with moderate endurance exercise, such as walking or cycling [56]. Observational studies have reported higher LTL in exercising individuals of different age groups and different activity levels [15,17,18,57,58]. However, available prospective and interventional studies provide conflicting results. In a 5-year longitudinal study by Soares-Miranda L et al., physical activity and physical performance were unrelated to LTL [59]. In contrast, Werner et al., showed increased LTL and an upregulation of telomerase and telomere repeat binding factor (TRF) 2 after 6 months of aerobic endurance training or high intensity training [17,18]. The results from animal studies are also inconclusive. While Ludlow et al. showed a preservation of TL in cardiomyocytes and hepatocytes of exercising mice [16,17,21,60], Werner et al. did not find differences between cardiac TL of exercising and sedentary mice [16,17,21,60]. Regardless of potential effects on TL, exercise seems to alter the expression telomere-regulating proteins, such as telomerase and shelterins [60].

Whether or not obesity and physical activity are causally related to telomere length and the expression of telomere-regulating proteins remains elusive. Furthermore, previous studies are limited to analyses of TL in leucocytes, myocardium, skeletal muscle, and liver. Additionally, potential interactions between the consumption of a hypercaloric diet and regular exercise has not been studied systematically. This aspect is of particular relevance as exercise is often used to compensate bad eating habits and to treat obesity. Therefore, the present study aimed to address this gap of knowledge by mapping TL in leucocytes and 9 solid tissues from aged sedentary rats that were fed for 10 months either a normal chow-based diet (ND) or a synthetic high-fat diet (HFD). In order to explore potential interactions between the consumption of HFD and exercise, half of the animals from both groups performed regular treadmill running with moderate intensity.

2. Materials and Methods

2.1. Animal Model

Ninety-six female Sprague Dawley (SD) rats were purchased from Janvier Labs (Le Genest-Saint-Isle, France) at four months of age. The animals were kept in groups of three animals per cage under constant conditions on a 12 h/12 h light/dark cycle at the core facility animal housing at the Medical University of Graz (Austria). Temperature was maintained between 22 and 25°C. Humidity ranged between 55 to 58%. After one week of acclimatization, the animals were randomly assigned to receive either a standard diet (ND) (Altromin, Lage, Germany) with 3230 kcal/kg and 11% fat or a custom-designed beef-tallow high-fat diet (HFD), rich in saturated fatty acids (SFA), in particular C16:0 and C18:0, with 5150 kcal/kg and 60% fat (Table 1; ssniff, Soest, Germany). Food and tap water were provided ad libitum.

Table 1. Organ weight in female SD rats after 10 months of treadmill exercise.

Organs	Measurement	ND		HFD	
		Sedentary $n = 22$	Exercising $n = 22$	Sedentary $n = 16$	Exercising $n = 12$
heart	average weight	1.31 ± 0.21	1.24 ± 0.11	1.40 ± 0.14	1.46 ± 0.19 ###
	normalized weight	0.28 ± 0.04	0.27 ± 0.03	0.30 ± 0.03	0.31 ± 0.03 ###
spleen	average weight	0.98 ± 0.16	0.97 ± 0.15	1.20 ± 0.16 ###	1.18 ± 0.23 ##
	normalized weight	0.21 ± 0.03	0.21 ± 0.03	0.24 ± 0.07	0.25 ± 0.04 ###
liver	average weight	12.53 ± 1.72	12.50 ± 1.80	14.03 ± 2.36 #	15.16 ± 4.40 #
	normalized weight	2.67 ± 0.31	2.56 ± 0.68	2.98 ± 0.52 #	3.28 ± 0.89 #
visceral fat	average weight	13.20 ± 5.26	10.46 ± 4.48	40.13 ± 12.81 ###	39.46 ± 23.20 ###
	normalized weight	0.03 ± 0.01	0.03 ± 0.01	0.08 ± 0.018 ###	0.07 ± 0.03 ###

Organ weight is given in grams. The weights of heart, spleen, and liver were normalized to total tibia length (cm), while visceral fat weight was normalized to body weight (g). Data are presented as mean ± SD; # $p < 0.05$, ## $p < 0.01$, ### $p < 0.001$ compared to the appropriate normal diet control group with the two-tailed Student's t-test for independent samples.

2.2. Experimental Design and Treatment

Animals were randomly allocated to following 4 groups, each consisting of 24 animals: coND, exeND, coHFD and exeHFD. coND and exeND animals were fed with ND for the entire study period, whereas coHFD and exeHFD received HFD. The animals from exeND and exeHFD groups performed a 10-month exercise program consisting of 30 min of forced running on a treadmill (Panlab, Barcelona, Spain) on five consecutive days followed by 2 days of rest. The running speed was constant and set at 30 cm/s. The training protocol was based on previous experimental studies [61–64]. The animals in the coND and coHFD groups did not exercise and had no access to a running wheel. These animals were used as sedentary controls.

2.3. Euthanasia and Sample Preparation

At the time of scarification, blood was collected by heart puncture into plasma-EDTA and serum tubes (Sarstedt, Nümbrecht, Germany) under deep isoflurane anaesthesia (Forane, Abbott, Austria). After centrifugation at 2000× g for 12 min at room temperature, plasma and serum samples were aliquoted and stored at −80°C until batched analysis. Immediately after blood collection, the following organs were explanted and snap frozen in liquid nitrogen: liver, skeletal muscle, heart, aorta, large intestine, spleen, kidney, brain, lung, visceral fat. Subsequently, all tissue samples were stored together deep-frozen at −80°C until analysis. Exclusion criteria were the development of illnesses or tumours during the intervention period.

2.4. Analysis of Relative Telomere Length (RTL) in PBMCs and Solid Organs

After diluting 100 µL of whole blood with 100 µL of dH_2O, DNA was isolated with the MagNA Pure LC instrument (Roche, Austria) using the Total Nucleic Isolation Kit (Roche, Austria). Subsequently, relative telomere length (RTL) of peripheral blood mononuclear cells (PBMCs) was measured by quantitative real-time PCR (qPCR) using a protocol developed by Cawthon [65]. This assay quantifies the ratio of average TL (T) to glyceraldehyde-3-phosphate dehydrogenase (GAPDH) as single copy reference gene (S). The single copy gene is used as amplification control for each sample and to determine the number of genome copies per sample. All qPCR analyses were performed on a Thermocycler CFX384 TouchTM (Biorad, Feldkirchen, Germany) instrument using the following primers:

1. Telomere For: 5′-CGGTTTGTTTGGGTTTGGGTTTGGGTTTGGGTTTGGGTT-3′;
2. Telomere Rev: 3′-GGCTTGCCTTACCCTTACCCTTACCCTTACCCTTACCCT-5′;
3. GAPDH For: 5′-CACCTAGACAAGGATGCAGAG-3′;
4. GAPDH Rev: 3′-GCATGACTGGAGGAATCACA-5′.

All primers have been purchased from Eurofins Genomics, Austria. Each run included a standard curve made by dilutions of isolated and pooled rat DNA from 21 different blood samples, to determine the quantity of the targeted templates. RTL has been calculated as the ratio of telomere quantity to single copy reference gene quantity (T/S ratio).

RTL in solid organs was analysed with the same method described before. For this purpose, approximately 10 mg of tissue were homogenised in 300 µL Magna Pure Lysis Buffer (Roche, Wien, Austria) using the MagnaLyser (Roche, Wien, Austria). Subsequently, the DNA was isolated and quantified using the same procedure as for blood leucocytes.

2.5. The mRNA Expression Analyses in Blood Cells and Solid Organs

TERT, TERF-1, and TERF-2 gene expression was analysed in RNA extracts of all solid organs. As blood leucocytes were used up for the measurement of RTL, they were not available for mRNA expression analyses. Therefore, mRNA expression in spleen was used as reference because the organ belongs to the lymphatic system and is rich in leucocytes. From each organ, 10 mg of tissue were homogenised in 300 µL Magna Pure Lysis Buffer (Roche, Wien, Austria) using the MagnaLyser (Roche, Wien, Austria). RNA was extracted from these homogenates with the Total Nucleic Isolation Kit (Roche, Wien, Austria) on a MagNA Pure LC instrument (Roche, Wien, Austria). Subsequently, the mRNA in these extracts was transcribed into cDNA using the QuantiTect Reverse Transcription kit (Qiagen, Hilden, Germany). Finally, mRNA expression of TERT, TERF-1, and TERF-2 was analysed by qPCR with TaqMan probes (Life Technologies dba Invitrogen, Waltham, MA, USA). The expression of each target gene expression was calculated with the $\Delta\Delta$CT method using β-actin as reference gene. The sequences of the probes used were as follows:

5. B-actin: 5′-CTTCCTTCCTGGGTATGGAATCCTG-3′;
6. Tert: 5′-ATCGAGCAGAGCATCTCCATGAATG-3′;
7. Terf-1: 5′-AAAACAGACATGGCTTTGGGAAGAA-3′;
8. Terf-2: 5′-GAGAAAATTTAGACTGTTCCTTTGA-3′.

2.6. Statistical Analyses

Results are shown as mean ± standard deviations (SD). Qualitative variables such as tumor abundance and type were assessed with the Fisher's exact test or the Chi-squared test. Group differences were assessed using the two-tailed Student's t test for dependent or independent samples or the Mann–Whitney U test depending on the distribution of the data. Group comparisons with three or more groups were analysed using the two-way ANOVA or the Kruskall–Wallis test for independent samples. Correlations between variables were determined by linear regression analysis according to Pearson (r, Pearson Correlation coefficient; p, univariate ANOVA). Data were plotted using Python programming language with Jupyter Notebook within the data science package Anaconda3 for Windows. IBM

SPSS v. 26 for Windows was used for explorative data analysis and a level of acceptance of the null hypothesis was set at $p = 0.05$.

3. Results

3.1. Characterization of the Animal Model

From the 96 rats, 6 were excluded prior to the end of the study due to general health issues. A total 18 animals developed benign tumours and, thus, were excluded from the final analysis. Tumours were more frequent in animals on HFD rather than on ND (16 vs. 2 rats, $p = 1.289 \times 10^{-4}$). The tumours in the HFD animals were of heterogeneous nature compared to the ND group ($p < 0.001$), as masses were found in breasts, ovaries, and abdomen of obese animals. Regular exercise did not significantly change tumor incidence in both diet groups (coND vs. exeND, $p = 0.975$; coHFD vs. exeHFD, $p = 0.347$) nor tumor diversity in the HFD group ($p = 0.197$). After exclusion of dropouts, 72 eligible animals were included in the final statistical analyses.

At study end, the animals in the two HFD groups had a significantly higher body weight than those in the ND groups (Figure 1). Median body weight between ND and HFD differed by 115 g in sedentary animals and by 90 g in exercising animals. In line with this finding, also the weight of individual organs and tissues, such as heart, liver, and visceral fat, was significantly higher in HFD animals (Table 1).

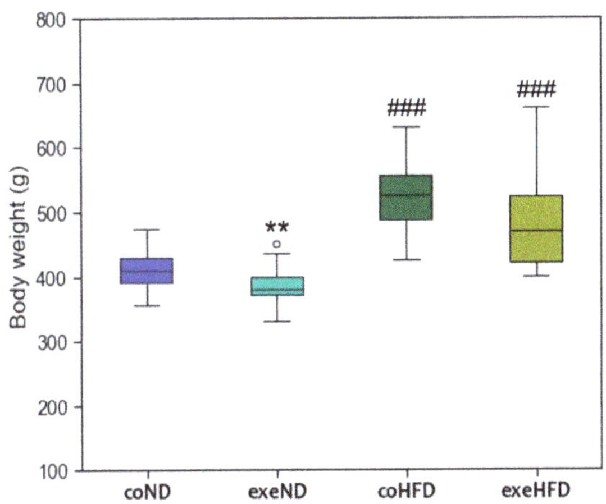

Figure 1. Box and Whisker plot of the body weight at the end of the 10 months study period. Outliers are shown as white circles above the box plots. The two-tailed Student's *t*-test was used for group comparison of independent samples. ** $p < 0.01$ compared to appropriate sedentary control group; ### $p < 0.001$ compared to appropriate normal diet control group.

The exercise protocol was well tolerated by the animals of both diet groups. Body weight of exeND animals was significantly lower than that of coND animals ($p < 0.01$), whereas coHFD and exeHFD animals showed no difference. In the factorial ANOVA, the main effects of diet and exercise on body weight were significant with F (1, 67) = 80.92, $p = 3.89 \times 10^{-13}$, and F (1, 67) = 8.29, $p = 0.005$, respectively. There was no significant interaction between diet and exercise, F (1, 67) = 0.138, $p = 0.712$. Regular exercise induced a higher organ weight of heart and liver in HFD animals, but not in ND animals (Table 1).

3.2. Influence of HFD on RTL and the Expression of Telomere-Regulating Genes in Different Tissues

When compared to ND, 10 months of HFD consumption had no systematic effect on RTL across different organs. In visceral fat RTL was significantly lower in coHFD animals than in coND animals. In contrast, renal RTL was slightly higher in coHFD than in coND animals. All other solid tissues and PBMCs showed comparable RTL between the two diet groups. (Figure 2).

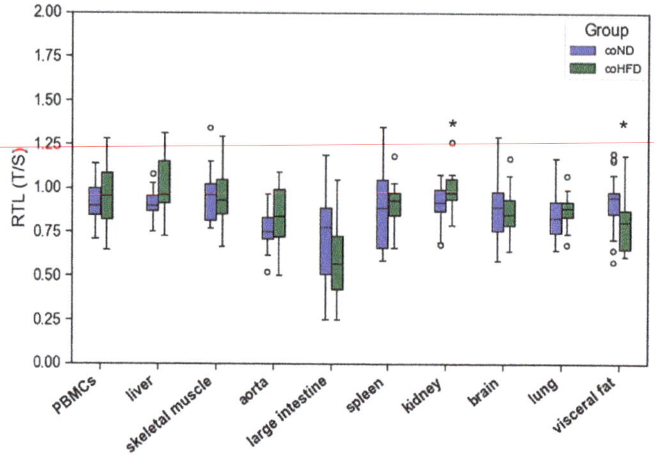

Figure 2. Distribution of RTL in PBMCs and nine solid organs isolated from lean (coND) and obese rats (coHFD). Outliers are shown as white circles above the box plots. RTL is expressed as ratio of average telomere length to the reference gene GAPDH. The two-tailed Student's *t*-test or the Mann–Whitney U-test were used for group comparison of independent samples. * $p < 0.05$ vs. coND.

TERT mRNA expression varied substantially between tissues with highest expression levels in liver and kidney. The consumption of HFD did not result in a systematic difference of TERT mRNA expression across different organs (Figure 3). Spleen, large intestine, and kidney showed higher TERT mRNA expression levels in coHFD than in coND animals, whereas in visceral fat a lower TERT mRNA expression was observed.

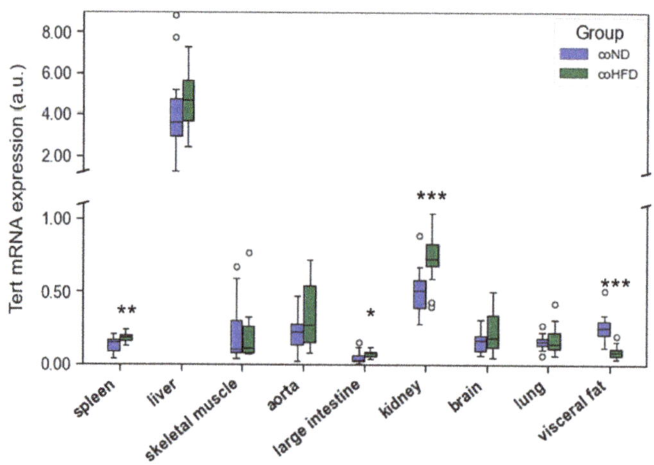

Figure 3. Differences in the TERT gene expression isolated from different solid organs of lean (coND) and obese rats (coHFD). Outliers are shown as white circles above the box plots. TERT mRNA expression is shown in arbitrary units. The two-tailed Student's *t*-test or the Mann–Whitney U-test were used for group comparison of independent samples. * $p < 0.05$; ** $p < 0.01$; *** $p < 0.001$ vs. coND.

Similar to TERT, also mRNA expression of the two shelterins, TERF-1 and TERF-2, varied substantially between tissues. The consumption of HFD upregulated TERF-1 and TERF-2 mRNA expression in 3 (Figure 4a) and 5 (Figure 4b) out of nine tissues, respectively. In contrast, a reduced mRNA expression of both shelterins was seen in only one and two tissues, respectively. A simultaneous upregulation of TERF-1 and TERF-2 in coHFD animals was found in skeletal muscle, aorta, and large intestine. In contrast, visceral fat showed a lower mRNA expression of both shelterins in these animals. Furthermore, TERF-1 was markedly reduced in the liver of coHFD rats, whereas TERF-2 was increased in spleen and kidney but decreased in lung tissue.

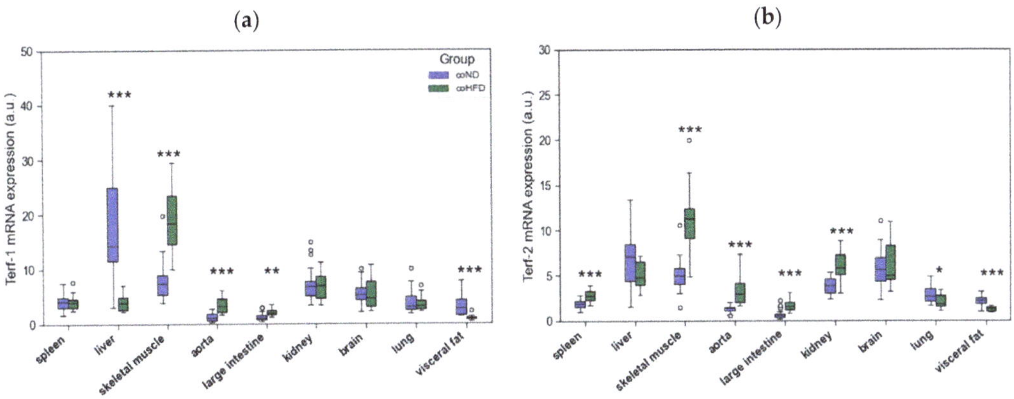

Figure 4. Differences in the gene expression of (**a**) TERF-1, and (**b**) TERF-2 isolated from different solid organs of lean (coND) and obese rats (coHFD). Outliers are shown as white circles above the box plots. TERF-1, and TERF-2 mRNA expression is shown in arbitrary units. The two-tailed Student's t-test or the Mann–Whitney U-test were used for group comparison of independent samples. * $p < 0.05$; ** $p < 0.01$; *** $p < 0.001$ vs. conD.

Combining all the differences described before in an effect matrix, it becomes apparent that only in visceral fat HFD consistently reduces RTL and the mRNA expression of telomere-regulating genes (Figure S1). Instead, in four out of nine tissues mRNA expression of one or more telomere-regulating genes was increased without a change in RTL.

3.3. *Influence of Exercise on RTL and the Expression of Telomere-Regulating Genes in Different Tissues*

Ten months of regular treadmill running had heterogeneous effects on RTL in different tissues with significantly longer telomeres in aorta and large intestine tissue, but shorter telomeres in PBMCs and liver RTL (Figure 5). In all other tissues, RTL did not significantly differ between sedentary and exercising animals. Of note, the simultaneous administration of HFD did not substantially change this pattern.

Exercise had vastly different effects on mRNA expression of TERT, TERF-1 and TERF-2 in different tissues. In some tissues, but not all, HFD altered the exercise-induced effects observed in ND animals. TERT mRNA expression was increased in spleen, liver, kidney, and lung of exeND animals compared to coND animals (Figure 6a). Conversely, in exeHFD animals, TERT expression in large intestine and kidney was significantly lower than in coHFD, whereas spleen, liver and lung showed no differences (Figure 6b).

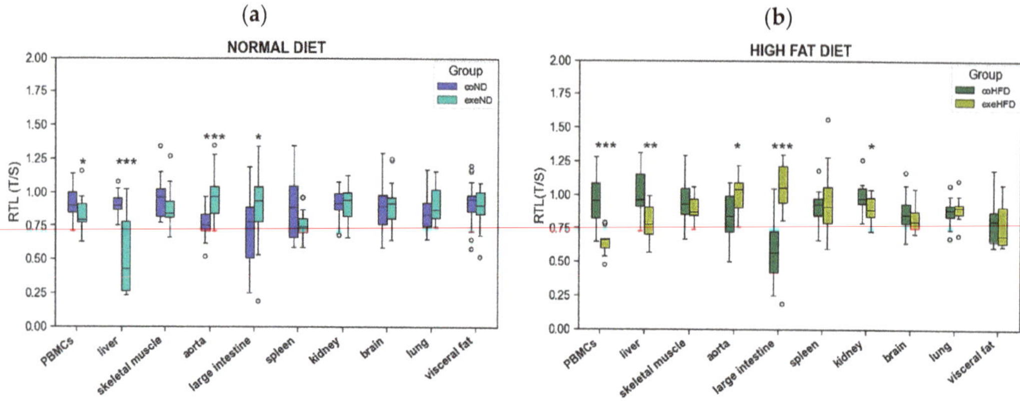

Figure 5. Comparison of RTL in PBMCs and nine solid organs isolated from exercising and sedentary SD rats that received normal diet or HFD for 10 months. (**a**) sedentary (coND) vs. exercising (exeND) animals on ND, (**b**) sedentary (coHFD) vs. exercising (exeHFD) animals on HFD. Outliers are shown as white circles above the box plots. RTL is expressed as ratio of average telomere length to the reference gene GAPDH. The two-tailed Student's t-test or the Mann–Whitney U-test were used for group comparison of independent samples. * $p < 0.05$, ** $p < 0.01$, *** $p < 0.001$ vs. respective sedentary controls.

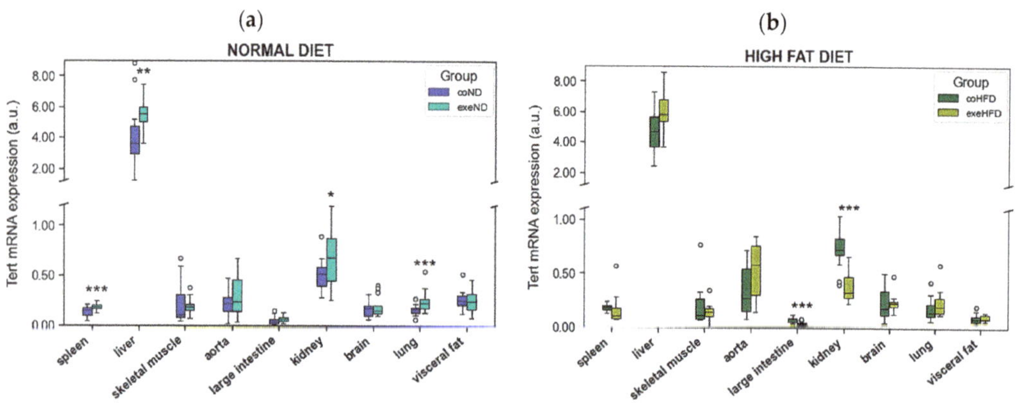

Figure 6. Comparison of TERT expression in nine solid organs from exercising and sedentary SD rats that received either normal diet or HFD for 10 months. (**a**) sedentary lean animals (coND) vs. exercising lean animals (exeND), (**b**) sedentary obese animals (coHFD) vs. exercising obese animals (exeHFD). Outliers are shown as white circles above the box plots. TERT mRNA expression is shown in arbitrary units. The two-tailed Student's *t*-test or the Mann–Whitney U-test were used for group comparison of independent samples. * $p < 0.05$; ** $p < 0.01$; *** $p < 0.001$ vs. respective sedentary controls.

TERF-1 mRNA expression was significantly reduced in liver, lung, and visceral fat, but increased in skeletal muscle, aorta, and large intestine of exeND rats when compared to coND animals (Figure 7a). In exeHFD animals instead, TERF-1 mRNA expression was increased in spleen, liver, large intestine, and kidney, but reduced in skeletal muscle when compared to coHFD animals (Figure 7c).

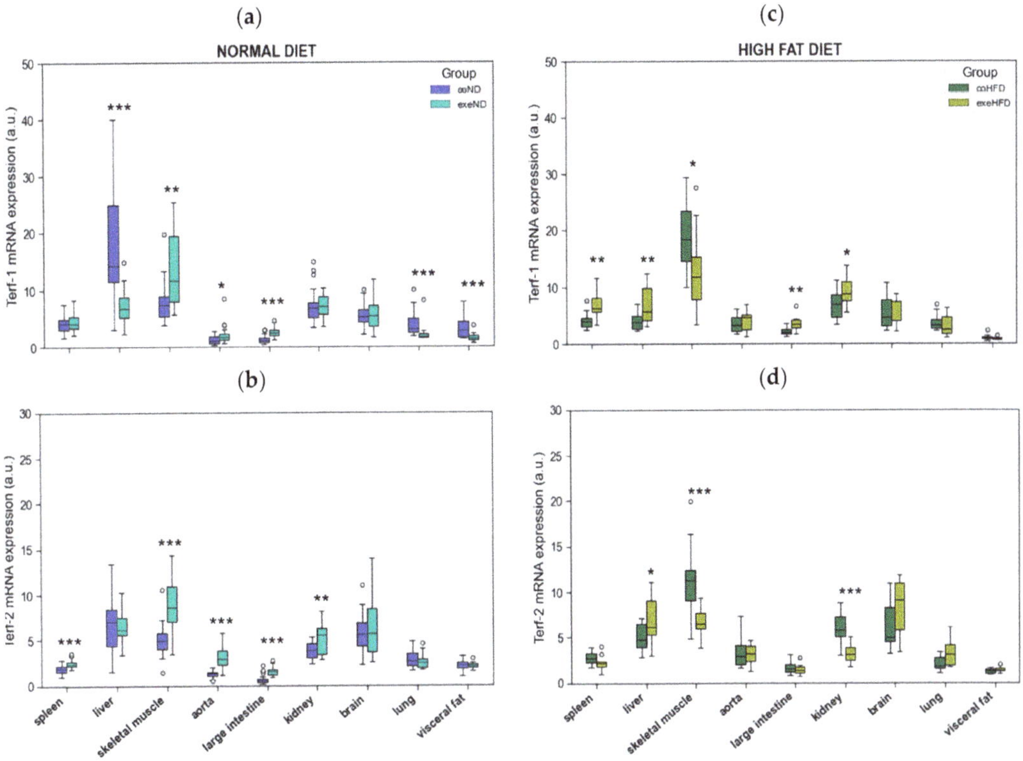

Figure 7. Comparison of TERF-1 and TERF-2 expression in nine solid organs from exercising and sedentary SD rats that received either normal diet or HFD for 10 months. (**a**) TERF-1 in sedentary lean animals (coND) vs. exercising lean animals (exeND), (**b**) TERF-2 in sedentary lean animals (coND) vs. exercising lean animals (exeND) (**c**) TERF-1 in sedentary obese animals (coHFD) vs. exercising obese animals (exeHFD), (**d**) TERF-2 in sedentary obese animals (coHFD) vs. exercising obese animals (exeHFD). Outliers are shown as white circles above the box plots. TERF-1 and TERF-2 mRNA expression is shown in arbitrary units. The two-tailed Student's *t*-test or the Mann–Whitney U-test were used for group comparison of independent samples. * $p < 0.05$; ** $p < 0.01$; *** $p < 0.001$ vs. respective sedentary controls.

TERF-2 mRNA expression was significantly increased in five out of nine solid organs of exeND animals when compared to coND rats (Figure 7b), namely spleen, skeletal muscle, aorta, large intestine, and kidney. In exercising obese animals instead, TERF-2 mRNA expression was profoundly different with a higher expression level in liver, but reduced expression levels in skeletal muscle and kidney in exeHFD rats when compared to coHFD animals (Figure 7d).

Summarizing all the results from sedentary and exercising lean animals in the effect matrix (left column), it becomes clear that only in aorta and large intestine an increase in one or more telomere-regulating genes was associated with an increase in RTL (Figure S1). All other differences in mRNA expression of telomere-regulating genes were unrelated to RTL. Similar to exercise, also HFD failed to induce systematic effects on RTL and telomere-regulating genes. Only in kidney and visceral fat of obese sedentary animals, did RTL and telomere-regulating genes show changes directed in the same way. However, the changes in both tissues pointed in opposite directions. An interaction between HFD and exercise was only observed in kidneys, where exercising obese rats exhibited a similar RTL to sedentary lean controls (Figure S1). Additionally, both exercising groups show reductions in hepatic RTL, but inconsistent changes in the hepatic expression of telomere-regulating genes.

3.4. Correlation Analysis

To further explore our hypothesis that lifestyle factors can modify RTL through regulatory effects on the expression of telomere-regulating genes, we performed correlation analyses that included the animals from all four groups. Figure 8 illustrates that RTL was not consistently correlated to any of the telomere-regulating genes. For example, in kidney ($R = 0.337$; $p = 0.004$) and visceral fat ($R = 0.337$; $p = 0.004$) RTL and TERT mRNA expression were positively correlated, whereas large intestine ($R = -0.252$; $p = 0.036$) and spleen ($R = -0.263$; $p = 0.028$) showed the opposite. Likewise, RTL and TERF-2 were positively correlated in aorta ($R = 0.373$; $p = 0.002$), kidney ($R = 0.318$; $p = 0.007$) and visceral fat ($R = 0.332$; $p = 0.007$), but negatively correlated in liver ($R = -0.247$; $p = 0.053$).

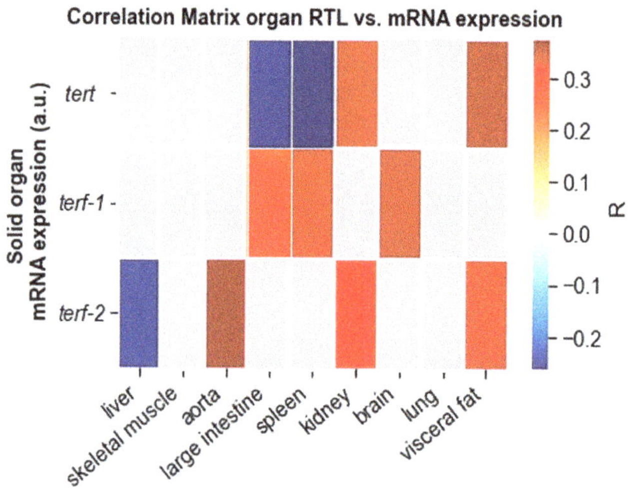

Figure 8. Correlation matrix for RTL and telomere-regulating genes in 9 solid tissue types. The colour bar on the right side of the figure shows the Pearson correlation coefficient as R. Only significant correlations are shown in colour ($p \leq 0.05$). Grey boxes indicate the lack of significant correlation.

4. Discussion

Here, we show for the first time that neither regular exercise nor the consumption of HFD have a systematic effect on RTL in solid tissues and PBMCs of SD rats. In fact, most tissues had comparable RTL in the respective intervention and control groups. Additionally, dual stimulation by feeding HFD to exercising animals did not change this result. Nevertheless, some tissues exhibited significantly higher RTL after 10 months of HFD (kidney) or exercise (aorta and small intestine), whereas other tissues showed reduced RTL upon HFD (visceral fat) or exercise (PBMCs and liver). These differences were not accompanied by a consistent mRNA expression pattern of the respective telomere-regulating genes *tert*, *terf-1* and *terf-2*. Therefore, the present results do not support the hypothesis that regular moderate endurance exercise or prolonged exposure to a diet rich in saturated lipids influence RTL through the expression of telomerase and shelterins.

The comprehensive mapping of RTL and related telomere-regulating genes after long-term exposure to HFD and exercise significantly expands existing knowledge on the influence of modifiable lifestyle factors on age-related telomere shortening. Previous in vivo studies have mostly focused on RTL in specific cells or tissue types, such as PBMCs or myocardium [16,17,21,60]. The results are rather inconsistent. Ludlow et al. showed in wild type derived short telomere mice (CAST/Ei) that 1 year of voluntary running in a running wheel preserved TL in myocardium and liver, but not in skeletal muscle [60]. Similar to the present study, these effects were not accompanied by consistent alterations of telomere-regulating genes that would explain these effects. In contrast, after 3 weeks of voluntary wheel running, Werner et al. reported an upregulation of telomerase activity (TA) in murine aorta and PBMCs, and an increased aortic gene expression of TERF-2. Additionally, senescence-associated genes, such as Chk2, p53, and p21, were lowered in the aorta of exercising animals. However, the increased expression of these telomere-regulating genes did not result in a significant difference of aortic TL after 6 months of exercise when compared to inactive controls [17]. The TL results reported by the two studies of Ludlow et al. and Werner et al. are not in line with our findings, where 10 months of regular moderate running exercise reduced RTL in PBMCs and liver, whereas aorta and large intestine showed a significant increase.

The inconsistencies between existing exercise studies in animals may, at least partly, be explained by differences in the animal models used. Our results were obtained in SD rats, whereas previous studies worked with CAST/Ei [60] and C57/Bl6 mice [17]. Additionally, the duration of exercise varied amongst existing studies between a few weeks and one year, which further limits comparability. In addition, a greater group size with 22 coND and 22 exeND animals provides robustness to our results. The studies from Ludlow et al. and Werner et al. were performed with no more than 10 animals per group, which limits statistical power and leaves more room for random effects. A major strength of the present study is strict standardization of the exercise intervention, which consisted in forced treadmill running for 30 min at fixed speed on 5 consecutive days per week. The efficacy of this intervention is evidenced by a significantly lower body weight at the time of scarification. In contrast, most previous studies used voluntary wheel running, which is not standardized.

Similar to the exercise studies discussed before, mouse models of obesity and metabolic syndrome also failed to show accelerated telomere shortening despite an upregulation of Chk2, p53, and p21 [22,23]. For example, feeding mice for 60 weeks with a high-fat/high-sucrose diet induced obesity and metabolic dysfunction, but did not accelerate LTL shortening [23]. With advancing obesity, the animals were physically less active, which should have amplified potentially adverse effects of obesity. Additionally, in genetically modified rats with metabolic syndrome, Takahashi et al. showed comparable myocardial TL than in wild-type controls [22]. At the same time, telomerase expression and TA were upregulated together with the senescence-associated genes Chk2, p53, and p21. These results are in line with our present study, showing similar RTL in PBMCs, liver, aorta, and skeletal muscle after 10 months of HFD or ND. Additionally, large intestine, spleen, brain, and lung showed comparable RTL in the two groups. In our model neither exercise nor HFD induced a consistent expression pattern of telomere-regulating genes, namely *tert*, *terf-1* and *terf-2*. Only kidney and visceral fat showed significant differences in RTL, but in opposite directions. Similar differences were detected for the expression of *tert* and *terf-2*. However, the relevance of these effects is questionable as TERT, TERF-1 and TERF-2 were also altered in several other tissues of HFD animals without affecting the respective RTL. Furthermore, most existing studies reported effects of exercise and obesity on telomerase and shelterins, but often this was not associated with changes in RTL. In line with existing data, correlation analyses in the present study showed inconsistent correlations between RTL and mRNA expression levels of the three telomere-regulating genes. Altogether, these results question the pathophysiological relevance of such observations.

A unique aspect of the present study is the combination of exercise and HFD. In modern societies, people often try to compensate adverse nutritional habits with exercise, but the efficacy of this approach is not well documented. Our results show that such an approach does produce a different outcome than exercise alone. Specifically, RTL was lower in PBMCs, liver and kidney of exercising animals on HFD, but higher in aorta and large intestine. Despite a comparable pattern of RTL in the different organs of exercising animals on normal diet and HFD, incongruent results were registered for the mRNA expression of *tert*, *terf-1* and *terf-2*. For example, *terf-1* and *terf-2* were both increased in skeletal muscle of exercising animals on ND but decreased in exercising animals on HFD. However, both groups showed comparable RTL in this tissue. In line with this argument, also correlation analyses that included all 72 animals did not show consistent correlations between RTL and the expression level of telomerase or shelterins. For example, an inverse correlation between RTL and *tert* was seen in spleen and large intestine, whereas kidney and visceral fat showed the opposite. In all other tissues both parameters were not correlated.

This present animal study does not support the results from human studies showing a reduced LTL in obese people [39,40] and a preservation of TL upon regular endurance exercise [17]. Although some studies do not support an inverse relationship between TL and obesity [41–43], a recent meta-analysis calculated a significantly lower LTL in obese individuals than in normal-weight individuals [49]. Moreover, LTL was inversely correlated with BMI, body fat content, waist circumference, waist-to-hip ratio, and nuchal fat thickness. However, the observational character of the studies included impedes any conclusion towards causality. Additional insights can be gained from longitudinal observation studies that assessed LTL in obese patients before and after bariatric surgery [66]. Available results indicate an improvement in LTL after >2 years, probably due to an improvement in inflammation and oxidative stress. However, only a small number such studies has been published, with rather heterogenous design and outcome. Human studies that investigated LTL in exercising and sedentary individuals are also inconsistent. Several observational studies have shown higher LTL in exercising individuals of different age groups and activity levels [15,17,18,57,58]. Additional support from prospective observation and intervention studies is strongly limited. Soares-Miranda L et al. performed serial blood collections over a 5-year period in 582 older US adults and found no significant association between physical activity, physical performance, and LTL [59]. In contrast, Werner et al. reported an increase in LTL, TA, and TERF-2 expression after 6 months of aerobic endurance training or high intensity training, which was not seen in controls [17,18].

A general downside of existing human studies is the limitation of TL analyses to blood leucocytes, which impedes conclusions about TL in solid tissues of obese and lean individuals. However, previous results from our group have shown that LTL does not provide reliable information on TL in other tissues [67]. While RTL in some tissues, exhibit a positive correlation with LTL, others show the opposite. Additionally, RTL in young and aged SD rats did not systematically change.

The present results should be interpreted with caution keeping in mind the strengths and limitation of this study. A rather large number of animals per group and a strictly standardized exercise intervention provide robustness to the results. In addition, the intervention period was quite long. However, results from Werner et al. suggest, that up to 18 months may be needed to observe a significant reduction in TL [16]. SD rats have an average life expectancy of 2 years so that our animals were sacrificed at advanced adult age, but they cannot be regarded old. The exercise protocol applied was rather moderate and a more intensive regimen might have produced different results. However, with this protocol we aimed to mimic a common recreational activity pattern in adults. Energy intake and energy expenditure may have varied between individual animals and different groups. The lacking information on both factors adds some uncertainty to the interpretation of our results. Another important limitation is the RT-qPCR method that has been used for the measurement of RTL. This method gives an average TL across all cells and chromosomes but does not provide information on the percentage of very short and long telomeres. There is some evidence that the percentage of very short telomeres rather than average TL is associated with aging and age-related disease [68]. However, determination of the shortest telomeres requires highly sophisticated and cumbersome methods, such as Telomere Shortest Length Assay (TeSLA) [69]. In addition, these methods are difficult to standardize and not suitable for high throughput analysis. As we had planned to analyze more than 1000 samples, these assays were deemed not feasible for our purpose. Lastly, telomere-regulating genes were only analysed by mRNA expression, but not at protein level. Although mRNA expression and protein analyses may give discordant results, we do not feel that this limits the overall meaning of our results. The absence of systemic effects on RTL in PBMCs and solid tissues and the highly inconsistent mRNA expression pattern of telomerase and shelterins limit the potential scope of these factors as relevant mediators of telomere effects induced by exercise and diet.

5. Conclusions

In summary, the present in vivo study does not provide evidence that modifiable lifestyle factors, such as obesity and exercise, have significant systemic effects on telomere shortening and the expression of telomere-regulating genes. Additionally, exercise and HFD do not show significant interaction. Any lifestyle-related effect on RTL and telomere-regulating genes in one tissue type does not allow conclusions on other tissues or cell types. Future research should address the impact of exercise and diet on the shortest telomeres and explore their role for aging and degenerative disease. Moreover, future studies on the effects of lifestyle factors on telomere length and telomere function should focus on advanced adult age, where degenerative disease most frequently occurs.

Supplementary Materials: The following supporting information can be downloaded at: https://www.mdpi.com/article/10.3390/cells11101605/s1, Figure S1: Effects matrix that visualizes the effects of exercise, HFD, and the interaction of both lifestyle factors on RTL and the mRNA expression of telomere-regulating genes.

Author Contributions: Conceptualization, M.H., H.-J.G. and G.A.; methodology, M.D.S., G.A. and H.-J.G.; software, H.-J.G. and M.D.S.; validation, W.R. and H.-J.G.; formal analysis, M.D.S., W.R. and H.-J.G.; investigation, M.H., H.-J.G., M.D.S. and G.A.; resources, M.D.S., H.-J.G. and G.A.; data curation, M.D.S.; writing—original draft preparation, M.D.S.; writing—review and editing, M.H., H.-J.G., W.R. and G.A.; visualization, M.D.S.; supervision, M.H. and H.-J.G.; project administration, M.H.; funding acquisition, M.H. All authors have read and agreed to the published version of the manuscript.

Funding: This research received no external funding.

Institutional Review Board Statement: The study was approved by the responsible National Ethics Committee (GZ: 66.010/0070-V/3b/2018) and conducted in accordance with the guidelines of the Animal Care and Use Committee of the Ministry of Science and Research, Vienna, Austria.

Data Availability Statement: All data underlying the findings reported in this manuscript are provided as part of the article. The raw data that are not already presented in the figures are available from the corresponding author upon reasonable request.

Acknowledgments: The authors would like to thank Sabine Pailer and Melanie Kaiser with the Medical University of Graz, and the Core Facility Experimental Biomodels, Division of Biomedical Research of the Medical University of Graz, 8036 Graz (Austria).

Conflicts of Interest: The authors declare no conflict of interest.

References

1. López-Otín, C.; Blasco, M.A.; Partridge, L.; Serrano, M.; Kroemer, G. The Hallmarks of Aging. *Cell* **2013**, *153*, 1194–1217. [CrossRef] [PubMed]
2. Shay, J.W. Role of Telomeres and Telomerase in Aging and Cancer. *Cancer Discov.* **2016**, *6*, 584. [CrossRef] [PubMed]
3. Benetos, A.; Gardner, J.P.; Zureik, M.; Labat, C.; Xiaobin, L.; Adamopoulos, C.; Temmar, M.; Bean, K.E.; Thomas, F.; Aviv, A. Short Telomeres Are Associated With Increased Carotid Atherosclerosis in Hypertensive Subjects. *Hypertension* **2004**, *43*, 182–185. [CrossRef] [PubMed]
4. Wong, L.S.; de Boer, R.A.; Samani, N.J.; van Veldhuisen, D.J.; van der Harst, P. Telomere biology in heart failure. *Eur. J. Heart Fail.* **2008**, *10*, 1049–1056. [CrossRef] [PubMed]
5. Farzaneh-Far, R.; Cawthon, R.M.; Na, B.; Browner, W.S.; Schiller, N.B.; Whooley, M.A. Prognostic value of leukocyte telomere length in patients with stable coronary artery disease: Data from the Heart and Soul Study. *Arter. Thromb Vasc. Biol.* **2008**, *28*, 1379–1384. [CrossRef]

6. Willeit, P.; Willeit, J.; Brandstätter, A.; Ehrlenbach, S.; Mayr, A.; Gasperi, A.; Weger, S.; Oberhollenzer, F.; Reindl, M.; Kronenberg, F.; et al. Cellular Aging Reflected by Leukocyte Telomere Length Predicts Advanced Atherosclerosis and Cardiovascular Disease Risk. *Arterioscler. Thromb. Vasc. Biol.* **2010**, *30*, 1649–1656. [CrossRef]
7. Testa, R.; Olivieri, F.; Sirolla, C.; Spazzafumo, L.; Rippo, M.R.; Marra, M.; Bonfigli, A.R.; Ceriello, A.; Antonicelli, R.; Franceschi, C.; et al. Leukocyte telomere length is associated with complications of Type 2 diabetes mellitus. *Diabet. Med.* **2011**, *28*, 1388–1394. [CrossRef]
8. Ehrlenbach, S.; Willeit, P.; Kiechl, S.; Willeit, J.; Reindl, M.; Schanda, K.; Kronenberg, F.; Brandstatter, A. Influences on the reduction of relative telomere length over 10 years in the population-based Bruneck Study: Introduction of a well-controlled high-throughput assay. *Int. J. Epidemiol.* **2009**, *38*, 1725–1734. [CrossRef]
9. Honig, L.S.; Kang, M.S.; Schupf, N.; Lee, J.H.; Mayeux, R. Association of Shorter Leukocyte Telomere Repeat Length With Dementia and Mortality. *Arch. Neurol.* **2012**, *69*, 1332–1339. [CrossRef]
10. Pusceddu, I.; Kleber, M.; Delgado, G.; Herrmann, W.; Marz, W.; Herrmann, M. Telomere length and mortality in the Ludwigshafen Risk and Cardiovascular Health study. *PLoS ONE* **2018**, *13*, e0198373. [CrossRef]
11. Wang, Q.; Zhan, Y.; Pedersen, N.L.; Fang, F.; Hägg, S. Telomere Length and All-Cause Mortality: A Meta-analysis. *Ageing Res. Rev.* **2018**, *48*, 11–20. [CrossRef] [PubMed]
12. Epel, E.S.; Blackburn, E.H.; Lin, J.; Dhabhar, F.S.; Adler, N.E.; Morrow, J.D.; Cawthon, R.M. Accelerated telomere shortening in response to life stress. *Proc. Natl. Acad. Sci. USA* **2004**, *101*, 17312. [CrossRef] [PubMed]
13. Freitas-Simoes, T.M.; Ros, E.; Sala-Vila, A. Nutrients, foods, dietary patterns and telomere length: Update of epidemiological studies and randomized trials. *Metabolism* **2016**, *65*, 406–415. [CrossRef] [PubMed]
14. Astuti, Y.; Wardhana, A.; Watkins, J.; Wulaningsih, W.; Network, P.R. Cigarette smoking and telomere length: A systematic review of 84 studies and meta-analysis. *Env. Res.* **2017**, *158*, 480–489. [CrossRef]
15. Semeraro, M.D.; Smith, C.; Kaiser, M.; Levinger, I.; Duque, G.; Gruber, H.-J.; Herrmann, M. Physical activity, a modulator of aging through effects on telomere biology. *Aging* **2020**, *12*, 13803–13823. [CrossRef]
16. Werner, C.; Hanhoun, M.; Widmann, T.; Kazakov, A.; Semenov, A.; Pöss, J.; Bauersachs, J.; Thum, T.; Pfreundschuh, M.; Müller, P.; et al. Effects of Physical Exercise on Myocardial Telomere-Regulating Proteins, Survival Pathways, and Apoptosis. *J. Am. Coll. Cardiol.* **2008**, *52*, 470. [CrossRef]
17. Werner, C.M.; Furster, T.; Widmann, T.; Poss, J.; Roggia, C.; Hanhoun, M.; Scharhag, J.; Buchner, N.; Meyer, T.; Kindermann, W.; et al. Physical exercise prevents cellular senescence in circulating leukocytes and in the vessel wall. *Circulation* **2009**, *120*, 2438–2447. [CrossRef]
18. Werner, C.M.; Hecksteden, A.; Morsch, A.; Zundler, J.; Wegmann, M.; Kratzsch, J.; Thiery, J.; Hohl, M.; Bittenbring, J.T.; Neumann, F.; et al. Differential effects of endurance, interval, and resistance training on telomerase activity and telomere length in a randomized, controlled study. *Eur. Heart J.* **2018**, *40*, 34–46. [CrossRef]
19. Ludlow, A.T.; Zimmerman, J.B.; Witkowski, S.; Hearn, J.W.; Hatfield, B.D.; Roth, S.M. Relationship between physical activity level, telomere length, and telomerase activity. *Med. Sci. Sports Exerc.* **2008**, *40*, 1764. [CrossRef]
20. Ludlow, A.T.; Lima, L.C.J.; Wang, J.; Hanson, E.D.; Guth, L.M.; Spangenburg, E.E.; Roth, S.M. Exercise alters mRNA expression of telomere-repeat binding factor 1 in skeletal muscle via p38 MAPK. *J. Appl. Physiol.* **2012**, *113*, 1737–1746. [CrossRef]
21. Ludlow, A.T.; Witkowski, S.; Marshall, M.R.; Wang, J.; Lima, L.C.; Guth, L.M.; Spangenburg, E.E.; Roth, S.M. Chronic exercise modifies age-related telomere dynamics in a tissue-specific fashion. *J. Gerontol. A Biol. Sci. Med. Sci.* **2012**, *67*, 911–926. [CrossRef] [PubMed]
22. Takahashi, K.; Takatsu, M.; Hattori, T.; Murase, T.; Ohura, S.; Takeshita, Y.; Watanabe, S.; Murohara, T.; Nagata, K. Premature cardiac senescence in DahlS.Z-Lepr(fa)/Lepr(fa) rats as a new animal model of metabolic syndrome. *Nagoya J. Med. Sci.* **2014**, *76*, 35–49. [PubMed]
23. Burchfield, J.G.; Kebede, M.A.; Meoli, C.C.; Stöckli, J.; Whitworth, P.T.; Wright, A.L.; Hoffman, N.J.; Minard, A.Y.; Ma, X.; Krycer, J.R.; et al. High dietary fat and sucrose results in an extensive and time-dependent deterioration in health of multiple physiological systems in mice. *J. Biol. Chem.* **2018**, *293*, 5731–5745. [CrossRef] [PubMed]
24. von Zglinicki, T. Oxidative stress shortens telomeres. *Trends Biochem. Sci.* **2002**, *27*, 339–344. [CrossRef]
25. Piepoli, M.F.; Hoes, A.W.; Agewall, S.; Albus, C.; Brotons, C.; Catapano, A.L.; Cooney, M.T.; Corra, U.; Cosyns, B.; Deaton, C.; et al. 2016 European Guidelines on cardiovascular disease prevention in clinical practice. *Rev. Esp. Cardiol.* **2016**, *69*, 939. [CrossRef]
26. Herrmann, M.; Pusceddu, I.; Marz, W.; Herrmann, W. Telomere biology and age-related diseases. *Clin. Chem. Lab. Med.* **2018**, *56*, 1210–1222. [CrossRef]
27. Ornish, D.; Lin, J.; Daubenmier, J.; Weidner, G.; Epel, E.; Kemp, C.; Magbanua, M.J.M.; Marlin, R.; Yglecias, L.; Carroll, P.R.; et al. Increased telomerase activity and comprehensive lifestyle changes: A pilot study. *Lancet Oncol.* **2008**, *9*, 1048–1057. [CrossRef]
28. Paul, L. Diet, nutrition and telomere length. *J. Nutr. Biochem.* **2011**, *22*, 895–901. [CrossRef]
29. Kiecolt-Glaser, J.K.; Epel, E.S.; Belury, M.A.; Andridge, R.; Lin, J.; Glaser, R.; Malarkey, W.B.; Hwang, B.S.; Blackburn, E. Omega-3 fatty acids, oxidative stress, and leukocyte telomere length: A randomized controlled trial. *Brain Behav. Immun.* **2013**, *28*, 16–24. [CrossRef]
30. Crous-Bou, M.; Fung, T.T.; Prescott, J.; Julin, B.; Du, M.; Sun, Q.; Rexrode, K.M.; Hu, F.B.; De Vivo, I. Mediterranean diet and telomere length in Nurses' Health Study: Population based cohort study. *BMJ* **2014**, *349*, g6674. [CrossRef]

31. Ludlow, A.T.; Ludlow, L.W.; Roth, S.M. Do Telomeres Adapt to Physiological Stress? Exploring the Effect of Exercise on Telomere Length and Telomere-Related Proteins. *BioMed Res. Int.* **2013**, *2013*, 15. [CrossRef] [PubMed]
32. Bandaria, J.N.; Qin, P.; Berk, V.; Chu, S.; Yildiz, A. Shelterin Protects Chromosome Ends by Compacting Telomeric Chromatin. *Cell* **2016**, *164*, 735–746. [CrossRef] [PubMed]
33. de Lange, T. Shelterin: The protein complex that shapes and safeguards human telomeres. *Genes Dev.* **2005**, *19*, 2100–2110. [CrossRef]
34. Morgan, R.G.; Donato, A.J.; Walker, A.E. Telomere uncapping and vascular aging. *Am. J. Physiol Heart Circ. Physiol.* **2018**, *315*, H1–H5. [CrossRef] [PubMed]
35. Chaput, J.-P.; Tremblay, A. Obesity and physical inactivity: The relevance of reconsidering the notion of sedentariness. *Obes Facts* **2009**, *2*, 249–254. [CrossRef] [PubMed]
36. Review, W.P. Most Obese Countries 2022. Available online: https://worldpopulationreview.com/country-rankings/most-obese-countries (accessed on 14 April 2022).
37. Division of Nutrition, Physical Activity and Obesity, National Center for Chronic Disease Prevention and Health Promotion. Available online: https://www.cdc.gov/physicalactivity/data/inactivity-prevalence-maps/index.html (accessed on 29 March 2022).
38. Reilly, J.J.; Kelly, J. Long-term impact of overweight and obesity in childhood and adolescence on morbidity and premature mortality in adulthood: Systematic review. *Int. J. Obes.* **2011**, *35*, 891–898. [CrossRef] [PubMed]
39. Sun, Q.; Shi, L.; Prescott, J.; Chiuve, S.E.; Hu, F.B.; De Vivo, I.; Stampfer, M.J.; Franks, P.W.; Manson, J.E.; Rexrode, K.M. Healthy lifestyle and leukocyte telomere length in U.S. women. *PLoS ONE* **2012**, *7*, e38374. [CrossRef]
40. Buxton, J.L.; Walters, R.G.; Visvikis-Siest, S.; Meyre, D.; Froguel, P.; Blakemore, A.I. Childhood obesity is associated with shorter leukocyte telomere length. *J. Clin. Endocrinol. Metab.* **2011**, *96*, 1500–1505. [CrossRef]
41. MacEneaney, O.J.; Kushner, E.J.; Westby, C.M.; Cech, J.N.; Greiner, J.J.; Stauffer, B.L.; DeSouza, C.A. Endothelial progenitor cell function, apoptosis, and telomere length in overweight/obese humans. *Obesity* **2010**, *18*, 1677–1682. [CrossRef]
42. Kim, S.; Parks, C.G.; DeRoo, L.A.; Chen, H.; Taylor, J.A.; Cawthon, R.M.; Sandler, D.P. Obesity and weight gain in adulthood and telomere length. *Cancer Epidemiol Biomark. Prev.* **2009**, *18*, 816–820. [CrossRef]
43. Cui, Y.; Gao, Y.T.; Cai, Q.; Qu, S.; Cai, H.; Li, H.L.; Wu, J.; Ji, B.T.; Yang, G.; Chow, W.H.; et al. Associations of leukocyte telomere length with body anthropometric indices and weight change in Chinese women. *Obesity* **2013**, *21*, 2582–2588. [CrossRef] [PubMed]
44. Nordfjall, K.; Eliasson, M.; Stegmayr, B.; Melander, O.; Nilsson, P.; Roos, G. Telomere length is associated with obesity parameters but with a gender difference. *Obesity* **2008**, *16*, 2682–2689. [CrossRef] [PubMed]
45. Moreno-Navarrete, J.M.; Ortega, F.; Sabater, M.; Ricart, W.; Fernandez-Real, J.M. Telomere length of subcutaneous adipose tissue cells is shorter in obese and formerly obese subjects. *Int. J. Obes.* **2010**, *34*, 1345–1348. [CrossRef] [PubMed]
46. Lee, M.; Martin, H.; Firpo, M.A.; Demerath, E.W. Inverse association between adiposity and telomere length: The Fels Longitudinal Study. *Am. J. Hum. Biol.* **2011**, *23*, 100–106. [CrossRef]
47. Njajou, O.T.; Cawthon, R.M.; Blackburn, E.H.; Harris, T.B.; Li, R.; Sanders, J.L.; Newman, A.B.; Nalls, M.; Cummings, S.R.; Hsueh, W.C. Shorter telomeres are associated with obesity and weight gain in the elderly. *Int. J. Obes.* **2012**, *36*, 1176–1179. [CrossRef]
48. Jones, D.A.; Prior, S.L.; Barry, J.D.; Caplin, S.; Baxter, J.N.; Stephens, J.W. Changes in markers of oxidative stress and DNA damage in human visceral adipose tissue from subjects with obesity and type 2 diabetes. *Diabetes Res. Clin. Pract.* **2014**, *106*, 627–633. [CrossRef]
49. Mundstock, E.; Sarria, E.E.; Zatti, H.; Mattos Louzada, F.; Kich Grun, L.; Herbert Jones, M.; Guma, F.T.; Mazzola, J.; Epifanio, M.; Stein, R.T.; et al. Effect of obesity on telomere length: Systematic review and meta-analysis. *Obesity* **2015**, *23*, 2165–2174. [CrossRef]
50. Mangge, H.; Herrmann, M.; Almer, G.; Zelzer, S.; Moeller, R.; Horejsi, R.; Renner, W. Telomere shortening associates with elevated insulin and nuchal fat accumulation. *Sci. Rep.* **2020**, *10*, 6863. [CrossRef]
51. Paffenbarger Jr, R.S.; Hyde, R.; Wing, A.L.; Hsieh, C.-c. Physical activity, all-cause mortality, and longevity of college alumni. *New Engl. J. Med.* **1986**, *314*, 605–613. [CrossRef]
52. Fraser, G.E.; Shavlik, D.J. Ten years of life: Is it a matter of choice? *Arch. Intern. Med.* **2001**, *161*, 1645–1652. [CrossRef]
53. Franco, O.H.; de Laet, C.; Peeters, A.; Jonker, J.; Mackenbach, J.; Nusselder, W. Effects of physical activity on life expectancy with cardiovascular disease. *Arch. Intern. Med.* **2005**, *165*, 2355–2360. [CrossRef] [PubMed]
54. Byberg, L.; Melhus, H.; Gedeborg, R.; Sundström, J.; Ahlbom, A.; Zethelius, B.; Berglund, L.G.; Wolk, A.; Michaëlsson, K. Total mortality after changes in leisure time physical activity in 50 year old men: 35 year follow-up of population based cohort. *BMJ* **2009**, *338*, b688. [CrossRef] [PubMed]
55. Wen, C.P.; Wai, J.P.M.; Tsai, M.K.; Yang, Y.C.; Cheng, T.Y.D.; Lee, M.-C.; Chan, H.T.; Tsao, C.K.; Tsai, S.P.; Wu, X. Minimum amount of physical activity for reduced mortality and extended life expectancy: A prospective cohort study. *Lancet* **2011**, *378*, 1244–1253. [CrossRef]
56. Barrow, D.R.; Abbate, L.M.; Paquette, M.R.; Driban, J.B.; Vincent, H.K.; Newman, C.; Messier, S.P.; Ambrose, K.R.; Shultz, S.P. Exercise prescription for weight management in obese adults at risk for osteoarthritis: Synthesis from a systematic review. *BMC Musculoskelet Disord* **2019**, *20*, 610. [CrossRef]

57. Cherkas, L.F.; Hunkin, J.L.; Kato, B.S.; Richards, J.B.; Gardner, J.P.; Surdulescu, G.L.; Kimura, M.; Lu, X.; Spector, T.D.; Aviv, A. The association between physical activity in leisure time and leukocyte telomere length. *Arch. Intern. Med.* **2008**, *168*, 154–158. [CrossRef]
58. Du, M.; Prescott, J.; Kraft, P.; Han, J.; Giovannucci, E.; Hankinson, S.E.; De Vivo, I. Physical activity, sedentary behavior, and leukocyte telomere length in women. *Am. J. Epidemiol.* **2012**, *175*, 414–422. [CrossRef]
59. Soares-Miranda, L.; Imamura, F.; Siscovick, D.; Jenny, N.S.; Fitzpatrick, A.L.; Mozaffarian, D. Physical activity, physical fitness and leukocyte telomere length: The cardiovascular health study. *Med. Sci. Sports Exerc.* **2015**, *47*, 2525. [CrossRef]
60. Ludlow, A.T.; Gratidao, L.; Ludlow, L.W.; Spangenburg, E.E.; Roth, S.M. Acute exercise activates p38 MAPK and increases the expression of telomere-protective genes in cardiac muscle. *Exp. Physiol.* **2017**, *102*, 397–410. [CrossRef]
61. Rocha-Rodrigues, S.; Rodríguez, A.; Gonçalves, I.O.; Moreira, A.; Maciel, E.; Santos, S.; Domingues, M.R.; Frühbeck, G.; Ascensão, A.; Magalhães, J. Impact of physical exercise on visceral adipose tissue fatty acid profile and inflammation in response to a high-fat diet regimen. *Int. J. Biochem Cell Biol.* **2017**, *87*, 114–124. [CrossRef]
62. Li, F.H.; Sun, L.; Zhu, M.; Li, T.; Gao, H.E.; Wu, D.S.; Zhu, L.; Duan, R.; Liu, T.C. Beneficial alterations in body composition, physical performance, oxidative stress, inflammatory markers, and adipocytokines induced by long-term high-intensity interval training in an aged rat model. *Exp. Gerontol.* **2018**, *113*, 150–162. [CrossRef]
63. Rao, Z.; Wang, S.; Bunner, W.P.; Chang, Y.; Shi, R. Exercise induced Right Ventricular Fibrosis is Associated with Myocardial Damage and Inflammation. *Korean Circ. J.* **2018**, *48*, 1014–1024. [CrossRef] [PubMed]
64. Ji, N.; Luan, J.; Hu, F.; Zhao, Y.; Lv, B.; Wang, W.; Xia, M.; Zhao, X.; Lao, K. Aerobic exercise-stimulated Klotho upregulation extends life span by attenuating the excess production of reactive oxygen species in the brain and kidney. *Exp. Med.* **2018**, *16*, 3511–3517. [CrossRef]
65. Cawthon, R.M. Telomere measurement by quantitative PCR. *Nucleic. Acids Res.* **2002**, *30*, e47. [CrossRef]
66. Peña, E.; León-Mengíbar, J.; Powell, T.R.; Caixàs, A.; Cardoner, N.; Rosa, A. Telomere length in patients with obesity submitted to bariatric surgery: A systematic review. *Eur. Eat. Disord Rev.* **2021**, *29*, 842–853. [CrossRef] [PubMed]
67. Semeraro, M.D.; Almer, G.; Renner, W.; Gruber, H.J.; Herrmann, M. Telomere length in leucocytes and solid tissues of young and aged rats. *Aging* **2022**, *14*, 1713–1728. [CrossRef] [PubMed]
68. Cherif, H.; Tarry, J.L.; Ozanne, S.E.; Hales, C.N. Ageing and telomeres: A study into organ- and gender-specific telomere shortening. *Nucleic Acids Res.* **2003**, *31*, 1576–1583. [CrossRef]
69. Lai, T.-P.; Zhang, N.; Noh, J.; Mender, I.; Tedone, E.; Huang, E.; Wright, W.E.; Danuser, G.; Shay, J.W. A method for measuring the distribution of the shortest telomeres in cells and tissues. *Nat. Commun.* **2017**, *8*, 1356. [CrossRef] [PubMed]

cells

MDPI

Article

Natural Product Library Screens Identify Sanguinarine Chloride as a Potent Inhibitor of Telomerase Expression and Activity

Siyu Yan [1], Song Lin [1], Kexin Chen [1], Shanshan Yin [1], Haoyue Peng [1], Nanshuo Cai [1], Wenbin Ma [1], Zhou Songyang [1,2,3] and Yan Huang [1,*]

1 MOE Key Laboratory of Gene Function and Regulation, Guangzhou Key Laboratory of Healthy Aging Research and State Key Laboratory of Biocontrol, School of Life Sciences, Sun Yat-Sen University, Guangzhou 510275, China; yansy3@mail2.sysu.edu.cn (S.Y.); lins3@mail2.sysu.edu.cn (S.L.); chenkx25@mail2.sysu.edu.cn (K.C.); yinshsh2@mail2.sysu.edu.cn (S.Y.); penghy8@mail2.sysu.edu.cn (H.P.); cainsh@mail2.sysu.edu.cn (N.C.); mawenbin@mail.sysu.edu.cn (W.M.); songyanz@mail.sysu.edu.cn (Z.S.)
2 Sun Yat-Sen Memorial Hospital, Sun Yat-Sen University, Guangzhou 510120, China
3 Bioland Laboratory (Guangzhou Regenerative Medicine and Health Guangdong Laboratory), Guangzhou 510005, China
* Correspondence: huangy336@mail.sysu.edu.cn

Abstract: Reverse transcriptase hTERT is essential to telomerase function in stem cells, as well as in 85–90% of human cancers. Its high expression in stem cells or cancer cells has made telomerase/hTERT an attractive therapeutic target for anti-aging and anti-tumor applications. In this study, we screened a natural product library containing 800 compounds using an endogenous hTERT reporter. Eight candidates have been identified, in which sanguinarine chloride (SC) and brazilin (Braz) were selected due to their leading inhibition. SC could induce an acute and strong suppressive effect on the expression of *hTERT* and telomerase activity in multiple cancer cells, whereas Braz selectively inhibited telomerase in certain types of cancer cells. Remarkably, SC long-term treatment could cause telomere attrition and cell growth retardation, which lead to senescence features in cancer cells, such as the accumulation of senescence-associated β-galactosidase (SA-β-gal)-positive cells, the upregulation of p16/p21/p53 pathways and telomere dysfunction-induced foci (TIFs). Additionally, SC exhibited excellent capabilities of anti-tumorigenesis, both in vitro and in vivo. In the mechanism, the compound down-regulated several active transcription factors including p65, a subunit of NF-κB complex, and reintroducing p65 could alleviate its suppression of the hTERT/telomerase. Moreover, SC could directly bind hTERT and inhibit telomerase activity in vitro. In conclusion, we identified that SC not only down-regulates the *hTERT* gene's expression, but also directly affects telomerase/hTERT. The dual function makes this compound an attractive drug candidate for anti-tumor therapy.

Keywords: telomerase/hTERT; anti-cancer; sanguinarine chloride; cellular senescence

Citation: Yan, S.; Lin, S.; Chen, K.; Yin, S.; Peng, H.; Cai, N.; Ma, W.; Songyang, Z.; Huang, Y. Natural Product Library Screens Identify Sanguinarine Chloride as a Potent Inhibitor of Telomerase Expression and Activity. *Cells* **2022**, *11*, 1485. https://doi.org/10.3390/cells11091485

Academic Editors: Nicole Wagner, Kay-Dietrich Wagner and Ursula Stochaj

Received: 19 March 2022
Accepted: 26 April 2022
Published: 28 April 2022

1. Introduction

Telomeres are continuously shortened during the process of DNA replication as DNA polymerase cannot synthesize chromosomal end sequences [1]. Mammalian telomeric DNA consists of TTAGGG hexanucleotide tandem DNA repeats forming loop structures, with the interactions of specialized telosome/shelterin proteins [2,3]. In order to maintain the genomic stability and protect cells from senescence, shortened telomeres could be elongated by two mechanisms: in a telomerase-dependent manner in most pluripotent stem cells and cancer cells, or the alternative lengthening of telomeres (ALT) in 10–15% cancer cells [4].

Human telomerase ribonucleoprotein contains the catalytic reverse transcriptase hTERT and RNA template hTERC associated with the accessory H/ACA proteins [5]. hTERT is highly conserved among species. Human TERT is composed of four main domains [6]. The

N-terminal extension (TEN) domain is connected with the 5′-terminal of the telomerase RNA-binding domain (TRBD) by a short linker sequence. The central reverse transcriptase (RT) domain and the C-terminal extension (CTE) domain make up a right-hand structure containing palm-, finger- and thumb-like subdomains. Together, the sequential TRBD-RT-CTE domains form a ring-like structure for telomere repeat addition [6].

Telomerase activity is strictly modulated at multiple levels, including the expression of telomerase subunits, the process of holoenzyme assembly and its recruitment to telomeres [7]. Compared with the ubiquitously expressed hTERC in cells, the restrictively expressed protein hTERT is the major limiting factor of telomerase activity regulation [8]. Suppressing telomerase activity by decreasing hTERT protein blocks telomere extension. Very short telomeres could trigger telomere dysfunction and induce a DNA damage response and cell senescence. Except for its involvement in telomeres elongation, hTERT turned out to participate in many non-telomeric biological events; for instance, the expressional regulation of aging-related genes or oncogenes [9]. Interestingly, c-myc can bind to the E-box motif in the *hTERT* promoter region to activate its transcription [10], while hTERT is also able to stabilize c-myc protein and modulates its binding to target promoters [11]. NF-κB p65 has been reported to modulate telomerase expression [12], and also mediate the nuclear translocation of the hTERT protein from cytoplasm via TNF-α in the cancer cell line [13].

In most cancer cells, stem/progenitor cells and certain somatic cells in special physiological states, such as activated T cells, hTERT is expressed to activate telomerase, making it attractive as a therapeutic target for cancer [14,15]. Imetelstat, also known as GRN163L, is chemically modified oligonucleotide, which can silence the telomerase assembly process [16]. BIBR1532 has been identified as a selective telomerase inhibitor that tightly binds to the FVYL motif near TRBD and can result in an impediment to the interaction of hTERT TRBD with the CR4/5 stem loop of telomerase RNA [17]. Yet, like most quinoline derivatives, BIBR1532 exhibits a certain degree of cytotoxicity, and causes apoptosis and senescence in high doses over 25 μM [18].

Cell senescence is a state of growth arrest caused by several factors, such as telomere loss and DNA damage [19]. Cancer cells show features such as senescence after exposure to certain chemotherapeutic compounds [20]. Therefore, the telomerase inhibitor, as a factor of accelerated cell senescence, is a double-edged sword on its applications, and is accepted in anti-cancer strategies via triggering senescence.

Successful clinical outcomes require prolonged treatment, which may lead to severe toxicity in patients. In comparison with synthetic chemicals, natural products are more acceptable and environmentally friendly. In this work, we used CRISPR/Cas9 to establish a cell-based platform aiming to screen out certain natural compounds that modulate the expression of endogenous *hTERT*, and found some compounds with potential applications, of which sanguinarine chloride (SC), a benzophenanthridine alkaloid extracted from the root of Sanguinaria canadensis, is very attractive.

SC exhibits clear-cut antitumor properties, with evidence of apoptotic cell death induction and anti-proliferation through generating reactive oxygen species [21], suppressing the NF-κB pathway [22], inhibiting cyclin-dependent kinases and cyclins [23] and blocking VEGF function in angiogenesis [24]. Besides this, sanguinarine is commonly used in toothpaste and oral health products because of its antibacterial and anti-inflammatory effects [25]. However, since 1999, a sanguinarine-added mouthwash product Viadent® was reported to be associated with age-related leukoplakia, indicating its pre-neoplastic adverse effects [26,27]. Notably, how sanguinarine induces leukoplakia remains unclear so far, and the underlying mechanism of the anti-tumor effect of sanguinarine remains elusive; thus, figuring out its biological target and detailed molecular mechanism is crucial for pharmacological usage.

2. Materials and Methods

2.1. Chemicals

Sanguinarine chloride hydrate (SC) was purchased from Aladdin (S101540; Shanghai, China) and Brazilin (Braz) was purchased from Sigma-Aldrich (SML2132; St. Louis, MO, USA). All the chemicals were dissolved in DMSO and stored at −20 °C.

2.2. Cell Culture and Transfection

The HEK293T cell line and cancer cell lines, including HTC75, HeLa, DLD1, MDA-MB-231, Hs578t and A549, were routinely cultured in Dulbecco's Modified Eagle's Medium (DMEM; Corning; New York, NY, USA) supplemented with 10% Fetal bovine serum (FBS; Excell Bio; Jiangsu, China). Human skin fibroblasts (HFs) and human umbilical vein smooth muscle cells (HUVSMCs) were cultured in Dulbecco's Modified Eagle Medium/F-12 Nutrition Mixture (DMEM/F12; Gibco; New York, NY, USA) containing 10% FBS (Hyclone; Logan, UT, USA). Peripheral blood mononuclear cells (PBMCs) were cultured in RPMI 1640 medium (Gibco; New York, NY, USA) with 10% FBS (Hyclone; Logan, UT, USA). Lipofectamine2000 reagents (Invitrogen; Carlsbad, CA, USA) were used for cell transfections of recombinant plasmids.

2.3. Flow Cytometry Screening

A total of 800 compounds of a natural product library (Natural Products Collection; Microsource; Gaylordsville, CT, USA) was applied and screened in a hTERT-P2A-GFP reporter cell line. Cells were treated by compound in 96wells for 48 h and were then harvested in PBS buffer. GFP and dsRed2 expressions were analyzed by using flow cytometry (Beckman CytoFLEX S; Brea, CA, USA). The mean fluorescence intensity (MFI) of collected cells was taken as the screening index indicating the expression of the target gene.

2.4. Quantitative Telomeric Repeat Amplification Protocol (Q-TRAP) and IP-TRAP

Q-TRAP assays were performed as described [28]. Briefly, 10^5 cells were lysed on ice for 30 min in 100 μL NP40 lysis buffer (10 mM Tris-HCl pH 8.0; 1 mM $MgCl_2$; 1 mM EDTA; 0.25 mM sodium deoxycholate; 150 mM NaCl; 1% NP-40; 10% glycerol; 1% fresh protease inhibitor cocktail) and then centrifuged at 13,200 rpm for 10 min at 4 °C. The supernatant was mixed with 100 ng/μL TS primer (5′-AATCCGTCGAGCAGAGTT-3′), 100 ng/μL ACX primer (5′-GCGCGGCTTACCCTTACCCTTACCCTAACC-3′), and 1mM EGTA in 2 × Real-Star Green Power Mixture (with ROX) (GeneStar; Beijing, China), and then incubated at 30 °C for 30 min for telomeric repeat extension and PCR amplification (40 cycles, 95 °C for 15 s and 60 °C for 60 s) using the Step One Plus^{TM} Real-Time PCR system (Applied Biosystems; Foster, CA, USA). As for IP-TRAP, the cell lysis was immunoprecipitated by anti-FLAG M2 beads (Sigma-Aldrich; St. Louis, MO, USA) at 4 °C for 3 h and eluted by 3 × FLAG peptides. The eluates were mixed with the compound and subjected to the TRAP assay.

2.5. Terminal Restriction Fragment (TRF)

The average telomere length was measured as described [29]. Briefly, genomic DNA was digested by Hinf I and Rsa I overnight at 37 °C, separated by agarose gel, then denatured and hybridized with a radio labeled telomeric probe $(TTAGGG)_4$. The dried gel was exposed to a phosphor screen and then scanned with Amersham Typhoon IP Phosphorimager (GE Healthcare; Torrington, CT, USA). The average telomere length was calculated using ImageJ (National Institutes of Health developed; Bethesda, MD, USA) and GraphPad Prism software (San Diego, CA, USA).

2.6. SA-β-Gal Staining

The assay was performed by using a Senescence β-Galactosidase Staining Kit (Beyotime; Shanghai, China). Briefly, cells were seeded in 12-well plates and fixed with 4% formaldehyde for 15 min at room temperature. The fixed cells were then washed with PBS

3 times and incubated with fresh SA-β-gal staining reagent mix containing 1.0 mg/mL X-galactosidase at 37 °C for 24 h to microscopically observe the staining.

2.7. GST Pull Down and Telomerase Activity Reconstitution In Vitro

The GST pull down assay for GST-hTERT purification was carried out as described previously [30]. Briefly, IPTG was added at 16 °C for 20 h to stimulate the hTERT fusion protein expression. The fusion protein was incubated with GST beads at 4 °C for 4 h and washed three times. The eluates were stored at −80 °C for further experiments.

The reconstitution of telomerase activity in vitro was performed as described [28]. Purified GST-tagged hTERT products (GST-opTERT) were incubated with in vitro transcribed hTERC in telomerase reconstruction buffer (25 mM Tris-HCl pH7.4; 2.6 mM KCl; 1 mM $MgCl_2$; 136 mM NaCl; 1 mM EGTA; 10% glycerol; 1 mM DTT; 1×proteinase inhibitor cocktail; 0.5 U/μL of RNase inhibitor) at 37 °C for 30 min. Compounds in serially diluted concentrations were added into the reconstructive products, followed by the TRAP assay.

2.8. Thioflavin T (ThT) Biochemical Assay

The experiments were conducted in 96-well microplates. In total, 1 μg genome of DNA sample was mixed with ThT at a final concentration of 2 μM in the buffer (20 mM Tris-HCl pH 7.0, 40 mM KCl) at room temperature. The fluorescence emission was collected at 491 nm in a multi-mode microplate reader (BioTek; Winooski, VT, USA).

2.9. Microscale Thermophoresis (MST) Assay

Human telomeric oligonucleotides (Telo24, 5′-Cy5-$(TTAGGG)_4$-3′) were annealed in the K+ buffer (10 mM K_2HPO_4/KH_2PO_4 pH 7.0, 100 mM KCl) by heating to 95 °C for 6 min, then cooled down to room temperature and store at 4 °C. The annealed telomeric G-quadruplex samples (1 μM) were incubated with compound SC at concentrations ranging from 0.15625 μM to 320 μM in the K^+ buffer for 30 min, followed by the MST assays. The MST assay was conducted using the Monolith NT.115 device (NanoTemper Technologies; Munich, Germany) according to the manufacturer's instructions. Data were analyzed using MO. Affinity Analysis software (NanoTemper Technologies; Munich, Germany).

2.10. Fluorescence Polarization Assay

SC at a high concentration over 10 μM exhibits autofluorescence. The equilibrium binding of the compound with hTERT TRBD protein was monitored by fluorescence polarization assay. All fluorescence polarization compound–protein binding assays were performed in 100 μL PBS buffer containing 10 μM SC and purified His tagged hTRBD in a serially diluted concentration from 2.5 μM to 40 μM in 96-well black polypropylene plates. Fluorescence polarization (FP) measurements were performed at room temperature using a Victor^{TM} X5 2030 Multiple Reader (PerkinElmer; Waltham, MA, USA). BSA was used as a negative control.

2.11. Soft-Agar Colony Formation Assay

The MDA-MB-231 cell suspension was mixed in 0.3% soft agar in DMEM containing 10% FBS and the compound, then layered on 0.6% solid agar in DMEM containing 10% FBS and the compound. In total, 1000 cells were seeded per well in a 6-well plate. After 14 days of culturing, colonies were observed under a microscope and the total numbers of colonies from ten random fields of view were counted for the statistical analysis.

2.12. In Vivo Cell Derived Xenograft (CDX)

MDA-MB-231 cells suspended in cold PBS buffer were inoculated subcutaneously into 6–8-week-old nude mice in situ. After the xenograft model was established, the mice were injected intravenously with the compound at a dosage of 1.1 μg/kg (the concentration of the compound in blood was 1 μM if the blood volume was estimated as 7% of the body weight). The compound was administrated every three days. The volume of tumor and the body weight were recorded before each injection. The tumor volume was calculated as $LW^2/2$, where L represents the long diameter and W represents the short diameter.

2.13. Statistical Analysis

Data are shown as mean ± SD. Experiments were carried out in three technical replicates. Student's t-test and one-way ANOVA test were used for statistical significance analyses with the software GraphPad Prism version 6.0 (San Diego, CA, USA). The fitting curves were depicted using Original version 9.0. A p value less than 0.05 is statistically significant (* $p < 0.05$, ** $p < 0.01$, *** $p < 0.001$, **** $p < 0.0001$).

3. Results

3.1. Natural Compound Screening in hTERT Promoter-Driven GFP Reporter Cell Line and Identification of Potential Inhibitor Candidates

In order to screen out *hTERT* regulatory molecules, we established a platform of *hTERT* reporter HEK293T cells using CRISPR/Cas9 to knock in the P2A-GFP fusion gene before the stop codon of the *hTERT* gene (Figure 1A). *hTERT* and *GFP* were transcribed together from the same promoter, and translated fusion proteins were self-cleaved by a small linker peptide, P2A. The red fluorescent protein dsRed2 was stably transfected into the reporter cell line as an internal control of the fluorescence-based FACs screening. Since the abundance of the hTERT protein was much lower than that of most of the other proteins in the cell, here, we took advantage of a monoclonal-derived cell line with a relatively low intensity of GFP and a high intensity of dsRed2 for the screening (Figure 1A and Figure S1A). The initial screening focused on the commercial natural compound library containing 800 small molecules (Supplementary Figure S1B); 69 compounds exhibiting the mean fluorescence intensity (MFI) of GFP, normalized by dsRed2 with at least 40% decline compared to the control (DMSO), were enriched for the second screening (Supplementary Figure S1B). These 69 natural products were conducted to three independent repetitive screens and 8 candidate compounds were repeatably obtained with a significant decrease in the MFI of GFP/dsRed2 (Figure 1B, Supplementary Table S1). These eight candidates were further verified, and SC and Braz were selected due to their outstanding inhibitory effects (Figure 1C and Figure S1C, Supplementary Table S1). Braz has been patented as a kind of natural telomerase inhibitor [31].

Based on the flow cytometry data, treating reporter cells with SC (1 μM) or Braz (10 μM) for 48 h decreased the MFI of GFP, but did not influence dsRed2 expression (Figure 1D and Figure S1D, Supplementary Table S1). The *hTERT* expression level and relative telomerase activity (RTA) were examined in reporter cells under treatment with SC or Braz. Indeed, both SC and Braz inhibited *hTERT* mRNA level and telomerase activity, which confirms our screening result (Figure 1E,F and Figure S1E,F).

3.2. Inhibitory Effects of SC on Telomerase Activity in HTC75 Cancer Cells

The characteristic of its higher expression in most cancer cells makes telomerase/hTERT a valuable predictive biomarker and drug target in malignant cells. To evaluate their inhibition of RTA in cancer cells, we treated HTC75 cells with the two candidate compounds respectively for 48 h, and then performed the Q-TRAP assay. HTC75 is a telomerase-positive fibrosarcoma cell line that can maintain a constant telomere length during in vitro passaging, and is commonly used in the telomere field [32]. The results suggested that SC suppresses the telomerase activity in a dose-dependent manner (Figure 2A). The CCK-8 cell proliferation assay showed the effect of SC at different dosages on the viability of HTC75 cell. The fitting curve indicated the median viable concentration was 2.18 μM (Figure 2B). To explore cancer cell proliferative inhibition induced by the compound, we carried out an analysis of cell cycle and apoptosis. SC-treated cells showed a subtle cell cycle arrest in the G2/M phase compared with control cells (Figure 2C,D). Cells incubated with 2 μM SC exhibited an acute increase in apoptotic cells (Figure 2E). The result was consistent with previously reported studies, wherein SC induced apoptosis through generating reactive oxygen species [33,34]. Furthermore, we wondered if SC effectively works in different types of cancer cells; the RTA of five other kinds of solid tumor cell lines were examined after 48 h treatment. SC exhibited a consistent suppressive effect on telomerase modulation in all tested cell lines, although with different degrees of inhibitory effects (Figure 2F).

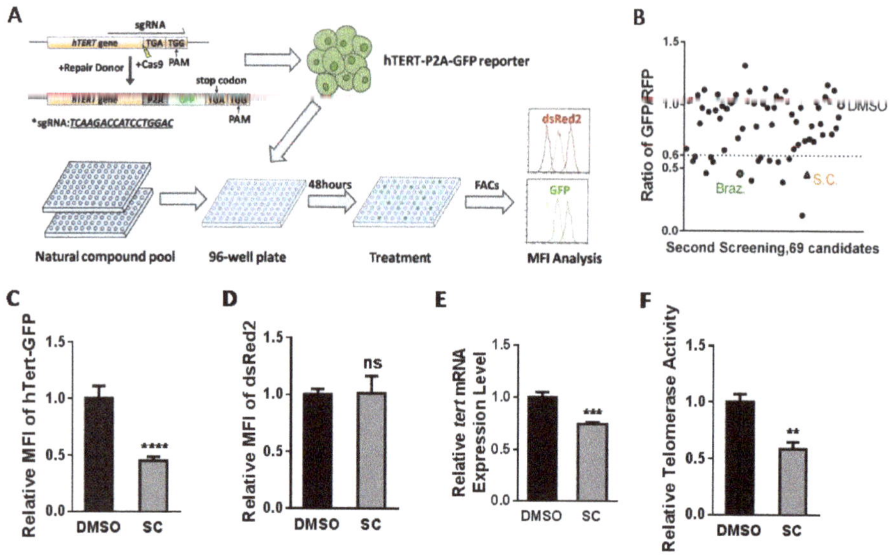

Figure 1. Screening of natural *hTERT* inhibitors and their verification in the endogenous hTERT-P2A-GFP knock-in HEK293Treporter cell line. (**A**) Schematics of the endogenous hTERT-P2A-GFP HEK293T reporter cell line construction and the screening strategy of a natural product pool for telomerase modulators. Two rounds of screening were carried out and compounds with GFP/RFP ratio ≤ 0.6 were selected as candidates. (**B**) Results from the second round of compound screening, with two candidates highlighted as potential inhibitors. The orange triangle indicates SC while the green rectangle indicates Braz. (**C**,**D**) Mean fluorescence intensity (MFI) quantification of endogenous hTERT-GFP (**C**) and internal reference dsRed2 (**D**) after 1 μM SC treatment for 48 h. (**E**) *hTERT* mRNA level by quantitative real-time PCR of reporter cell line upon the treatment of SC (1 μM) for 24 h. DMSO served as the control group. (**F**) Real-time quantitative telomeric repeat amplification protocol (Q-TRAP) assay in reporter cells treated with SC (1 μM). DMSO served as the control group. The screening was performed three independent times. Real-time PCR and Q-TRAP assays are from triplicate samples (* $p < 0.05$, ** $p < 0.01$, *** $p < 0.001$, **** $p < 0.0001$).

Compared with SC, Braz could also induce cell cycle arrest in the G2/M phase (Supplementary Figure S2A), whereas no exacerbation of apoptosis events was observed at the concentrations that inhibit telomerase activities in HTC75 cells (Supplementary Figure S2B). In terms of the potential use as a natural telomerase inhibitor, Braz could only inhibit telomerase in certain types of cancer cells, which indicates its restricted applicability in multiple kinds of tumors, compared to the broad-spectrum anti-telomerase property of the compound SC (Supplementary Figure S2C). Furthermore, we assessed the cytotoxicity of Braz in different cancer cells. The compound showed proliferative inhibition effects in HeLa, DLD1 and HTC75 cells, which also inhibit RTA in these three cell lines (Supplementary Figure S2D).

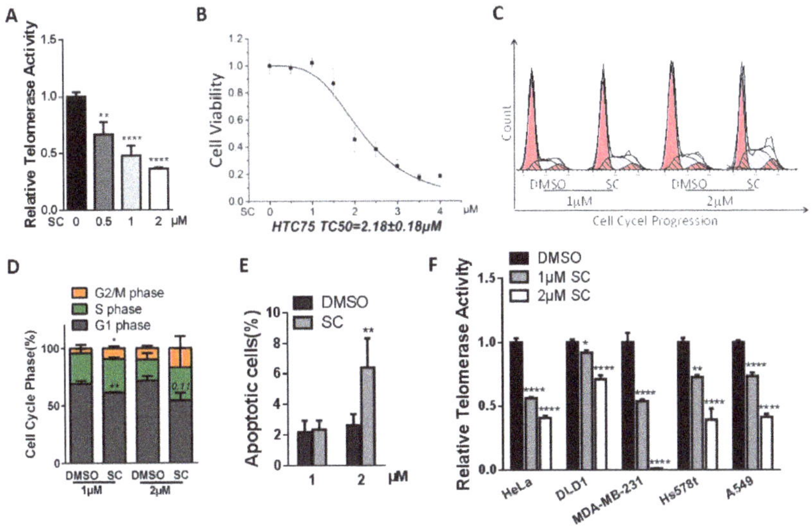

Figure 2. Effects of SC treatment on different cancer cells. (**A**) The inhibitory effect of SC on telomerase activity in HTC75 cells was evaluated. The relative telomerase activity (RTA) level suggested a dose-dependent suppressive effect. (**B**) The CCK-8 assay showed the cell viability curve of SC in HTC75 cells. The median toxic concentration (TC50) was 2.18 μM. (**C**) Cell cycle analysis through PI staining and the following flow cytometry for HTC75 cell treated with SC. (**D**) Quantification of cell cycle populations measured in (**C**). (**E**) Quantification of apoptotic cells (Annexin-V FITC positive cells). (**F**) The 1 μM SC treatment for 48 h hindered RTA in multiple cancer cells, normalized by RTA of the DMSO group. All the analyses were performed on triplicate samples (* $p < 0.05$, ** $p < 0.01$, **** $p < 0.0001$).

3.3. Effects of SC on Cancer Cell Senescence and Telomere Length through Prolonged Treatment

To identify the optimal anti-telomerase dosage of SC, we evaluated RTA in HTC75 tumor cells treated with SC at different concentrations for 48 h. The Q-TRAP results showed the IC50 (half maximal inhibitory concentration) of SC to telomerase was 1.21 μM (Figure 3A). Cell cycle arrest and apoptosis events may indirectly down-regulate the telomerase. These indirect negative cellular events should be avoided whenever possible when the telomerase inhibitor is applied in anti-tumor therapy. Based on these considerations and the data mentioned above, 1 μM SC could substantially inhibit telomerase activity with no obvious apoptosis induction in cancer cells. Therefore, we chose this concentration for further experiments.

Upon continuous treatment with the specified dose of compound SC, the HTC75 cells retained unchanged morphological characteristics compared to the DMSO-treated control cells in a short period; however, prolonged SC-treated cells became shrunken and irregular (Figure 3B). A curve of cumulative population doubling was plotted to assess cancer cell growth. The proliferation of SC-treated cells was much slower than the DMSO-treated cells (Figure 3C). The persistent inhibition of hTERT protein level caused by prolonged compound treatment (Figure 3D) may cause the accumulation of telomere attrition. Telomere length was measured by the terminal restriction fragment (TRF) assay. The average telomere length of HTC75 cells treated with SC for a long time was obviously shortened, compared to the DMSO-treated HTC75 cells (Figure 3E,F). The shortened telomeres can manifest a DNA damage response and drive the cells into senescence. SA-β-gal staining showed more senescent cells in the SC-treated group (Figure 3G). The senescent markers p16/p21/p53 were all up-regulated at the protein levels (Figure 3H). Prolonged SC treatment also induced more telomere dysfunction-induced foci (TIFs) in HTC75 cells (Figure 3I).

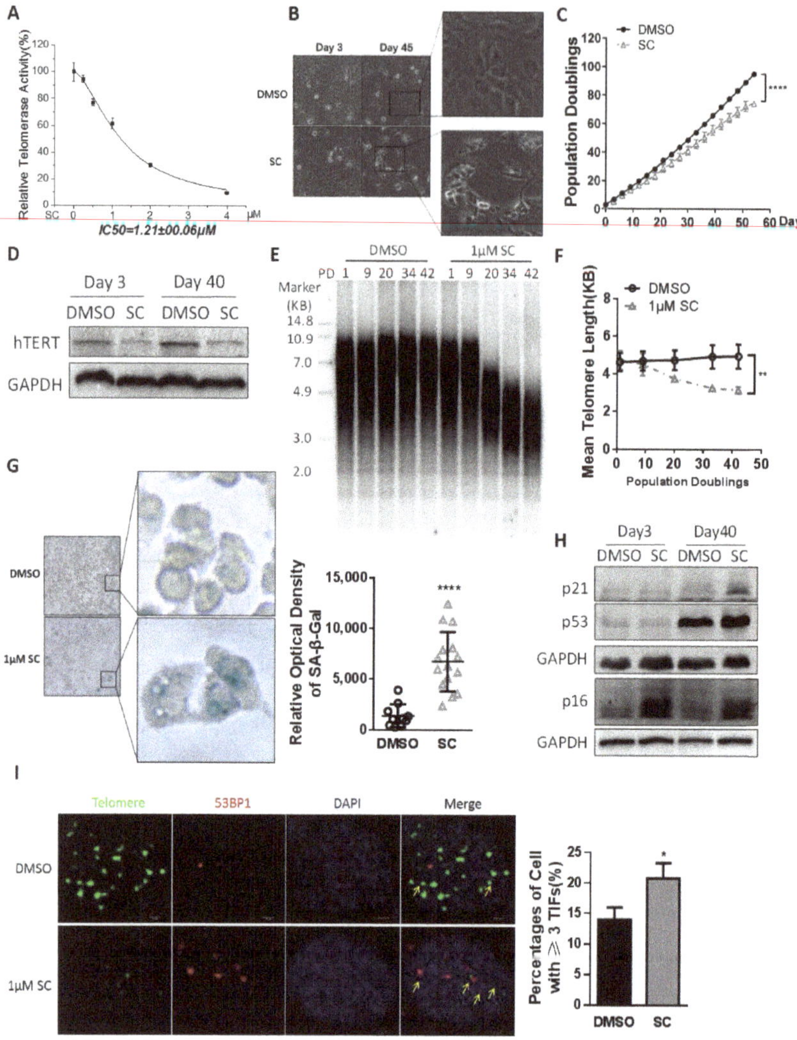

Figure 3. Continuous treatment with SC led to cell senescence in cancer cells. (**A**) Fitting curve of inhibitory effect of SC on RTA in HTC75 cells. IC50 = 1.21 μM. (**B**) Morphology of HTC75 cells treated with 1 μM SC versus DMSO between day 3 and day 45. (**C**) The cell growth curve of the HTC75 cells continuously treated with 1 μM SC, comparing to DMSO-treated groups. (**D**) Western blotting analysis to confirm the inhibitory effect of SC on hTERT protein level. GAPDH served as an internal reference. (**E**) HTC75 cells continuously treated with 1 μM SC were analyzed by terminal restriction fragment (TRF) assay. (**F**) Quantitative average telomere length from (**E**). (**G**) SA-β-gal staining assay to identify cell senescence in HTC75 cells continuously treated with 1 μM SC. (**H**) The senescence markers p16/p21/p53 were up-regulated in cancer cells after chronic treatment of the compound. (**I**) Telomere dysfunction-induced foci (TIFs) were analyzed using anti-53BP1 antibody (red) and PNA-conjugated telomere C strand probe (green) when HTC75 cells were treated with the compound for 40 days. Cells with TIFs ≥3 were counted for the significance test. The experiments were performed in 3 independent cell lines and the results are shown as mean ± SD, $n = 3$ (* $p < 0.05$, ** $p < 0.01$, **** $p < 0.0001$).

As a moderate telomerase inhibitor in cancer cells, Braz suppressed HTC75 cell growth as well as RTA in a long-period treatment (Supplementary Figure S3A,B). However, the average telomere length of Braz-treated cells remained unchanged compared to that of the control cells (Supplementary Figure S3C).

3.4. SC Inhibits Telomerase Depending on p65 Expression

We have confirmed that SC could reduce the mRNA level of *hTERT* gene; the dual luciferase reporter assay of *hTERT* promoter (−1200 bp) suggested the suppressive effect of SC on transcriptional activity (Figure 4A). In addition, we also treated the cancer cells with the transcription blocking reagent Actinomycin D and SC to assess the mRNA stability of *hTERT*, and found that SC does not affect *hTERT* mRNA stability (Supplementary Figure S4A). So far, we have speculated that SC suppresses *hTERT* expression by modulating *hTERT* promoter activity, but not mRNA stability.

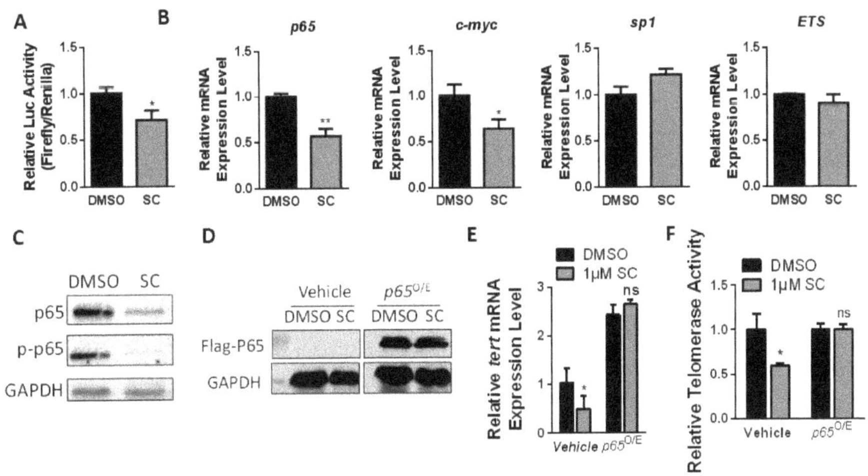

Figure 4. SC inhibited telomerase activity depending on *p65* expression in cancer cells. (**A**) The dual luciferase reporter assay suggested the compound has suppressive effects on *hTERT* transcriptional activities. (**B**) The relative mRNA levels of partial common transcription factors, which have been reported to regulate *hTERT* transcription. (**C**) The protein levels of pan-p65 and pi-p65 in cancer cells after 1 μM SC treatment for 48 h. DMSO served as a control. (**D**) Western blot was carried out to detect the protein level of overexpressed p65 in sanguinarine chloride-treated cells. (**E**) The mRNA level of *hTERT* was rescued in the SC-treated group by overexpressing p65. (**F**) Telomerase activity assay in p65 re-introduced HTC75 cells with SC treatment for 48 h. All the analyses were from triplicate samples (* $p < 0.05$, ** $p < 0.01$, ns means no significance).

The *hTERT* promoter contains many transcription factor-binding sites, including GC-motifs and E-boxes, which can directly modulate telomerase transcription in response to physiological processes, including tumorigenesis. The transcriptional regulation of telomerase is complicated in different cancer cells due to diversified mutations and multi-layered networks [35]. For cells chronically exposed to the compound SC, we detected the mRNA levels of 10 previously reported classic transcription factors and found a significant decrease in *p65*, *c-MYC*, and *MXD1* levels. Among the positively correlated transcription factor genes, *p65* was observed to be down-regulated by SC to the most significant extent (Figure 4B and Figure S4B). Moreover, Western blotting confirmed a decreased level of the p65 protein in the cells treated with the compound (Figure 4C).

To investigate the mechanism by which SC inhibits hTERT/telomerase, we transiently transfected p65 and c-myc plasmids respectively into cancer cells treated with the compound (Figure 4D and Figure S4C). The reintroduction of p65, rather than c-myc, alleviated the inhibitory effect of SC on telomerase at both the *hTERT* mRNA level (Figure 4E and Figure S4D) and the RTA level (Figure 4F and Figure S4E). These results indicate that SC inhibits hTERT/telomerase in a p65-dependent manner.

3.5. SC Directly Modulates Telomerase In Vitro

The rapid attrition of telomere length in SC-treated cells suggests that SC might also directly inhibit telomerase activity, besides decreasing *hTERT* expression. Previously, it was found that the addition of sanguinarine at 10 μM has strong affinity for human telomere repeats and *c-MYC* promoter sequence, enabling the forming of a G-quadruplex structure in vitro [36]. Isoquinoline alkaloids, represented by sanguinarine, were found to selectively recognize the telomeric G-quadruplexes in vitro and inhibit telomerase in MCF-7 cells [37]. In brief, the G-quadruplex is a common target for telomerase or other reverse transcriptases. Moreover, c-myc is an essential transcription factor in cell growth, acting via regulating the expression of related genes, including *hTERT*. The promoter region of *c-MYC* also contains abundant G-quadruplex motifs [38]. Our work suggested that SC retards HTC75 cell growth and represses the expression of hTERT and *c-MYC* (Figures 3D and 4B). In contrast to the reported 10 μM concentration of sanguinarine that recognizes the G-quadruplex in vitro, the effective dosage of SC as a telomerase inhibitor in our system was lowered to 1 μM in various cells. Notably, cells were unable to survive at a dosage/concentration of over 4 μM. Nevertheless, we wondered whether telomerase could be inhibited in our system by SC via G-quadruplex binding or not.

Thus, we performed IP-TRAP and telomerase reconstitution assays. Firstly, we want to evaluate the inhibitory effects of the compound in vitro. The schematic experimental processes of IP-TRAP are depicted in Figure 5A. The immunoprecipitated telomerase complex from hTERT-overexpressed cells was incubated with serial dilutions of the compound for the telomere extension reaction. SC displayed a dose-dependent inhibitory effect on the activity of immunoprecipitated telomerase holoenzyme. The IC50 to immunoprecipitated telomerase in vitro was 1.40 μM (Figure 5A), which is close to the IC50 value when in cell culture (1.21 μM in Figure 3A). This IC50 was much lower than the concentration necessary for SC to bind the G-quadruplex.

We next purified the GST-tagged hTERT protein and in vitro transcribed hTERC, and then incubated them in a water bath. Subsequently, the addition of the compound significantly impeded the reconstituted telomerase activity in a concentration-dependent manner. Surprisingly, the suppressive effect of the compound on the reconstituted telomerase activity in vitro was exhibited at the nanomole level (Figure 5B). In conclusion, we identified that sanguinarine chloride directly inhibits telomerase at a concentration much lower than the 10 μM reported in vitro.

To investigate whether the cellular telomerase inhibition by the compound depends on telomeric G-quadruplex formation, we carried out a series of biochemical assays. Based on a previous work [36], we synthesized a human telomeric oligonucleotide (Telo24) labeled with Cy5 fluorophores. A Microscale Thermophoresis (MST) assay showed the dose-dependent binding of the compound SC to the telomeric G-quadruplex DNA. Unexpectedly, the EC50 (half maximal effective concentration) to telomeric G-quadruplexes was 100 times more than the IC50 to telomerase in cancer cells (120 μM in Figure 5C vs. 1.21 μM in Figure 3A). Thioflavin T (ThT) is a fluorescent dye used to sense G-quadruplex structures, especially in human telomeric DNA [39]. The fluorescence signal of ThT showed no difference after treatment with 1 or 2 μM of the compound or DMSO, while Pyridostatin, a G-quadruplex DNA-stabilizing agent, significantly enhanced the cellular ThT signal intensity (Figure 5D). BG4 is an antibody specific to the G-quadruplex structure. Immunofluorescence was used to visualize and quantify the cellular co-localization of G-quadruplex motifs and telomeres (indicated by an antibody against the telomeric repeat binding factor 2). Following treatment with 1 μM SC, the cell were comparable to in the DMSO control, both in the BG4 foci and in the colocalized foci of BG4 and TRF2 (Figure 5E). Furthermore,

SC at high concentrations (more than 10 μM) could emit an autofluorescence signal, thus fluorescence polarization assays were carried out. The binding curve implied that SC could directly interact with TRBD of the hTERT protein in vitro (Figure 5F).

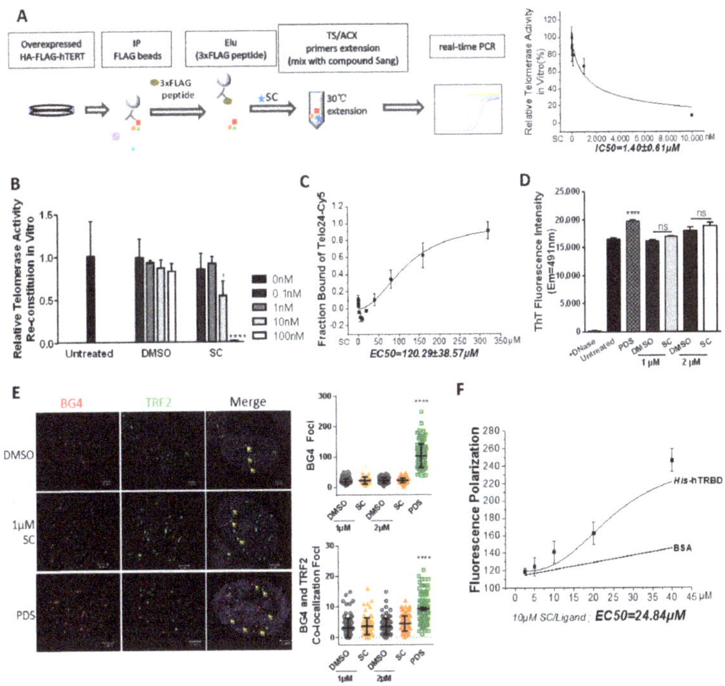

Figure 5. SC directly suppressed telomeric repeat extension in vitro without inducing a G-quadruplex motif. (**A**) Schematic representation of immunoprecipitation-based TRAP experiment. FLAG-hTERT-overexpressing HEK293T cells were subjected to immunoprecipitation. The elutes mixed with SC were used to perform the TRAP assay in vitro. The inhibition of natural telomerase activity by SC was presented in a dose-dependent manner, and the fitting curve showed that its IC50 to natural telomerase in vitro was 1.4 μM. (**B**) Purified GST-opTERT was incubated with in vitro-transcribed hTERC for 30 min, then mixed with SC to perform the TRAP assay. SC directly suppressed the activity of reconstituted telomerase at the nanomole level in vitro. (**C**) MST analysis of the interaction of the telomeric G-quadruplex with SC. The EC50 was 120 μM. (**D**) Detection of 2 μM Thioflavin T (ThT) fluorescence intensity at 491 nm for the whole genomic DNA in a K^+ Tris-HCl buffer. Pyridostatin (PDS) was the positive compound used to induce G-quadruplex structure. (**E**) Representative immunofluorescence images of the G-quadruplex (recognized by BG4 antibody, red) and TRF2 (green) foci in HTC75 cells treated with SC or DMSO for 48 h. Quantification of the number of G-quadruplex foci (recognized by BG4 antibody) per nucleus in compound-treated HTC75 cells (right upper) and quantification of the number of colocalized G-quadruplex foci (recognized by BG4 antibody) and TRF2 in the nucleus (right bottom). In total, 100 nuclei were counted and statistically analyzed. (**F**) A fluorescence polarization binding assay with His-tagged hTRBD and SC was performed, and the EC50 was 24.84 μM; the BSA protein served as a negative control. All the analyses were performed on triplicate samples (* $p < 0.05$, **** $p < 0.0001$, ns means no significance).

Taken together, the results showed that exposing cells to 1 μM SC does not change the formation of the G-quadruplex, and indicated that the compound at the concentration of 1 μM suppresses the telomerase activity in cancer cells by directly binding to the hTERT protein.

3.6. Assessment of Safety and Antitumor Efficacy of SC

As a matter of fact, sanguinarine has shown potential antitumor value in animal models [40,41]. In our system, we also needed to evaluate its safety performance and antitumor efficacy in vitro and in vivo. Firstly, we detected the cell viability of three SC-treated human primary cells with no telomerase expression. The growth of human skin fibroblasts (HFs) and HUVSMCs was analyzed via the CCK-8 assay kit, and the viability of PBMCs was traced based on CFSE labeling. All the results pointed to the safe and non-poisonous characteristics of SC in relation to primary somatic cells at a low dosage (Figure 6A,B). The cell cycle and apoptosis assays performed in the HFs suggested no increased apoptosis or cell cycle arrest was induced (Figure 6C,D). In the tested cells, 1 or 2 μM of SC had no proliferative inhibition effect.

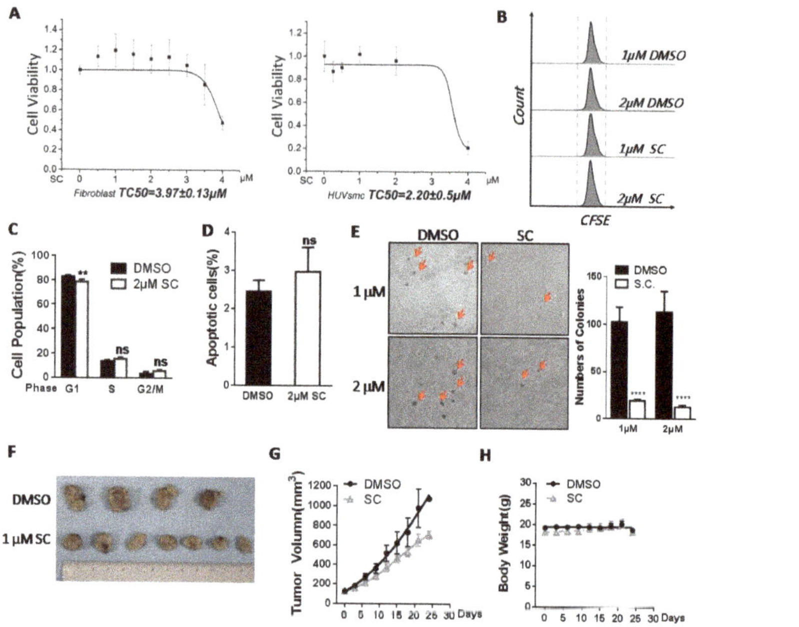

Figure 6. Application of SC to the inhibition of tumor formation. (**A**) the CCK-8 assay showed no proliferative inhibitory effect of SC in human primary skin fibroblast cells (**left**) and HUVSMC (**right**) at the concentration that was effective in inhibiting telomerase activity in cancer cells (data shown in Figure 3, IC50 to telomerase = 1.21 μM). (**B**) Human PBMCs were measured by CFSE labeling and then treated with 2 μM SC for 72 h. The similar characteristics of the histograms indicate that SC does not affect the proliferation of PBMCs. (**C**) Cell cycle PI staining assay was performed in fibroblast with 2 μM of compound. (**D**) Apoptosis analysis for human fibroblasts following SC treatment. (**E**) Representative photographs of the colonies from the soft agar colony formation assay in MDA-MB-231 cells treated with SC or DMSO (control). The red arrows indicate the formed cell colonies. The statistical number of colonies formed in ten randomly visual fields was quantified. (**F**,**H**) MDA-MB-231 cells were used to establish an orthotopic xenograft model in nude mice. Here, 1 μM blood concentration of SC or DMSO (vehicle) was injected into the tail vein. The total blood volume of a mouse was estimated as 7% of the body weight. A representative picture of the developed tumors of each group (**F**); tumor volume was measured every 3 days (**G**), and the body weights of xenograft nude mice was measured before every injection (**H**). All cellular assays were performed on triplicate samples, and the animal experiment was carried out in 4–7 mice ($n = 4$ in control group; $n = 7$ in SC treatment group; four sets of data were used in the volume and body weight assays, ** $p < 0.01$, **** $p < 0.0001$, ns means no significance).

MDA-MB-231 is a triple-negative breast cancer cell line commonly used to represent one kind of advanced breast cancer. Triple-negative breast cancer is considered to be the most dangerous because of its aggressive behavior, the lack of an effective therapy, and the high mortality in clinic. Since we found that SC works effectively in breast cancer cell lines (Figure 2F), we next wondered about the anti-tumor effect of SC. MDA-MB-231 cells were seeded into soft agar medium with drug treatment. After 14 days of culture, in the presence of 1 or 2 μM SC, a dramatic reduction in MDA-MB-231 cell-derived colonies was observed in comparison with the control (Figure 6E). Moreover, MDA-MB-231 cells were transplanted into nude mice in situ to evaluate the compound's capability of suppressing tumorigenesis. The compound (final concentration ~1 μM in blood) was injected intravenously into the xenograft model every 3 days throughout the experimental period, and the same proportion of DMSO was used as the control. The volumes of tumors and the mouse body weight were monitored before each injection. The tumors in SC-treated mice reduced 40% versus those in control mice after 24 days of administration (Figure 6F,G), whereas the body weights remained at a constant level (Figure 6H). Taken together, we see that 1 μM SC exhibited strong antitumor efficacy both in vitro and in vivo.

4. Discussion

Natural products and traditional Chinese medicine have been reported to exhibit various anti-cancer capabilities. The discovery of natural compounds that inhibit telomerase can lead to advancements in tumor therapy. Here, we utilized a telomerase reporter cell line indicating the expression level of endogenous *hTERT*, which can sensitively reflect telomerase modulation under physiological conditions. After stably expressing dsRed2 as an internal control, this reporter can indicate the relative expression of *hTERT* based on the ratio of MFI. The screening approach is simple, rapid, and low-cost. Given the advantages of the reporter, we carried out a high-throughput screening of the natural product library. We succeeded in finding a few small molecules that can modulate telomerase positively (data not shown) or negatively, which are promising for use in anti-aging or anti-cancer applications.

Among the eight identified candidate inhibitors, we found that Braz has been patented as a natural telomerase inhibitor previously [31], while few related experimental studies about its suppressive effect on telomerase have been published. Here, we identified that Braz could inhibit RTA by down-regulating the *hTERT* gene and retarded the cell growth via G2/M phase arrest in cancer cells. However, only certain types of cancer cells sensitively responded to Braz. Q-TRAP data and CCK8 assay demonstrated reduced telomerase activities and cell proliferation in HTC75, HeLa and DLD1 cells upon treatment with the compound, while MDA-MB-231, Hs578t and A549 cells have no response to 8 μM Braz (Supplementary Figure S2C). The effective inhibition of telomerase caused by Braz may be associated with the cytotoxicity of this compound, implying that the telomerase repression by Braz is likely operated via an indirect or complex mechanism. Furthermore, telomerase activity was inhibited in cancer cells following a long-term treatment with Braz, whereas telomere shortening was not observed (Supplementary Figure S3C). Isolated from Caesalpinia sappan L. [42], Braz displayed antitumor abilities by inducing apoptosis and cell cycle arrest [43]. How Braz suppresses telomerase and its anti-carcinogenesis role in vivo remain to be further studied in the future.

SC was another telomerase inhibitor candidate with leading inhibition that we identified from the library. SC exhibited a broad effective range in cancer cell types and an acute inhibitory effect on telomerase and cancer cell growth. This compound down-regulated *hTERT* expression, which may be mediated by directly changing the *hTERT* transcriptional activity because of the decreased expression of several *hTERT* regulatory transcription factors, including the c-myc family, p65 et al. The oncogene *c-MYC* is a common and essential factor in the modulation of hTERT/telomerase. P65 is a subunit of the NF-κB complex and can target the remote region of the *hTERT* promoter (near −600 bp) to regulate its transcription [44]. Moreover, p65 could interact with hTERT to facilitate it transporting into the nucleus [13]. Reintroducing c-myc or p65 into the compound-treated cells indicated that the telomerase inhibition could only be rescued by p65, suggesting SC suppresses telomerase depending on p65 expression but not c-myc.

In addition, overexpressing FLAG-hTERT in the *hTERT* promoter reporter cell line caused an increase in GFP fluorescence intensity (Supplementary Figure S5A). This result implied that TERT transcription may enable positive feedback via its encoded protein, i.e., hTERT may act as a transcription coactivator to regulate its expression; consistently, c-myc could target the *hTERT* promoter region to activate its transcription, and hTERT could also regulate and stabilize *c-MYC* at the transcriptional level [11].

Apart from that, we also found that low dosages of SC could directly suppress telomerase activity both in vitro and in vivo, while not affecting telomeric G-quadruplex formation. Moreover, this inhibition in vitro was achieved at the nanomole level, much lower than the EC50 to the telomeric G-quadruplex in K^+ solution and the IC50 to telomerase at the cellular level. Thus, SC may also directly bind to and suppress hTERT/telomerase. We performed a fluorescence polarization binding assay with TRBD and SC. The preliminary result showed a dose-dependent interaction in vitro (Figure 5F). Furthermore, we also found that this compound specifically decreases exogenous hTERT protein levels compared to the control, possibly by modulating its stability (Supplementary Figure S5B).

Sanguinarine has been used against different tumor or chronic diseases via different mechanisms [45]. Specifically, in breast cancer and cervical cancer, sanguinarine generates reactive oxygen species to induce apoptosis, and suppresses the NF-κB pathway to prevent metastasis [22,46,47]. In prostate cancer, the compound arrests the cell cycle by inhibiting cycle kinases and cyclins [23]. In myeloid cells, it targets the stability and phosphorylation of the IκB protein, and in certain cancers, this compound inhibits VEGF function in angiogenesis [24,48–50]. hTERT has also been reported to participate in the modulation of angiogenesis. Considering the complicated function of sanguinarine, researchers have designed different therapeutic approaches depending on specific cancers. For instance, low concentrations of sanguinarine could synergistically enhance the therapeutic efficacy of the chemotherapeutic agent doxorubicin in drug-resistant leukemia cells [21,51–53].

Different from other anti-tumorigenesis studies of sanguinarine, which were carried out by inducing apoptosis and cell cycle arrest at high concentrations [54], the administration dosage of SC was much lower in our system. Cancer cells, chronically exposed to the compound at a lower concentration for a long time, can show remarkably shortened telomeres, consequently inducing cancer cell senescence.

In conclusion, SC displayed an inhibitory effect on *hTERT* expression and telomerase activity that slowed down cell growth. Long-term treatment with SC induced changes of cell morphology and triggered senescence events, including an increase in SA-β-gal activity, the up-regulation of the expression of p16/p21/p53 pathways, progressive telomere dysfunctions (TIFs), and telomere shortening. Together, all these events triggered by SC led to senescence in cancer cells, thus blocking their progression. SC inhibits telomerase by dual functions, and its antitumor effect is potent and safe.

Previously, the sanguinarine-added mouthwash product Viadent® has been reported to be associated with age-related leukoplakia, indicating its pre-neoplastic adverse effects. Oral leukoplakia is a classic symptom in dyskeratosis congenita patients [55]. Our findings show that long-term SC treatment will decrease telomere length, suggesting human adult stem cells may also be affected by long-term treatment with sanguinarine. This may explain the mechanism of the adverse effects of the Viadent® mouthwash product. More precise works judging the appropriate dosage, duration and drug delivery system will improve its application in pharmacological contexts.

5. Conclusions

Robust telomerase activity is a common feature shared by 90% cancers. Our study identified SC at the precise dosage as an effective telomerase inhibitor, with anticancer applications. Historically, telomerase inhibitors have shown unsatisfactory performances in clinic, although they have exhibited strong suppressive effects at the cell or animal level. Therefore, as a natural telomerase inhibitor with dual functions (regulation on the mRNA and protein levels) and little proliferative inhibition effects on somatic cells, SC is potent and safe, providing a potential therapeutic approach for human malignancies. Our study proposes a prolonged treatment approach using SC to induce cancer cell senescence. Anti-tumor drugs such as SC may be synergistically used with senolytics that kill senescent cells to improve the efficacy. Additionally, the precise dosage and duration of SC application in cancer therapeutics need to be considered in future research.

Supplementary Materials: The following supporting information can be downloaded at: https://www.mdpi.com/article/10.3390/cells11091485/s1, Figure S1: Compound screening for endogenous hTERT inhibitors; Figure S2: Effects of Braz treatment on cancer cells; Figure S3: Analysis of HTC75 cell treated with Braz; Figure S4: The mechanism of SC regulates hTERT transcription; Figure S5: Discussion supplement; Table S1: The result of the second screening.

Author Contributions: S.Y. (Siyu Yan) performed the experiments, data analysis and draft writing; S.L. provided technique support and research resources; K.C. helped in the cell culture experiments; S.Y. (Shanshan Yin) helped in the nude mice experiment; H.P. helped in the qPCR experiment; N.C. helped in the Western blot experiments; W.M. provided suggestions for the study design; Z.S. revised the manuscript; Y.H. designed the study, arranged the data and revised the manuscript. All authors approved the final manuscript. All authors have read and agreed to the published version of the manuscript.

Funding: This work was funded by the National Key Research and Development Program of China (2018YFA0107003), the National Natural Science Foundation (81871109, 82071587, 31930058, 32170757, 31871479), and Guang Dong Basic and Applied Basic Research Foundation (2020A1515010462).

Institutional Review Board Statement: All animal studies involved were approved by the Animal Care and Use Committee of Sun Yat-sen University.

Informed Consent Statement: Not applicable.

Data Availability Statement: The data published in this study are available from the corresponding authors upon reasonable request.

Acknowledgments: The authors would like to thank Qinong Ye for providing the optimized GST-hTERT plasmid.

Conflicts of Interest: All authors declare no conflict of interest.

References

1. Blackburn, E.H. Switching and signaling at the telomere. *Cell* **2001**, *106*, 661–673. [CrossRef]
2. Liu, D.; O'Connor, M.S.; Qin, J.; Songyang, Z. Telosome, a mammalian telomere-associated complex formed by multiple telomeric proteins. *J. Biol. Chem.* **2004**, *279*, 51338–51342. [CrossRef]
3. Palm, W.; de Lange, T. How shelterin protects mammalian telomeres. *Annu. Rev. Genet.* **2008**, *42*, 301–334. [CrossRef]
4. Heaphy, C.M.; Subhawong, A.P.; Hong, S.-M.; Goggins, M.G.; Montgomery, E.A.; Gabrielson, E.; Netto, G.J.; Epstein, J.I.; Lotan, T.; Westra, W.H.; et al. Prevalence of the alternative lengthening of telomeres telomere maintenance mechanism in human cancer subtypes. *Am. J. Pathol.* **2011**, *179*, 1608–1615. [CrossRef]
5. Nguyen, T.H.D.; Tam, J.; Wu, R.A.; Greber, B.J.; Toso, D.; Nogales, E.; Collins, K. Cryo-EM structure of substrate-bound human telomerase holoenzyme. *Nature* **2018**, *557*, 190–195. [CrossRef]
6. Gillis, J.A.; Schuller, A.P.; Skordalakes, E. Structure of the Tribolium castaneum telomerase catalytic subunit TERT. *Nature* **2008**, *455*, 633–637. [CrossRef]
7. Schmidt, J.C.; Cech, T.R. Human telomerase: Biogenesis, trafficking, recruitment, and activation. *Genes Dev.* **2015**, *29*, 1095–1105. [CrossRef]
8. MacKenzie, K.L.; Franco, S.; May, C.; Sadelain, M.; Moore, M.A. Mass cultured human fibroblasts overexpressing hTERT encounter a growth crisis following an extended period of proliferation. *Exp. Cell Res.* **2000**, *259*, 336–350. [CrossRef]
9. Zhou, L.; Zheng, D.; Wang, M.; Cong, Y.-S. Telomerase reverse transcriptase activates the expression of vascular endothelial growth factor independent of telomerase activity. *Biochem. Biophys. Res. Commun.* **2009**, *386*, 739–743. [CrossRef]
10. Wu, K.-J.; Grandori, C.; Amacker, M.; Simon-Vermot, N.; Polack, A.; Lingner, J.; Dalla-Favera, R. Direct activation of TERT transcription by c-MYC. *Nat. Genet.* **1999**, *21*, 220–224. [CrossRef]

11. Koh, C.M.; Khattar, E.; Leow, S.C.; Liu, C.Y.; Muller, J.; Ang, W.X.; Li, Y.; Franzoso, G.; Li, S.; Guccione, E.; et al. Telomerase regulates MYC-driven oncogenesis independent of its reverse transcriptase activity. *J. Clin. Investig.* **2015**, *125*, 2109–2122. [CrossRef]
12. Zuo, Q.P.; Liu, S.K.; Li, Z.J.; Li, B.; Zhou, Y.L.; Guo, R.; Huang, L.H. NF-kappaB p65 modulates the telomerase reverse transcriptase in the HepG$_2$ hepatoma cell line. *Eur. J. Pharmacol.* **2011**, *672*, 113–120. [CrossRef]
13. Akiyama, M.; Hideshima, T.; Hayashi, T.; Tai, Y.-T.; Mitsiades, C.S.; Mitsiades, N.; Chauhan, D.; Richardson, P.; Munshi, N.C.; Anderson, K.C. Nuclear factor-kappaB p65 mediates tumor necrosis factor alpha-induced nuclear translocation of telomerase reverse transcriptase protein. *Cancer Res.* **2003**, *63*, 18–21.
14. Huang, E.; Tedone, E.; O'Hara, R.; Cornelius, C.; Lai, T.-P.; Ludlow, A.; Wright, W.; Shay, J.W. The maintenance of telomere length in CD28+ T cells during T lymphocyte stimulation. *Sci. Rep.* **2017**, *7*, 6785. [CrossRef]
15. Zou, Y.; Cong, Y.S.; Zhou, J. Implications of telomerase reverse transcriptase in tumor metastasis. *BMB Rep.* **2020**, *53*, 458–465. [CrossRef]
16. Herbert, B.S.; Gellert, G.C.; Hochreiter, A.; Pongracz, K.; Wright, W.E.; Zielinska, D.; Chin, A.C.; Harley, C.B.; Shay, J.W.; Gryaznov, S.M. Lipid modification of GRN163, an N3'–>P5' thio-phosphoramidate oligonucleotide, enhances the potency of telomerase inhibition. *Oncogene* **2005**, *24*, 5262–5268. [CrossRef]
17. Bryan, C.; Rice, C.; Hoffman, H.; Harkisheimer, M.; Sweeney, M.; Skordalakes, E. Structural basis of telomerase inhibition by the highly specific BIBR1532. *Structure* **2015**, *23*, 1934–1942. [CrossRef]
18. Doğan, F.; Özateş, N.P.; Bağca, B.G.; Abbaszadeh, Z.; Söğütlü, F.; Gasimli, R.; Gündüz, C.; Avcı Çığır, B. Investigation of the effect of telomerase inhibitor BIBR1532 on breast cancer and breast cancer stem cells. *J. Cell. Biochem.* **2019**, *120*, 1282–1293. [CrossRef]
19. Collado, M.; Blasco, M.A.; Serrano, M. Cellular senescence in cancer and aging. *Cell* **2007**, *130*, 223–233. [CrossRef]
20. Chang, B.D.; Broude, E.V.; Dokmanovic, M.; Zhu, H.; Ruth, A.; Xuan, Y.; Kandel, E.S.; Lausch, E.; Christov, K.; Roninson, I.B. A senescence-like phenotype distinguishes tumor cells that undergo terminal proliferation arrest after exposure to anticancer agents. *Cancer Res.* **1999**, *59*, 3761–3767.
21. Fu, C.; Guan, G.; Wang, H. The Anticancer Effect of Sanguinarine: A Review. *Curr. Pharm. Des.* **2018**, *24*, 2760–2764. [CrossRef] [PubMed]
22. Chaturvedi, M.M.; Kumar, A.; Darnay, B.G.; Chainy, G.B.; Agarwal, S.; Aggarwal, B.B. Sanguinarine (pseudochelerythrine) is a potent inhibitor of NF-kappaB activation, IkappaBalpha phosphorylation, and degradation. *J. Biol. Chem.* **1997**, *272*, 30129–30134. [CrossRef] [PubMed]
23. Adhami, V.M.; Aziz, M.H.; Reagan-Shaw, S.R.; Nihal, M.; Mukhtar, H.; Ahmad, N. Sanguinarine causes cell cycle blockade and apoptosis of human prostate carcinoma cells via modulation of cyclin kinase inhibitor-cyclin-cyclin-dependent kinase machinery. *Mol. Cancer Ther.* **2004**, *3*, 933–940. [PubMed]
24. Xu, J.-Y.; Meng, Q.-H.; Chong, Y.; Jiao, Y.; Zhao, L.; Rosen, E.M.; Fan, S. Sanguinarine is a novel VEGF inhibitor involved in the suppression of angiogenesis and cell migration. *Mol. Clin. Oncol.* **2013**, *1*, 331–336. [CrossRef] [PubMed]
25. Mackraj, I.; Govender, T.; Gathiram, P. Sanguinarine. *Cardiovasc. Ther.* **2008**, *26*, 75–83.
26. Damm, D.D.; Curran, A.; White, D.K.; Drummond, J.F. Leukoplakia of the maxillary vestibule–an association with Viadent? *Oral Surg. Oral Med. Oral Pathol. Oral Radiol. Endodontol.* **1999**, *87*, 61–66. [CrossRef]
27. Mascarenhas, A.K.; Allen, C.M.; Loudon, J. The association between Viadent use and oral leukoplakia. *Epidemiology* **2001**, *12*, 741–743. [CrossRef]
28. Deng, T.; Huang, Y.; Weng, K.; Lin, S.; Li, Y.; Shi, G.; Chen, Y.; Huang, J.; Liu, D.; Ma, W.; et al. TOE1 acts as a 3′ exonuclease for telomerase RNA and regulates telomere maintenance. *Nucleic Acids Res.* **2019**, *47*, 391–405. [CrossRef]
29. Xin, H.; Liu, D.; Wan, M.; Safari, A.; Kim, H.; Sun, W.; O'Connor, M.S.; Zhou, S. TPP1 is a homologue of ciliate TEBP-beta and interacts with POT1 to recruit telomerase. *Nature* **2007**, *445*, 559–562. [CrossRef]
30. Cheng, L.; Yuan, B.; Ying, S.; Niu, C.; Mai, H.; Guan, X.; Yang, X.; Teng, Y.; Lin, J.; Huang, J.; et al. PES1 is a critical component of telomerase assembly and regulates cellular senescence. *Sci. Adv.* **2019**, *5*, eaav1090. [CrossRef]
31. Chin, R.L.; Tolman, A.C. Telomerase Inhibitors and Methods of Their Use. WO Patent 0,193,864, 13 December 2001.
32. Van Steensel, B.; De Lange, T. Control of telomere length by the human telomeric protein TRF1. *Nature* **1997**, *385*, 740–743. [CrossRef]
33. Choi, W.Y.; Kim, G.-Y.; Lee, W.H.; Choi, Y.H. Sanguinarine, a benzophenanthridine alkaloid, induces apoptosis in MDA-MB-231 human breast carcinoma cells through a reactive oxygen species-mediated mitochondrial pathway. *Chemotherapy* **2008**, *54*, 279–287. [CrossRef] [PubMed]
34. Kim, S.; Lee, T.-J.; Leem, J.; Choi, K.S.; Park, J.-W.; Kwon, T.K. Sanguinarine-induced apoptosis: Generation of ROS, down-regulation of Bcl-2, c-FLIP, and synergy with TRAIL. *J. Cell. Biochem.* **2008**, *104*, 895–907. [CrossRef] [PubMed]
35. Eitsuka, T.; Nakagawa, K.; Kato, S.; Ito, J.; Otoki, Y.; Takasu, S.; Shimizu, N.; Takahashi, T.; Miyazawa, T. Modulation of telomerase activity in cancer cells by dietary compounds: A review. *Int. J. Mol. Sci.* **2018**, *19*, 478. [CrossRef] [PubMed]
36. Ghosh, S.; Pradhan, S.K.; Kar, A.; Chowdhury, S.; Dasgupta, D. Molecular basis of recognition of quadruplexes human telomere and c-myc promoter by the putative anticancer agent sanguinarine. *Biochim. Biophys. Acta BBA—Gen. Subj.* **2013**, *1830*, 4189–4201. [CrossRef]
37. Noureini, S.K.; Esmaeili, H.; Abachi, F.; Khiali, S.; Islam, B.; Kuta, M.; Saboury, A.A.; Hoffmann, M.; Sponer, J.; Parkinson, G.; et al. Selectivity of major isoquinoline alkaloids from Chelidonium majus towards telomeric G-quadruplex: A study using a transition-FRET (t-FRET) assay. *Biochim. Biophys. Acta BBA—Gen. Subj.* **2017**, *1861*, 2020–2030. [CrossRef]

38. Siddiqui-Jain, A.; Grand, C.L.; Bearss, D.J.; Hurley, L.H. Direct evidence for a G-quadruplex in a promoter region and its targeting with a small molecule to repress c-MYC transcription. *Proc. Natl. Acad. Sci. USA* **2002**, *99*, 11593–11598. [CrossRef]
39. Mohanty, J.; Barooah, N.; Dhamodharan, V.; Harikrishna, S.; Pradeepkumar, P.I.; Bhasikuttan, A.C. Thioflavin T as an efficient inducer and selective fluorescent sensor for the human telomeric G-quadruplex DNA. *J. Am. Chem. Soc.* **2013**, *135*, 367–376. [CrossRef]
40. Pica, F.; Balestrieri, E.; Serafino, A.; Sorrentino, R.; Gaziano, R.; Moroni, G.; Moroni, N.; Palmieri, G.; Mattei, M.; Garaci, E.; et al. Antitumor effects of the benzophenanthridine alkaloid sanguinarine in a rat syngeneic model of colorectal cancer. *Anti-Cancer Drugs* **2012**, *23*, 32–42. [CrossRef]
41. Wei, G.; Xu, Y.; Peng, T.; Yan, J.; Wang, Z.; Sun, Z. Sanguinarine exhibits antitumor activity via up-regulation of Fas-associated factor 1 in non-small cell lung cancer. *J. Biochem. Mol. Toxicol.* **2017**, *31*, e21914. [CrossRef]
42. Baek, N.-I.; Jeon, S.G.; Ahn, E.-M.; Hahn, J.-T.; Bahn, J.H.; Jang, J.S.; Cho, S.-W.; Park, J.K.; Choi, S.Y. Anticonvulsant compounds from the wood of *Caesalpinia sappan* L. *Arch. Pharmacal Res.* **2000**, *23*, 344–348. [CrossRef] [PubMed]
43. Kim, B.; Kim, S.H.; Jeong, S.J.; Sohn, E.J.; Jung, J.H.; Lee, M.H.; Kim, S.H. Brazilin induces apoptosis and G2/M arrest via inactivation of histone deacetylase in multiple myeloma U266 cells. *J. Agric. Food Chem.* **2012**, *60*, 9882–9889. [CrossRef] [PubMed]
44. Gizard, F.; Heywood, E.B.; Findeisen, H.M.; Zhao, Y.; Jones, K.L.; Cudejko, C.; Post, G.R.; Staels, B.; Bruemmer, D. Telomerase activation in atherosclerosis and induction of telomerase reverse transcriptase expression by inflammatory stimuli in macrophages. *Arter. Thromb. Vasc. Biol.* **2011**, *31*, 245–252. [CrossRef] [PubMed]
45. Basu, P.; Kumar, G.S. Sanguinarine and Its Role in Chronic Diseases. *Adv. Exp. Med. Biol.* **2016**, *928*, 155–172. [PubMed]
46. Ahsan, H.; Reagan-Shaw, S.; Breur, J.; Ahmad, N. Sanguinarine induces apoptosis of human pancreatic carcinoma AsPC-1 and BxPC-3 cells via modulations in Bcl-2 family proteins. *Cancer Lett.* **2007**, *249*, 198–208. [CrossRef]
47. Park, S.Y.; Jin, M.L.; Kim, Y.H.; Lee, S.-J.; Park, G. Sanguinarine inhibits invasiveness and the MMP-9 and COX-2 expression in TPA-induced breast cancer cells by inducing HO-1 expression. *Oncol. Rep.* **2014**, *31*, 497–504. [CrossRef]
48. Eun, J.-P.; Koh, G.Y. Suppression of angiogenesis by the plant alkaloid, sanguinarine. *Biochem. Biophys. Res. Commun.* **2004**, *317*, 618–624. [CrossRef]
49. Basini, G.; Bussolati, S.; Santini, S.E.; Grasselli, F. Sanguinarine inhibits VEGF-induced angiogenesis in a fibrin gel matrix. *Biofactors* **2007**, *29*, 11–18. [CrossRef]
50. De Stefano, I.; Raspaglio, G.; Zannoni, G.F.; Travaglia, D.; Prisco, M.G.; Mosca, M.; Ferlini, C.; Scambia, G.; Gallo, D. Antiproliferative and antiangiogenic effects of the benzophenanthridine alkaloid sanguinarine in melanoma. *Biochem. Pharmacol.* **2009**, *78*, 1374–1381. [CrossRef]
51. Eid, S.Y.; El-Readi, M.Z.; Wink, M. Synergism of three-drug combinations of sanguinarine and other plant secondary metabolites with digitonin and doxorubicin in multi-drug resistant cancer cells. *Phytomedicine* **2012**, *19*, 1288–1297. [CrossRef]
52. Gaziano, R.; Moroni, G.; Buè, C.; Miele, M.T.; Sinibaldi-Vallebona, P.; Pica, F. Antitumor effects of the benzophenanthridine alkaloid sanguinarine: Evidence and perspectives. *World J. Gastrointest. Oncol.* **2016**, *8*, 30–39. [CrossRef] [PubMed]
53. Achkar, I.W.; Mraiche, F.; Mohammad, R.M.; Uddin, S. Anticancer potential of sanguinarine for various human malignancies. *Future Med. Chem.* **2017**, *9*, 933–950. [CrossRef] [PubMed]
54. Fan, H.N.; Chen, W.; Peng, S.Q.; Chen, X.Y.; Zhang, R.; Liang, R.; Liu, H.; Zhu, J.-S.; Zhang, J. Sanguinarine inhibits the tumorigenesis of gastric cancer by regulating the TOX/DNA-PKcs/ KU70/80 pathway. *Pathol.-Res. Pract.* **2019**, *215*, 152677. [CrossRef]
55. Savage, S.A.; Alter, B.P. Dyskeratosis congenita. *Hematol. Oncol. Clin. N. Am.* **2009**, *23*, 215–231. [CrossRef] [PubMed]

cells MDPI

Article

Detection of Cellular Senescence in Human Primary Melanocytes and Malignant Melanoma Cells In Vitro

Tom Zimmermann [1], Michaela Pommer [1], Viola Kluge [1], Chafia Chiheb [1], Susanne Muehlich [2] and Anja-Katrin Bosserhoff [1,3,*]

1 Institute of Biochemistry, Friedrich-Alexander-Universität Erlangen-Nürnberg (FAU), 91052 Erlangen, Germany; tom.zimmermann@fau.de (T.Z.); michaela.pommer@fau.de (M.P.); viola.kluge@fau.de (V.K.); chafia.chiheb@fau.de (C.C.)
2 Department of Chemistry and Pharmacy, Friedrich-Alexander-Universität Erlangen-Nürnberg (FAU), 91052 Erlangen, Germany; susanne.muehlich@fau.de
3 Comprehensive Cancer Center (CCC) Erlangen-EMN, 91054 Erlangen, Germany
* Correspondence: anja.bosserhoff@fau.de

Abstract: Detection and quantification of senescent cells remain difficult due to variable phenotypes and the absence of highly specific and reliable biomarkers. It is therefore widely accepted to use a combination of multiple markers and cellular characteristics to define senescent cells in vitro. The exact choice of these markers is a subject of ongoing discussion and usually depends on objective reasons such as cell type and treatment conditions, as well as subjective considerations including feasibility and personal experience. This study aims to provide a comprehensive comparison of biomarkers and cellular characteristics used to detect senescence in melanocytic systems. Each marker was assessed in primary human melanocytes that overexpress mutant BRAFV600E, as it is commonly found in melanocytic nevi, and melanoma cells after treatment with the chemotherapeutic agent etoposide. The combined use of these two experimental settings is thought to allow profound conclusions on the choice of senescence biomarkers when working with melanocytic systems. Further, this study supports the development of standardized senescence detection and quantification by providing a comparative analysis that might also be helpful for other cell types and experimental conditions.

Keywords: senescence; melanocyte; melanoma; beta-galactosidase

Citation: Zimmermann, T.; Pommer, M.; Kluge, V.; Chiheb, C.; Muehlich, S.; Bosserhoff, A.-K. Detection of Cellular Senescence in Human Primary Melanocytes and Malignant Melanoma Cells In Vitro. *Cells* **2022**, *11*, 1489. https://doi.org/10.3390/cells11091489

Academic Editors: Nicole Wagner and Kay-Dietrich Wagner

Received: 24 March 2022
Accepted: 26 April 2022
Published: 28 April 2022

Publisher's Note: MDPI stays neutral with regard to jurisdictional claims in published maps and institutional affiliations.

1. Introduction

Cellular senescence describes a stable state of growth arrest, commonly accompanied by ample molecular and phenotypical changes. In the decades since its first description by Hayflick and Moorhead in 1961 [1], cellular senescence has been linked to numerous physiological and pathological conditions, ranging from developmental processes to neurodegenerative diseases and cancer [2]. In each of these conditions, establishment of senescence is caused by one of three major mechanisms: telomere shortening, oncogene activation, or extensive DNA damage [3]. When it comes to melanocytes and malignant melanoma, two of these mechanisms are of special importance: first, oncogene-induced senescence (OIS) as a central feature of melanocytic nevi that prevents further oncogenesis and malignant transformation of such benign lesions [4]. As melanocytic nevi have been causally linked to development of malignant melanoma [5,6], stabilization of OIS or clearing of senescent melanocytes remain an important and promising approach in preventing tumorigenesis [7,8]. Second, DNA damage-induced senescence is an important part of therapeutic treatment of malignant melanoma. The majority of cytotoxic treatments cause DNA damage to induce apoptosis and senescence, thereby halting tumor growth [9]. However, growing evidence was found that senescent cancer cells might become therapy-resistant, resulting in residual tumor masses and potentially recurrent malignancies [10,11]. Consequently, there is a need for efficient and reliable detection and targeting of senescent cells.

Although fundamental hallmarks of cellular senescence are conserved in most experimental and clinical conditions, the exact phenotype is often variable and affects the reliability of biomarkers [12]. In addition, the majority of these markers are not completely specific for cellular senescence, e.g., growth arrest and activation of DNA damage response [12,13]. Melanocytes introduce another potential problem since they physiologically show high levels of lysosomal beta-galactosidase activity [14]. This potentially interferes with detection of senescence-associated beta-galactosidase (SA-β-Gal) [15], one of the most widely used markers of cellular senescence [16], and needs to be taken into consideration. The aim of this study is to comprehensively evaluate biomarkers of cellular senescence for their use in primary melanocytes and melanoma cells. It is thought to support the ongoing discussion on the choice of the best markers, especially in the complex field of melanoma research, and thereby improving the reliability and reproducibility of senescence detection.

2. Materials and Methods

2.1. Cell Culture

Normal human melanocytes (NHEM, neonatal) were obtained from Lonza and cultivated in MGM-4 BulletKit medium (Lonza, Basel, Switzerland) with 1% penicillin/streptomycin. NHEM from different donors were used between passages 6 and 8. Cells from the same donor, but at a different passage, were considered biological replicates. HEK293T cells for transduction were a generous gift from Prof. Stephan Hahn (Ruhr-Universität Bochum, Germany). Their cultivation required high-glucose Dulbecco's modified Eagle's medium (DMEM) with 10% FCS and 1% penicillin/streptomycin. Both NHEM and HEK293T cells were incubated at 37 °C and 5% CO_2 in a humified atmosphere. Melanoma cell line Mel Juso was cultivated in RPMI 1640 medium with 2% sodium bicarbonate, 10% FCS and 1% penicillin/streptomycin at 37 °C, and 8% CO_2. Mycoplasma contamination was regularly excluded for all primary cells and cell lines. When reaching approximately 80% confluence, cells were washed with PBS and detached using a solution of 0.05% trypsin and 0.02% EDTA in PBS. After centrifugation and removal of the trypsin solution, cells were either passaged or counted using a Neubauer counting chamber. Melanoma cell line Mel Im was cultured as described in Section 2.3. Unless otherwise stated, cell culture chemicals and media were obtained from Sigma Aldrich (Steinheim, Germany).

2.2. Lentiviral Transduction of Melanocytes

Lentiviral transduction using a third-generation vector system was described elsewhere [17]. Briefly, HEK293T cells were seeded in 10 cm plates at a density of 2×10^6 cells/plate. On the next day, three vectors were introduced simultaneously using transfection with Lipofectamine® LTX (Thermo Fisher, Waltham, MA, USA): an envelope plasmid pHIT-G, a packaging plasmid pCMV ΔR8.2, and a target plasmid with the DNA of interest (either copGFP or B-RAF^{V600E}). Cells were incubated 16 h before the medium was changed to MGM-4 BulletKit medium. After additional incubation for 24 h, supernatants were collected, filtered, and applied to NHEM. Polybrene® (Santa Cruz, Dallas, TX, USA) was added to a final concentration of 1 μg/mL to increase the efficiency of viral uptake. Due to high sensitivity of primary cells, lentiviral supernatants were removed after approximately 6 h. Cells were washed three times with PBS and cultivated in regular MGM-4 BulletKit medium. All experiments using transduced melanocytes started exactly 7 days after transduction to allow establishment of a senescent phenotype. All data in this study are derived from samples that are either untransfected or transfected with scrambled siRNAs. Different transductions are referred to as Mock (copGFP control plasmid) or BRAFm (B-RAFV600E).

2.3. Induction of Senescence in Melanoma Cells

Etoposide treatment started 24 h after approximately 200,000 cells/well were seeded in 6-well plates. Etoposide (R&D Systems, Minneapolis, MN, USA) was dissolved in DMSO to achieve a stock solution of 50 mM, which was then diluted in culture medium to a final concentration of 100 μM and applied to the cells. Control cells were treated with a similar amount of DMSO to exclude effects of the solvent. After an incubation period of 48 h, cells were detached as described in Section 2.1 and either collected for further processing or counted using a Neubauer counting chamber.

Treatment with acidified nitrite was described recently [18]. After a treatment period of 5 min, cells were incubated for 48 h in culture medium and eventually detached and collected for further processing. For the analysis of long-time acidosis effects, melanoma metastasis cell line Mel Im was cultured in medium at pH 6.7 for at least 2 months prior to analyzation. Therefore, low-glucose DMEM was supplemented with 10% FCS and 1% penicillin/streptomycin and 0.2% sodium bicarbonate as buffer. Cells were then incubated at 37 °C and 8% CO_2 to set the desired pH value. Control cells were cultured conventionally at pH 7.4 in low-glucose DMEM including 3.7% sodium bicarbonate, supplemented with 10% FCS and 1% penicillin/streptomycin at 37 °C, and 8% CO_2.

2.4. Analysis of mRNA Expression Using Real-Time PCR

Total RNA isolation was achieved using E.Z.N.A.® Total RNA Kit II (Omega Bio-Tek, Norcross, GA, USA) according to manufacturer's instructions. Generation of cDNA was performed as previously described [19]. For real-time PCR, LightCycler® 480 II devices (Roche, Basel, Switzerland) were used with forward and reverse primers from Sigma-Aldrich. Primer sequences can be found in Table 1.

Table 1. Oligonucleotides used for real-time PCR.

Gene	Forward Primer	Reverse Primer
CDKN2A	GGAGCAGCATGGAGCCTTCGGC	CCACCAGCGTGTCCAGGAAGC
CDKN1A	CGAGGCACCGAGGCACTCAGAGG	CCTGCCTCCTCCCAACTCATCCC
TP53	AAGTCTAGAGCCACCGTCCA	AGTCTGGCTGCCAATCCA
CXCL2	ATCAATGTGACGGCAGGGAAA	CGAAACCTCTCTGCTCTAACAC
CCL8	CCCAGGTGCAGTGTGACATTA	GGGAGGACCCCACAACACTA
18s	TCTGTGATGCCCTTAGATGTCC	CCATCCAATCGGTAGTAGCG

2.5. Western Blot Protein Analysis

Total protein isolation was realized using radio-immunoprecipitation assay buffer (Roche) as described previously [17]. A total of 20 μg protein were loaded on 12.75% SDS polyacrylamide gels for electrophoresis and immediately blotted onto a PVDF membrane (Bio-Rad, Hercules, CA, USA). Protein load of each sample was quantified using Ponceau S staining. After a washing step using double distilled water, membranes were blocked for 1 h using 5% non-fat dried milk/TBS-T. Primary antibodies against γH2AX (1:1000 in 5% NFDM, Cell Signaling, 9718), p16 (1:500 in TBS-T, R&D Systems, AF5779, previously used in [20]), p21 (1:1000 in 5% NFDM, Abcam, ab109199), p53 (1:2000 in 5% BSA, Santa Cruz, sc-126), pERK and ERK (1:1000 in 5% BSA, Cell signaling, 4370 and 9102) were incubated overnight, shaking at 4 °C. The primary antibody against β-actin (1:5000 in 5% BSA, Sigma Aldrich, A5441) was incubated for 1 h at room temperature. Secondary antibodies conjugated to horseradish peroxidase (HRP, Cell Signaling 7074 and 7076, and Dako P0449) were applied for 1 h at room temperature. Afterwards, Clarity™ Western ECL Substrate (Bio-Rad) was added to the membranes for visualization with a Chemostar chemiluminescence imager (Intas, Goettingen, Germany). Quantification of signal intensity was achieved using LabImage software (Version 4.2.3, Kapelan Bio-Imaging GmbH, Germany). One sample of NHEM was excluded from analysis of p16 protein levels due to artifacts interfering with signal quantification.

2.6. Immunofluorescent Stainings

Approximately 20,000 cells were seeded on 18 mm round coverslips and incubated overnight. On the next day, cells were washed twice with PBS and subsequently fixed with

4% PFA for 10 min. Staining procedure started immediately after this fixation step, since even short-term storage was found to interfere with nuclear PML signal in our experiments. Permeabilization using 0.1% Triton-X100 in PBS for 3 min was followed by 30 min blocking with 10% BSA in PBS. The primary antibody against PML (1:200 in 1.5% BSA/PBS, Santa Cruz, sc-966) was added and incubated overnight at 4 °C. The secondary antibody (1:400 in 1.5% BSA/PBS, Thermo Fisher, A32727) was incubated for 1 h at room temperature. Cells were then stained with DAPI (1:10,000 in PBS, Sigma Aldrich) and mounted on microscope slides using Aqua-Poly/Mount (Polysciences, Warrington, PA, USA). Final stainings were analyzed using an Olympus IX83 inverted microscope in combination with Olympus CellSens Dimension software (Version 2.3, Olympus, Tokyo, Japan). DAPI staining of the stainings was used to analyze heterochromatin formation. Brightness and contrast of representative images were adjusted evenly to increase visibility of the staining.

2.7. *Real-Time Cell Proliferation Analysis (RTCA)*

Proliferation was measured using the xCELLigence System (Roche) as described elsewhere [21]. In short, approximately 3000 cells/well were seeded on specific plates and loaded into the device. Proliferation was monitored for five (Mel Juso) or nine days (NHEM) without replacing culture medium. The parameter *slope* describes the steepness of each curve during proliferation and was normalized to control treatment.

2.8. *XTT Cell Viability Assay*

Approximately 3000 cells/well were seeded in a 96-well plate. Cell viability was assessed after an incubation period of 7 days (NHEM) or 48 h (Mel Juso) using the Cell Proliferation Kit II (Roche) according to the manufacturer's instructions. Due to low cell density, it was not necessary to replace culture medium at any time during the incubation period. A Clariostar Plus Multiplate reader (BMG Labtech, Ortenberg, Germany) was used for photometric detection. Absorbance values were normalized to control treatment.

2.9. *Staining of Senescence-Associated Beta-Galactosidase Activity*

Quantification of β-galactosidase activity was done 7 days post transduction in NHEM and 48 h post treatment in Mel Juso. Fixation and staining were performed using the senescence β-galactosidase staining kit (Cell Signaling, Danvers, MA, USA) according to the manufacturer's instructions. After staining, cells were washed twice with PBS and stored at 4 °C for up to two weeks. An Olympus IX83 inverted microscope in combination with Olympus CellSens Dimension Software (Olympus) was used to acquire images of the stainings, which were then quantified manually using ImageJ. The same images were used to manually assess and quantify changes in cellular morphology. Brightness and contrast of representative images were adjusted evenly to increase visibility of the staining.

2.10. *Flow Cytometry of Fluorescent Beta-Galactosidase Substrates*

Activity of β-galactosidase was also quantified using fluorescent substrates in combination with flow cytometry. After treatment and adequate incubation times (see Sections 2.2 and 2.3), approximately 350,000 cells were seeded in 6-well plates exactly 12 h prior staining. The ImaGene Red™ C_{12}RG lacZ Gene Expression Kit (Molecular Probes) was used in accordance to the manufacturer's instructions. In short, staining began by addition of 300 μM chloroquine reagent in 1 mL prewarmed cell medium. After incubation for 30 min at 37 °C, 6.67 μL substrate reagent were added directly to the supernatant to achieve a final concentration of 33 μM. Cells were incubated for another 1 h at 37 °C before they were detached and collected. Following centrifugation, samples were resuspended in 1 mM PETG reagent in 1% BSA/PBS and transferred to FACS tubes. For staining with DDAO, a solution of 20 μM DDAO galactoside (Thermo Fisher) and 0.1 μM Bafilomycin A1 (Sigma Aldrich) cell culture medium was prepared and applied to the cells. Plates were then sealed with parafilm and incubated for 90 min at 37 °C. Cells were eventually washed with PBS and detached, followed by centrifugation and resuspension in 1% BSA/PBS. All samples were measured using a BD LSRFortessa™ flow cytometer in combination with BD FACSDiva™ software (Version 8.0, BD Biosciences, San Jose, CA, USA).

2.11. Statistical Analysis

Analysis and visualization of experimental results was done using GraphPad Prism 9 software (Version 9.1.2, GraphPad Software Inc., San Diego, CA, USA). If not otherwise stated, at least three biological replicated were measured and statistical analysis was performed by Student's unpaired *t*-test. All results are normalized to the respective control treatment and shown as mean ± SEM. A critical value of $p < 0.05$ was considered statistically significant.

3. Results

3.1. RNA Markers of Senescence

Quantification of gene expression displays an easy and reliable approach to assess cellular conditions, including senescence. We here used quantitative RT-PCR to detect specific mRNAs regulated in senescence of normal human melanocytes (NHEM) and melanoma cells. NHEM received lentiviral transduction of mutated BRAF^{V600E}, which leads to oncogene-induced senescence (OIS) as initially described by Michaloglou et al. [4]. Melanoma cell line Mel Juso was treated with 100 μM etoposide, an inhibitor of topoisomerase II, to induce DNA damage and thereby trigger cellular senescence. Traditional mRNA markers of senescence include cell cycle inhibitors and members of the senescence-associated secretory phenotype (SASP). We started with assessing cell cycle inhibitors p21$^{CIP1/WAF1}$ and p16^{INK4A}, which are encoded by *CDKN1A* and *CDKN2A* genes, respectively. While CDKN2A was significantly induced in both experimental settings, we detected a significant increase of CDKN1A only in melanoma cells treated with etoposide, while there was no significant effect in senescent NHEM (Figure 1A,B). A third cell cycle inhibitor, p53 encoded by the TP53 gene, did not show any regulation on mRNA level in both systems. We further assessed gene expression of CXCL2 and CCL8, which are associated with the SASP [22]. While both markers were increased in senescent NHEM and melanoma cells, a statistical significance could only be detected for CXCL2.

3.2. Protein Markers of Senescence

Since mRNA markers have a number of limitations, mostly due to the possibility of translational and posttranslational regulation, we next tested for different protein markers of cellular senescence. Interestingly, aforementioned cell cycle inhibitors p16^{INK4A}, p21$^{CIP1/WAF1}$ and p53 are among the most common proteins for detection and quantification of cellular senescence [23]. We, therefore, used Western blot analysis to assess and compare their regulation in different experimental settings. Protein levels of all three cell cycle inhibitors doubled during OIS in NHEM (Figure 2A). Melanoma cells treated with etoposide showed a stronger increase of p21$^{CIP1/WAF1}$, while upregulation of p53 turned out to be rather variable (Figure 2B). Cell cycle inhibitor p16^{INK4A} protein, however, was absent in melanoma cells (data not shown). As several studies found an association of ERK1/2 activation with cellular senescence [18,24], we assessed this marker next. A significant increase of phosphorylated ERK1/2 was found in both senescent NHEM and melanoma cells. It is important to note that the effect in NHEM is potentially caused by overexpression of mutant BRAF^{V600E}, an upstream kinase of ERK1/2. The melanoma cell line used in our experiments also carries a mutation upstream of ERK1/2, affecting HRAS and NRAS genes. In contrast to NHEM, however, both control and treatment cells bear these mutations, indicating that they do not attribute for the increased phosphorylation of ERK1/2 after etoposide treatment. Furthermore, we detected γH2AX as an indicator of DNA damage in melanoma cells, and found a strong but not reliable upregulation when assessing biological replicates, thereby preventing statistical significance. The same marker could not be detected in senescent NHEM (data not shown). Another molecular marker of DNA damage showed increased levels of nuclear promyelocytic leukemia protein (PML), which is best assessed using immunocytochemistry. While we could detect a significant increase of nuclear PML staining in NHEM, there was only a tendency toward an upregulation in melanoma cells treated with etoposide (Figure 2C,D).

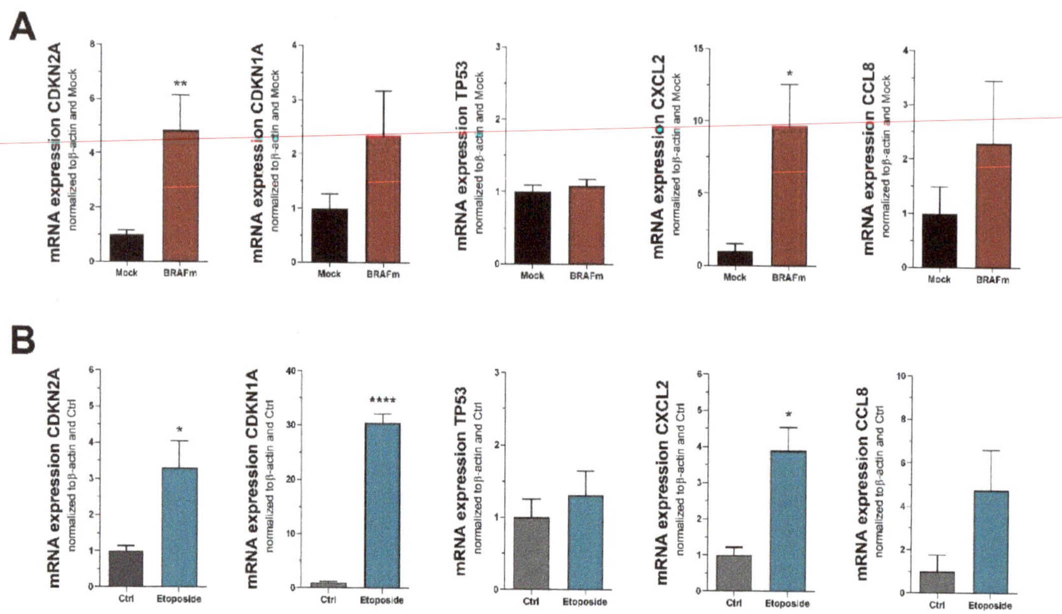

Figure 1. RNA markers of cellular senescence in (**A**) NHEM transduced with mutated BRAF^{V600E} (n = 7) and (**B**) melanoma cell line Mel Juso treated with 100 µM etoposide (n = 3). Bars are shown as mean ± SEM (Student's *t*-test). (*: $p < 0.05$, **: $p < 0.01$, ****: $p < 0.0001$).

3.3. Functional and Morphological Markers of Senescence

A central hallmark of cellular senescence is the discontinuation of cell division and thereby proliferation. We here used real-time cell proliferation analysis (RTCA) to track cell growth over time, and found a significant decrease in both experimental settings (Figure 3A,B). In addition, proliferative activity can be measured indirectly by incubating cells for an appropriate time period followed by cell viability analysis. Since incubation times are largely dependent on the proliferation rate of control cells, we used 7 days (NHEM) and 48 h (Mel Juso) in our experiments. An XTT assay revealed significantly reduced cell viability in BRAF^{V600E}-transduced NHEM and etoposide-treated Mel Juso cells, hence indicating reduced proliferation (Figure 3C,D).

Some functional consequences of cellular senescence lead to morphological changes, including heterochromatin formation, flattening, and multinucleation of cells. Senescence-associated heterochromatin foci (SAHF) were detected using DAPI staining, revealing a significant increase in both primary and melanoma cells (Figure 3E,F). Next, we assessed flattening and multinucleation, which were both significantly elevated after treatment of Mel Juso with etoposide (Figure 3H). NHEM, however, already had high levels of flattened cells in the control treatment, with a small but significant increase after entering OIS (Figure 3G). Interestingly, multinucleation in NHEM was negligible, as only very few cells with two or more nuclei were found.

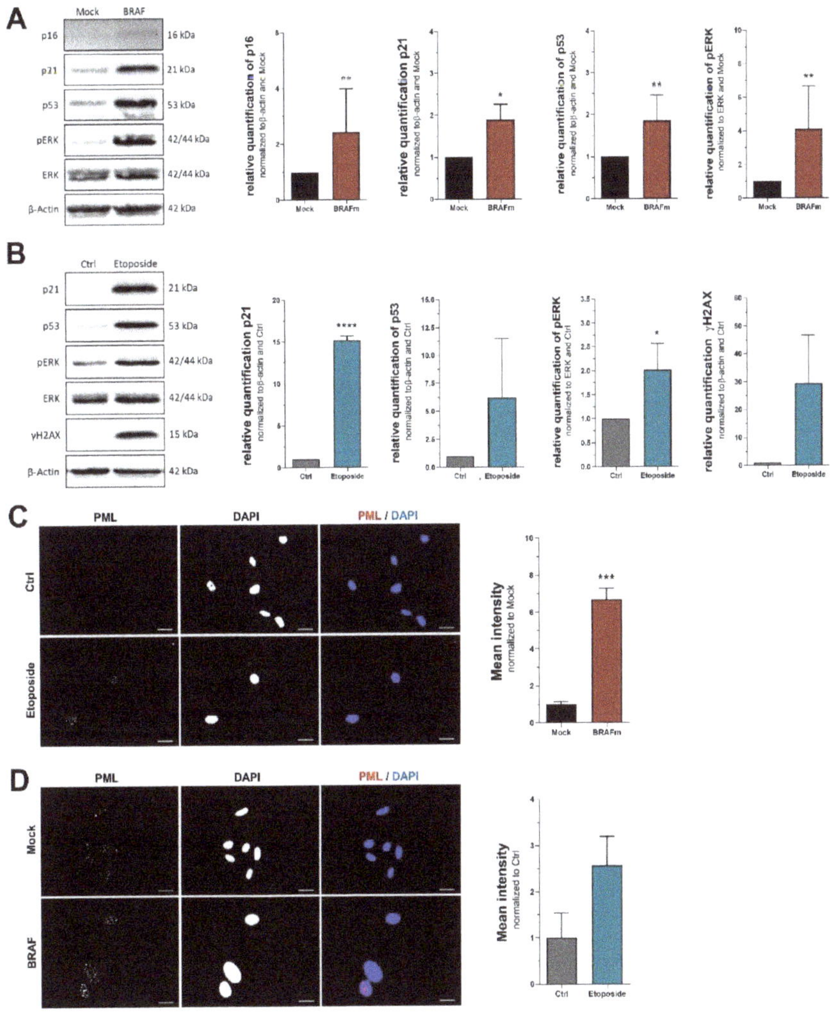

Figure 2. Protein markers of cellular senescence. Western blot analysis of (**A**) NHEM transduced with mutated BRAF^{V600E} (all n = 7, p16 n = 6) and (**B**) melanoma cell line Mel Juso treated with 100 μM etoposide (n = 3). (**C**,**D**) Immunofluorescent stainings of PML and DAPI. Scale bars equal 20 μm. Bars are shown as mean ± SEM (Student's *t*-test). (*: $p < 0.05$, **: $p < 0.01$, ***: $p < 0.001$, ****: $p < 0.0001$).

3.4. Quantification of Senescence-Associated Beta-Galactosidase Activity

Next, we quantified activity of senescence-associated beta-galactosidase (SA-β-Gal), which is stated to be the gold standard when it comes to detection of cellular senescence [16]. Multiple experimental approaches have been developed to reliably measure activity of beta-galactosidase in vitro, with the method of Dimri et al. [14] being the most widespread. It is based on the cleavage of X-Gal to yield a blue and insoluble dye, which can be easily detected using bright field microscopy. We used this method to quantify senescent primary and melanoma cells, and detected a significant increase compared to the respective controls (Figure 4A,B). Next, two fluorescent substrates of beta-galactosidase were used in combination with flow cytometry, namely $C_{12}RG$ (Figure 4C,D) and DDAO galactoside (Figure 4E,F). While both substrates share the main feature of producing a fluorescent molecule upon hydrolysis by beta-galactosidase, their chemical and functional properties differ notably. The resorufin-based $C_{12}RG$ is not fluorescent in its inactive state and carries a lipophilic tail that integrates in the cellular membrane to anchor the fluorescent product within the cell and ensure signal stability. DDAO galactoside, on the other hand, is an intrinsically fluorescent molecule that drastically changes its excitation and emission spectra after hydrolysis. In our experiments, both molecules successfully detected senescent cells similar to the traditional X-Gal assay. The percentage of positively stained cells was comparable among all three detection methods, in both control and treatment settings.

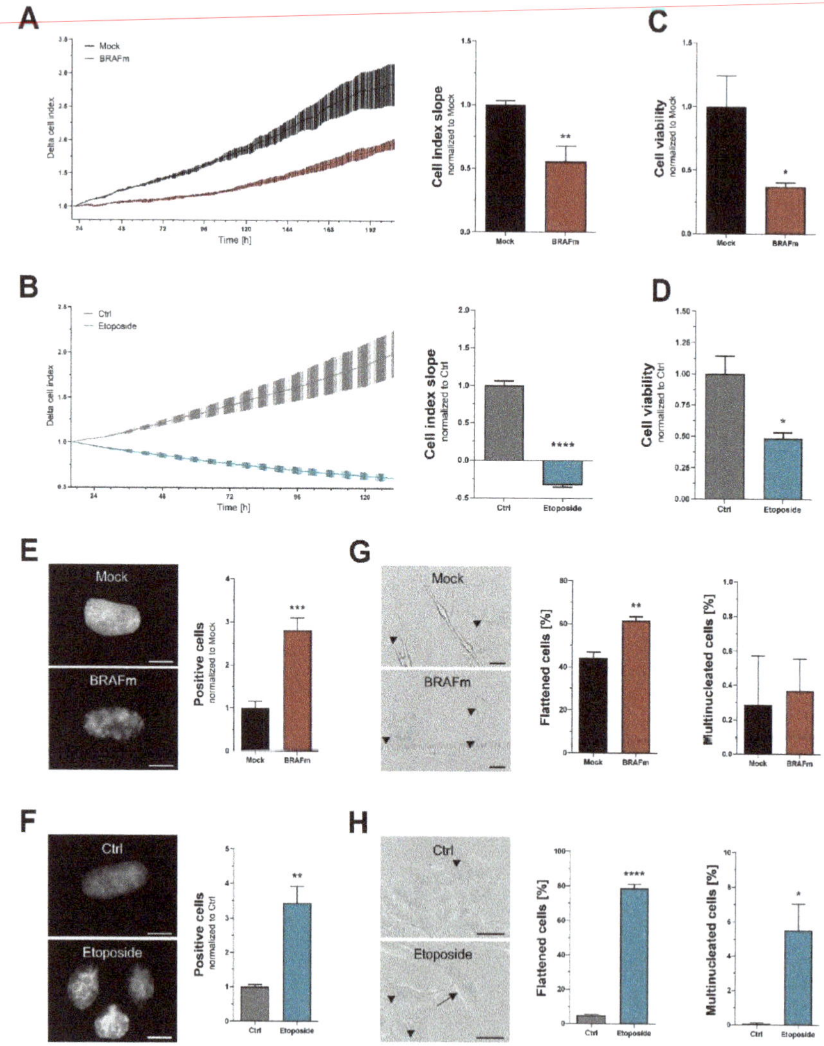

Figure 3. Functional and morphological markers of senescence. RTCA assay using (**A**) NHEM (n = 7) and (**B**) melanoma cell line Mel Juso (n = 3). A representative example is shown on the left. (**C**,**D**) Cell viability analysis 7 days (NHEM, n = 7) or 48 h (Mel Juso, n = 3) after treatment. (**E**,**F**) Detection of cells with visible SAHF using DAPI staining. Scale bars equal 10 μm. (**G**,**H**) Quantification of flattened and multinucleated cells. Arrowheads indicate flattened cells, arrows mark multinucleation. Scale bars equal 20 μm. Bars are shown as mean ± SEM (Student's *t*-test). (*: $p < 0.05$, **: $p < 0.01$, ***: $p < 0.001$, ****: $p < 0.0001$).

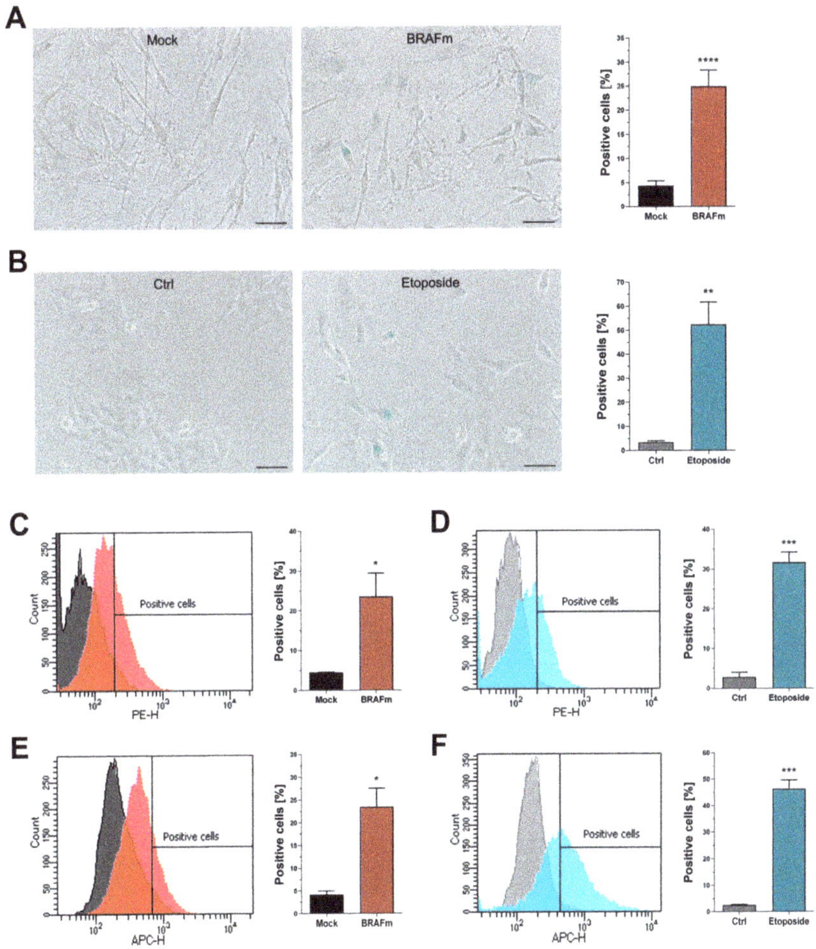

Figure 4. Quantification of SA-β-Gal activity. (**A**,**B**) Traditional method using X-gal and bright field microscopy (NHEM n = 7, Mel Juso n = 3). (**C**,**D**) Flow cytometry after staining with C_{12}RG. (**E**,**F**) Flow cytometry after staining with DDAO galactoside. Histograms are representative examples. Scale bars equal 50 µm. Bars are shown as mean ± SEM (Student's *t*-test). (*: $p < 0.05$, **: $p < 0.01$, ***: $p < 0.001$, ****: $p < 0.0001$).

3.5. Validating Selected Markers for Detection of Cellular Senescence in Melanoma Cells

While primary melanocytes are well characterized even in their senescent state, malignant melanoma is one of the most highly mutated cancers and thereby comes with a high grade of heterogeneity [25]. Further, induction of senescence in therapeutic or physiological settings might be due to a broad variety of stimuli, indicating that further validation is necessary for this cell type. Based on the data described in this study, we selected a set of three markers that showed the best results in melanoma cells treated with etoposide: SA-β-Gal, p21, and morphological changes, including flattening and multinucleation. In a first step, Mel Juso cells were treated with acidified nitrite, a novel antitumor treatment previously described [18]. A significant increase of SA-β-Gal, detected via X-Gal assay, was revealed (Figure 5A). Induction of p21 was present on both mRNA and protein level (Figure 5B). Since acidified nitrite interfered with β-actin expression (data not shown), we used 18s mRNA as reference, as well as Ponceau S staining during protein quantification.

When assessing morphological changes, we could detect a significant increase of flattened cells, while multinucleation remained scarce (Figure 5C). To account for mutational heterogeneity of malignant melanoma, cell line Mel Im was used in addition. In contrast to Mel Juso cells, which bear a NRAS^{Q61L} mutation but wild-type BRAF, Mel Im cells carry wild-type NRAS but mutated BRAF^{V600E}. Further, we introduced a physiological stimulus of cellular senescence, long-term acidosis, as described recently [26]. During long-term acidosis, Mel Im cells exhibited a strong increase of SA-β-Gal staining (Figure 5D), combined with increased p21 mRNA and protein (Figure 5E). Similar to aforementioned treatment with acidified nitrite, we detected elevated levels of flattened cells without any relevant effect on multinucleation (Figure 5F).

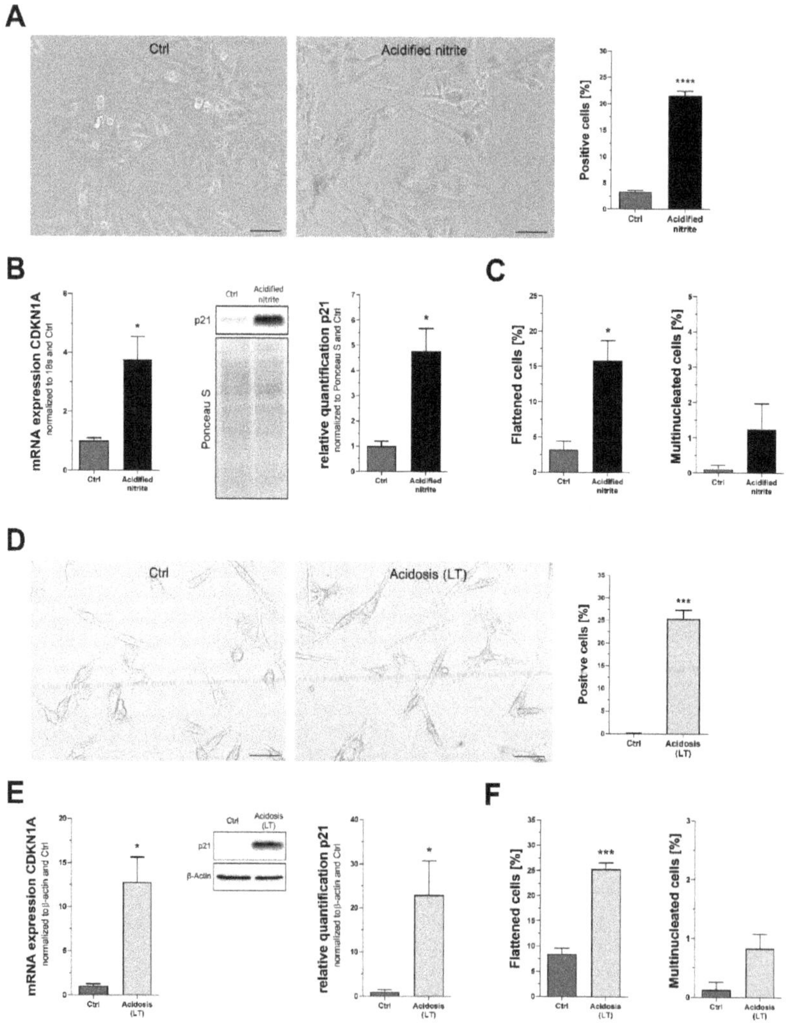

Figure 5. Validation of selected markers in malignant melanoma cells. (**A**–**C**) Mel Juso cells treated with acidified nitrite showed increased SA-β-Gal staining (**A**), p21 mRNA and protein (**B**), and some effects on morphology (**C**). (**D**–**F**) Mel Im cells during long term (LT) acidosis were found to have strong SA-β-Gal activity too (**D**), accompanied by p21 mRNA and protein (**E**) as well as morphological alterations (**F**). Scale bars equal 50 µm. Bars are shown as mean ± SEM (Student's *t*-test). (*: $p < 0.05$, ***: $p < 0.001$, ****: $p < 0.0001$).

4. Discussion

Reliable detection and quantification of senescence have been major challenges ever since its first description decades ago. Heterogeneous cell populations, combined with a strong dependency on cell type and senescence trigger [27], increased the difficulty of developing universal molecular markers. To date, the only generally valid marker of cellular senescence is thought to be SA-β-Gal [16], while the vast majority of remaining molecules have to be carefully assessed and validated for each experimental setting. This study focusses on melanocytic systems by comparing two clinically relevant in vitro models: Primary melanocytes bearing mutant BRAF^{V600E} to enter OIS, as commonly found in melanocytic nevi [4], and melanoma cells after treatment with chemotherapeutic agent etoposide. It thereby combines a physiological setting, in which the difference between proliferating and senescent cells is relatively small, with a therapeutic approach that compares highly proliferative cancer cells with severely damaged cells after treatment. The expected distance of control and treatment conditions, in terms of the senescent phenotype, is an important consideration to be made before selection of senescence biomarkers, as it defines the appropriate sensitivity required for the experiment. A second consideration is the quantity of molecular markers. While diagnostic and therapeutic approaches require sophisticated sets of biomarkers to detect senescent cells with great sensitivity and specificity, a reduced and simplified selection might be sufficient for most research applications. The latter will be addressed at the end of this section, as we will propose a condensed set of senescence biomarkers designated for research on melanocytes or melanoma cells.

When assessing cellular senescence, cell cycle inhibitors are commonly used as mRNA and protein markers. Their importance has been extensively reviewed elsewhere [28–30]. Briefly, two different pathways are induced, namely Arf/p53/p21 and p16/pRb, both resulting in cell cycle arrest as reviewed by Larsson [31]. However, several limitations have to be taken into account: first, activation of cell cycle inhibitors is dynamic and depends on the cellular state. As indicated by several studies, p21^{CIP1} and p53 are commonly found during the initiation of senescence, but may decline afterwards, while p16^{INK4A} shows delayed upregulation to stabilize senescence [32,33]. Consequently, assessing single cell cycle inhibitors might not be sufficient to detect senescent cells reliably. A second limitation is introduced by their low specificity, especially p21^{CIP1} and p53, as other cellular conditions can cause a similar upregulation. This includes quiescence [34,35], apoptosis [36,37], and cellular dormancy [38]. Finally, cancer cells commonly bear mutations of cell cycle inhibitors [39], potentially interfering with their re-activation and limiting their use as a molecular marker of senescence. As pathways for induction of senescence are far from being fully understood, establishment of a senescent phenotype without activation of major cell cycle inhibitors seems reasonable. This is of special importance for malignant melanoma, as it belongs to the most highly mutated cancers [25].

The SASP is an important hallmark of cellular senescence and can be detected either by qPCR to measure mRNA levels of SASP components or via enzyme-linked immunosorbent assay (ELISA). We here used qPCR to detect levels of CXCL2 and CCL8, while many more SASP components might be equally useful [40,41]. Most of these markers, however are easily affected by cell type, senescence trigger, and cellular microenvironment [42]. Further, SASP components are reportedly increased during quiescence and even apoptosis [43,44]. Moreover, DNA damage is considered to a central mediator of cellular senescence, as it was previously linked to replicative as well as premature senescence caused by oncogene activation, cytotoxic therapy, or other triggers [45]. We assessed two members of the DNA repair response (DRR), gamma-H2AX and PML, which both turned out to be rather unreliable. A possible explanation is found in the experimental settings used in this study: gamma-H2AX is one of the first steps for recruitment and localization of DRR proteins [46], which possibly explains why it could not be detected in primary melanocytes seven days after oncogene overexpression in full senescence. The exact function and regulation of PML during DRR remains unclear, but it was found to be more stable in our experiments. However, it is important to note that PML has widespread cellular functions and is involved in several physiological and pathological processes [47,48]. Another, yet uncommon marker of senescence is phospho-ERK1/2 as an indicator of MAPK activity. Although it seems counterintuitive at first, recent studies have reported significant evidence that MAPK signaling contributes not only to proliferation, but also to cellular senescence [49,50]. Since pathways regulating ERK activation are affected by mutations in many cancers and the majority of melanomas [51], special caution should be exercised when using phospho-ERK1/2 for detection of senescence.

Discontinuation of proliferation displays the most important functional consequence of cellular senescence. A common and feasible method to detect this is metabolic assays based on mitochondrial activity, including XTT assay used in this study. Such indirect measurement of proliferation has certain drawbacks, as mitochondrial activity might be affected without any further consequences on proliferation, leading to false positive results. On the other hand, senescence itself was shown to affect mitochondria [52], thereby impairing the necessary correlation of mitochondrial activity and cell count. Consequently, metabolic assays have only limited reliability when measuring proliferation rates in the context of cellular senescence. Impedance-based systems like RTCA bypass these problems by directly quantifying the coverage of specific cell culture plates, which is a result of cell count and cell size. Although senescent cells commonly show changes in morphology and size [53], such effects can easily be excluded by analyzing the slope of the proliferation curve rather than raw impedance values. Morphological changes might serve as markers of senescence on their own, with the main advantage that they do not require any processing or staining. Senescent cells are usually flattened [53], which was supported by our experiments. Multinucleation could not be found in melanocytes, but melanoma cells treated with etoposide, the reason for which is unknown. Further, there was no relevant increase of multinucleation when testing different senescence inducers and a second melanoma cell line, indicating that this parameter is not reliable. Since research on morphological changes during senescence and underlying mechanisms is sparse, critical evaluation of their use as markers of senescence is barely possible.

Finally, we assessed the gold standard of senescence detection, SA-β-Gal. Its main advantages include easy detection and comparatively high, but not perfect specificity for senescent cells [54]. Melanocytes are among the very few cell types that physiologically express high levels of lysosomal β-galactosidase, the same enzyme referred to as SA-β-Gal in senescent cells [14,15]. Since its expression increases over time, it is generally advisable to use neonatal melanocytes for in vitro experiments, as it was done in this study. Furthermore, experiments including SA-β-Gal should be conducted at low passage numbers. From our experience, primary neonatal melanocytes start to show increased β-galactosidase activity after approximately 15 population doublings (equals 10 passages), which is why all experiments in this study were performed before cells had doubled 12 times (equals 8 passages). We then used two fluorescent substrates in combination with flow cytometry to measure SA-β-Gal and got results comparable to the traditional X-Gal assay, thereby confirming their validity. Flow cytometry has a number of advantages, including the possibility to analyze full samples with a consistent threshold, instead of manually analyzing a small percentage of cells and individually defining positive and negative cells. Pigmented melanocytes introduce another challenge during analysis of X-Gal stainings, as it may be difficult to distinguish between brown pigment and blue dye. The combination of fluorescent substrates and automated analysis using flow cytometry displays an easy solution to overcome such difficulties. Finally, neither staining with fluorescent substrates nor flow cytometry require fixation or preprocessing of cells. This introduces the possibility of flow cytometric sorting of cells with increased SA-β-Gal activity, as it was already done in recent studies [55,56].

After evaluating the molecular markers used in this study, it becomes evident that although the majority of them reliably detects senescence, there is no single molecule or cellular property with sufficient specificity to discriminate between senescence and other, possibly related cellular states. A combination of several markers, each with its own advantages and limitations, might display an adequate solution. As described initially, such a set of markers should be adjusted and validated for each cell type. Based on our data, we suggest two different sets of molecular markers for primary melanocytes and melanoma cells: when working with NHEM, increased SA-β-Gal activity should be the first marker and is preferentially assessed using flow cytometry. Cell cycle inhibitor $p16^{INK4A}$ was found to be strongly and reliably induced, thereby rendering it the second best molecule to detect when investigating cellular senescence. As morphological and functional characteristics were somewhat variable in NHEM, we suggest to add either CXCL2 as a marker of the SASP, or PML immunofluorescence for detection of DNA damage as a third marker. In melanoma cells, SA-β-Gal activity also represents the main marker of cellular senescence, with the advantage that detection via traditional X-Gal assay is sufficient. Beside this, morphological flattening and induction of cell cycle inhibitor $p21^{CIP1/WAF1}$ should be used as additional and reliable markers.

In summary, this study assessed a variety of senescence markers in two different melanocytic systems. We found most of them working reliably, but critical evaluation of their capabilities and drawbacks highlighted the importance of elaborate combined solutions. Finally, we proposed a set of up to three molecular markers for primary melanocytes and malignant melanoma cells to ensure reliable detection of cellular senescence in vitro. Our data supports the ongoing discussion and potentially improves senescence detection, until novel and sophisticated molecular markers are found.

Author Contributions: Conceptualization, T.Z. and A.-K.B.; methodology, T.Z. and A.-K.B.; formal analysis, T.Z.; investigation, T.Z., M.P., V.K. and C.C.; resources, A.-K.B.; writing—original draft, T.Z. and A.-K.B.; writing—review and editing, T.Z., M.P., V.K., C.C., S.M. and A.-K.B.; visualization: T.Z.; supervision, A.-K.B.; funding acquisition, A.-K.B. All authors have read and agreed to the published version of the manuscript.

Funding: The project is funded by the IZKF Erlangen (D31) and the Wilhelm Sander Foundation.

Institutional Review Board Statement: Not applicable.

Informed Consent Statement: Not applicable.

Data Availability Statement: Not applicable.

Acknowledgments: We thank Ingmar Henz for technical and methodological assistance and discussions.

Conflicts of Interest: The authors declare no conflict of interest.

References

1. Hayflick, L.; Moorhead, P.S. The serial cultivation of human diploid cell strains. *Exp. Cell Res.* **1961**, *25*, 585–621. [CrossRef]
2. Rhinn, M.; Ritschka, B.; Keyes, W.M. Cellular senescence in development, regeneration and disease. *Development* **2019**, *146*, dev151837. [CrossRef]
3. McHugh, D.; Gil, J. Senescence and Aging: Causes, Consequences, and Therapeutic Avenues. *J. Cell Biol.* **2017**, *217*, 65–77. [CrossRef]
4. Michaloglou, C.; Vredeveld, L.C.W.; Soengas, M.S.; Denoyelle, C.; Kuilman, T.; Van Der Horst, C.M.A.M.; Majoor, D.M.; Shay, J.W.; Mooi, W.J.; Peeper, D.S. BRAFE600-Associated Senescence-like Cell Cycle Arrest of Human Naevi. *Nature* **2005**, *436*, 720–724. [CrossRef]
5. Shain, A.H.; Bastian, B.C. From Melanocytes to Melanomas. *Nat. Rev. Cancer* **2016**, *16*, 345–358. [CrossRef]
6. Leclerc, J.; Ballotti, R.; Bertolotto, C. Pathways from senescence to melanoma: Focus on MITF sumoylation. *Oncogene* **2017**, *36*, 6659–6667. [CrossRef]
7. Saleh, T.; Carpenter, V.J. Potential Use of Senolytics for Pharmacological Targeting of Precancerous Lesions. *Mol. Pharmacol.* **2021**, *100*, 580–587. [CrossRef]
8. L'Hôte, V.; Courbeyrette, R.; Pinna, G.; Cintrat, J.C.; Le Pavec, G.; Delaunay-Moisan, A.; Mann, C.; Thuret, J.Y. Ouabain and Chloroquine Trigger Senolysis of BRAF-V600E-Induced Senescent Cells by Targeting Autophagy. *Aging Cell* **2021**, *20*, e13447. [CrossRef]
9. Hosoya, N.; Miyagawa, K. Targeting DNA Damage Response in Cancer Therapy. *Cancer Sci.* **2014**, *105*, 370–388. [CrossRef]
10. Gordon, R.R.; Nelson, P.S. Cellular senescence and cancer chemotherapy resistance. *Drug Resist. Updat.* **2012**, *15*, 123–131. [CrossRef]

11. Guillon, J.; Petit, C.; Toutain, B.; Guette, C.; Lelievre, E.; Coqueret, O. Chemotherapy-induced senescence, an adaptive mechanism driving resistance and tumor heterogeneity. *Cell Cycle* **2019**, *18*, 2385–2397. [CrossRef] [PubMed]
12. Hernandez-Segura, A.; Nehme, J.; Demaria, M. Hallmarks of Cellular Senescence. *Trends Cell Biol.* **2018**, *28*, 436–453. [CrossRef] [PubMed]
13. Matt, S.; Hofmann, T.G. The DNA damage-induced cell death response: A roadmap to kill cancer cells. *Experientia* **2016**, *73*, 2829–2850. [CrossRef] [PubMed]
14. Dimri, G.P.; Lee, X.; Basile, G.; Acosta, M.; Scott, G.; Roskelley, C.; Medrano, E.E.; Linskens, M.; Rubelj, I.; Pereira-Smith, O. A Biomarker That Identifies Senescent Human Cells in Culture and in Aging Skin in vivo. *Proc. Natl. Acad. Sci. USA* **1995**, *92*, 9363–9367. [CrossRef]
15. Lee, B.Y.; Han, J.A.; Im, J.S.; Morrone, A.; Johung, K.; Goodwin, E.C.; Kleijer, W.J.; DiMaio, D.; Hwang, E.S. Senescence-Associated Beta-Galactosidase Is Lysosomal Beta-Galactosidase. *Aging Cell* **2006**, *5*, 187–195. [CrossRef]
16. Itahana, K.; Campisi, J.; Dimri, G.P. Methods to Detect Biomarkers of Cellular Senescence: The Senescence-Associated Beta-Galactosidase Assay. *Methods Mol. Biol.* **2007**, *371*, 21–31. [CrossRef]
17. Feuerer, L.; Lamm, S.; Henz, I.; Kappelmann-Fenzl, M.; Haferkamp, S.; Meierjohann, S.; Hellerbrand, C.; Kuphal, S.; Bosserhoff, A.K. Role of Melanoma Inhibitory Activity in Melanocyte Senescence. *Pigment. Cell Melanoma Res.* **2019**, *32*, 777–791. [CrossRef]
18. Zimmermann, T.; Gebhardt, L.; Kreiss, L.; Schneider, C.; Arndt, S.; Karrer, S.; Friedrich, O.; Fischer, M.; Bosserhoff, A.-K. Acidified Nitrite Contributes to the Antitumor Effect of Cold Atmospheric Plasma on Melanoma Cells. *Int. J. Mol. Sci.* **2021**, *22*, 3757. [CrossRef]
19. Schiffner, S.; Braunger, B.M.; de Jel, M.M.; Coupland, S.E.; Tamm, E.R.; Bosserhoff, A.K. Tg(Grm1) Transgenic Mice: A Murine Model That Mimics Spontaneous Uveal Melanoma in Humans? *Exp. Eye Res.* **2014**, *127*, 59–68. [CrossRef]
20. Schreyer, L.; Mittermeier, C.; Franz, M.J.; Meier, M.A.; Martin, D.E.; Maier, K.C.; Huebner, K.; Schneider-Stock, R.; Singer, S.; Holzer, K.; et al. Tetraspanin 5 (TSPAN5), a Novel Gatekeeper of the Tumor Suppressor DLC1 and Myocardin-Related Transcription Factors (MRTFs), Controls HCC Growth and Senescence. *Cancers* **2021**, *13*, 5373. [CrossRef]
21. Bosserhoff, A.K.; Ellmann, L.; Kuphal, S. Melanoblasts in Culture as an in Vitro System to Determine Molecular Changes in Melanoma. *Exp. Dermatol.* **2011**, *20*, 435–440. [CrossRef] [PubMed]
22. Coppé, J.-P.; Desprez, P.-Y.; Krtolica, A.; Campisi, J. The Senescence-Associated Secretory Phenotype: The Dark Side of Tumor Suppression. *Annu. Rev. Pathol. Mech. Dis.* **2010**, *5*, 99–118. [CrossRef] [PubMed]
23. Wang, A.S.; Dreesen, O. Biomarkers of Cellular Senescence and Skin Aging. *Front. Genet.* **2018**, *9*, 247. [CrossRef] [PubMed]
24. Zou, J.; Lei, T.; Guo, P.; Yu, J.; Xu, Q.; Luo, Y.; Ke, R.; Huang, D. Mechanisms Shaping the Role of ERK1/2 in Cellular Senescence. *Mol. Med. Rep.* **2018**, *19*, 759–770. [CrossRef] [PubMed]
25. Davis, E.J.; Johnson, D.B.; Sosman, J.A.; Chandra, S. Melanoma: What Do All the Mutations Mean? *Cancer* **2018**, *124*, 3490–3499. [CrossRef]
26. Böhme, I.; Bosserhoff, A. Extracellular Acidosis Triggers a Senescence-like Phenotype in Human Melanoma Cells. *Pigment. Cell Melanoma Res.* **2019**, *33*, 41–51. [CrossRef]
27. Jochems, F.; Thijssen, B.; De Conti, G.; Jansen, R.; Pogacar, Z.; Groot, K.; Wang, L.; Schepers, A.; Wang, C.; Jin, H.; et al. The Cancer SENESCopedia: A Delineation of Cancer Cell Senescence. *Cell Rep.* **2021**, *36*, 109441. [CrossRef]
28. Rayess, H.; Wang, M.B.; Srivatsan, E.S. Cellular Senescence and Tumor Suppressor Gene P16. *Int. J. Cancer* **2012**, *130*, 1715–1725. [CrossRef]
29. Kumari, R.; Jat, P. Mechanisms of Cellular Senescence: Cell Cycle Arrest and Senescence Associated Secretory Phenotype. *Front. Cell Dev. Biol.* **2021**, *9*, 485. [CrossRef]
30. Mijit, M.; Caracciolo, V.; Melillo, A.; Amicarelli, F.; Giordano, A. Role of P53 in the Regulation of Cellular Senescence. *Biomolecules* **2020**, *10*, 420. [CrossRef]
31. Larsson, L.G. Oncogene- and Tumor Suppressor Gene-Mediated Suppression of Cellular Senescence. *Semin. Cancer Biol.* **2011**, *21*, 367–376. [CrossRef] [PubMed]
32. Stein, G.H.; Drullinger, L.F.; Soulard, A.; Dulić, V. Differential Roles for Cyclin-Dependent Kinase Inhibitors P21 and P16 in the Mechanisms of Senescence and Differentiation in Human Fibroblasts. *Mol. Cell. Biol.* **1999**, *19*, 2109–2117. [CrossRef] [PubMed]
33. Te Poele, R.H.; Okorokov, A.L.; Jardine, L.; Cummings, J.; Joel, S.P. DNA Damage Is Able to Induce Senescence in Tumor Cells In Vitro and In Vivo 1. *Cancer Res.* **2002**, *62*, 1876–1883. [PubMed]
34. Brien, G.L.; Bracken, A.P. The PCL1-p53 axis promotes cellular quiescence. *Cell Cycle* **2016**, *15*, 305–306. [CrossRef]
35. Li, L.; Bhatia, R. Stem Cell Quiescence. *Clin. Cancer Res.* **2011**, *17*, 4936–4941. [CrossRef]
36. Lee, S.B.; Lee, S.; Park, J.Y.; Lee, S.Y.; Kim, H.S. Induction of P53-Dependent Apoptosis by Prostaglandin A2. *Biomolecules* **2020**, *10*, 492. [CrossRef]
37. Wang, X.; Simpson, E.R.; Brown, K.A. P53: Protection against Tumor Growth beyond Effects on Cell Cycle and Apoptosis. *Cancer Res.* **2015**, *75*, 5001–5007. [CrossRef]
38. Dai, Y.; Wang, L.; Tang, J.; Cao, P.; Luo, Z.; Sun, J.; Kiflu, A.; Sai, B.; Zhang, M.; Wang, F.; et al. Activation of Anaphase-Promoting Complex by P53 Induces a State of Dormancy in Cancer Cells against Chemotherapeutic Stress. *Oncotarget* **2016**, *7*, 25478–25492. [CrossRef]
39. He, S.; Sharpless, N.E. Senescence in Health and Disease. *Cell* **2017**, *169*, 1000–1011. [CrossRef]
40. Birch, J.; Gil, J. Senescence and the SASP: Many Therapeutic Avenues. *Genes Dev.* **2020**, *34*, 1565–1576. [CrossRef]

41. Basisty, N.; Kale, A.; Jeon, O.H.; Kuehnemann, C.; Payne, T.; Rao, C.; Holtz, A.; Shah, S.; Sharma, V.; Ferrucci, L.; et al. A Proteomic Atlas of Senescence-Associated Secretomes for Aging Biomarker Development. *PLoS Biol.* **2020**, *18*, e3000599. [CrossRef] [PubMed]
42. Chambers, C.R.; Ritchie, S.; Pereira, B.A.; Timpson, P. Overcoming the Senescence associated Secretory Phenotype (SASP): A Complex Mechanism of Resistance in the Treatment of Cancer. *Mol. Oncol.* **2021**, *15*, 3242–3255. [CrossRef]
43. Tang, W.; Wang, W.; Zhang, Y.; Liu, S.; Liu, Y.; Zheng, D. Tumour Necrosis Factor-Related Apoptosis-Inducing Ligand (TRAIL)-Induced Chemokine Release in Both TRAIL-Resistant and TRAIL-Sensitive Cells via Nuclear Factor Kappa B. *FEBS J.* **2008**, *276*, 581–593. [CrossRef] [PubMed]
44. Hernandez-Segura, A.; de Jong, T.V.; Melov, S.; Guryev, V.; Campisi, J.; Demaria, M. Unmasking Transcriptional Heterogeneity in Senescent Cells. *Curr. Biol.* **2017**, *27*, 2652–2660.e4. [CrossRef] [PubMed]
45. Chen, J.H.; Hales, C.N.; Ozanne, S.E. DNA Damage, Cellular Senescence and Organismal Ageing: Causal or Correlative? *Nucleic Acids Res.* **2007**, *35*, 7417–7428. [CrossRef]
46. Kuo, L.J.; Yang, L.-X. γ-H2AX-A Novel Biomarker for DNA Double-Strand Breaks. *In Vivo* **2008**, *22*, 305–309. [PubMed]
47. Previati, M.; Missiroli, S.; Perrone, M.; Caroccia, N.; Paliotto, F.; Milani, D.; Giorgi, C. Functions and dys-functions of promyelocytic leukemia protein PML. *Rend. Lince* **2018**, *29*, 411–420. [CrossRef]
48. Bernardi, R.; Pandolfi, P.P. Structure, dynamics and functions of promyelocytic leukaemia nuclear bodies. *Nat. Rev. Mol. Cell Biol.* **2007**, *8*, 1006–1016. [CrossRef]
49. Boucher, M.-J.; Jean, D.; Vézina, A.; Rivard, N. Dual role of MEK/ERK signaling in senescence and transformation of intestinal epithelial cells. *Am. J. Physiol. Liver Physiol.* **2004**, *286*, G736–G746. [CrossRef]
50. Anerillas, C.; Abdelmohsen, K.; Gorospe, M. Regulation of senescence traits by MAPKs. *GeroScience* **2020**, *42*, 397–408. [CrossRef]
51. Inamdar, G.S.; Madhunapantula, S.V.; Robertson, G.P. Targeting the MAPK pathway in melanoma: Why some approaches succeed and other fail. *Biochem. Pharmacol.* **2010**, *80*, 624–637. [CrossRef] [PubMed]
52. Chapman, J.; Fielder, E.; Passos, J.F. Mitochondrial Dysfunction and Cell Senescence: Deciphering a Complex Relationship. *FEBS Lett.* **2019**, *593*, 1566–1579. [CrossRef] [PubMed]
53. Herranz, N.; Gil, J. Mechanisms and functions of cellular senescence. *J. Clin. Investig.* **2018**, *128*, 1238–1246. [CrossRef] [PubMed]
54. Debacq-Chainiaux, F.; Erusalimsky, J.D.; Campisi, J.; Toussaint, O. Protocols to detect senescence-associated beta-galactosidase (SA-βgal) activity, a biomarker of senescent cells in culture and in vivo. *Nat. Protoc.* **2009**, *4*, 1798–1806. [CrossRef]
55. Macedo, J.C.; Vaz, S.; Bakker, B.; Ribeiro, R.; Bakker, P.L.; Escandell, J.M.; Ferreira, M.G.; Medema, R.; Foijer, F.; Logarinho, E. FoxM1 repression during human aging leads to mitotic decline and aneuploidy-driven full senescence. *Nat. Commun.* **2018**, *9*, 2834. [CrossRef]
56. Flor, A.C.; Wolfgeher, D.; Wu, D.; Kron, S.J. A signature of enhanced lipid metabolism, lipid peroxidation and aldehyde stress in therapy-induced senescence. *Cell Death Discov.* **2017**, *3*, 17075. [CrossRef]

cells

Article

Dynamic Spatiotemporal Expression Pattern of the Senescence-Associated Factor p16Ink4a in Development and Aging

Hasan Safwan-Zaiter [1,†], Nicole Wagner [1,*,†], Jean-François Michiels [2] and Kay-Dietrich Wagner [1,*]

1 The National Center for Scientific Research (CNRS), The National Institute of Health and Medical Research (INSERM), iBV, Université Côte d'Azur, 06107 Nice, France; Hasan.Safwan-Zaiter@unice.fr
2 Department of Pathology, CHU Nice, 06107 Nice, France; michiels.jf@chu-nice.fr
* Correspondence: nwagner@unice.fr (N.W.); kwagner@unice.fr (K.-D.W.)
† These authors contributed equally to this work.

Abstract: A plethora of factors have been attributed to underly aging, including oxidative stress, telomere shortening and cellular senescence. Several studies have shown a significant role of the cyclin-dependent kinase inhibitor p16ink4a in senescence and aging. However, its expression in development has been less well documented. Therefore, to further clarify a potential role of p16 in development and aging, we conducted a developmental expression study of p16, as well as of p19ARF and p21, and investigated their expression on the RNA level in brain, heart, liver, and kidney of mice at embryonic, postnatal, adult, and old ages. P16 expression was further assessed on the protein level by immunohistochemistry. Expression of p16 was highly dynamic in all organs in embryonic and postnatal stages and increased dramatically in old mice. Expression of p19 and p21 was less variable and increased to a moderate extent at old age. In addition, we observed a predominant expression of p16 mRNA and protein in liver endothelial cells versus non-endothelial cells of old mice, which suggests a functional role specifically in liver endothelium of old subjects. Thus, p16 dynamic spatiotemporal expression might implicate p16 in developmental and physiological processes in addition to its well-known function in the build-up of senescence.

Keywords: aging; endothelial cells; development; liver; heart; brain; kidney; senescence; SASP

Citation: Safwan-Zaiter, H.; Wagner, N.; Michiels, J.-F.; Wagner, K.-D. Dynamic Spatiotemporal Expression Pattern of the Senescence-Associated Factor p16Ink4a in Development and Aging. *Cells* **2022**, *11*, 541. https://doi.org/10.3390/cells11030541

Academic Editor: Antonio Paolo Beltrami

Received: 13 January 2022
Accepted: 2 February 2022
Published: 4 February 2022

1. Introduction

Aging is characterized by the gradual continuous decline of functions of cells, tissues, and the whole organism [1]. This age-related functional degeneration affects each organism that passes through developmental phases up to aging, as it is experienced by single cellular and multicellular organisms [2]. In mammals, aging is associated with a variety of pathologies and has been classified as the leading predictive factor of many chronic diseases that account for the majority of morbidity and mortality worldwide [3]. These diseases include neurodegenerative (Alzheimer's and Parkinson), cardiovascular, pulmonary, renal, and bone disorders, and cancers [4–9]. What makes aging a common risk factor is the fact that it arises from molecular mechanisms and pathological pathways that are cornerstones for the development of all these diseases. This includes oxidative stress and overproduction of reactive oxygen species, overproduction of inflammatory cytokines, activation of oncogenes, DNA damage, telomere shortening, and, consequently, accumulation of senescent cells [10–15].

Cellular senescence is a stress response defined as an irreversible arrest of cellular proliferation that results from experiencing potentially oncogenic stress [16]. Senescence was first discovered in primary cell culture in which cells exhibited a replicative senescence after extended period of growth which was termed the Hayflick's limit [17,18]. Senescent cells are usually characterized by phenotypic changes, morphological and biochemical, and adopt a secretory phenotype known as the senescence-associated secretory phenotype (SASP) [3,19–21]. Morphologically, senescent cells are usually larger than normal ones and exhibit a flattened shape, sometimes with multi-nuclei. However biochemically, these cells show a differential expression profile especially for some genes which rendered them as senescence fingerprints. Senescence-associated β-galactosidase, is an enzyme that is upregulated in senescent cells, and which acts as senescence biomarker [22]. Moreover, ectopic expression or upregulation of several genes has been identified, which includes augmented secretion of proinflammatory cytokines, proteases, and growth factors, which are all together termed the SASP [23–25]. A variety of causes underly the induction of cellular senescence; this includes oncogenic stress, telomere shortening, mitogenic signals, genomic DNA damage, epigenomic modifications, and tumor suppressor gene dysregulation [26–33].

Two major pathways have been identified to generate and maintain senescence, representing the intrinsic arm of cellular senescence. The key regulatory proteins of these pathways are the cell cycle regulators p16Ink4a (afterwards termed p16), p19Arf (afterwards p19), and p21 in addition to p53 and retinoblastoma protein (pRB). p21 acts mainly as a downstream effector of p53, and p16 is an upstream regulator of pRB via inhibition of cyclin-dependent kinases Cdk4 and Cdk6 [34–38]. Based on their action in regulating the cell cycle, p16, p19, and p21 were associated with cancer, aging, senescence, regeneration, and tumor suppression [21,35,39]. Expression of p19and p21 in embryonic development has been described [21,40–43], while little is known about the expression of p16 during development [44–46]. Therefore, we investigated p16, p19, and p21 RNA expression and p16 protein localization in several organs during embryonic and postnatal development as well as in adult and old mice.

2. Materials and Methods

2.1. Mice and Tissue Preparation

All animal work was conducted according to national and international guidelines and was approved by the local ethics committee (PEA-NCE/2013/106).

Timed pregnant mice (NMRI and C57BL/6) were purchased from Janvier Labs (Le Genest-Saint-Isle, France). The day of vaginal plug was considered embryonic day (E) 0.5. Pregnant mice were sacrificed by cervical dislocation at the indicated time points. Embryonic tissues were dissected, and tissues were used to prepare RNA. The day of birth was considered postnatal day (P) 0.

2.2. Mouse Tissue Samples, Histology, and Immunohistology

For immunohistochemistry, collections of paraffin-embedded whole embryos were used up to E18.5; for later stages, hearts, livers, kidneys, and brains were dissected. Samples from at least three different animals per time point were analyzed. Three-micrometer paraffin sections were used for histological and immunohistological procedures. For p16 immunohistology, after heat-mediated antigen retrieval and quenching of endogenous peroxidase activity, the antigen was detected after antibody application (1:500 dilution, p16 mouse monoclonal antibody, clone 2D9A12; ab54210, Abcam, Cambridge, UK,; additionally for some samples, a p16 mouse monoclonal antibody, clone 1E12E10, MA5-17142, Thermo Scientific, Courtaboeuf, France) using the M.O.M peroxidase kit from Vector (Vector Laboratories, PK-2200, Burlingame, CA, USA.) following the manufacturer's instructions. Avidin/Biotin blocking was performed using a kit from Vector (SP-2001). Diaminobenzidine (DAB) served as substrate (Sigma, Saint-Quentin-Fallavier, France). Sections were counterstained with hematoxylin (Dako, Trappes, France) [47,48]. Omission of the first antibody served as a negative control, and additional controls were livers from p16 knockout mice. Slides were photographed using a slide scanner (Leica Microsystems, Nanterre, France) or an epifluorescence microscope (DMLB, Leica, Germany) connected to a digital camera (Spot RT Slider, Diagnostic Instruments, Sterling Heights, MI, USA). For immunofluorescence double-labelling of mouse livers, anti-CD31 rabbit monoclonal antibody (1:2000 dilution, clone EPR17259, Ref: ab225883) from Abcam was combined with the mouse monoclonal anti-p16 antibody (Abcam) using Alexa Fluor 594 donkey anti rabbit and Alexa-Fluor 488 donkey anti mouse secondary antibodies (Jackson ImmunoResearch, Newmarket, Suffolk, UK) [49]. Negative controls were obtained by omission of first antibodies. Images were taken using a confocal ZEISS LSM Exciter microscope (Zeiss, Jena, Germany).

2.3. RNA Isolation, Reverse Transcription, and Quantitative PCR

Using the Trizol reagent (Thermo Scientific, Courtaboeuf, France), total RNA was isolated from brain, heart, liver, and kidneys of four different samples each at different stages of development (embryonic day 10.5, 12.5, 14.5, 16.5, and 18.5; postnatal days 1, 7, 21, 3 months, and 16–18 months) [50]. For E10.5 and E12.5, tissues from 7 embryos each were pooled per sample. For E14.5 and E16.5, organs from 4 embryos were used per sample. First-strand cDNA synthesis was performed with 500 ng of total RNA using the Thermo Scientific Maxima First Strand cDNA Synthesis Kit (#K1672, Thermo Scientific, Courtaboeuf, France), which contains DNase I, RNase inhibitor, oligo (DT) and random hexamer primers. The cDNAs were diluted 10 times in nuclease free water. Two microliters of the diluted reaction product were taken for real-time RT-PCR amplification which was performed using a StepOne Plus thermocycler (Thermo Scientific) and the PowerUp SYBR® Green Master Mix (#A25742, Thermo Scientific) or EurobioGreen Mix (GAEMMX02H, Eurobio, Les Ulis, France). For each sample, expression of the housekeeping genes *Gapdh, Rplp0, and β-actin* was determined. Three independent housekeeping genes were used as expression for each gene might vary under different experimental conditions [51,52]. Expression for each sample was calculated by subtracting the mean value of housekeeping gene Ct's from the gene of interest Ct using the ΔCt method [47,48,50,52–58]. Afterward, relative gene expression values were obtained by normalization of each sample against the mean value of all samples at E10.5 to determine differences between the organs and time points investigated. The mean value of all samples at E10.5 was set to 1 for easier illustration as described [50]. Primer sequences are listed in Table 1.

Table 1. Primers used for quantitative RT-PCR.

Gene of Interest	Oligonucleotide Sequences	References
p16ink4	F: AGGGCCGTGTGCATGACGTG R: GCACCGGGCGGGAGAAGGTA	[59]
p19arf	F: CGCTCTGGCTTTCGTGAAC R: GTGCGGCCCTCTTCTCAA	[60]
p21	F: AATTGGAGTCAGGCGCAGAT R: CATGAGCGCATCGCAATCAC	[61]
Tgf-b1	F: AGCTGGTGAAACGGAAGCG R: GCGAGCCTTAGTTTGGACAGG	This study
Vegfa	F: CTCACCAAAGCCAGCACATA R: AATGCTTTCTCCGCTCTGAA	[54]
Il-6	F: CACTTCACAAGTCGGAGGCT R: TGCCATTGCACAACTCTTTTCT	[54]
Mmp9	F: CCATGCACTGGGCTTAGATCA B: GGCCTTGGGTCAGGCTTAGA	[54]
Gapdh	F: AGGTCGGTGTGAACGGATTTG R: TGTAGACCATGTAGTTGAGGTCA	[47,48,54,58]
β-actin	F: CTTCCTCCCTGGAGAAGAGC R: ATGCCACAGGATTCCATACC	[47,48,54,58]
Rplp0	F: CACTGGTCTAGGACCCGAGAAG R: GGTGCCTCTGGAGATTTTCG	[47,48,54,58]

2.4. Endothelial Cell Magnetic-Activated Cell Sorting (MACS)

Kidneys, livers, hearts, and brains were isolated from four adult (3 months) mice and four old (18 months) mice each. Organs were minced and afterward digested with 0.1 mg/mL of DNase I (10104159001, Roche Diagnostics, Mannheim, Germany) and 1 mg/mL of Collagenase A (11088793001, Roche) in 10 mL of DMEM culture media (ThermoScientific) for 1 h at 37 °C. Digested samples were passed through 70-μm filters (SmartStrainers, 130-098-462, Miltenyi Biotec, Paris, France), centrifuged, and washed twice with PBS containing 2% fetal calf serum (FCS) and 0.5 mM of EDTA (ThermoScientific). Cells were re-suspended in 90 μL of the same buffer (PBS + FCS + EDTA)/10^7 cells. Endothelial cells were labelled by adding 10 μL/10^7 cells of magnetic microbead-associated anti-CD31 antibody (130-097-418, Miltenyi) at 4 °C for 15–30 min. Cells were separated via LS column (130-042-401, Miltenyi) pre-washed with 3 mL of PBS + FCS + EDTA and attached to a MidiMACS separator magnet (130-042-302, Miltenyi). Non-endothelial cells were eluted by washes with 3x 3 mL of PBS + FCS + EDTA. Afterward, endothelial cells were eluted by removing the LS columns from the magnetic field and flushing with 6 mL of PBS + FCS + EDTA. Eluted cells were separated as 1/3 for RNA extraction (see above) and 2/3 for protein extraction and quantification.

2.5. Protein Isolation, Quantification, and Western Blot

After endothelial cell sorting as described above, 2/3 of each endothelial and organ cells were taken from the total cell suspension. Cells where centrifuged at 3000 rpm for 10 min at 4 °C. Then, cells were incubated with 100 μL and 150 μL of RIPA buffer (Sigma) for endothelial and organ cells, respectively, and kept on ice for 30 min. Afterwards, samples were agitated overnight at 4 °C. The next day, the tubes were centrifuged at 16,000 rpm for 30 min at 4 °C. The total protein containing supernatant was recovered and stored at −80 °C.

Proteins were quantified by colorimetric BCA assay according to manufacturer's instructions (Uptima, Montluçon, France). Samples were diluted 20 times in distilled water and loaded in triplicates of 10 μL each, in transparent 96-well plates. In addition, BSA standards ranging from 0 to 2 mg were loaded in triplicates (10 μL). Absorbance was measured at a wavelength of 560 nm in a plate spectrophotometer (Biorad, Marnes-la-Coquette, France).

For Western blotting, 60 μg of protein in Laemmli buffer was denatured at 95 °C for 5–10 min. Samples were loaded on acrylamide gels (acrylamide/bisacrylamide 37.5/1) and set for electrophoresis. Afterward, proteins were transferred to PVDF membranes (162-0177, Biorad), and the membranes were blocked with 5% milk for 1 h (232100, Difco Skim Milk). p16 was detected using a rabbit monoclonal anti-p16 (Abcam; ab211542) diluted 1:2000 in PBS + 0.05% Tween 20 + 2.5% milk powder (overnight, 4 °C), followed by anti-rabbit peroxidase-labeled secondary antibody addition (Vector Laboratories) diluted 1:2000 in PBS + 0.05% Tween 20 + 2.5% milk powder for 1 h. Then, the chemiluminescence signal was obtained by incubation with the enzyme-specific substrate (RPN2235, Amersham, ECL Select Western blotting detection reagent). Afterward, the membrane was stripped by application of 10 mL of stripping buffer for 15 min (ST010, Gene Bio-Application L.T.D., Kfar-Hanagid, Israel) and washed 5x 5 min with distilled water before a second identical blocking step with milk for the detection of Gapdh as housekeeping protein. A rabbit monoclonal anti-Gapdh antibody (Abcam; ab181602) was used, and the signal was generated with same secondary antibody and substrate mentioned above.

2.6. Statistics

Data are expressed as means ± standard error of the mean (S.E.M.). Statistical differences were assessed by analysis of variance (ANOVA) followed by the Bonferroni post-hoc test (Graph Pad Instat, GraphPad Software, Inc., San Diego, CA, USA). A p-value < 0.05 was considered to reflect statistical significance.

3. Results

3.1. p16Ink4a, p19, and p21 mRNA Expression during Embryonic Development and Postnatal Stages in Different Organs

Expression of the mRNAs of the three genes *p16*, *p19* and *p21* was assessed at different ages (E10.5, E12.5, E14.5, E16.5, E18.5, P1, P7, P21, 3 months (adult), and 16–18 months (old)). Experiments were conducted on brain, heart, kidney, and liver tissues from which RNA samples were extracted and quantified by reverse transcription-quantitative PCR, normalized to the respective means of *Rplp0*, *Gapdh*, and *β-actin* housekeeping genes. The results below show the comparison of relative expression levels at all investigated ages in each organ for the three genes of interest (Figure 1) and the comparison of the expression levels of each gene in the different organs at each age (Figure 2).

In the brain, we observed a significant upregulation of *p16* expression beginning at E14.5 until P7 compared to E10.5. Surprisingly, *p16* expression dropped significantly at P21 compared to P7 ($p < 0.05$) to reach the highest levels in old animals. *p21* levels increased significantly around E16.5 during embryonic development and remained at stable levels during further development, increasing less than *p16* in brains from old mice. *p19* expression became upregulated around E14.5 and remained more or less stable during further brain development, showing an increase only in brains of old subjects.

Also in the heart, kidney, and liver *p16* expression increased constantly over time with higher expression levels than *p19* and *p21*, which both showed rather low, fluctuating expression during embryonic and postnatal development. Interestingly, in the heart, *p16* tended to drop between P7 and P21 ($p = 0.070$) comparable to the time course in the developing brain. In old stages, *p16* expression in brain, heart, kidney, and liver was largely increased. Also, *p19* and *p21* levels were upregulated in the respective organs, but to a much lesser extent than p16 (Figure 1).

To further analyze the relative mRNA expression data for *p16*, *p19*, and *p21* in embryonic development and in postnatal stages, we compared the expression of each gene in the brain, heart, kidneys, and liver at each time point (Figure 2). Expression levels of *p16* were significantly higher at E10.5 in the brain compared to the developing heart, kidney, and liver. At E14.5, E18.5, and P21, the liver displayed the highest *p16* expression compared to the other investigated organs. At adult and old life stages, p16 expression was high, but not significantly different in the four organ systems studied. *P19* expression did not vary much between brain, heart, kidneys, and liver. An increase of *p19* could be observed in the kidney during development at embryonic day E12.5. Therefore, although p16 and *p19* are situated in the same chromosomal region, spatiotemporal expression patterns seem to be unrelated. Expression of *p21* increased mostly in the brain during embryonic and postnatal development beginning at E16.5 compared to E10.5, while in the other organs, only temporary very restricted significant alterations were observed. Only in old animals, in all organs a significant increase in *p21* expression was noted (Figure 1). At embryonic day 12.5, *p21* expression was highest in the kidneys, while at E16.5 and P21, it was highest in the brain, compared to heart, kidneys, and liver. Even in old animals, *p21* mRNA levels were elevated in brains compared to the kidneys (Figure 2).

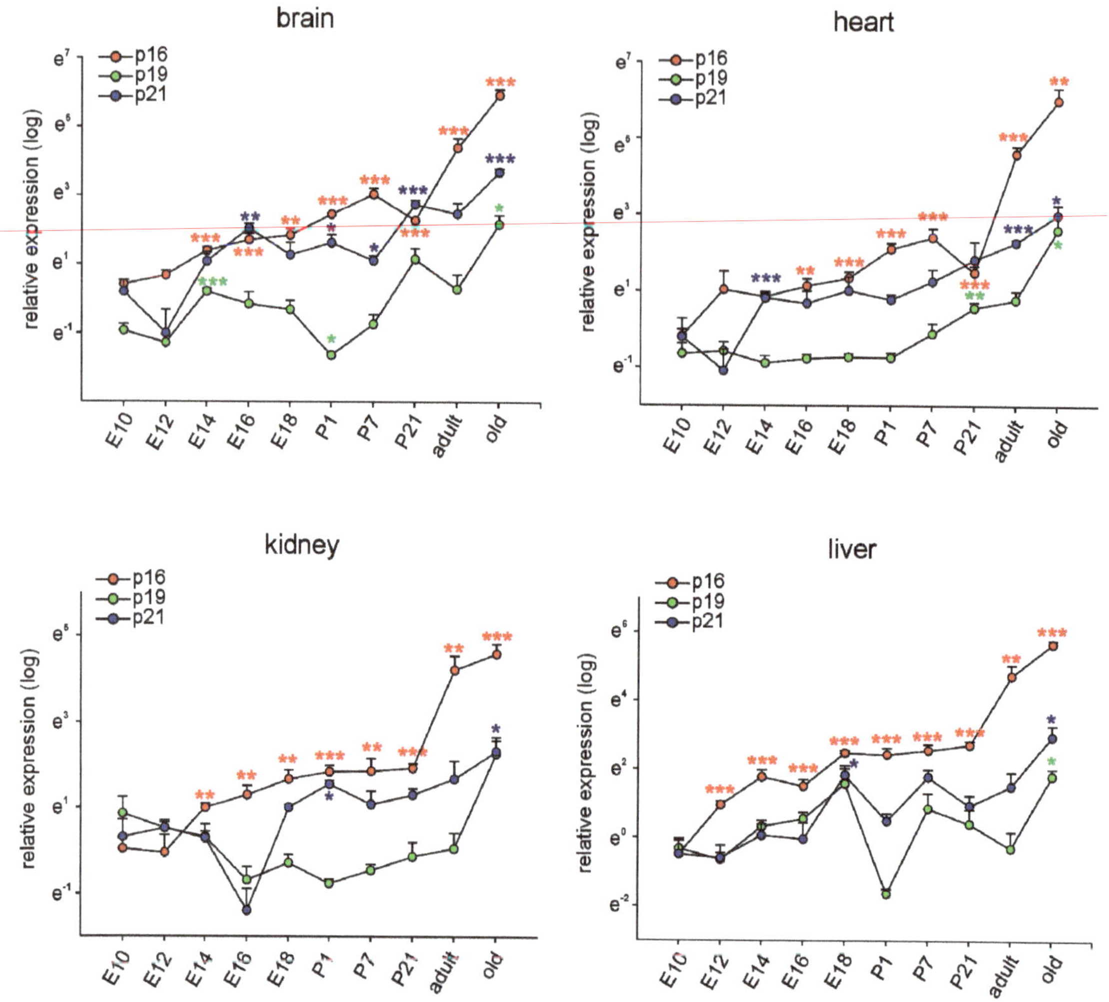

Figure 1. p16, p19, and p21 are differentially expressed during development and adulthood. Quantitative RT-PCRs for p16, p19, and p21 in mouse brains, hearts, kidneys, and livers at different time-points of development and in adulthood (n = 4 each, the four samples for E10.5 were each pooled from 7 organs, at E12.5, and 14.5 the four samples were pooled from four organs each). E: embryonic day, P: postnatal day, adult: 3 months of age, old: 16–18 months of age. Expression of each gene was normalized to the respective *Gapdh*, *actin*, and *Rplp0* expression. Next, the average of all organs and samples at E10.5 was calculated. Individual samples were then normalized against this average value (see Materials and Methods for details). Significance was tested for all time points between E10.5 and 18 months. Data are mean ± SEM. * $p < 0.05$, ** $p < 0.01$, *** $p < 0.001$.

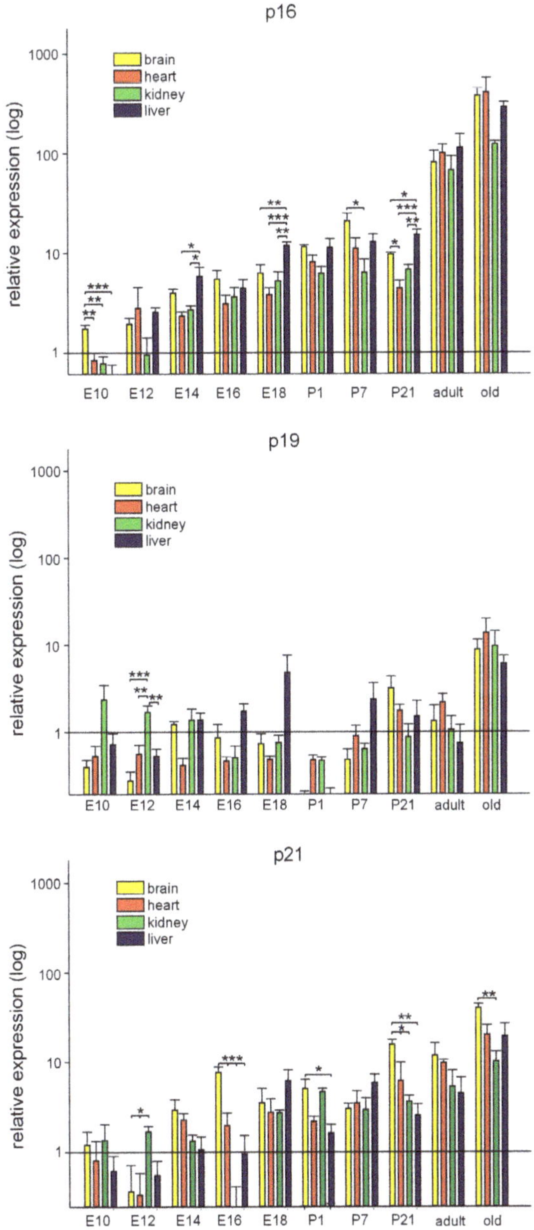

Figure 2. Differential spatiotemporal expression of p16, p19, and p21. Quantitative RT-PCRs for p16, p19, and p21 in mouse brains, hearts, kidneys, and livers at different time points of development and in adulthood (n = 4 each, the four samples for E10.5 were each pooled from 7 organs, at E12.5 and 14.5 the four samples were pooled from four organs each). E: embryonic day, P: postnatal day, adult: 3 months of age, old: 16–18 months of age. Expression of each gene was normalized to the respective *Gapdh*, *actin*, and *Rplp0* expression. The average of all organs and samples at E10.5 was calculated and set to 1. Individual samples were then normalized against this average value (see Materials and Methods for details). Significance was tested between the different organs for each time point. Data are mean ± SEM. * $p < 0.05$, ** $p < 0.01$, *** $p < 0.001$.

3.2. Immunohistochemical Investigation of p16 Expression

In addition to quantitative *p16* assessment on the mRNA level, we investigated its expression in the brain, heart, kidneys, and liver at the different time points by immunohistochemistry. In the developing brain, we detected p16 in neuronal cells of the cephalic mesenchyme (E10.5) and the neopallial cortex (E12.5–E18.5). The number of p16-positive neurons increased with differentiation of the brain up to E18.5. (Figures 3 and 4). Neurons of the cortex of old animals displayed a high p16 reactivity. Endothelial cells of the cortex occasionally showed a faint p16 signal (Figure 5), which increased at adult (3 months) and old (16–18 months) stages (Figure 6). Some cardiomyocytes showed p16 expression at early embryonic stages E10.5–E12.5 (Figure 3). With compaction of the myocardium, the number of p16 expressing cardiomyocytes increased from E14.5 to P1. From P7 on, the frequency of p16 expressing cardiomyocytes decreased (Figures 3–5). Endothelial cardiac cells frequently showed p16 expression, with a strong increase in old animals (Figure 6). Similarly to the brain and the heart, more p16-positive cells were found upon differentiation of the kidney. Whereas only faint expression of p16 could be detected in the ureteric bud at E10.5 and E12.5, during formation of the metanephric nephrons and interstitial mesenchyme, a high number of cells expressed p16. The number of p16 expressing cells in the kidney decreased postnatally (Figures 3–5). In old mice, p16 was highly expressed in glomerular structures, composed of podocytes, fibroblasts, and endothelial cells, and in vessels of the kidney (Figure 6). In the hepatic primordium (E10.5–E12.5), very few cells exhibited p16 expression. With the onset of hepatic hematopoiesis, the number of p16 expressing cells in the embryonic liver increased (Figure 3). From P1 on, when the bone marrow becomes the dominant hematopoietic organ, very few cells in the liver expressed p16 (Figure 5). In livers of old mice, we detected a strong signal in endothelial cells compared to hepatocytes (Figure 6). Liver sections with omission of the primary antibody and sections from p16 knockout mice were used as negative controls for the immunostaining (Figure S1). Embryonic p16 expression was confirmed using a different monoclonal antibody (Figure S2).

3.3. Selected SASP Factor Expression

To gain additional insights into the potential relevance and function of *p16* expression during embryonic and postnatal development and in adult and old mice, we measured mRNA expression of *Il-6*, *Mmp9*, *Tgfb1*, and *Vegfa* as selected SASP (senescence associated secretory phenotype) factors [21,62,63] during the time points and in the organs mentioned before. As each of these genes has individual functions at different time points and in different organs, we considered only a concomitant modification of the four genes as indicative of SASP. In agreement with the literature, we observed an increase in the measured SASP factors in all organs in old mice (Figure 7). *Mmp9* was transiently upregulated during late embryonic and early postnatal development in the liver, but as the other investigated genes did not follow the same time course, it might not be indicative for SASP, and the increase in *p16* expression during embryonic and postnatal development alone is not indicative for senescence.

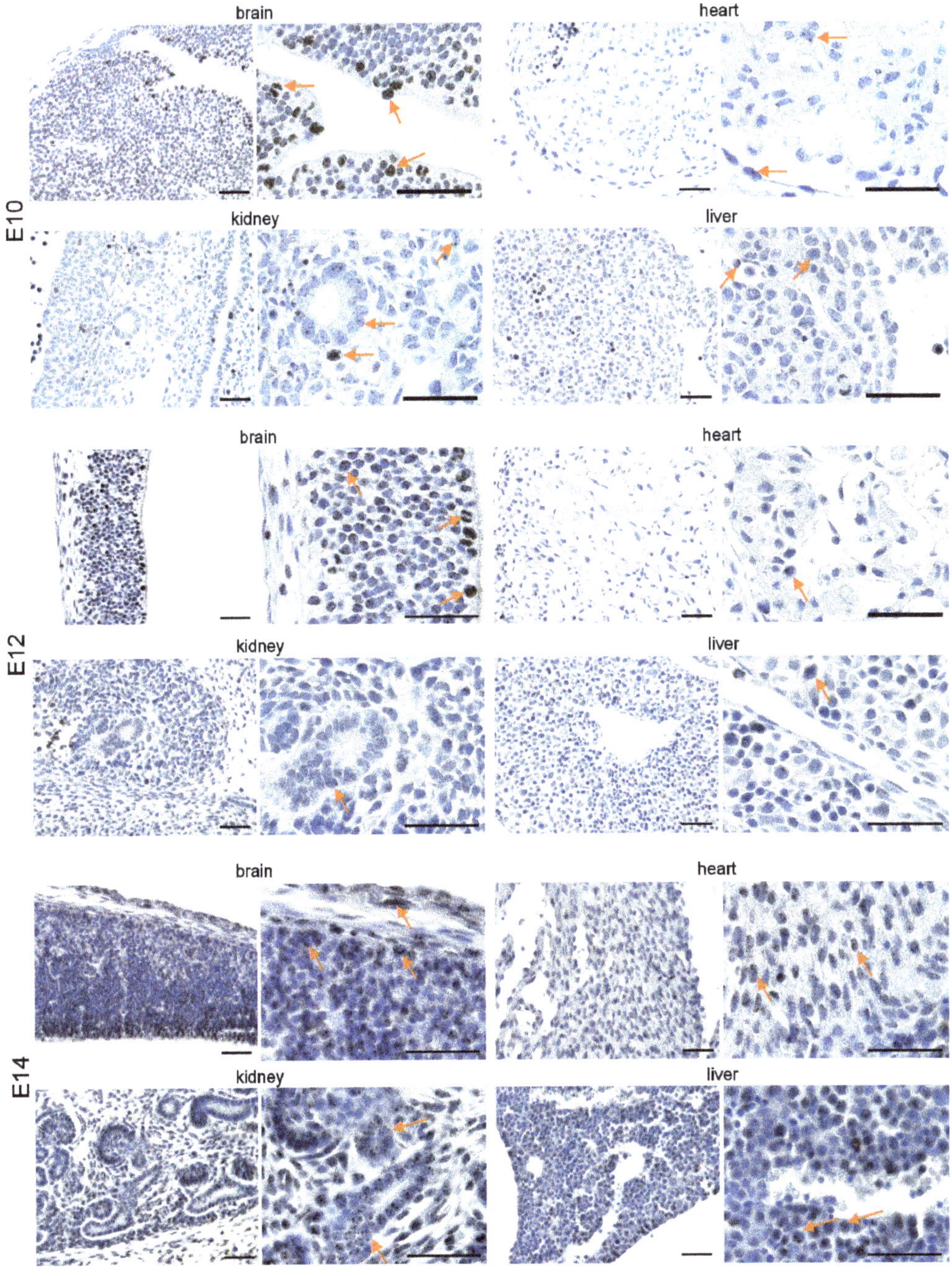

Figure 3. p16 is expressed in the brain, heart, kidneys, and liver during embryonic development (E10–E14). Representative photomicrographs of p16 immunostaining on sections of mouse embryos (3,3′ diaminobenzidine (DAB) substrate, brown, hematoxylin counterstaining) at different stages before birth. Arrows indicate exemplary p16-positive cells. Scale bars represent 50 µm.

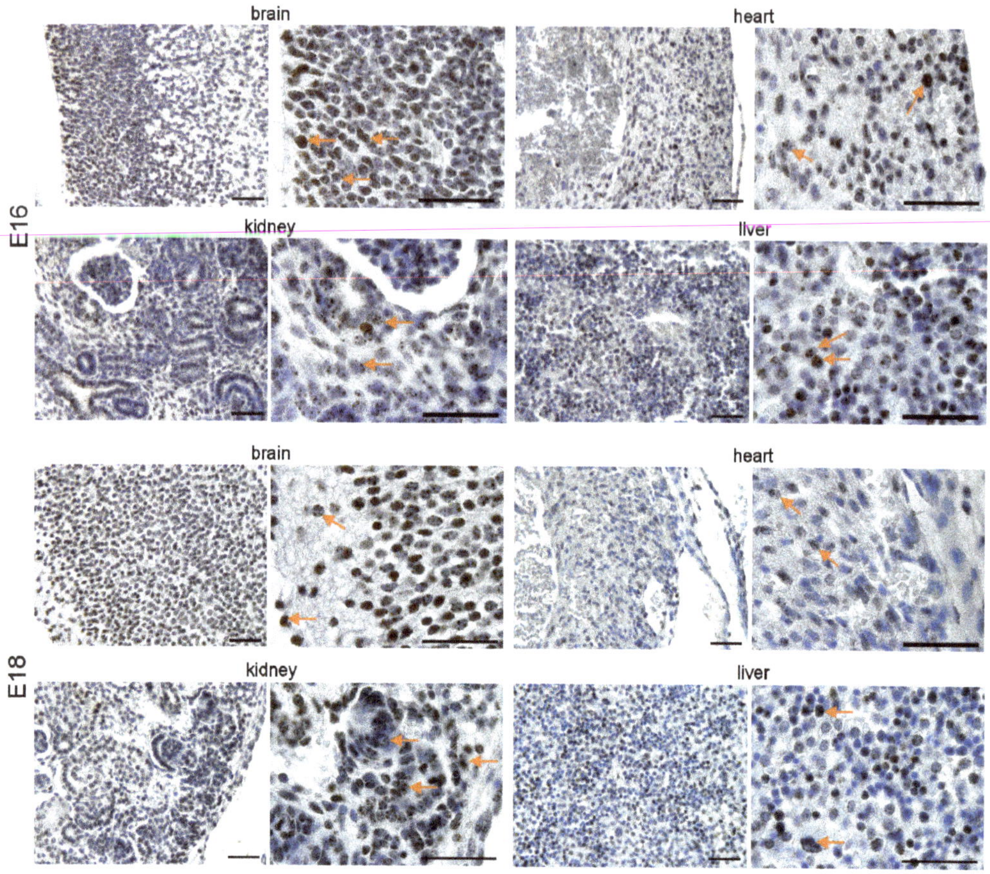

Figure 4. p16 is expressed in the brain, heart, kidneys, and liver during embryonic development (E16–E18). Representative photomicrographs of p16 immunostaining on sections of mouse embryos (3,3′ diaminobenzidine (DAB) substrate, brown, hematoxylin counterstaining) at different stages before birth. Arrows indicate exemplary p16-positive cells. Scale bars represent 50 µm.

3.4. Higher p16 Expression in Endothelial versus Non-Endothelial Cells in the Liver

As our immunohistochemistry approach suggested higher p16 expression in endothelial versus non-endothelial cells in the liver, we confirmed colocalization of p16 with Cd31 by double-labelling and confocal imaging of wild-type mice livers at 3 and 18 months of age (Figure 8a). For quantitative determinations, endothelial and organ liver cells were isolated from adult (3 months) and old (18 months) mice. In adult livers, endothelial vs. non-endothelial p16 mRNA levels tended to be higher, which became highly significant in old livers (Figure 8b). A comparable result was obtained in Western Blot analyses, showing slightly higher p16 expression in endothelial vs. non-endothelial cells in adult livers, and a dramatic increase of p16 expression in endothelial cells from livers of old animals (Figure 8c). Expression levels for p16 mRNA in endothelial versus non-endothelial cells did not differ for the other adult and old organs, except for the hearts of old mice, where organ cells showed higher p16 expression than endothelial cells. Interestingly, although we observed high p16 expression in old liver endothelial cells versus non-endothelial cells, this was not correlated with an increase in the expression of SASP genes except for *Tgfb1*, while *Vegfa* expression was even lower in endothelial cells (Figure S3). This is consistent with the previous observation that liver endothelial cells in aged mice are highly metabolic active and functional despite high p16 expression [64].

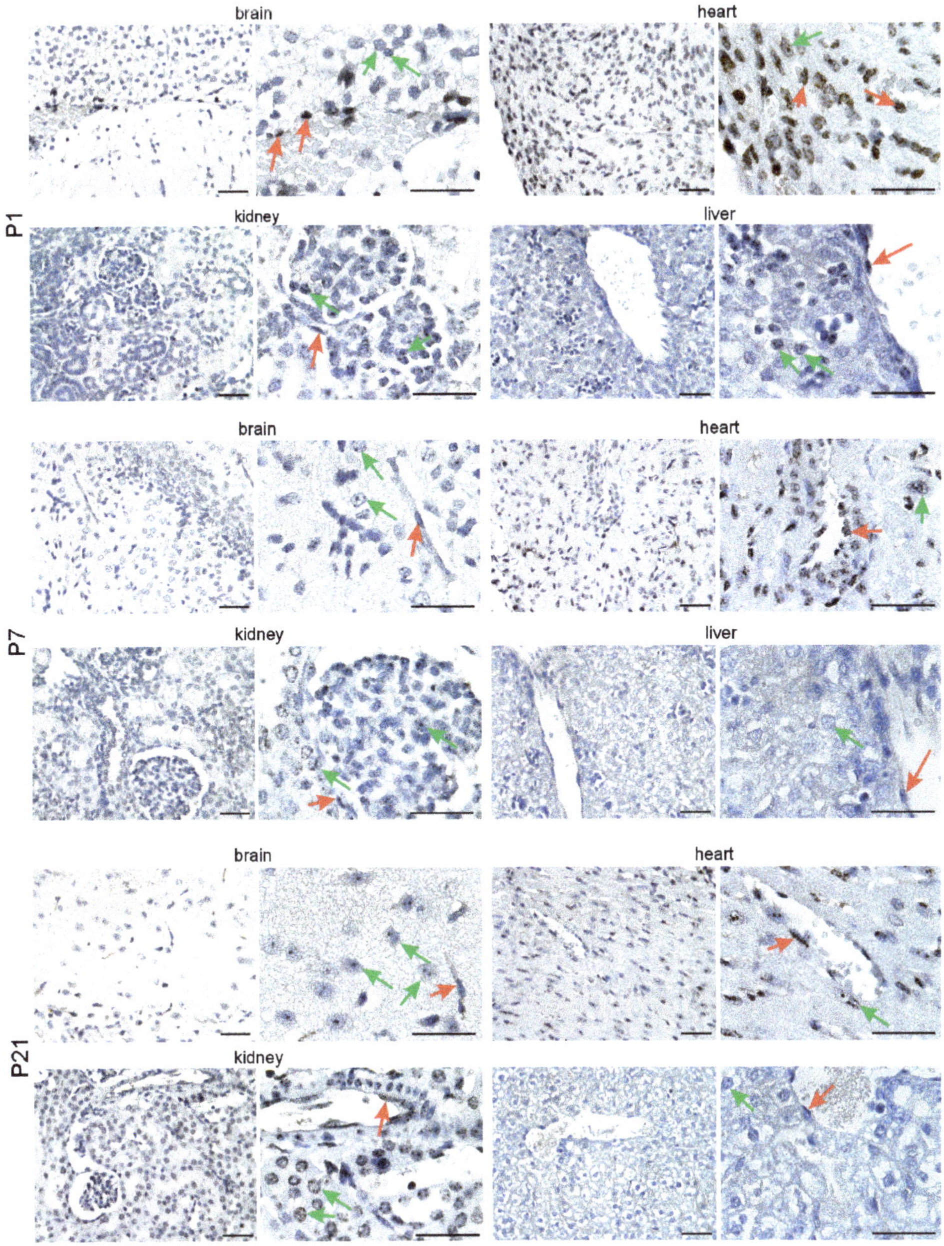

Figure 5. p16 is continuously expressed after birth in vascular and organ cells. Representative photomicrographs of p16 immunostaining for the brain, heart, kidneys, and liver (3,3′ diaminobenzidine (DAB) substrate, brown, hematoxylin counterstaining) at different stages after birth. Note the persistent expression of p16 in neuronal cells of the brain, cardiomyocytes, tubular and glomerular kidney cells, and hepatocytes (green arrows) and endothelial cells (red arrows). Scale bars indicate 50 μm.

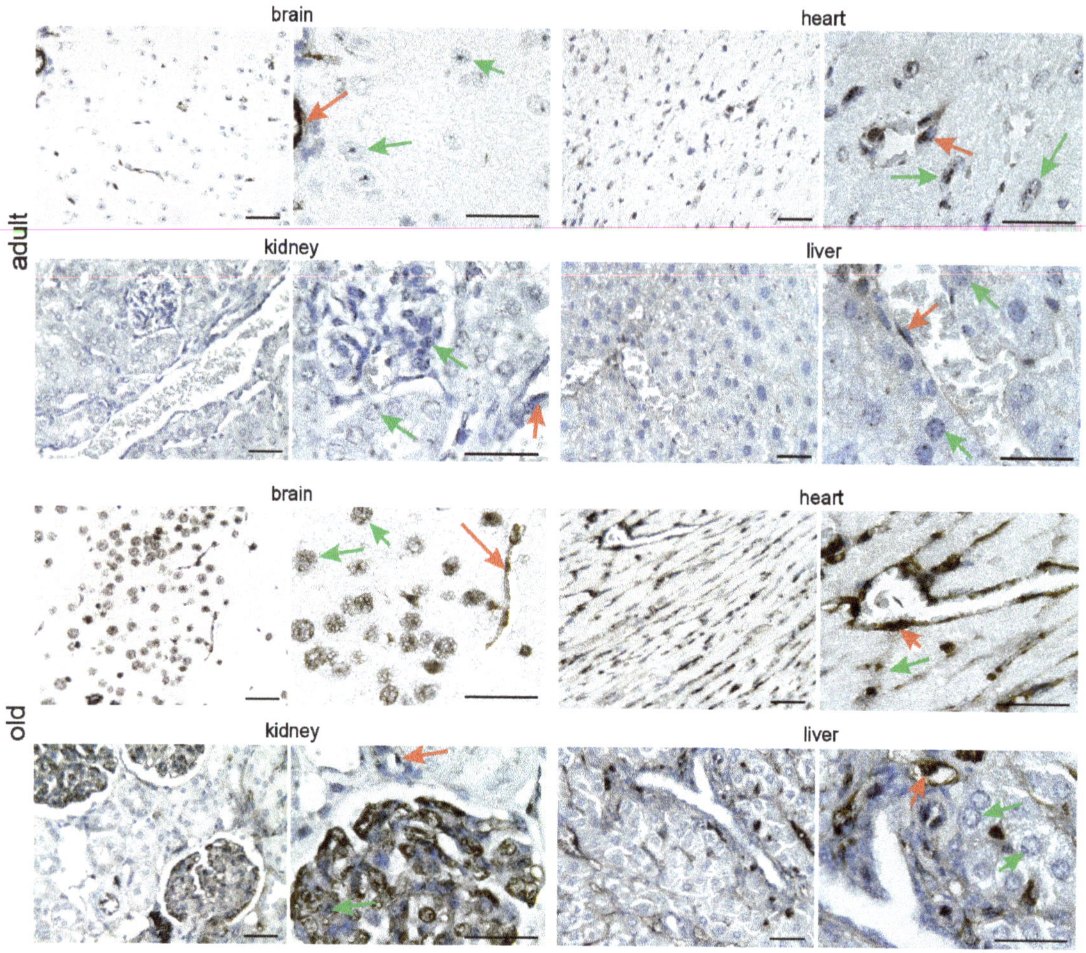

Figure 6. p16 is continuously expressed in adults and increases in old animals. Representative photomicrographs of p16 immunostaining for the brain, heart, kidneys, and liver (3,3′ diaminobenzidine (DAB) substrate, brown, hematoxylin counterstaining). Note the persistent expression of p16 in neuronal cells of the brain, cardiomyocytes, tubular and glomerular kidney cells, and hepatocytes (green arrows) and endothelial cells (red arrows), which increases with age. Scale bars indicate 50 μm.

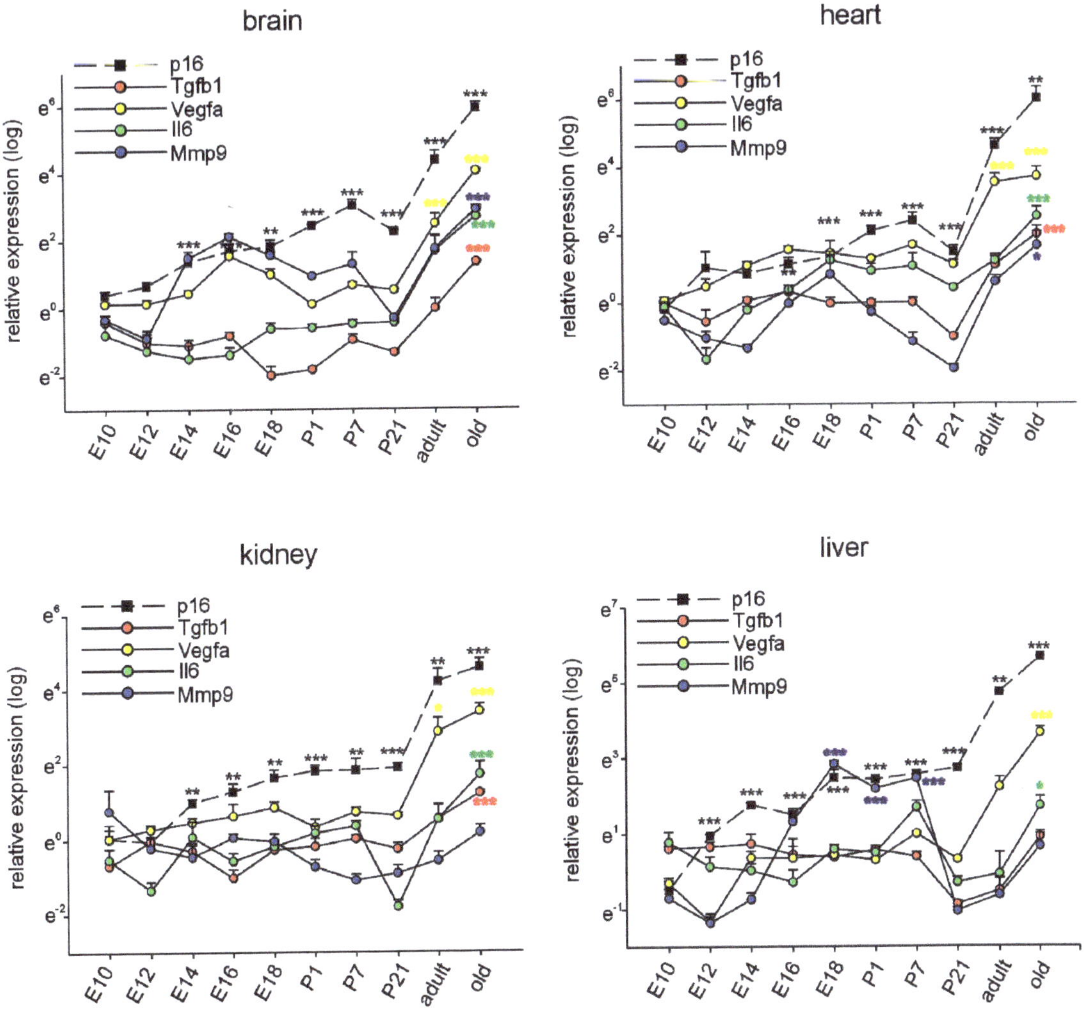

Figure 7. Differential spatiotemporal expression of SASP factors in comparison to *p16*. Quantitative RT-PCRs for *p16*, *Tgfb*, *Vegfa*, *Il6*, and *Mmp9* in mouse brains, hearts, kidneys, and livers at different time points of development and in adulthood ($n = 4$ each, the four samples for E10.5 were each pooled from 7 organs, at E12.5 and 14.5 the four samples were pooled from four organs each). Expression of each gene was normalized to the respective *Gapdh, actin*, and *Rplp0* expression. Next, the average of all organs and samples at E10.5 was calculated. Individual samples were then normalized against this average value. Significance was tested for all time points between E10.5 and 18 months. E: embryonic day, P: postnatal day, adult: 3 months of age, old: 16–18 months of age. Data are mean ± SEM. * $p < 0.05$, ** $p < 0.01$, *** $p < 0.001$.

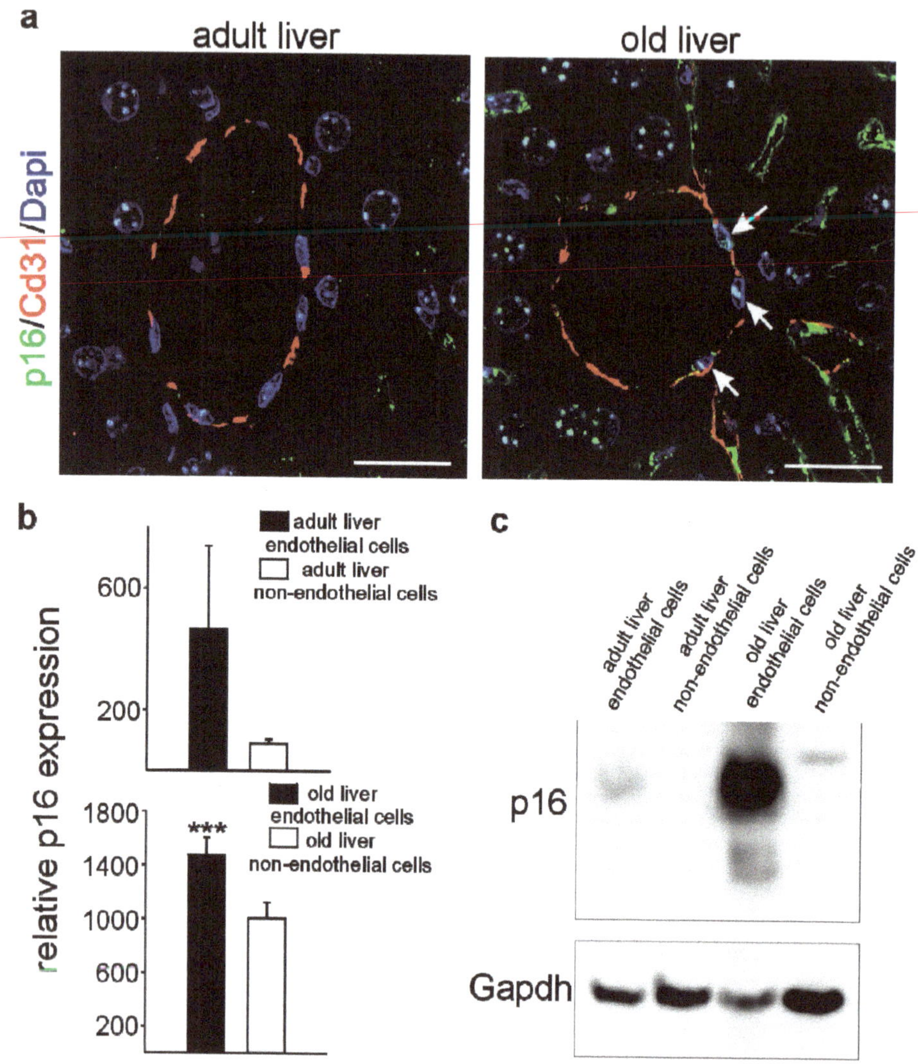

Figure 8. Liver vascular cells express higher levels of p16 than liver cells with aging. (**a**) Confocal images of Cd31 (red)/p16 (green) double-labeling on adult (3 months) (left image) and old (18 months) (right image) liver tissues. Arrows indicate p16/Cd31-positive vascular cells. Scale bars represent 50 µm. (**b**) Quantitative RT-PCRs for *p16* of sorted liver endothelial cells (black bar) and liver organ cells (white bar) at 3 months (upper panel) and 18 months (lower panel). Expression of p16 was normalized to the respective *Gapdh, actin*, and *Rplp0* expression. Data are mean ± SEM. *** $p < 0.001$. (**c**) Western Blot for p16 in 3- or 18-month-old liver endothelial cells and 3- or 18-month-old liver cells. Gapdh served as standard.

4. Discussion

Our results have shown dynamic and differential expression of *p16* during embryonic and postnatal development as well as in adult and old mice in the brain, heart, kidneys, and liver. Expression of *p16* was varying significantly within each organ during embryonic development in a matter of days. At the same time, *p19* and *p21* did not show such a remarkable variation of expression. We limited the current study to the investigation of brain, heart, kidney, and liver as these organs already develop at the embryonic time points chosen [65–68] and are relatively easy to isolate. Nevertheless, it is possible that *p16*, *p19*, and *p21* might be expressed in a variety of developing organs. For example, *p21* expression has already been described during embryonic development, i.e., in muscle, nasal epithelium, tongue muscles, hair follicles, epidermis, and cartilage, and was related in part to growth arrest and senescence [42,69,70]. Expression of *p19* has been described in the developing nervous system [44], while p16 has not been detected during embryonic development in earlier studies [45]. However, the authors of this study did not exclude that p16INK4a mRNA might be expressed at low levels or restricted sites in embryos. The authors reported an upregulation of p16 transcripts in organs from 15 month-old mice; however, the original PCR data do not show a specific p16 signal [45]. Using highly sensitive quantitative RT-PCR and antibody staining methods [48,54–58,71], our finding of a relatively high *p16* expression during development and especially at old stages might be more accurate.

Upregulation of *p16*/*p19* and *p21* is widely accepted as a marker of aging and senescence [16,72–74]. In human tissue samples, P16 was detected in endocrine and exocrine pancreas, skin, kidneys, liver, intestine, spleen, brain, and lung. Its expression increased in all investigated organs except for the lung with increasing age [75]. We demonstrate here that murine *p16* expression highly increased in all organs investigated between 3 months and 16 months of age. We observed a less pronounced increase in *p19* and *p21* compared to *p16* in old versus adult mice, which is consistent with previous reports in mice and humans [76]. We could not detect organ-specific differences in *p16* expression at 16 months of age, which contrasts with a recent study from Yousefzadeh et al. [77]. This might be explained by the age difference of the animals used in their study which compares mice aged from 15–19 weeks with 120-week-old subjects. A study comparing P16 protein expression by immunohistochemistry in human tissues from young, middle-aged, and old donors confirmed a significant increase of P16 in the liver, kidneys, and brain in old subjects. However, no P16 expression could be detected at all ages investigated in the heart, which might be due to species differences [75].

Regarding expression of *p16*, *p19*, and *p21* and the role of senescence during embryonic development, the literature is more controversial. Unlike *p21*, *p16* and *p19* were reported to be absent in early studies as discussed above [45,69]. Senescence, however, has been detected based on SA-β-galactosidase staining during embryonic development, which seems to depend on *p21* expression [42,43,78]. Interestingly, absence of p21 was compensated by apoptosis, but still slight developmental abnormalities were detectable [43]. Although these studies focused mainly on p21, *p16* loss has also been shown to result in developmental defects in the eye [79]; inactivation of *p16* and *p19* induced cardiomyocyte proliferation [80]; p16 has been detected in the ventricular and subventricular zones at embryonic and early postnatal stages in the rat brain; SA-β-galactosidase activity and *p16* expression has been detected in regressing mesonephros of quails [81]; *p16* expression in mouse embryos has been detected in motoneurons and the senolytic ABT-263 decreased the number of these cells [82]. Nevertheless, not all highly p16-positive cells are necessarily senescent [83,84]. For example, overexpression of *p16* slowed cell cycle progression in the G0/G1 phase and induced erythroid lineage differentiation [85], which might correspond to the early p16 expression in embryonic mouse livers [86]. Lack of *p16* is linked to increased cardiomyocyte proliferation [80], while lower cardiomyocyte proliferation, differentiation, and specification are required for myocardial compaction [87,88], which coincides with our observed cardiac p16 expression.

Furthermore, the notion of senescence as an irreversible form of cell cycle arrest, leading to death of the cell [18] has been recently questioned by the observation that cancer cells can escape from the senescence induced cell cycle arrest and gain a highly aggressive growth potential [89]. Highly interesting, it has also been demonstrated that embryonic senescent cells re-enter the cell cycle and contribute later to tissue formation [40]. We observed organ specific variations of *p16* expression, especially by immunohistochemical localization of p16 protein during development. Expression of p16 in development might reflect its function of slowing down cell cycle progression, a process essential for cell type specific differentiation. Knockout mice for *p16/p19* and selectively for *p16* are prone to tumor development [90–92], but potential developmental defects have not been investigated as the mice are viable and fertile. Thus, re-evaluation of potential developmental defects in mice with inactivation of *p16* or elimination of p16 expressing cells remains an interesting challenge for further studies.

In postnatal livers, p16 has been intensively studied. p16 has protective effects in non-alcoholic steatohepatitis and liver fibrosis through the regulation of reactive oxygen species (ROS) and oxidative stress [93,94]. Specific removal of liver endothelial cells expressing high levels of p16 resulted in fibrosis and liver deterioration, indicating that these cells are required for the maintenance of liver physiology [64]. However, detailed future studies using conditional cell type-specific knockout approaches will be needed to determine the specific function of p16 in liver endothelial cells.

5. Conclusions

Taken together, p16 expression in embryonic stages might reflect an implication in developmental differentiation processes. Further elucidation of the characteristics of p16 expressing cells, using embryos with inactivation or specific elimination of p16 expressing cells will hopefully shed light on the possible functions of p16 in differentiation, in addition to its implication in senescence and aging. Moreover, in aged mice, the significant upregulation of p16 expression in liver endothelial cells points to a selective role in liver endothelial physiology.

Supplementary Materials: The following are available online at https://www.mdpi.com/article/10.3390/cells11030541/s1, Figure S1: Representative photomicrographs of p16 immunostaining on sections of mouse livers, Figure S2: Representative photomicrographs of p16 immunostaining using a different p16 antibody (clone 1E12E10) on sections of mouse embryos, Figure S3: Expression of selected senescence-associated secretory phenotype (SASP) factors and p16 in endothelial and non-endothelial cells.

Author Contributions: Conceptualization, N.W. and K.-D.W.; methodology, H.S.-Z., N.W. and K.-D.W.; formal analysis, H.S.-Z., N.W., J.-F.M. and K.-D.W.; investigation, H.S.-Z., N.W. and K.-D.W.; resources, N.W., J.-F.M. and K.-D.W.; writing—original draft preparation, H.S.-Z., N.W. and K.-D.W.; writing—review and editing, N.W. and K.-D.W.; visualization, H.S.-Z. and N.W.; supervision, N.W. and K.-D.W.; project administration, N.W. and K.-D.W.; funding acquisition, N.W. and K.-D.W. All authors have read and agreed to the published version of the manuscript.

Funding: This research was funded by Fondation pour la Recherche Medicale, grant number FRM DPC20170139474 (K.-D.W.), Fondation ARC pour la recherche sur le cancer", grant number n°PJA 20161204650 (N.W.), Gemluc (N.W.), Plan Cancer INSERM (K.-D.W.), and Agence Nationale de la Recherche, grant R19125AA "Senage" (K.-D.W.).

Institutional Review Board Statement: All animal work was conducted according to national and international guidelines and was approved by the local ethics committee (Ciepal Cote d'Azur, PEA-NCE/2013/106).

Informed Consent Statement: Not applicable.

Acknowledgments: The authors thank A. Biancardini, A. Martres, A. Borderie, M. Goracci, and M. Cutajar-Bossert for technical assistance.

Conflicts of Interest: The authors declare no conflict of interest.

References

1. Campisi, J. Aging, Cellular Senescence, and Cancer. *Annu. Rev. Physiol.* **2013**, *75*, 685–705. [CrossRef]
2. Rose, M.R.; Charlesworth, B. A test of evolutionary theories of senescence. *Nature* **1980**, *287*, 141–142. [CrossRef]
3. Tchkonia, T.; Kirkland, J.L. Aging, Cell Senescence, and Chronic Disease. *J. Am. Med. Assoc.* **2018**, *320*, 1319–1320. [CrossRef]

4. Hou, Y.; Dan, X.; Babbar, M.; Wei, Y.; Hasselbalch, S.G.; Croteau, D.L.; Bohr, V.A. Ageing as a risk factor for neurodegenerative disease. *Nat. Rev. Neurol.* **2019**, *15*, 565–581. [CrossRef]
5. North, B.J.; Sinclair, D.A. The Intersection Between Aging and Cardiovascular Disease. *Circ. Res.* **2012**, *110*, 1097–1108. [CrossRef]
6. Ramly, E.; Kaafarani, H.M.; Velmahos, G.C. The Effect of Aging on Pulmonary Function. *Surg. Clin. N. Am.* **2015**, *95*, 53–69. [CrossRef]
7. Meyer, B.R. Renal Function in Aging. *J. Am. Geriatr. Soc.* **1989**, *37*, 791–800. [CrossRef]
8. Tung, S.; Iqbal, J. Evolution, Aging, and Osteoporosis. *Ann. N. Y. Acad. Sci.* **2007**, *1116*, 499–506. [CrossRef]
9. Misra, D.; Seo, P.H.; Cohen, H.J. Aging and cancer. Clin. *Adv. Hematol. Oncol.* **2004**, *2*, 457–465.
10. Liguori, I.; Russo, G.; Curcio, F.; Bulli, G.; Aran, L.; DELLA-Morte, D.; Gargiulo, G.; Testa, G.; Cacciatore, F.; Bonaduce, D.; et al. Oxidative stress, aging, and diseases. *Clin. Interv. Aging* **2018**, *13*, 757–772. [CrossRef]
11. Michaud, M.; Balardy, L.; Moulis, G.; Gaudin, C.; Peyrot, C.; Vellas, B.; Cesari, M.; Nourhashemi, F. Proinflammatory Cytokines, Aging, and Age-Related Diseases. *J. Am. Med. Dir. Assoc.* **2013**, *14*, 877–882. [CrossRef]
12. Burhans, W.C.; Weinberger, M. DNA replication stress, genome instability and aging. *Nucleic Acids Res.* **2007**, *35*, 7545–7556. [CrossRef]
13. Best, B.P. Nuclear DNA Damage as a Direct Cause of Aging. *Rejuvenation Res.* **2009**, *12*, 199–208. [CrossRef]
14. Freitas, A.; de Magalhaes, J.P. A review and appraisal of the DNA damage theory of ageing. *Mutat. Res. Mutat. Res.* **2011**, *728*, 12–22. [CrossRef]
15. Ohtani, N.; Mann, D.J.; Hara, E. Cellular senescence: Its role in tumor suppression and aging. *Cancer Sci.* **2009**, *100*, 792–797. [CrossRef]
16. Campisi, J.; d'Adda di Fagagna, F. Cellular senescence: When bad things happen to good cells. *Nat. Rev. Mol. Cell Biol.* **2007**, *8*, 729–740. [CrossRef]
17. Hayflick, L.; Moorhead, P.S. The serial cultivation of human diploid cell strains. *Exp. Cell Res.* **1961**, *25*, 585–621. [CrossRef]
18. Hayflick, L. The limited in vitro lifetime of human diploid cell strains. *Exp. Cell Res.* **1965**, *37*, 614–636. [CrossRef]
19. Tchkonia, T.; Zhu, Y.; Van Deursen, J.; Campisi, J.; Kirkland, J.L. Cellular senescence and the senescent secretory phenotype: Therapeutic opportunities. *J. Clin. Investig.* **2013**, *123*, 966–972. [CrossRef]
20. Di Micco, R.; Krizhanovsky, V.; Baker, D.; di Fagagna, F.D. Cellular senescence in ageing: From mechanisms to therapeutic opportunities. *Nat. Rev. Mol. Cell Biol.* **2021**, *22*, 75–95. [CrossRef]
21. Rhinn, M.; Ritschka, B.; Keyes, W.M. Cellular senescence in development, regeneration and disease. *Development* **2019**, *146*, dev151837. [CrossRef]
22. Dimri, G.P.; Lee, X.; Basile, G.; Acosta, M.; Scott, G.; Roskelley, C.; Medrano, E.E.; Linskens, M.; Rubelj, I.; Pereira-Smith, O.; et al. A biomarker that identifies senescent human cells in culture and in aging skin in vivo. *Proc. Natl. Acad. Sci. USA* **1995**, *92*, 9363–9367. [CrossRef]
23. Acosta, J.C.; Banito, A.; Wuestefeld, T.; Georgilis, A.; Janich, P.; Morton, J.P.; Athineos, D.; Kang, T.-W.; Lasitschka, F.; Andrulis, M.; et al. A complex secretory program orchestrated by the inflammasome controls paracrine senescence. *Nat. Cell Biol.* **2013**, *15*, 978–990. [CrossRef]
24. Coppé, J.-P.; Patil, C.K.; Rodier, F.; Sun, Y.; Muñoz, D.P.; Goldstein, J.; Nelson, P.S.; Desprez, P.-Y.; Campisi, J. Senescence-Associated Secretory Phenotypes Reveal Cell-Nonautonomous Functions of Oncogenic RAS and the p53 Tumor Suppressor. *PLoS Biol.* **2008**, *6*, e301. [CrossRef]
25. Coppé, J.-P.; Desprez, P.-Y.; Krtolica, A.; Campisi, J. The Senescence-Associated Secretory Phenotype: The Dark Side of Tumor Suppression. *Annu. Rev. Pathol. Mech. Dis.* **2010**, *5*, 99–118. [CrossRef]
26. Levy, M.Z.; Allsopp, R.C.; Futcher, A.; Greider, C.; Harley, C.B. Telomere end-replication problem and cell aging. *J. Mol. Biol.* **1992**, *225*, 951–960. [CrossRef]
27. Allsopp, R.C.; Chang, E.; Kashefi-Aazam, M.; Rogaev, E.I.; Piatyszek, M.A.; Shay, J.W.; Harley, C.B. Telomere Shortening Is Associated with Cell Division in Vitro and in Vivo. *Exp. Cell Res.* **1995**, *220*, 194–200. [CrossRef]
28. McEachern, M.J.; Krauskopf, A.; Blackburn, E.H. TELOMERES AND THEIR CONTROL. *Annu. Rev. Genet.* **2000**, *34*, 331–358. [CrossRef]
29. Serrano, M.; Lin, A.W.; McCurrach, M.E.; Beach, D.; Lowe, S.W. Oncogenic ras Provokes Premature Cell Senescence Associated with Accumulation of p53 and p16INK4a. *Cell* **1997**, *88*, 593–602. [CrossRef]
30. Nakamura, A.J.; Chiang, Y.J.; Hathcock, K.S.; Horikawa, I.; A Sedelnikova, O.; Hodes, R.J.; Bonner, W.M. Both telomeric and non-telomeric DNA damage are determinants of mammalian cellular senescence. *Epigenetics Chromatin* **2008**, *1*, 6–12. [CrossRef]
31. Narita, M.; Nuñez, S.; Heard, E.; Narita, M.; Lin, A.W.; Hearn, S.A.; Spector, D.L.; Hannon, G.J.; Lowe, S.W. Rb-Mediated Heterochromatin Formation and Silencing of E2F Target Genes during Cellular Senescence. *Cell* **2003**, *113*, 703–716. [CrossRef]
32. Adams, P.D. Healing and Hurting: Molecular Mechanisms, Functions, and Pathologies of Cellular Senescence. *Mol. Cell* **2009**, *36*, 2–14. [CrossRef] [PubMed]
33. Munro, J.; I Barr, N.; Ireland, H.; Morrison, V.; Parkinson, E. Histone deacetylase inhibitors induce a senescence-like state in human cells by a p16-dependent mechanism that is independent of a mitotic clock. *Exp. Cell Res.* **2004**, *295*, 525–538. [CrossRef]
34. Collins, C.J.; Sedivy, J.M. Involvement of the INK4a/Arf gene locus in senescence. *Aging Cell* **2003**, *2*, 145–150. [CrossRef] [PubMed]

35. Takeuchi, S.; Takahashi, A.; Motoi, N.; Yoshimoto, S.; Tajima, T.; Yamakoshi, K.; Hirao, A.; Yanagi, S.; Fukami, K.; Ishikawa, Y.; et al. Intrinsic Cooperation between p16INK4a and p21Waf1/Cip1 in the Onset of Cellular Senescence and Tumor Suppression In vivo. *Cancer Res.* **2010**, *70*, 9381–9390. [CrossRef]
36. A Weinberg, R. The Cat and Mouse Games That Genes, Viruses, and Cells Play. *Cell* **1997**, *88*, 573–575. [CrossRef]
37. Serrano, M.; Hannon, G.J.; Beach, D.C. A new regulatory motif in cell-cycle control causing specific inhibition of cyclin D/CDK4. *Nature* **1993**, *366*, 704–707. [CrossRef]
38. Parry, D.; Bates, S.; Mann, D.J.; Peters, G. Lack of cyclin D-Cdk complexes in Rb-negative cells correlates with high levels of p16INK4/MTS1 tumour suppressor gene product. *EMBO J.* **1995**, *14*, 503–511. [CrossRef]
39. Shamloo, B.; Usluer, S. p21 in Cancer Research. *Cancers* **2019**, *11*, 1178. [CrossRef]
40. Li, Y.; Zhao, H.; Huang, X.; Tang, J.; Zhang, S.; Li, Y.; Liu, X.; He, L.; Ju, Z.; Lui, K.O.; et al. Embryonic senescent cells re-enter cell cycle and contribute to tissues after birth. *Cell Res.* **2018**, *28*, 775–778. [CrossRef]
41. Hosako, H.; Francisco, L.E.; Martin, G.S.; Mirkes, P.E. The roles of p53 and p21 in normal development and hyperthermia-induced malformations. *Birth Defects Res. Part B Dev. Reprod. Toxicol.* **2009**, *86*, 40–47. [CrossRef]
42. Storer, M.; Mas, A.; Robert-Moreno, À.; Pecoraro, M.; Ortells, M.C.; Di Giacomo, V.; Yosef, R.; Pilpel, N.; Krizhanovsky, V.; Sharpe, J.; et al. Senescence Is a Developmental Mechanism that Contributes to Embryonic Growth and Patterning. *Cell* **2013**, *155*, 1119–1130. [CrossRef]
43. Muñoz-Espín, D.; Cañamero, M.; Maraver, A.; López, G.G.; Contreras, J.; Murillo-Cuesta, S.; Rodríguez-Baeza, A.; Varela-Nieto, I.; Ruberte, J.; Collado, M.; et al. Programmed Cell Senescence during Mammalian Embryonic Development. *Cell* **2013**, *155*, 1104–1118. [CrossRef]
44. Zindy, F.; Soares, H.; Herzog, K.H.; Morgan, J.; Sherr, C.J.; Roussel, M.F. Expression of INK4 inhibitors of cyclin D-dependent kinases during mouse brain development. *Cell Growth Differ. Mol. Biol. J. Am. Assoc. Cancer Res.* **1997**, *8*, 1139–1150.
45. Zindy, F.; Quelle, D.; Roussel, M.F.; Sherr, C.J. Expression of the p16INK4a tumor suppressor versus other INK4 family members during mouse development and aging. *Oncogene* **1997**, *15*, 203–211. [CrossRef]
46. Ouelle, D.E.; Zindy, F.; Ashmun, R.A.; Sherr, C.J. Alternative reading frames of the INK4a tumor suppressor gene encode two unrelated proteins capable of inducing cell cycle arrest. *Cell* **1995**, *83*, 993–1000. [CrossRef]
47. Wagner, K.-D.; Cherfils-Vicini, J.; Hosen, N.; Hohenstein, P.; Gilson, E.; Hastie, N.D.; Michiels, J.-F.; Wagner, N. The Wilms' tumour suppressor Wt1 is a major regulator of tumour angiogenesis and progression. *Nat. Commun.* **2014**, *5*, 5852. [CrossRef]
48. Wagner, N.; Ninkov, M.; Vukolic, A.; Deniz, G.C.; Rassoulzadegan, M.; Michiels, J.-F.; Wagner, K.-D. Implications of the Wilms' Tumor Suppressor Wt1 in Cardiomyocyte Differentiation. *Int. J. Mol. Sci.* **2021**, *22*, 4346. [CrossRef]
49. Wagner, N.; Michiels, J.F.; Schedl, A.; Wagner, K.-D. The Wilms' tumour suppressor WT1 is involved in endothelial cell proliferation and migration: Expression in tumour vessels in vivo. *Oncogene* **2008**, *27*, 3662–3672. [CrossRef]
50. Wagner, K.-D.; Ying, Y.; Leong, W.; Jiang, J.; Hu, X.; Chen, Y.; Michiels, J.-F.; Lu, Y.; Gilson, E.; Wagner, N.; et al. The differential spatiotemporal expression pattern of shelterin genes throughout lifespan. *Aging* **2017**, *9*, 1219–1232. [CrossRef]
51. Silver, N.; Best, S.; Jiang, J.; Thein, S.L. Selection of housekeeping genes for gene expression studies in human reticulocytes using real-time PCR. *BMC Mol. Biol.* **2006**, *7*, 33. [CrossRef]
52. Keber, R.; Motaln, H.; Wagner, K.D.; Debeljak, N.; Rassoulzadegan, M.; Ačimovič, J.; Rozman, D.; Horvat, S. Mouse Knockout of the Cholesterogenic Cytochrome P450 Lanosterol 14α-Demethylase (Cyp51) Resembles Antley-Bixler Syndrome. *J. Biol. Chem.* **2011**, *286*, 29086–29097. [CrossRef]
53. Faulkner, A.; Lynam, E.; Purcell, R.; Jones, C.; Lopez, C.; Board, M.; Wagner, K.-D.; Wagner, N.; Carr, C.; Wheeler-Jones, C. Context-dependent regulation of endothelial cell metabolism: Differential effects of the PPARβ/δ agonist GW0742 and VEGF-A. *Sci. Rep.* **2020**, *10*, 7849. [CrossRef]
54. Wagner, K.D.; Du, S.; Martin, L.; Leccia, N.; Michiels, J.-F.; Wagner, N. Vascular PPARβ/δ Promotes Tumor Angiogenesis and Progression. *Cells* **2019**, *8*, 1623. [CrossRef]
55. Wagner, K.-D.; El Maï, M.; Ladomery, M.; Belali, T.; Leccia, N.; Michiels, J.-F.; Wagner, N. Altered VEGF Splicing Isoform Balance in Tumor Endothelium Involves Activation of Splicing Factors Srpk1 and Srsf1 by the Wilms' Tumor Suppressor Wt1. *Cells* **2019**, *8*, 41. [CrossRef]
56. Wagner, K.-D.; Vukolic, A.; Baudouy, D.; Michiels, J.-F.; Wagner, N. Inducible Conditional Vascular-Specific Overexpression of Peroxisome Proliferator-Activated Receptor Beta/Delta Leads to Rapid Cardiac Hypertrophy. *PPAR Res.* **2016**, *2016*, 7631085. [CrossRef]
57. EL Maï, M.; Wagner, K.-D.; Michiels, J.-F.; Ambrosetti, D.; Borderie, A.; Destree, S.; Renault, V.; Djerbi, N.; Giraud-Panis, M.-J.; Gilson, E.; et al. The Telomeric Protein TRF2 Regulates Angiogenesis by Binding and Activating the PDGFRβ Promoter. *Cell Rep.* **2014**, *9*, 1047–1060. [CrossRef]
58. Wagner, N.; Morrison, H.; Pagnotta, S.; Michiels, J.-F.; Schwab, Y.; Tryggvason, K.; Schedl, A.; Wagner, K.-D. The podocyte protein nephrin is required for cardiac vessel formation. *Hum. Mol. Genet.* **2011**, *20*, 2182–2194. [CrossRef]
59. Sato, S.; Kawamata, Y.; Takahashi, A.; Imai, Y.; Hanyu, A.; Okuma, A.; Takasugi, M.; Yamakoshi, K.; Sorimachi, H.; Kanda, H.; et al. Ablation of the p16INK4a tumour suppressor reverses ageing phenotypes of klotho mice. *Nat. Commun.* **2015**, *6*, 7035. [CrossRef]

60. Bruggeman, S.W.; Valk-Lingbeek, M.E.; van der Stoop, P.P.; Jacobs, J.J.; Kieboom, K.; Tanger, E.; Hulsman, D.; Leung, C.; Arsenijevic, Y.; Marino, S.; et al. Ink4a and Arf differentially affect cell proliferation and neural stem cell self-renewal in Bmi1-deficient mice. *Genes Dev.* **2005**, *19*, 1438–1443. [CrossRef]
61. Sladky, V.C.; Knapp, K.; Soratroi, C.; Heppke, J.; Eichin, F.; Rocamora-Reverte, L.; Szabó, T.G.; Bongiovanni, L.; Westendorp, B.; Moreno, E.; et al. E2F-Family Members Engage the PIDDosome to Limit Hepatocyte Ploidy in Liver Development and Regeneration. *Dev. Cell* **2020**, *52*, 335–349.e7. [CrossRef]
62. Pulido, T.; Velarde, M.C.; Alimirah, F. The senescence-associated secretory phenotype: Fueling a wound that never heals. *Mech. Ageing Dev.* **2021**, *199*, 111561. [CrossRef]
63. Chambers, C.R.; Ritchie, S.; Pereira, B.A.; Timpson, P. Overcoming the senescence-associated secretory phenotype (SASP): A complex mechanism of resistance in the treatment of cancer. *Mol. Oncol.* **2021**. [CrossRef]
64. Grosse, L.; Wagner, N.; Emelyanov, A.; Molina, C.; Lacas-Gervais, S.; Wagner, K.-D.; Bulavin, D.V. Defined p16High Senescent Cell Types Are Indispensable for Mouse Healthspan. *Cell Metab.* **2020**, *32*, 87–99.e6. [CrossRef]
65. Costantini, F.; Kopan, R. Patterning a Complex Organ: Branching Morphogenesis and Nephron Segmentation in Kidney Development. *Dev. Cell* **2010**, *18*, 698–712. [CrossRef]
66. Wagner, N.; Wagner, K.-D. Every Beat You Take—The Wilms′ Tumor Suppressor WT1 and the Heart. *Int. J. Mol. Sci.* **2021**, *22*, 7675. [CrossRef]
67. Henry, A.M.; Hohmann, J.G. High-resolution gene expression atlases for adult and developing mouse brain and spinal cord. *Mamm. Genome* **2012**, *23*, 539–549. [CrossRef]
68. Zhao, R.; Duncan, S.A. Embryonic development of the liver. *Hepatology* **2005**, *41*, 956–967. [CrossRef]
69. Vasey, D.B.; Wolf, C.R.; Brown, K.; Whitelaw, C.B.A. Spatial p21 expression profile in the mid-term mouse embryo. *Transgenic Res.* **2010**, *20*, 23–28. [CrossRef]
70. Parker, S.B.; Eichele, G.; Zhang, P.; Rawls, A.; Sands, A.T.; Bradley, A.; Olson, E.N.; Harper, J.W.; Elledge, S.J. p53-Independent Expression of p21 Cip1 in Muscle and Other Terminally Differentiating Cells. *Science* **1995**, *267*, 1024–1027. [CrossRef]
71. Wagner, N.; Wagner, K.-D.; Scholz, H.; Kirschner, K.; Schedl, A. Intermediate filament protein nestin is expressed in developing kidney and heart and might be regulated by the Wilms' tumor suppressor Wt1. *Am. J. Physiol. Integr. Comp. Physiol.* **2006**, *291*, R779–R787. [CrossRef]
72. Krishnamurthy, J.; Torrice, C.; Ramsey, M.R.; Kovalev, G.I.; Al-Regaiey, K.; Su, L.; Sharpless, N.E. Ink4a/Arf expression is a biomarker of aging. *J. Clin. Investig.* **2004**, *114*, 1299–1307. [CrossRef]
73. Sharpless, N.E.; Sherr, C.J. Forging a signature of in vivo senescence. *Nat. Rev. Cancer* **2015**, *15*, 397–408. [CrossRef]
74. López-Domínguez, J.A.; Rodríguez-López, S.; Ahumada-Castro, U.; Desprez, P.-Y.; Konovalenko, M.; Laberge, R.-M.; Cárdenas, C.; Villalba, J.M.; Campisi, J. Cdkn1a transcript variant 2 is a marker of aging and cellular senescence. *Aging* **2021**, *13*, 13380–13392. [CrossRef]
75. Idda, M.L.; McClusky, W.G.; Lodde, V.; Munk, R.; Abdelmohsen, K.; Rossi, M.; Gorospe, M. Survey of senescent cell markers with age in human tissues. *Aging* **2020**, *12*, 4052–4066. [CrossRef]
76. Hudgins, A.D.; Tazearslan, C.; Tare, A.; Zhu, Y.; Huffman, D.; Suh, Y. Age- and Tissue-Specific Expression of Senescence Biomarkers in Mice. *Front. Genet.* **2018**, *9*, 59. [CrossRef]
77. Yousefzadeh, M.J.; Zhao, J.; Bukata, C.; Wade, E.A.; McGowan, S.J.; Angelini, L.A.; Bank, M.P.; Gurkar, A.; McGuckian, C.A.; Calubag, M.F.; et al. Tissue specificity of senescent cell accumulation during physiologic and accelerated aging of mice. *Aging Cell* **2020**, *19*, e13094. [CrossRef]
78. Wanner, E.; Thoppil, H.; Riabowol, K. Senescence and Apoptosis: Architects of Mammalian Development. *Front. Cell Dev. Biol.* **2021**, *8*, 620089. [CrossRef]
79. Cheong, C.; Sung, Y.H.; Lee, J.; Choi, Y.S.; Song, J.; Kee, C.; Lee, H.-W. Role of INK4a locus in normal eye development and cataract genesis. *Mech. Ageing Dev.* **2006**, *127*, 633–638. [CrossRef]
80. An, S.; Chen, Y.; Gao, C.; Qin, B.; Du, X.; Meng, F.; Qi, Y. Inactivation of INK4a and ARF induces myocardial proliferation and improves cardiac repair following ischemia-reperfusion. *Mol. Med. Rep.* **2015**, *12*, 5911–5916. [CrossRef]
81. Nacher, V.; Carretero, A.; Navarro, M.; Ayuso, E.; Ramos, D.; Luppo, M.; Rodríguez, A.; Mendes, L.; Herrero-Fresneda, I.; Ruberte, J. Endothelial Cell Transduction in Primary Cultures from Regressing Mesonephros. *Cells Tissues Organs* **2010**, *191*, 84–95. [CrossRef]
82. Dominguez-Bautista, J.A.; Acevo-Rodríguez, P.S.; Castro-Obregón, S. Programmed Cell Senescence in the Mouse Developing Spinal Cord and Notochord. *Front. Cell Dev. Biol.* **2021**, *9*, 587096. [CrossRef]
83. Frescas, D.; Hall, B.M.; Strom, E.; Virtuoso, L.P.; Gupta, M.; Gleiberman, A.S.; Rydkina, E.; Balan, V.; Vujcic, S.; Chernova, O.B.; et al. Murine mesenchymal cells that express elevated levels of the CDK inhibitor p16(Ink4a) in vivo are not necessarily senescent. *Cell Cycle* **2017**, *16*, 1526–1533. [CrossRef]
84. Hall, B.M.; Balan, V.; Gleiberman, A.S.; Strom, E.; Krasnov, P.; Virtuoso, L.P.; Rydkina, E.; Vujcic, S.; Balan, K.; Gitlin, I.; et al. Aging of mice is associated with p16(Ink4a)- and β-galactosidase-positive macrophage accumulation that can be induced in young mice by senescent cells. *Aging* **2016**, *8*, 1294–1315. [CrossRef]
85. Minami, R.; Muta, K.; Umemura, T.; Motomura, S.; Abe, Y.; Nishimura, J.; Nawata, H. p16INK4a induces differentiation and apoptosis in erythroid lineage cells. *Exp. Hematol.* **2003**, *31*, 355–362. [CrossRef]

86. Wang, X.; Yang, L.; Wang, Y.-C.; Xu, Z.-R.; Feng, Y.; Zhang, J.; Wang, Y.; Xu, C.-R. Comparative analysis of cell lineage differentiation during hepatogenesis in humans and mice at the single-cell transcriptome level. *Cell Res.* **2020**, *30*, 1109–1126. [CrossRef]
87. Wu, T.; Liang, Z.; Zhang, Z.; Liu, C.; Zhang, L.; Gu, Y.; Peterson, K.L.; Evans, S.M.; Fu, X.-D.; Chen, J. PRDM16 Is a Compact Myocardium-Enriched Transcription Factor Required to Maintain Compact Myocardial Cardiomyocyte Identity in Left Ventricle. *Circulation* **2021**. [CrossRef]
88. Sedmera, D.; Pexieder, T.; Vuillemin, M.; Thompson, R.P.; Anderson, R.H. Developmental patterning of the myocardium. *Anat. Rec.* **2000**, *258*, 319–337. [CrossRef]
89. Milanovic, M.; Fan, D.N.Y.; Belenki, D.; Däbritz, J.H.M.; Zhao, Z.; Yu, Y.; Dörr, J.R.; Dimitrova, L.; Lenze, D.; Barbosa, I.A.M.; et al. Senescence-associated reprogramming promotes cancer stemness. *Nature* **2017**, *553*, 96–100. [CrossRef]
90. E Sharpless, N.; Ramsey, M.; Balasubramanian, P.; Castrillon, D.H.; DePinho, R. The differential impact of p16INK4a or p19ARF deficiency on cell growth and tumorigenesis. *Oncogene* **2004**, *23*, 379–385. [CrossRef]
91. Sharpless, N.; Bardeesy, N.; Lee, K.-H.; Carrasco, D.; Castrillon, D.H.; Aguirre, A.J.; Wu, E.A.; Horner, J.W.; DePinho, R. Loss of p16Ink4a with retention of p19Arf predisposes mice to tumorigenesis. *Nature* **2001**, *413*, 86–91. [CrossRef] [PubMed]
92. Serrano, M.; Lee, H.-W.; Chin, L.; Cordon-Cardo, C.; Beach, D.; DePinho, R. Role of the INK4a Locus in Tumor Suppression and Cell Mortality. *Cell* **1996**, *85*, 27–37. [CrossRef]
93. Lv, F.; Li, N.; Kong, M.; Wu, J.; Fan, Z.; Miao, D.; Xu, Y.; Ye, Q.; Wang, Y. CDKN2a/p16 Antagonizes Hepatic Stellate Cell Activation and Liver Fibrosis by Modulating ROS Levels. *Front. Cell Dev. Biol.* **2020**, *8*, 176. [CrossRef] [PubMed]
94. Lv, F.; Wu, J.; Miao, D.; An, W.; Wang, Y. p16 deficiency promotes nonalcoholic steatohepatitis via regulation of hepatic oxidative stress. *Biochem. Biophys. Res. Commun.* **2017**, *486*, 264–269. [CrossRef] [PubMed]

cells MDPI

Review

The Senescence Markers p16INK4A, p14ARF/p19ARF, and p21 in Organ Development and Homeostasis

Kay-Dietrich Wagner * and **Nicole Wagner** *

CNRS, INSERM, iBV, Université Côte d'Azur, 06107 Nice, France
* Correspondence: kwagner@unice.fr (K.-D.W.); nwagner@unice.fr (N.W.)

Abstract: It is widely accepted that senescent cells accumulate with aging. They are characterized by replicative arrest and the release of a myriad of factors commonly called the senescence-associated secretory phenotype. Despite the replicative cell cycle arrest, these cells are metabolically active and functional. The release of SASP factors is mostly thought to cause tissue dysfunction and to induce senescence in surrounding cells. As major markers for aging and senescence, p16INK4, p14ARF/p19ARF, and p21 are established. Importantly, senescence is also implicated in development, cancer, and tissue homeostasis. While many markers of senescence have been identified, none are able to unambiguously identify all senescent cells. However, increased levels of the cyclin-dependent kinase inhibitors p16INK4A and p21 are often used to identify cells with senescence-associated phenotypes. We review here the knowledge of senescence, p16INK4A, p14ARF/p19ARF, and p21 in embryonic and postnatal development and potential functions in pathophysiology and homeostasis. The establishment of senolytic therapies with the ultimate goal to improve healthy aging requires care and detailed knowledge about the involvement of senescence and senescence-associated proteins in developmental processes and homeostatic mechanism. The review contributes to these topics, summarizes open questions, and provides some directions for future research.

Keywords: development; aging; endothelial cells; senescence; SASP; metabolic function; stem cells

Citation: Wagner, K.-D.; Wagner, N. The Senescence Markers p16INK4A, p14ARF/p19ARF, and p21 in Organ Development and Homeostasis. *Cells* **2022**, *11*, 1966. https://doi.org/10.3390/cells11121966

Academic Editor: Christoph Englert

Received: 11 May 2022
Accepted: 15 June 2022
Published: 19 June 2022

1. Introduction

Senescence was first described by Hayflick in isolated fibroblasts in culture [1,2]. In response to repeated replication, DNA damage, metabolic alterations, reactive oxygen species or cytotoxic drugs, cells enter permanent cell cycle arrest, change their morphology to more flat and large cells, express and secrete cytokines, chemokines, growth factors, bioactive lipids, and pro-apoptotic factors—the so-called senescence-associated secretory phenotype (SASP) and become positive for senescence-associated beta-galactosidase (SAβG) [3–11]. Although the morphological features are easy to follow in cultured cells, the identification in vivo or on histological sections is more problematic. SAβG staining is also not uniform in all old cells or in response to typical inducers of senescence, e.g., doxorubicin [12]. During embryonic development, even some co-localization of SAβG staining with proliferation markers was detectable [13]. Thus, recently the use of combinations of different markers and expression of SASP factors was suggested from the International Cell Senescence Association to correctly identify senescent cells [5]. In addition, the expression of SASP factors varies depending on different cell types [14]. Whether different cell types are to the same extent susceptible to age-related senescence is equally unclear. The conventional view in agreement with the Hayflick experiments would suggest that replicative cells are prone to senescence with increasing age. Nevertheless, senescence-like features were also observed in terminally differentiated non-cycling cells [15–18] and in macrophages and T-cells [19–21]. As typical markers for aging and senescence p16INK4A, p14ARF/p19ARFArf, and p21 are accepted [3–7,10,11,22–28]. These proteins were originally identified as cell cycle inhibitors (for details see below). Thus, senescence could also be viewed as an extreme case of cell cycle inhibition except for the case of postmitotic cells. p16INK4A is one of the most attractive and intensively investigated marker of aging and senescence as expression has been initially reported to be absent during embryonic development [29,30] and it is highly expressed in advanced age and senescence [24–28,31–37]. We and others provided recent evidence that p16INK4A is expressed during development in several organs [38]. The elimination of p16INK4A-expressing cells in aged animals did not only have the expected positive effects, but also negatively impacted the health span, caused liver fibrosis [39] and interfered with normal wound healing [40,41]. Thus, it seems timely to review knowledge of senescence, p16INK4A, p14ARF/19ASRF, and p21 in embryonic and postnatal development, in disease and homeostasis.

2. p16INK4A, p14ARF/p19ARF, and p21—Basic Molecular Mechanisms

p16INK4A was originally identified as a tumor suppressor gene [42,43]. Initially, different names, i.e., multiple tumor suppressor-1 (MTS-1), inhibitor of cyclin dependent kinase 4a (INK4a), cyclin dependent kinase inhibitor 2a (CDKN2A), have been used, *CDKN2A* now being the official gene symbol. The human *p16INK4A* gene is located on the short arm of chromosome 9 (9p21.3) while the mouse gene is located on chromosome 4. The use of different open reading frames on the locus generates in both species' alternative proteins (p14p14ARF in humans and p19ARF in mice). In comparison to p16INK4A, they differ in the first exon while they share the second exon, resulting in the translation of different reading frames [44,45] (reviewed in [46]). The *p21* gene (*CDKN1A*) is completely independent and localized on chromosomes 6 and 17 in humans and mice, respectively. p16INK4A acts as a specific inhibitor of the cyclin-dependent kinases CDK4 and CDK6 that is mainly active in the G1 phase of the cell cycle to prevent the cell transition from the G1 to S phase and subsequent proliferation arrest by rendering retinoblastoma protein (pRB) in a hypo-phosphorylated state. CDK 4/6 bind cyclin D to form a complex that phosphorylates retinoblastoma protein. When phosphorylated, pRB dissociates from E2F transcription factors which translocate to the nucleus and activate transcription of S phase genes which results in a cellular proliferation [47–49]. p16INK4A expression is tightly regulated via a negative feedback loop with pRB. pRB phosphorylation promotes E2F translocation and induces p16INK4A expression, which in turn inhibits CDK 4/6 and increases hypo-phosphorylated pRB, leading to the downregulation of p16INK4A [50]. Alternatively, elevated p16INK4A transcription in pRB negative cells has also been reported, indicating alternative mechanisms for p16INK4A upregulation [25]. Furthermore, differences in p16INK4A RNA expression did not correlate well with the pRB status of the cells [25]. p16INK4A and p19p14ARF/p19ARF are suppressed by promoter hypermethylation via PRC1 and PRC2 complexes involving BMI-1, EZH2, ZFP 277, Mel18, CXB7, and CXB8 proteins [51–61]. Interestingly, pRB seems to also be involved in this regulatory loop as a lack of pRB results in loss of histone H3K27 trimethylation and less recruitment of BMI-1 and repression of the p16INK4A locus [62]. Activators of the p16INK4A locus include AP-1 [63], JDP-2 [64–66], CTCF [67], Tcf-1 [68], p300 with Sp-1 [69], Meis1 [70], and PPAR gamma [71]. These in vitro molecular studies should be interpreted with care. For example, multiple beneficial effects were attributed to removal of p16INK4A-expressing senescent cells in mice [17,72–82]. PPAR gamma stimulation induces p16INK4A-expression and might result in senescent cell-based multi-organ failure. However, glitazones (PPAR gamma activators, e.g., rosiglitazone) have been in clinical use as antidiabetic drugs for more than 20 years [83].

Combined in vivo and in vitro studies using knockout mouse models, chromatin immunoprecipitation (CHIP), and RNA sequencing showed that non-cleaved general transcription factor TFIIA acts as a repressor of the p16INK4A, p14ARF/p19ARF, and p21 loci. Taspase1-mediated (TASP1-mediated) cleavage of TFIIA ensures rapid cell proliferation and morphogenesis by reducing transcription of p16INK4A and p14ARF/p19ARF. Consequently, Tasp1 knockout mice showed variable degrees of micro-ophthalmia, anophthalmia, agnathia, general growth retardation, and defects of development of forebrain neurons, which were partially rescued by combined knockout of p16INK4A and p14ARF/p19ARF [84].

Elegant in vivo studies showed that a common variant of a 58 kb non-coding sequence in humans (70 kb in mice) flanking the p16INK4A/p14ARF/p19ARF locus is associated with an increased risk of coronary artery disease [85–87]. The removal of this sequence resulted in a low expression of p16INK4A/p14ARF/p19ARF and excessive proliferation of aortic smooth muscle cells indicating that this region has a pivotal role in the regulation of p16INK4A/p14ARF/p19ARF expression and control of cell proliferation [88].

Coordinated suppression/activation of the p16INK4A/p14ARF/p19ARF locus would further implicate that p16INK4A and p14ARF/p19ARF expression patterns are related. Our recent study on several organs during development and aging showed that this is not the case [38]. Moreover, p14ARF/p19ARF shows different downstream signaling from p16INK4A. p14ARF/p19ARF acts as a cell cycle inhibitor by interfering with the activation of the P53 pathway, through binding to MDM2 and blocking MDM2-mediated P53 degradation [89], although p14ARF/p19ARF might also have some P53-independent actions [90]. p14ARF/p19ARF might induce apoptosis via Bax in a P53 independent manner [91]. p14ARF/p19ARF is activated by Myc [92] and in a feedback mechanism seems to physically interact with Myc protein and prevent its function as a transcriptional activator. In addition, this action is independent of P53 [93,94]. p21 is another cyclin-dependent kinase inhibitor and has been shown to fulfill anti-proliferative functions in a mechanism that is P53-dependent. p16INK4A might activate p21, which acts by inhibiting CDK2-cyclin E active complex formation, such as p16INK4A inhibition of CDK4/6 cyclin D. The consequence is also hypo-phosphorylation of pRB and cell cycle arrest [95]. Interesting, low p21 levels promote CDK-cyclin complex formation, while higher expression of p21 inhibits the activity of the complex [96]. This might explain to some extent the diverse effects of altering the levels in vivo described below.

pRB interacts through various cellular proteins, among which E2F transcription factors are the best characterized [97–99]. While transient E2F overexpression promotes cell cycle progression and hyperplasia, sustained E2F3 overexpression promotes a senescence-like phenotype in a p16INK4A-pRB-p14ARF/p19ARF pathway-dependent manner [100] points again to the different outcomes dependent on timing and cellular model. E2F3 in addition occupies the p14ARF/p19ARF promoter in mouse embryonic fibroblasts and E2f3 loss is sufficient to derepress p14ARF/p19ARF, which in turn triggers activation of p53 and expression of p21 [101]. The combined loss of all E2F transcription factors also results in overexpression of p21, leading to a decrease in cyclin-dependent kinase activity and Rb phosphorylation [98,99]. p21 is furthermore transcriptionally inhibited by a Myc-Miz complex [102,103] and activated by Smad/FoxO complexes in response to TGF beta stimulation [104]. The regulation of p16INK4A, p14/p14ARF/p19ARFArf, and p21 are reviewed in detail elsewhere [105–110].

3. p16INK4A, p14ARF/p19ARF, and p21 in Organ Development

Earlier studies were not able to detect p16INK4A expression during mouse embryonic development [29,30]. However, the authors did not exclude the possibility that p16INK4A might be expressed in different developing organs and time points, but the lack of p16INK4A detection might rather represent technical limits [29]. We used recently sensitive quantitative RT-PCR and immunohistochemistry techniques [111–115] to re-evaluate p16INK4A expression during mouse embryonic development, in the adult, and in old mice [38]. We determined p16INK4A expression between embryonic day (E10) and birth, at postnatal day seven (P7), postnatal day 21, which corresponds to weaning, in adults, and 16–18-month-old mice. We focused on the heart, brain, liver, and kidney as these organs or progenitors are already present at the first time point chosen [116–119]. p16INK4A, p14ARF/p19ARF, and p21 were detectable at all investigated embryonic and postnatal time points. Compared to p14ARF/p19ARF and p21, p16INK4A expression continued to increase during development, remained then stable in adulthood and became dramatically upregulated in the organs of old animals. This high rise of p16INK4A expression with old age is in principle in agreement with the literature defining p16INK4A as a marker of aging and senescence [5,72,120–124]. In agreement with this, we also detected a significant increase in SASP markers in all investigated organs of old animals. Interestingly, in the organs of old mice, we observed the highest p16INK4A expression in vascular structures, especially in the liver and the heart. During embryonic development, high p16INK4A expression did not correspond with increased SASP expression and was observed in the respective parenchymal cells and not in vessels, which coincided with the corresponding time points of differentiation in the organs investigated [38], suggesting that in this instance, p16INK4A might act in a classical way as cell cycle inhibitor as pre-requisite for differentiation. Although we did not yet identify potential molecular regulators of p16INK4A expression during embryonic development, it is interesting to note that p16INK4A and p14ARF/p19ARF displayed a differential developmental expression pattern indicating that not the genomic locus, but more specific regulatory elements of p16INK4A might be activated.

In contrast to the early reports of absent p16INK4A expression during mouse development [29,30], expression during rat brain development was described only slightly afterward. In agreement with our results, p16INK4A expression colocalized with p53 in the ventricular and subventricular zones at embryonic and early postnatal stages and p53 was mainly found in postmitotic cells of the cerebral cortex and hippocampus [125]. In the olfactory epithelium, p16INK4A and p21 were detectable around birth, with p16INK4A marking differentiating and p21 mature neurons [126]. p16INK4A expression was also detected in bone marrow derived hematopoietic progenitor cells of adults [127] and in epiphyseal growth plate chondrocytes and bone lining osteoblasts in growing mice [128]. In these cases, higher p16INK4A expression was associated with reduced cell proliferation, but senescence of these cells had not been reported. Increasing p16INK4A and p21 expression has been also observed in male germ cells coinciding with the timing of mitotic arrest, but not with senescence [129]. These male germ cells enter meiosis during post-natal life [130]. Increased p16INK4A expression was noted already in the endometrium between days 2 to 5 of pregnancy in mice. Injection of a p16INK4A antibody decreased the number of implanted blastocysts compared with that of a saline-injected group suggesting a role of p16INK4A in blastocyst implantation [131]. This observation seems to be in contrast to normal Mendelian frequencies of birth in p16INK4A knockout mice [132], but slight deviations from Mendelian inheritance might become obvious only when analyzing large numbers of pups [115] and implantation defects would be only detectable if the female mice in mating are p16INK4A knockout instead of heterozygotes. p16INK4A expression was also described in human endometrium during pregnancy [133].

During mouse embryonic development, p16INK4A was further detected in the gut in intestinal stem cells and progenitor compartments. Loss of Bmi1 resulted in accumulation of p16INK4A and p14ARF/p19ARF and reduced intestinal stem cell proliferation, which was accompanied by increased differentiation to the post-mitotic goblet cell lineage. Bmi1 expression in intestinal stem cells was co-regulated by Notch and beta-catenin [134]. Bmi1 plays also important roles for maintenance of neural stem cell self-renewal [135–139], for mesenchymal stem cell renewal and bone formation [140], for immature retinal progenitor/stem cells and retinal development [141], and for hepatic stem cell expansion [142] via negative regulation of p16INK4A, p14ARF/p19ARF, and p21.

Already in three-month-old mice, a significant number of p16INK4A-expressing cardiomyocytes, mostly bi- and multinucleated cells, had been described [143]. We investigated expression much earlier during embryonic development and found increased expression coinciding with cardiomyocyte differentiation [38]. As isolated cardiomyocytes were immunostained in the previously mentioned report, potential expression in endothelial cells at this time point was not detected. The authors considered p16INK4A expression as a marker of senescence without further approaches to identify the cells as senescent [143]. Another study investigated the proliferation of cardiomyocytes by PCNA staining ex vivo in p16INK4A/p14ARF/p19ARF knockout mice. Surprisingly, the authors report 70% of proliferating cardiomyocytes from 8 weeks old mice [144], which is in obvious contrast to all data in the literature.

Specific p16INK4A knockout mice which retained p14ARF/p19ARF function displayed an increased incidence of spontaneous and carcinogen-induced cancers [132] and melanomas [145] and thymus hyperplasia [132]. Thymus hyperplasia was associated with increased numbers of CD4 and CD8 lymphocytes, which was surprisingly not due to increased proliferation, but to reduced apoptosis of lymphocytes [146]. Mice lacking p16INK4A and p14ARF/p19ARF on an FVBN genetic background develop cataracts and micro-ophthalmia. They showed beginning from E15.5 defects in the developmental regression of the hyaloid vascular system, retinal dysplasia, abnormal differentiation of the lens, and cataracts [147]. Interestingly, the micro-ophthalmia phenotype in Task1 knockout mice was partially rescued by the lack of p16INK4A and p14ARF/p19ARF suggesting a fine-tuning of neuronal and eye development by the two proteins [84].

In addition, p14ARF/p19ARF knockout mice are prone to spontaneous and carcinogen-induced cancers [148]. p14ARF/p19ARF is involved in perivascular cell accumulation postnatally in the mouse eye before eye development is completed [147,149–151]. p14ARF/p19ARF decreased Pdgfr beta expression and blocked Pdgf-B-driven proliferation independently of Mdm2 and p53, which prevents the accumulation of perivascular cells and allows regression of the hyaloid vascular system of the developing eye [152,153]. Tgfbeta2 is required for p14ARF/p19ARF transcription in the hyaloid vascular system as well as in the cornea and the umbilical arteries [154,155].

p14ARF/p19ARF is detectable in developing hepatoblasts [156], which agrees with our recent report. Lack of the Tbx3 member of the T-box family of transcription factors results in upregulation of p14ARF/p19ARF and p21 in the developing liver, which is associated with severe defects in proliferation and in hepatobiliary lineage segregation, including the promotion of cholangiocyte differentiation and abnormal liver development [156]. Whether Tbx3 might directly regulate p14ARF/p19ARF and p21 expression was not determined in this study.

p21 knockout mice were reported initially to develop normally despite defective G1 checkpoint control in isolated knockout embryonic fibroblasts [157]. Interestingly, p21 expression was detected by Western Blot in human fetal atrial tissue, but not in adult hearts [158]. p21 was also found in developing rat ventricular myocytes [159]. In both studies, no comparison with old ages was made. Some p21-expressing cardiomyocytes were detected in E15.5 developing mouse embryos [160] and in trabecular myocardium at E18.5 [161]. The number was largely increased in Foxm1 knockout embryos as well as in Tbx20 overexpressing hearts at the early stages of development, which correlated with reduced proliferation and cardiac hypoplasia [160,162,163]. Fog-2 was described as a direct transcriptional repressor of p21 in cardiac development. Fog-2 mutant embryos showed multiple cardiac malformations, upregulation of p21, and thin-walled myocardium [164]. p21 expression had also been described in developing skeletal muscle, bones, lung, and spinal cord [165–169]. p21 has been also implicated in the mitotic arrest in male mouse germ cells during embryonic development [170]. An elegant study analyzing p21 and P57 double-mutant mice showed that both proteins redundantly control differentiation of skeletal muscle, bones, and alveoli in the lungs. Mice lacking both p21 and p57 failed to form myotubes, and displayed enhanced proliferation and apoptotic rates of myoblasts clearly indicating a role of p21 and P57 in normal muscle development [171]. Skeletal defects were more pronounced in embryos lacking p21 [171]. This report clearly shows the redundancy of the different proteins in cell cycle control and might explain the only few phenotypes observed in single knockout animals despite the importance of the cell cycle regulators for embryonic development.

Besides these studies implicating mostly p21 in embryonic development and differentiation, several reports also pointed to senescence as a potential mechanism for normal embryonic development. Munoz–Espin and colleagues performed whole-mount senescence-associated β-galactosidase SaβG) staining in mouse embryos. They detected SaβG activity in endolymphatic sacs of the developing ear, in the closing neural tube, the apical ectodermal ridge (AER) of the limbs, and later in regressing interdigital webs, around the vibrissae, and in the mesonephros of dissected gonad-mesonephros complexes [13]. However, in the dissected gonad-mesonephros picture of the manuscript, some SAβG staining also seems to be visible in the gonad and the Wolffian/Muellerian duct system. In further analyses, the authors focused on the endolymphatic sac and the mesonephros. SAβG activity in regressing mesonephros had been reported already earlier in chicken embryos [172]. SAβG activity was also detected in mesonephros and endolymphatic sacs of human embryos around 9 weeks of development [13]. As a marker of proliferation, they used Ki67 staining and found low proliferation in cells with SAβG activity. Nevertheless, during several developmental time points, some Ki67-positive cells were still detectable in SAbG-positive structures. As a major actor in developmental senescence, the authors suggested p21 based on immunostainings for several markers of senescence in endolymphatic sacs and mesonephros. Interestingly, the authors detected high p16INK4A expression in the gonad, which was not further commented upon. SAβG-positive cells were surrounded by macrophages and disappeared during ongoing development while the macrophage infiltration persisted longer. The attraction of macrophages was attributed to the SASP of SAβG-positive cells, which resulted in the now widely accepted concept that senescent cells secrete a cocktail of molecules, which beside other effects attract macrophages that finally clear the senescent cells [13,173–176]. A subset of p16INK4A expressing macrophages, which are SAβG-positive and might mediate this effect was identified recently in mouse tissues [177]. However, as Munz–Espin and colleagues immunostained the embryos also for p16INK4A, the macrophages in their model might represent a distinct population. Also, in tumor-bearing mice, doxorubicin induced senescence and a SASP in the skin, independent of macrophages and neutrophils [178], suggesting a certain variability in the events of senescent cell clearance. Finally, Munz–Espin investigated potential developmental defects in p21-deficient embryos. p21 knockout embryos had abnormal endolymphatic sacs with infoldings at late stages of development (E18.5), which disappeared after birth most likely due to macrophage clearance. Also in this case, the developmental program to remove the abnormal cells was independent of SAβG-positive cells or p21. In the uterus, which partially derives from the regressing Wolffian duct, the authors observed frequent septation and consequently a lower number of pups in p21 knockout mice [13], a phenotype, which had been missed in the first global analyses of these animals. Storer et al. used in parallel a similar approach and detected SAβG-positive cells in the AER, otic vesicle, the eye, branchial arches, gut endoderm, neural tube, tail, gall bladder, and interdigital tissue [179]. Similarly, in this report, p16INK4A and p14ARF/p19ARF seemed not to be involved in embryonic senescence, but p21 knockout embryos displayed less SAβG-positive cells. Instead of becoming senescent, cells underwent apoptotic cell death and were cleared by macrophages. Interestingly, the mesenchyme directly below the AER showed reduced proliferation indicating that developmental senescence is directly linked to cell proliferation and patterning of neighboring structures [179]. As additional sites of SAβG-staining, the developing bones, placental trophoblast cells [180], and the visceral endoderm [181] were identified. In the case of the visceral endoderm, SAβG-staining was not indicative of senescence [181]. Senescent cells have been described in a variety of developing organisms including birds, zebrafish, axolotl, naked mole rats, xenopus, mouse, and humans [13,172,179,182–188], mostly on the basis of SAbG-staining. More recently, the utility of SAβG-staining for the detection of developmental senescence has been questioned as also apoptotic cells, e.g., in the interdigital tissue and postmitotic neurons are stained independent of senescence [189–191]. Additionally, SAβG and p16INK4A expression have been shown to be induced in macrophages in response to physiological stimuli, without the cells being senescent [192]. Furthermore, we described recently p16INK4A expression at different developmental time points and several organs, which did not correspond to the known sites of SAβG expression. Only in old animals, but not during development, was a significant correlation between p16INK4A expression and SASP factors detectable. Of interest is also the detection of senescence cells and significant SASP activation in the development and response to stress in naked mole rats, which are considered a model of cancer-free longevity [186]. Reported sites of SAβG-staining, p16INK4A, p19p14ARF/p19ARF, and p21 expression during development are briefly summarized in Table 1 and illustrated in Figure 1.

Table 1. Detection of senescence markers during development.

Localization	Detected Signal	Species	Ref.
Heart, kidney, brain, liver	p16INK4A, p14ARF/p19ARF, p21 mRNA, p16INK4A protein	mouse	[38]
Brain	p16INK4A mRNA	rat	[125]
Olfactory epithelium	p16INK4A, p14ARF/p19ARF, p21 protein	mouse	[126]
Hematopoietic stem cells	p16INK4A, p14ARF/p19ARF mRNA	mouse	[127]
Chondrocytes, osteoblasts	p16INK4A, p21 protein	mouse	[128]
Male germ cells	p16INK4A, p21 mRNA	mouse	[129,170]
Endometrium in early pregnancy	p16INK4A mRNA, p16INK4A protein	mouse	[131]
Endometrium in pregnancy	p16INK4A protein	human	[133]
Syncytiotrophoblast	p16INK4A, p21 protein	human	[182]
Intestinal stem cells	p16INK4A protein	mouse	[134]
Perivascular ocular cells	p14ARF/p19ARF protein	mouse	[147,149–152]
Hepatoblasts	p14ARF/p19ARF, p21 protein	mouse	[156]
Heart	p21 protein	human, rat, mouse	[158–161]
Muscle, cartilage, skin, nasal epithelium, hair follicles	p21 mRNA, p21 protein	mouse	[165–167,171]
Mesonephros	SAβG	bird	[172]
Endolymphatic sacs, mesonephros	SAβG	mouse, human	[13]
Neural tube, AER, vibrissae	SAβG	mouse	[193]
AER, otic vesicle, eye, branchial arches, gut endoderm, neural tube, tail, gall bladder, and interdigital tissue	SAβG	mouse	[179]
Bones, placenta trophoblast cells	SAβG	mouse	[180]
Visceral endoderm	SAβG	mouse	[181]
Inner ear	SAβG	birds	[183]
Pronephros, cement gland, oral cavity, olfactory epithelium, lateral organs, gums	SAβG	axolotl	[184,185]
Yolk sac epithelium, lower part of the gut	SAβG	zebrafish	[185]
Nail bed, dermis, hair follicle, nasopharyngeal cavity	SAβG	Naked mole rat	[186]

Abbreviations: Ref.—Reference, AER—apical ectodermal ridge, SaβG—senescence-associated beta galactosidase

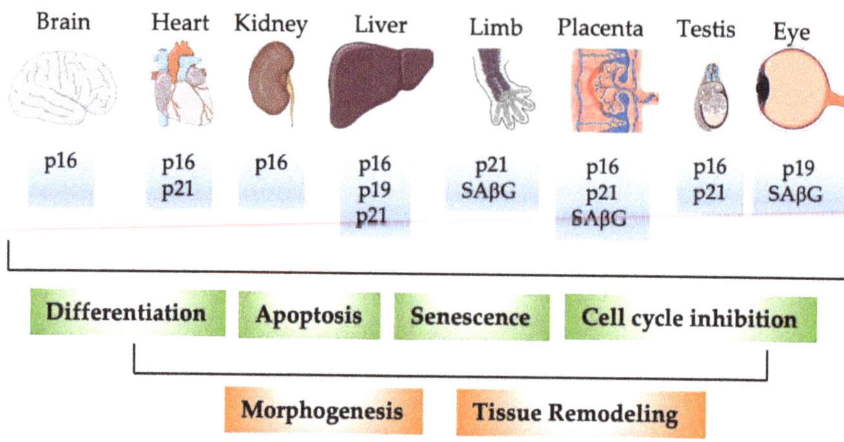

Figure 1. Schematic illustration of detection of p16INK4A, p14ARF/p19ARF, p21, and SAβG in selected murine organs during development. P16: p16INK4A; p19: p14ARF/p19ARF. During development, p16INK4A, p14ARF/p19ARF, p21, and SAβG not only mark senescent cells. p16INK4A, p14ARF/p19ARF, and p21 proteins are expressed in distinct cell types during different developmental stages. Expression of the individual proteins correlates with lower expression of proliferation markers in agreement with their classical function as cell cycle inhibitors, with apoptosis, and cellular differentiation. These fundamental processes contribute dynamically to tissue remodeling and morphogenesis during embryonic development.

4. p16INK4A, p14ARF/p19ARF, and p21 in Homeostasis

The implications of p16INK4A, p14ARF/p19ARF, and p21 in senescence and aging are extensively investigated and reviewed elsewhere [4–6,11,56,194–199]. Senescence has long been considered an important mechanism to prevent tumorigenesis, thus acting as a guardian of homeostasis, which agrees with p16INK4A, p14ARF/p19ARF, and p21 knock-out mouse models. However, more recent data allow to draw a more differentiated picture of senescence and the SASP in tumor initiation and progression (reviewed in [200–204]). Organ and tissue homeostasis, however, do not only play a role in cancer prevention, but represent the central organizing principle of physiology and pathophysiology [205]. Major homeostatic and pathophysiological processes involving p16INK4A, p14ARF/p19ARF, and p21 are summarized in Table 2 and described below.

Table 2. Major phenotypes associated with p16INK4A, p14ARF/p19ARF, or p21 modifications in homeostasis and pathophysiology.

Pathophysiology/Homeostatic Mechanism	Intervention/Model	Outcome	Ref.
Physiology			
Age-related cardiomyocyte hypertrophy	INK-ATTAC mouse	Cardiac cell size↓	[73]
Age-related lipodystrophy	INK-ATTAC mouse	Adipose tissue mass ↑	[73]
Health-span	INK-ATTAC mouse	Survival ↑	[73]
Health-span	p16INK4ACre; DTA	Survival ↓	[39]
Age-related bone loss	p16INK4A-3MR mouse	=	[206]

Table 2. *Cont.*

Pathophysiology/Homeostatic Mechanism	Intervention/Model	Outcome	Ref.
Aging-related intervertebral disc degeneration	p16INK4A-3MR mouse	Histological disc morphology improved	[23]
Aging features	p16INK4A overexpression	Accelerated	[207]
Adipocyte formation	p16INK4A−/−	Adipogenesis ↑	[208]
Longevity	p16INK4A−/−, p14ARF/p19ARF−/−, P53−/−	Lifespan ↑	[209]
Longevity Male fertility	p16INK4A/p14ARF/p19ARF overexpression	Lifespan ↑ Absence of sperm	[210]
Lifespan	INK-ATTAC mouse, BubR1$^{H/H}$ background	=	[72]
Physical fitness	INK-ATTAC mouse BubR1$^{H/H}$ background	Endurance ↑	[72]
Aging-associated liver fibrosis	p16INK4ACre; DTA	Fibrosis ↑	[39]
Aging-associated hepatic steatosis	INK-ATACC mouse	Fat accumulation ↓	[17]
Wound healing	p16INK4A-3MR mouse	Wound closure ↓	[41]
Wound healing	CCN1-dependent p16INK4A induction	Fibrosis ↓	[211]
Aging-associated glomerulosclerosis	INK-ATTAC mouse	Glomerulosclerosis ↓	[73]
Aging-related physical activity loss	p21Cre;DTA	Physical fitness ↑	[212]
Sarcopenia	INK-ATTAC mouse, BubR1$^{H/H}$ background	Sarcopenia delayed	[72]
Glaucoma	INK-ATTAC mouse, BubR1$^{H/H}$ background	Glaucoma onset delayed	[72]
Pathophysiology			
Myocardial infarction	INK-ATTAC mouse, senolytics	Cardiomyocyte proliferation ↑	[78]
Myocardial infarction	p16INK4A overexpression	Cardiac function ↑ Fibrosis ↓	[213]
Myocardial infarction	p16INK4A−/−, p14ARF/p19ARF−/−	Cardiomyocyte proliferation ↑ Cardiac function ↑	[144]
Obesity	INK-ATACC mouse	Insulin sensitivity ↑ Metabolic dysfunction ↓	[214]

Table 2. *Cont.*

Pathophysiology/Homeostatic Mechanism	Intervention/Model	Outcome	Ref.
Adipocyte conversion	p16INK4A−/−	White to brown ↑	[215]
Diabetes	p16INK4A overexpression	Insulin secretion ↑	[216]
Glucose homeostasis	Human p16INK4A inactivating mutations	Insulin secretion ↑ Insulin sensitivity ↓	[217]
Glucose homeostasis Insulin sensitivity in obese mice	p21Cre;DTA	GTT ↑ ITT ↑	[218]
Diabetes	p16INK4A overexpression	Insulin sensitivity ↑ Metabolic dysfunction ↓	[219]
Pancreatic beta cell regeneration	p16INK4A overexpression	Islet proliferation ↓	[220]
Pancreatic beta cell regeneration	p16INK4A−/−	Islet proliferation ↑	[220]
Liver fibrosis	p53−/−; p16INK4A/p14ARF/ p19ARF−/−	Fibrosis ↑	[221]
Ionizing radiation-induced reduction of neurogenesis	p16INK4A−/−	partial restoration	[222]
Radiation-induced impairment of cognitive function	p16INK4A-3MR mouse	Cognitive function ↑	[223]
Cisplatin-induced peripheral neuropathy	p16INK4A-3MR mouse, senolytics	Neuropathy ↓	[224]
Post-traumatic osteoarthritis	p16INK4A-3MR mouse	Osteoarthritis ↓	[225]
Radiation-induced osteoporosis	p21INK-ATTAC mouse	Osteoporosis ↓	[226]
Macrophage polarization	p16INK4A−/−	Anti-inflammatory phenotype ↑	[227]
Macrophage polarization	Human p16INK4A silencing	Anti-inflammatory phenotype ↑	[228]
Irradiation-induced immune dysfunction	p16INK4A-3MR mouse	T-cell proliferation ↑ Macrophage phagocytosis ↑	[229,230]

↑: Higher, ↓: Lower, =: not significantly different, −/−: knockout mouse model, BubR1$^{H/H}$: mouse model of accelerated aging with multiple age-related pathologies, INK-ATTAC mouse: allows deletion of p16INK4A expressing cells, p16INK4ACre;DTA: mouse model allows deletion of p16INK4A expressing cells, p16INK4A-3MR mouse: allows deletion of p16INK4A expressing cells, GTT: glucose tolerance test, ITT: insulin tolerance test

4.1. *p16INK4A*

Maintenance of cardiac function during aging and cardiac remodeling had to some extent been attributed to the expansion and differentiation of cardiac-resident stem cells (reviewed in [197]). To which extent cardiac stem and progenitor cells contribute to myocytes, endothelium, smooth muscle cells, etc., in cardiac repair is still a matter of debate [111,117,197,231–234]. In contrast to earlier publications, it is now widely accepted that cardiac, but not hematopoietic-derived progenitor cells are implicated in the cardiac repair [235]. With increasing age, the fraction of p16INK4A-expressing cardiac stem cells and expression of SASP factors increased in human biopsies [78]. A fraction of SAβG-negative cardiac stem cells improved cardiac function after experimental myocardial infarction in immunosuppressed mice while the fraction of SAβG-positive cells did not [78]. Notably, injection of the SAβG-positive cells did not worsen cardiac function after experimental myocardial infarction, which contrasts with the title of the manuscript [78]. The combination of the senolytic drugs dasatinib and quercetin as well as the elimination of p16INK4A-positive cells in the INK-ATTAC mouse model improved some cardiac parameters [78]. Unfortunately, neither the number of p16INK4A-positive cells nor cardiac function was determined in this set of experiments. As the values in INK-ATTAC mice and dasatinib and quercetin-treated animals differed for most parameters, it is possible that the cocktail of senolytic drugs has additional effects besides the elimination of p16INK4A-expressing cells. Of note, the original paper describing the generation and characterization of INK-ATTAC mice [72] reported a lack of INK-ATTAC induction in the heart, liver, and aorta, making it likely that the observed beneficial effects are due to secondary paracrine (SASP) induced events. In this original mouse model, time course studies showed that the elimination of p16INK4A expressing cells reflects the attenuated progression of age-related declines rather than a reversal of aging [72]. This seems to be somehow in contrast to the study mentioned before [78]. Most of the original investigations were done in the BubR1$^{H/H}$ progeroid mouse genetic background, which might be slightly different from aged mice. In a following manuscript, the same group detected increasing p16INK4A expression in aged mice in all organs, but induction of the transgene with AP20187 did not affect the colon or liver expression of senescence markers [73]. However, heart and kidney morphological and expression parameters were to some extent normalized in aging INK-ATTAC mice treated with AP20187 and healthy lifespan extended. The shortest survival was measured in C57 wild-type mice treated with AP20187 [73]. In the heart, cardiomyocyte diameters were reduced in aging INK-ATTAC mice treated with AP20187, while the left ventricular wall thickness as an alternative measure of hypertrophy was unaffected suggesting that the clearance in INK-ATTAC mice is partial and tissue-selective [73]. This transgenic mouse model under the control of a 2.6 kB p16INK4A-promoter fragment might not completely reflect endogenous p16INK4A expression and regulation as we detected p16INK4A expression in the heart and liver [38,39] and elimination of p16INK4A expressing cells in the p16INK4ACre;DTA model caused cardiac and liver fibrosis and reduced health span [39], which is in agreement with the notion that senescent cells contribute to tissue repair and maintenance [211,221].

Elevated expression of endogenous p16INK4A has been recently demonstrated in a myocardial infarction (MI) model in mice. Forced overexpression of p16INK4A improved cardiac function while silencing of p16INK4A deteriorated cardiac function. As a possible underlying mechanism, reduced fibroblast proliferation, and collagen accumulation and less cardiac fibrosis was attributed to the classical cell-cycle inhibitory function of p16INK4A [213]. Increased cardiomyocyte proliferation and better functional recovery after MI has been reported in p16INK4A knockout mice [144]. This discrepancy remains currently unresolved.

Genome-wide association studies have implicated the human p16INK4AInk4a/Arf locus in the risk for cardiovascular and metabolic diseases and type 2 diabetes mellitus [236–238]. Deletion of a homologous region in mice caused reduced expression of p16INK4A and Cdkn2b, increased tumor incidence, and increased body weights and mortality in the animals [88]. Knockdown of *p16INK4A* enhanced adipogenesis in vitro, and adipose tissue formation especially in the pericardial fat was enhanced in *p16INK4A* knockout mice [208]. The role of p16INK4A in adipogenesis seems to be related via several molecular mechanisms to PPAR gamma (reviewed in [214]). Senolytic drug treatment or the use of INK-ATACC mice has been shown to alleviate metabolic and adipose tissue dysfunction, improve glucose tolerance, enhance insulin sensitivity, lower circulating inflammatory mediators, and promote adipogenesis in obese mice [239]. p16INK4A regulates adipogenesis and adipose tissue insulin sensitivity mainly via CDK4 [208,240,241]. Part of the action of p16INK4A in adipose tissue is related to obesity-induced inflammation and immune cell polarization [228,242]. Bone marrow-derived macrophages from *p16INK4A* knockout mice show polarization towards an anti-inflammatory M2 phenotype and silencing of p16INK4A in macrophages from obese patients equally shifts the phenotype towards M2 macrophages [227,228]. These effects seem to be independent of proliferation and senescence [214], although earlier data indicated a critical role of the p16INK4A locus in proliferation and programming of progenitor cell populations [243]. Besides the effects of p16INK4A on macrophage polarization in adipose tissue, also increased white-to-brown adipocyte conversion associated with enhanced energy expenditure and insulin sensitivity has been reported in *p16INK4A* knockout mice [215]. Whether this is due to enhanced direct conversion from white to brown adipocytes or it results from enhanced differentiation of progenitor cells remains an open question.

In contrast to the results described above for the INK-ATACC model, which eliminates p16INK4A expressing cells, a transgenic "Super-Ink4/Arf" mouse model with slightly increased p16INK4A RNA expression in the liver has been described [219]. Despite one extra copy of p16INK4A, the animals showed no significant increase in p16INK4A protein expression in the liver, heart, muscle, or pancreatic islets. Nevertheless, they did not develop glucose intolerance with age and showed a higher insulin sensitivity. The authors argued that the small increases in p16INK4A are causing this protective effect against the development of age-related diabetes mellitus [219]. Increasing p16INK4A expression with age in pancreatic islets has been described. Forced overexpression of p16INK4A reduced islet proliferation, while old mice lacking p16INK4A in pancreatic islets demonstrated enhanced islet proliferation and survival after beta-cell ablation, which agrees with the "classical" antiproliferative effect of p16INK4A [220]. Several additional publications implicated p16INK4A in insulin secretion and beta-cell proliferation [79,216,244,245]. In addition, p16INK4A deficiency enhances fasting-induced hepatic glucose production via activation of PKA-CREB-PGC1α signaling [246]. Accumulation of senescent cells during aging promotes hepatic fat accumulation and steatosis via reduced capabilities of mitochondria to metabolize fatty acids. Elimination of senescent cells in INK-ATTAC mice or by treatment with a combination of the senolytic drugs dasatinib and quercetin reduces hepatic steatosis [17], while specific elimination of p16INK4A-expressing liver sinusoidal endothelial cells induces hepatic fibrosis and premature death [39]. In humans with loss-of-function mutations in CDKN2A encoding p16INK4A and p14ARF, carriers showed increased insulin secretion, impaired insulin sensitivity, and reduced hepatic insulin clearance. There were no significant differences between patients with mutations affecting both p16INK4A and p14ARF and subjects with mutations affecting p16INK4A only suggesting that these effects are indeed due to the p16INK4A loss of function [217]. Taken together, the different reports from mice and humans suggest that p16INK4A acts at multiple levels of glucose homeostasis and metabolism especially in older individuals. Potential developments of therapeutic strategies for type 2 diabetes mellitus by modifying p16INK4A should be considered with care given the potential cancer risk.

Besides the described implications of p16INK4A in the cardiovascular system, adipose tissue, and metabolism, several publications also identified potential functions in the circadian clock [247], neurogenesis, neuronal trans-differentiation, and axon regeneration [222,248–250], most of them in agreement with cell cycle control by p16INK4A.

In an elegant study, Demaria and colleagues identified senescence as a potential adaptative mechanism for tissue repair. They generated a bacterial artificial chromosome (BAC)-transgenic mouse line containing 50 kb of the genomic region of the p16INK4A locus, a luciferase and red fluorescent protein (RFP) reporter, and a truncated herpes simplex virus 1 (HSV-1) thymidine kinase (HSV-TK) cassette allowing the elimination of cells with activated p16INK4A locus upon treatment with ganciclovir [41]. RFP-positive cells showed increased SAβG staining and increased levels of mRNAs encoding p16INK4A, p21, and the SASP factors IL-6, MMP-3, and VEGF, but not IL-5, suggesting that the RFP-marked cells are indeed senescent. The elimination of these cells caused delayed cutaneous wound healing. A similar phenomenon was also observed in p16INK4A/p21 double knockout mice, which do not show senescence [251] but not in single p16INK4A or p21 knockout animals, which are able to compensate the lack of one protein by the other in terms of senescence [41,251]. As major p16INK4A-positive cell types in the cutaneous injury model, endothelial cells and fibroblasts were identified [41], which agrees with our recent observations [38,39]. Senescent endothelial cells and fibroblasts appear early after injury and accelerate wound closure by inducing myofibroblast differentiation through the secretion of platelet-derived growth factor AA [41]. Using the same mouse model, several reports indicated that the removal of p16INK4A-expressing cells attenuated post-traumatic osteoarthritis [225], had no effect on age-related bone loss [206], prevented age-related intervertebral disc degeneration [23], improved irradiation-induced immune cell functional decline [229,230], protected cognitive function [223], and alleviated cisplatin-induced peripheral neuropathy in mice [224]. Senescent cells might also contribute to the release of hemostasis-related factors, which in excess might contribute to thromboembolic events in the old [252]. Most recently, the mouse model was used to study cellular senescence in cigarette smoke-induced lung injuries in adult and old mice [253]. Cigarette smoke induced senescence, p16INK4A, and p21 expression in adult animals, though surprisingly the opposite was observed in old animals [253].

In line with the role of p16INK4A in cardiovascular progenitor cells mentioned above, a potential function was postulated in skin stem and progenitor cells [254] and a higher colony-forming ability and replating efficiency measured in bone marrow-derived progenitor cells from p16INK4A knockout mice [255], which has been reviewed elsewhere [105,256,257]. In aged p16INK4A knockout mice, superior repopulating ability in bone marrow transplantation experiments compared with wild-type animals was noted, while only tiny differences were detectable under baseline conditions [258]. In mice with tetracycline-inducible overexpression of a human p16INK4A transgene, proliferation of intestinal stem cells was diminished, and animals showed signs of accelerated aging, which were mostly reversible upon withdrawal of tetracycline [207]. In this model, p16INK4A overexpression was not associated with senescence as evidenced by lack of SAβG staining. In contrast to these mouse models, to the best of our knowledge, neither major skin nor hematopoietic nor intestinal stem cell abnormalities were reported in patients with p16INK4A mutations.

4.2. *p14ARF/p19ARF*

Although p16INK4A and p14ARF/p19ARFArf are transcribed from the same locus, the proteins have some overlapping as well as distinct functions. Mice with an extra copy of Ink4/Arf or the downstream effector P53 showed resistance against cancer, which is in line with the general cell cycle and tumor suppressor function [259–262]. Intercrosses of both mouse lines showed additional resistance to cancer and extended longevity [209]. It is likely that the extended longevity in this model is at least in part due to the preservation of the stem cell pool in different organs [209,263–266]. Extra copies of Ink4/Arf in homozygous mice induced delayed aging, reduced the cancer incidence, improved longevity, diminished kidney lesions, and DNA damage, but also caused male infertility [210]. Different mouse models with activated P53 signaling also showed resistance to cancer development, but decreased the lifespan and premature onset of age-related diseases such as osteoporosis and tissue atrophy [267,268]. In line with this, these mouse models present reduced hematopoietic, mammary gland, neuronal, and pancreatic stem and progenitor cells with impaired hematopoiesis, mammary atrophy, decreased olfaction, and disturbed glucose homeostasis [269–272]. Whether the discrepancies in the longevity of the various mouse models are due to different levels of activation of the Arf-P53 pathway remains elusive. Taken together, the p14ARF/p19ARFArf-p53 pathway seems to be mostly responsible to maintain the stem cell pool and promote homeostasis, while data mostly from the transgenic p16INK4A-INK-ATTAC and p16INK4A-3MR [41,239] mouse models suggest that elimination of p16INK4A might be beneficial for homeostasis and healthy aging although this view was challenged recently [39].

4.3. p21

Recently, two mouse models were established to specifically address the role of p21 in senescence and tissue homeostasis. The first consists of an inducible p21-Cre model (CreERT2), which allows after crossing with different floxed mice monitoring or elimination of p21 expressing cells [212]. The second mouse strain is comparable to the p16INK4A-INK-ATTAC mouse model but uses a 3.2 kb p21 promoter fragment driving expression of the FKBP–Caspase-8 fusion suicide protein. The construct was inserted in the Rosa26 locus [226]. The p21-CreERT2 mice were crossed with a luciferase reporter, and luminescence was measured in vivo after doxorubicin treatment or a high-fat diet as known inducers of senescence. Next, p21-CreERT2 animals were crossed with floxed knock-in tdTomato mice confirming the expected increase in fluorescent cells in several organs in old mice. Finally, the p21-CreERT2 line was crossed with a DTA ablator line, and physical fitness was measured in old mice treated with Tamoxifen and controls. The elimination of p21-positive senescent cells increased walking speed, grip strength, hanging endurance, daily food intake, and daily activity indicating a rejuvenation phenotype in response to the elimination of p21-expressing cells [212]. Surprisingly, p16INK4A- and p21-expressing cell populations seem to be different [212], which is contrasting with the lack of senescence in p16INK4A/p21 double knockout animals [251]. Also in the p21–ATTAC model, the clearance of p21- but not p16INK4A-positive senescent cells prevented radiation-induced osteoporosis and bone marrow adiposity [226], supporting the view that p16INK4A- and p21-dependent senescence comprise different and independent pathways [3,5,22,273]. A high number of p21- but not p16INK4A-expressing cells was detected in visceral adipose tissue of obese mice, mostly preadipocytes, endothelial cells, and macrophages [218]. In contrast to visceral adipose tissue, the heart, kidney, liver, and brain of old mice express high levels of p16INK4A in endothelial cells [38,39]. Elimination of p21-expressing cells using the p21-CreERT2 line crossed with the DTA ablator line worked in preadipocytes, macrophages, and leukocytes, but not in the endothelial compartment. Functionally, it improved glucose homeostasis and insulin sensitivity in obese mice. Interestingly, the removal of p21-positive cells had less metabolic benefits in female than male mice [218] adding one more layer of complexity to potential translational approaches. Of note, the senolytic cocktail of dasatinib plus quercetin was able to remove p21-positive senescent adipocytes but not endothelial cells and macrophages [218]. Nevertheless, it improved glucose homeostasis and insulin sensitivity and reduced pro-inflammatory SASP secretion [218]. Although this elegant study clearly supports the idea of senolytic drugs as a therapeutic strategy for obesity-induced metabolic dysfunction, it also raises new questions about the mode of action of the senolytic drug cocktail, which seems to target one specific senescent cell type.

A recent elegant study showed that in response to cellular stress, p21 and p16INK4A are upregulated. Both induce cell cycle arrest and SASP expression, but the SASPs are different [274]. The p21-induced secretome is characterized by the release of additional immunosurveillance factors, in particular Cxcl14, which are lacking in the p16INK4A-induced SASP. Consequently, the p21-induced secretome attracts macrophages. At later stages, the macrophages polarize into a M1 phenotype, and the p21-exressing cells are cleared via T cells. Most importantly, the authors showed that the p21-induced SASP places the cells under immunosurveillance and establishes a timer mechanism for the cell fate. In the case of p21, the expression normalizes within 4 days in mice, macrophages withdraw, and the cells are not cleared. Thus, the specific p21-induced SASP sets the time frame for the switch between surveillance and cell clearance mode of the immune system [274]. This mechanism might contribute to the developmental decisions described above, where individual cells are mostly characterized by transient expression of p21.

5. Open Questions and Perspectives

The establishment of several p16INK4A- and p21-deleter mouse lines mentioned above contributes largely to our understanding of senescence and aging phenotypes. As both proteins are expressed in different cell types and ablation has diverse effects, senescence is not one biological entity, but comprises different cellular events and divergent SASPs. The picture might be even more complex considering that in a given cell type aging is heterogenous [275] and tissues are in different stages of senescence [276,277]. The observation of beneficial effects in organs where the transgene is not expressed in p16INK4A-INK-ATTAC mice suggests a major role of SASP normalization instead of direct elimination of senescent cells. This is further supported by the recent p21-Cre line data [218] and the fact that the SASP from a small number of cells is sufficient to induce senescence in young mice and senolytic drugs induced a rejuvenation phenotype [278]. The next complicating issue is that the SASP is also not a homogenous cocktail of released factors but might highly differ in the composition of immunomodulatory factors and thus determine more physiological aging versus pro-inflammatory deteriorating phenotype (reviewed in [3,279]). Interestingly, different p16INK4A-positive cell elimination mouse models showed diverse phenotypes with the p16INK4A-INK-ATTAC model delaying aging phenotypes and increasing lifespan [72], while in the p16INK4A-3MR model wound healing was disturbed [41], and in p16INK4ACre;DTA mice liver fibrosis and reduced health-span were observed [39]. Thus, it would be important to determine whether p16INK4A-expressing cells are the same in the three models under baseline conditions. For this purpose, our recently established and knockout-validated immunohistochemistry protocol could be a useful tool [38]. As p16INK4A expression is not an off–on phenomenon, but increases from embryonic stages until old age [38], in the next step it would be interesting to determine whether p16INK4A-expressing cells in the mouse models are eliminated at different levels of p16INK4A expression. If this is the case, sorting of the cells and secretome analysis could define the secretory phenotype of protective versus detrimental p16INK4A expressing cells which finally may serve as a rejuvenation approach in aged patients without the need and limitations of overexpression of reprogramming factors [279].

Author Contributions: Conceptualization, K.-D.W. and N.W.; formal analysis, K.-D.W. and N.W.; investigation, K.-D.W. and N.W.; writing—original draft preparation, K.-D.W. and N.W.; writing—review and editing, K.-D.W. and N.W.; supervision, K.-D.W. and N.W.; project administration, K.-D.W. and N.W.; funding acquisition, K.-D.W. and N.W. All authors have read and agreed to the published version of the manuscript.

Funding: This research was funded by Fondation pour la Recherche Medicale, grant number FRM DPC20170139474 (K.-D.W.), Fondation ARC pour la recherche sur le cancer", grant number n°PJA 20161204650 (N.W.), Gemluc (N.W.), Plan Cancer INSERM (K.-D.W.), and Agence Nationale de la Recherche, grant R19125AA "Senage" (K.-D.W.), Fondation ARC pour la recherche sur le cancer", grant number n°PJA 20161204650 (K.-D.W.).

Institutional Review Board Statement: Not applicable.

Informed Consent Statement: Not applicable.

Acknowledgments: Parts of the figure were drawn by using pictures from Servier Medical Art. Servier Medical Art by Servier is licensed under a Creative Commons Attribution 3.0 Unported License (https://creativecommons.org/licenses/by/3.0/).

Conflicts of Interest: The authors declare no conflict of interest.

References

1. Hayflick, L. The limited in vitro lifetime of human diploid cell strains. *Exp. Cell Res.* **1965**, *37*, 614–636. [CrossRef]
2. Hayflick, L.; Moorhead, P.S. The serial cultivation of human diploid cell strains. *Exp. Cell Res.* **1961**, *25*, 585–621. [CrossRef]
3. Tripathi, U.; Misra, A.; Tchkonia, T.; Kirkland, J.L. Impact of Senescent Cell Subtypes on Tissue Dysfunction and Repair: Importance and Research Questions. *Mech. Ageing Dev.* **2021**, *198*, 111548. [CrossRef] [PubMed]

4. Wiley, C.D.; Campisi, J. The metabolic roots of senescence: Mechanisms and opportunities for intervention. *Nat. Metab.* **2021**, *3*, 1290–1301. [CrossRef] [PubMed]
5. Gorgoulis, V.; Adams, P.D.; Alimonti, A.; Bennett, D.C.; Bischof, O.; Bishop, C.; Campisi, J.; Collado, M.; Evangelou, K.; Ferbeyre, G.; et al. Cellular Senescence: Defining a Path Forward. *Cell* **2019**, *179*, 813–827. [CrossRef]
6. Campisi, J.; Kapahi, P.; Lithgow, G.J.; Melov, S.; Newman, J.C.; Verdin, E. From discoveries in ageing research to therapeutics for healthy ageing. *Nature* **2019**, *571*, 183–192. [CrossRef]
7. Campisi, J. Aging, Cellular Senescence, and Cancer. *Annu. Rev. Physiol.* **2013**, *75*, 685–705. [CrossRef]
8. Coppé, J.P.; Desprez, P.Y.; Krtolica, A.; Campisi, J. The senescence-associated secretory phenotype: The dark side of tumor suppression. *Annu. Rev. Pathol.* **2010**, *5*, 99–118. [CrossRef]
9. Coppé, J.P.; Patil, C.K.; Rodier, F.; Sun, Y.; Muñoz, D.P.; Goldstein, J.; Nelson, P.S.; Desprez, P.-Y.; Campisi, J. Senescence-Associated Secretory Phenotypes Reveal Cell-Nonautonomous Functions of Oncogenic RAS and the p53 Tumor Suppressor. *PLoS Biol.* **2008**, *6*, e301. [CrossRef]
10. Campisi, J.; d'Adda di Fagagna, F. Cellular senescence: When bad things happen to good cells. *Nat. Rev. Mol. Cell Biol.* **2007**, *8*, 729–740. [CrossRef]
11. Tchkonia, T.; Kirkland, J.L. Aging, Cell Senescence, and Chronic Disease: Emerging Therapeutic Strategies. *JAMA* **2018**, *320*, 1319–1320. [CrossRef]
12. Fitzner, B.; Müller, S.; Walther, M.; Fischer, M.; Engelmann, R.; Müller-Hilke, B.; Pützer, B.M.; Kreutzer, M.; Nizze, H.; Jaster, R. Senescence determines the fate of activated rat pancreatic stellate cells. *J. Cell Mol. Med.* **2012**, *16*, 2620–2630. [CrossRef]
13. Muñoz-Espín, D.; Cañamero, M.; Maraver, A.; Gómez-López, G.; Contreras, J.; Murillo-Cuesta, S.; Rodríguez-Baeza, A.; Varela-Nieto, I.; Ruberte, J.; Collado, M.; et al. Programmed cell senescence during mammalian embryonic development. *Cell* **2013**, *155*, 1104–1118. [CrossRef]
14. Schafer, M.J.; Zhang, X.; Kumar, A.; Atkinson, E.J.; Zhu, Y.; Jachim, S.; Mazula, D.L.; Brown, A.K.; Berning, M.; Aversa, Z.; et al. The senescence-associated secretome as an indicator of age and medical risk. *JCI Insight* **2020**, *5*, e133668. [CrossRef]
15. Von Zglinicki, T.; Wan, T.; Miwa, S. Senescence in Post-Mitotic Cells: A Driver of Aging? *Antioxid. Redox Signal.* **2021**, *34*, 308–323. [CrossRef]
16. Jurk, D.; Wang, C.; Miwa, S.; Maddick, M.; Korolchuk, V.; Tsolou, A.; Gonos, E.S.; Thrasivoulou, C.; Saffrey, M.J.; Cameron, K.; et al. Postmitotic neurons develop a p21-dependent senescence-like phenotype driven by a DNA damage response. *Aging Cell* **2012**, *11*, 996–1004. [CrossRef]
17. Ogrodnik, M.; Miwa, S.; Tchkonia, T.; Tiniakos, D.; Wilson, C.L.; Lahat, A.; Day, C.P.; Burt, A.; Palmer, A.; Anstee, Q.M.; et al. Cellular senescence drives age-dependent hepatic steatosis. *Nat. Commun.* **2017**, *8*, 15691. [CrossRef]
18. Anderson, R.; Lagnado, A.; Maggiorani, D.; Walaszczyk, A.; Dookun, E.; Chapman, J.; Birch, J.; Salmonowicz, H.; Ogrodnik, M.; Jurk, D.; et al. Length-independent telomere damage drives post-mitotic cardiomyocyte senescence. *EMBO J.* **2019**, *38*, e100492. [CrossRef]
19. Childs, B.G.; Baker, D.J.; Wijshake, T.; Conover, C.A.; Campisi, J.; van Deursen, J.M. Senescent intimal foam cells are deleterious at all stages of atherosclerosis. *Science* **2016**, *354*, 472–477. [CrossRef]
20. Covre, L.P.; De Maeyer, R.P.H.; Gomes, D.C.O.; Akbar, A.N. The role of senescent T cells in immunopathology. *Aging Cell* **2020**, *19*, e13272. [CrossRef]
21. Tallerico, R.; Garofalo, C.; Carbone, E. A New Biological Feature of Natural Killer Cells: The Recognition of Solid Tumor-Derived Cancer Stem Cells. *Front. Immunol.* **2016**, *7*, 179. [CrossRef]
22. Gasek, N.S.; Kuchel, G.A.; Kirkland, J.L.; Xu, M. Strategies for Targeting Senescent Cells in Human Disease. *Nat. Aging* **2021**, *1*, 870–879. [CrossRef]
23. Patil, P.; Dong, Q.; Wang, D.; Chang, J.; Wiley, C.; Demaria, M.; Lee, J.; Kang, J.; Niedernhofer, L.J.; Robbins, P.D.; et al. Systemic clearance of p16. *Aging Cell* **2019**, *18*, e12927. [CrossRef]
24. Wong, H.; Riabowol, K. Differential CDK-inhibitor gene expression in aging human diploid fibroblasts. *Exp. Gerontol.* **1996**, *31*, 311–325. [CrossRef]
25. Hara, E.; Smith, R.; Parry, D.; Tahara, H.; Stone, S.; Peters, G. Regulation of p16CDKN2 expression and its implications for cell immortalization and senescence. *Mol. Cell Biol.* **1996**, *16*, 859–867. [CrossRef] [PubMed]
26. Reznikoff, C.A.; Yeager, T.R.; Belair, C.D.; Savelieva, E.; Puthenveettil, J.A.; Stadler, W.M. Elevated p16 at senescence and loss of p16 at immortalization in human papillomavirus 16 E6, but not E7, transformed human uroepithelial cells. *Cancer Res.* **1996**, *56*, 2886–2890.
27. Loughran, O.; Malliri, A.; Owens, D.; Gallimore, P.H.; Stanley, M.A.; Ozanne, B.; Frame, M.C.; Parkinson, E.K. Association of CDKN2A/p16INK4A with human head and neck keratinocyte replicative senescence: Relationship of dysfunction to immortality and neoplasia. *Oncogene* **1996**, *13*, 561–568. [PubMed]
28. Alcorta, D.A.; Xiong, Y.; Phelps, D.; Hannon, G.; Beach, D.; Barrett, J.C. Involvement of the cyclin-dependent kinase inhibitor p16 (INK4a) in replicative senescence of normal human fibroblasts. *Proc. Natl. Acad. Sci. USA* **1996**, *93*, 13742–13747. [CrossRef] [PubMed]
29. Zindy, F.; Quelle, D.E.; Roussel, M.F.; Sherr, C.J. Expression of the p16INK4a tumor suppressor versus other INK4 family members during mouse development and aging. *Oncogene* **1997**, *15*, 203–211. [CrossRef] [PubMed]
30. Zindy, F.; Soares, H.; Herzog, K.H.; Morgan, J.; Sherr, C.J.; Roussel, M.F. Expression of INK4 inhibitors of cyclin D-dependent kinases during mouse brain development. *Cell Growth Differ.* **1997**, *8*, 1139–1150. [PubMed]

31. Beauséjour, C.M.; Krtolica, A.; Galimi, F.; Narita, M.; Lowe, S.W.; Yaswen, P.; Campisi, J. Reversal of human cellular senescence: Roles of the p53 and p16 pathways. *EMBO J.* **2003**, *22*, 4212–4222. [CrossRef]
32. Erickson, S.; Sangfelt, O.; Heyman, M.; Castro, J.; Einhorn, S.; Grandér, D. Involvement of the Ink4 proteins p16 and p15 in T-lymphocyte senescence. *Oncogene* **1998**, *17*, 595–602. [CrossRef]
33. Brenner, A.J.; Stampfer, M.R.; Aldaz, C.M. Increased p16 expression with first senescence arrest in human mammary epithelial cells and extended growth capacity with p16 inactivation. *Oncogene* **1998**, *17*, 199–205. [CrossRef]
34. Stein, G.H.; Drullinger, L.F.; Soulard, A.; Dulić, V. Differential roles for cyclin-dependent kinase inhibitors p21 and p16 in the mechanisms of senescence and differentiation in human fibroblasts. *Mol. Cell Biol.* **1999**, *19*, 2109–2117. [CrossRef]
35. Coppé, J.P.; Rodier, F.; Patil, C.K.; Freund, A.; Desprez, P.Y.; Campisi, J. Tumor suppressor and aging biomarker p16(INK4a) induces cellular senescence without the associated inflammatory secretory phenotype. *J. Biol. Chem.* **2011**, *286*, 36396–36403. [CrossRef]
36. Di Micco, R.; Krizhanovsky, V.; Baker, D.; d'Adda di Fagagna, F. Cellular senescence in ageing: From mechanisms to therapeutic opportunities. *Nat. Rev. Mol. Cell Biol.* **2021**, *22*, 75–95. [CrossRef]
37. Dai, C.Y.; Enders, G.H. p16 INK4a can initiate an autonomous senescence program. *Oncogene* **2000**, *19*, 1613–1622. [CrossRef]
38. Safwan-Zaiter, H.; Wagner, N.; Michiels, J.-F.; Wagner, K.-D. Dynamic Spatiotemporal Expression Pattern of the Senescence-Associated Factor p16Ink4a in Development and Aging. *Cells* **2022**, *11*, 541. [CrossRef]
39. Grosse, L.; Wagner, N.; Emelyanov, A.; Molina, C.; Lacas-Gervais, S.; Wagner, K.-D.; Bulavin, D.V. Defined p16High Senescent Cell Types Are Indispensable for Mouse Healthspan. *Cell Metab.* **2020**, *32*, 87–99. [CrossRef]
40. Natarajan, E.; Omobono, J.D.; Jones, J.C.; Rheinwald, J.G. Co-expression of p16INK4A and laminin 5 by keratinocytes: A wound-healing response coupling hypermotility with growth arrest that goes awry during epithelial neoplastic progression. *J. Investig. Dermatology Symp. Proc.* **2005**, *10*, 72–85. [CrossRef]
41. Demaria, M.; Ohtani, N.; Youssef, S.A.; Rodier, F.; Toussaint, W.; Mitchell, J.R.; Laberge, R.-M.; Vijg, J.; Van Steeg, H.; Dollé, M.E.T.; et al. An Essential Role for Senescent Cells in Optimal Wound Healing through Secretion of PDGF-AA. *Dev. Cell* **2014**, *31*, 722–733. [CrossRef]
42. Xiong, Y.; Zhang, H.; Beach, D. Subunit rearrangement of the cyclin-dependent kinases is associated with cellular transformation. *Genes Dev.* **1993**, *7*, 1572–1583. [CrossRef]
43. Kamb, A.; Gruis, N.A.; Weaver-Feldhaus, J.; Liu, Q.; Harshman, K.; Tavtigian, S.V.; Stockert, E.; Day, R.S.; Johnson, B.E.; Skolnick, M.H. A cell cycle regulator potentially involved in genesis of many tumor types. *Science* **1994**, *264*, 436–440. [CrossRef]
44. Stone, S.; Jiang, P.; Dayananth, P.; Tavtigian, S.V.; Katcher, H.; Parry, D.; Peters, G.; Kamb, A. Complex structure and regulation of the P16 (MTS1) locus. *Cancer Res.* **1995**, *55*, 2988–2994.
45. Mao, L.; Merlo, A.; Bedi, G.; Shapiro, G.I.; Edwards, C.D.; Rollins, B.J.; Sidransky, D. A novel p16INK4A transcript. *Cancer Res.* **1995**, *55*, 2995–2997.
46. Leon, K.E.; Tangudu, N.K.; Aird, K.M.; Buj, R. Loss of p16: A Bouncer of the Immunological Surveillance? *Life* **2021**, *11*, 309. [CrossRef]
47. Serrano, M.; Hannon, G.J.; Beach, D. A new regulatory motif in cell-cycle control causing specific inhibition of cyclin D/CDK4. *Nature* **1993**, *366*, 704–707. [CrossRef]
48. Parry, D.; Bates, S.; Mann, D.J.; Peters, G. Lack of cyclin D-Cdk complexes in Rb-negative cells correlates with high levels of p16INK4/MTS1 tumour suppressor gene product. *EMBO J.* **1995**, *14*, 503–511. [CrossRef]
49. Weinberg, R.A. The cat and mouse games that genes, viruses, and cells play. *Cell* **1997**, *88*, 573–575. [CrossRef]
50. Li, Y.; Nichols, M.A.; Shay, J.W.; Xiong, Y. Transcriptional repression of the D-type cyclin-dependent kinase inhibitor p16 by the retinoblastoma susceptibility gene product pRb. *Cancer Res.* **1994**, *54*, 6078–6082. [PubMed]
51. Sherr, C.J.; Roberts, J.M. Living with or without cyclins and cyclin-dependent kinases. *Genes Dev.* **2004**, *18*, 2699–2711. [CrossRef] [PubMed]
52. Negishi, M.; Saraya, A.; Mochizuki, S.; Helin, K.; Koseki, H.; Iwama, A. A novel zinc finger protein Zfp277 mediates transcriptional repression of the Ink4a/arf locus through polycomb repressive complex 1. *PLoS ONE* **2010**, *5*, e12373. [CrossRef] [PubMed]
53. Sparmann, A.; van Lohuizen, M. Polycomb silencers control cell fate, development and cancer. *Nat. Rev. Cancer* **2006**, *6*, 846–856. [CrossRef]
54. Schwartz, Y.B.; Pirrotta, V. Polycomb silencing mechanisms and the management of genomic programmes. *Nat. Rev. Genet.* **2007**, *8*, 9–22. [CrossRef]
55. Bracken, A.P.; Kleine-Kohlbrecher, D.; Dietrich, N.; Pasini, D.; Gargiulo, G.; Beekman, C.; Theilgaard-Mönch, K.; Minucci, S.; Porse, B.T.; Marine, J.-C.; et al. The Polycomb group proteins bind throughout the INK4A-ARF locus and are disassociated in senescent cells. *Genes Dev.* **2007**, *21*, 525–530. [CrossRef]
56. Rayess, H.; Wang, M.B.; Srivatsan, E.S. Cellular senescence and tumor suppressor gene p16. *Int. J. Cancer* **2012**, *130*, 1715–1725. [CrossRef]
57. Barradas, M.; Anderton, E.; Acosta, J.C.; Li, S.; Banito, A.; Rodriguez-Niedenführ, M.; Maertens, G.; Banck, M.; Zhou, M.-M.; Walsh, M.J.; et al. Histone demethylase JMJD3 contributes to epigenetic control of INK4a/ARF by oncogenic RAS. *Genes Dev.* **2009**, *23*, 1177–1182. [CrossRef]
58. Maertens, G.N.; El Messaoudi-Aubert, S.; Racek, T.; Stock, J.K.; Nicholls, J.; Rodriguez-Niedenführ, M.; Gil, J.; Peters, G. Several distinct polycomb complexes regulate and co-localize on the INK4a tumor suppressor locus. *PLoS ONE* **2009**, *4*, e6380. [CrossRef]
59. Wang, Y.; Guan, Y.; Wang, F.; Huang, A.; Wang, S.; Zhang, Y.A. Bmi-1 regulates self-renewal, proliferation and senescence of human fetal neural stem cells in vitro. *Neurosci. Lett.* **2010**, *476*, 74–78. [CrossRef]

60. Guo, W.J.; Datta, S.; Band, V.; Dimri, G.P. Mel-18, a polycomb group protein, regulates cell proliferation and senescence via transcriptional repression of Bmi-1 and c-Myc oncoproteins. *Mol. Biol. Cell* **2007**, *18*, 536–546. [CrossRef]
61. Jacobs, J.J.; Kieboom, K.; Marino, S.; DePinho, R.A.; van Lohuizen, M. The oncogene and Polycomb-group gene bmi-1 regulates cell proliferation and senescence through the ink4a locus. *Nature* **1999**, *397*, 164–168. [CrossRef]
62. Kotake, Y.; Cao, R.; Viatour, P.; Sage, J.; Zhang, Y.; Xiong, Y. pRB family proteins are required for H3K27 trimethylation and Polycomb repression complexes binding to and silencing p16INK4alpha tumor suppressor gene. *Genes Dev.* **2007**, *21*, 49–54. [CrossRef]
63. Gil, J.; Peters, G. Regulation of the INK4b-ARF-INK4a tumour suppressor locus: All for one or one for all. *Nat. Rev. Mol. Cell Biol.* **2006**, *7*, 667–677. [CrossRef]
64. Huang, Y.C.; Saito, S.; Yokoyama, K.K. Histone chaperone Jun dimerization protein 2 (JDP2): Role in cellular senescence and aging. *Kaohsiung J. Med. Sci.* **2010**, *26*, 515–531. [CrossRef]
65. Nakade, K.; Pan, J.; Yamasaki, T.; Murata, T.; Wasylyk, B.; Yokoyama, K.K. JDP2 (Jun Dimerization Protein 2)-deficient mouse embryonic fibroblasts are resistant to replicative senescence. *J. Biol. Chem.* **2009**, *284*, 10808–10817. [CrossRef]
66. Nakade, K.; Lin, C.S.; Chen, X.Y.; Tsai, M.H.; Wuputra, K.; Zhu, Z.W.; Pan, J.Z.; Yokoyama, K.K. Jun dimerization protein 2 controls hypoxia-induced replicative senescence via both the p16. *FEBS Open Bio* **2017**, *7*, 1793–1804. [CrossRef]
67. Witcher, M.; Emerson, B.M. Epigenetic silencing of the p16(INK4a) tumor suppressor is associated with loss of CTCF binding and a chromatin boundary. *Mol. Cell* **2009**, *34*, 271–284. [CrossRef]
68. De Jaime-Soguero, A.; Aulicino, F.; Ertaylan, G.; Griego, A.; Cerrato, A.; Tallam, A.; Del Sol, A.; Cosma, M.P.; Lluis, F. Wnt/Tcf1 pathway restricts embryonic stem cell cycle through activation of the Ink4/Arf locus. *PLoS Genet.* **2017**, *13*, e1006682. [CrossRef]
69. Wang, X.; Pan, L.; Feng, Y.; Wang, Y.; Han, Q.; Han, L.; Han, S.; Guo, J.; Huang, B.; Lu, J. P300 plays a role in p16(INK4a) expression and cell cycle arrest. *Oncogene* **2008**, *27*, 1894–1904. [CrossRef]
70. Mahmoud, A.I.; Kocabas, F.; Muralidhar, S.A.; Kimura, W.; Koura, A.S.; Thet, S.; Porrello, E.R.; Sadek, H.A. Meis1 regulates postnatal cardiomyocyte cell cycle arrest. *Nature* **2013**, *497*, 249–253. [CrossRef]
71. Gan, Q.; Huang, J.; Zhou, R.; Niu, J.; Zhu, X.; Wang, J.; Zhang, Z.; Tong, T. PPAR{gamma} accelerates cellular senescence by inducing p16INK4{alpha} expression in human diploid fibroblasts. *J. Cell Sci.* **2008**, *121*, 2235–2245. [CrossRef] [PubMed]
72. Baker, D.J.; Wijshake, T.; Tchkonia, T.; LeBrasseur, N.K.; Childs, B.G.; van de Sluis, B.; Kirkland, J.L.; van Deursen, J.M. Clearance of p16Ink4a-positive senescent cells delays ageing-associated disorders. *Nature* **2011**, *479*, 232–236. [CrossRef]
73. Baker, D.J.; Childs, B.G.; Durik, M.; Wijers, M.E.; Sieben, C.J.; Zhong, J.; Saltness, R.A.; Jeganathan, K.B.; Verzosa, G.C.; Pezeshki, A.; et al. Naturally occurring p16(Ink4a)-positive cells shorten healthy lifespan. *Nature* **2016**, *530*, 184–189. [CrossRef] [PubMed]
74. Roos, C.M.; Zhang, B.; Palmer, A.K.; Ogrodnik, M.B.; Pirtskhalava, T.; Thalji, N.M.; Hagler, M.; Jurk, D.; Smith, L.A.; Casaclang-Verzosa, G.; et al. Chronic senolytic treatment alleviates established vasomotor dysfunction in aged or atherosclerotic mice. *Aging Cell* **2016**, *15*, 973–977. [CrossRef]
75. Farr, J.N.; Xu, M.; Weivoda, M.M.; Monroe, D.G.; Fraser, D.G.; Onken, J.L.; Negley, B.A.; Sfeir, J.G.; Ogrodnik, M.B.; Hachfeld, C.M.; et al. Targeting cellular senescence prevents age-related bone loss in mice. *Nat. Med.* **2017**, *23*, 1072–1079. [CrossRef]
76. Bussian, T.J.; Aziz, A.; Meyer, C.F.; Swenson, B.L.; van Deursen, J.M.; Baker, D.J. Clearance of senescent glial cells prevents tau-dependent pathology and cognitive decline. *Nature* **2018**, *562*, 578–582. [CrossRef]
77. Ogrodnik, M.; Zhu, Y.; Langhi, L.G.P.; Tchkonia, T.; Krüger, P.; Fielder, E.; Victorelli, S.; Ruswhandi, R.A.; Giorgadze, N.; Pirtskhalava, T.; et al. Obesity-Induced Cellular Senescence Drives Anxiety and Impairs Neurogenesis. *Cell Metab.* **2019**, *29*, 1233. [CrossRef]
78. Lewis-McDougall, F.C.; Ruchaya, P.J.; Domenjo-Vila, E.; Shin Teoh, T.; Prata, L.; Cottle, B.J.; Clark, J.E.; Punjabi, P.P.; Awad, W.; Torella, D.; et al. Aged-senescent cells contribute to impaired heart regeneration. *Aging Cell* **2019**, *18*, e12931. [CrossRef]
79. Aguayo-Mazzucato, C.; Andle, J.; Lee, T.B.; Midha, A.; Talemal, L.; Chipashvili, V.; Hollister-Lock, J.; van Deursen, J.; Weir, G.; Bonner-Weir, S. Acceleration of β Cell Aging Determines Diabetes and Senolysis Improves Disease Outcomes. *Cell Metab.* **2019**, *30*, 129–142. [CrossRef]
80. Ogrodnik, M.; Evans, S.A.; Fielder, E.; Victorelli, S.; Kruger, P.; Salmonowicz, H.; Weigand, B.M.; Patel, A.D.; Pirtskhalava, T.; Inman, C.L.; et al. Whole-body senescent cell clearance alleviates age-related brain inflammation and cognitive impairment in mice. *Aging Cell* **2021**, *20*, e13296. [CrossRef]
81. Kim, S.R.; Puranik, A.S.; Jiang, K.; Chen, X.; Zhu, X.Y.; Taylor, I.; Khodadadi-Jamayran, A.; Lerman, A.; Hickson, L.J.; Childs, B.G.; et al. Progressive Cellular Senescence Mediates Renal Dysfunction in Ischemic Nephropathy. *J. Am. Soc. Nephrol.* **2021**, *32*, 1987–2004. [CrossRef] [PubMed]
82. Cohen, C.; Le Goff, O.; Soysouvanh, F.; Vasseur, F.; Tanou, M.; Nguyen, C.; Amrouche, L.; Le Guen, J.; Saltel-Fulero, O.; Meunier, T.; et al. Glomerular endothelial cell senescence drives age-related kidney disease through PAI-1. *EMBO Mol. Med.* **2021**, *13*, e14146. [CrossRef] [PubMed]
83. Goldstein, B.J. Rosiglitazone. *Int J. Clin. Pract* **2000**, *54*, 333–337. [PubMed]
84. Takeda, S.; Sasagawa, S.; Oyama, T.; Searleman, A.C.; Westergard, T.D.; Cheng, E.H.; Hsieh, J.J. Taspase1-dependent TFIIA cleavage coordinates head morphogenesis by limiting Cdkn2a locus transcription. *J. Clin. Invest.* **2015**, *125*, 1203–1214. [CrossRef]
85. Chen, Z.; Qian, Q.; Ma, G.; Wang, J.; Zhang, X.; Feng, Y.; Shen, C.; Yao, Y. A common variant on chromosome 9p21 affects the risk of early-onset coronary artery disease. *Mol. Biol. Rep.* **2009**, *36*, 889–893. [CrossRef]

86. Helgadottir, A.; Thorleifsson, G.; Manolescu, A.; Gretarsdottir, S.; Blondal, T.; Jonasdottir, A.; Sigurdsson, A.; Baker, A.; Palsson, A.; Masson, G.; et al. A common variant on chromosome 9p21 affects the risk of myocardial infarction. *Science* **2007**, *316*, 1491–1493. [CrossRef]
87. McPherson, R.; Pertsemlidis, A.; Kavaslar, N.; Stewart, A.; Roberts, R.; Cox, D.R.; Hinds, D.A.; Pennacchio, L.A.; Tybjaerg-Hansen, A.; Folsom, A.R.; et al. A common allele on chromosome 9 associated with coronary heart disease. *Science* **2007**, *316*, 1488–1491. [CrossRef]
88. Visel, A.; Zhu, Y.; May, D.; Afzal, V.; Gong, E.; Attanasio, C.; Blow, M.J.; Cohen, J.C.; Rubin, E.M.; Pennacchio, L.A. Targeted deletion of the 9p21 non-coding coronary artery disease risk interval in mice. *Nature* **2010**, *464*, 409–412. [CrossRef]
89. Palmero, I.; Pantoja, C.; Serrano, M. p19ARF links the tumour suppressor p53 to Ras. *Nature* **1998**, *395*, 125–126. [CrossRef]
90. Weber, J.D.; Jeffers, J.R.; Rehg, J.E.; Randle, D.H.; Lozano, G.; Roussel, M.F.; Sherr, C.J.; Zambetti, G.P. p53-independent functions of the p19(ARF) tumor suppressor. *Genes Dev.* **2000**, *14*, 2358–2365. [CrossRef]
91. Suzuki, H.; Kurita, M.; Mizumoto, K.; Nishimoto, I.; Ogata, E.; Matsuoka, M. p19ARF-induced p53-independent apoptosis largely occurs through BAX. *Biochem. Biophys. Res. Commun.* **2003**, *312*, 1273–1277. [CrossRef]
92. Zindy, F.; Eischen, C.M.; Randle, D.H.; Kamijo, T.; Cleveland, J.L.; Sherr, C.J.; Roussel, M.F. Myc signaling via the ARF tumor suppressor regulates p53-dependent apoptosis and immortalization. *Genes Dev.* **1998**, *12*, 2424–2433. [CrossRef]
93. Qi, Y.; Gregory, M.A.; Li, Z.; Brousal, J.P.; West, K.; Hann, S.R. p19ARF directly and differentially controls the functions of c-Myc independently of p53. *Nature* **2004**, *431*, 712–717. [CrossRef]
94. Cleveland, J.L.; Sherr, C.J. Antagonism of Myc functions by Arf. *Cancer Cell* **2004**, *6*, 309–311. [CrossRef]
95. Mitra, J.; Dai, C.Y.; Somasundaram, K.; El-Deiry, W.S.; Satyamoorthy, K.; Herlyn, M.; Enders, G.H. Induction of p21WAF1/CIP1and Inhibition of Cdk2 Mediated by the Tumor Suppressor p16INK4a. *Mol. Cell. Biol.* **1999**, *19*, 3916–3928. [CrossRef]
96. LaBaer, J.; Garrett, M.D.; Stevenson, L.F.; Slingerland, J.M.; Sandhu, C.; Chou, H.S.; Fattaey, A.; Harlow, E. New functional activities for the p21 family of CDK inhibitors. *Genes Dev.* **1997**, *11*, 847–862. [CrossRef]
97. Yamasaki, L.; Bronson, R.; Williams, B.O.; Dyson, N.J.; Harlow, E.; Jacks, T. Loss of E2F-1 reduces tumorigenesis and extends the lifespan of Rb1(+/-)mice. *Nat. Genet.* **1998**, *18*, 360–364. [CrossRef]
98. Sharma, N.; Timmers, C.; Trikha, P.; Saavedra, H.I.; Obery, A.; Leone, G. Control of the p53-p21CIP1 Axis by E2f1, E2f2, and E2f3 is essential for G1/S progression and cellular transformation. *J. Biol. Chem.* **2006**, *281*, 36124–36131. [CrossRef]
99. Wu, L.; Timmers, C.; Maiti, B.; Saavedra, H.I.; Sang, L.; Chong, G.T.; Nuckolls, F.; Giangrande, P.; Wright, F.A.; Field, S.J.; et al. The E2F1-3 transcription factors are essential for cellular proliferation. *Nature* **2001**, *414*, 457–462. [CrossRef]
100. Denchi, E.L.; Attwooll, C.; Pasini, D.; Helin, K. Deregulated E2F Activity Induces Hyperplasia and Senescence-Like Features in the Mouse Pituitary Gland. *Mol. Cell. Biol.* **2005**, *25*, 2660–2672. [CrossRef]
101. Aslanian, A.; Iaquinta, P.J.; Verona, R.; Lees, J.A. Repression of the Arf tumor suppressor by E2F3 is required for normal cell cycle kinetics. *Genes Dev.* **2004**, *18*, 1413–1422. [CrossRef] [PubMed]
102. Herold, S.; Wanzel, M.; Beuger, V.; Frohme, C.; Beul, D.; Hillukkala, T.; Syvaoja, J.; Saluz, H.P.; Haenel, F.; Eilers, M. Negative regulation of the mammalian UV response by Myc through association with Miz-1. *Mol. Cell* **2002**, *10*, 509–521. [CrossRef]
103. Seoane, J.; Le, H.V.; Massagué, J. Myc suppression of the p21(Cip1) Cdk inhibitor influences the outcome of the p53 response to DNA damage. *Nature* **2002**, *419*, 729–734. [CrossRef] [PubMed]
104. Seoane, J.; Le, H.V.; Shen, L.; Anderson, S.A.; Massagué, J. Integration of Smad and forkhead pathways in the control of neuroepithelial and glioblastoma cell proliferation. *Cell* **2004**, *117*, 211–223. [CrossRef]
105. LaPak, K.M.; Burd, C.E. The molecular balancing act of p16(INK4a) in cancer and aging. *Mol. Cancer Res.* **2014**, *12*, 167–183. [CrossRef]
106. Lagopati, N.; Belogiannis, K.; Angelopoulou, A.; Papaspyropoulos, A.; Gorgoulis, V. Non-Canonical Functions of the ARF Tumor Suppressor in Development and Tumorigenesis. *Biomolecules* **2021**, *11*, 86. [CrossRef]
107. Gallagher, S.J.; Kefford, R.F.; Rizos, H. The ARF tumour suppressor. *Int J. Biochem. Cell Biol.* **2006**, *38*, 1637–1641. [CrossRef]
108. Engeland, K. Cell cycle regulation: p53-p21-RB signaling. *Cell Death Differ.* **2022**, *29*, 946–960. [CrossRef]
109. Lai, L.; Shin, G.Y.; Qiu, H. The Role of Cell Cycle Regulators in Cell Survival-Dual Functions of Cyclin-Dependent Kinase 20 and p21. *Int. J. Mol. Sci.* **2020**, *21*, 8504. [CrossRef]
110. Warfel, N.A.; El-Deiry, W.S. p21WAF1 and tumourigenesis: 20 years after. *Curr. Opin. Oncol.* **2013**, *25*, 52–58. [CrossRef]
111. Wagner, N.; Ninkov, M.; Vukolic, A.; Cubukcuoglu Deniz, G.; Rassoulzadegan, M.; Michiels, J.F.; Wagner, K.D. Implications of the Wilms' Tumor Suppressor Wt1 in Cardiomyocyte Differentiation. *Int. J. Mol. Sci.* **2021**, *22*, 4346. [CrossRef]
112. Wagner, K.D.; El Maï, M.; Ladomery, M.; Belali, T.; Leccia, N.; Michiels, J.F.; Wagner, N. Altered VEGF Splicing Isoform Balance in Tumor Endothelium Involves Activation of Splicing Factors Srpk1 and Srsf1 by the Wilms' Tumor Suppressor Wt1. *Cells* **2019**, *8*, 41. [CrossRef]
113. El Maï, M.; Wagner, K.D.; Michiels, J.F.; Ambrosetti, D.; Borderie, A.; Destree, S.; Renault, V.; Djerbi, N.; Giraud-Panis, M.J.; Gilson, E.; et al. The Telomeric Protein TRF2 Regulates Angiogenesis by Binding and Activating the PDGFRβ Promoter. *Cell Rep.* **2014**, *9*, 1047–1060. [CrossRef]
114. Wagner, K.D.; Cherfils-Vicini, J.; Hosen, N.; Hohenstein, P.; Gilson, E.; Hastie, N.D.; Michiels, J.F.; Wagner, N. The Wilms' tumour suppressor Wt1 is a major regulator of tumour angiogenesis and progression. *Nat. Commun.* **2014**, *5*, 5852. [CrossRef]
115. Wagner, N.; Morrison, H.; Pagnotta, S.; Michiels, J.F.; Schwab, Y.; Tryggvason, K.; Schedl, A.; Wagner, K.D. The podocyte protein nephrin is required for cardiac vessel formation. *Hum. Mol. Genet.* **2011**, *20*, 2182–2194. [CrossRef]

116. Costantini, F.; Kopan, R. Patterning a complex organ: Branching morphogenesis and nephron segmentation in kidney development. *Dev. Cell* **2010**, *18*, 698–712. [CrossRef]
117. Wagner, N.; Wagner, K.D. Every Beat You Take-The Wilms' Tumor Suppressor WT1 and the Heart. *Int. J. Mol. Sci.* **2021**, *22*, 7675. [CrossRef]
118. Zhao, R.; Duncan, S.A. Embryonic development of the liver. *Hepatology* **2005**, *41*, 956–967. [CrossRef]
119. Henry, A.M.; Hohmann, J.G. High-resolution gene expression atlases for adult and developing mouse brain and spinal cord. *Mamm. Genome* **2012**, *23*, 539–549. [CrossRef]
120. Dimri, G.P. The search for biomarkers of aging: Next stop INK4a/ARF locus. *Sci. Aging Knowl. Environ.* **2004**, *2004*, pe40. [CrossRef]
121. Krishnamurthy, J.; Torrice, C.; Ramsey, M.R.; Kovalev, G.I.; Al-Regaiey, K.; Su, L.; Sharpless, N.E. Ink4a/Arf expression is a biomarker of aging. *J. Clin. Invest.* **2004**, *114*, 1299–1307. [CrossRef]
122. Sharpless, N.E.; Sherr, C.J. Forging a signature of in vivo senescence. *Nat. Rev. Cancer* **2015**, *15*, 397–408. [CrossRef]
123. González-Gualda, E.; Baker, A.G.; Fruk, L.; Muñoz-Espín, D. A guide to assessing cellular senescence in vitro and in vivo. *FEBS J.* **2021**, *288*, 56–80. [CrossRef]
124. Basisty, N.; Kale, A.; Jeon, O.H.; Kuehnemann, C.; Payne, T.; Rao, C.; Holtz, A.; Shah, S.; Sharma, V.; Ferrucci, L.; et al. A proteomic atlas of senescence-associated secretomes for aging biomarker development. *PLoS Biol.* **2020**, *18*, e3000599. [CrossRef]
125. Van Lookeren Campagne, M.; Gill, R. Tumor-suppressor p53 is expressed in proliferating and newly formed neurons of the embryonic and postnatal rat brain: Comparison with expression of the cell cycle regulators p21Waf1/Cip1, p27Kip1, p57Kip2, p16Ink4a, cyclin G1, and the proto-oncogene Bax. *J. Comp. Neurol.* **1998**, *397*, 181–198. [CrossRef]
126. Legrier, M.E.; Ducray, A.; Propper, A.; Chao, M.; Kastner, A. Cell cycle regulation during mouse olfactory neurogenesis. *Cell Growth Differ.* **2001**, *12*, 591–601.
127. Park, I.K.; Qian, D.; Kiel, M.; Becker, M.W.; Pihalja, M.; Weissman, I.L.; Morrison, S.J.; Clarke, M.F. Bmi-1 is required for maintenance of adult self-renewing haematopoietic stem cells. *Nature* **2003**, *423*, 302–305. [CrossRef]
128. Hess, J.; Hartenstein, B.; Teurich, S.; Schmidt, D.; Schorpp-Kistner, M.; Angel, P. Defective endochondral ossification in mice with strongly compromised expression of JunB. *J. Cell Sci.* **2003**, *116*, 4587–4596. [CrossRef] [PubMed]
129. Western, P.S.; Miles, D.C.; van den Bergen, J.A.; Burton, M.; Sinclair, A.H. Dynamic regulation of mitotic arrest in fetal male germ cells. *Stem Cells* **2008**, *26*, 339–347. [CrossRef] [PubMed]
130. Wolgemuth, D.J.; Roberts, S.S. Regulating mitosis and meiosis in the male germ line: Critical functions for cyclins. *Philos. Trans. R. Soc. Lond. B. Biol. Sci.* **2010**, *365*, 1653–1662. [CrossRef] [PubMed]
131. Yang, H.; Xie, Y.; Yang, R.; Wei, S.L.; Xi, Q. Expression of p16INK4a in mouse endometrium and its effect during blastocyst implantation. *Sheng Li Xue Bao* **2008**, *60*, 547–552.
132. Sharpless, N.E.; Bardeesy, N.; Lee, K.H.; Carrasco, D.; Castrillon, D.H.; Aguirre, A.J.; Wu, E.A.; Horner, J.W.; DePinho, R.A. Loss of p16Ink4a with retention of p19Arf predisposes mice to tumorigenesis. *Nature* **2001**, *413*, 86–91. [CrossRef]
133. Parvanov, D.; Ganeva, R.; Vidolova, N.; Stamenov, G. Decreased number of p16-positive senescent cells in human endometrium as a marker of miscarriage. *J. Assist. Reprod. Genet.* **2021**, *38*, 2087–2095. [CrossRef]
134. López-Arribillaga, E.; Rodilla, V.; Pellegrinet, L.; Guiu, J.; Iglesias, M.; Roman, A.C.; Gutarra, S.; González, S.; Muñoz-Cánoves, P.; Fernández-Salguero, P.; et al. Bmi1 regulates murine intestinal stem cell proliferation and self-renewal downstream of Notch. *Development* **2015**, *142*, 41–50. [CrossRef]
135. Fasano, C.A.; Dimos, J.T.; Ivanova, N.B.; Lowry, N.; Lemischka, I.R.; Temple, S. shRNA knockdown of Bmi-1 reveals a critical role for p21-Rb pathway in NSC self-renewal during development. *Cell Stem Cell* **2007**, *1*, 87–99. [CrossRef]
136. Moon, J.H.; Yoon, B.S.; Kim, B.; Park, G.; Jung, H.Y.; Maeng, I.; Jun, E.K.; Yoo, S.J.; Kim, A.; Oh, S.; et al. Induction of neural stem cell-like cells (NSCLCs) from mouse astrocytes by Bmi1. *Biochem. Biophys. Res. Commun.* **2008**, *371*, 267–272. [CrossRef]
137. Nishino, J.; Kim, I.; Chada, K.; Morrison, S.J. Hmga2 promotes neural stem cell self-renewal in young but not old mice by reducing p16Ink4a and p19Arf Expression. *Cell* **2008**, *135*, 227–239. [CrossRef]
138. Fasano, C.A.; Phoenix, T.N.; Kokovay, E.; Lowry, N.; Elkabetz, Y.; Dimos, J.T.; Lemischka, I.R.; Studer, L.; Temple, S. Bmi-1 cooperates with Foxg1 to maintain neural stem cell self-renewal in the forebrain. *Genes Dev.* **2009**, *23*, 561–574. [CrossRef]
139. He, S.; Iwashita, T.; Buchstaller, J.; Molofsky, A.V.; Thomas, D.; Morrison, S.J. Bmi-1 over-expression in neural stem/progenitor cells increases proliferation and neurogenesis in culture but has little effect on these functions in vivo. *Dev. Biol.* **2009**, *328*, 257–272. [CrossRef]
140. Zhang, H.W.; Ding, J.; Jin, J.L.; Guo, J.; Liu, J.N.; Karaplis, A.; Goltzman, D.; Miao, D. Defects in mesenchymal stem cell self-renewal and cell fate determination lead to an osteopenic phenotype in Bmi-1 null mice. *J. Bone Miner. Res.* **2010**, *25*, 640–652. [CrossRef]
141. Chatoo, W.; Abdouh, M.; Duparc, R.H.; Bernier, G. Bmi1 distinguishes immature retinal progenitor/stem cells from the main progenitor cell population and is required for normal retinal development. *Stem Cells* **2010**, *28*, 1412–1423. [CrossRef] [PubMed]
142. Chiba, T.; Seki, A.; Aoki, R.; Ichikawa, H.; Negishi, M.; Miyagi, S.; Oguro, H.; Saraya, A.; Kamiya, A.; Nakauchi, H.; et al. Bmi1 promotes hepatic stem cell expansion and tumorigenicity in both Ink4a/Arf-dependent and -independent manners in mice. *Hepatology* **2010**, *52*, 1111–1123. [CrossRef] [PubMed]
143. Rota, M.; Hosoda, T.; De Angelis, A.; Arcarese, M.L.; Esposito, G.; Rizzi, R.; Tillmanns, J.; Tugal, D.; Musso, E.; Rimoldi, O.; et al. The young mouse heart is composed of myocytes heterogeneous in age and function. *Circ. Res.* **2007**, *101*, 387–399. [CrossRef] [PubMed]

144. An, S.; Chen, Y.; Gao, C.; Qin, B.; Du, X.; Meng, F.; Qi, Y. Inactivation of INK4a and ARF induces myocardial proliferation and improves cardiac repair following ischemia–reperfusion. *Mol. Med. Rep.* **2015**, *12*, 5911–5916. [CrossRef]
145. Krimpenfort, P.; Quon, K.C.; Mooi, W.J.; Loonstra, A.; Berns, A. Loss of p16Ink4a confers susceptibility to metastatic melanoma in mice. *Nature* **2001**, *413*, 83–86. [CrossRef]
146. Bianchi, T.; Rufer, N.; MacDonald, H.R.; Migliaccio, M. The tumor suppressor p16Ink4a regulates T lymphocytesurvival. *Oncogene* **2006**, *25*, 4110–4115. [CrossRef]
147. Cheong, C.; Sung, Y.H.; Lee, J.; Choi, Y.S.; Song, J.; Kee, C.; Lee, H.W. Role of INK4a locus in normal eye development and cataract genesis. *Mech. Ageing Dev.* **2006**, *127*, 633–638. [CrossRef]
148. Kamijo, T.; Bodner, S.; van de Kamp, E.; Randle, D.H.; Sherr, C.J. Tumor spectrum in ARF-deficient mice. *Cancer Res.* **1999**, *59*, 2217–2222.
149. McKeller, R.N.; Fowler, J.L.; Cunningham, J.J.; Warner, N.; Smeyne, R.J.; Zindy, F.; Skapek, S.X. The Arf tumor suppressor gene promotes hyaloid vascular regression during mouse eye development. *Proc. Natl. Acad. Sci. USA* **2002**, *99*, 3848–3853. [CrossRef]
150. Martin, A.C.; Thornton, J.D.; Liu, J.; Wang, X.; Zuo, J.; Jablonski, M.M.; Chaum, E.; Zindy, F.; Skapek, S.X. Pathogenesis of persistent hyperplastic primary vitreous in mice lacking the arf tumor suppressor gene. *Invest. Ophthalmol. Vis. Sci.* **2004**, *45*, 3387–3396. [CrossRef]
151. Thornton, J.D.; Swanson, D.J.; Mary, M.N.; Pei, D.; Martin, A.C.; Pounds, S.; Goldowitz, D.; Skapek, S.X. Persistent hyperplastic primary vitreous due to somatic mosaic deletion of the arf tumor suppressor. *Invest. Ophthalmol. Vis. Sci.* **2007**, *48*, 491–499. [CrossRef]
152. Silva, R.L.; Thornton, J.D.; Martin, A.C.; Rehg, J.E.; Bertwistle, D.; Zindy, F.; Skapek, S.X. Arf-dependent regulation of Pdgf signaling in perivascular cells in the developing mouse eye. *EMBO J.* **2005**, *24*, 2803–2814. [CrossRef]
153. Widau, R.C.; Zheng, Y.; Sung, C.Y.; Zelivianskaia, A.; Roach, L.E.; Bachmeyer, K.M.; Abramova, T.; Desgardin, A.; Rosner, A.; Cunningham, J.M.; et al. p19Arf represses platelet-derived growth factor receptor β by transcriptional and posttranscriptional mechanisms. *Mol. Cell Biol.* **2012**, *32*, 4270–4282. [CrossRef]
154. Freeman-Anderson, N.E.; Zheng, Y.; McCalla-Martin, A.C.; Treanor, L.M.; Zhao, Y.D.; Garfin, P.M.; He, T.C.; Mary, M.N.; Thornton, J.D.; Anderson, C.; et al. Expression of the Arf tumor suppressor gene is controlled by Tgfbeta2 during development. *Development* **2009**, *136*, 2081–2089. [CrossRef]
155. Zheng, Y.; Devitt, C.; Liu, J.; Mei, J.; Skapek, S.X. A distant, cis-acting enhancer drives induction of Arf by Tgfβ in the developing eye. *Dev. Biol.* **2013**, *380*, 49–57. [CrossRef]
156. Suzuki, A.; Sekiya, S.; Büscher, D.; Izpisúa Belmonte, J.C.; Taniguchi, H. Tbx3 controls the fate of hepatic progenitor cells in liver development by suppressing p19ARF expression. *Development* **2008**, *135*, 1589–1595. [CrossRef]
157. Deng, C.; Zhang, P.; Harper, J.W.; Elledge, S.J.; Leder, P. Mice lacking p21CIP1/WAF1 undergo normal development, but are defective in G1 checkpoint control. *Cell* **1995**, *82*, 675–684. [CrossRef]
158. Kim, W.H.; Joo, C.U.; Ku, J.H.; Ryu, C.H.; Koh, K.N.; Koh, G.Y.; Ko, J.K. Cell cycle regulators during human atrial development. *Korean J. Intern. Med.* **1998**, *13*, 77–82. [CrossRef]
159. Poolman, R.A.; Gilchrist, R.; Brooks, G. Cell cycle profiles and expressions of p21CIP1 AND P27KIP1 during myocyte development. *Int. J. Cardiol.* **1998**, *67*, 133–142. [CrossRef]
160. Ramakrishna, S.; Kim, I.M.; Petrovic, V.; Malin, D.; Wang, I.C.; Kalin, T.V.; Meliton, L.; Zhao, Y.Y.; Ackerson, T.; Qin, Y.; et al. Myocardium defects and ventricular hypoplasia in mice homozygous null for the Forkhead Box M1 transcription factor. *Dev. Dyn.* **2007**, *236*, 1000–1013. [CrossRef]
161. Evans-Anderson, H.J.; Alfieri, C.M.; Yutzey, K.E. Regulation of cardiomyocyte proliferation and myocardial growth during development by FOXO transcription factors. *Circ. Res.* **2008**, *102*, 686–694. [CrossRef]
162. Bolte, C.; Zhang, Y.; Wang, I.C.; Kalin, T.V.; Molkentin, J.D.; Kalinichenko, V.V. Expression of Foxm1 transcription factor in cardiomyocytes is required for myocardial development. *PLoS ONE* **2011**, *6*, e22217. [CrossRef]
163. Chakraborty, S.; Yutzey, K.E. Tbx20 regulation of cardiac cell proliferation and lineage specialization during embryonic and fetal development in vivo. *Dev. Biol.* **2012**, *363*, 234–246. [CrossRef]
164. Garnatz, A.S.; Gao, Z.; Broman, M.; Martens, S.; Earley, J.U.; Svensson, E.C. FOG-2 mediated recruitment of the NuRD complex regulates cardiomyocyte proliferation during heart development. *Dev. Biol.* **2014**, *395*, 50–61. [CrossRef]
165. Parker, S.B.; Eichele, G.; Zhang, P.; Rawls, A.; Sands, A.T.; Bradley, A.; Olson, E.N.; Harper, J.W.; Elledge, S.J. p53-independent expression of p21Cip1 in muscle and other terminally differentiating cells. *Science* **1995**, *267*, 1024–1027. [CrossRef]
166. Guo, K.; Wang, J.; Andrés, V.; Smith, R.C.; Walsh, K. MyoD-induced expression of p21 inhibits cyclin-dependent kinase activity upon myocyte terminal differentiation. *Mol. Cell Biol.* **1995**, *15*, 3823–3829. [CrossRef] [PubMed]
167. Halevy, O.; Novitch, B.G.; Spicer, D.B.; Skapek, S.X.; Rhee, J.; Hannon, G.J.; Beach, D.; Lassar, A.B. Correlation of terminal cell cycle arrest of skeletal muscle with induction of p21 by MyoD. *Science* **1995**, *267*, 1018–1021. [CrossRef]
168. Ikoma, T.; Ito, T.; Okudela, K.; Hayashi, H.; Yazawa, T.; Kitamura, H. Modulation of the expression of the Cip/Kip family of cyclin-dependent kinase inhibitors in foetal developing lungs of hamsters. *Cell Prolif.* **2001**, *34*, 233–241. [CrossRef]
169. Gui, H.; Li, S.; Matise, M.P. A cell-autonomous requirement for Cip/Kip cyclin-kinase inhibitors in regulating neuronal cell cycle exit but not differentiation in the developing spinal cord. *Dev. Biol.* **2007**, *301*, 14–26. [CrossRef]

170. Moniot, B.; Ujjan, S.; Champagne, J.; Hirai, H.; Aritake, K.; Nagata, K.; Dubois, E.; Nidelet, S.; Nakamura, M.; Urade, Y.; et al. Prostaglandin D2 acts through the Dp2 receptor to influence male germ cell differentiation in the foetal mouse testis. *Development* **2014**, *141*, 3561–3571. [CrossRef]
171. Zhang, P.; Wong, C.; Liu, D.; Finegold, M.; Harper, J.W.; Elledge, S.J. p21(CIP1) and p57(KIP2) control muscle differentiation at the myogenin step. *Genes Dev.* **1999**, *13*, 213–224. [CrossRef] [PubMed]
172. Nacher, V.; Carretero, A.; Navarro, M.; Armengol, C.; Llombart, C.; Rodríguez, A.; Herrero-Fresneda, I.; Ayuso, E.; Ruberte, J. The quail mesonephros: A new model for renal senescence? *J. Vasc. Res.* **2006**, *43*, 581–586. [CrossRef] [PubMed]
173. Kang, T.W.; Yevsa, T.; Woller, N.; Hoenicke, L.; Wuestefeld, T.; Dauch, D.; Hohmeyer, A.; Gereke, M.; Rudalska, R.; Potapova, A.; et al. Senescence surveillance of pre-malignant hepatocytes limits liver cancer development. *Nature* **2011**, *479*, 547–551. [CrossRef] [PubMed]
174. Xue, W.; Zender, L.; Miething, C.; Dickins, R.A.; Hernando, E.; Krizhanovsky, V.; Cordon-Cardo, C.; Lowe, S.W. Senescence and tumour clearance is triggered by p53 restoration in murine liver carcinomas. *Nature* **2007**, *445*, 656–660. [CrossRef]
175. Hoenicke, L.; Zender, L. Immune surveillance of senescent cells–biological significance in cancer- and non-cancer pathologies. *Carcinogenesis* **2012**, *33*, 1123–1126. [CrossRef]
176. Flavell, R.A.; Sanjabi, S.; Wrzesinski, S.H.; Licona-Limón, P. The polarization of immune cells in the tumour environment by TGFbeta. *Nat. Rev. Immunol.* **2010**, *10*, 554–567. [CrossRef]
177. Hall, B.M.; Balan, V.; Gleiberman, A.S.; Strom, E.; Krasnov, P.; Virtuoso, L.P.; Rydkina, E.; Vujcic, S.; Balan, K.; Gitlin, I.; et al. Aging of mice is associated with p16(Ink4a)- and β-galactosidase-positive macrophage accumulation that can be induced in young mice by senescent cells. *Aging* **2016**, *8*, 1294–1315. [CrossRef]
178. Alimirah, F.; Pulido, T.; Valdovinos, A.; Alptekin, S.; Chang, E.; Jones, E.; Diaz, D.A.; Flores, J.; Velarde, M.C.; Demaria, M.; et al. Cellular Senescence Promotes Skin Carcinogenesis through p38MAPK and p44/42MAPK Signaling. *Cancer Res.* **2020**, *80*, 3606–3619. [CrossRef]
179. Storer, M.; Mas, A.; Robert-Moreno, A.; Pecoraro, M.; Ortells, M.C.; Di Giacomo, V.; Yosef, R.; Pilpel, N.; Krizhanovsky, V.; Sharpe, J.; et al. Senescence Is a Developmental Mechanism that Contributes to Embryonic Growth and Patterning. *Cell* **2013**, *155*, 1119–1130. [CrossRef]
180. Zhang, K.; Chen, C.; Liu, Y.; Chen, H.; Liu, J.P. Cellular senescence occurred widespread to multiple selective sites in the fetal tissues and organs of mice. *Clin. Exp. Pharmacol. Physiol.* **2014**, *41*, 965–975. [CrossRef]
181. Huang, T.; Rivera-Pérez, J.A. Senescence-associated β-galactosidase activity marks the visceral endoderm of mouse embryos but is not indicative of senescence. *Genesis* **2014**, *52*, 300–308. [CrossRef]
182. Chuprin, A.; Gal, H.; Biron-Shental, T.; Biran, A.; Amiel, A.; Rozenblatt, S.; Krizhanovsky, V. Cell fusion induced by ERVWE1 or measles virus causes cellular senescence. *Genes Dev.* **2013**, *27*, 2356–2366. [CrossRef]
183. Gibaja, A.; Aburto, M.R.; Pulido, S.; Collado, M.; Hurle, J.M.; Varela-Nieto, I.; Magariños, M. TGFβ2-induced senescence during early inner ear development. *Sci. Rep.* **2019**, *9*, 5912. [CrossRef]
184. Davaapil, H.; Brockes, J.P.; Yun, M.H. Conserved and novel functions of programmed cellular senescence during vertebrate development. *Development* **2017**, *144*, 106–114. [CrossRef]
185. Villiard, É.; Denis, J.F.; Hashemi, F.S.; Igelmann, S.; Ferbeyre, G.; Roy, S. Senescence gives insights into the morphogenetic evolution of anamniotes. *Biol. Open* **2017**, *6*, 891–896. [CrossRef]
186. Zhao, Y.; Tyshkovskiy, A.; Muñoz-Espín, D.; Tian, X.; Serrano, M.; de Magalhaes, J.P.; Nevo, E.; Gladyshev, V.N.; Seluanov, A.; Gorbunova, V. Naked mole rats can undergo developmental, oncogene-induced and DNA damage-induced cellular senescence. *Proc. Natl. Acad. Sci. USA* **2018**, *115*, 1801–1806. [CrossRef]
187. Rhinn, M.; Zapata-Bodalo, I.; Klein, A.; Plassat, J.-L.; Knauer-Meyer, T.; Keyes, W.M. Aberrant induction of p19Arf-mediated cellular senescence contributes to neurodevelopmental defects. *PloS Biol.* **2021**, *20*, e3001664. [CrossRef]
188. Rhinn, M.; Ritschka, B.; Keyes, W.M. Cellular senescence in development, regeneration and disease. *Development* **2019**, *146*, dev151837. [CrossRef]
189. De Mera-Rodríguez, J.A.; Álvarez-Hernán, G.; Gañán, Y.; Martín-Partido, G.; Rodríguez-León, J.; Francisco-Morcillo, J. Is Senescence-Associated β-Galactosidase a Reliable. *Front. Cell Dev. Biol.* **2021**, *9*, 623175. [CrossRef]
190. De Mera-Rodríguez, J.A.; Álvarez-Hernán, G.; Gañán, Y.; Santos-Almeida, A.; Martín-Partido, G.; Rodríguez-León, J.; Francisco-Morcillo, J. Endogenous pH 6.0 β-Galactosidase Activity Is Linked to Neuronal Differentiation in the Olfactory Epithelium. *Cells* **2022**, *11*, 298. [CrossRef]
191. De Mera-Rodríguez, J.A.; Álvarez-Hernán, G.; Gañán, Y.; Martín-Partido, G.; Rodríguez-León, J.; Francisco-Morcillo, J. Senescence-associated β-galactosidase activity in the developing avian retina. *Dev. Dyn.* **2019**, *248*, 850–865. [CrossRef]
192. Hall, B.M.; Balan, V.; Gleiberman, A.S.; Strom, E.; Krasnov, P.; Virtuoso, L.P.; Rydkina, E.; Vujcic, S.; Balan, K.; Gitlin, I.I.; et al. p16(Ink4a) and senescence-associated β-galactosidase can be induced in macrophages as part of a reversible response to physiological stimuli. *Aging* **2017**, *9*, 1867–1884. [CrossRef]
193. Jurk, D. Chapter 6—Cellular senescence during aging and chronic liver diseases: Mechanisms and therapeutic opportunities. In *Cellular Senescence in Disease*; Serrano, M., Muñoz-Espín, D., Eds.; Academic Press: Cambridge, MA, USA, 2022; pp. 155–178.
194. He, S.; Sharpless, N.E. Senescence in Health and Disease. *Cell* **2017**, *169*, 1000–1011. [CrossRef]
195. Childs, B.G.; Durik, M.; Baker, D.J.; van Deursen, J.M. Cellular senescence in aging and age-related disease: From mechanisms to therapy. *Nat. Med.* **2015**, *21*, 1424–1435. [CrossRef]

196. Salama, R.; Sadaie, M.; Hoare, M.; Narita, M. Cellular senescence and its effector programs. *Genes Dev.* **2014**, *28*, 99–114. [CrossRef]
197. Cianflone, E.; Torella, M.; Biamonte, F.; De Angelis, A.; Urbanek, K.; Costanzo, F.S.; Rota, M.; Ellison-Hughes, G.M.; Torella, D. Targeting Cardiac Stem Cell Senescence to Treat Cardiac Aging and Disease. *Cells* **2020**, *9*, 1558. [CrossRef]
198. Sperka, T.; Wang, J.; Rudolph, K.L. DNA damage checkpoints in stem cells, ageing and cancer. *Nat. Rev. Mol. Cell Biol* **2012**, *13*, 579–590. [CrossRef]
199. Shamloo, B.; Usluer, S. p21 in Cancer Research. *Cancers* **2019**, *11*, 1178. [CrossRef]
200. Ohtani, N. The roles and mechanisms of senescence-associated secretory phenotype (SASP): Can it be controlled by senolysis? *Inflamm Regen* **2022**, *42*, 11. [CrossRef]
201. Ohtani, N.; Mann, D.J.; Hara, E. Cellular senescence: Its role in tumor suppression and aging. *Cancer Sci.* **2009**, *100*, 792–797. [CrossRef]
202. Sánchez-Díaz, L.; Espinosa-Sánchez, A.; Blanco, J.R.; Carnero, A. Senotherapeutics in Cancer and HIV. *Cells* **2022**, *11*, 1222. [CrossRef] [PubMed]
203. Cortesi, M.; Zanoni, M.; Pirini, F.; Tumedei, M.M.; Ravaioli, S.; Rapposelli, I.G.; Frassineti, G.L.; Bravaccini, S. Pancreatic Cancer and Cellular Senescence: Tumor Microenvironment under the Spotlight. *Int. J. Mol. Sci.* **2021**, *23*, 254. [CrossRef] [PubMed]
204. Niklander, S.E.; Lambert, D.W.; Hunter, K.D. Senescent Cells in Cancer: Wanted or Unwanted Citizens. *Cells* **2021**, *10*, 3315. [CrossRef] [PubMed]
205. Billman, G.E. Homeostasis: The Underappreciated and Far Too Often Ignored Central Organizing Principle of Physiology. *Front. Physiol.* **2020**, *11*, 200. [CrossRef] [PubMed]
206. Kim, H.N.; Chang, J.; Iyer, S.; Han, L.; Campisi, J.; Manolagas, S.C.; Zhou, D.; Almeida, M. Elimination of senescent osteoclast progenitors has no effect on the age-associated loss of bone mass in mice. *Aging Cell* **2019**, *18*, e12923. [CrossRef]
207. Boquoi, A.; Arora, S.; Chen, T.; Litwin, S.; Koh, J.; Enders, G.H. Reversible cell cycle inhibition and premature aging features imposed by conditional expression of p16Ink4a. *Aging Cell* **2015**, *14*, 139–147. [CrossRef] [PubMed]
208. Wouters, K.; Deleye, Y.; Hannou, S.A.; Vanhoutte, J.; Maréchal, X.; Coisne, A.; Tagzirt, M.; Derudas, B.; Bouchaert, E.; Duhem, C.; et al. The tumour suppressor CDKN2A/p16. *Diab. Vasc. Dis. Res.* **2017**, *14*, 516–524. [CrossRef]
209. Matheu, A.; Maraver, A.; Klatt, P.; Flores, I.; Garcia-Cao, I.; Borras, C.; Flores, J.M.; Viña, J.; Blasco, M.A.; Serrano, M. Delayed ageing through damage protection by the Arf/p53 pathway. *Nature* **2007**, *448*, 375–379. [CrossRef]
210. Matheu, A.; Maraver, A.; Collado, M.; Garcia-Cao, I.; Cañamero, M.; Borras, C.; Flores, J.M.; Klatt, P.; Viña, J.; Serrano, M. Anti-aging activity of the Ink4/Arf locus. *Aging Cell* **2009**, *8*, 152–161. [CrossRef]
211. Jun, J.I.; Lau, L.F. The matricellular protein CCN1 induces fibroblast senescence and restricts fibrosis in cutaneous wound healing. *Nat. Cell Biol.* **2010**, *12*, 676–685. [CrossRef]
212. Wang, B.; Wang, L.; Gasek, N.S.; Zhou, Y.; Kim, T.; Guo, C.; Jellison, E.R.; Haynes, L.; Yadav, S.; Tchkonia, T.; et al. An inducible. *Nat. Aging* **2021**, *1*, 962–973. [CrossRef]
213. Shi, J.; Sun, J.; Liu, L.; Shan, T.; Meng, H.; Yang, T.; Wang, S.; Wei, T.; Chen, B.; Ma, Y.; et al. P16ink4a overexpression ameliorates cardiac remodeling of mouse following myocardial infarction via CDK4/pRb pathway. *Biochem. Biophys. Res. Commun.* **2022**, *595*, 62–68. [CrossRef]
214. Kahoul, Y.; Oger, F.; Montaigne, J.; Froguel, P.; Breton, C.; Annicotte, J.S. Emerging Roles for the INK4a/ARF (CDKN2A) Locus in Adipose Tissue: Implications for Obesity and Type 2 Diabetes. *Biomolecules* **2020**, *10*, 1350. [CrossRef]
215. Rabhi, N.; Hannou, S.A.; Gromada, X.; Salas, E.; Yao, X.; Oger, F.; Carney, C.; Lopez-Mejia, I.C.; Durand, E.; Rabearivelo, I.; et al. Cdkn2a deficiency promotes adipose tissue browning. *Mol. Metab.* **2018**, *8*, 65–76. [CrossRef]
216. Helman, A.; Klochendler, A.; Azazmeh, N.; Gabai, Y.; Horwitz, E.; Anzi, S.; Swisa, A.; Condiotti, R.; Granit, R.Z.; Nevo, Y.; et al. p16(Ink4a)-induced senescence of pancreatic beta cells enhances insulin secretion. *Nat. Med.* **2016**, *22*, 412–420. [CrossRef]
217. Pal, A.; Potjer, T.P.; Thomsen, S.K.; Ng, H.J.; Barrett, A.; Scharfmann, R.; James, T.J.; Bishop, D.T.; Karpe, F.; Godsland, I.F.; et al. Loss-of-Function Mutations in the Cell-Cycle Control Gene CDKN2A Impact on Glucose Homeostasis in Humans. *Diabetes* **2016**, *65*, 527–533. [CrossRef]
218. Wang, L.; Wang, B.; Gasek, N.S.; Zhou, Y.; Cohn, R.L.; Martin, D.E.; Zuo, W.; Flynn, W.F.; Guo, C.; Jellison, E.R.; et al. Targeting p21Cip1 highly expressing cells in adipose tissue alleviates insulin resistance in obesity. *Cell Metab.* **2022**, *34*, 75–89. [CrossRef]
219. González-Navarro, H.; Vinué, Á.; Sanz, M.J.; Delgado, M.; Pozo, M.A.; Serrano, M.; Burks, D.J.; Andrés, V. Increased dosage of Ink4/Arf protects against glucose intolerance and insulin resistance associated with aging. *Aging Cell* **2013**, *12*, 102–111. [CrossRef]
220. Krishnamurthy, J.; Ramsey, M.R.; Ligon, K.L.; Torrice, C.; Koh, A.; Bonner-Weir, S.; Sharpless, N.E. p16INK4a induces an age-dependent decline in islet regenerative potential. *Nature* **2006**, *443*, 453–457. [CrossRef]
221. Krizhanovsky, V.; Yon, M.; Dickins, R.A.; Hearn, S.; Simon, J.; Miething, C.; Yee, H.; Zender, L.; Lowe, S.W. Senescence of Activated Stellate Cells Limits Liver Fibrosis. *Cell* **2008**, *134*, 657–667. [CrossRef]
222. Le, O.; Palacio, L.; Bernier, G.; Batinic-Haberle, I.; Hickson, G.; Beauséjour, C. INK4a/ARF Expression Impairs Neurogenesis in the Brain of Irradiated Mice. *Stem Cell Rep.* **2018**, *10*, 1721–1733. [CrossRef] [PubMed]
223. Yabluchanskiy, A.; Tarantini, S.; Balasubramanian, P.; Kiss, T.; Csipo, T.; Fülöp, G.A.; Lipecz, A.; Ahire, C.; DelFavero, J.; Nyul-Toth, A.; et al. Pharmacological or genetic depletion of senescent astrocytes prevents whole brain irradiation-induced impairment of neurovascular coupling responses protecting cognitive function in mice. *Geroscience* **2020**, *42*, 409–428. [CrossRef] [PubMed]

224. Acklin, S.; Zhang, M.; Du, W.; Zhao, X.; Plotkin, M.; Chang, J.; Campisi, J.; Zhou, D.; Xia, F. Depletion of senescent-like neuronal cells alleviates cisplatin-induced peripheral neuropathy in mice. *Sci. Rep.* **2020**, *10*, 14170. [CrossRef] [PubMed]
225. Jeon, O.H.; Kim, C.; Laberge, R.M.; Demaria, M.; Rathod, S.; Vasserot, A.P.; Chung, J.W.; Kim, D.H.; Poon, Y.; David, N.; et al. Local clearance of senescent cells attenuates the development of post-traumatic osteoarthritis and creates a pro-regenerative environment. *Nat. Med.* **2017**, *23*, 775–781. [CrossRef]
226. Chandra, A.; Lagnado, A.B.; Farr, J.N.; Doolittle, M.; Tchkonia, T.; Kirkland, J.L.; LeBrasseur, N.K.; Robbins, P.D.; Niedernhofer, L.J.; Ikeno, Y.; et al. Targeted clearance of p21- but not p16-positive senescent cells prevents radiation-induced osteoporosis and increased marrow adiposity. *Aging Cell* **2022**, *21*, e13602. [CrossRef]
227. Cudejko, C.; Wouters, K.; Fuentes, L.; Hannou, S.A.; Paquet, C.; Bantubungi, K.; Bouchaert, E.; Vanhoutte, J.; Fleury, S.; Remy, P.; et al. p16INK4a deficiency promotes IL-4-induced polarization and inhibits proinflammatory signaling in macrophages. *Blood* **2011**, *118*, 2556–2566. [CrossRef]
228. Fuentes, L.; Wouters, K.; Hannou, S.A.; Cudejko, C.; Rigamonti, E.; Mayi, T.H.; Derudas, B.; Pattou, F.; Chinetti-Gbaguidi, G.; Staels, B.; et al. Downregulation of the tumour suppressor p16INK4A contributes to the polarisation of human macrophages toward an adipose tissue macrophage (ATM)-like phenotype. *Diabetologia* **2011**, *54*, 3150–3156. [CrossRef]
229. Palacio, L.; Goyer, M.L.; Maggiorani, D.; Espinosa, A.; Villeneuve, N.; Bourbonnais, S.; Moquin-Beaudry, G.; Le, O.; Demaria, M.; Davalos, A.R.; et al. Restored immune cell functions upon clearance of senescence in the irradiated splenic environment. *Aging Cell* **2019**, *18*, e12971. [CrossRef]
230. Sadhu, S.; Decker, C.; Sansbury, B.E.; Marinello, M.; Seyfried, A.; Howard, J.; Mori, M.; Hosseini, Z.; Arunachalam, T.; Finn, A.V.; et al. Radiation-Induced Macrophage Senescence Impairs Resolution Programs and Drives Cardiovascular Inflammation. *J. Immunol.* **2021**, *207*, 1812–1823. [CrossRef]
231. Cano, E.; Carmona, R.; Ruiz-Villalba, A.; Rojas, A.; Chau, Y.Y.; Wagner, K.D.; Wagner, N.; Hastie, N.D.; Muñoz-Chápuli, R.; Pérez-Pomares, J.M. Extracardiac septum transversum/proepicardial endothelial cells pattern embryonic coronary arterio-venous connections. *Proc. Natl. Acad. Sci. USA* **2016**, *113*, 656–661. [CrossRef]
232. Wagner, K.D.; Wagner, N.; Bondke, A.; Nafz, B.; Flemming, B.; Theres, H.; Scholz, H. The Wilms' tumor suppressor Wt1 is expressed in the coronary vasculature after myocardial infarction. *FASEB J.* **2002**, *16*, 1117–1119. [CrossRef]
233. Cianflone, E.; Aquila, I.; Scalise, M.; Marotta, P.; Torella, M.; Nadal-Ginard, B.; Torella, D. Molecular basis of functional myogenic specification of Bona Fide multipotent adult cardiac stem cells. *Cell Cycle* **2018**, *17*, 927–946. [CrossRef]
234. Vicinanza, C.; Aquila, I.; Scalise, M.; Cristiano, F.; Marino, F.; Cianflone, E.; Mancuso, T.; Marotta, P.; Sacco, W.; Lewis, F.C.; et al. Adult cardiac stem cells are multipotent and robustly myogenic: C-kit expression is necessary but not sufficient for their identification. *Cell Death Differ.* **2017**, *24*, 2101–2116. [CrossRef]
235. Epstein, J.A. A Time to Press Reset and Regenerate Cardiac Stem Cell Biology. *JAMA Cardiol.* **2019**, *4*, 95–96. [CrossRef] [PubMed]
236. Hannou, S.A.; Wouters, K.; Paumelle, R.; Staels, B. Functional genomics of the CDKN2A/B locus in cardiovascular and metabolic disease: What have we learned from GWASs? *Trends Endocrinol. Metab.* **2015**, *26*, 176–184. [CrossRef] [PubMed]
237. Kong, Y.; Sharma, R.B.; Nwosu, B.U.; Alonso, L.C. Islet biology, the CDKN2A/B locus and type 2 diabetes risk. *Diabetologia* **2016**, *59*, 1579–1593. [CrossRef]
238. Morris, A.P.; Voight, B.F.; Teslovich, T.M.; Ferreira, T.; Segrè, A.V.; Steinthorsdottir, V.; Strawbridge, R.J.; Khan, H.; Grallert, H.; Mahajan, A.; et al. Large-scale association analysis provides insights into the genetic architecture and pathophysiology of type 2 diabetes. *Nat. Genet.* **2012**, *44*, 981–990. [CrossRef] [PubMed]
239. Palmer, A.K.; Xu, M.; Zhu, Y.; Pirtskhalava, T.; Weivoda, M.M.; Hachfeld, C.M.; Prata, L.G.; van Dijk, T.H.; Verkade, E.; Casaclang-Verzosa, G.; et al. Targeting senescent cells alleviates obesity-induced metabolic dysfunction. *Aging Cell* **2019**, *18*, e12950. [CrossRef]
240. Lagarrigue, S.; Lopez-Mejia, I.C.; Denechaud, P.D.; Escoté, X.; Castillo-Armengol, J.; Jimenez, V.; Chavey, C.; Giralt, A.; Lai, Q.; Zhang, L.; et al. CDK4 is an essential insulin effector in adipocytes. *J. Clin. Invest.* **2016**, *126*, 335–348. [CrossRef]
241. Rane, S.G.; Dubus, P.; Mettus, R.V.; Galbreath, E.J.; Boden, G.; Reddy, E.P.; Barbacid, M. Loss of Cdk4 expression causes insulin-deficient diabetes and Cdk4 activation results in beta-islet cell hyperplasia. *Nat. Genet.* **1999**, *22*, 44–52. [CrossRef]
242. Kim, J.A.; Hong, S.; Lee, B.; Hong, J.W.; Kwak, J.Y.; Cho, S.; Kim, C.C. The inhibition of T-cells proliferation by mouse mesenchymal stem cells through the induction of p16INK4A-cyclin D1/cdk4 and p21waf1, p27kip1-cyclin E/cdk2 pathways. *Cell Immunol.* **2007**, *245*, 16–23. [CrossRef]
243. Li, H.; Collado, M.; Villasante, A.; Strati, K.; Ortega, S.; Cañamero, M.; Blasco, M.A.; Serrano, M. The Ink4/Arf locus is a barrier for iPS cell reprogramming. *Nature* **2009**, *460*, 1136–1139. [CrossRef]
244. Aguayo-Mazzucato, C.; van Haaren, M.; Mruk, M.; Lee, T.B.; Crawford, C.; Hollister-Lock, J.; Sullivan, B.A.; Johnson, J.W.; Ebrahimi, A.; Dreyfuss, J.M.; et al. β Cell Aging Markers Have Heterogeneous Distribution and Are Induced by Insulin Resistance. *Cell Metab.* **2017**, *25*, 898–910. [CrossRef]
245. Kong, Y.; Sharma, R.B.; Ly, S.; Stamateris, R.E.; Jesdale, W.M.; Alonso, L.C. T2D Genome-Wide Association Study Risk SNPs Impact Locus Gene Expression and Proliferation in Human Islets. *Diabetes* **2018**, *67*, 872–884. [CrossRef]
246. Bantubungi, K.; Hannou, S.A.; Caron-Houde, S.; Vallez, E.; Baron, M.; Lucas, A.; Bouchaert, E.; Paumelle, R.; Tailleux, A.; Staels, B. Cdkn2a/p16Ink4a regulates fasting-induced hepatic gluconeogenesis through the PKA-CREB-PGC1α pathway. *Diabetes* **2014**, *63*, 3199–3209. [CrossRef]

247. El-Athman, R.; Genov, N.N.; Mazuch, J.; Zhang, K.; Yu, Y.; Fuhr, L.; Abreu, M.; Li, Y.; Wallach, T.; Kramer, A.; et al. The Ink4a/Arf locus operates as a regulator of the circadian clock modulating RAS activity. *PLoS Biol.* **2017**, *15*, e2002940. [CrossRef]
248. Price, J.D.; Park, K.Y.; Chen, J.; Salinas, R.D.; Cho, M.J.; Kriegstein, A.R.; Lim, D.A. The Ink4a/Arf locus is a barrier to direct neuronal transdifferentiation. *J. Neurosci.* **2014**, *34*, 12560–12567. [CrossRef]
249. Ma, K.H.; Duong, P.; Moran, J.J.; Junaidi, N.; Svaren, J. Polycomb repression regulates Schwann cell proliferation and axon regeneration after nerve injury. *Glia* **2018**, *66*, 2487–2502. [CrossRef]
250. Gomez-Sanchez, J.A.; Gomis-Coloma, C.; Morenilla-Palao, C.; Peiro, G.; Serra, E.; Serrano, M.; Cabedo, H. Epigenetic induction of the Ink4a/Arf locus prevents Schwann cell overproliferation during nerve regeneration and after tumorigenic challenge. *Brain* **2013**, *136*, 2262–2278. [CrossRef]
251. Takeuchi, S.; Takahashi, A.; Motoi, N.; Yoshimoto, S.; Tajima, T.; Yamakoshi, K.; Hirao, A.; Yanagi, S.; Fukami, K.; Ishikawa, Y.; et al. Intrinsic cooperation between p16INK4a and p21Waf1/Cip1 in the onset of cellular senescence and tumor suppression in vivo. *Cancer Res.* **2010**, *70*, 9381–9390. [CrossRef]
252. Wiley, C.D.; Liu, S.; Limbad, C.; Zawadzka, A.M.; Beck, J.; Demaria, M.; Artwood, R.; Alimirah, F.; Lopez-Dominguez, J.A.; Kuehnemann, C.; et al. SILAC Analysis Reveals Increased Secretion of Hemostasis-Related Factors by Senescent Cells. *Cell Rep.* **2019**, *28*, 3329–3337. [CrossRef] [PubMed]
253. Kaur, G.; Sundar, I.K.; Rahman, I. p16-3MR: A Novel Model to Study Cellular Senescence in Cigarette Smoke-Induced Lung Injuries. *Int. J. Mol. Sci.* **2021**, *22*, 4834. [CrossRef] [PubMed]
254. Dellambra, E.; Golisano, O.; Bondanza, S.; Siviero, E.; Lacal, P.; Molinari, M.; D'Atri, S.; De Luca, M. Downregulation of 14-3-3sigma prevents clonal evolution and leads to immortalization of primary human keratinocytes. *J. Cell Biol* **2000**, *149*, 1117–1130. [CrossRef] [PubMed]
255. Lewis, J.L.; Chinswangwatanakul, W.; Zheng, B.; Marley, S.B.; Nguyen, D.X.; Cross, N.C.; Banerji, L.; Glassford, J.; Thomas, N.S.; Goldman, J.M.; et al. The influence of INK4 proteins on growth and self-renewal kinetics of hematopoietic progenitor cells. *Blood* **2001**, *97*, 2604–2610. [CrossRef] [PubMed]
256. Sharpless, N.E.; DePinho, R.A. Telomeres, stem cells, senescence, and cancer. *J. Clin. Invest.* **2004**, *113*, 160–168. [CrossRef] [PubMed]
257. D'Arcangelo, D.; Tinaburri, L.; Dellambra, E. The Role of p16INK4a Pathway in Human Epidermal Stem Cell Self-Renewal, Aging and Cancer. *Int J. Mol. Sci.* **2017**, *18*, 1591. [CrossRef]
258. Janzen, V.; Forkert, R.; Fleming, H.E.; Saito, Y.; Waring, M.T.; Dombkowski, D.M.; Cheng, T.; DePinho, R.A.; Sharpless, N.E.; Scadden, D.T. Stem-cell ageing modified by the cyclin-dependent kinase inhibitor p16INK4a. *Nature* **2006**, *443*, 421–426. [CrossRef]
259. García-Cao, I.; García-Cao, M.; Martín-Caballero, J.; Criado, L.M.; Klatt, P.; Flores, J.M.; Weill, J.C.; Blasco, M.A.; Serrano, M. "Super p53" mice exhibit enhanced DNA damage response, are tumor resistant and age normally. *EMBO J.* **2002**, *21*, 6225–6235. [CrossRef]
260. Matheu, A.; Pantoja, C.; Efeyan, A.; Criado, L.M.; Martín-Caballero, J.; Flores, J.M.; Klatt, P.; Serrano, M. Increased gene dosage of Ink4a/Arf results in cancer resistance and normal aging. *Genes Dev.* **2004**, *18*, 2736–2746. [CrossRef]
261. García-Cao, I.; García-Cao, M.; Tomás-Loba, A.; Martín-Caballero, J.; Flores, J.M.; Klatt, P.; Blasco, M.A.; Serrano, M. Increased p53 activity does not accelerate telomere-driven ageing. *EMBO Rep.* **2006**, *7*, 546–552. [CrossRef]
262. Menendez, S.; Camus, S.; Herreria, A.; Paramonov, I.; Morera, L.B.; Collado, M.; Pekarik, V.; Maceda, I.; Edel, M.; Consiglio, A.; et al. Increased dosage of tumor suppressors limits the tumorigenicity of iPS cells without affecting their pluripotency. *Aging Cell* **2012**, *11*, 41–50. [CrossRef]
263. Carrasco-Garcia, E.; Arrizabalaga, O.; Serrano, M.; Lovell-Badge, R.; Matheu, A. Increased gene dosage of Ink4/Arf and p53 delays age-associated central nervous system functional decline. *Aging Cell* **2015**, *14*, 710–714. [CrossRef]
264. Li, Y.; Liu, J.; Li, W.; Brown, A.; Baddoo, M.; Li, M.; Carroll, T.; Oxburgh, L.; Feng, Y.; Saifudeen, Z. p53 Enables metabolic fitness and self-renewal of nephron progenitor cells. *Development* **2015**, *142*, 1228–1241. [CrossRef]
265. Tomás-Loba, A.; Flores, I.; Fernández-Marcos, P.J.; Cayuela, M.L.; Maraver, A.; Tejera, A.; Borrás, C.; Matheu, A.; Klatt, P.; Flores, J.M.; et al. Telomerase reverse transcriptase delays aging in cancer-resistant mice. *Cell* **2008**, *135*, 609–622. [CrossRef]
266. Carrasco-Garcia, E.; Moreno, M.; Moreno-Cugnon, L.; Matheu, A. Increased Arf/p53 activity in stem cells, aging and cancer. *Aging Cell* **2017**, *16*, 219–225. [CrossRef]
267. Tyner, S.D.; Venkatachalam, S.; Choi, J.; Jones, S.; Ghebranious, N.; Igelmann, H.; Lu, X.; Soron, G.; Cooper, B.; Brayton, C.; et al. p53 mutant mice that display early ageing-associated phenotypes. *Nature* **2002**, *415*, 45–53. [CrossRef]
268. Maier, B.; Gluba, W.; Bernier, B.; Turner, T.; Mohammad, K.; Guise, T.; Sutherland, A.; Thorner, M.; Scrable, H. Modulation of mammalian life span by the short isoform of p53. *Genes Dev.* **2004**, *18*, 306–319. [CrossRef]
269. Dumble, M.; Moore, L.; Chambers, S.M.; Geiger, H.; Van Zant, G.; Goodell, M.A.; Donehower, L.A. The impact of altered p53 dosage on hematopoietic stem cell dynamics during aging. *Blood* **2007**, *109*, 1736–1742. [CrossRef]
270. Gatza, C.E.; Dumble, M.; Kittrell, F.; Edwards, D.G.; Dearth, R.K.; Lee, A.V.; Xu, J.; Medina, D.; Donehower, L.A. Altered mammary gland development in the p53+/m mouse, a model of accelerated aging. *Dev. Biol.* **2008**, *313*, 130–141. [CrossRef]
271. Medrano, S.; Burns-Cusato, M.; Atienza, M.B.; Rahimi, D.; Scrable, H. Regenerative capacity of neural precursors in the adult mammalian brain is under the control of p53. *Neurobiol. Aging* **2009**, *30*, 483–497. [CrossRef]
272. Hinault, C.; Kawamori, D.; Liew, C.W.; Maier, B.; Hu, J.; Keller, S.R.; Mirmira, R.G.; Scrable, H.; Kulkarni, R.N. Δ40 Isoform of p53 controls β-cell proliferation and glucose homeostasis in mice. *Diabetes* **2011**, *60*, 1210–1222. [CrossRef] [PubMed]

273. Zhang, Y.; Shao, C.; Li, H.; Wu, K.; Gong, L.; Zheng, Q.; Dan, J.; Jia, S.; Tang, X.; Wu, X.; et al. The Distinct Function of p21. *Front. Genet.* **2021**, *12*, 597566. [CrossRef] [PubMed]
274. Sturmlechner, I.; Zhang, C.; Sine, C.C.; van Deursen, E.J.; Jeganathan, K.B.; Hamada, N.; Grasic, J.; Friedman, D.; Stutchman, J.T.; Can, I.; et al. p21 produces a bioactive secretome that places stressed cells under immunosurveillance. *Science* **2021**, *374*, eabb3420. [CrossRef] [PubMed]
275. Mahmoudi, S.; Mancini, E.; Xu, L.; Moore, A.; Jahanbani, F.; Hebestreit, K.; Srinivasan, R.; Li, X.; Devarajan, K.; Prélot, L.; et al. Heterogeneity in old fibroblasts is linked to variability in reprogramming and wound healing. *Nature* **2019**, *574*, 553–558. [CrossRef]
276. Consortium, T.M. A single-cell transcriptomic atlas characterizes ageing tissues in the mouse. *Nature* **2020**, *583*, 590–595. [CrossRef]
277. Schaum, N.; Lehallier, B.; Hahn, O.; Pálovics, R.; Hosseinzadeh, S.; Lee, S.E.; Sit, R.; Lee, D.P.; Losada, P.M.; Zardeneta, M.E.; et al. Ageing hallmarks exhibit organ-specific temporal signatures. *Nature* **2020**, *583*, 596–602. [CrossRef]
278. Xu, M.; Pirtskhalava, T.; Farr, J.N.; Weigand, B.M.; Palmer, A.K.; Weivoda, M.M.; Inman, C.L.; Ogrodnik, M.B.; Hachfeld, C.M.; Fraser, D.G.; et al. Senolytics improve physical function and increase lifespan in old age. *Nat. Med.* **2018**, *24*, 1246–1256. [CrossRef]
279. Chen, R.; Skutella, T. Synergistic Anti-Ageing through Senescent Cells Specific Reprogramming. *Cells* **2022**, *11*, 830. [CrossRef]

cells

MDPI

Review

Synergistic Anti-Ageing through Senescent Cells Specific Reprogramming

Rui Chen and Thomas Skutella *

Group for Regeneration and Reprogramming, Medical Faculty, Department of Neuroanatomy, Institute for Anatomy and Cell Biology, Heidelberg University, 69120 Heidelberg, Germany; rui.chen@uni-heidelberg.de
* Correspondence: thomas.skutella@uni-heidelberg.de

Abstract: In this review, we seek a novel strategy for establishing a rejuvenating microenvironment through senescent cells specific reprogramming. We suggest that partial reprogramming can produce a secretory phenotype that facilitates cellular rejuvenation. This strategy is desired for specific partial reprogramming under control to avoid tumour risk and organ failure due to loss of cellular identity. It also alleviates the chronic inflammatory state associated with ageing and secondary senescence in adjacent cells by improving the senescence-associated secretory phenotype. This manuscript also hopes to explore whether intervening in cellular senescence can improve ageing and promote damage repair, in general, to increase people's healthy lifespan and reduce frailty. Feasible and safe clinical translational protocols are critical in rejuvenation by controlled reprogramming advances. This review discusses the limitations and controversies of these advances' application (while organizing the manuscript according to potential clinical translation schemes) to explore directions and hypotheses that have translational value for subsequent research.

Keywords: ageing; senescence; senolytics/senostatics; p16^{Ink4a}; p19^{Arf}; p21$^{Waf1/Cip1}$; SASP

Citation: Chen, R.; Skutella, T. Synergistic Anti-Ageing through Senescent Cells Specific Reprogramming. *Cells* **2022**, *11*, 830. https://doi.org/10.3390/cells11050830

Academic Editors: Nicole Wagner and Kay-Dietrich Wagner

Received: 10 January 2022
Accepted: 24 February 2022
Published: 28 February 2022

1. Introduction

Ageing can be defined as a time-dependent decline in the functionality of the body. At the cellular level, its essence can be seen as a gradual loss of normal cell function accompanied by a series of ageing phenotypes [1,2]. Stress factors such as telomere dysfunction, DNA damage, oncogene activation, and organelle dysfunction accelerate the progression of senescence at the cellular level, spreading through the cellular microenvironment and accelerating organ dysfunction throughout the tissues, culminating in the loss of all vital functions of the body (organ failure).

Interestingly, there is a self-balancing of repair after injury and renewal after ageing in the organism itself. These balancing mechanisms include the conversion of resting stem cells to progenitor cells in various tissues within dynamic homeostasis. Alternative or anti-dysfunctional mechanisms act at the organelle level of dysfunction (e.g., alternative mitochondrial glycolytic and glutamine metabolic pathways, activation of nuclear repair through nuclear damage) [3–5]. At the tissue level, the senescence-associated secretory phenotype (SASP) has some anti-ageing potential in addition to being a senescence-promoting factor (e.g., interleukin-6 (IL-6) can enhance tissue repair by promoting reprogramming [6,7]; interleukin-1 (IL-1) can promote the clearance of senescent cells by immunopurified NK cells [8–10]). It is important to note that there is no evidence that anti-ageing can be achieved by exploiting the beneficial aspects of ageing itself. This review aims to show that reprogramming interventions on senescent cells have the potential to retain these valuable components compared with senescent cell removal. There is also the possibility that reprogramming can be self-modified by the above result, in which senescence promotes reprogramming: the same expression level of reprogramming factors may be more effective in the senescent environment. In contrast, the efficiency of reprogramming is reduced when the senescent environment is mitigated or reversed, thus avoiding the harm caused by over-induction.

To explore the anti-ageing potential of this balance, this review will discuss the possibility and feasibility of attenuating senescence through intercellular interactions as follows:

a. To propose synergistic anti-ageing in order to clarify the idea of combined anti-ageing with multiple rejuvenating factors and to serve the next research on new key pathways that can be applied in combination with known anti-ageing pathways.
b. The hypothesis of a youthful secretory phenotype is proposed to generalise the anti-ageing factors (particularly, NAD^+, eNAMPT, GSTM2, etc. are involved in whole body anti-ageing by regulating the circulating NAD^+/NADH balance) found in the secretome of young blood and young cells and to serve for future clinical translation and co-application.
c. To propose the hypothesis that controlled reprogramming (defined as the induction of Yamanaka factors expression to reverse the ageing phenotype of cells but without iPSCs-induced pluripotent stem cell formation) may synergistically anti-age by a youthful secretory phenotype.

2. The Characteristics of Ageing and their Potential for Translation

2.1. *"Asynchronous Effect" in Ageing*

As we age, the ageing of different organs and tissues is not synchronized, and the parenchymal cells that perform the biological functions of organs/tissues are in different stages of senescence [1].

It has been shown in mice that plasma cells and the antibodies they secrete infiltrate various organs, appearing in the kidney, heart, liver, muscle, fat, lung, and thymus [2]. This implies that ageing in one organ may trigger or accelerate ageing-related chronic inflammation and dysfunction throughout the body via the systemic circulation. Furthermore, when mice reach middle age, immune cells (T and B cells [2], M1 macrophages [11,12]) are extensively activated in adipose tissue [2]. These studies suggest that ageing and immunity are inextricably linked, and ageing is an "asynchronous" process. On the other hand, adipose tissue is one of the first areas of the body to show senescence-related phenotypes (inflammatory cell infiltration and the appearance of senescence-related secretory phenotypes) [1,2].

In fact, the removal of senescent cells can reduce the adverse effects of pre-senescent fractions (e.g., senescent cell ablation-senolytics) [13]. As cellular senescence is not synchronized, the senescence microenvironments originating from pre-senescent cells can cause a vicious cycle (the senescence of one organ promoting the decline throughout the body) at the tissue level [1,2]. The "old factors" that promote ageing may be diluted or suppressed by another "rejuvenating factor (existing in young blood as well as being secreted by a heterogeneous subgroup of ageing cells)", thus acting as a rejuvenating agent [14,15]. Through transplanting pre-senescent adipocytes, it was found that the senescence of a small number of adipose precursor cells was sufficient to induce organ senescence in juvenile mice. The removal of transplanted senescent cells from young mice and naturally senescent cells from naturally senescent mice by intermittent oral senolytics improved ageing (organ function enhanced, survival increased by 36%, and the risk of death reduced by 65%) [13].

The heterogeneity of ageing is reflected on the one hand in the fact that chronic inflammation differs in different tissues, specifically in the fact that SASP has different levels in different tissues [2]. On the other hand, it is reflected in the different sequences of appearance of senescent cells and the different rates of accumulation [1].

For example, in the kidney, where senescent cells increase significantly with age, Cu/Zn-superoxide dismutase (Sod1) knockout mice result in high levels of oxidative cellular senescence [16]. Senescence-related secretory phenotypes (particularly IL-6 and IL-1β) are also significantly increased [16]. The higher levels of circulating cytokines suggest that the accelerated senescence phenotype may be due to increased inflammation caused by the accelerated accumulation of senescent cells [17]. The accumulation of senescent cells, in turn, caused an increase in chronic inflammation [16].

Remarkably, asynchrony of senescence persists even in the same ageing cells (fibroblasts in aged mice) [14]. Different subpopulations with different secretory phenotypes influence the rate of wound healing in vivo by affecting reprogramming efficiency [14].

Taken together, it is feasible to intervene in pre-senescent tissues for overall benefit by taking advantage of the asynchrony in ageing [18]. It is also known that the microenvironment has a key influence on the state of cell ageing (cells exposed to the secretome of senescent cells will age faster) [13]. Rather, improving the cellular micro-environment (increasing the secretory phenotype of young cells) to combat ageing is a promising direction [19]. In particular, adipose tissue is one of the first to be affected by ageing (also having a key influence on the inflammatory state associated with ageing) [1,2].

2.2. "Synergistic Effect" in Anti-Ageing

It is shown that blood therapy involves several different anti-ageing factors (GDF11 [20], GPLD1 [21], clusterin [22], Klotho [23], etc.). On the one hand, it leads to controversy [19], but on the other hand, it also suggests that greater benefits may be achieved by synchronizing multiple factors to combat ageing. Synergistic anti-ageing is a phenomenon in which the combined modulation of multiple anti-ageing factors produces a higher effect than the sum of the effects of modulating any of them individually. Therefore, studies of ageing patterns are instructive when discovering critical pathway of anti-ageing with synergistic potential (e.g., in simple eukaryotes).

Some yeast cells show significant nuclear stability changes during cell ageing and exhibit ribosomal senescence, while others develop mitochondrial dysfunction. In yeast with a ribosomal senescence pattern, overexpression of Sir2 (a lysine deacetylase that contributes to ribosomal DNA silencing) would extend the average lifespan of the yeast [24,25]. Overexpression of Sir2 and Hap4 prolongs lifespan by producing synergistic rather than additive effects [26]. A similar synergistic effect is observed when the fob1Δ longevity mutant enhances rDNA stability combined with Hap4 overexpression [25]. This model also explains the anti-ageing synergy between caloric restriction, promoting heme activator protein (HAP) and Sir2 [26]. These two seemingly independent lifespan factors can be understood as two key anti-ageing nodes and targets with synergistic anti-ageing effects. Both could be considered critical anti-ageing nodes that need to be regulated simultaneously.

In another study using *Caenorhabditis elegans* (*C. elegans*) as a model, an essential regulator gene called CYC-2.1 (a nematode cytochrome C ortholog), a cytochrome strongly associated with mitochondrial ageing was identified. Reducing CYC-2.1 expression activated the "unfolded protein response" in mitochondria, promoting their division and thus significantly extending nematode lifespan [27]. The rsks-1 (the *C. elegans* ribosomal S6K ortholog) mutation increased average lifespan by 20%, the daf-2 (a nematode insulin growth factor 1 receptor ortholog) mutation increased the average lifespan by 169%, and the daf-2 and rsks-1 double mutation increased the average lifespan by 454% over the wild type; thus, the increased longevity of the daf-2 and risk-1 double mutants is not simply additive but has a synergistic effect on longevity [28]. On the other hand, TOR (target of rapamycin) regulates mRNA translation levels via ribosomal S6 kinase (S6K) [29]; therefore, it demonstrates a significant synergistic anti-mitochondrial ageing effect of the IIS (insulin/insulin-like signalling) and TOR [27]. This study also suggests that synergistic regulation of ribosomal protein genes and mitochondrial function could increase the synergistic effects of key anti-ageing factors to a greater extent. It also implies that mitochondrial function regulation could be achieved through metabolic reprogramming induced by remote intercellular regulation and could elicit a broader response from various immune-metabolic cells and organs via an anti-ageing secretory phenotype.

Reactive oxygen species (ROS)-induced DNA damage response (DDR) activates mTORC1 through direct phosphorylation of protein kinase B (PKB/Akt) by ATM, and activated Akt directly phosphorylates the TSC1/TSC2 complex, in this way activating mTORC1.

Activation of mTORC1 promotes ROS-dependent DDR, and, through the mitochondrial biogenesis transcriptional co-activator peroxisome proliferator-activated receptor-gamma coactivator-1beta (PGC-1β), promotes ageing phenotypes (e.g., ASAP), ultimately resulting in ROS-mediated DDR activation (upregulation of DDR protein γH2A.X) and cell cycle arrest (with reduced expression of $p21^{Waf1/Cip1}$ and $p16^{INK4a}$) [4]. This apparent vicious circle, if not broken, results in increasing senescence. Thus, improvements in mitochondrial function and altered redox status are key factors in breaking this impasse.

Mitochondrial dysfunction-associated senescence (MiDAS) leads to a reduction in the $NAD^+/NADH$ ratio, which leads to activation of AMPK and p53, which leads to both growth arrest of senescent cells (caused by p53 activation, with pyruvate preventing MiDAS growth arrest but restoring NF-κB activity) and AMPK-mediated p53 activation reducing IL-1 secretion [3]. This implies that there are multiple and intricate patterns of ASAP such that reprogramming strategies dependent on IL-6 boosting efficiency can break the mitochondrial dysfunction (MiD)-ROS-dependent DDR vicious cycle by modulating the $NAD^+/NADH$ ratio (possibly in parallel with the pyruvate response) while responding to the promotion of the senescence microenvironment. It implies that the emergence and persistence of rejuvenation microenvironments in the blood [30] (e.g., endocrine rejuvenation microenvironment), as well as immune rejuvenation microenvironments, are crucial because of their systemic nature (affecting almost all cells) and the breadth of their effects (participating in nearly all rejuvenation-related pathways).

Short-term exposure to Oct4, Sox2, Klf4, and c-Myc (OSKM) (also called "Yamanaka factors") reverses the ageing phenotype of cells [31], demonstrating that senescence is reversible [32]. This means that rejuvenating senescent cells is a new strategy for disrupting the vicious cycle of ageing by creating dynamic rejuvenation homeostasis in multiple pathways together. However, it is important to note that premature termination of reprogramming can lead to failure in the rejuvenation of MSCs [33]. Thus, partial reprogramming (defined as inducing Yamanaka factors expression to reverse the ageing phenotype of cells but without iPSCs-induced pluripotent stem cells forming) is a potential anti-ageing intervention [31] (Figure 1).

3. Strategies for Reversing Senescence and the Potential Underlying Mechanisms

3.1. Reprogramming-Based Therapies to Reverse Senescence

Partial reprogramming simultaneously lengthens telomeres, inhibits p53, and restores mitochondrial function [31]. Interestingly, the telomerase reverse transcriptase overexpression in transgenic mice (Sp53/Sp16/SArf/Tg Tert mice) showed improved tumour resistance and was found to prevent ageing-related degeneration (mainly atrophy) and inflammatory processes, higher blood levels of IGF1, and a reduction in γ-H2AX foci. Increased glucose tolerance and neuromuscular coordination cause a longer average lifespan [40]. The telomere–p53–PGC pathway and its downstream gene network regulate the functional state of multiple organs and ageing: increased levels of p53 (Trp53) lead to inhibition of peroxisome proliferator-activated receptor-gamma coactivator-1 alpha (PGC-1α). The germline deletion of p53 fully restores PGC network expression; PGC-1α expression restores mitochondrial respiration, cardiac function, and glucose allosterism [41]. Furthermore, reducing peroxisome proliferator-activated receptor-gamma coactivator-1beta (PGC-1b) attenuates cellular senescence-related phenotypes [4]. This implies that short-term cyclic expression of OSKM can rejuvenate senescent cells' epigenome in vivo, reduce $p16^{Ink4a}$ and SASP, and affect various senescence-related regulatory pathways (such as mitochondria dysfunction, DNA damage, impaired protein folding, telomere shortening, and inflammation [31]), thus exerting a synergistic anti-ageing effect.

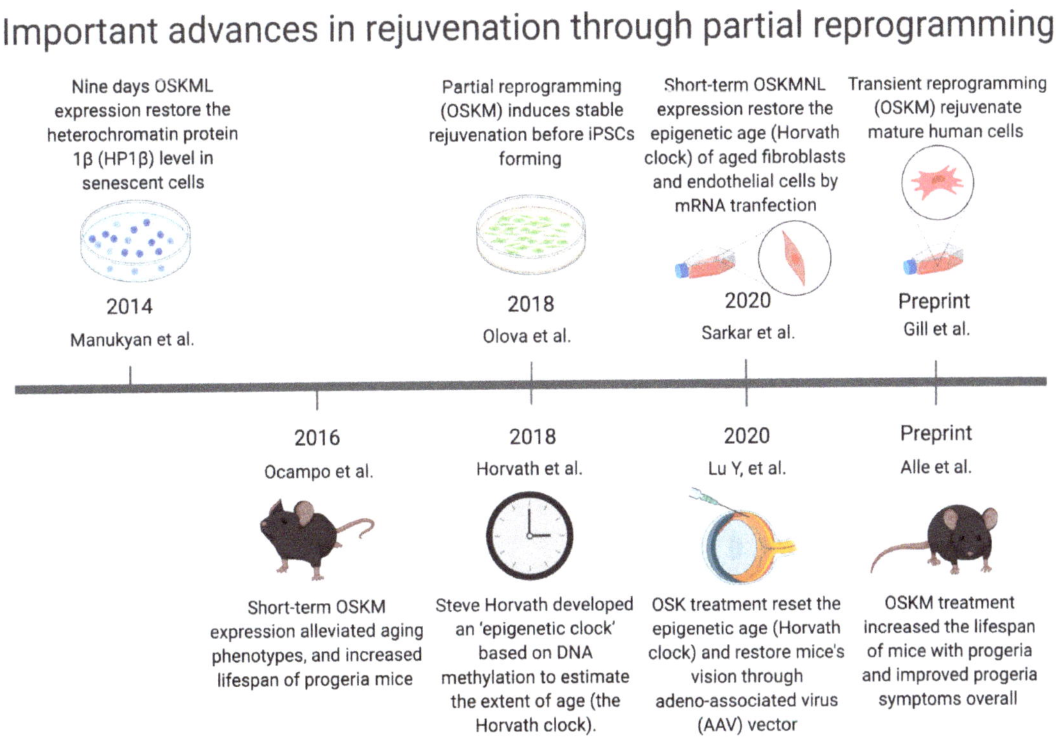

Figure 1. Important advances in rejuvenation through partial reprogramming. Manukyan et al.: Nine days OSKML expression restored the heterochromatin protein 1β (HP1β) level in senescent human fibroblasts [32]. Ocampo et al.: Short-term OSKM expression alleviated ageing phenotypes and increased lifespan of the progeria mice (LAKI 4F mice) [31]. Olova et al.: Partial reprogramming (OSKM) induced stable rejuvenation of adult human fibroblasts before iPSCs forming [34]. Horvath et al.: Steve Horvath developed an "epigenetic clock" based on DNA methylation to estimate the extent of age (the Horvath clock) [35]. Sarkar et al.: Short-term OSKMNL expression restored the epigenetic age (Horvath clock) of aged human fibroblasts and endothelial cells by mRNA transfection [36]. Lu Y et al.: OSK treatment reset the epigenetic age (Horvath clock) and restored mice's vision through adeno-associated virus (AAV) vector [37]. Gill et al.: Transient reprogramming (OSKM) rejuvenated mature human cells [38]. Alle et al.: OSKM treatment increased the lifespan and improved premature phenotypes in the progeria mice [39].

Due to the "asynchronous" character of ageing, senescent cells reprogramming preferentially affects the tissues that are first influenced by ageing (e.g., adipose tissue, the immune system, and fibroblasts [1,2]). We, therefore, start our discussion with adipose tissue (Figure 2). Ageing is often accompanied by a decline in subcutaneous adipocytes marked by the depletion of adipose precursor cells [42], which in turn causes a change in fat tissue distribution—i.e., more visceral white fat and less brown fat [43,44] as well as ectopic fat deposits [45]. This transformation leads to a vicious circle of producing an ageing microenvironment through an imbalance in the inflammatory state and cellular metabolic state associated with ageing and, consequently, a disruption of cellular homeostasis (proteostasis) [46].

Senescence of adipose precursor cells (caused by sirtuin 1 reduction) leads to the accumulation of senescent adipocytes [43], which secrete pro-inflammatory factors that constitute the first part of the senescent microenvironment and cause chronic inflammatory infiltration of adipose tissue [47]. As ageing redistributes fat (visceral fat increases), senescent adipose tissue carries the chronic inflammatory state associated with senescence (Mcp-1 and Il-6) throughout the body and gradually accumulates.

Increased white adipose tissue causes a decrease in glutamine levels in adipose tissue, leading to increased macrophage glycolysis in adipose tissue, increased pro-inflammatory transcription, and secretion of large amounts of SASP into the peripheral microcirculation, generating a second part of the senescent microenvironment [48].

M1 macrophages in senescent white adipose tissue consume large amounts of NAD+ [11,12], and adipocytes secrete less eNAMPT due to senescence [49,50], resulting in a systemic $NAD^+/NADH$ ratio imbalance (lower), which accelerates mitochondrial dysfunction-related senescence in cells throughout the body [3], resulting in an imbalance in energy metabolic status (glycolysis increase) and creating the third part of the senescent microenvironment.

Mitochondrial metabolic disorders cause enhanced glycolytic pathways and cellular redox disorders, resulting in systemic redox disorders [3,4]. Systemic fibroblasts under the influence of the first three parts of senescence and their own senescence, decrease GST secretion, exacerbating systemic peroxidation and creating the fourth part of the senescence microenvironment [51].

The redox disorder strongly affects genomic stability [5], generating many misconfigured proteins, which form aggregates that are expelled from the cells and adipocytes, which also discharge aged mitochondria, forming the fifth part of the senescent microenvironment [52–54].

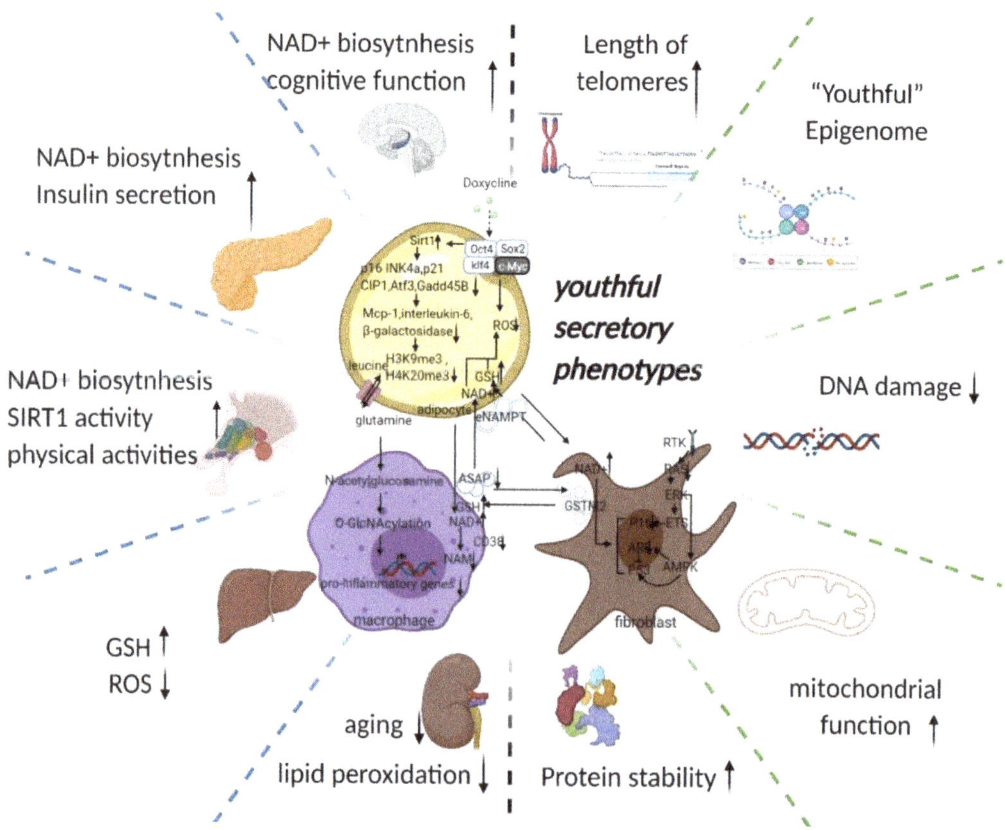

Figure 2. Potential intercellular mechanisms related to senescent cells specific reprogramming. Senescence in adipose precursor cells can be improved directly or indirectly (via reduced p21 and p16

pathways by overexpression of Sirt1 [43,55]) by doxycycline-induced overexpression of OSKM. The reversal of senescence by reprogramming can comprehensively improve senescence indicators (decrease in p16, p21, senescence-associated β-galactosidase, etc.) and can, at the same time, ameliorate senescence-associated secretory phenotypes (decreased Mcp-1 and Il-6, MMP13) and even improve histone methylation status (decrease in H3K9me3, H4K20me3) [31]. With the rejuvenation of adipose tissue (telomere lengthening, phenotypic rejuvenation remodelling, and promotion of gene damage repair), the upregulation of adipocyte glutaminase 1 [56] is reversed and the tissue is therefore rescued from the glutamine depleted state caused by ageing. Increased levels of glutamine will improve the chronic inflammatory state associated with ageing on a systemic scale by reducing the transcription of pro-inflammatory genes in macrophages in adipose tissue [48]. This means that the production of senescence-associated secretory phenotypes is reduced, thereby favouring the maintenance of a youthful state in surrounding fibroblasts, adipocytes, and themselves. The reprogramming also promotes the production of secretory eNAMPT in extracellular vesicles. By altering the NAD+ content of cells to regulate their mitochondrial metabolic state and redox homeostasis, eNAMPT promotes the rejuvenation of various cells throughout the body (improves pancreatic and hypothalamic secretion phenotypes, thereby amplifying anti-ageing effects via hormones) [49,57]. Macrophage rejuvenation not only improves the rejuvenation of the systemic secretory phenotype but also attenuates NAD^+ degradation through reduced CD38 expression [11]. This may have a synergistic anti-senescence effect with eNAMPT. NAD^+ and a rejuvenated secretory phenotype (possibly through metabolic reprogramming or cell rejuvenation via ERK–AMPK regulation of P16 and P53) improve the GST secretory capacity of fibroblasts. Delivery of GST to organs throughout the body via extracellular vesicles improved cellular redox homeostasis, resulting in a promising anti-ageing effect (improves liver redox status and kidney ageing) [51]. Taken together, local reprogramming through systemic cellular communication (eNAMPT, YSAP, and GST, etc.) produces synergistic anti-ageing effects (improvement in redox and metabolic imbalances caused by mitochondrial senescence and protein instability caused by ribosomal senescence). However, it is worth noting that further studies are needed to determine whether reprogramming can produce sufficient alterations in the secretory phenotype and whether intercellular communication can alter the secretory phenotype of adjacent cells. (black arrow: direct stimulatory, round arrow: cycle, dotted arrow: tentative stimulatory, down faded arrow: decrease, up faded arrow: increase; the grey dotted lines depict macro-level improvements on the left and micro-level improvements on the right, both separated by green dotted lines).

Protein mismatches reach the upper limit of cellular discharge and continue to accumulate, damaging the cell's genetic repair mechanism, which in turn disrupts all cellular functions into a state of irreversible death, releasing waste after its death and thus creating the sixth part of the senescent microenvironment. (Therefore, the simple removal of senescent cells might not avoid the senescence signal release process during senescent cell death).

3.2. Potential Key Mechanisms Related to Reprogramming-Based Therapies

3.2.1. Cyclin-Dependent Kinase Inhibitors ($p16^{INK4A}$)

Multiple lines of evidence suggest $p16^{INK4A}$ extensive involvement in the ageing process, which could serve as an alternative regulatory centre in anti-ageing strategies. Mice with low levels of cell cycle checkpoint kinase BubR1 expression suffer from an acceleration in ageing as well as high levels of $p16^{INK4A}$ in tissues with age-related histopathology [58]. Targeted mutation of $p16^{INK4A}$ caused an ageing delay in the BubR1 mice. This delay went on together with reduced levels of senescent cells. This reveals a connection between biological ageing and cellular senescence [59]. In INK-ATTAC mice, where senescent cells expressing $p16^{INK4A}$ are specifically killed, the loss of senescent cells increases the lifespan and health span [60]. It has been observed that $p16^{INK4A}$ blocks E2F function and thus inhibits α-klotho promoter activity to accelerated senescence [61]. The $p16^{INK4a}$ prevents inactivation of retinoblastoma (Rb) phosphorylation by inhibiting cyclin D-dependent kinases. Then Rb represses E2F transcription factors expression by recruiting histone deacetylases to its promoter. The activated retinoblastoma (Rb) pathway simultaneously promotes formation of senescence-associated heterochromatic foci (SAHF), similarly refining the senescence-promoting mechanism of $p16^{INK4a}$ [62].

It is controversial whether there are side effects of $p16^{INK4a}$ -positive cells ablation. Removal of $p16^{INK4a}$ positive cells can lead to the side effect of fibrosis in the liver and perivascular tissue, which in turn reduces life expectancy [63]. Partial reprogramming of senescent cells is therefore one of the possible solutions to this problem.

3.2.2. Senescence-Associated Secretory Phenotype (SASP)

On the whole SASP is detrimental. For example, less than 1% of senescent preadipocytes can cause extensive physical dysfunction in young mice [13]. The killing of adjacent normal cells by SASP affects organ function [64], causing secondary senescence and increasing the accumulation of senescent cells that cause a variety of chronic inflammation/diseases [37] (senescent cells themselves are stalled in replication, so their primary cause of increased senescence is secondary senescence [64]). The anti-apoptotic capacity of senescent cells also increases the accumulation of senescent cells, and this property also protects these cells from SASP (creating a vicious circle) [65]. Targeted reprogramming of these cells may kill cells by breaking the anti-apoptotic capacity of senescent cells (it has been demonstrated in vivo in acute myeloid leukaemia cells, where short-term activation of OSKM expression induces apoptosis in leukaemic cells with little effect on normal haematopoietic stem and progenitor cells [66]). However, another possibility is to retain the beneficial components while eliminating the harmful ones (senescent cells are heterogeneous, and one subpopulation is beneficial for reprogramming and regeneration [14]). The validation of this hypothesis is one of the valuable directions for future research, so this section will comment on the beneficial potential of SASP.

Cells with $p16^{INK4a}$ promoter activation were monitored in vitro and in vivo to accumulate senescence and inflammation. They showed senescence features such as reduced cell proliferation and activation of senescence-associated β-galactosidase (SA-β-gal). Additionally, they augmented the expression of genes related to the SASP [67]. The biological conditions associated with ageing, $p16^{Ink4a}$, create a relaxed tissue environment by producing the cytokine interleukin 6, which supports reprogramming of OSKM in vivo [6].

Skeletal muscle and (white) adipose tissue are two tissues that develop phenotypes associated with early senescence in response to BubR1 dysfunction, and they are high in $p16^{INK4A}$ and $p19^{Arf}$ [58]. $p16^{INK4A}$ inactivation in BubR1-deficient mice attenuated cellular senescence and premature senescence in these tissues. In contrast, $p19^{Arf}$ inactivation exacerbated senescence and senescence in BubR1 mutant mice. Thus, BubR1 functional incompetence triggers Cdkn2a locus activation in some mouse tissues [58]. $p16^{INK4A}/p19^{Arf}$-free tissue attenuates cellular senescence and reduces IL6 production and reprogramming efficiency. Tissues without p53, on the other hand, are extensively damaged and senescent, create high levels of IL6, and are efficiently reprogrammed. Thus $p16^{INK4A}$, but not $p19^{Arf}$, is required for OSKM-induced senescence and paracrine stimulation [68].

Specific removal of senescent cells, instead, reduces reprogramming effectiveness, and the outcome of senescence on reprogramming is mediated in part by interleukin-6 (IL-6) [7]. Thus, SASP promotes reprogramming, but reprogramming decreases SASP and thus can create a weak negative feedback regulation. On the plus side, it may prevent loss of organ function and teratomas caused by excessive reprogramming. However, it may also result in less efficient reprogramming. In addition to selecting safe tissues for reprogramming, one should also consider the extensive linkage of the selected tissue to the whole body and the simultaneous regulation of factors with synergistic effects in anti-ageing.

In addition to its contribution to reprogramming, SASP, as a major dynamic component of the senescence microenvironment, also assumes a role in regulating the cellular senescence state. Senescent cells can transfer proteins directly to neighbouring cells, and this cellular communication enhances the immune surveillance of cell senescence by natural killer (NK) cells [52]. A direct attempt to exploit this mechanism is to liberate NK cells from inhibition to target senescent cells for killing. This strategy shares a feature with two other strategies (i.e., first, targeting senescent cells for killing by means of chimeric antigen receptor T (CAR T) cells and NK cells [69], and second, 2-BCL-2 inhibition to induce apoptosis to kill senescent cells [70]).

Senescence transmission has been found to be transmitted via soluble factors and extracellular vesicles (sEVs) that makeup SASP [71]. In addition to SASP, Ras (rat sarcoma viral oncogene)-induced senescence through a juxtacrine NOTCH1 (Notch Receptor 1)–JAG1 (Jagged1) pathway contributes to senescence in adjacent cells, defined as "secondary senescence" [72,73]. This shows the intercellular transmission of the senescence state (NOTCH1/JAG1) and the potential of SASP secreted by senescent cells as a key node in anti-ageing strategies. After reprogramming deeply aged cells with OSKM, the senescent microenvironment is transformed into a rejuvenating microenvironment by altering SASP; multiple substances in the rejuvenating microenvironment metabolically remodel moderately aged cells (improved mitochondrial function) and then remodel mitochondrial nucleosome interactions such as ROS-DDR. The rejuvenation microenvironment is characterized by a wide range of substances that reshape the metabolism of moderately senescent cells (improved mitochondrial function). SASP is thus an essential part of the microenvironmental remodelling and intercellular communication; another part of SASP's role is to link the endocrine (e.g., eNAMPT) and immune systems (e.g., glutamine) in this strategy to amplify the effects and scope of anti-ageing, which is described below (Figure 2).

3.2.3. DNA Methylation Level (Epigenetic Clock)

Epigenetics, characterized by acetylation and methylation (especially methylation of histone and the cytosines of CpG dinucleotides [35,74]), plays an essential role in cellular ageing. Thus, the "epigenetic clock (using the key age-related CpGs in a weighted linear model to predict chronological age)" might be indicative of biological age [37]. In addition, multiple studies have shown that epigenetic rejuvenation is possible through partial reprogramming, as reflected by age-deceleration in epigenetic clocks [37]. Therefore, epigenetic remodelling might be one of the most important ways to achieve a synergistic reversal of ageing. A genome-wide knockdown screen of human embryonic stem cells carrying a premature ageing mutation (CRISPR-Cas9-based) revealed that inactivation of the histone acetyltransferase KAT7 could inhibit $p15^{INK4b}$ transcription by reducing acetylation of histone H3 lysine 14 (H3K14) and is anti-ageing [75]. Reprogramming resets telomeres in supercentenarian cells, implying its massive role in cell rejuvenation [76]. Even with extensive epigenetic defects, reprogramming can still reset the epigenetic pattern to a revitalized pluripotent state [77].

3.2.4. Telomeres

In addition to remodelling the epigenetic landscape, partial reprogramming also prolongs telomeres in senescent cells [31]. Telomeres are repetitive nucleotide sequences located at the ends of chromosomes, are directly linked to cellular senescence, and are regulated by telomerase. Telomerase is a ribonucleoprotein complex that in humans consists of an enzyme, telomerase reverse transcriptase (TERT), plus a non-coding human telomerase RNA (hTR). The latter acts as a template for the prolonging of telomer length at the ends of chromosomes [78].

Tert overexpression significantly delayed ageing in mice by slowing telomere wear and preserving stem cell proliferative potential, but this required an increase in tumour suppression to counteract the pro-tumorigenic effects telomerase [40]. Abnormal telomere function inhibits PGC-1α and its downstream gene network via the p53-PGC pathway, thereby affecting cellular metabolism, causing organ dysfunction and leading to ageing [41]. Transient activation of telomerase restores neurogenesis in the subventricular zone and improves odour detection, suggesting that telomere lengthening reverses neural ageing and enhances its regenerative capacity, broadly improving organ function [79]. Therefore, tissue-specific transient telomere activation appears to be beneficial. For example, studies of telomerase gene therapy in mice by expressing pancreatic TERT with a broadly targeted adeno-associated virus (AAV) have also achieved beneficial effects, including increased pancreatic lifespan with reduced insulin sensitivity, osteoporosis, neuromuscular coordination, and molecular markers of ageing but no additional cancers occurred, which may suggest that the known oncogenic activity of telomerase is reduced when expressed in adult or aged organisms using an AAV vector [80].

In vitro assays with fibroblasts obtained from Tert knockout mice showed that $mTert^{-/-}$ cells are more susceptible to senescence and malignancy than $mTert^{+/+}$ cells. Telomerase reverse transcriptase (TERT) expression is upregulated by $mTert^{+/+}$ cells prior to senescence. In addition, knockdown or downregulation of TERT by CRISPR/Cas9 or shRNA reproduced the $mTert^{-/-}$ phenotype, while overexpression of TERT in $mTert^{-/-}$ cells was rescued [81]. In summary, whether transient induction of telomerase expression is beneficial should also be investigated in different tissues in vivo, but the vast differences in the oncogenic capacity of human and mouse limit the potential application of telomerase.

The activation of yes-associated protein 1 (YAP1), which upregulates the pro-inflammatory factor interleukin-18, can be rescued by mTert reactivation in mice with telomere dysfunction. In contrast, conventional SASP (IL-1, IL-6, IL8) did not show much change [82]. Thus, reprogramming strategies (which can regulate more inflammatory factors while lengthening telomeres) may have a synergistic effect on reducing the ageing-related chronic inflammation.

3.2.5. Youthful Secretory Phenotype (YSP)

The youthful secretory phenotype is a newly proposed hypothesis. It aims to generalize the anti-ageing factors (including GDF11, GPLD1, clusterin, Klotho, NAD^+, eNAMPT, GSTM2, exosomes, et al.) found in young blood and in the secretome of young cells [20–23,49,51,57]. These "young factors (existing in young blood as well as being secreted by a heterogeneous subgroup of ageing cells)" may dilute or inhibit "old factors" promoting ageing, thus playing a rejuvenating role [14,15].

Plasma proteome alteration can also interfere with senescence through intercellular and organ–organ communication [19]. Exposure of aged mice to young serum improved regeneration of senescent satellite cells (through Notch signalling activation), increased senescent hepatocytes' proliferation, and restored the cEBP-α complex to youthful levels [83]. It is shown that blood input from young donors to elderly recipients improves the latter's senescence-related phenotype. It is not unique that the soluble factors and extracellular vesicles (sEVs) that make up SASP can influence other cells' senescence state and even transmit senescence by secretion [71]. Therefore, it is tempting to think that the "youthful secretory phenotype (YSP)" (secreted by young cells) could also convey youth across the whole body. For example, neonatal umbilical cord (UC)-derived mesenchymal stem cell extracellular vesicles (MSC-EV), which are rich in anti-ageing rejuvenation signals, rejuvenate senescent adult bone marrow-derived mesenchymal stem cells (AB-MSC) [84]. Exposure of neonatal umbilical cord-derived MSC extracellular vesicles (UC-EV) increased telomere length in AB-MSC with a significant improvement in SASP. It improved age-related degeneration of mouse bones and kidneys at the organ level.

After blood alteration, cardiac hypertrophy and cardiomyocyte size decreased significantly in old mice with concomitant molecular remodelling [85]. Growth and differentiation factor 11 (GDF11) could be one of the "rejuvenating" factors in young blood. Restoring circulating GDF11 levels reverses functional and genetic damage in aged muscle stem cells [86]. However, it remains controversial whether blood exchange therapy works by restoring GDF11 in aged mice to youthful levels [86,87]. For example, whether young blood improves synaptic plasticity and cognitive function through the activation of cyclic AMP response element-binding protein (Creb) in dentate gyrus neurons [88] or by enhancing neurogenesis in ageing mice via GDF11 [20].

In addition, the mechanism of young blood combating senescence remains controversial. The autophagic activity of aged livers can be restored by exposure to plasma from juvenile donors. Conversely, inhibition of autophagic activity eliminates the anti-ageing effect of plasmapheresis on the liver [89]. Exosomes from young serum significantly downregulated senescence-related genes (cyclin-dependent kinase inhibitor 2A, mechanistic target of rapamycin, and insulin-like growth factor 1 receptor) and upregulated telomerase related genes (e.g., Men1, Mre11a, Tep1, Terf2, Tert, and Tnks) in lung and liver by reversing mmu-miR-126b-5p levels of aged mice [90]. Thus, circulating anti-ageing factors are not unique and may work together through different pathways to exert anti-ageing effects. Exercise stimulates the liver to produce glycosylphosphatidylinositol-specific phospholipase D (GPLD1) [21]. It cannot cross the blood–brain barrier; instead, it improves age-related cognitive decline by reducing inflammation and increasing blood supply to the brain [21]. Thus, remote mediating mechanisms between organs can be both directly and indirectly anti-ageing. Not limited to organs such as the liver, kidney and brain, the cytokines MCP-1 and IL-6 (pro-inflammatory) were found to be reduced in visceral adipose tissue (VAT) of aged (18 months) mice by exposure to young plasma (from 3-month-old mice). Ageing adipose tissue-derived stromovascular fraction cells showed a decrease in the expression of the senescence markers ($p16^{Ink4a}$ and $p21^{Waf1/Cip1}$) [91]. Thus, amelioration of ageing-related hypofunction by providing a rejuvenating microenvironment (plasma proteome alteration) for senescent cells is widely applicable in a wide range of tissues [18].

In addition to blood-based evidence, this "youthful secretory phenotype (YSP)" (secreted by young cells) anti-ageing phenomenon is also widely observed in a variety of tissues (e.g., muscle, adipose, etc.). Cardiac stem cells are absent from the adult myocardium, but paracrine effects derived from young cardiomyocytes lengthen the telomeres not restricted to senescent cardiomyocytes. In aged rats treated with cardiac sphere-derived cells (CDCs) from young donors, circulating levels of the inflammatory cytokines interleukin-1β and interleukin-6 were reduced, along with elevated anti-inflammatory interleukin-10 levels, correlated with the observed improvements in exercise capacity, muscle reduction, hair regeneration, and renal function [92]. This research shows that young source tissue cells deliberately produce systemic benefits through systemic improvements rather than local improvements in single-organ ageing alone. Allogenic CDC intracoronary infusion in patients with heart attack increased left ventricle (LV) volume and N-terminal prob-type natriuretic peptide (NT-proBNP) compared with placebo but did not reduce scarring [93].

Further studies have shown that the ageing bone marrow can be reconstituted by tail vein transplantation of juvenile-derived antigen 1 positive (Sca-1+) bone marrow (BM) stem cells. It also promotes the rejuvenation of the heart by activating the c-x-c motif chemokine ligand 12 c-x-c chemokine receptor type 4 (Cxcl12/Cxcr4) pathway in cardiac endothelial cells [94]. CDC transplanted into the rat heart after myocardial infarction can derive exosomes into the bloodstream to promote myocardial repair via microRNAs associated with myocardial recovery [95].

Nicotinamide adenine dinucleotide (NAD) is a metabolic cofactor that decreases with age and supplementation with its precursors, most commonly nicotinamide (NAM), is thought to be anti-ageing. Although NAM is a NAD^+ precursor, NAM inhibits the NAM salvage pathway without producing a net elevation in the NAD metabolome. NAM treatment increases global protein acetylation, indicating that total sirtuins are inhibited. Combined with the reduction in nicotinamide phosphoribosyltransferase (NAMPT) levels it causes [96], the upper limit of the effect of NAD^+ precursor supplementation alone may not be sufficient to meet anti-ageing requirements (enzymatic reaction balance may have an inhibitory effect on enzymes and enzyme depletion and inactivation due to ageing). Therefore, the reduction in NAD depletion caused by heavy braiding, enzyme renewal, and mitochondrial rejuvenation may be more advantageous. The NAD^+-dependent deacetylase (SIRT1) in mammalian adipocytes deacetylates intracellular nicotinamide phosphoribosyltransferase (iNAMPT) to facilitate secretion to form extracellular nicotinamide phosphoribosyltransferase (eNAMPT) and enhance eNAMPT activity. eNAMPT in adipose tissue enhances hypothalamic NAD^+/SIRT1 signalling and physical activity [57]. On the other hand, the hypothalamus is thought to be the superior control centre for ageing in mammals. By influencing endocrine regulatory centres, eNAMPT links adipose reprogramming anti-ageing strategies to endocrine regulation. The levels of eNAMPT in extracellular vesicles (EVs) decreased significantly with age and increased the secretion of eNAMPT-containing EVs from adipocytes; it significantly improved the behavioural phenotype of ageing and prolonged lifespan in mice [49]. In addition to the direct effects on cellular energy metabolism and redox, we should also consider the indirect effects caused by eNAMPT, such as acetylation, which is widely involved in a variety of metabolic and chromatin regulations (histone modifications) and thus extensively affects mitochondrial metabolic remodelling and epigenetic remodelling in reprogramming strategies. Some easily overlooked mechanisms of gene damage also need to be investigated in an integrated manner to complete the network of crucial factors for reversal of ageing; e.g., in addition to ROS-dependent DDR, DNA-protein cross-linking (DPC) is also one of the triggers of nuclear senescence. Gene damage induces ATM/ATR activation, which activates the deubiquitinating enzyme VCPIP1/VCIP135 by phosphorylation, and VCPIP1 deubiquitinates SPRTN(a DNA-dependent metalloproteinase) to create the conditions for its subsequent acetylation, which eventually localizes SPRTN to chromatin damage sites to cleave DPCs proteins and protect genomic stability from senescence [97]. Thus eNAMPT and the response induced by changes in NAD^+/NADH ratios (and even changes in sirtuin1,3) are by no means limited to the previously described crosstalk mechanisms between ROS-induced DNA damage response (DDR) and mitochondrial dysfunction-associated senescence (MiDAS) [3,4] but are instead more broad. Reduced glutamine levels in lipid cells reduce uridine diphosphate N-acetylglucosamine (UDP-GlcNAc) levels via glycolysis, reducing O-linked β-N-acetylglucosamine (O-GlcNAc). This ultimately leads to chronic inflammation development by promoting a pro-inflammatory transcriptional response through O-GlcNAcy of chromatin-binding proteins near inflammatory genes [48] (Figure 3).

Extracellular vesicles (EVs) secreted by young donor fibroblasts are rich in GST-active GSTM2, increasing the level of GSH in mice and humans and reducing ROS accumulation and lipid oxidation in aged mice and humans. This modulates brown adipose tissue (BAT) and kidney and lung senescence markers and influences the SASP in serum, thereby systematically improving the cellular and physiological characteristics of ageing [51]. Therefore the potential of sEVs in young individuals to ameliorate ageing-associated cell damage was investigated [51]. The secretion of sEVs by old individual cells ("old cells") can induce paracrine senescence characteristics to the proliferating cells of young individuals ("young cells") to facilitate their reprogramming [7]. Thus, the above two relationships can be partially reprogrammed to create a self-regulating virtuous cycle in the body. White adipose tissue infiltrated by $CD38^+$ M1 macrophages in senescent animals leads to a decrease in the levels of NAD^+ and its precursor, NMN. Knockdown of the CD38 gene prevented a decline in NAD^+. In vitro co-culture of senescent cells and macrophages resulted in elevated CD38 expression and activity. Inhibition of SASP reduced CD38 in vivo, thereby reversing the decline in NAD^+ levels. It was shown that senescence-induced infiltration of $CD38^+$ macrophages in white adipose tissue overexpressing CD38 leads to an overall decrease in NAD^+. Small amounts of CD38 present in cells are sufficient to regulate their NAD^+ levels, but CD38 can also reduce NAD^+ levels by blocking their use of extracellular NMN. Exposure to CD38 antibody reversed the decline in NAD^+ levels in white adipose tissue of aged mice, and the effect was superimposed on NMN supplementation [11,12]. Thus, the introduction of adipose reprogramming can also broadly regulate multiple organs throughout the body (vital metabolic organs such as the hypothalamus, liver, and pancreas) by modulating $CD38^+$ macrophages and circulating eNMPT and GST in both immune and cellular metabolic pathways.

The mmu-miR-126b-5p in young serum-derived exosomes downregulates senescence-related genes in ($p16^{Ink4A}$, MTOR, and IGF1R) in lung and liver tissues of aged mice and upregulates telomerase-related genes (e.g., Men1, Mre11a, Tep1, Terf2, Tert, and Tnks) in the liver to show anti-ageing potential [90]. An earlier study showed that transfection of miR-21a-5p, miR-103-3p, and miR-30c-5p decreased Mre11a, p16^{INK4a}, and Mtor in the liver of aged mice [98]. Mmu-miR-291a-3p in mouse ESC exosomes decreased the activity of β-galactosidase, improved proliferation, and decreased expression and translation of TGF-β receptors 2, p53, and p21 in senescent cells. The human homologs of mma-miR-291a-3p, hsa-miR-371a-3p, and hsa-miR-520e, exercise comparable anti-ageing activity [99]. On the other hand, increased miR-29b-3p in exosomes released by BM-MSC in ageing mice, and consequently inhibition of sirtuin 1 (Sirt1), lead to increased senescence-associated insulin resistance [100].

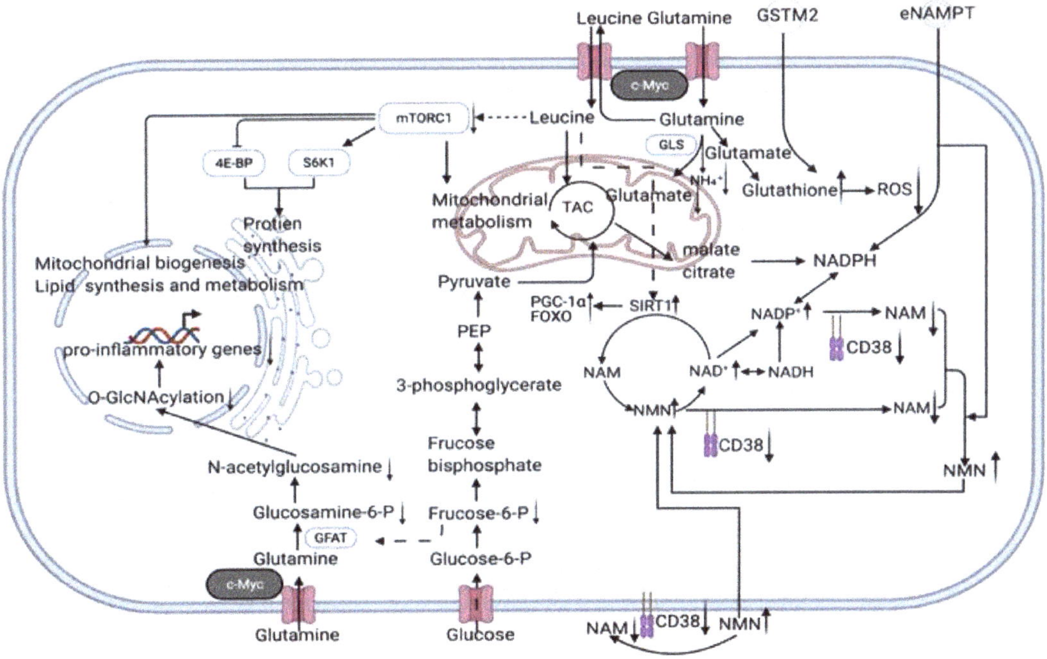

Figure 3. Potential intracellular mechanisms related to senescent cells specific reprogramming. Youthful secretory phenotype reverses the upregulation of cellular glutaminase 1 [56], so more leucine is transported into adipocytes with glutamine export. Leucine activation of mTORC1 and p70 ribosomal protein S6 kinase 1 (S6K1) promotes lipolysis and inhibits fatty acid synthesis [101]. However, its overall gains are obtained by increasing leptin and lipocalin secretion and/or synthesis in adipocytes, activating AMPK/SIRT1/PGC-1α signalling to regulate mitochondrial metabolism, and inhibiting the detrimental factors associated with mTORC1 activation to promote browning and fatty acid oxidation [102]. Therefore, in addition to amplifying the anti-ageing effect, the use of synergistic effects should also be considered to offset each other's side effects to achieve an overall benefit to highlight the superiority of this strategy. For example, direct intake of leucine causes inhibition of Sestrin2, and thus activation of the rapamycin complex 1 (mTORC1) pathway, which shortens lifespan [103]. However, by improving adipose tissue function, rejuvenated adipose precursor cells produce more functional adipocytes, as we know that branched chain amino acids (BCAAs) promote adipose precursor cell differentiation, which in turn reduces acetyl coenzyme A (AcCoA) production from sugar and glutamine (thus increasing glutamine cycle levels and facilitating leucine entry into adipocytes) and increase branched chain amino acid (BCAA) catabolic fluxes [104]. BCAA catabolism, which reduced leucine levels in other tissues (thereby derepressing Sestrin2 inhibition and inactivating the rapamycin complex 1 (mTORC1) pathway). This leucine distribution

(adipocyte enrichment) facilitates the activation of AMPK/SIRT1/PGC-1α signalling to regulate mitochondrial metabolism while suppressing the detrimental effects of mTORC1 activation in other tissues. Just as senescent adipocytes maintain survival by passing mitochondria to macrophages [54,105], increased glutamine catabolism in senescent cells to lower intracellular pH (glutamine-glutamate+$NH_4{}^+$) is a way to save themselves [56], and the susceptibility of adipose to senescence leads to high amide consumption, which reminds us that macrophages secrete large amounts of SASP due to elevated glucolysis in a low glutamine environment [48]. Thus, increased glutamine levels inhibit glucolysis and thus improve inflammation, while GSTM2 in exosomes provides more GSH to maintain redox homeostasis [51]. In combination with the promotion of glycine adipocyte enrichment and the reduction in mTORC1 activation in other tissues, this is a synergistic anti-ageing effect. Furthermore, due to eNAMPT and cell rejuvenation, excess branched-chain amino acids are consumed via the tricarboxylic acid cycle, thereby retaining their role in promoting leptin secretion and Sirt1 activity while preventing excessive mTORC1 activation. CD38 expression increases with macrophage senescence, which can lead to a significant depletion of NAD^+ [11,12], combined with a decrease in Sirt1 [43] function and eNAMPT [49,50] secretion triggered by adipocyte senescence and a decrease in the NAD^+/NADH ratio [3]. This series of imbalances leads to mitochondrial dysfunction and an imbalance in redox status. The removal of these cells will delay senescence by reducing the above disorders, but the rejuvenation of these senescent cells may be achieved through the tricarboxylic acid cycle and the rearrangement of the disordered metabolic and transcriptional state through de novo synthesis and salvage production pathway (NAM–NMN–NAD^+). The activation of Sirt1 has an anti-ageing effect via the PGC-1α and FOXO pathways [106–108]. (black arrow: direct stimulatory, round arrow: cycle, dotted arrow: tentative stimulatory, down faded arrow: decrease, up faded arrow: increase).

4. Potential Candidates of Clinical Translation Scheme to Achieve Senescent Cells Specific Reprogramming

Partial reprogramming can be achieved by the transient expression of nuclear reprogramming factors transfected by mRNAs [36] or by AAV delivery [37]. The other methods used for complete reprogramming are also potent for applying partial reprogramming by drug control, for example, by injecting plasmids encoding OSKM by the single hydrodynamic tail vein (HTV) [109]. The effective cell permeability of recombinant transcription factors was established by fusing it with the minimal transduction domain of ZEBRA protein, triggering human fibroblasts' regeneration and CD34+ cells without genetic interference and becoming one of the candidate approaches for a senescent cell-specific partial reprogramming strategy [110]. Even reprogramming senescent cells with a low-dose OKSM delivery vector of pseudo typic adeno-associated virus (AAV) using AAV-DJ-coated shells is also an alternative approach [111]. However, since the amount of protein overexpressed by AAV is itself low, it does not indicate its long-term safety, and the same AAV-mediated reprogramming of OSK is more advantageous due to the removal of the oncogene c-Myc. It would also be interesting to investigate whether there is a synergistic anti-ageing effect of simultaneous overexpression of TERT and OSK (OSKM lengthens telomeres, the effect of OSK on telomere length is currently unknown [37]). It is also suggesting that specific induction of overexpression in pioneer senescent cells is a more favourable strategy than uncontrollable direct overexpression. Customized solutions and a safe organization's choice are the directions that it will take to improve this strategy.

Pioneer transcription factors are defined as a class of transcription factors that can differentially bind DNA-binding domains (DBDs) to silence chromatin targets and initiate cell fate trajectory shifts [112]. In a reprogramming strategy, the TFs OSK act as pioneer transcription factors in reprogramming to target the LIN28B motif, a key site for reprogramming and pluripotency on nucleosomes, and binding to Myc, which cannot target nucleosomes alone, to direct its targeting to the condensed E-box on nucleosomes and together regulate cell fate [112]. Some studies have shown that OSK combinations alone still have an anti-ageing effect [37,113], but Myc, as an efficiency enhancer, greatly enhances reprogramming efficiency [114]. There is a critical role of Myc for glutamate metabolism. Overexpression of c-Myc increases glutamine levels in lipocytes, thereby decreasing glycolysis, and the glycolytic pathway produces less uridine diphosphate N-acetylamino glucose (UDP-GlcNAc), further decreasing nuclear O-linked β-N-acetylamino glucose (O-GlcNAc)). The reduced level of O-GlcNAc in the nucleus restricts the O-GlcNAcylation of inflammatory gene-associated chromatin-binding proteins, thereby inhibiting pro-inflammatory transcription [48]. This will lead to attenuated chronic inflammation caused by enriched immune cells in the vicinity of white adipocytes and through macrophages in tissues, attenuating NAD^+ depletion in vivo [11,12]. This cascade breaks the ROS-induced DNA damage response by increasing the $NAD^+/NADH$ ratio (DDR) and mitochondrial dysfunction-associated senescence (MiDAS) crosstalk [3,4] while reshaping the rejuvenation pattern at the epigenetic, metabolic, and redox levels and mitochondrial metabolism. Mitochondria regulation also goes beyond energy metabolism and redox; as described above, the function of mitochondrial DNA should also be considered. For example, Humanin, a peptide encoded by a short open reading frame within the mitochondrial genome, can extend the lifespan through the daf-16/Foxo pathway [115]. Therefore, comparing the efficiency and side effects of OSKM and OSK schemes in anti-ageing is also a way to resolve the controversy (which is superior in anti-ageing).

5. Conclusions

Senescence-specific phenotypes are manifested by increased expression of senescence-associated genes (senescence transcriptional program, epigenetic alterations, telomerase shortening, senescence-associated chronic inflammatory state) and altered metabolic state (altered mitochondrial function, redox imbalance, disturbances in the metabolism of key molecules of senescence, such as NAD^+), while cell cycle (cell cycle withdrawal) and protein synthesis (ribosomal ageing phenotype, senescence-associated endocrine phenotype) also appear to be characteristically altered. Of these, the senescence-associated endocrine phenotype (which encompasses the indirect cellular communication of all senescent cells with immune cells, young cells, and organ-specific functional cells) is an essential component of the senescence microenvironment. The multiple cytokines, enzymes, and extracellular vesicles (EVs) [71] that make up the SASP can interact with young cells through the senescence microenvironment, a balance that generally promotes senescence. Still, the rejuvenating microenvironment of immature cells can also improve the metabolic state of senescent cells at the tissue level and thus break the senescence signature within senescent cells through the remodelling of protein synthesis and gene expression. It is possible that the vicious cycle of senescence within senescent cells can be broken through the remodelling of protein synthesis and gene expression patterns. For example, blood-derived factors and extracellular vesicles of young origin have been shown to rejuvenate senescent mice [71,83], implying that breaking the dominance of the senescent microenvironment in the senescent organism and changing this balance to one dominated by the rejuvenating microenvironment has the opportunity to reprogram the metabolism of senescent cells and thus break the characteristic cycle of senescence within senescent cells.

Additionally, the senescent microenvironment itself promotes cellular reprogramming, which may be an opportunity left by evolution to combat senescence with controlled reprogramming of local tissues (based on the OSKM Yamanaka factor, which essentially creates a persistent young environment in a controlled manner), in turn, radically improving the overall senescence homeostasis of senescent cells through metabolic reprogramming and epigenetic remodelling [51], and this deadlock-breaking anti-ageing strategy is autonomously regulated by the ageing microenvironment, depending on the degree of senescence (the more the microenvironment is inclined to senescence, the easier the local reprogramming, metabolic reprogramming, and epigenetic remodelling).

In summary, the phenomena we expect to see in future research and clinical translation are as follows: As rejuvenation becomes more pronounced, local reprogramming loses the promotion from SASP and combines with a controlled induction system to avoid tumours and loss of cellular function. It must be stressed that this strategy must be implemented in non-fatal tissues (such as adipose, fascia rather than myocardium, and nerves).

Author Contributions: Conceptualization, R.C. and T.S.; software, R.C.; writing—original draft preparation, R.C.; writing—review and editing, R.C.; visualization, R.C.; supervision, T.S. All authors have read and agreed to the published version of the manuscript.

Funding: The research project was financed by the University Clinic Heidelberg and the European Wellness Academy (ASIA PACIFIC). R.C. is supported by a Chinese Scholar Council fellowship.

Acknowledgments: All figures were created using BioRender by authors of this manucript, https://biorender.com (accessed on 25 January 2022). The authors acknowledge the help of the University Clinic Heidelberg and the European Wellness Academy (ASIA PACIFIC). R.C. acknowledge the support of Chinese Scholar Council fellowship.

Conflicts of Interest: The authors declare no conflict of interest.

References

1. Consortium, T.T.M. A single-cell transcriptomic atlas characterizes ageing tissues in the mouse. *Nature* **2020**, *583*, 590–595. [CrossRef] [PubMed]
2. Schaum, N.; Lehallier, B.; Hahn, O.; Pálovics, R.; Hosseinzadeh, S.; Lee, S.E.; Sit, R.; Lee, D.P.; Losada, P.M.; Zardeneta, M.E.; et al. Ageing hallmarks exhibit organ-specific temporal signatures. *Nature* **2020**, *583*, 596–602. [CrossRef] [PubMed]
3. Wiley, C.D.; Velarde, M.C.; Lecot, P.; Liu, S.; Sarnoski, E.A.; Freund, A.; Shirakawa, K.; Lim, H.W.; Davis, S.S.; Ramanathan, A.; et al. Mitochondrial Dysfunction Induces Senescence with a Distinct Secretory Phenotype. *Cell Metab.* **2016**, *23*, 303–314. [CrossRef] [PubMed]
4. Correia-Melo, C.; Marques, F.D.; Anderson, R.; Hewitt, G.; Hewitt, R.; Cole, J.; Carroll, B.M.; Miwa, S.; Birch, J.; Merz, A.; et al. Mitochondria are required for pro-ageing features of the senescent phenotype. *EMBO J.* **2016**, *35*, 724–742. [CrossRef]
5. Vizioli, M.G.; Liu, T.; Miller, K.N.; Robertson, N.A.; Gilroy, K.; Lagnado, A.B.; Perez-Garcia, A.; Kiourtis, C.; Dasgupta, N.; Lei, X.; et al. Mitochondria-to-nucleus retrograde signaling drives formation of cytoplasmic chromatin and inflammation in senescence. *Genes Dev.* **2020**, *34*, 428–445. [CrossRef]
6. Mosteiro, L.; Pantoja, C.; Alcazar, N.; Marión, R.M.; Chondronasiou, D.; Rovira, M.; Fernandez-Marcos, P.J.; Muñoz-Martin, M.; Blanco-Aparicio, C.; Pastor, J.; et al. Tissue damage and senescence provide critical signals for cellular reprogramming in vivo. *Science* **2016**, *354*, aaf4445. [CrossRef]
7. Chiche, A.; Le Roux, I.; von Joest, M.; Sakai, H.; Aguín, S.B.; Cazin, C.; Salam, R.; Fiette, L.; Alegria, O.; Flamant, P.; et al. Injury-Induced Senescence Enables In Vivo Reprogramming in Skeletal Muscle. *Cell Stem Cell* **2017**, *20*, 407–414. [CrossRef]
8. Orjalo, A.V.; Bhaumik, D.; Gengler, B.K.; Scott, G.K.; Campisi, J. Cell surface-bound IL-1alpha is an upstream regulator of the senescence-associated IL-6/IL-8 cytokine network. *Proc. Natl. Acad. Sci. USA* **2009**, *106*, 17031–17036. [CrossRef]
9. Krizhanovsky, V.; Yon, M.; Dickins, R.A.; Hearn, S.; Simon, J.; Miething, C.; Yee, H.; Zender, L.; Lowe, S.W. Senescence of activated stellate cells limits liver fibrosis. *Cell* **2008**, *134*, 657–667. [CrossRef]
10. Sagiv, A.; Burton, D.G.; Moshayev, Z.; Vadai, E.; Wensveen, F.; Ben-Dor, S.; Golani, O.; Polic, B.; Krizhanovsky, V. NKG2D ligands mediate immunosurveillance of senescent cells. *Aging (Albany NY)* **2016**, *8*, 328–344. [CrossRef]
11. Chini, C.C.S.; Peclat, T.R.; Warner, G.M.; Kashyap, S.; Espindola-Netto, J.M.; de Oliveira, G.C.; Gomez, L.S.; Hogan, K.A.; Tarragó, M.G.; Puranik, A.S.; et al. CD38 ecto-enzyme in immune cells is induced during aging and regulates NAD(+) and NMN levels. *Nat. Metab.* **2020**, *2*, 1284–1304. [CrossRef] [PubMed]
12. Covarrubias, A.J.; Kale, A.; Perrone, R.; Lopez-Dominguez, J.A.; Pisco, A.O.; Kasler, H.G.; Schmidt, M.S.; Heckenbach, I.; Kwok, R.; Wiley, C.D.; et al. Senescent cells promote tissue NAD(+) decline during ageing via the activation of CD38(+) macrophages. *Nat. Metab.* **2020**, *2*, 1265–1283. [CrossRef]
13. Xu, M.; Pirtskhalava, T.; Farr, J.N.; Weigand, B.M.; Palmer, A.K.; Weivoda, M.M.; Inman, C.L.; Ogrodnik, M.B.; Hachfeld, C.M.; Fraser, D.G.; et al. Senolytics improve physical function and increase lifespan in old age. *Nat. Med.* **2018**, *24*, 1246–1256. [CrossRef]
14. Mahmoudi, S.; Mancini, E.; Xu, L.; Moore, A.; Jahanbani, F.; Hebestreit, K.; Srinivasan, R.; Li, X.; Devarajan, K.; Prélot, L.; et al. Heterogeneity in old fibroblasts is linked to variability in reprogramming and wound healing. *Nature* **2019**, *574*, 553–558. [CrossRef] [PubMed]
15. Mahmoudi, S.; Xu, L.; Brunet, A. Turning back time with emerging rejuvenation strategies. *Nat. Cell Biol.* **2019**, *21*, 32–43. [CrossRef] [PubMed]
16. Zhang, Y.; Unnikrishnan, A.; Deepa, S.S.; Liu, Y.; Li, Y.; Ikeno, Y.; Sosnowska, D.; Van Remmen, H.; Richardson, A. A new role for oxidative stress in aging: The accelerated aging phenotype in Sod1(-/)(-) mice is correlated to increased cellular senescence. *Redox Biol.* **2017**, *11*, 30–37. [CrossRef]
17. Ovadya, Y.; Landsberger, T.; Leins, H.; Vadai, E.; Gal, H.; Biran, A.; Yosef, R.; Sagiv, A.; Agrawal, A.; Shapira, A.; et al. Impaired immune surveillance accelerates accumulation of senescent cells and aging. *Nat. Commun.* **2018**, *9*, 5435. [CrossRef]

18. Yousefzadeh, M.J.; Wilkinson, J.E.; Hughes, B.; Gadela, N.; Ladiges, W.C.; Vo, N.; Niedernhofer, L.J.; Huffman, D.M.; Robbins, P.D. Heterochronic parabiosis regulates the extent of cellular senescence in multiple tissues. *Geroscience* **2020**, *42*, 951–961. [CrossRef]
19. Castellano, J.M. Blood-Based Therapies to Combat Aging. *Gerontology* **2019**, *65*, 84–89. [CrossRef]
20. Katsimpardi, L.; Litterman, N.K.; Schein, P.A.; Miller, C.M.; Loffredo, F.S.; Wojtkiewicz, G.R.; Chen, J.W.; Lee, R.T.; Wagers, A.J.; Rubin, L.L. Vascular and neurogenic rejuvenation of the aging mouse brain by young systemic factors. *Science* **2014**, *344*, 630–634. [CrossRef]
21. Horowitz, A.M.; Fan, X.; Bieri, G.; Smith, L.K.; Sanchez-Diaz, C.I.; Schroer, A.B.; Gontier, G.; Casaletto, K.B.; Kramer, J.H.; Williams, K.E.; et al. Blood factors transfer beneficial effects of exercise on neurogenesis and cognition to the aged brain. *Science* **2020**, *369*, 167–173. [CrossRef] [PubMed]
22. De Miguel, Z.; Khoury, N.; Betley, M.J.; Lehallier, B.; Willoughby, D.; Olsson, N.; Yang, A.C.; Hahn, O.; Lu, N.; Vest, R.T.; et al. Exercise plasma boosts memory and dampens brain inflammation via clusterin. *Nature* **2021**, *600*, 494–499. [CrossRef] [PubMed]
23. Sahu, A.; Clemens, Z.J.; Shinde, S.N.; Sivakumar, S.; Pius, A.; Bhatia, A.; Picciolini, S.; Carlomagno, C.; Gualerzi, A.; Bedoni, M.; et al. Regulation of aged skeletal muscle regeneration by circulating extracellular vesicles. *Nat. Aging* **2021**, *1*, 1148–1161. [CrossRef]
24. Lombard, D.B.; Chua, K.F.; Mostoslavsky, R.; Franco, S.; Gostissa, M.; Alt, F.W. DNA repair, genome stability, and aging. *Cell* **2005**, *120*, 497–512. [CrossRef]
25. Kaeberlein, M.; Kirkland, K.T.; Fields, S.; Kennedy, B.K. Sir2-independent life span extension by calorie restriction in yeast. *PLoS Biol.* **2004**, *2*, 296. [CrossRef]
26. Li, Y.; Jiang, Y.; Paxman, J.; O'Laughlin, R.; Klepin, S.; Zhu, Y.; Pillus, L.; Tsimring, L.S.; Hasty, J.; Hao, N. A programmable fate decision landscape underlies single-cell aging in yeast. *Science* **2020**, *369*, 325–329. [CrossRef]
27. Lan, J.; Rollins, J.A.; Zang, X.; Wu, D.; Zou, L.; Wang, Z.; Ye, C.; Wu, Z.; Kapahi, P.; Rogers, A.N.; et al. Translational Regulation of Non-autonomous Mitochondrial Stress Response Promotes Longevity. *Cell Rep.* **2019**, *28*, 1050–1062.e1056. [CrossRef]
28. Chen, D.; Li, P.W.; Goldstein, B.A.; Cai, W.; Thomas, E.L.; Chen, F.; Hubbard, A.E.; Melov, S.; Kapahi, P. Germline signaling mediates the synergistically prolonged longevity produced by double mutations in daf-2 and rsks-1 in C. elegans. *Cell Rep.* **2013**, *5*, 1600–1610. [CrossRef]
29. Zid, B.M.; Rogers, A.N.; Katewa, S.D.; Vargas, M.A.; Kolipinski, M.C.; Lu, T.A.; Benzer, S.; Kapahi, P. 4E-BP extends lifespan upon dietary restriction by enhancing mitochondrial activity in Drosophila. *Cell* **2009**, *139*, 149–160. [CrossRef]
30. Tanaka, T.; Biancotto, A.; Moaddel, R.; Moore, A.Z.; Gonzalez-Freire, M.; Aon, M.A.; Candia, J.; Zhang, P.; Cheung, F.; Fantoni, G.; et al. Plasma proteomic signature of age in healthy humans. *Aging Cell* **2018**, *17*, 12799. [CrossRef]
31. Ocampo, A.; Reddy, P.; Martinez-Redondo, P.; Platero-Luengo, A.; Hatanaka, F.; Hishida, T.; Li, M.; Lam, D.; Kurita, M.; Beyret, E.; et al. In Vivo Amelioration of Age-Associated Hallmarks by Partial Reprogramming. *Cell* **2016**, *167*, 1719–1733.e1712. [CrossRef]
32. Manukyan, M.; Singh, P.B. Epigenome rejuvenation: HP1β mobility as a measure of pluripotent and senescent chromatin ground states. *Sci. Rep.* **2014**, *4*, 4789. [CrossRef]
33. Göbel, C.; Goetzke, R.; Eggermann, T.; Wagner, W. Interrupted reprogramming into induced pluripotent stem cells does not rejuvenate human mesenchymal stromal cells. *Sci. Rep.* **2018**, *8*, 11676. [CrossRef]
34. Olova, N.; Simpson, D.J.; Marioni, R.E.; Chandra, T. Partial reprogramming induces a steady decline in epigenetic age before loss of somatic identity. *Aging Cell* **2019**, *18*, 12877. [CrossRef]
35. Horvath, S.; Raj, K. DNA methylation-based biomarkers and the epigenetic clock theory of ageing. *Nat. Rev. Genet.* **2018**, *19*, 371–384. [CrossRef]
36. Sarkar, T.J.; Quarta, M.; Mukherjee, S.; Colville, A.; Paine, P.; Doan, L.; Tran, C.M.; Chu, C.R.; Horvath, S.; Qi, L.S.; et al. Transient non-integrative expression of nuclear reprogramming factors promotes multifaceted amelioration of aging in human cells. *Nat. Commun.* **2020**, *11*, 1545. [CrossRef]
37. Lu, Y.; Brommer, B.; Tian, X.; Krishnan, A.; Meer, M.; Wang, C.; Vera, D.L.; Zeng, Q.; Yu, D.; Bonkowski, M.S.; et al. Reprogramming to recover youthful epigenetic information and restore vision. *Nature* **2020**, *588*, 124–129. [CrossRef]
38. Gill, D.; Parry, A.; Santos, F.; Hernando-Herraez, I.; Stubbs, T.M.; Milagre, I.; Reik, W. Multi-omic rejuvenation of human cells by maturation phase transient reprogramming. *bioRxiv* **2021**, 426786. [CrossRef]
39. Alle, Q.; Le Borgne, E.; Bensadoun, P.; Lemey, C.; Béchir, N.; Gabanou, M.; Estermann, F.; Bertrand-Gaday, C.; Pessemesse, L.; Toupet, K.; et al. A single short reprogramming early in life improves fitness and increases lifespan in old age. *bioRxiv* **2021**, 443979. [CrossRef]
40. Tomás-Loba, A.; Flores, I.; Fernández-Marcos, P.J.; Cayuela, M.L.; Maraver, A.; Tejera, A.; Borrás, C.; Matheu, A.; Klatt, P.; Flores, J.M.; et al. Telomerase reverse transcriptase delays aging in cancer-resistant mice. *Cell* **2008**, *135*, 609–622. [CrossRef]
41. Sahin, E.; Colla, S.; Liesa, M.; Moslehi, J.; Müller, F.L.; Guo, M.; Cooper, M.; Kotton, D.; Fabian, A.J.; Walkey, C.; et al. Telomere dysfunction induces metabolic and mitochondrial compromise. *Nature* **2011**, *470*, 359–365. [CrossRef]
42. Caso, G.; McNurlan, M.A.; Mileva, I.; Zemlyak, A.; Mynarcik, D.C.; Gelato, M.C. Peripheral fat loss and decline in adipogenesis in older humans. *Metabolism* **2013**, *62*, 337–340. [CrossRef]
43. Khanh, V.C.; Zulkifli, A.F.; Tokunaga, C.; Yamashita, T.; Hiramatsu, Y.; Ohneda, O. Aging impairs beige adipocyte differentiation of mesenchymal stem cells via the reduced expression of Sirtuin 1. *Biochem. Biophys. Res. Commun.* **2018**, *500*, 682–690. [CrossRef]
44. Berry, D.C.; Jiang, Y.; Arpke, R.W.; Close, E.L.; Uchida, A.; Reading, D.; Berglund, E.D.; Kyba, M.; Graff, J.M. Cellular Aging Contributes to Failure of Cold-Induced Beige Adipocyte Formation in Old Mice and Humans. *Cell Metab.* **2017**, *25*, 481. [CrossRef]

45. Palikaras, K.; Mari, M.; Petanidou, B.; Pasparaki, A.; Filippidis, G.; Tavernarakis, N. Ectopic fat deposition contributes to age-associated pathology in Caenorhabditis elegans. *J. Lipid Res.* **2017**, *58*, 72–80. [CrossRef]
46. Tchkonia, T.; Pirtskhalava, T.; Thomou, T.; Cartwright, M.J.; Wise, B.; Karagiannides, I.; Shpilman, A.; Lash, T.L.; Becherer, J.D.; Kirkland, J.L. Increased TNFalpha and CCAAT/enhancer-binding protein homologous protein with aging predispose preadipocytes to resist adipogenesis. *Am. J. Physiol. Endocrinol. Metab.* **2007**, *293*, 1810–1819. [CrossRef]
47. Graja, A.; Gohlke, S.; Schulz, T.J. Aging of Brown and Beige/Brite Adipose Tissue. *Handb. Exp. Pharm.* **2019**, *251*, 55–72. [CrossRef]
48. Petrus, P.; Lecoutre, S.; Dollet, L.; Wiel, C.; Sulen, A.; Gao, H.; Tavira, B.; Laurencikiene, J.; Rooyackers, O.; Checa, A.; et al. Glutamine Links Obesity to Inflammation in Human White Adipose Tissue. *Cell Metab.* **2020**, *31*, 375–390. [CrossRef]
49. Yoshida, M.; Satoh, A.; Lin, J.B.; Mills, K.F.; Sasaki, Y.; Rensing, N.; Wong, M.; Apte, R.S.; Imai, S.-I. Extracellular Vesicle-Contained eNAMPT Delays Aging and Extends Lifespan in Mice. *Cell Metab.* **2019**, *30*, 329–342. [CrossRef]
50. Ma, C.; Pi, C.; Yang, Y.; Lin, L.; Shi, Y.; Li, Y.; Li, Y.; He, X. Nampt Expression Decreases Age-Related Senescence in Rat Bone Marrow Mesenchymal Stem Cells by Targeting Sirt1. *PLoS ONE* **2017**, *12*, e0170930. [CrossRef]
51. Fafián-Labora, J.A.; Rodríguez-Navarro, J.A.; O'Loghlen, A. Small Extracellular Vesicles Have GST Activity and Ameliorate Senescence-Related Tissue Damage. *Cell Metab.* **2020**, *32*, 71–86. [CrossRef] [PubMed]
52. Biran, A.; Perelmutter, M.; Gal, H.; Burton, D.G.; Ovadya, Y.; Vadai, E.; Geiger, T.; Krizhanovsky, V. Senescent cells communicate via intercellular protein transfer. *Genes Dev.* **2015**, *29*, 791–802. [CrossRef]
53. Takahashi, A.; Okada, R.; Nagao, K.; Kawamata, Y.; Hanyu, A.; Yoshimoto, S.; Takasugi, M.; Watanabe, S.; Kanemaki, M.T.; Obuse, C.; et al. Exosomes maintain cellular homeostasis by excreting harmful DNA from cells. *Nat. Commun.* **2017**, *8*, 15287. [CrossRef] [PubMed]
54. Altshuler-Keylin, S.; Shinoda, K.; Hasegawa, Y.; Ikeda, K.; Hong, H.; Kang, Q.; Yang, Y.; Perera, R.M.; Debnath, J.; Kajimura, S. Beige Adipocyte Maintenance Is Regulated by Autophagy-Induced Mitochondrial Clearance. *Cell Metab.* **2016**, *24*, 402–419. [CrossRef]
55. Vassallo, P.F.; Simoncini, S.; Ligi, I.; Chateau, A.L.; Bachelier, R.; Robert, S.; Morere, J.; Fernandez, S.; Guillet, B.; Marcelli, M.; et al. Accelerated senescence of cord blood endothelial progenitor cells in premature neonates is driven by SIRT1 decreased expression. *Blood* **2014**, *123*, 2116–2126. [CrossRef] [PubMed]
56. Johmura, Y.; Yamanaka, T.; Omori, S.; Wang, T.W.; Sugiura, Y.; Matsumoto, M.; Suzuki, N.; Kumamoto, S.; Yamaguchi, K.; Hatakeyama, S.; et al. Senolysis by glutaminolysis inhibition ameliorates various age-associated disorders. *Science* **2021**, *371*, 265–270. [CrossRef] [PubMed]
57. Yoon, M.J.; Yoshida, M.; Johnson, S.; Takikawa, A.; Usui, I.; Tobe, K.; Nakagawa, T.; Yoshino, J.; Imai, S.-i. SIRT1-Mediated eNAMPT Secretion from Adipose Tissue Regulates Hypothalamic NAD+ and Function in Mice. *Cell Metab.* **2015**, *21*, 706–717. [CrossRef]
58. Baker, D.J.; Perez-Terzic, C.; Jin, F.; Pitel, K.S.; Pitel, K.; Niederländer, N.J.; Jeganathan, K.; Yamada, S.; Reyes, S.; Rowe, L.; et al. Opposing roles for p16Ink4a and p19Arf in senescence and ageing caused by BubR1 insufficiency. *Nat. Cell Biol.* **2008**, *10*, 825–836. [CrossRef]
59. Baker, D.J.; Childs, B.G.; Durik, M.; Wijers, M.E.; Sieben, C.J.; Zhong, J.; Saltness, R.A.; Jeganathan, K.B.; Verzosa, G.C.; Pezeshki, A.; et al. Naturally occurring p16(Ink4a)-positive cells shorten healthy lifespan. *Nature* **2016**, *530*, 184–189. [CrossRef]
60. Baker, D.J.; Wijshake, T.; Tchkonia, T.; LeBrasseur, N.K.; Childs, B.G.; van de Sluis, B.; Kirkland, J.L.; van Deursen, J.M. Clearance of p16Ink4a-positive senescent cells delays ageing-associated disorders. *Nature* **2011**, *479*, 232–236. [CrossRef]
61. Sato, S.; Kawamata, Y.; Takahashi, A.; Imai, Y.; Hanyu, A.; Okuma, A.; Takasugi, M.; Yamakoshi, K.; Sorimachi, H.; Kanda, H.; et al. Ablation of the p16(INK4a) tumour suppressor reverses ageing phenotypes of klotho mice. *Nat. Commun.* **2015**, *6*, 7035. [CrossRef] [PubMed]
62. Narita, M.; Nũnez, S.; Heard, E.; Narita, M.; Lin, A.W.; Hearn, S.A.; Spector, D.L.; Hannon, G.J.; Lowe, S.W. Rb-mediated heterochromatin formation and silencing of E2F target genes during cellular senescence. *Cell* **2003**, *113*, 703–716. [CrossRef]
63. Grosse, L.; Wagner, N.; Emelyanov, A.; Molina, C.; Lacas-Gervais, S.; Wagner, K.-D.; Bulavin, D.V. Defined p16High Senescent Cell Types Are Indispensable for Mouse Healthspan. *Cell Metab.* **2020**, *32*, 87–99.e86. [CrossRef] [PubMed]
64. López-Otín, C.; Blasco, M.A.; Partridge, L.; Serrano, M.; Kroemer, G. The Hallmarks of Aging. *Cell* **2013**, *153*, 1194–1217. [CrossRef] [PubMed]
65. Kirkland, J.L.; Tchkonia, T. Senolytic drugs: From discovery to translation. *J. Intern. Med.* **2020**, *288*, 518–536. [CrossRef] [PubMed]
66. Wang, Y.; Lu, T.; Sun, G.; Zheng, Y.; Yang, S.; Zhang, H.; Hao, S.; Liu, Y.; Ma, S.; Zhang, H.; et al. Targeting of apoptosis gene loci by reprogramming factors leads to selective eradication of leukemia cells. *Nat. Commun.* **2019**, *10*, 5594. [CrossRef] [PubMed]
67. Liu, J.-Y.; Souroullas, G.P.; Diekman, B.O.; Krishnamurthy, J.; Hall, B.M.; Sorrentino, J.A.; Parker, J.S.; Sessions, G.A.; Gudkov, A.V.; Sharpless, N.E. Cells exhibiting strong promoter activation in vivo display features of senescence. *Proc. Natl. Acad. Sci. USA* **2019**, *116*, 2603–2611. [CrossRef]
68. Mosteiro, L.; Pantoja, C.; de Martino, A.; Serrano, M. Senescence promotes in vivo reprogramming through p16 and IL-6. *Aging Cell* **2018**, *17*. [CrossRef]
69. Amor, C.; Feucht, J.; Leibold, J.; Ho, Y.J.; Zhu, C.; Alonso-Curbelo, D.; Mansilla-Soto, J.; Boyer, J.A.; Li, X.; Giavridis, T.; et al. Senolytic CAR T cells reverse senescence-associated pathologies. *Nature* **2020**, *583*, 127–132. [CrossRef]
70. Yosef, R.; Pilpel, N.; Tokarsky-Amiel, R.; Biran, A.; Ovadya, Y.; Cohen, S.; Vadai, E.; Dassa, L.; Shahar, E.; Condiotti, R.; et al. Directed elimination of senescent cells by inhibition of BCL-W and BCL-XL. *Nat. Commun.* **2016**, *7*, 11190. [CrossRef]

71. Borghesan, M.; Fafián-Labora, J.; Eleftheriadou, O.; Carpintero-Fernández, P.; Paez-Ribes, M.; Vizcay-Barrena, G.; Swisa, A.; Kolodkin-Gal, D.; Ximénez-Embún, P.; Lowe, R.; et al. Small Extracellular Vesicles Are Key Regulators of Non-cell Autonomous Intercellular Communication in Senescence via the Interferon Protein IFITM3. *Cell Rep.* **2019**, *27*, 3956–3971.e3956. [CrossRef]
72. Hoare, M.; Ito, Y.; Kang, T.W.; Weekes, M.P.; Matheson, N.J.; Patten, D.A.; Shetty, S.; Parry, A.J.; Menon, S.; Salama, R.; et al. NOTCH1 mediates a switch between two distinct secretomes during senescence. *Nat. Cell Biol.* **2016**, *18*, 979–992. [CrossRef]
73. Teo, Y.V.; Rattanavirotkul, N.; Olova, N.; Salzano, A.; Quintanilla, A.; Tarrats, N.; Kiourtis, C.; Müller, M.; Green, A.R.; Adams, P.D.; et al. Notch Signaling Mediates Secondary Senescence. *Cell. Rep.* **2019**, *27*, 997–1007. [CrossRef]
74. Benayoun, B.A.; Pollina, E.A.; Brunet, A. Epigenetic regulation of ageing: Linking environmental inputs to genomic stability. *Nat. Rev. Mol. Cell Biol.* **2015**, *16*, 593–610. [CrossRef]
75. Wang, W.; Zheng, Y.; Sun, S.; Li, W.; Song, M.; Ji, Q.; Wu, Z.; Liu, Z.; Fan, Y.; Liu, F.; et al. A genome-wide CRISPR-based screen identifies KAT7 as a driver of cellular senescence. *Sci. Transl. Med.* **2021**, *13*, eabd2655. [CrossRef]
76. Lee, J.; Bignone, P.A.; Coles, L.S.; Liu, Y.; Snyder, E.; Larocca, D. Induced pluripotency and spontaneous reversal of cellular aging in supercentenarian donor cells. *Biochem. Biophys. Res. Commun.* **2020**, *525*, 563–569. [CrossRef]
77. Chen, Z.; Chang, W.Y.; Etheridge, A.; Strickfaden, H.; Jin, Z.; Palidwor, G.; Cho, J.-H.; Wang, K.; Kwon, S.Y.; Doré, C.; et al. Reprogramming progeria fibroblasts re-establishes a normal epigenetic landscape. *Aging Cell* **2017**, *16*, 870–887. [CrossRef]
78. Roake, C.M.; Artandi, S.E. Regulation of human telomerase in homeostasis and disease. *Nat. Rev. Mol. Cell Biol.* **2020**, *21*, 384–397. [CrossRef]
79. Jaskelioff, M.; Muller, F.L.; Paik, J.-H.; Thomas, E.; Jiang, S.; Adams, A.C.; Sahin, E.; Kost-Alimova, M.; Protopopov, A.; Cadiñanos, J.; et al. Telomerase reactivation reverses tissue degeneration in aged telomerase-deficient mice. *Nature* **2011**, *469*, 102–106. [CrossRef]
80. Bernardes de Jesus, B.; Vera, E.; Schneeberger, K.; Tejera, A.M.; Ayuso, E.; Bosch, F.; Blasco, M.A. Telomerase gene therapy in adult and old mice delays aging and increases longevity without increasing cancer. *EMBO Mol. Med.* **2012**, *4*, 691–704. [CrossRef]
81. Sun, L.; Chiang, J.Y.; Choi, J.Y.; Xiong, Z.-M.; Mao, X.; Collins, F.S.; Hodes, R.J.; Cao, K. Transient induction of telomerase expression mediates senescence and reduces tumorigenesis in primary fibroblasts. *Proc. Natl. Acad. Sci. USA* **2019**, *116*, 18983–18993. [CrossRef]
82. Chakravarti, D.; Hu, B.; Mao, X.; Rashid, A.; Li, J.; Li, J.; Liao, W.-T.; Whitley, E.M.; Dey, P.; Hou, P.; et al. Telomere dysfunction activates YAP1 to drive tissue inflammation. *Nat. Commun.* **2020**, *11*, 4766. [CrossRef]
83. Conboy, I.M.; Conboy, M.J.; Wagers, A.J.; Girma, E.R.; Weissman, I.L.; Rando, T.A. Rejuvenation of aged progenitor cells by exposure to a young systemic environment. *Nature* **2005**, *433*, 760–764. [CrossRef]
84. Lei, Q.; Gao, F.; Liu, T.; Ren, W.; Chen, L.; Cao, Y.; Chen, W.; Guo, S.; Zhang, Q.; Chen, W.; et al. Extracellular vesicles deposit PCNA to rejuvenate aged bone marrow-derived mesenchymal stem cells and slow age-related degeneration. *Sci. Transl. Med.* **2021**, *13*, eaaz8697. [CrossRef]
85. Loffredo, F.S.; Steinhauser, M.L.; Jay, S.M.; Gannon, J.; Pancoast, J.R.; Yalamanchi, P.; Sinha, M.; Dall'Osso, C.; Khong, D.; Shadrach, J.L.; et al. Growth differentiation factor 11 is a circulating factor that reverses age-related cardiac hypertrophy. *Cell* **2013**, *153*, 828–839. [CrossRef]
86. Sinha, M.; Jang, Y.C.; Oh, J.; Khong, D.; Wu, E.Y.; Manohar, R.; Miller, C.; Regalado, S.G.; Loffredo, F.S.; Pancoast, J.R.; et al. Restoring systemic GDF11 levels reverses age-related dysfunction in mouse skeletal muscle. *Science* **2014**, *344*, 649–652. [CrossRef]
87. Harper, S.C.; Brack, A.; MacDonnell, S.; Franti, M.; Olwin, B.B.; Bailey, B.A.; Rudnicki, M.A.; Houser, S.R. Is Growth Differentiation Factor 11 a Realistic Therapeutic for Aging-Dependent Muscle Defects? *Circ. Res.* **2016**, *118*, 1143–1150. [CrossRef]
88. Villeda, S.A.; Plambeck, K.E.; Middeldorp, J.; Castellano, J.M.; Mosher, K.I.; Luo, J.; Smith, L.K.; Bieri, G.; Lin, K.; Berdnik, D.; et al. Young blood reverses age-related impairments in cognitive function and synaptic plasticity in mice. *Nat. Med.* **2014**, *20*, 659–663. [CrossRef]
89. Liu, A.; Guo, E.; Yang, J.; Yang, Y.; Liu, S.; Jiang, X.; Hu, Q.; Dirsch, O.; Dahmen, U.; Zhang, C.; et al. Young plasma reverses age-dependent alterations in hepatic function through the restoration of autophagy. *Aging Cell* **2018**, *17*. [CrossRef]
90. Lee, B.-R.; Kim, J.-H.; Choi, E.-S.; Cho, J.-H.; Kim, E. Effect of young exosomes injected in aged mice. *Int. J. Nanomed.* **2018**, *13*, 5335–5345. [CrossRef]
91. Ghosh, A.K.; O'Brien, M.; Mau, T.; Qi, N.; Yung, R. Adipose Tissue Senescence and Inflammation in Aging is Reversed by the Young Milieu. *J. Gerontol. A Biol. Sci. Med. Sci.* **2019**, *74*, 1709–1715. [CrossRef] [PubMed]
92. Grigorian-Shamagian, L.; Liu, W.; Fereydooni, S.; Middleton, R.C.; Valle, J.; Cho, J.H.; Marbán, E. Cardiac and systemic rejuvenation after cardiosphere-derived cell therapy in senescent rats. *Eur. Heart J.* **2017**, *38*, 2957–2967. [CrossRef] [PubMed]
93. Makkar, R.R.; Kereiakes, D.J.; Aguirre, F.; Kowalchuk, G.; Chakravarty, T.; Malliaras, K.; Francis, G.S.; Povsic, T.J.; Schatz, R.; Traverse, J.H.; et al. Intracoronary ALLogeneic heart STem cells to Achieve myocardial Regeneration (ALLSTAR): A randomized, placebo-controlled, double-blinded trial. *Eur. Heart J.* **2020**, *41*, 3451–3458. [CrossRef]
94. Li, J.; Li, S.-H.; Dong, J.; Alibhai, F.J.; Zhang, C.; Shao, Z.-B.; Song, H.-F.; He, S.; Yin, W.-J.; Wu, J.; et al. Long-term repopulation of aged bone marrow stem cells using young Sca-1 cells promotes aged heart rejuvenation. *Aging Cell* **2019**, *18*, e13026. [CrossRef]
95. Saha, P.; Sharma, S.; Korutla, L.; Datla, S.R.; Shoja-Taheri, F.; Mishra, R.; Bigham, G.E.; Sarkar, M.; Morales, D.; Bittle, G.; et al. Circulating exosomes derived from transplanted progenitor cells aid the functional recovery of ischemic myocardium. *Sci. Transl. Med.* **2019**, *11*, eaau1168. [CrossRef]

96. Mitchell, S.J.; Bernier, M.; Aon, M.A.; Cortassa, S.; Kim, E.Y.; Fang, E.F.; Palacios, H.H.; Ali, A.; Navas-Enamorado, I. Di Francesco, A.; et al. Nicotinamide Improves Aspects of Healthspan, but Not Lifespan, in Mice. *Cell Metab.* **2018**, *27*, 667–676. [CrossRef] [PubMed]
97. Huang, J.; Zhou, Q.; Gao, M.; Nowsheen, S.; Zhao, F.; Kim, W.; Zhu, Q.; Kojima, Y.; Yin, P.; Zhang, Y.; et al. Tandem Deubiquitination and Acetylation of SPRTN Promotes DNA-Protein Crosslink Repair and Protects against Aging. *Mol. Cell* **2020**, *79*, 824–835. [CrossRef]
98. Kim, J.-H.; Lee, B.-R.; Choi, E.-S.; Lee, K.-M.; Choi, S.-K.; Cho, J.H.; Jeon, W.B.; Kim, E. Reverse Expression of Aging-Associated Molecules through Transfection of miRNAs to Aged Mice. *Mol. Ther. Nucleic Acids* **2017**, *6*, 106–115. [CrossRef]
99. Bae, Y.-U.; Son, Y.; Kim, C.-H.; Kim, K.S.; Hyun, S.H.; Woo, H.G.; Jee, B.A.; Choi, J.-H.; Sung, H.-K.; Choi, H.-C.; et al. Embryonic Stem Cell-Derived mmu-miR-291a-3p Inhibits Cellular Senescence in Human Dermal Fibroblasts Through the TGF-β Receptor 2 Pathway. *J. Gerontol. A Biol. Sci. Med. Sci.* **2019**, *74*, 1359–1367. [CrossRef]
100. Su, T.; Xiao, Y.; Xiao, Y.; Guo, Q.; Li, C.; Huang, Y.; Deng, Q.; Wen, J.; Zhou, F.; Luo, X.-H. Bone Marrow Mesenchymal Stem Cells-Derived Exosomal MiR-29b-3p Regulates Aging-Associated Insulin Resistance. *ACS Nano* **2019**, *13*, 2450–2462. [CrossRef]
101. Arif, A.; Terenzi, F.; Potdar, A.A.; Jia, J.; Sacks, J.; China, A.; Halawani, D.; Vasu, K.; Li, X.; Brown, J.M.; et al. EPRS is a critical mTORC1-S6K1 effector that influences adiposity in mice. *Nature* **2017**, *542*, 357–361. [CrossRef] [PubMed]
102. Zhang, L.; Li, F.; Guo, Q.; Duan, Y.; Wang, W.; Zhong, Y.; Yang, Y.; Yin, Y. Leucine Supplementation: A Novel Strategy for Modulating Lipid Metabolism and Energy Homeostasis. *Nutrients* **2020**, *12*, 1299. [CrossRef] [PubMed]
103. Richardson, N.E.; Konon, E.N.; Schuster, H.S.; Mitchell, A.T.; Boyle, C.; Rodgers, A.C.; Finke, M.; Haider, L.R.; Yu, D.; Flores, V.; et al. Lifelong restriction of dietary branched-chain amino acids has sex-specific benefits for frailty and life span in mice. *Nat. Aging* **2021**, *1*, 73–86. [CrossRef]
104. Green, C.R.; Wallace, M.; Divakaruni, A.S.; Phillips, S.A.; Murphy, A.N.; Ciaraldi, T.P.; Metallo, C.M. Branched-chain amino acid catabolism fuels adipocyte differentiation and lipogenesis. *Nat. Chem. Biol* **2016**, *12*, 15–21. [CrossRef]
105. Brestoff, J.R.; Wilen, C.B.; Moley, J.R.; Li, Y.; Zou, W.; Malvin, N.P.; Rowen, M.N.; Saunders, B.T.; Ma, H.; Mack, M.R.; et al. Intercellular Mitochondria Transfer to Macrophages Regulates White Adipose Tissue Homeostasis and Is Impaired in Obesity. *Cell Metab* **2021**, *33*, 270–282.e278. [CrossRef]
106. Baur, J.A.; Pearson, K.J.; Price, N.L.; Jamieson, H.A.; Lerin, C.; Kalra, A.; Prabhu, V.V.; Allard, J.S.; Lopez-Lluch, G.; Lewis, K.; et al. Resveratrol improves health and survival of mice on a high-calorie diet. *Nature* **2006**, *444*, 337–342. [CrossRef]
107. Zhao, W.; Kruse, J.P.; Tang, Y.; Jung, S.Y.; Qin, J.; Gu, W. Negative regulation of the deacetylase SIRT1 by DBC1. *Nature* **2008**, *451*, 587–590. [CrossRef]
108. Mouchiroud, L.; Houtkooper, R.H.; Moullan, N.; Katsyuba, E.; Ryu, D.; Cantó, C.; Mottis, A.; Jo, Y.S.; Viswanathan, M.; Schoonjans, K.; et al. The NAD(+)/Sirtuin Pathway Modulates Longevity through Activation of Mitochondrial UPR and FOXO Signaling. *Cell* **2013**, *154*, 430–441. [CrossRef]
109. Yilmazer, A.; de Lázaro, I.; Bussy, C.; Kostarelos, K. In vivo cell reprogramming towards pluripotency by virus-free overexpression of defined factors. *PLoS ONE* **2013**, *8*, e54754. [CrossRef]
110. Caulier, B.; Berthoin, L.; Coradin, H.; Garban, F.; Dagher, M.C.; Polack, B.; Toussaint, B.; Lenormand, J.L.; Laurin, D. Targeted release of transcription factors for human cell reprogramming by ZEBRA cell-penetrating peptide. *Int. J. Pharm.* **2017**, *529*, 65–74. [CrossRef]
111. Senís, E.; Mosteiro, L.; Wilkening, S.; Wiedtke, E.; Nowrouzi, A.; Afzal, S.; Fronza, R.; Landerer, H.; Abad, M.; Niopek, D.; et al. AAVvector-mediated in vivo reprogramming into pluripotency. *Nat. Commun.* **2018**, *9*, 2651. [CrossRef]
112. Soufi, A.; Garcia, M.F.; Jaroszewicz, A.; Osman, N.; Pellegrini, M.; Zaret, K.S. Pioneer transcription factors target partial DNA motifs on nucleosomes to initiate reprogramming. *Cell* **2015**, *161*, 555–568. [CrossRef]
113. Roux, A.; Zhang, C.; Paw, J.; Zavala-Solorio, J.; Vijay, T.; Kolumam, G.; Kenyon, C.; Kimmel, J.C. Partial reprogramming restores youthful gene expression through transient suppression of cell identity. *bioRxiv* **2021**, 444–556. [CrossRef]
114. Park, I.H.; Zhao, R.; West, J.A.; Yabuuchi, A.; Huo, H.; Ince, T.A.; Lerou, P.H.; Lensch, M.W.; Daley, G.Q. Reprogramming of human somatic cells to pluripotency with defined factors. *Nature* **2008**, *451*, 141–146. [CrossRef]
115. Yen, K.; Mehta, H.H.; Kim, S.-J.; Lue, Y.; Hoang, J.; Guerrero, N.; Port, J.; Bi, Q.; Navarrete, G.; Brandhorst, S.; et al. The mitochondrial derived peptide humanin is a regulator of lifespan and healthspan. *Aging (Albany NY)* **2020**, *12*, 11185–11199. [CrossRef]

cells

Review

Molecular Mechanisms of Alveolar Epithelial Stem Cell Senescence and Senescence-Associated Differentiation Disorders in Pulmonary Fibrosis

Xiaojing Hong [1], Lihui Wang [1], Kexiong Zhang [1], Jun Liu [1] and Jun-Ping Liu [1,2,3,*]

1 Institute of Ageing Research, Hangzhou Normal University School of Medicine, Hangzhou 311121, China; xiaojing.hong@hznu.edu.cn (X.H.); wanglihui@hznu.edu.cn (L.W.); kxzhang@hznu.edu.cn (K.Z.); junliu262@hznu.edu.cn (J.L.)
2 Department of Immunology and Pathology, Monash University Faculty of Medicine, Prahran, VIC 3181, Australia
3 Hudson Institute of Medical Research, Monash University Department of Molecular and Translational Science, Clayton, VIC 3168, Australia
* Correspondence: jun-ping.liu@monash.edu

Abstract: Pulmonary senescence is accelerated by unresolved DNA damage response, underpinning susceptibility to pulmonary fibrosis. Recently it was reported that the SARS-Cov-2 viral infection induces acute pulmonary epithelial senescence followed by fibrosis, although the mechanism remains unclear. Here, we examine roles of alveolar epithelial stem cell senescence and senescence-associated differentiation disorders in pulmonary fibrosis, exploring the mechanisms mediating and preventing pulmonary fibrogenic crisis. Notably, the TGF-β signalling pathway mediates alveolar epithelial stem cell senescence by mechanisms involving suppression of the *telomerase reverse transcriptase* gene in pulmonary fibrosis. Alternatively, telomere uncapping caused by stress-induced telomeric shelterin protein TPP1 degradation mediates DNA damage response, pulmonary senescence and fibrosis. However, targeted intervention of cellular senescence disrupts pulmonary remodelling and fibrosis by clearing senescent cells using senolytics or preventing senescence using telomere dysfunction inhibitor (TELODIN). Studies indicate that the development of senescence-associated differentiation disorders is reprogrammable and reversible by inhibiting stem cell replicative senescence in pulmonary fibrosis, providing a framework for targeted intervention of the molecular mechanisms of alveolar stem cell senescence and pulmonary fibrosis. **Abbreviations:** DPS, developmental programmed senescence; IPF, idiopathic pulmonary fibrosis; OIS, oncogene-induced replicative senescence; SADD, senescence-associated differentiation disorder; SALI, senescence-associated low-grade inflammation; SIPS, stress-induced premature senescence; TERC, telomerase RNA component; TERT, telomerase reverse transcriptase; TIFs, telomere dysfunction-induced foci; TIS, therapy-induced senescence; VIS, virus-induced senescence.

Keywords: replicative senescence; DNA damage response; telomerase and telomeres; TGF-β signalling; pulmonary fibrosis; COVID-19

Citation: Hong, X.; Wang, L.; Zhang, K.; Liu, J.; Liu, J.-P. Molecular Mechanisms of Alveolar Epithelial Stem Cell Senescence and Senescence-Associated Differentiation Disorders in Pulmonary Fibrosis. *Cells* **2022**, *11*, 877. https://doi.org/10.3390/cells11050877

Academic Editors: Nicole Wagner and Kay-Dietrich Wagner

Received: 10 February 2022
Accepted: 2 March 2022
Published: 3 March 2022

1. Introduction

Ageing, characterized by the gradual loss of the integrity of the body's physiological function over time, occurs at a different pace as different tissue stem cells experience different stress pressures and damage accumulations. In the lung, pulmonary epithelial senescence represents the highest risk of idiopathic pulmonary fibrosis (IPF) [1,2]. Recently it was reported that SARS-Cov-2 viruses cause acute pulmonary virus-induced senescence (VIS), and subsequently fibrosis, illustrating a major mechanism of coronavirus disease 2019 (COVID-19) [3–5]. VIS features not only alveolar stem cell exhaustion, but also epithelial sloughing, dishevelled repair and fibrogenesis, ultimately resulting in recurrent dyspnea and respiratory failure similar to IPF [1,3–8]. As the alveolar epithelia contain predominantly alveolar monolayer squamous epithelial type 1 (AEC1) cells (80%) that are terminally differentiated without replicative capacity and the round cubic AEC2 stem cells (~15%) undertaking self-renewal, proliferation and differentiation to ACE1 [9,10], VIS exemplifies an acutely accelerated form of alveolar senescence in the wake of the cell cycle arrest [3]. However, the mechanism of the accelerated cellular replicative senescence remains to be elucidated in association with widespread alveolar epithelial interstitial damages, epithelial cell repair failure, myofibroblast fibrogenic proliferation and extracellular matrix deposition [3–5].

Differing from such pulmonary stem cells at the distal end of the bronchus as the clublike stem cells [11], distal airway stem cells (DASC) [11,12] and bronchioalveolar stem cells (BASC) [13], AEC2 stem cells reside within the alveolar epithelium with a significant heterogeneity with variable levels of the marker CD44, telomerase activity, and proliferative and differentiation potentials in response to a variety of cellular signalling in ageing [14,15]. Whereas chemically defined EGF/NOGGIN medium supports the basal proliferation of clonal AEC2 cells in the organoids [16], potentially through a constitutive activity of the single Wnt5a-signalling stromal fibroblast niche, an injury-associated activation of AEC2 stem cell expansion involves autocrine Wnt signalling associated with a potential checkpoint regulation in lieu of the niche juxtacrine Wnt signalling [17]. Responding to a transient interstitial macrophage-derived IL-1beta, the AEC2 stem cell population undergoes a reshuffle of a distinct subpopulation (damage-associated transient progenitors, DATPs) expressing Il1r1 for alveolar regeneration via a HIF1alpha-mediated glycolysis pathway [18]. However, in contrast to a full differentiation to AEC1 cells [16,18,19], AEC2 cell differentiation to AEC1 cells is hampered at an intermediate, partially differentiated status of AEC2 stem cells (called pre-alveolar type-1 transitional cell state, PATS or alveolar differentiation intermediate, ADI) under persistent inflammatory stress conditions, showing cellular replicative senescence in pulmonary fibrosis of both animal models and patients [4,18–23]. These findings of AEC2 stem cell senescence-associated differentiation disorder (SADD) may have significant implications in not only the impaired AEC1 cell replenishment during alveolar epithelial repair, but also aberrant trans-differentiation during the pathogenic development of pulmonary fibrosis (Figure 1).

Cellular replicative senescence occurs in various forms depending on the mechanisms leading to replicative senescence. Such forms have been described as stress-induced premature senescence (SIPS), oncogene-induced replicative senescence (OIS), developmental programmed senescence (DPS) and therapy-induced senescence (TIS), which display a general underlying mechanism of the permanent cell cycle arrest in cellular mitotic divisions. Initially referred to as the irreversible cell proliferation ability after a limited number of continuous population doublings for human diploid fibroblasts (HDFs), cell senescence occurs as a general feature of the different types of cell replicative senescence to the resistance to both cell proliferation and cell death signalling, e.g., without responding of mitosis to growth factor induction [24,25]. Thus, the primary cellular mechanism of cell replicative senescence converges on the cell cycle permanent arrest provoked by significant stress insults along with sustained DNA damage in response to a variety of physical, chemical and biological stimuli (Figure 1). Although a variety of diverse stressors accelerate premature terminations of the cell cycle, including chemical toxins, antibiotics and oxidative stress in the airways [24,26–29], cell replicative senescence appears irreversible as demonstrated in SIPS and OIS [30]. With the enhanced mitogenic signalling such as activation of the Ras/MAPK pathway, and the epigenetic modifications such as histone-3 lysine-9 trimethylation (H3K9Me3) in the senescence-associated heterochromatin foci [30,31], OIS occurs to serve as an oncogenic checkpoint to prevent tumor cell clonal expansion and micro-evolutionary transformation with increased oncogenic proteins (such as c-myc [32] and mTOR [33]). In contrast to OIS, SIPS and VIS occur with excessive damages to serve as a proliferative and differentiative checkpoint to prevent cellular damage from passing onto daughter cells [3,34,35]. However, it remains unclear how the cell cycle arrest is specifically regulated in SIPS, OIS and VIS. Previous studies suggest that cellular replicative senescence involves the activation of retinoblastoma protein (RB) signal transduction pathways of tumor suppressor genes *p53-p21^{CIP1}/CDKN1A* and *p16^{INK4A}/CDKN2A* [3,36–38]. Additionally, DPS may occur via the transforming growth factor-β (TGF-β), PI3K and ERK1/2 pathways to regulate tissue remodelling [39]. Moreover, the ER unfolded protein response (ER^{UPR}) may participate in TIS with increased ER associated degradation (ERAD), resulting in ER-related senescence-associated apoptosis in a caspase-12 and caspase-3 dependent manner [40]. While it remains unclear if ER-mediated TIS entails DNA damage response (DDR), recent studies show that TIS reprograms cancerous cells to acquire the cellular phenotypic features of cellular stemness [41].

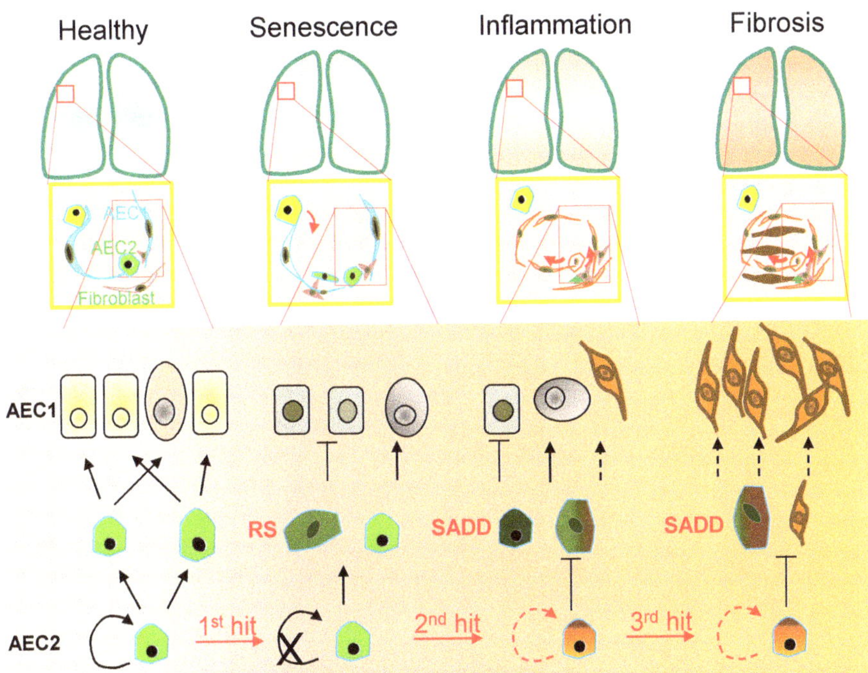

Figure 1. Alveolar monolayer squamous epithelial type 2 (AEC2) stem cell differentiation arrest and transdifferentiation disorder in pulmonary fibrosis. During pulmonary fibrogenesis, AEC2 stem cells are susceptible to stress assaults triggering telomeric DNA damage response (DDR) and replicative senescence and senescence-associated cease of the directional differentiation to alveolar monolayer squamous epithelial type 1 (AEC1) cells. Chronic stress induces senescent AEC2 stem cells to undergo transdifferentiation. The senescence-associated differentiation disorders (SADDs) contribute to myofibroblast proliferation under the condition of senescence-associated low grade inflammation (SALI).

Intriguingly, the intrinsic reprograming mechanism of a potential cell transition from ceasing replication to entering senescence may involve the transcriptional repression of the *telomerase reverse transcriptase* (*TERT*) gene expression, conferring cell replicative senescence on oncogene-stimulated cell division [42]. Additionally, TP53 phosphorylation at S15 residue [38] and TP53 fluctuation in oscillatory dynamics [43] may also be involved in the pathway switching from cell replicative senescence to a cell proliferative state. Furthermore, DNA methylation of the Oct4A enhancers may regulate the TP53-dependent TIS through tuning alternative splicing [44]. With an instrumental non-cell-autonomous role of SASP [41,45,46], cell replicative senescence plasticity represents a hotspot of further investigation in terms of autocrine/juxtacrine/paracrine actions of growth factors, proteases, cytokines, chemokines and extracellular matrix (ECM) components [47–51]. In this regard, studies showed that the extracellular vesicles (EVs)—exosomes, ectosomes, microvessels and microparticles—are increased in cell replicative senescence responding to ionizing radiation, chemical reagents or overexpression of oncogenes in a number of cell types including epithelial cells [52,53], and that TP53 activation increases EV release and mediates prostate cancer cell senescence induced by telomere shortening or radiation [53], suggesting that extracellular vesicular signalling represents a potential mechanism shaping cellular responses to and from replicative senescence. In the following sections, we focus on the phenotypic characteristics of SADD in the development of pathological fibrogenic remodelling and explore the molecular mechanisms mediating telomere maintenance, homeostatic disruption and prophylactic/therapeutic interventions as well in stress-induced cell replicative senescence and pulmonary fibrosis.

2. Senescence-Associated Differentiation Disorder (SADD)

As the lung peripheral tissue stem cells that are essentially required for alveolar epithelial damage repair and regeneration [10–13], AEC2 stem cells are compromised in replenishing AEC1 cells in pulmonary fibrosis [23,54,55]. In a long-term process of AEC2 cell self-renewal and directional differentiation to form AEC1 cells by pedigree tracing [10], AEC2 cells respond to chronic stress with the phenotypes of permanent termination of the cell cycle and differentiation to AEC1 [9,18,19,22] by an unclear molecular mechanism [18–23]. In the case of diffuse alveolar damage, in addition to discontinued differentiation to AEC1 cells, AEC2 stem cells arrested at G_2/M of the cell cycle can express a high level of KRT8, and exhibit the cubic and partial spreading morphologies [18–21,23,56,57]. Moreover, the AEC2 senescent cells appeared to be undergoing a program of transdifferentiation to a mesenchymal state in association with increased levels of α-SMA and collagen [35,58–63]. Collectively, these studies implicate AEC2 senescent stem cells as behaving with a SADD characteristic of the loss of differentiation to AEC1 cells and the gain of aberrant transdifferentiation during pulmonary pathogenic fibrogenesis. The features of stress-induced AEC2 stem cell replicative senescence and SADD in mice are consistent with recent clinical findings that AEC2 stem cells express increased senescence markers, including $p16^{INK4A}$/CDKN2A, $p14^{INK4B}$/CDKN2B, TP53 and $p21^{CIP1}$/CDKN1A [3,4,7], and that AEC2 stem cell SADD existed throughout the course studied in the pneumonia-induced acute respiratory distress syndrome and pulmonary fibrogenesis [4,64,65]. Importantly, AEC2 stem cell SADD manifested in association with pulmonary interstitial deposition in both COVID-19 [64] and IPF [66]. It is thereby probable that AEC2 stem cell replicative senescence mediates the differentiation failure of AEC2 stem cell replenishing damaged AEC1 cells by a mechanism at least involving the compromised proliferative potential [3]. Inducing AEC2 stem cell senescence or otherwise depletion demonstrated that AEC2 senescence, rather than AEC2 cell loss, promotes progressive fibrosis [38], consistent with a gain of malfunction of senescent AEC2 cells as a mechanism in causing pathogenic fibrogenesis. Moreover, preventing AEC2 stem cells from entering replicative senescence [35], or eliminating replicative senescence cells in COVID-19 [3], rescues pulmonary fibrosis in the animal models, underscoring that AEC2 stem cell replicative senescence is a prerequisite driver of pulmonary fibrosis [38].The mechanism of AEC2 stem cell SADD remains to be investigated extensively, although dysregulation of several intracellular signalling pathways (including the TGF-β family) participates in the deregulation of AEC2 differentiation [19,67]. Notably, serving as a regulatory switch, BMP4 was reported to inhibit AEC2 stem cell proliferation while stimulating AEC2 stem cell differentiation to AEC1 cells [67]. Conversely, the antagonists, follistatin and noggin, promote AEC2 stem cell self-renewal and inhibit AEC2 stem cell differentiation to AEC1 cells [67]. Moreover, a number of other extracellular factors including TGF-β, TGF-α, IL-13, TNF-α, IL-1β, IL-1R1, retinoic acid and platelet-derived growth factor are likewise implicated in shutting down AEC2-to-AEC1 cell differentiation in the milieu of the alveolar interstitium [18,68,69]. It is suggested that TGF-β and follistatin-like 1 form a positive feedback loop in accelerating the pathogenic differentiation of fibrogenesis in IPF [70]. While the downstream effectors of TGF-β may engage in regulating the telomerase *TERT* gene and telomere maintenance negatively, it has been shown recently that protecting telomeres from dysfunction eliminates stress-induced AEC2 stem cell senescence and subsequent pulmonary fibrosis (see below for details).

In addition to the discontinued AEC2-to-AEC1 cell differentiation in SADD of IPF, AEC2 cell transdifferentiation in the pathological fibrogenic condition [35,58–63] may involve epithelial-mesenchymal transition (EMT) [62,63,71–76]. Evidence of AEC2 cell contributing to fibrogenesis via transdifferentiation includes co-localization of SPC with α-SMA, Col1α1 and hydroxyproline in the mouse models of pulmonary fibrosis [35,61–63]. In addition, recent single-cell RNA sequencing studies revealed that a cell population of replicative senescence is associated with transdifferentiation in the profiles of $KRT5^-/KRT17^+$ marks and increased gene expressions of a subset of mesenchymal genes such as Col1α1 in the fibrotic lung of patients [56,57]. Lineage-tracing studies in mice show that AEC2 stem cells undergo aberrant cellular remodeling with stretching morphological alterations in association with replicative senescence characteristic of SADD, by a mechanism involving increased TGF-*β* and TP53 signaling [19,38,77]. Thus, it is conceivable that SADD underlies the discontinuation of AEC2 stem cell differentiation to AEC1 cells and initiations of myofibroblast activation and matrix deposition under the conditions of alveolar senescence-associated low-grade inflammation (SALI) [78–81] with inflammatory cell and cytokine signaling [19,62,63,71–76,82,83]. Further studies are required to investigate if cell replicative senescence may dictate SADD through intracellular mechanisms prior to SALI (Figure 2), so inhibiting cell replicative senescence may impede SADD as to prevent fibrogenesis in the animal models of telomere dysfunction-induced pulmonary fibrosis [35].

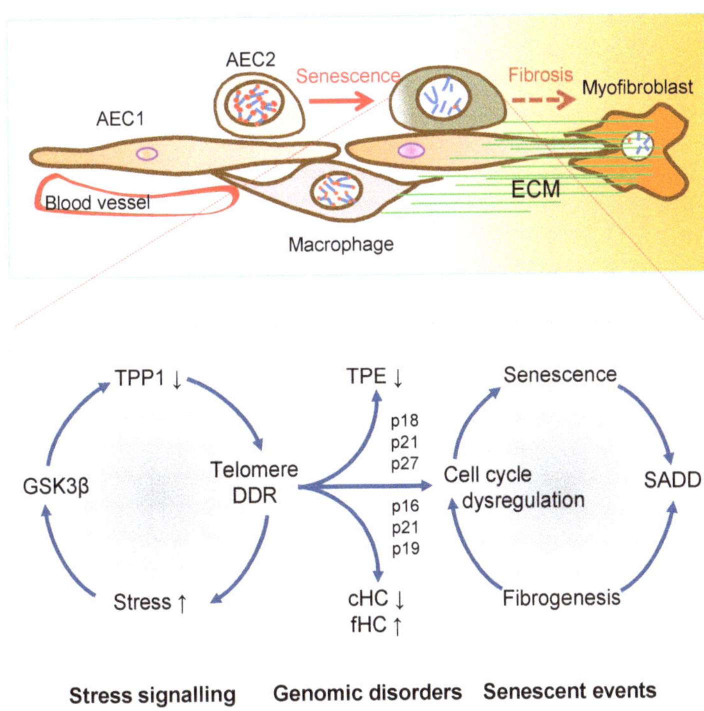

Figure 2. Mechanisms of AEC2 stem cell senescence and SADD. Cellular stress signalling triggers GSK3β-targeting of telomere shelterin complex, inducing the telomerase recruitment protein TPP1 phosphorylation,

subjecting phosphorylated TPP1 multisite polyubiquitination and degradation, resulting in telomere uncapping. The telomere uncapping triggers telomeric DDR, resulting in activation of the cyclin-dependent protein kinase inhibitors and cell cycle deregulation through telomere position effect (TPE) and altered constitutive and facultative heterochromatins (cHC and fHC). Unresolved telomeric DNA repair and cell cycle arrest result in stem cell senescence and subsequent SADD including differentiation arrest and trans-differentiation underlying pulmonary fibrosis.

3. Telomere Dysfunction Mediates Pulmonary Senescence and Fibrosis

AEC2 stem cell replicative senescence has been shown to involve DDR in SADD and IPF [35,84–87] under SALI, which is subserved by senescence-associated secretory phenotype (SASP) and infiltrations of inflammatory and immune cells [78–81] (Figures 1 and 2). Genetically, loss of function mutations of the genes encoding *TERT*, *telomerase RNA component* (*TERC*), *poly-(A)-specific ribonuclease* (*PARN*) and *regulator of telomere length 1* (*RTEL1*) underpins patient familial recessive hereditary IPF [2,18,88–93]. Notably, over a third of IPF cases with hereditary susceptibility of autosomal recessive gene damages could be related to telomere-related gene mutations including telomerase catalytic subunit *TERT* and RNA subunit *TERC* genes [6,35,93–97], and some could be related to mutations of the genes coding SPC or SPA of alveolar stem cells [98–100]. With the hypothesis that genetic susceptibility involves vulnerable genetic elements such as telomeres predisposing increased pulmonary sensitivity to environmental hazard-accelerated damages [35], it has been shown that stress-induced DDR occurs rapidly to telomeres, in a more abundant scale in telomeres as compared with non-telomeric regions in the genome [35,101,102]. The telomeric DDR takes place irrespective of telomerase activity, resistant to DNA repair, rendering irreparable DDR to be persistent in causing cellular replicative senescence [103–106]. Thus, preventing stress-induced telomere damages may protect the cells from entering replicative senescence and fibrogenesis.

Environmental factors triggering pulmonary senescence and fibrotic disease onset have been a hot topic in aiming to prevent the incurable disease IPF without yet effective therapy [91,96]. Since the 1980s, environmental stress factors that drive pulmonary fibrotic onset include smoking, bacterial toxin bleomycin [107,108], ionizing radiation (IR) [109–111], and oxidative metabolite reactive oxygen species [84,112–115]. Such environmental stress cues as bleomycin [58,116,117], IR [105,118,119] or H_2O_2 [102,120–122] have been shown to induce telomere DNA damage, resulting in telomere shortening. Consistent with cigarette smoking as causing pulmonary lesions by a mechanism of telomeric DDR [123–125], a significant number of smokers with IPF and chronic obstructive pulmonary disease (COPD) demonstrated acceleratory shortening of telomeres [47,123,126]. We demonstrated that radiation exposure, oxidative stress (such as H_2O_2) or bacterial product bleomycin each trigger telomere shelterin protein TPP1 degradation in AEC2 stem cells, provoking telomere uncapping, DDR, stem cell exhaustion, fibrogenic gene expressions, and pulmonary fibrosis [35]. These findings indicate a primary mechanism through which environmental stress induces AEC1 cell damage and coerces AEC2 stem cells to succumb to replicative senescence and SADD in pulmonary fibrosis.

In the telomere syndrome traits, IPF appeared more common than bone marrow diseases [58,95,127] (such as aplastic anemia [128] or dyskeratosis congenita (DC) [129–132]). The incidence of IPF is related to the aggravation of lung epithelial injuries, alveolar stem cell replicative senescence and SADD induced by environmental stress factors such as smoking, infection, radiation injury, oxidative stress [16,35,93,117]. It is thought that once in replicative senescence, AEC2 stem cells are entangled in the milieu of SALI in relaying bidirectional signal transductions via a variety of inflammatory factors to and from senescent cells, thereby driving the development of fibrogenesis [78,133–135]. Animal studies have demonstrated that environmental stresses, such as viral infection, induce extensive inflammatory responses, interstitial hyperplasia and edema of the lung tissue, resulting in fibrosis and dyspnea [35,136,137], characteristic of chronic pulmonary epithelial reduction, diffuse alveolitis and interstitial myofibroblast proliferation [78,138]. Investigation of the key molecular mechanisms mediating stress-induced pulmonary cellular senescence may unveil molecular targets pivotal for intervention in pulmonary fibrosis as in IPF. What remains unclear for decades includes how telomere shortening or replicative senescence is triggered and accelerated to provoke IPF by various environmental stressors. Telomeres are composed of the DNA repeats, $(TTAGGG)_n$, and their binding protein complex shelterin at chromosome ends, underpinning chromosome integrity and stability in the regulation of genome homeostasis in cell development, proliferation and differentiation [139–142]. Shelterin contains six proteins, telomere repeat binding factor 1 (TRF1), TRF2, RAP1, TIN2, POT1 and TPP1, and covers both double-stranded and single-stranded telomeric DNA, preventing telomeres from inadvertently activating DDR and replicative senescence in stress responses [143,144]. Under physiological conditions, telomere maintenance, or shortening as it occurs with cell replication, is regulated by the levels of shelterin and operates as the mitotic clock that results in their exit from the cell cycle in replicative senescence [145,146]. However, telomere dysfunction occurs as a function of sustained abundance of environmental stress stimulations, preceding non-telomeric DNA damages which ensue to telomere dysfunction [35]. Critical stress pressures [6,35,119,147,148], and chronic insufficiency of telomeric DNA repair, such as repressions of telomerase genes, repressions [35,78,149,150], together result in telomere shortening, ultimately premature senescence, and ageing-related diseases. Telomere dysfunction as a potential cause of ageing-related diseases [127,151,152] results from mutations of telomere related genes, including those encoding shelterin proteins [143], and telomerase [78,153].

Shelterin and telomerase function to protect telomeres by preventing cells from telomere DDR [29,154–157] and entering senescence [27,28,78,152,153]. Genetic mutagenesis of TRF2 or telomerase genes induces pulmonary fibrosis [58,78], whereas transgenically expressing *TERT* holds a therapeutic potential in alleviating pulmonary fibrosis [158]. In addition, the *ACD* gene coding for TPP1, which caps the telomere ends and recruits telomerase [159–165], was associated with such stem cell diseases as aplastic anemia with bone marrow failure [166] and Hoyeraal–Hreidarsson syndrome [167], and the expression levels of TPP1 were significantly reduced in the lung tissue of patients with chronic obstructive pulmonary disease (COPD) [168]. These studies suggest that qualitative and quantitative alterations in telomere shelterin and telomerase components accelerate telomere shortening, stem cell replicative senescence and related disorders.

3.1. Telomerase Gene Deficiency in Ageing-Related Disorders

Telomerase replenishes telomeres to counteract shortening by *TERC* template reverse transcription [169,170]. Telomerase catalytic subunit TERT determines telomerase activity in lengthening telomeres, conferring the ability of continuous proliferation and survival on stem cells and tissue precursor cells [171–176]. In addition to elongating telomeres, TERT is involved in maintaining telomere heterochromatin and synthesizing double-stranded RNA through RNA-dependent RNA polymerase activity (RdRP) [177–179]. Telomerase deficiency due to *TERT* mutations results in stem cell replicative senescence and exhaustion, tissue fibrosis, aplastic anemia, and skin diseases of premature ageing [35,78,132,180,181]. While *TERT* gene expressions are fundamental to telomerase activity and stem cell renewal, tissue regeneration, the molecular mechanisms underlying *TERT* gene expression remain largely elusive [182]. Multiple lines of evidence indicate that several transcription factors play fundamental roles in regulating *TERT* transcription [183] (Figures 3 and 4). Our laboratory established for the first time that downstream MAP kinase signalling [184], ETS transcription factors bind to the *TERT* gene promoter ETS binding motif (*CCTT*) directly, functioning as an essentially key transcription factor for *TERT* gene expression in cancer [184,185]. Moreover, ETS activates *TERT* transcription not only directly by binding to the *CCTT* element and in cooperation with the proto-oncogene c-myc that binds to the E-box (*CACGTG*) at the *TERT* promoter in the regulation of telomerase activity and cell proliferation [185–188] (Figures 3 and 4). The role of ETS in *TERT* gene expression has been confirmed independently in cancer patients with C→T nucleotide mutations creating additional ETS binding sites [189,190] and mechanistically [191]. Furthermore, Wu et al. showed first that c-myc activates *TERT* by binding to the E-box at the *TERT* gene promoter [192]. However, the action of c-myc at the E-box of *TERT* promoter is regulated negatively by Max/Mad1 interaction [193] and by TGF-β signalling via Smad3 interactions with c-myc and the *TERT* gene promoter [186,188] (Figures 3 and 4). Our laboratory demonstrated that estrogen upregulates *TERT* gene expression and telomerase activity in the endocrine organs of ovary and adrenal gland for tissue homeostasis [149,150,194,195]. Deficiency of estrogen in mice by targeted disruption of the aromatase gene responsible for estrogen synthesis results in compromised cell proliferation and stem cell exhaustion [149,150]. Interestingly, hormone replacement therapy increases the activities of telomerase and stem cell renewal positive for Ki67 staining and the stem cell markers stem-cell antigen-1 (Sca-1) and c-kit expression, attenuating premature ageing and tissue atrophy at both adrenal and ovarian glands [149,150]. These studies are consistent with the current notion that telomerase TERT is required for the maintenance of telomeres and stem cell functionalities in the endocrine, epithelial and endothelial tissues against tissue degenerative changes and proliferative disorders [196–198].

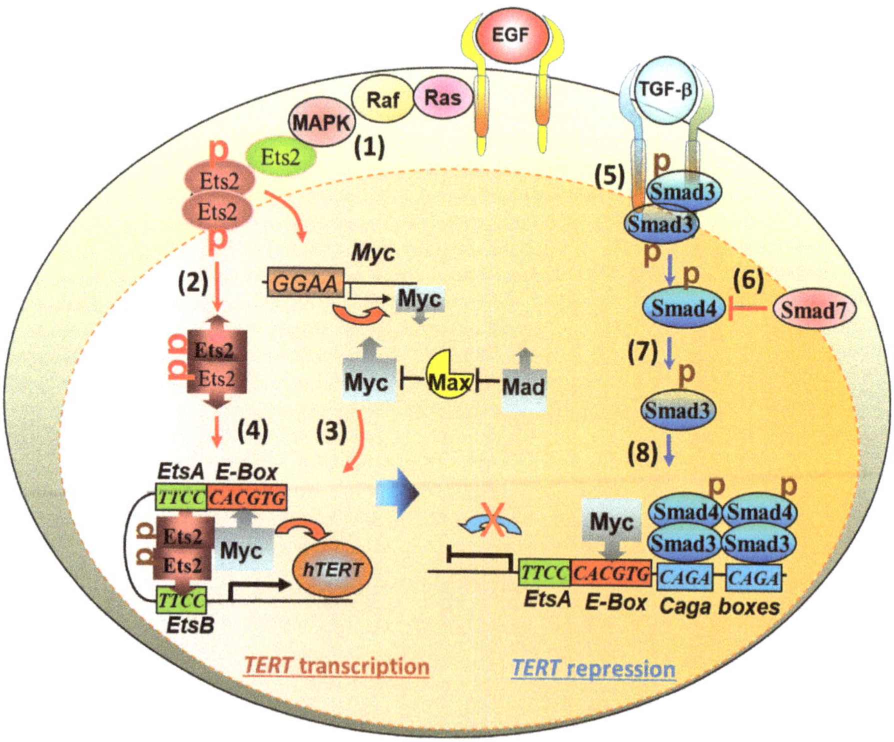

Figure 3. Intracellular signalling pathways of growth factors to the transcription factors and repressors in the regulation of telomerase reverse transcriptase (TERT) gene transcription. The TERT gene promoter assumes both active and repressive conformations under the molecular regulation of the MAP kinase and TGF-β signalling pathways, respectively. Epithelial growth factor (EGF) stimulated mitogenic signalling induces MAP kinase-mediated Ets2 transcription factor phosphorylation, nuclear retentions and dimerisation (**1**–**2**). Ets2 binds the CCTT element in the TERT and c-myc gene promoters, driving c-myc (**3**) and TERT (**4**) transcriptions in the upregulation of cell proliferative immortality. On the other hand, TGF-β activates TGF-β RII receptors by auto- and trans-phosphorylation, resulting in Smad3 phosphorylation and mobilization (**5**), which is regulated

natively by Smad7 (**6**) and positively by Smad4 (**7**) in the Smad3 nuclear localization and action. Smad3 binds to the CAGA element in the TERT promoter to repress TERT gene transcription in pathological fibrogenesis such as pulmonary fibrosis (**8**).

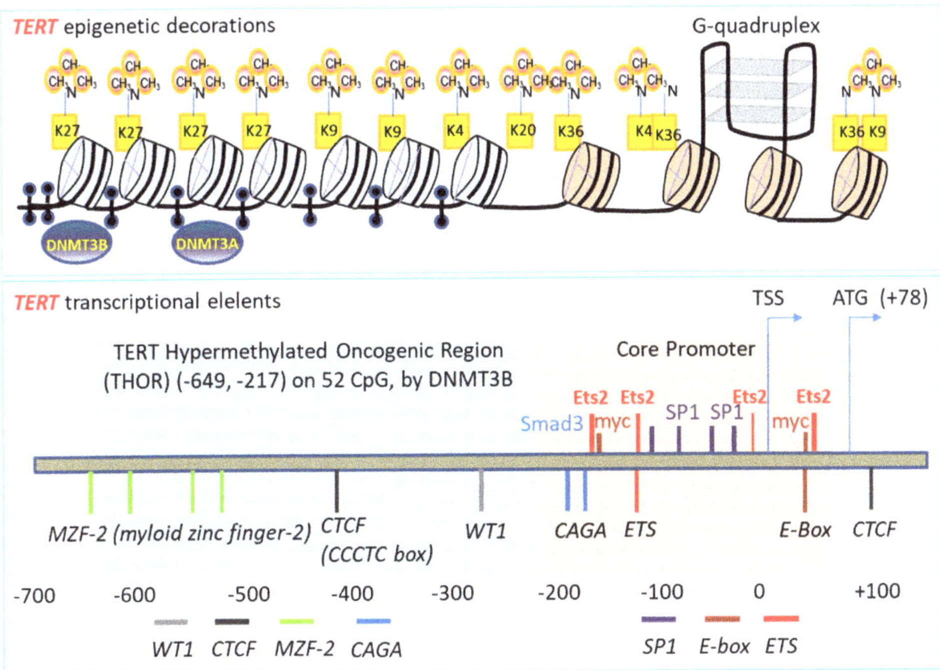

Figure 4. The regulatory network of TERT gene transcription. Top panel: Epigenetic regulatory organization of the TERT gene promoter repression involves trimethylation of the various lysine residues (K) on the nucleosome histone tails, methylation of C5 cytosine of CpG dinucleotides by DNA methyltransferase 3 alpha and beta (DNMT3A and 3B), and G-quadruplex. Bottom panel: TERT promoter DNA regulatory elements (italic), and engaged transcription factors and repressors (non-italic) in the positive (red) and negative (blue) regulations of TERT gene transcription. The scale labelled is for relative positions of promoter upstream regulatory DNA elements, but not in proportion.

3.2. TGF-β Signaling to Telomerase TERT Gene Repression

TGF-β signaling is closely involved in pulmonary fibrosis [19,20,22,23,56,57,82,199–201]. The mechanisms of how TGF-β family cytokines regulate the intracellular events mediating pulmonary fibrogenesis remain unclear [202]. Studies indicate that mediating the interactions between AEC2 stem cells and fibroblasts, TGF-β released from cells [203] induces Sonic Hedgehog (SHH) pathway in alveolar epithelial cells by autocrine mechanisms, whereas SHH induces TGF-β in lung fibroblasts stimulating myofibrosis by paracrine mechanisms, suggesting that re-emergence of SHH in epithelial cells mediates TGF-β signaling and induces myofibroblast differentiation in a Smoothened receptor-dependent manner with subsequent transcription factor Gli1 activation of the *α-SMA* promoter [204]. In addition, elevated extracellular mechanical tension between myoblasts and AEC2 stem cells activates TGF-β RII receptor signaling in AEC2 stem cells, resulting in AEC2 stem cells with increased TGF-β1 and becoming gradually unable to differentiate to AEC1 cells [23]. Deletion of the TGF-β RII receptor specifically in lung epithelium protects mice from bleomycin-induced pulmonary fibrosis, further supporting a central role TGF-β signaling of alveolar epithelium in fibrogenesis [82,205]. Reflecting SADD, moreover, recent studies show that in radiation-induced pulmonary fibrosis, both TGF-β and GSK3β are increased in AEC2 cells undergoing transdifferentiation with increased AEC1 cells and mesenchymal markers such as α-SMA [206]. These studies suggest that TGF-β in the intercellular milieu stimulates AEC2 cell transdifferentiation by mechanisms involving GSK3β in association with SHH cellular signaling.

The mechanism of intracellular transduction initiated by TGF-β in the induction of pulmonary fibrosis entails receptor-mediated Smad3 signaling. Firstly, epithelium-specific disruption of TGF-β RII receptor increases Smad2 phosphorylation and decreases Smad3 phosphorylation, and protects mice from bleomycin-induced pulmonary fibrosis with increased survival [82] (Figure 3). In addition, knockout of Smad3 [207], or inhibition of phosphorylated Smad3 into the nucleus by polypeptide [208], inhibits TGF-β1-induced pulmonary fibrosis in mice. Moreover, in an attempt to determine the downstream mechanisms of TGF-β activation of Smad3 co-transcription factor, our laboratory demonstrated that by acting on different specific RII receptors, TGF-β1 induces Smad3 translocation from the cytoplasm to the nucleus where Smad3 binds the *CAGA* box in the *TERT* gene promoter interfering c-myc binding and repressing *TERT* transcription [188,209–211] (Figures 3 and 4). Furthermore, the TGF-β family member BMP7 induces the *TERT* gene repression in a BMP RII receptor- and Smad3-dependent manner [211,212]. Chronic exposure to BMP7 results in telomere shortening, cell replicative senescence and apoptosis, but mutation of the BMP RII receptor (but not TGFbRII, ACTRIIA or ACTRIIB receptors) inhibits BMP7-induced *TERT* repression, leading to increased telomerase activity, lengthened telomeres and continued cell proliferation [211]. The effect of BMP7-induced cell replicative senescence and apoptosis is *TERT* repression-dependent as overexpression of *TERT* reverses BMP7-induced cell replicative senescence [211]. These data together suggest that Smad3-mediated repression of the *TERT* gene and shortening of telomeres are critical to TGF-β induced pulmonary senescence, SADD, and pathological fibrogenesis.

3.3. Stress-Induced TPP1 Degradation and Telomere Uncapping

In the regulation of telomerase lengthening of telomeres, recruiting telomerase to telomeres represents a most effective step in telomere maintenance. Our recent studies show an intertwined relationship of TPP1 capping of telomeres, recruitment of telomerase, deficiency-induced telomere uncapping, pulmonary senescence, and fibrosis [35,213]. By targeted protection of telomere uncapping, we demonstrate that pulmonary senescence and fibrosis with dyspnea are altogether prevented in mice under persistent fibrogenic stresses including whole body ionizing radiation or pulmonary bacterial toxin bleomycin [35]. The stress-induced pulmonary senescence and fibrotic onset are targetable through animal inhalation of either recombinant cDNA coding for the telomere protein TPP1 or a small 8-mer peptide (telomere dysfunction inhibitor or TELODIN) to prevent telomere uncapping, indicating an early effective intervention strategy on the mechanisms of telomere dysfunction and subsequent cellular senescence prior to differentiation disorder of pulmonary fibrosis [35,213]. These findings dovetail with a large body of literature indicating that telomere DDR underpins IPF [35,58,93,95,103–106,214], providing an important strategy for intervention of cell replicative senescence-associated diseases.

TPP1 is a subunit of shelterin that is a multi-subunit protein complex capping telomeric DNA [143,144]. We showed that pulmonary fibrosis initiated by chronic stress in mouse models is mediated by the tumor suppressor protein FBW7-mediated degradation of TPP1 [35]. FBW7 E3 ubiquitin ligase binds TPP1 and mediates stress-stimulated TPP1 multisite polyubiquitination at K299, K453 and K459, causing TPP1 degradation in proteasomes and telomere uncapping [35]. The multisite polyubiquitination of TPP1 requires FBW7 interaction with a Cdc4 phosphodegron (CPD) site [215,216] in the serine/threonine (S/T) region (aa 341–482) of TPP1, with the CPD being phosphorylated by GSK3β at the S354 and S358 residues [35]. We showed that the binding to the phosphorylated CPD of TPP1 by FBW7 is mediated by the seventh β-strand blade containing the ^{689}R residue frequently mutated in cancers at the WD40 propeller domain of the C-terminal region of FBW7 [35]. Consistent with stress-mediated TPP1 degradation, thereby telomere shortening and pulmonary fibrosis, we found that under-expression of TPP1 is sufficient to induce pulmonary senescence and fibrosis, indicating that TPP1 polyubiquitination at multiple sites triggered by stress incurs telomere uncapping, DDR and shortening, resulting in pulmonary senescence and fibrosis [35]. Moreover, overexpression of TPP1 enhances respiratory physiological function with increased AEC2 stem cell population with lengthened telomeres, and confers pulmonary resistance to stress-induced onset of pulmonary fibrosis [35].

4. Targeted Intervention of Cellular Senescence and Tissue Fibrosis

Since cellular replicative senescence plays a causal role in tissue remodelling, it appears particularly appealing to determine if elimination or prevention of cellular replicative senescence may provide beneficial outcomes to mitigate pathogenic tissue remodelling and ageing-related disorders under various disease conditions. Considerable evidence suggests that targeted removal or prevention of cell replicative senescence alleviates SADD in fibrosis, including VIS or stress-induced fibrosis [3,35].

4.1. Targeting Anti-Apoptotic Gene Bcl-2 to Clear Senescent Cells

Since replicative senescence cells have upregulated anti-apoptotic proteins Bcl-w and Bcl XL which underpin senescent cells' antiapoptotic ability with long-term survival, drugs that inhibit Bcl survival proteins have been studied and named as senolytics for removing replicative senescence cells [217,218] (Table 1). Although the U.S. Food and Drug Administration (FDA) approved the selective Bcl-2 inhibitor venetoclax (abt-199) used in leukemia, it has not had a significant effect on anti-ageing tests in vitro. Compound screening-identified that its homologue navitoclax (abt-263) effectively inhibits the effects of Bcl-2, Bcl XL and Bcl-w, and induces apoptosis and thus clearance of replicative senescence cells [217]. In addition to abt-263, siRNAs and chemical inhibitor abt-737 simultaneously inhibit Bcl-w and Bcl XL, induce replicative senescence cell specific apoptosis, reduce RC in the populations of hematopoietic stem cells, muscle stem cells, and pulmonary and epidermal tissues of mice, with cleared replicative senescence cells harboring DNA damage and p14ARF-activated TP53 [218]. Moreover, a combination of dasatinib and quercetin (D + Q) reduces senescent cells, improving mouse idiopathic pulmonary fibrosis [219], and nerve regeneration and obesity-related anxiety behavior [220].

Recently, the use of navitoclax and D+Q selectively eliminated VIS cells and mitigated COVID-19-reminiscent lung disease with reduced inflammation in SARS-CoV-2-driven hamster and mouse models [3]. Treatment with navitoclax for less than a week to the SARS-CoV-2 infected Syrian golden hamster model with increased $p16^{INK4A}$ led to a profound decrease in senescent cells and SALI [3]. On another hamster model of the Roborovski dwarf animals, the navitoclax or D + Q regimen led to a substantial reduction of the cell senescence markers H3K9me3 and $p16^{INK4A}$ in the respiratory epithelium and increased lifespan [3]. Furthermore, in two randomized clinical trials (NCT04578158 and NCT04861298), quercetin showed significant effects of symptom improvements with significant risk reductions regarding the needs of hospitalization and oxygen therapy in a total of 194 COVID-19 patients [3]. The senolytic targeting of VIS as a novel regimen option against COVID-19 indicates that cellular replicative senescence is causative to severe stress-induced acute SADD, tissue fibrogenic remodelling, and phenotypic onset, and that eliminating cellular replicative senescence is beneficial in managing pulmonary fibrosis.

4.2. Targeting TP53 and $p16^{INK4A}$ Tumor Suppressor Genes to Clear Senescent Cells

Recent studies have consistently shown that senescent cells can be eliminated by targeting the cells with high levels of $p16^{INK4A}$ and TP53 tumor suppressor gene expressions. Because $p16^{INK4a}$ is significantly increased in senescent cells, AP20187 that targets FK506 using a minimal $p16^{INK4a}$ promoter element triggers a dimer formation of FK506 binding protein and caspase-8 to effectively induce senescent cell apoptosis, resulting in a significant reduction of the incidence and mortality of cardiovascular diseases in aged mice [221]. This regimen of p16^{INK4a} sensitive caspase-8-mediated apoptosis also delays osteoporosis and persistent intervertebral disc proteoglycans [217], restores nerve regeneration in obese induced by high-fat feeding or leptin receptor deficiency, and reduces anxiety-related behaviors in mice [220]. In another model of p16^{INK4A}-3MR transgenic mice, the expression of a reporter protein (3MR) under the control of p16^{INK4A} to selectively remove senescent cells through using the small molecule compound UBX0101 reduced post-traumatic osteoarthritis and extended lifespan [222]. Furthermore, studies show that TP53 is crucial in AEC2 stem cell senescence preceding pulmonary fibrosis, especially with potential accumulation of phosphorylated TP53 at the S15 residue [38,77,223]. Therefore, the use of a FOXO4 polypeptide—harboring 60 amino acids to interfere with the interaction between FOXO4 and TP53—may be fundamental in triggering TP53-dependent apoptosis of senescent cells, delaying fibrogenesis and accelerating the rehabilitation of pulmonary ageing [223].

Table 1. Senescence-Targeting Senolytics and Other Compounds.

Class	Senolytic Agent	Mechanism of Action	Reference
BCL-2 Family Inhibitors	Dasatinib + Quercetin ABT-737 ABT-263 A-1331852 A-1155463	PI3K/Akt inhibition/DNA intercalation Inhibiting BCL-2, BCL-XL, BCL-W Inhibiting BCL-2, BCL-XL, BCL-W Inhibiting BCL-XL Inhibiting BCL-XL	[3,217,218,220,224]
Targeting p53	FOXO4-DRI UBX0101 RG7112 (RO5045337) P5091 P22077 NOX1 and NOX2 dual inhibitor PAI-1 inhibitor (TM5275)	Disrupting FOXO4-p53 interaction Disrupting MDM2-p53 interaction Disrupting MDM2-p53 interaction USP7 inhibitor USP7 inhibitor Activating p53 and apoptosis Activating p53 and apoptosis	[222,223,225,226]
HSP90 Inhibitors	17-DMAG (alvespimycin) Geldanamycin 17-AAG (tanespimycin) Ganetespib	Disrupting HSP90-AKT interaction Attenuating HSP90 Attenuating HSP90 Attenuating HSP90	[227,228]
Natural Products and their Analogues	Fisetin Curcumin *O*-Vanillin (curcumin metabolite) EF-24 (curcumin analogue) Piperlongumine and its analogues (compounds 47–49) GL-V9	BCL-2, PI3K/AKT, p53, NF-κB and more Down-regulating Nrf2 and NF-κB pathways Down-regulating Nrf2 and NF-κB pathways Attenuating the BCL-2 family Unclear, OXR1, NF-κB - Unclear, increasing ROS, alkalizing lysosome	[224,229–232]
Cardiac Glycosides	Proscillaridin A Ouabain Ouabagenin Digoxin Bufalin K-Stropanthin Strophanthidin	Inhibiting Na^+/K^+-ATPase Inhibiting Na^+/K^+-ATPase, increasing BCL2-Family protein NOXA Inhibiting Na^+/K^+-ATPase Inhibiting Na^+/K^+-ATPase, increasing BCL2-Family protein NOXA Inhibiting Na^+/K^+-ATPase Inhibiting Na^+/K^+-ATPase Inhibiting Na^+/K^+-ATPase	[233,234]
Galactose Modified Prodrugs	SSK1 Pro-drug A (JHB75B) Nav-Gal 5FURGa	Targeting SA-β-galactosidase Targeting SA-β-galactosidase Targeting SA-β-galactosidase Targeting SA-β-galactosidase	[235–237]

Table 1. *Cont.*

Class	Senolytic Agent	Mechanism of Action	Reference
PROTACs	PZ15227 ARV825	Degrading BCL-XL Degrading BRD4	[238,239]
Miscellaneous	Fenofibrate Azithromycin Roxithromycin Tamatinib (R406) MitoTam Panobinostat AT-406 Rapamycin Metformin	PPARα agonist Inducing autophagy and glycolysis NOX4 Syk inhibitor, FAK and p38MAPK Reducing mitochondrial membrane potential, Inhibiting OXPHOS Histone deacetylase inhibitor Inhibitor of c-IAP1, c-IAP2 and XIAP Inhibiting mTOR, p16 and p21 Inhibiting NF-κB pathways/AKT	[240–247]

4.3. Preventing Telomere Dysfunction and Pulmonary Fibrosis by TELODIN

By screening a peptide library, we discovered an 8-mer peptide (TELODIN) that significantly inhibited telomere dysfunction [213]. Corresponding to the β-turn hairpin-like blade 7 of FBW7 E3 ubiquitin ligase WD40 domain, as synthesized in a native or retro-inverso (reversed, inversed in dextral amino acids) configuration, TELODIN competitively inhibited stress-induced TPP1 accelerated turnover, telomere shortening and pulmonary fibrosis once applied through intratracheal instillation in mice [35,213]. TELODIN inhalation through the respiratory airway promoted alveolar stem cell proliferation and enhanced pulmonary resistance to stress-induced pulmonary fibrogenic onset induced by different chronic stresses including ionizing radiation, or by overexpression of FBW7 in causing TPP1 deficiency, telomere uncapping and DDR [35]. The effect of TELODIN on alveolar stem cell proliferation is due to increased TPP1 and subsequently protection of telomeres against GSK3β-primed FBW7-mediated TPP1 degradation [35,213]. Thus, to the emerged molecular target of TPP1 accelerated turnover, TELODIN prevents chronic stress induced premature pulmonary ageing and fibrosis, highlighting an effective protection of TPP1 and thus telomere dysfunction as a novel approach for intervention into pulmonary fibrosis.

5. Conclusions and Perspectives

Pulmonary ageing and fibrosis occur in association with AEC2 stem cell senescence and SADD. AEC2 stem cell failure of differentiation to AEC1 cells in epithelial damage repair, and trans-differentiation to mesenchymal fibrogenic remodeling, are the two critical cellular processes mediated by telomere dysfunction. Downstream of stress-induced uncapping of telomeres or TGF-β-signaling repression of the telomerase *TERT* gene, AEC2 stem cell senescence drives alveolar epithelial fibrogenesis, representing a key molecular target for intervention. Preventing AEC2 stem cell senescence by promoting telomere capping integrity using TELODIN, or removing senescent cells by promoting apoptosis using senolytics, is emerging as a promising strategy in the studies of pulmonary fibrosis intervention.

More investigations are underway to detect and intervene in stress-induced premature pulmonary senescence and the early stage of fibrosis, informing molecular targeting strategies of prophylactic and therapeutic intervention. The contemporary strategies and technologies—such as sequencing single cell populations to differentiate gene profiling of cell senescence and differentiation statues, deciphering structural modifications post gene transcription, defining disordered molecular networks, and decoding denatured macromolecules and compromised protective complexes—will lead to more studies of interventions of alveolar stem cell senescence and SADD in pulmonary fibrogenesis. Dissecting cellular and molecular interplays will likely produce further insights into the development of premature cellular senescence and fibrosis for some prophylactic and therapeutic outcomes.

Author Contributions: X.H. conceived and wrote the manuscript. L.W., K.Z. and J.L. participated in writing. J.-P.L. conceived, supervised and wrote the manuscript. All authors contributed to the discussions and presentations. All authors have read and agreed to the published version of the manuscript.

Funding: This research was funded by National Key Research and Development Program of China (2018YFC2000100, and 2021ZD0202402), the National Natural Science Foundation of China (91949207, 82130044, 92149302, 91849124, and 81871112), and the Victorian Government's Operational Infrastructure Support Program of Australia.

Acknowledgments: This work was supported by grants from the National Key Research and Development Program of China (2018YFC2000100, and 2021ZD0202402), the National Natural Science Foundation of China (91949207, 82130044, 92149302, 91849124, and 81871112), and the Victorian Government's Operational Infrastructure Support Program of Australia.

Conflicts of Interest: The authors declare no conflict of interest.

References

1. Cookson, W.O.; Moffatt, M.F. Bedside to gene and back in idiopathic pulmonary fibrosis. *N. Engl. J. Med.* **2013**, *368*, 2228–2230. [CrossRef] [PubMed]
2. Armanios, M.Y. Telomerase mutations in families with idiopathic pulmonary fibrosis. *N. Engl. J. Med.* **2007**, *356*, 1370–1372. [CrossRef] [PubMed]
3. Lee, S.; Yu, Y.; Trimpert, J.; Benthani, F.; Mairhofer, M.; Richter-Pechanska, P.; Wyler, E.; Belenki, D.; Kaltenbrunner, S.; Pammer, M.; et al. Virus-induced senescence is driver and therapeutic target in COVID-19. *Nature* **2021**, *599*, 283–289. [CrossRef] [PubMed]
4. Wang, S.; Yao, X.; Ma, S.; Ping, Y.; Fan, Y.; Sun, S.; He, Z.; Shi, Y.; Sun, L.; Xiao, S.; et al. A single-cell transcriptomic landscape of the lungs of patients with COVID-19. *Nat. Cell Biol.* **2021**, *23*, 1314–1328. [CrossRef]
5. Nie, X.; Qian, L.; Sun, R.; Huang, B.; Dong, X.; Xiao, Q.; Zhang, Q.; Lu, T.; Yue, L.; Chen, S.; et al. Multi-organ proteomic landscape of COVID-19 autopsies. *Cell* **2021**, *184*, 775–791.e14. [CrossRef]
6. Armanios, M.Y.; Chen, J.J.; Cogan, J.D.; Alder, J.K.; Ingersoll, R.G.; Markin, C.; Lawson, W.E.; Xie, M.; Vulto, I.; Phillips, J.A., 3rd; et al. Telomerase mutations in families with idiopathic pulmonary fibrosis. *N. Engl. J. Med.* **2007**, *356*, 1317–1326. [CrossRef]
7. Ting, C.; Aspal, M.; Vaishampayan, N.; Huang, S.K.; Wang, F.; Farver, C.; Zemans, R.L. Ineffectual AEC1 differentiation from KRT8 (hi) transitional state without fibrosis is associated with fatal COVID-19 ARDS. *bioRxiv* **2021**. [CrossRef]
8. Drake, T.M.; Docherty, A.B.; Harrison, E.M.; Quint, J.K.; Adamali, H.; Agnew, S.; Babu, S.; Barber, C.M.; Barratt, S.; Bendstrup, E.; et al. Outcome of Hospitalization for COVID-19 in Patients with Interstitial Lung Disease. An International Multicenter Study. *Am. J. Respir. Crit. Care Med.* **2020**, *202*, 1656–1665. [CrossRef]
9. Jansing, N.L.; McClendon, J.; Henson, P.M.; Tuder, R.M.; Hyde, D.M.; Zemans, R.L. Unbiased Quantitation of Alveolar Type II to Alveolar Type I Cell Transdifferentiation during Repair after Lung Injury in Mice. *Am. J. Respir. Cell Mol. Biol.* **2017**, *57*, 519–526. [CrossRef]
10. Barkauskas, C.E.; Cronce, M.J.; Rackley, C.R.; Bowie, E.J.; Keene, D.R.; Stripp, B.R.; Randell, S.H.; Noble, P.W.; Hogan, B.L. Type 2 alveolar cells are stem cells in adult lung. *J. Clin. Investig.* **2013**, *123*, 3025–3036. [CrossRef]
11. Kathiriya, J.J.; Brumwell, A.N.; Jackson, J.R.; Tang, X.; Chapman, H.A. Distinct Airway Epithelial Stem Cells Hide among Club Cells but Mobilize to Promote Alveolar Regeneration. *Cell Stem Cell* **2020**, *26*, 346–358.e4. [CrossRef] [PubMed]
12. Rao, W.; Wang, S.; Duleba, M.; Niroula, S.; Goller, K.; Xie, J.; Mahalingam, R.; Neupane, R.; Liew, A.A.; Vincent, M.; et al. Regenerative Metaplastic Clones in COPD Lung Drive Inflammation and Fibrosis. *Cell* **2020**, *181*, 848–864.e18. [CrossRef]
13. Lee, J.H.; Bhang, D.H.; Beede, A.; Huang, T.L.; Stripp, B.R.; Bloch, K.D.; Wagers, A.J.; Tseng, Y.H.; Ryeom, S.; Kim, C.F. Lung stem cell differentiation in mice directed by endothelial cells via a BMP4-NFATc1-thrombospondin-1 axis. *Cell* **2014**, *156*, 440–455. [CrossRef]
14. Chen, Q.; Suresh Kumar, V.; Finn, J.; Jiang, D.; Liang, J.; Zhao, Y.Y.; Liu, Y. CD44(high) alveolar type II cells show stem cell properties during steady-state alveolar homeostasis. *Am. J. Physiol. Lung Cell. Mol. Physiol.* **2017**, *313*, L41–L51. [CrossRef] [PubMed]
15. Reddy, R.; Buckley, S.; Doerken, M.; Barsky, L.; Weinberg, K.; Anderson, K.D.; Warburton, D.; Driscoll, B. Isolation of a putative progenitor subpopulation of alveolar epithelial type 2 cells. *Am. J. Physiol. Lung Cell. Mol. Physiol.* **2004**, *286*, 658–667. [CrossRef] [PubMed]
16. Salahudeen, A.A.; Choi, S.S.; Rustagi, A.; Zhu, J.; van Unen, V.; de la O, S.M.; Flynn, R.A.; Margalef-Catala, M.; Santos, A.J.M.; Ju, J.; et al. Progenitor identification and SARS-CoV-2 infection in human distal lung organoids. *Nature* **2020**, *588*, 670–675. [CrossRef] [PubMed]
17. Nabhan, A.N.; Brownfield, D.G.; Harbury, P.B.; Krasnow, M.A.; Desai, T.J. Single-cell Wnt signaling niches maintain stemness of alveolar type 2 cells. *Science* **2018**, *359*, 1118–1123. [CrossRef]
18. Choi, J.; Park, J.E.; Tsagkogeorga, G.; Yanagita, M.; Koo, B.K.; Han, N.; Lee, J.H. Inflammatory Signals Induce AT2 Cell-Derived Damage-Associated Transient Progenitors that Mediate Alveolar Regeneration. *Cell Stem Cell* **2020**, *27*, 366–382.e7. [CrossRef]

19. Kobayashi, Y.; Tata, A.; Konkimalla, A.; Katsura, H.; Lee, R.F.; Ou, J.; Banovich, N.E.; Kropski, J.A.; Tata, P.R. Persistence of a regeneration-associated, transitional alveolar epithelial cell state in pulmonary fibrosis. *Nat. Cell Biol.* **2020**, *22*, 934–946. [CrossRef]
20. Strunz, M.; Simon, L.M.; Ansari, M.; Kathiriya, J.J.; Angelidis, I.; Mayr, C.H.; Tsidiridis, G.; Lange, M.; Mattner, L.F.; Yee, M.; et al. Alveolar regeneration through a Krt8+ transitional stem cell state that persists in human lung fibrosis. *Nat. Commun.* **2020**, *11*, 3559. [CrossRef]
21. Jiang, P.; Gil de Rubio, R.; Hrycaj, S.M.; Gurczynski, S.J.; Riemondy, K.A.; Moore, B.B.; Omary, M.B.; Ridge, K.M.; Zemans, R.L. Ineffectual Type 2-to-Type 1 Alveolar Epithelial Cell Differentiation in Idiopathic Pulmonary Fibrosis: Persistence of the KRT8(hi) Transitional State. *Am. J. Respir. Crit. Care Med.* **2020**, *201*, 1443–1447. [CrossRef] [PubMed]
22. Riemondy, K.A.; Jansing, N.L.; Jiang, P.; Redente, E.F.; Gillen, A.E.; Fu, R.; Miller, A.J.; Spence, J.R.; Gerber, A.N.; Hesselberth, J.R.; et al. Single cell RNA sequencing identifies TGFbeta as a key regenerative cue following LPS-induced lung injury. *JCI Insight* **2019**, *4*, e123637. [CrossRef] [PubMed]
23. Wu, H.; Yu, Y.; Huang, H.; Hu, Y.; Fu, S.; Wang, Z.; Shi, M.; Zhao, X.; Yuan, J.; Li, J.; et al. Progressive Pulmonary Fibrosis Is Caused by Elevated Mechanical Tension on Alveolar Stem Cells. *Cell* **2020**, *180*, 107–121.e4. [CrossRef] [PubMed]
24. Salama, R.; Sadaie, M.; Hoare, M.; Narita, M. Cellular senescence and its effector programs. *Genes Dev.* **2014**, *28*, 99–114. [CrossRef]
25. Hayflick, L.; Moorhead, P.S. The serial cultivation of human diploid cell strains. *Exp. Cell Res.* **1961**, *25*, 585–621. [CrossRef]
26. Munoz-Espin, D.; Serrano, M. Cellular senescence: From physiology to pathology. *Nat. Rev. Mol. Cell Biol* **2014**, *15*, 482–496. [CrossRef]
27. Maciel-Baron, L.A.; Morales-Rosales, S.L.; Aquino-Cruz, A.A.; Triana-Martinez, F.; Galvan-Arzate, S.; Luna-Lopez, A.; Gonzalez-Puertos, V.Y.; Lopez-Diazguerrero, N.E.; Torres, C.; Konigsberg, M. Senescence associated secretory phenotype profile from primary lung mice fibroblasts depends on the senescence induction stimuli. *Age* **2016**, *38*, 26. [CrossRef]
28. Balaban, R.S.; Nemoto, S.; Finkel, T. Mitochondria, oxidants, and aging. *Cell* **2005**, *120*, 483–495. [CrossRef]
29. Rivera, T.; Haggblom, C.; Cosconati, S.; Karlseder, J. A balance between elongation and trimming regulates telomere stability in stem cells. *Nat. Struct. Mol. Biol.* **2017**, *24*, 30–39. [CrossRef]
30. Paluvai, H.; Di Giorgio, E.; Brancolini, C. The Histone Code of Senescence. *Cells* **2020**, *9*, 466. [CrossRef]
31. Sati, S.; Bonev, B.; Szabo, Q.; Jost, D.; Bensadoun, P.; Serra, F.; Loubiere, V.; Papadopoulos, G.L.; Rivera-Mulia, J.C.; Fritsch, L.; et al. 4D Genome Rewiring during Oncogene-Induced and Replicative Senescence. *Mol. Cell* **2020**, *78*, 522–538.e9. [CrossRef] [PubMed]
32. Campaner, S.; Doni, M.; Hydbring, P.; Verrecchia, A.; Bianchi, L.; Sardella, D.; Schleker, T.; Perna, D.; Tronnersjo, S.; Murga, M.; et al. Cdk2 suppresses cellular senescence induced by the c-myc oncogene. *Nat. Cell Biol.* **2010**, *12*, 54–59. [CrossRef] [PubMed]
33. Laberge, R.M.; Sun, Y.; Orjalo, A.V.; Patil, C.K.; Freund, A.; Zhou, L.; Curran, S.C.; Davalos, A.R.; Wilson-Edell, K.A.; Liu, S.; et al. MTOR regulates the pro-tumorigenic senescence-associated secretory phenotype by promoting IL1A translation. *Nat. Cell Biol.* **2015**, *17*, 1049–1061. [CrossRef]
34. Sieben, C.J.; Sturmlechner, I.; van de Sluis, B.; van Deursen, J.M. Two-Step Senescence-Focused Cancer Therapies. *Trends Cell Biol.* **2018**, *28*, 723–737. [CrossRef]
35. Wang, L.; Chen, R.; Li, G.; Wang, Z.; Liu, J.; Liang, Y.; Liu, J.P. FBW7 Mediates Senescence and Pulmonary Fibrosis through Telomere Uncapping. *Cell Metab.* **2020**, *32*, 860–877.e9. [CrossRef] [PubMed]
36. Sharpless, N.E.; Sherr, C.J. Forging a signature of In Vivo senescence. *Nat. Rev. Cancer* **2015**, *15*, 397–408. [CrossRef] [PubMed]
37. Ito, Y.; Hoare, M.; Narita, M. Spatial and Temporal Control of Senescence. *Trends Cell Biol.* **2017**, *27*, 820–832. [CrossRef] [PubMed]
38. Yao, C.; Guan, X.; Carraro, G.; Parimon, T.; Liu, X.; Huang, G.; Mulay, A.; Soukiasian, H.J.; David, G.; Weigt, S.S.; et al. Senescence of Alveolar Type 2 Cells Drives Progressive Pulmonary Fibrosis. *Am. J. Respir. Crit. Care Med.* **2021**, *203*, 707–717. [CrossRef]
39. Zhang, K.; Chen, C.; Liu, Y.; Chen, H.; Liu, J.P. Cellular senescence occurred widespread to multiple selective sites in the fetal tissues and organs of mice. *Clin. Exp. Pharmacol. Physiol.* **2014**, *41*, 965–975. [CrossRef]
40. Dorr, J.R.; Yu, Y.; Milanovic, M.; Beuster, G.; Zasada, C.; Dabritz, J.H.; Lisec, J.; Lenze, D.; Gerhardt, A.; Schleicher, K.; et al. Synthetic lethal metabolic targeting of cellular senescence in cancer therapy. *Nature* **2013**, *501*, 421–425. [CrossRef]
41. Milanovic, M.; Fan, D.N.Y.; Belenki, D.; Däbritz, J.H.M.; Zhao, Z.; Yu, Y.; Dörr, J.R.; Dimitrova, L.; Lenze, D.; Monteiro Barbosa, I.A.; et al. Senescence-associated reprogramming promotes cancer stemness. *Nature* **2018**, *553*, 96–100. [CrossRef] [PubMed]
42. Patel, P.L.; Suram, A.; Mirani, N.; Bischof, O.; Herbig, U. Derepression of hTERT gene expression promotes escape from oncogene-induced cellular senescence. *Proc. Natl. Acad. Sci. USA* **2016**, *113*, 5024–5033. [CrossRef] [PubMed]
43. Reyes, J.; Chen, J.Y.; Stewart-Ornstein, J.; Karhohs, K.W.; Mock, C.S.; Lahav, G. Fluctuations in p53 Signaling Allow Escape from Cell-Cycle Arrest. *Mol. Cell* **2018**, *71*, 581–591.e5. [CrossRef] [PubMed]
44. Baryshev, M.; Inashkina, I.; Salmina, K.; Huna, A.; Jackson, T.R.; Erenpreisa, J. DNA methylation of the Oct4A enhancers in embryonal carcinoma cells after etoposide treatment is associated with alternative splicing and altered pluripotency in reversibly senescent cells. *Cell Cycle* **2018**, *17*, 362–366. [CrossRef]
45. Mosteiro, L.; Pantoja, C.; Alcazar, N.; Marión, R.M.; Chondronasiou, D.; Rovira, M.; Fernandez-Marcos, P.J.; Muñoz-Martin, M.; Blanco-Aparicio, C.; Pastor, J.; et al. Tissue damage and senescence provide critical signals for cellular reprogramming In Vivo. *Science* **2016**, *354*, aaf4445. [CrossRef]
46. Ritschka, B.; Storer, M.; Mas, A.; Heinzmann, F.; Ortells, M.C.; Morton, J.P.; Sansom, O.J.; Zender, L.; Keyes, W.M. The senescence-associated secretory phenotype induces cellular plasticity and tissue regeneration. *Genes Dev.* **2017**, *31*, 172–183. [CrossRef]

47. Demaria, M.; O'Leary, M.N.; Chang, J.; Shao, L.; Liu, S.; Alimirah, F.; Koenig, K.; Le, C.; Mitin, N.; Deal, A.M.; et al. Cellular Senescence Promotes Adverse Effects of Chemotherapy and Cancer Relapse. *Cancer Discov.* **2017**, *7*, 165–176. [CrossRef]
48. Campisi, J. Senescent cells, tumor suppression, and organismal aging: Good citizens, bad neighbors. *Cell* **2005**, *120*, 513–522. [CrossRef]
49. Coppe, J.P.; Desprez, P.Y.; Krtolica, A.; Campisi, J. The senescence-associated secretory phenotype: The dark side of tumor suppression. *Annu. Rev. Pathol.* **2010**, *5*, 99–118. [CrossRef]
50. Hernandez-Segura, A.; de Jong, T.V.; Melov, S.; Guryev, V.; Campisi, J.; Demaria, M. Unmasking Transcriptional Heterogeneity in Senescent Cells. *Curr. Biol. CB* **2017**, *27*, 2652–2660.e4. [CrossRef]
51. Takasugi, M.; Okada, R.; Takahashi, A.; Virya Chen, D.; Watanabe, S.; Hara, E. Small extracellular vesicles secreted from senescent cells promote cancer cell proliferation through EphA2. *Nat. Commun.* **2017**, *8*, 15729. [CrossRef] [PubMed]
52. Takahashi, A.; Okada, R.; Nagao, K.; Kawamata, Y.; Hanyu, A.; Yoshimoto, S.; Takasugi, M.; Watanabe, S.; Kanemaki, M.T.; Obuse, C.; et al. Exosomes maintain cellular homeostasis by excreting harmful DNA from cells. *Nat. Commun.* **2017**, *8*, 15287. [CrossRef] [PubMed]
53. Lehmann, B.D.; Paine, M.S.; Brooks, A.M.; McCubrey, J.A.; Renegar, R.H.; Wang, R.; Terrian, D.M. Senescence-associated exosome release from human prostate cancer cells. *Cancer Res.* **2008**, *68*, 7864–7871. [CrossRef] [PubMed]
54. Konopka, K.E.; Nguyen, T.; Jentzen, J.M.; Rayes, O.; Schmidt, C.J.; Wilson, A.M.; Farver, C.F.; Myers, J.L. Diffuse alveolar damage (DAD) resulting from coronavirus disease 2019 Infection is Morphologically Indistinguishable from Other Causes of DAD. *Histopathology* **2020**, *77*, 570–578. [CrossRef]
55. Carsana, L.; Sonzogni, A.; Nasr, A.; Rossi, R.S.; Pellegrinelli, A.; Zerbi, P.; Rech, R.; Colombo, R.; Antinori, S.; Corbellino, M.; et al. Pulmonary post-mortem findings in a series of COVID-19 cases from northern Italy: A two-centre descriptive study. *Lancet Infect. Dis.* **2020**, *20*, 1135–1140. [CrossRef]
56. Adams, T.S.; Schupp, J.C.; Poli, S.; Ayaub, E.A.; Neumark, N.; Ahangari, F.; Chu, S.G.; Raby, B.A.; DeIuliis, G.; Januszyk, M.; et al. Single-cell RNA-seq reveals ectopic and aberrant lung-resident cell populations in idiopathic pulmonary fibrosis. *Sci. Adv.* **2020**, *6*, eaba1983. [CrossRef] [PubMed]
57. Habermann, A.C.; Gutierrez, A.J.; Bui, L.T.; Yahn, S.L.; Winters, N.I.; Calvi, C.L.; Peter, L.; Chung, M.I.; Taylor, C.J.; Jetter, C.; et al. Single-cell RNA sequencing reveals profibrotic roles of distinct epithelial and mesenchymal lineages in pulmonary fibrosis. *Sci. Adv.* **2020**, *6*, eaba1972. [CrossRef]
58. Alder, J.K.; Barkauskas, C.E.; Limjunyawong, N.; Stanley, S.E.; Kembou, F.; Tuder, R.M.; Hogan, B.L.; Mitzner, W.; Armanios, M. Telomere dysfunction causes alveolar stem cell failure. *Proc. Natl. Acad. Sci. USA* **2015**, *112*, 5099–5104. [CrossRef]
59. Razdan, N.; Vasilopoulos, T.; Herbig, U. Telomere dysfunction promotes transdifferentiation of human fibroblasts into myofibroblasts. *Aging Cell* **2018**, *17*, e12838. [CrossRef]
60. Lee, J.; Reddy, R.; Barsky, L.; Scholes, J.; Chen, H.; Shi, W.; Driscoll, B. Lung alveolar integrity is compromised by telomere shortening in telomerase-null mice. *Am. J. Physiol. Lung Cell. Mol. Physiol.* **2009**, *296*, 57–70. [CrossRef]
61. Willis, B.C.; Liebler, J.M.; Luby-Phelps, K.; Nicholson, A.G.; Crandall, E.D.; du Bois, R.M.; Borok, Z. Induction of epithelial-mesenchymal transition in alveolar epithelial cells by transforming growth factor-beta1: Potential role in idiopathic pulmonary fibrosis. *Am. J. Pathol.* **2005**, *166*, 1321–1332. [CrossRef]
62. Marmai, C.; Sutherland, R.E.; Kim, K.K.; Dolganov, G.M.; Fang, X.; Kim, S.S.; Jiang, S.; Golden, J.A.; Hoopes, C.W.; Matthay, M.A.; et al. Alveolar epithelial cells express mesenchymal proteins in patients with idiopathic pulmonary fibrosis. *Am. J. Physiol. Lung Cell. Mol. Physiol.* **2011**, *301*, 71–78. [CrossRef] [PubMed]
63. Kim, K.K.; Kugler, M.C.; Wolters, P.J.; Robillard, L.; Galvez, M.G.; Brumwell, A.N.; Sheppard, D.; Chapman, H.A. Alveolar epithelial cell mesenchymal transition develops In Vivo during pulmonary fibrosis and is regulated by the extracellular matrix. *Proc. Natl. Acad. Sci. USA* **2006**, *103*, 13180–13185. [CrossRef] [PubMed]
64. Bharat, A.; Querrey, M.; Markov, N.S.; Kim, S.; Kurihara, C.; Garza-Castillon, R.; Manerikar, A.; Shilatifard, A.; Tomic, R.; Politanska, Y.; et al. Lung transplantation for patients with severe COVID-19. *Sci. Transl. Med.* **2020**, *12*, eabe4282. [CrossRef]
65. Chen, J.; Wu, H.; Yu, Y.; Tang, N. Pulmonary alveolar regeneration in adult COVID-19 patients. *Cell Res.* **2020**, *30*, 708–710. [CrossRef]
66. Kulkarni, T.; de Andrade, J.; Zhou, Y.; Luckhardt, T.; Thannickal, V.J. Alveolar epithelial disintegrity in pulmonary fibrosis. *Am. J. Physiol. Lung Cell. Mol. Physiol.* **2016**, *311*, 185–191. [CrossRef]
67. Chung, M.I.; Bujnis, M.; Barkauskas, C.E.; Kobayashi, Y.; Hogan, B.L.M. Niche-mediated BMP/SMAD signaling regulates lung alveolar stem cell proliferation and differentiation. *Development* **2018**, *145*, dev163014. [CrossRef]
68. Schruf, E.; Schroeder, V.; Le, H.Q.; Schonberger, T.; Raedel, D.; Stewart, E.L.; Fundel-Clemens, K.; Bluhmki, T.; Weigle, S.; Schuler, M.; et al. Recapitulating idiopathic pulmonary fibrosis related alveolar epithelial dysfunction in a human iPSC-derived air-liquid interface model. *FASEB J.* **2020**, *34*, 7825–7846. [CrossRef]
69. Gokey, J.J.; Snowball, J.; Green, J.; Waltamath, M.; Spinney, J.J.; Black, K.E.; Hariri, L.P.; Xu, Y.; Perl, A.K. Pretreatment of aged mice with retinoic acid supports alveolar regeneration via upregulation of reciprocal PDGFA signalling. *Thorax* **2021**, *76*, 456–467. [CrossRef]
70. Zheng, X.; Qi, C.; Zhang, S.; Fang, Y.; Ning, W. TGF-beta1 induces Fstl1 via the Smad3-c-Jun pathway in lung fibroblasts. *Am. J. Physiol. Lung Cell. Mol. Physiol.* **2017**, *313*, 240–251. [CrossRef]

71. Kim, K.K.; Wei, Y.; Szekeres, C.; Kugler, M.C.; Wolters, P.J.; Hill, M.L.; Frank, J.A.; Brumwell, A.N.; Wheeler, S.E.; Kreidberg, J.A.; et al. Epithelial cell alpha3beta1 integrin links beta-catenin and Smad signaling to promote myofibroblast formation and pulmonary fibrosis. *J. Clin. Investig.* **2009**, *119*, 213–224. [CrossRef] [PubMed]
72. Namba, T.; Tanaka, K.I.; Ito, Y.; Hoshino, T.; Matoyama, M.; Yamakawa, N.; Isohama, Y.; Azuma, A.; Mizushima, T. Induction of EMT-like phenotypes by an active metabolite of leflunomide and its contribution to pulmonary fibrosis. *Cell Death Differ.* **2010**, *17*, 1882–1895. [CrossRef] [PubMed]
73. Schneider, D.J.; Wu, M.; Le, T.T.; Cho, S.H.; Brenner, M.B.; Blackburn, M.R.; Agarwal, S.K. Cadherin-11 contributes to pulmonary fibrosis: Potential role in TGF-beta production and epithelial to mesenchymal transition. *FASEB J.* **2012**, *26*, 503–512. [CrossRef] [PubMed]
74. Tanjore, H.; Cheng, D.S.; Degryse, A.L.; Zoz, D.F.; Abdolrasulnia, R.; Lawson, W.E.; Blackwell, T.S. Alveolar epithelial cells undergo epithelial-to-mesenchymal transition in response to endoplasmic reticulum stress. *J. Biol. Chem.* **2011**, *286*, 30972–30980. [CrossRef]
75. Tanjore, H.; Xu, X.C.; Polosukhin, V.V.; Degryse, A.L.; Li, B.; Han, W.; Sherrill, T.P.; Plieth, D.; Neilson, E.G.; Blackwell, T.S.; et al. Contribution of epithelial-derived fibroblasts to bleomycin-induced lung fibrosis. *Am. J. Respir. Crit. Care Med.* **2009**, *180*, 657–665. [CrossRef]
76. Zhou, B.; Liu, Y.; Kahn, M.; Ann, D.K.; Han, A.; Wang, H.; Nguyen, C.; Flodby, P.; Zhong, Q.; Krishnaveni, M.S.; et al. Interactions between β-catenin and transforming growth factor-β signaling pathways mediate epithelial-mesenchymal transition and are dependent on the transcriptional co-activator cAMP-response element-binding protein (CREB)-binding protein (CBP). *J. Biol. Chem.* **2012**, *287*, 7026–7038. [CrossRef]
77. Zhang, K.; Wang, L.; Hong, X.; Chen, H.; Shi, Y.; Liu, Y.; Liu, J.; Liu, J.P. Pulmonary Alveolar Stem Cell Senescence, Apoptosis, and Differentiation by p53-Dependent and -Independent Mechanisms in Telomerase-Deficient Mice. *Cells* **2021**, *10*, 2892. [CrossRef]
78. Chen, R.; Zhang, K.; Chen, H.; Zhao, X.; Wang, J.; Li, L.; Cong, Y.; Ju, Z.; Xu, D.; Williams, B.R.; et al. Telomerase deficiency causes alveolar stem cell senescence-associated low-grade inflammation in lungs. *J. Biol. Chem.* **2015**, *290*, 30813–30829. [CrossRef]
79. Wang, L.; Yu, X.; Liu, J.P. Telomere Damage Response and Low-Grade Inflammation. *Adv. Exp. Med. Biol* **2017**, *1024*, 213–224. [CrossRef]
80. Birnhuber, A.; Egemnazarov, B.; Biasin, V.; Bonyadi Rad, E.; Wygrecka, M.; Olschewski, H.; Kwapiszewska, G.; Marsh, L.M. CDK4/6 inhibition enhances pulmonary inflammatory infiltration in bleomycin-induced lung fibrosis. *Respir. Res.* **2020**, *21*, 167. [CrossRef]
81. Sala, M.A.; Balderas-Martinez, Y.I.; Buendia-Roldan, I.; Abdala-Valencia, H.; Nam, K.; Jain, M.; Bhorade, S.; Bharat, A.; Reyfman, P.A.; Ridge, K.M.; et al. Inflammatory pathways are upregulated in the nasal epithelium in patients with idiopathic pulmonary fibrosis. *Respir. Res.* **2018**, *19*, 233. [CrossRef] [PubMed]
82. Li, M.; Krishnaveni, M.S.; Li, C.; Zhou, B.; Xing, Y.; Banfalvi, A.; Li, A.; Lombardi, V.; Akbari, O.; Borok, Z.; et al. Epithelium-specific deletion of TGF-beta receptor type II protects mice from bleomycin-induced pulmonary fibrosis. *J. Clin. Investig.* **2011**, *121*, 277–287. [CrossRef] [PubMed]
83. Aumiller, V.; Balsara, N.; Wilhelm, J.; Gunther, A.; Konigshoff, M. WNT/beta-catenin signaling induces IL-1beta expression by alveolar epithelial cells in pulmonary fibrosis. *Am. J. Respir. Cell Mol. Biol.* **2013**, *49*, 96–104. [CrossRef] [PubMed]
84. Yu, G.; Tzouvelekis, A.; Wang, R.; Herazo-Maya, J.D.; Ibarra, G.H.; Srivastava, A.; de Castro, J.P.W.; DeIuliis, G.; Ahangari, F.; Woolard, T.; et al. Thyroid hormone inhibits lung fibrosis in mice by improving epithelial mitochondrial function. *Nat. Med.* **2018**, *24*, 39–49. [CrossRef]
85. Cheresh, P.; Kim, S.J.; Tulasiram, S.; Kamp, D.W. Oxidative stress and pulmonary fibrosis. *Biochim. Biophys. Acta* **2013**, *1832*, 1028–1040. [CrossRef] [PubMed]
86. Otoupalova, E.; Smith, S.; Cheng, G.; Thannickal, V.J. Oxidative Stress in Pulmonary Fibrosis. *Compr. Physiol.* **2020**, *10*, 509–547. [CrossRef]
87. Korthagen, N.M.; van Moorsel, C.H.; Barlo, N.P.; Kazemier, K.M.; Ruven, H.J.; Grutters, J.C. Association between variations in cell cycle genes and idiopathic pulmonary fibrosis. *PLoS ONE* **2012**, *7*, e30442. [CrossRef]
88. Dressen, A.; Abbas, A.R.; Cabanski, C.; Reeder, J.; Ramalingam, T.R.; Neighbors, M.; Bhangale, T.R.; Brauer, M.J.; Hunkapiller, J.; Reeder, J.; et al. Analysis of protein-altering variants in telomerase genes and their association with MUC5B common variant status in patients with idiopathic pulmonary fibrosis: A candidate gene sequencing study. *Lancet Respir. Med.* **2018**, *6*, 603–614. [CrossRef]
89. King, T.E., Jr.; Pardo, A.; Selman, M. Idiopathic pulmonary fibrosis. *Lancet* **2011**, *378*, 1949–1961. [CrossRef]
90. Tsakiri, K.D.; Cronkhite, J.T.; Kuan, P.J.; Xing, C.; Raghu, G.; Weissler, J.C.; Rosenblatt, R.L.; Shay, J.W.; Garcia, C.K. Adult-onset pulmonary fibrosis caused by mutations in telomerase. *Proc. Natl. Acad. Sci. USA* **2007**, *104*, 7552–7557. [CrossRef]
91. Stuart, B.D.; Choi, J.; Zaidi, S.; Xing, C.; Holohan, B.; Chen, R.; Choi, M.; Dharwadkar, P.; Torres, F.; Girod, C.E.; et al. Exome sequencing links mutations in PARN and RTEL1 with familial pulmonary fibrosis and telomere shortening. *Nat. Genet.* **2015**, *47*, 512–517. [CrossRef] [PubMed]
92. Sousa, S.R.; Caetano Mota, P.; Melo, N.; Bastos, H.N.; Padrao, E.; Pereira, J.M.; Cunha, R.; Souto Moura, C.; Guimaraes, S.; Morais, A. Heterozygous TERT gene mutation associated with familial idiopathic pulmonary fibrosis. *Respir. Med. Case Rep.* **2019**, *26*, 118–122. [CrossRef] [PubMed]

93. Alder, J.K.; Chen, J.J.; Lancaster, L.; Danoff, S.; Su, S.C.; Cogan, J.D.; Vulto, I.; Xie, M.; Qi, X.; Tuder, R.M.; et al. Short telomeres are a risk factor for idiopathic pulmonary fibrosis. *Proc. Natl. Acad. Sci. USA* **2008**, *105*, 13051–13056. [CrossRef] [PubMed]
94. Alder, J.K.; Cogan, J.D.; Brown, A.F.; Anderson, C.J.; Lawson, W.E.; Lansdorp, P.M.; Phillips, J.A., 3rd; Loyd, J.E.; Chen, J.J.; Armanios, M. Ancestral mutation in telomerase causes defects in repeat addition processivity and manifests as familial pulmonary fibrosis. *PLoS Genet.* **2011**, *7*, e1001352. [CrossRef]
95. Armanios, M. Syndromes of Telomere Shortening. *Annu Rev. Genom. Hum. Genet.* **2009**, *62*, 6405–6409. [CrossRef]
96. Petrovski, S.; Todd, J.L.; Durheim, M.T.; Wang, Q.; Chien, J.W.; Kelly, F.L.; Frankel, C.; Mebane, C.M.; Ren, Z.; Bridgers, J.; et al. An Exome Sequencing Study to Assess the Role of Rare Genetic Variation in Pulmonary Fibrosis. *Am. J. Respir. Crit. Care Med.* **2017**, *196*, 82–93. [CrossRef]
97. Chilosi, M.; Poletti, V.; Rossi, A. The pathogenesis of COPD and IPF: Distinct horns of the same devil? *Respir. Res.* **2012**, *13*, 3. [CrossRef]
98. Nogee, L.M.; Dunbar, A.E., 3rd; Wert, S.E.; Askin, F.; Hamvas, A.; Whitsett, J.A. A mutation in the surfactant protein C gene associated with familial interstitial lung disease. *N. Engl. J. Med.* **2001**, *344*, 573–579. [CrossRef]
99. Cottin, V.; Reix, P.; Khouatra, C.; Thivolet-Bejui, F.; Feldmann, D.; Cordier, J.F. Combined pulmonary fibrosis and emphysema syndrome associated with familial SFTPC mutation. *Thorax* **2011**, *66*, 918–919. [CrossRef]
100. Nathan, N.; Giraud, V.; Picard, C.; Nunes, H.; Dastot-Le Moal, F.; Copin, B.; Galeron, L.; De Ligniville, A.; Kuziner, N.; Reynaud-Gaubert, M.; et al. Germline SFTPA1 mutation in familial idiopathic interstitial pneumonia and lung cancer. *Hum. Mol. Genet.* **2016**, *25*, 1457–1467. [CrossRef]
101. Wu, J.; McKeague, M.; Sturla, S.J. Nucleotide-Resolution Genome-Wide Mapping of Oxidative DNA Damage by Click-Code-Seq. *J. Am. Chem Soc.* **2018**, *140*, 9783–9787. [CrossRef] [PubMed]
102. Qian, W.; Kumar, N.; Roginskaya, V.; Fouquerel, E.; Opresko, P.L.; Shiva, S.; Watkins, S.C.; Kolodieznyi, D.; Bruchez, M.P.; Van Houten, B. Chemoptogenetic damage to mitochondria causes rapid telomere dysfunction. *Proc. Natl. Acad. Sci. USA* **2019**, *116*, 18435–18444. [CrossRef] [PubMed]
103. Hewitt, G.; Jurk, D.; Marques, F.D.; Correia-Melo, C.; Hardy, T.; Gackowska, A.; Anderson, R.; Taschuk, M.; Mann, J.; Passos, J.F. Telomeres are favoured targets of a persistent DNA damage response in ageing and stress-induced senescence. *Nat. Commun.* **2012**, *3*, 708. [CrossRef] [PubMed]
104. Degryse, A.L.; Xu, X.C.; Newman, J.L.; Mitchell, D.B.; Tanjore, H.; Polosukhin, V.V.; Jones, B.R.; McMahon, F.B.; Gleaves, L.A.; Phillips, J.A., 3rd; et al. Telomerase deficiency does not alter bleomycin-induced fibrosis in mice. *Exp. Lung Res.* **2012**, *38*, 124–134. [CrossRef] [PubMed]
105. Fumagalli, M.; Rossiello, F.; Clerici, M.; Barozzi, S.; Cittaro, D.; Kaplunov, J.M.; Bucci, G.; Dobreva, M.; Matti, V.; Beausejour, C.M.; et al. Telomeric DNA damage is irreparable and causes persistent DNA-damage-response activation. *Nat. Cell Biol.* **2012**, *14*, 355–365. [CrossRef]
106. Coluzzi, E.; Colamartino, M.; Cozzi, R.; Leone, S.; Meneghini, C.; O'Callaghan, N.; Sgura, A. Oxidative stress induces persistent telomeric DNA damage responsible for nuclear morphology change in mammalian cells. *PLoS ONE* **2014**, *9*, e110963. [CrossRef]
107. Aitken, M.L.; Dugowson, C.; Schmidt, R.A.; Fer, M. Bleomycin-induced pulmonary fibrosis in a patient with rheumatoid arthritis. A possible synergistic effect? *West. J. Med.* **1989**, *150*, 344–346.
108. Santrach, P.J.; Askin, F.B.; Wells, R.J.; Azizkhan, R.G.; Merten, D.F. Nodular form of bleomycin-related pulmonary injury in patients with osteogenic sarcoma. *Cancer* **1989**, *64*, 806–811. [CrossRef]
109. Parfrey, H.; Babar, J.; Fiddler, C.A.; Chilvers, E.R. Idiopathic pulmonary fibrosis in a Christmas Island nuclear test veteran. *BMJ Case Rep.* **2010**, *2010*, bcr0620103102. [CrossRef]
110. Desai, M.Y.; Karunakaravel, K.; Wu, W.; Agarwal, S.; Smedira, N.G.; Lytle, B.W.; Griffin, B.P. Pulmonary fibrosis on multidetector computed tomography and mortality in patients with radiation-associated cardiac disease undergoing cardiac surgery. *J. Thorac. Cardiovasc. Surg.* **2014**, *148*, 475–481.e3. [CrossRef]
111. Gross, N.J. Pulmonary effects of radiation therapy. *Ann. Intern. Med.* **1977**, *86*, 81–92. [CrossRef] [PubMed]
112. Chung, K.P.; Hsu, C.L.; Fan, L.C.; Huang, Z.; Bhatia, D.; Chen, Y.J.; Hisata, S.; Cho, S.J.; Nakahira, K.; Imamura, M.; et al. Mitofusins regulate lipid metabolism to mediate the development of lung fibrosis. *Nat. Commun.* **2019**, *10*, 3390. [CrossRef]
113. Anathy, V.; Lahue, K.G.; Chapman, D.G.; Chia, S.B.; Casey, D.T.; Aboushousha, R.; van der Velden, J.L.J.; Elko, E.; Hoffman, S.M.; McMillan, D.H.; et al. Reducing protein oxidation reverses lung fibrosis. *Nat. Med.* **2018**, *24*, 1128–1135. [CrossRef] [PubMed]
114. Demedts, M.; Behr, J.; Buhl, R.; Costabel, U.; Dekhuijzen, R.; Jansen, H.M.; MacNee, W.; Thomeer, M.; Wallaert, B.; Laurent, F.; et al. High-dose acetylcysteine in idiopathic pulmonary fibrosis. *N. Engl. J. Med.* **2005**, *353*, 2229–2242. [CrossRef] [PubMed]
115. Borok, Z.; Buhl, R.; Grimes, G.J.; Bokser, A.D.; Hubbard, R.C.; Holroyd, K.J.; Roum, J.H.; Czerski, D.B.; Cantin, A.M.; Crystal, R.G. Effect of glutathione aerosol on oxidant-antioxidant imbalance in idiopathic pulmonary fibrosis. *Lancet* **1991**, *338*, 215–216. [CrossRef]
116. Guan, L.L.; Kuwahara, J.; Sugiura, Y. Guanine-specific binding by bleomycin-nickel(III) complex and its reactivity for guanine-quartet telomeric DNA. *Biochemistry* **1993**, *32*, 6141–6145. [CrossRef] [PubMed]
117. Povedano, J.M.; Martinez, P.; Flores, J.M.; Mulero, F.; Blasco, M.A. Mice with Pulmonary Fibrosis Driven by Telomere Dysfunction. *Cell Rep.* **2015**, *12*, 286–299. [CrossRef]
118. Cornforth, M.N.; Meyne, J.; Littlefield, L.G.; Bailey, S.M.; Moyzis, R.K. Telomere staining of human chromosomes and the mechanism of radiation-induced dicentric formation. *Radiat. Res.* **1989**, *120*, 205–212. [CrossRef]

119. Metcalfe, J.A.; Parkhill, J.; Campbell, L.; Stacey, M.; Biggs, P.; Byrd, P.J.; Taylor, A.M. Accelerated telomere shortening in ataxia telangiectasia. *Nat. Genet.* **1996**, *13*, 350–353. [CrossRef]
120. Gorbunova, V.; Seluanov, A.; Pereira-Smith, O.M. Expression of hTERT does not prevent stress-induced senescence in normal human fibroblasts, but protects the cells from stress-induced apoptosis and necrosis. *J. Biol. Chem.* **2002**, *24*, 24. [CrossRef]
121. Petersen, S.; Saretzki, G.; von Zglinicki, T. Preferential accumulation of single-stranded regions in telomeres of human fibroblasts. *Exp. Cell Res.* **1998**, *239*, 152–160. [CrossRef] [PubMed]
122. von Zglinicki, T. Oxidative stress shortens telomeres. *Trends Biochem. Sci.* **2002**, *27*, 339–344. [CrossRef]
123. Halu, A.; Liu, S.; Baek, S.H.; Hobbs, B.D.; Hunninghake, G.M.; Cho, M.H.; Silverman, E.K.; Sharma, A. Exploring the cross-phenotype network region of disease modules reveals concordant and discordant pathways between chronic obstructive pulmonary disease and idiopathic pulmonary fibrosis. *Hum. Mol. Genet.* **2019**, *28*, 2352–2364. [CrossRef]
124. Kumar, M.; Seeger, W.; Voswinckel, R. Senescence-associated secretory phenotype and its possible role in chronic obstructive pulmonary disease. *Am. J. Respir. Cell Mol. Biol.* **2014**, *51*, 323–333. [CrossRef] [PubMed]
125. Mirabello, L.; Huang, W.Y.; Wong, J.Y.; Chatterjee, N.; Reding, D.; Crawford, E.D.; De Vivo, I.; Hayes, R.B.; Savage, S.A. The association between leukocyte telomere length and cigarette smoking, dietary and physical variables, and risk of prostate cancer. *Aging Cell* **2009**, *8*, 405–413. [CrossRef]
126. Stanley, S.E.; Chen, J.J.; Podlevsky, J.D.; Alder, J.K.; Hansel, N.N.; Mathias, R.A.; Qi, X.; Rafaels, N.M.; Wise, R.A.; Silverman, E.K.; et al. Telomerase mutations in smokers with severe emphysema. *J. Clin. Investig.* **2015**, *125*, 563–570. [CrossRef]
127. Blackburn, E.H.; Epel, E.S.; Lin, J. Human telomere biology: A contributory and interactive factor in aging, disease risks, and protection. *Science* **2015**, *350*, 1193–1198. [CrossRef]
128. Aalbers, A.M.; Kajigaya, S.; van den Heuvel-Eibrink, M.M.; van der Velden, V.H.; Calado, R.T.; Young, N.S. Human telomere disease due to disruption of the CCAAT box of the TERC promoter. *Blood* **2012**, *119*, 3060–3063. [CrossRef]
129. Hao, L.Y.; Armanios, M.; Strong, M.A.; Karim, B.; Feldser, D.M.; Huso, D.; Greider, C.W. Short telomeres, even in the presence of telomerase, limit tissue renewal capacity. *Cell* **2005**, *123*, 1121–1131. [CrossRef]
130. Fogarty, P.F.; Yamaguchi, H.; Wiestner, A.; Baerlocher, G.M.; Sloand, E.; Zeng, W.S.; Read, E.J.; Lansdorp, P.M.; Young, N.S. Late presentation of dyskeratosis congenita as apparently acquired aplastic anaemia due to mutations in telomerase RNA. *Lancet* **2003**, *362*, 1628–1630. [CrossRef]
131. Cawthon, R.M.; Smith, K.R.; O'Brien, E.; Sivatchenko, A.; Kerber, R.A. Association between telomere length in blood and mortality in people aged 60 years or older. *Lancet* **2003**, *361*, 393–395. [CrossRef]
132. Mitchell, J.R.; Wood, E.; Collins, K. A telomerase component is defective in the human disease dyskeratosis congenita. *Nature* **1999**, *402*, 551–555. [CrossRef] [PubMed]
133. Kadota, T.; Yoshioka, Y.; Fujita, Y.; Araya, J.; Minagawa, S.; Hara, H.; Miyamoto, A.; Suzuki, S.; Fujimori, S.; Kohno, T.; et al. Extracellular Vesicles from Fibroblasts Induce Epithelial-Cell Senescence in Pulmonary Fibrosis. *Am. J. Respir. Cell Mol. Biol.* **2020**, *63*, 623–636. [CrossRef] [PubMed]
134. Khayrullin, A.; Krishnan, P.; Martinez-Nater, L.; Mendhe, B.; Fulzele, S.; Liu, Y.; Mattison, J.A.; Hamrick, M.W. Very Long-Chain C24:1 Ceramide Is Increased in Serum Extracellular Vesicles with Aging and Can Induce Senescence in Bone-Derived Mesenchymal Stem Cells. *Cells* **2019**, *8*, 37. [CrossRef] [PubMed]
135. Misawa, T.; Tanaka, Y.; Okada, R.; Takahashi, A. Biology of extracellular vesicles secreted from senescent cells as senescence-associated secretory phenotype factors. *Geriatr. Gerontol. Int.* **2020**, *20*, 539–546. [CrossRef]
136. Pardo, A.; Selman, M. Lung Fibroblasts, Aging, and Idiopathic Pulmonary Fibrosis. *Ann. Am. Thorac Soc.* **2016**, *13* (Suppl. 5), S417–S421. [CrossRef]
137. Kelley, W.J.; Zemans, R.L.; Goldstein, D.R. Cellular senescence: Friend or foe to respiratory viral infections? *Eur. Respir. J.* **2020**, *56*, 2002708. [CrossRef]
138. Ley, B.; Newton, C.A.; Arnould, I.; Elicker, B.M.; Henry, T.S.; Vittinghoff, E.; Golden, J.A.; Jones, K.D.; Batra, K.; Torrealba, J.; et al. The MUC5B promoter polymorphism and telomere length in patients with chronic hypersensitivity pneumonitis: An observational cohort-control study. *Lancet Respir. Med.* **2017**, *5*, 639–647. [CrossRef]
139. Blackburn, E.H.; Gall, J.G. A tandemly repeated sequence at the termini of the extrachromosomal ribosomal RNA genes in Tetrahymena. *J. Mol. Biol.* **1978**, *120*, 33–53. [CrossRef]
140. Shampay, J.; Szostak, J.W.; Blackburn, E.H. DNA sequences of telomeres maintained in yeast. *Nature* **1984**, *310*, 154–157. [CrossRef]
141. Greider, C.W.; Blackburn, E.H. Identification of a specific telomere terminal transferase enzyme with two kinds of primer specificity. *Cell* **1985**, *51*, 405–413. [CrossRef]
142. de Lange, T.; Shiue, L.; Myers, R.M.; Cox, D.R.; Naylor, S.L.; Killery, A.M.; Varmus, H.E. Structure and variability of human chromosome ends. *Mol. Cell. Biol.* **1990**, *10*, 518–527. [PubMed]
143. de Lange, T. Shelterin: The protein complex that shapes and safeguards human telomeres. *Genes Dev.* **2005**, *19*, 2100–2110. [CrossRef] [PubMed]
144. de Lange, T. Shelterin-Mediated Telomere Protection. *Annu. Rev. Genet.* **2018**, *52*, 223–247. [CrossRef] [PubMed]
145. Harley, C.B.; Futcher, A.B.; Greider, C.W. Telomeres shorten during ageing of human fibroblasts. *Nature* **1990**, *345*, 458–460. [CrossRef] [PubMed]
146. Hastie, N.D.; Dempster, M.; Dunlop, M.G.; Thompson, A.M.; Green, D.K.; Allshire, R.C. Telomere reduction in human colorectal carcinoma and with ageing. *Nature* **1990**, *346*, 866–868. [CrossRef]

147. Oexle, K.; Zwirner, A. Advanced telomere shortening in respiratory chain disorders. *Hum. Mol. Genet.* **1997**, *6*, 905–908. [CrossRef]
148. Vaziri, H.; West, M.D.; Allsopp, R.C.; Davison, T.S.; Wu, Y.S.; Arrowsmith, C.H.; Poirier, G.G.; Benchimol, S. ATM-dependent telomere loss in aging human diploid fibroblasts and DNA damage lead to the post-translational activation of p53 protein involving poly(ADP-ribose) polymerase. *EMBO J.* **1997**, *16*, 6018–6033. [CrossRef]
149. Bayne, S.; Li, H.; Jones, M.E.; Pinto, A.R.; van Sinderen, M.; Drummond, A.; Simpson, E.R.; Liu, J.P. Estrogen deficiency reversibly induces telomere shortening in mouse granulosa cells and ovarian aging In Vivo. *Protein Cell* **2011**, *2*, 333–346. [CrossRef]
150. Bayne, S.; Jones, M.E.; Li, H.; Pinto, A.R.; Simpson, E.R.; Liu, J.P. Estrogen deficiency leads to telomerase inhibition, telomere shortening and reduced cell proliferation in the adrenal gland of mice. *Cell Res.* **2008**, *18*, 1141–1150. [CrossRef]
151. Greider, C.W. Telomeres, telomerase and senescence. *BioEssays News Rev. Mol. Cell. Dev. Biol.* **1990**, *12*, 363–368. [CrossRef] [PubMed]
152. Greider, C.W. Telomeres and senescence: The history, the experiment, the future. *Curr. Biol. CB* **1998**, *8*, 178–181. [CrossRef]
153. Xie, Z.; Jay, K.A.; Smith, D.L.; Zhang, Y.; Liu, Z.; Zheng, J.; Tian, R.; Li, H.; Blackburn, E.H. Early telomerase inactivation accelerates aging independently of telomere length. *Cell* **2015**, *160*, 928–939. [CrossRef] [PubMed]
154. Karlseder, J.; Smogorzewska, A.; de Lange, T. Senescence induced by altered telomere state, not telomere loss. *Science* **2002**, *295*, 2446–2449. [CrossRef]
155. Denchi, E.L.; de Lange, T. Protection of telomeres through independent control of ATM and ATR by TRF2 and POT1. *Nature* **2007**, *448*, 1068–1071. [CrossRef]
156. d'Adda di Fagagna, F.; Reaper, P.M.; Clay-Farrace, L.; Fiegler, H.; Carr, P.; Von Zglinicki, T.; Saretzki, G.; Carter, N.P.; Jackson, S.P. A DNA damage checkpoint response in telomere-initiated senescence. *Nature* **2003**, *426*, 194–198. [CrossRef] [PubMed]
157. Deng, Y.; Guo, X.; Ferguson, D.O.; Chang, S. Multiple roles for MRE11 at uncapped telomeres. *Nature* **2009**, *460*, 914–918. [CrossRef]
158. Povedano, J.M.; Martinez, P.; Serrano, R.; Tejera, A.; Gomez-Lopez, G.; Bobadilla, M.; Flores, J.M.; Bosch, F.; Blasco, M.A. Therapeutic effects of telomerase in mice with pulmonary fibrosis induced by damage to the lungs and short telomeres. *eLife* **2018**, *7*, e31299. [CrossRef]
159. Xin, H.; Liu, D.; Wan, M.; Safari, A.; Kim, H.; Sun, W.; O'Connor, M.S.; Songyang, Z. TPP1 is a homologue of ciliate TEBP-beta and interacts with POT1 to recruit telomerase. *Nature* **2007**, *445*, 559–562. [CrossRef]
160. Wang, F.; Podell, E.R.; Zaug, A.J.; Yang, Y.; Baciu, P.; Cech, T.R.; Lei, M. The POT1-TPP1 telomere complex is a telomerase processivity factor. *Nature* **2007**, *445*, 506–510. [CrossRef]
161. Miyoshi, T.; Kanoh, J.; Saito, M.; Ishikawa, F. Fission yeast Pot1-Tpp1 protects telomeres and regulates telomere length. *Science* **2008**, *320*, 1341–1344. [CrossRef] [PubMed]
162. Zhong, F.L.; Batista, L.F.; Freund, A.; Pech, M.F.; Venteicher, A.S.; Artandi, S.E. TPP1 OB-fold domain controls telomere maintenance by recruiting telomerase to chromosome ends. *Cell* **2012**, *150*, 481–494. [CrossRef] [PubMed]
163. Nandakumar, J.; Bell, C.F.; Weidenfeld, I.; Zaug, A.J.; Leinwand, L.A.; Cech, T.R. The TEL patch of telomere protein TPP1 mediates telomerase recruitment and processivity. *Nature* **2012**, *492*, 285–289. [CrossRef] [PubMed]
164. Schmidt, J.C.; Zaug, A.J.; Cech, T.R. Live Cell Imaging Reveals the Dynamics of Telomerase Recruitment to Telomeres. *Cell* **2016**, *166*, 1188–1197.e9. [CrossRef]
165. Palm, W.; de Lange, T. How shelterin protects mammalian telomeres. *Annu. Rev. Genet.* **2008**, *42*, 301–334. [CrossRef]
166. Guo, Y.; Kartawinata, M.; Li, J.; Pickett, H.A.; Teo, J.; Kilo, T.; Barbaro, P.M.; Keating, B.; Chen, Y.; Tian, L.; et al. Inherited bone marrow failure associated with germline mutation of ACD, the gene encoding telomere protein TPP1. *Blood* **2014**, *124*, 2767–2774. [CrossRef]
167. Kocak, H.; Ballew, B.J.; Bisht, K.; Eggebeen, R.; Hicks, B.D.; Suman, S.; O'Neil, A.; Giri, N.; Laboratory, N.D.C.G.R.; Group, N.D.C.S.W.; et al. Hoyeraal-Hreidarsson syndrome caused by a germline mutation in the TEL patch of the telomere protein TPP1. *Genes Dev.* **2014**, *28*, 2090–2102. [CrossRef]
168. Ahmad, T.; Sundar, I.K.; Tormos, A.M.; Lerner, C.A.; Gerloff, J.; Yao, H.; Rahman, I. Shelterin Telomere Protection Protein 1 Reduction Causes Telomere Attrition and Cellular Senescence via Sirtuin 1 Deacetylase in Chronic Obstructive Pulmonary Disease. *Am. J. Respir. Cell Mol. Biol.* **2017**, *56*, 38–49. [CrossRef]
169. Greider, C.W.; Blackburn, E.H. A telomeric sequence in the RNA of Tetrahymena telomerase required for telomere repeat synthesis. *Nature* **1989**, *337*, 331–337. [CrossRef]
170. Blackburn, E.H.; Greider, C.W.; Henderson, E.; Lee, M.S.; Shampay, J.; Shippen-Lentz, D. Recognition and elongation of telomeres by telomerase. *Genome* **1989**, *31*, 553–560. [CrossRef]
171. Bodnar, A.G.; Ouellette, M.; Frolkis, M.; Holt, S.E.; Chiu, C.P.; Morin, G.B.; Harley, C.B.; Shay, J.W.; Lichtsteiner, S.; Wright, W.E. Extension of life-span by introduction of telomerase into normal human cells. *Science* **1998**, *279*, 349–352. [CrossRef]
172. Harle-Bachor, C.; Boukamp, P. Telomerase activity in the regenerative basal layer of the epidermis inhuman skin and in immortal and carcinoma-derived skin keratinocytes. *Proc. Natl. Acad. Sci. USA* **1996**, *93*, 6476–6481. [CrossRef]
173. Counter, C.M.; Hahn, W.C.; Wei, W.; Caddle, S.D.; Beijersbergen, R.L.; Lansdorp, P.M.; Sedivy, J.M.; Weinberg, R.A. Dissociation among In Vitro telomerase activity, telomere maintenance, and cellular immortalization. *Proc. Natl. Acad. Sci. USA* **1998**, *95*, 14723–14728. [CrossRef] [PubMed]
174. Lee, H.W.; Blasco, M.A.; Gottlieb, G.J.; Horner, J.W., 2nd; Greider, C.W.; DePinho, R.A. Essential role of mouse telomerase in highly proliferative organs. *Nature* **1998**, *392*, 569–574. [CrossRef] [PubMed]

175. Collins, K. Mammalian telomeres and telomerase. *Curr. Opin. Cell Biol.* **2000**, *12*, 378–383. [CrossRef]
176. Mitchell, J.R.; Collins, K. Human telomerase activation requires two independent interactions between telomerase RNA and telomerase reverse transcriptase. *Mol. Cell* **2000**, *6*, 361–371. [CrossRef]
177. Maida, Y.; Yasukawa, M.; Furuuchi, M.; Lassmann, T.; Possemato, R.; Okamoto, N.; Kasim, V.; Hayashizaki, Y.; Hahn, W.C.; Masutomi, K. An RNA-dependent RNA polymerase formed by TERT and the RMRP RNA. *Nature* **2009**, *461*, 230–235. [CrossRef]
178. Maida, Y.; Yasukawa, M.; Masutomi, K. De Novo RNA Synthesis by RNA-Dependent RNA Polymerase Activity of Telomerase Reverse Transcriptase. *Mol. Cell. Biol.* **2016**, *36*, 1248–1259. [CrossRef]
179. Yasukawa, M.; Ando, Y.; Yamashita, T.; Matsuda, Y.; Shoji, S.; Morioka, M.S.; Kawaji, H.; Shiozawa, K.; Machitani, M.; Abe, T.; et al. CDK1 dependent phosphorylation of hTERT contributes to cancer progression. *Nat. Commun.* **2020**, *11*, 1557. [CrossRef]
180. Rudolph, K.L.; Chang, S.; Millard, M.; Schreiber-Agus, N.; DePinho, R.A. Inhibition of experimental liver cirrhosis in mice by telomerase gene delivery. *Science* **2000**, *287*, 1253–1258. [CrossRef]
181. Lin, S.; Nascimento, E.M.; Gajera, C.R.; Chen, L.; Neuhofer, P.; Garbuzov, A.; Wang, S.; Artandi, S.E. Distributed hepatocytes expressing telomerase repopulate the liver in homeostasis and injury. *Nature* **2018**, *556*, 244–248. [CrossRef] [PubMed]
182. Nicholls, C.; Li, H.; Wang, J.Q.; Liu, J.P. Molecular regulation of telomerase activity in aging. *Protein Cell* **2012**, *2*, 726–738. [CrossRef] [PubMed]
183. Yuan, X.; Larsson, C.; Xu, D. Mechanisms underlying the activation of TERT transcription and telomerase activity in human cancer: Old actors and new players. *Oncogene* **2019**, *38*, 6172–6183. [CrossRef]
184. Xu, D.; Li, H.; Liu, J.P. Inhibition of telomerase by targeting MAP kinase signaling. *Methods Mol. Biol.* **2007**, *405*, 147–165. [CrossRef]
185. Xu, D.; Dwyer, J.; Li, H.; Duan, W.; Liu, J.P. Ets2 maintains hTERT gene expression and breast cancer cell proliferation by interacting with c-Myc. *J. Biol. Chem.* **2008**, *283*, 23567–23580. [CrossRef] [PubMed]
186. Dwyer, J.; Li, H.; Xu, D.; Liu, J.P. Transcriptional regulation of telomerase activity: Roles of the the ets transcription factor family. *Ann. N. Y. Acad. Sci.* **2007**, *1114*, 36–47. [CrossRef]
187. Dwyer, J.M.; Liu, J.P. Ets2 Transcription Factor, Telomerase Activity and Breast Cancer. *Clin. Exp. Pharmacol. Physiol.* **2009**, *37*, 83–87. [CrossRef] [PubMed]
188. Li, H.; Xu, D.; Li, J.; Berndt, M.C.; Liu, J.P. Transforming growth factor beta suppresses human telomerase reverse transcriptase (hTERT) by Smad3 interactions with c-Myc and the hTERT gene. *J. Biol. Chem.* **2006**, *281*, 25588–25600. [CrossRef]
189. Horn, S.; Figl, A.; Rachakonda, P.S.; Fischer, C.; Sucker, A.; Gast, A.; Kadel, S.; Moll, I.; Nagore, E.; Hemminki, K.; et al. TERT promoter mutations in familial and sporadic melanoma. *Science* **2013**, *339*, 959–961. [CrossRef]
190. Huang, F.W.; Hodis, E.; Xu, M.J.; Kryukov, G.V.; Chin, L.; Garraway, L.A. Highly recurrent TERT promoter mutations in human melanoma. *Science* **2013**, *339*, 957–959. [CrossRef]
191. Zhang, F.; Wang, S.; Zhu, J. ETS variant transcription factor 5 and c-Myc cooperate in derepressing the human telomerase gene promoter via composite ETS/E-box motifs. *J. Biol. Chem.* **2020**, *295*, 10062–10075. [CrossRef]
192. Wu, K.J.; Grandori, C.; Amacker, M.; Simon-Vermot, N.; Polack, A.; Lingner, J.; Dalla-Favera, R. Direct activation of TERT transcription by c-MYC. *Nat. Genet.* **1999**, *21*, 220–224. [CrossRef] [PubMed]
193. Xu, D.; Popov, N.; Hou, M.; Wang, Q.; Bjorkholm, M.; Gruber, A.; Menkel, A.R.; Henriksson, M. Switch from Myc/Max to Mad1/Max binding and decrease in histone acetylation at the telomerase reverse transcriptase promoter during differentiation of HL60 cells. *Proc. Natl. Acad. Sci. USA* **2001**, *98*, 3826–3831. [CrossRef] [PubMed]
194. Bayne, S.; Jones, M.E.; Li, H.; Liu, J.P. Potential roles for estrogen regulation of telomerase activity in aging. *Ann. N. Y. Acad. Sci.* **2007**, *1114*, 48–55. [CrossRef] [PubMed]
195. Bayne, S.; Liu, J.P. Hormones and growth factors regulate telomerase activity in ageing and cancer. *Mol. Cell. Endocrinol.* **2005**, *240*, 11–22. [CrossRef] [PubMed]
196. Liu, J.; Wang, L.; Wang, Z.; Liu, J.P. Roles of Telomere Biology in Cell Senescence, Replicative and Chronological Ageing. *Cells* **2019**, *8*, 54. [CrossRef] [PubMed]
197. Liu, N.; Ding, D.; Hao, W.; Yang, F.; Wu, X.; Wang, M.; Xu, X.; Ju, Z.; Liu, J.P.; Song, Z.; et al. hTERT promotes tumor angiogenesis by activating VEGF via interactions with the Sp1 transcription factor. *Nucleic Acids Res.* **2016**, *44*, 8693–8703. [CrossRef]
198. Liu, N.; Guo, X.H.; Liu, J.P.; Cong, Y.S. Role of telomerase in the tumour microenvironment. *Clin. Exp. Pharmacol. Physiol.* **2020**, *47*, 357–364. [CrossRef]
199. Xu, J.; Xu, X.; Jiang, L.; Dua, K.; Hansbro, P.M.; Liu, G. SARS-CoV-2 induces transcriptional signatures in human lung epithelial cells that promote lung fibrosis. *Respir. Res.* **2020**, *21*, 182. [CrossRef]
200. Wicik, Z.; Eyileten, C.; Jakubik, D.; Simoes, S.N.; Martins, D.C., Jr.; Pavao, R.; Siller-Matula, J.M.; Postula, M. ACE2 Interaction Networks in COVID-19: A Physiological Framework for Prediction of Outcome in Patients with Cardiovascular Risk Factors. *J. Clin. Med.* **2020**, *9*, 3743. [CrossRef]
201. Acosta, J.C.; Banito, A.; Wuestefeld, T.; Georgilis, A.; Janich, P.; Morton, J.P.; Athineos, D.; Kang, T.W.; Lasitschka, F.; Andrulis, M.; et al. A complex secretory program orchestrated by the inflammasome controls paracrine senescence. *Nat. Cell Biol.* **2013**, *15*, 978–990. [CrossRef] [PubMed]
202. Nieto, M.A.; Huang, R.Y.; Jackson, R.A.; Thiery, J.P. EMT: 2016. *Cell* **2016**, *166*, 21–45. [CrossRef] [PubMed]
203. Xu, Y.D.; Hua, J.; Mui, A.; O'Connor, R.; Grotendorst, G.; Khalil, N. Release of biologically active TGF-beta1 by alveolar epithelial cells results in pulmonary fibrosis. *Am. J. Physiol. Lung Cell. Mol. Physiol.* **2003**, *285*, 527–539. [CrossRef] [PubMed]

204. Hu, B.; Liu, J.; Wu, Z.; Liu, T.; Ullenbruch, M.R.; Ding, L.; Henke, C.A.; Bitterman, P.B.; Phan, S.H. Reemergence of hedgehog mediates epithelial-mesenchymal crosstalk in pulmonary fibrosis. *Am. J. Respir. Cell Mol. Biol.* **2015**, *52*, 418–428. [CrossRef] [PubMed]
205. Degryse, A.L.; Tanjore, H.; Xu, X.C.; Polosukhin, V.V.; Jones, B.R.; Boomershine, C.S.; Ortiz, C.; Sherrill, T.P.; McMahon, F.B.; Gleaves, L.A.; et al. TGFbeta signaling in lung epithelium regulates bleomycin-induced alveolar injury and fibroblast recruitment. *Am. J. Physiol. Lung Cell. Mol. Physiol.* **2011**, *300*, 887–897. [CrossRef]
206. Zhang, T.; Zhou, J.; Yue, H.; Du, C.; Xiao, Z.; Zhao, W.; Li, N.; Wang, X.; Liu, X.; Li, Y.; et al. Glycogen synthase kinase-3beta promotes radiation-induced lung fibrosis by regulating beta-catenin/lin28 signaling network to determine type II alveolar stem cell transdifferentiation state. *FASEB J.* **2020**, *34*, 12466–12480. [CrossRef]
207. Bonniaud, P.; Kolb, M.; Galt, T.; Robertson, J.; Robbins, C.; Stampfli, M.; Lavery, C.; Margetts, P.J.; Roberts, A.B.; Gauldie, J. Smad3 null mice develop airspace enlargement and are resistant to TGF-beta-mediated pulmonary fibrosis. *J. Immunol.* **2004**, *173*, 2099–2108. [CrossRef]
208. Kang, J.H.; Jung, M.Y.; Yin, X.; Andrianifahanana, M.; Hernandez, D.M.; Leof, E.B. Cell-penetrating peptides selectively targeting SMAD3 inhibit profibrotic TGF-beta signaling. *J. Clin. Investig.* **2017**, *127*, 2541–2554. [CrossRef]
209. Cassar, L.; Li, H.; Pinto, A.R.; Nicholls, C.; Bayne, S.; Liu, J.P. Bone morphogenetic protein-7 inhibits telomerase activity, telomere maintenance, and cervical tumor growth. *Cancer Res.* **2008**, *68*, 9157–9166. [CrossRef]
210. Cassar, L.; Li, H.; Jiang, F.X.; Liu, J.P. TGF-beta induces telomerase-dependent pancreatic tumor cell cycle arrest. *Mol. Cell. Endocrinol.* **2010**, *320*, 97–105. [CrossRef]
211. Cassar, L.; Nicholls, C.; Pinto, A.R.; Chen, R.; Wang, L.; Li, H.; Liu, J.P. TGF-beta receptor mediated telomerase inhibition, telomere shortening and breast cancer cell senescence. *Protein Cell* **2017**, *8*, 39–54. [CrossRef] [PubMed]
212. Cassar, L.; Nicholls, C.; Pinto, A.R.; Li, H.; Liu, J.P. Bone morphogenetic protein-7 induces telomerase inhibition, telomere shortening, breast cancer cell senescence, and death via Smad3. *FASEB J.* **2009**, *23*, 1880–1892. [CrossRef] [PubMed]
213. Wang, L.; Wang, Z.; Liu, J.P. Identification of peptidomimetic telomere dysfunction inhibitor (TELODIN) through telomere dysfunction-induced foci (TIF) assay. *STAR Protoc.* **2021**, *2*, 100620. [CrossRef]
214. Prowse, K.R.; Greider, C.W. Developmental and tissue-specific regulation of mouse telomerase and telomere length. *Proc. Natl. Acad. Sci. USA* **1995**, *92*, 4818–4822. [CrossRef]
215. Davis, R.J.; Welcker, M.; Clurman, B.E. Tumor suppression by the Fbw7 ubiquitin ligase: Mechanisms and opportunities. *Cancer Cell* **2014**, *26*, 455–464. [CrossRef] [PubMed]
216. Gallo, L.H.; Ko, J.; Donoghue, D.J. The importance of regulatory ubiquitination in cancer and metastasis. *Cell Cycle* **2017**, *16*, 634–648. [CrossRef] [PubMed]
217. Chang, J.; Wang, Y.; Shao, L.; Laberge, R.M.; Demaria, M.; Campisi, J.; Janakiraman, K.; Sharpless, N.E.; Ding, S.; Feng, W.; et al. Clearance of senescent cells by ABT263 rejuvenates aged hematopoietic stem cells in mice. *Nat. Med.* **2016**, *22*, 78–83. [CrossRef]
218. Yosef, R.; Pilpel, N.; Tokarsky-Amiel, R.; Biran, A.; Ovadya, Y.; Cohen, S.; Vadai, E.; Dassa, L.; Shahar, E.; Condiotti, R.; et al. Directed elimination of senescent cells by inhibition of BCL-W and BCL-XL. *Nat. Commun* **2016**, *7*, 11190. [CrossRef]
219. Schafer, M.J.; White, T.A.; Iijima, K.; Haak, A.J.; Ligresti, G.; Atkinson, E.J.; Oberg, A.L.; Birch, J.; Salmonowicz, H.; Zhu, Y.; et al. Cellular senescence mediates fibrotic pulmonary disease. *Nat. Commun.* **2017**, *8*, 14532. [CrossRef]
220. Ogrodnik, M.; Zhu, Y.; Langhi, L.G.P.; Tchkonia, T.; Kruger, P.; Fielder, E.; Victorelli, S.; Ruswhandi, R.A.; Giorgadze, N.; Pirtskhalava, T.; et al. Obesity-Induced Cellular Senescence Drives Anxiety and Impairs Neurogenesis. *Cell Metab.* **2019**, *29*, 1061–1077.e8. [CrossRef]
221. Baker, D.J.; Childs, B.G.; Durik, M.; Wijers, M.E.; Sieben, C.J.; Zhong, J.; Saltness, R.A.; Jeganathan, K.B.; Verzosa, G.C.; Pezeshki, A.; et al. Naturally occurring p16(Ink4a)-positive cells shorten healthy lifespan. *Nature* **2016**, *530*, 184–189. [CrossRef] [PubMed]
222. Jeon, O.H.; Kim, C.; Laberge, R.M.; Demaria, M.; Rathod, S.; Vasserot, A.P.; Chung, J.W.; Kim, D.H.; Poon, Y.; David, N.; et al. Local clearance of senescent cells attenuates the development of post-traumatic osteoarthritis and creates a pro-regenerative environment. *Nat. Med.* **2017**, *23*, 775–781. [CrossRef] [PubMed]
223. Baar, M.P.; Brandt, R.M.C.; Putavet, D.A.; Klein, J.D.D.; Derks, K.W.J.; Bourgeois, B.R.M.; Stryeck, S.; Rijksen, Y.; van Willigenburg, H.; Feijtel, D.A.; et al. Targeted Apoptosis of Senescent Cells Restores Tissue Homeostasis in Response to Chemotoxicity and Aging. *Cell* **2017**, *169*, 132–147.e16. [CrossRef] [PubMed]
224. Zhu, Y.; Doornebal, E.J.; Pirtskhalava, T.; Giorgadze, N.; Wentworth, M.; Fuhrmann-Stroissnigg, H.; Niedernhofer, L.J.; Robbins, P.D.; Tchkonia, T.; Kirkland, J.L. New agents that target senescent cells: The flavone, fisetin, and the BCL-XL inhibitors, A1331852 and A1155463. *Aging* **2017**, *9*, 955–963. [CrossRef]
225. He, Y.; Li, W.; Lv, D.; Zhang, X.; Zhang, X.; Ortiz, Y.T.; Budamagunta, V.; Campisi, J.; Zheng, G.; Zhou, D. Inhibition of USP7 activity selectively eliminates senescent cells in part via restoration of p53 activity. *Aging Cell* **2020**, *19*, e13117. [CrossRef]
226. Weber, L. Patented inhibitors of p53-Mdm2 interaction (2006–2008). *Expert Opin. Ther. Pat.* **2010**, *20*, 179–191. [CrossRef]
227. Fuhrmann-Stroissnigg, H.; Ling, Y.Y.; Zhao, J.; McGowan, S.J.; Zhu, Y.; Brooks, R.W.; Grassi, D.; Gregg, S.Q.; Stripay, J.L.; Dorronsoro, A.; et al. Identification of HSP90 inhibitors as a novel class of senolytics. *Nat. Commun.* **2017**, *8*, 422. [CrossRef]
228. Miyata, Y. Hsp90 inhibitor geldanamycin and its derivatives as novel cancer chemotherapeutic agents. *Curr. Pharm. Des.* **2005**, *11*, 1131–1138. [CrossRef]
229. Cherif, H.; Bisson, D.G.; Jarzem, P.; Weber, M.; Ouellet, J.A.; Haglund, L. Curcumin and o-Vanillin Exhibit Evidence of Senolytic Activity in Human IVD Cells In Vitro. *J. Clin. Med.* **2019**, *8*, 433. [CrossRef]

230. Li, W.; He, Y.; Zhang, R.; Zheng, G.; Zhou, D. The curcumin analog EF24 is a novel senolytic agent. *Aging* **2019**, *11*, 771–782. [CrossRef]
231. Wang, Y.; Chang, J.; Liu, X.; Zhang, X.; Zhang, S.; Zhang, X.; Zhou, D.; Zheng, G. Discovery of piperlongumine as a potential novel lead for the development of senolytic agents. *Aging* **2016**, *8*, 2915–2926. [CrossRef]
232. Yang, D.; Tian, X.; Ye, Y.; Liang, Y.; Zhao, J.; Wu, T.; Lu, N. Identification of GL-V9 as a novel senolytic agent against senescent breast cancer cells. *Life Sci.* **2021**, *272*, 119196. [CrossRef] [PubMed]
233. Guerrero, A.; Herranz, N.; Sun, B.; Wagner, V.; Gallage, S.; Guiho, R.; Wolter, K.; Pombo, J.; Irvine, E.E.; Innes, A.J.; et al. Cardiac glycosides are broad-spectrum senolytics. *Nat. Metab* **2019**, *1*, 1074–1088. [CrossRef]
234. Triana-Martinez, F.; Picallos-Rabina, P.; Da Silva-Alvarez, S.; Pietrocola, F.; Llanos, S.; Rodilla, V.; Soprano, E.; Pedrosa, P.; Ferreiros, A.; Barradas, M.; et al. Identification and characterization of Cardiac Glycosides as senolytic compounds. *Nat. Commun.* **2019**, *10*, 4731. [CrossRef] [PubMed]
235. Gonzalez-Gualda, E.; Paez-Ribes, M.; Lozano-Torres, B.; Macias, D.; Wilson, J.R., 3rd; Gonzalez-Lopez, C.; Ou, H.L.; Miron-Barroso, S.; Zhang, Z.; Lerida-Viso, A.; et al. Galacto-conjugation of Navitoclax as an efficient strategy to increase senolytic specificity and reduce platelet toxicity. *Aging Cell* **2020**, *19*, e13142. [CrossRef]
236. Cai, Y.; Zhou, H.; Zhu, Y.; Sun, Q.; Ji, Y.; Xue, A.; Wang, Y.; Chen, W.; Yu, X.; Wang, L.; et al. Elimination of senescent cells by beta-galactosidase-targeted prodrug attenuates inflammation and restores physical function in aged mice. *Cell Res.* **2020**, *30*, 574–589. [CrossRef] [PubMed]
237. Guerrero, A.; Guiho, R.; Herranz, N.; Uren, A.; Withers, D.J.; Martinez-Barbera, J.P.; Tietze, L.F.; Gil, J. Galactose-modified duocarmycin prodrugs as senolytics. *Aging Cell* **2020**, *19*, e13133. [CrossRef] [PubMed]
238. He, Y.; Zhang, X.; Chang, J.; Kim, H.N.; Zhang, P.; Wang, Y.; Khan, S.; Liu, X.; Zhang, X.; Lv, D.; et al. Using proteolysis-targeting chimera technology to reduce navitoclax platelet toxicity and improve its senolytic activity. *Nat. Commun.* **2020**, *11*, 1996. [CrossRef]
239. Wakita, M.; Takahashi, A.; Sano, O.; Loo, T.M.; Imai, Y.; Narukawa, M.; Iwata, H.; Matsudaira, T.; Kawamoto, S.; Ohtani, N.; et al. A BET family protein degrader provokes senolysis by targeting NHEJ and autophagy in senescent cells. *Nat. Commun.* **2020**, *11*, 1935. [CrossRef] [PubMed]
240. Nogueira-Recalde, U.; Lorenzo-Gomez, I.; Blanco, F.J.; Loza, M.I.; Grassi, D.; Shirinsky, V.; Shirinsky, I.; Lotz, M.; Robbins, P.D.; Dominguez, E.; et al. Fibrates as drugs with senolytic and autophagic activity for osteoarthritis therapy. *EBioMedicine* **2019**, *45*, 588–605. [CrossRef]
241. Ozsvari, B.; Nuttall, J.R.; Sotgia, F.; Lisanti, M.P. Azithromycin and Roxithromycin define a new family of "senolytic" drugs that target senescent human fibroblasts. *Aging* **2018**, *10*, 3294–3307. [CrossRef] [PubMed]
242. Cho, H.J.; Yang, E.J.; Park, J.T.; Kim, J.R.; Kim, E.C.; Jung, K.J.; Park, S.C.; Lee, Y.S. Identification of SYK inhibitor, R406 as a novel senolytic agent. *Aging* **2020**, *12*, 8221–8240. [CrossRef] [PubMed]
243. Demidenko, Z.N.; Zubova, S.G.; Bukreeva, E.I.; Pospelov, V.A.; Pospelova, T.V.; Blagosklonny, M.V. Rapamycin decelerates cellular senescence. *Cell Cycle* **2009**, *8*, 1888–1895. [CrossRef] [PubMed]
244. Hubackova, S.; Davidova, E.; Rohlenova, K.; Stursa, J.; Werner, L.; Andera, L.; Dong, L.; Terp, M.G.; Hodny, Z.; Ditzel, H.J.; et al. Selective elimination of senescent cells by mitochondrial targeting is regulated by ANT2. *Cell Death Differ.* **2019**, *26*, 276–290. [CrossRef]
245. Moiseeva, O.; Deschenes-Simard, X.; St-Germain, E.; Igelmann, S.; Huot, G.; Cadar, A.E.; Bourdeau, V.; Pollak, M.N.; Ferbeyre, G. Metformin inhibits the senescence-associated secretory phenotype by interfering with IKK/NF-kappaB activation. *Aging Cell* **2013**, *12*, 489–498. [CrossRef]
246. Peilin, W.; Songsong, T.; Chengyu, Z.; Zhi, C.; Chunhui, M.; Yinxian, Y.; Lei, Z.; Min, M.; Zongyi, W.; Mengkai, Y.; et al. Directed elimination of senescent cells attenuates development of osteoarthritis by inhibition of c-IAP and XIAP. *Biochim. Biophys. Acta Mol. Basis Dis.* **2019**, *1865*, 2618–2632. [CrossRef]
247. Samaraweera, L.; Adomako, A.; Rodriguez-Gabin, A.; McDaid, H.M. A Novel Indication for Panobinostat as a Senolytic Drug in NSCLC and HNSCC. *Sci. Rep.* **2017**, *7*, 1900. [CrossRef]

cells

MDPI

Review

Cellular Senescence in Aging Lungs and Diseases

Arbi Aghali [1], Maunick Lefin Koloko Ngassie [2,3], Christina M. Pabelick [1,4] and Y. S. Prakash [1,4,*]

1 Department of Physiology and Biomedical Engineering, Mayo Clinic, Rochester, MN 55905, USA; aghali.arbi@mayo.edu (A.A.); pabelick.christina@mayo.edu (C.M.P.)
2 Department of Pathology and Medical Biology, University Medical Center Groningen, University of Groningen, 9713 GZ Groningen, The Netherlands; kolokongassie.maunicklefin@mayo.edu
3 Groningen Research Institute for Asthma and COPD, University Medical Center Groningen, University of Groningen, 9700 RB Groningen, The Netherlands
4 Department of Anesthesiology and Perioperative Medicine, Mayo Clinic, Rochester, MN 55905, USA
* Correspondence: prakash.ys@mayo.edu

Abstract: Cellular senescence represents a state of irreversible cell cycle arrest occurring naturally or in response to exogenous stressors. Following the initial arrest, progressive phenotypic changes define conditions of cellular senescence. Understanding molecular mechanisms that drive senescence can help to recognize the importance of such pathways in lung health and disease. There is increasing interest in the role of cellular senescence in conditions such as chronic obstructive pulmonary disease (COPD) and idiopathic pulmonary fibrosis (IPF) in the context of understanding pathophysiology and identification of novel therapies. Herein, we discuss the current knowledge of molecular mechanisms and mitochondrial dysfunction regulating different aspects of cellular senescence-related to chronic lung diseases to develop rational strategies for modulating the senescent cell phenotype in the lung for therapeutic benefit.

Keywords: aging; senescence; lung diseases; asthma; COPD; pulmonary fibrosis; remodeling; mitochondrial dysfunction

Citation: Aghali, A.; Koloko Ngassie, M.L.; Pabelick, C.M.; Prakash, Y.S. Cellular Senescence in Aging Lungs and Diseases. *Cells* **2022**, *11*, 1781. https://doi.org/10.3390/cells11111781

Academic Editor: Karima Djabali

Received: 24 April 2022
Accepted: 26 May 2022
Published: 29 May 2022

Publisher's Note: MDPI stays neutral with regard to jurisdictional claims in published maps and institutional affiliations.

1. Introduction

Cellular senescence is characterized by a permanent cell-cycle arrest triggered by various stimuli, including DNA damage to telomere shortening, genomic instability, epigenetic alterations, loss of proteostasis, and mitochondrial dysfunction (Figure 1) [1]. Despite being in cell-cycle arrest, senescent cells are resistant to apoptosis due to activation of anti-apoptotic signaling. Senescent cells remain metabolically active, secreting inflammatory cytokines, growth factors, chemokines (CXCs), and extracellular matrix (ECM) proteins, collectively known as senescence-associated secretory phenotype (SASP) [2,3].

Senescent cells are thought to have beneficial effects on repairing injured tissue and maintaining organismal integrity. The role of senescent cells in tumor suppression is also recognized. Under normal conditions, senescent cell burden is limited by removing excessive senescent cells via the immune system. However, with aging, impairment of the immune response results in accumulation of senescent cells that can exacerbate their effects leading to detrimental consequences, i.e., diseases of aging. Furthermore, there is now increasing evidence for different senescent cell phenotypes such that a shift towards detrimental, pro-inflammatory, pro-fibrotic senescent cells and SASP can occur with aging, contributing to disease.

Compared to other organ systems where senescence, SASP and contributions to aging and diseases of aging have been substantially explored, there is relatively less data on the aging lung, and senescent cells in aging-associated lung diseases such as COPD, pulmonary fibrosis, and even asthma. Senescent cells do accumulate in aging lungs and can exacerbate lung diseases [4–6] (Figure 1). However, the mechanisms by which senescent cells, via their SASP, can induce paracrine signaling to activate neighboring naïve cells to induce remodeling (altered ECM deposition and/or cell proliferation) or modulate cell–cell interactions to promote disease are still under investigation.

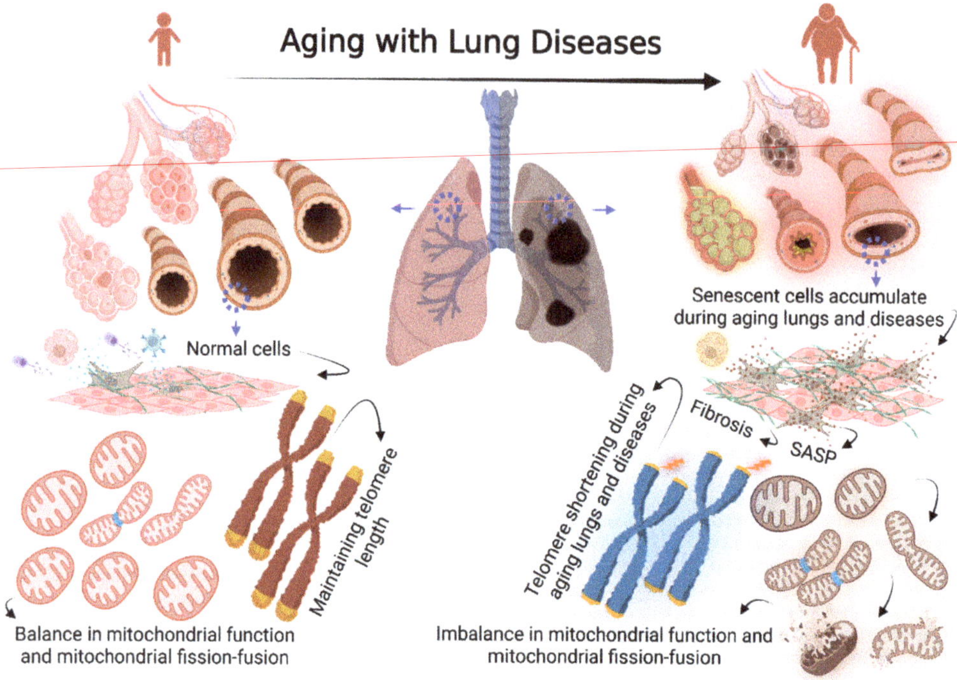

Figure 1. Left schematic figure shows a normal lung of young age with a low number of senescent cells rapidly cleared by immune cells, normal cells maintain a baseline of telomere length and mitochondrial homeostasis. Right schematic figure shows aged and diseased lung associated with increased fibrosis, higher numbers of senescent cells, and slow response of immune cells to clear senescent cells. Senescent cells are characterized by telomere shortening, secreting high rates of SASP, mitochondrial dysfunction, and an imbalance in mitochondrial fission and fusion. Figure 1 was created with BioRender.com accessed on 24 April 2022.

One factor relevant to cellular senescence and aging that may be of importance to the lung is cellular stress, which promotes mitochondrial dysfunction, including mitochondrial oxidative stress, mitochondrial DNA (mtDNA) mutation, imbalance in mitochondrial fission and fusion, and alterations in mitochondrial respiration [7]. Mitochondrial oxidative stress has been thought to be involved in accelerating aging effects. Separately, mitochondrial oxidative stress has been associated with lung diseases such as COPD and IPF and could thus play a role in stimulating as well as maintaining cellular senescence towards impaired lung function [8–10].

In this review, we discuss mechanisms of cellular senescence relevant to different aspects of the structure and function of aging lungs and to lung diseases, focusing on COPD and IPF. We review the influence of mitochondrial dysfunction in the context of cellular senescence and lung diseases. Finally, we summarize promising methods currently used to target senescent cells as a potential therapy to improve healthspan in the context of normal aging lung, and counteract lung diseases associated with aging. We appreciate that cellular senescence and SASP signaling is complex, and likely cell- and context-dependent. Accordingly, a review of these topics is necessarily brief and perhaps simple, but is relevant to the specific topic of aging lung and associated diseases.

2. Overview of Cellular Senescence

Cellular senescence was originally described by Hayflick and Moorhead [11], where they demonstrated that human fetal fibroblasts lose their ability to divide after a certain number of subcultures (i.e., replicative senescence), associated with changes in cellular morphology such as flattening and increased cell size. Several studies have since described a similar phenomenon of replicative senescence in other cell types from different organs [12,13]. It is also now clear that other factors can accelerate cells towards losing their ability to proliferate in vitro, including the age and donor health condition [14], as well as environmental and genotoxic stresses. Examples include telomere shortening, inflammatory signaling, mitochondrial dysfunction and oxidative stress, oncogene-induced senescence (OIS), cell differentiation [15,16], chemotherapeutic drugs such as etoposide [17,18], exposure to UV radiation, and DNA damage stress.

Among cellular stressors, telomere shortening is now recognized as a hallmark of aging and senescence. Telomeres contain a repetitive nucleotide sequence of complementary double-strand DNA (5′-AGGGT-3′ and 3′-TCCCA-5) and wind up with a tail of a single-stranded DNA (5′-TTAGGG-3′) [19–21]. Telomeres protect chromosomal ends from recombination and fusion, and maintain DNA stability. Without telomeres, the DNA damage response is initiated [21]. In replicative senescence, the telomere shortens due to the inability of DNA polymerase to complete DNA replication. When a short length of telomeres is reached, a damage signal is initiated in the coiled DNA [22,23]. Therefore, telomere shortening has been used as a hallmark of aged and senescent cells [23–25].

Senescent cells are thought to support physiological functions during embryonic and postnatal development, tissue regeneration, and wound healing [26–29]. For instance, upon wound closure, activated myofibroblasts limit excessive fibrosis at the injury site [26–29]. The effects of senescent cells are kept in check by immune monitoring and clearance of senescent cells. Indeed, it is thought that during development, senescent cells avoid elimination from their microenvironment by altering their SASP components to avoid the immune system [26–29]. However, with aging, the efficiency of the immune system to clear senescent cells is impaired [30]. Consequently, senescent cells accumulate, secreting SASP factors that may become detrimental to naïve/neighboring cells by virtue of the quantity of such factors or an altered phenotype involved in more inflammatory and fibrotic elements [29–33].

3. Cellular Senescence Signaling Pathways

Cellular senescence is regulated by two signaling pathways that interact but are also independent: p53-p21^{CIP1} and p16^{Ink4a}-Rb [18]. Permanent arrest of cell cycle occurs at the G_1/S transitional phase distinguishing it from the quiescent phase, G_0 [24,34–36]. The DNA damage response (DDR) regulates tumor suppressor of transcriptional factor p53 and downstream signaling p21^{CIP1}, to result in permanent arrest in the cell cycle [18,24,34–36].

In the nuclei, DDR foci originate in response to DNA double-strand breaks (DSB). A subnuclear focus and accumulation of DDR proteins such as p53-binding protein 1 (53BP1), histone variant H2AX phosphorylated at serine-139 residue (γ-H2AX), and Ataxia Telangiectasia Mutated (ATM) in the vicinity of chromosomal DNA double-strand reflect early molecular events of cellular responses to DSB [34,37,38]. DDR then initiates a series of molecular events to repair DSB and to prevent potential DNA mutations. Phosphorylated at serine-139 in H2AX is mediated by ATM and Ataxia Telangiectasia and Rad3 related protein (ATR) kinases, which lead to visible DNA damage foci within the chromatin [18]. p53 binding protein is a key modulator rapidly localized to DNA damage foci after, for instance, ionizing radiation that causes DSB [37–41]. Although the key functions of p53 have not been fully understood, accumulated evidence suggests that the roles of p53 binding protein are engaging DSB proteins, such as interferon regulatory factor 4 (also known as MUM1) [35,38] and RAP1-interacting factor 1 (RIF1) [38,42], amplifying ATM activity, and promoting checkpoint signaling in response to low levels of DNA damage signals [38–40,43–45].

ATM and ATR stabilize p53 by activating cyclin-dependent kinase inhibitor $p21^{CIP1}$, which in turn inhibits cyclin-dependent kinases-2 (CDK2) [33,39,40,43,46]. CDK2 triggers family members of tumor suppressors, retinoblastoma proteins (Rb), stopping the cell cycle in the S phase, and subsequently preventing DNA replication [29,31,43,47]. The signaling pathway of ATM-p53-$p21^{CIP1}$/Rb results in a permanent arrest in the cell cycle [29,31,35,43].

Another tumor suppressor that influences key roles during cessation of cell division is the *INK4a-ARF-INK4b* locus [29,41,45,46,48,49]. The *INK4a* and *INK4b* locus encode for two cyclin-dependent kinase inhibitors, $p16^{ink4a}$ and $p15^{ink4b}$, while *ARF* is associated with $p14^{ARF}$ in humans ($p19^{ARF}$ in mice) [32,45,46,48,49]. *INK4/ARF* activates cyclin-dependent kinase inhibitor $p16^{ink4a}$ that selectively inhibits cyclin-dependent kinases-4 (CDK4) and cyclin-dependent kinases-6 (CDK6) [29,31,32,43]. Upon activation, CDK4/6 phosphorylates retinoblastoma protein (Rb). As a result, transcriptional factor E2F3 is upregulated and leads to cell cycle arrest in the S phase [29,31,32,43,50,51]. Although upregulation of $p16^{ink4a}$ is meditated by the downstream signaling of p53-$p21^{CIP1}$ [29,31,32,43,52], it is believed that the transcriptional factor $p21^{CIP1}$ upregulates earlier than p16 ^{ink4a} [47], giving a chance for cultured cells to go for another division cycle before making it to a complete cell cycle arrest [29,31,32,43,51,53]. Thus, the expressions of p53-p21 and p16 ^{ink4a} appear to demonstrate a non-linear functional relationship.

4. Biomarkers of Cellular Senescence

Accumulation of senescent cells can be recognized by utilizing various methods in vitro and in vivo. For instance, upregulation of the transcriptional factors p53, p21, and p16, and SASP elements such as IL-6 and IL-8, are well-validated markers [31–33,41]. Senescence-associated β-galactosidase (SA-βgal) is another technique that is widely used to identify senescent cells in vitro and in vivo [28,50], where due to increased levels of lysosomal enzyme, the enzymatic activity of SA-βgal results in blue color at a pH of 6.0 [50,52,54]. However, SA-βgal is not the most sensitive or specific marker of senescence. Fluorescence in situ hybridization (FISH) of telomerase combined with immunofluorescence staining of γ-H2AX results in localization of telomere-associated foci (TAF) and has more recently been used to identify senescent cells [22,24,28,55]. SASP and SASP regulators are also used to characterize senescent cells, including (1) proinflammatory factors such as IL-1α, and IL1ß, IL-6 and IL-8; (2) signaling pathway such as Akt and MAPK; (3) NF-kB [51]; (4) growth factors such as TGF-β1 and matrix-degrading enzymes, metalloproteinases; (5) extracellular matrix proteins such as fibronectin [47]. However, it should be noted that the SASP profile is highly cell and context dependent, and it is not unusual for the profile to change with time, making it difficult to identify a unique, stable, and broadly applicable set of senescence markers. In this regard, while RNA sequencing and whole-genome analysis have been widely utilized to identify senescence-associated genes [47,53,56–58], there is substantial interest in the use of fluorescence-activated cell sorting (FACS) and particularly cytometry by time of flight (CyTOF) using antibodies that recognize antigens selectively expressed in senescent cells and can distinguish between detrimental and beneficial phenotypes based on expression of p16 and p21 (generalized markers) and that of NF-kB (detrimental) [55,59].

5. Senescence Signaling in Lung Diseases

Given the clinical impact of aging per se, and that of aging-associated lung diseases, it is important to identify biomarkers and signaling pathways in the context of senescence and its contributions to COPD [24] and IPF [60,61], and even asthma [18,62]. The importance of this area is reflected by the increasing number of research and review papers published on samples from human COPD and IPF patients and in animal models (Figure 2).

Lung tissues from patients with COPD and IPF show hallmarks of senescent cells [24,35,63,64]. Key biomarkers of senescence in aging adults are upregulation of p53, $p21^{CIP1}$, $p16^{ink4a}$, a robust release of SASP, positive staining for SA-βgal, TAF, and upregulation of anti-apoptotic signaling networks [24,25,35,63,65–69]. Increased expression of proinflammatory cytokines such as MCP-1, KC, MIP-1α, IL-12p40, and G-CSF have been observed in a mouse model of COPD [64,70]. In ROS-induced human senescent fibroblasts, IL-6 and IL-8 are increased following 14 and 25 days in culture [24,26].

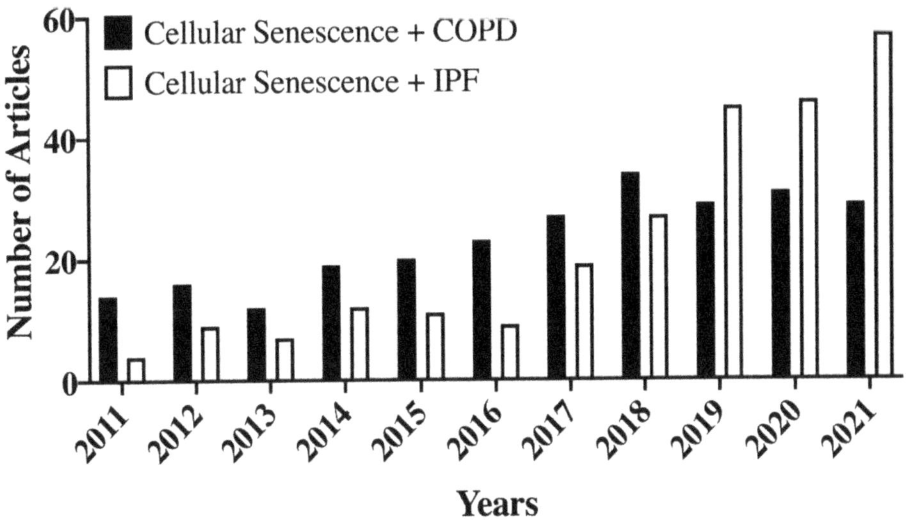

Figure 2. An increased number of articles related to Cellular Senescence and COPD or IPF have been published in PubMed-indexed journals during the past 10 years. Keywords used in PubMed search engines are "Cellular Senescence Chronic Obstructive Pulmonary Disease" or "Cellular Senescence Idiopathic Pulmonary Fibrosis". Figure 2 was created from National Library of Medicine (https://pubmed.ncbi.nlm.nih.gov accessed on 24 April 2022).

6. Cellular Senescence in COPD

COPD is a major healthcare issue with a high morbidity and mortality rate [71]. COPD is characterized by obstruction in small airways (bronchiolitis), alveolar emphysema, and airway remodeling. Although tobacco smoke is the leading cause of COPD [63,72], air pollution, genetic disorders (alpha-1 deficiency), and respiratory infections are also risks for COPD. While there is no known cure, COPD is managed via lifestyle changes and medications, but these therapies have limitations, necessitating exploration of novel therapies.

Studies have shown that endothelial colony-forming cells (ECFC) derived from COPD patients have increased expression of SA-βgal, p16, p21, and γ-H2AX compared to ECFC isolated from control group patients [73]. In addition, lung fibroblasts derived from COPD patients show greater release of IL-6 and IL-8 and a higher percentage of SA-βGal staining [74]. Increased p21 and p16-positive epithelial cells have also been reported in COPD lung tissues compared to control groups [75]. More recently, COPD lung fibroblasts have been found to show senescence and 42 SASP secretome elements, which are implicated in chronic inflammation of COPD [76].

Tobacco smoke can trigger cellular senescence via oxidative stress-mediated DNA damage. Conversely, targeting p16-positive cells can inhibit tobacco smoke-induced emphysema in mouse models [63]. Furthermore, tobacco smoke accelerates telomere erosion and causes oxidative damage in cells [77]. Increased production of ROS associated with oxidative stress and changes in mitochondrial complex II, III, and V expression enhance cellular senescence [20]. Increased senescence in airway epithelial cells of severe COPD patients along with increased SASP has also been observed [3,78,79], and is relevant given the role of inflammation in COPD.

Multiple senescence signaling pathways may be involved in COPD, and activated in patients with a history of tobacco smoking and/or E-cigarette vaping [9,24,63,80,81]. For example, in older COPD patients, phosphoinositide-3-kinase (PI3K)-Akt and p38 MAPK cascades are activated [82,83]. Oxidative stress in COPD inhibits PTEN phosphatase activity, which in turn activates downstream signaling of PI3K-Akt and of mammalian target rapamycin complex 1 (mTORC1) protein kinase, which is a key player in cellular senescence. mTORC1 can also be activated by AMP-activated protein kinase (AMPK), an energy sensor that responds to an imbalance between AMP:ATP and ADP:ATP ratios [84–88]. Although AMPK is best known for its roles in cellular metabolism [78], its signaling is also important in the regulation of mitochondrial biogenesis and mitophagy [79,85–88]. Sirtuins, proteins involved in metabolic activity, have been implicated in aging and COPD lungs [80]. For instance, activation of mTOR upregulates microRNA-34a (miR-34a), and in return inhibits sirtuin-1 (SIRT1) and sirtuin-6 (SIRT6) activities [81,89,90]. Inhibition of SIRT1 dysregulates oxidative energy metabolism and influences NF-kB activity [91]. Activation of NF-kB stimulates SASP expressions found in many age-related diseases.

Another signaling pathway that plays a role in cellular senescence and COPD is p38 MAPK [92,93]. Increased p38 MAPK phosphorylation has been found in bronchial epithelial cells of COPD and asthmatic patients [92–95]. p38 MAPK signaling is known to enhance senescence burden in the lung [82,93]. SASP secretomes and oxidative stress stimuli such as tobacco smoke as well as respiratory pathogens can drive p38 MAPK phosphorylation in COPD lungs [90,93]. Increased p38 MAPK upregulates c-Jun proteins and activator protein-1 (AP-1), resulting in upregulation of microRNA-570 (miR-570), which inhibits SIRT1 and enhances NF-kB activity, leading to downstream activation of p53 and enhanced SASP expression [20,84].

Overall, these data provide evidence of senescence in COPD, and the potential involvement of multiple signaling pathways that could contribute to at least the inflammatory aspects of this disease. Of note, these signaling pathways are also well known to contribute to cell proliferation and remodeling and may thus be relevant to these aspects of COPD as well.

7. Cellular Senescence in IPF

IPF is a life-threatening chronic lung disease with poor prognosis and survival. IPF is characterized by scarred lungs associated with hyperproduction of ECM proteins [96,97]. In the past decade, there has been increasing interest in understanding the contributions of senescence to IPF (Figure 2). Several studies have shown that higher senescence markers are detected in IPF-derived cells and IPF tissues harvested from humans or in animal models. For example, upregulation of senescence-related pathways in alveolar type 2 (AT2) cells has been noted in a mouse model of IPF where AT2 Sin3a has been knocked out to induce senescence [61]. Conversely, targeting p53 signaling in alveoli reduces fibrosis [61]. p21 and p16-positive cells have also been shown to accumulate in IPF lung tissues [61,75]. Furthermore, SASP, such as matrix metalloproteinases MMP2 and MMP9 and collagen type I alpha 1 (COL1A1), show higher expression in IPF lungs [60]. Increased expression of p16 along with increased pro-fibrotic SASP has been reported in bleomycin-induced pulmonary fibrosis mouse models [60]. Sirtuins also play an important role during IPF as shown in fibroblast–myofibroblast differentiation (FMD), a process often triggered by TGF-β1. Reduced expression of SIRT-3 has been observed in IPF lung tissue, and inhibiting SIRT-3 has been associated with increased FMD in a murine IPF model after exposure to TGF-β1 [98]. Overexpression of SIRT-3 prevents TGF-β1-mediated FMD [98]. Thus, these limited data highlight the importance of senescence and associated signaling pathways in IPF.

8. Mitochondria in Senescence and Aging

Mitochondria are essential in eukaryotic cells for maintaining cellular homeostasis and function. Mitochondria regulate numerous cellular activities such as metabolism, replication, differentiation, senescence, and apoptosis [99]. Mitochondria produce energy for cells to perform essential functions by metabolized sources of macromolecules, such as glucose, amino acids, monosaccharides, and monoacylglycerols [99]. Several enzymes participate in the mitochondrial respiratory chain, a multistep process required to convert macronutrients into high-level energy. Mitochondrial respiration of glycolysis and the electron transport chain has been discussed in the literature extensively [100,101]. Herein, we will discuss mitochondrial roles in the context of senescence, aging lungs, and lung-related diseases.

Mitochondrial dysfunction can contribute to cellular senescence. For instance, gradual alterations of mitochondrial DNA (mtDNA mutations), variations in mitochondrial fission and fusion, elevated mitochondrial ROS production, and changes related to mitochondrial morphology (increased mitochondrial mass and elongation) can all play a role [35,102–104]. Senescent fibroblasts (replicative senescence) show dynamic changes in mitochondrial mass [26,28], while other studies show a dynamic feedback loop between damaged DNA and mitochondria [26,28,105,106].

Induction of senescence via disruption of mitochondrial function results in a distinctive SASP portfolio compared to senescence induced by genotoxic stress [107]. This mitochondrial dysfunction-associated senescence has been termed MiDAS and has been further shown to be a low NAD+/NADH ratio that drives it through AMPK-mediated p53 activation. Specifically, MiDAS secretomes are distinguished by higher levels of IL-10, CCL27, and TNF-α than core components of SASP such as IL-6 and IL-8 [107].

Overproduction of mitochondrial ROS is also an important player that causes DNA damage and results in DDR. The circle of ROS-DNA damage involves phosphorylation of DDR kinase ATM and Akt [108]. ATM activation initiates a series of phosphorylation events through ATM, NEMO, and IKK, ultimately activating nuclear transcription factor NF-kB, which enhances inflammation [108,109]. However, it is important to note that NF-kB activity is also affected by several factors, including metabolic activity and ROS production. For instance, reduced NAD+ and an alteration in AMP:ATP and ADP:ATP ratios affect SASP through NF-kB regulation [108]. Furthermore, activation of sirtuins such as SIRT1 and SIRT6 has been shown to inhibit NF-kB activity, affecting multiple SASP genes [100,110,111]. Activation of inhibitor sirtuins requires cofactor NAD+ [107,112]. Therefore, a reduction in the sirtuin cofactor NAD+ can increase NF-kB activity, and ultimately SASP responses.

Patients with COPD show changes in proteins that influence oxidative stress. PTEN-induced protein kinase-1 (PINK1), a mitochondrial stress protein marker that accumulates on the outer membrane of damaged mitochondria, is found to be elevated in COPD [72,113]. On the other hand, excessive production of mitochondrial catalase, an enzyme that protects cells from oxidative damage catalyzing hydrogen peroxide to oxygen and water, extends lifespan in the mouse [114]. Conversely, a reduction in prohibitin genes, such as PHB1 in the inner mitochondrial membrane that maintain mitochondrial function, has been observed in COPD and in smokers with no history of COPD [115,116]. Hydrogen peroxide can promote mitochondrial dysfunction in airway smooth muscle (ASM) cells [117–120]. ASM cells from patients with COPD show higher ROS associated with (1) increased IL-8 release, (2) decreased mitochondrial complex enzyme expression, and (3) reduced mitochondrial membrane potential [18,119].

9. Mitochondrial DNA Mutation in Aging Lungs and Diseases

Unlike the nuclear genome, the mitochondrial genome is a ~16.6 kb circular DNA molecule encoding subunits of polypeptides [99,105,106,121]. Diseases associated with mitochondria are driven by a variety of genetic mutations encoded by either the mitochondrial genome or nuclear genome [110,122,123]. Mammalian cells have multiple mitochondria, each having ~10 copies of DNA [106,121]. Mutation in mtDNA can be heteroplasmic or homoplasmic [106,121]. mtDNA is maternally inherited during embryonic development [111,124,125]. However, mtDNA mutations often occur during aging [82,126,127], where mutation rates are much higher in mtDNA than in nuclear DNA (nDNA) [111,128].

Emerging evidence illustrates that alterations of mtDNA are associated with electron transport efficiency. These changes are due to mutations in encoded subunits of polypeptides making up mitochondrial respiratory complexes that serve as primary sources of ROS.

Depletion of mtDNA has been associated with premature aging and multiple chronic diseases [129]. Using a murine model, inducing mutation in mtDNA (by depletion) diminishes mitochondrial respiratory complexes I, III, and IV and ATP synthase [129]. These changes are associated with accelerated aging, skin hair loss, and increased inflammation [129]. Furthermore, introducing a deficiency in proofreading of mitochondrial DNA polymerase (POLG) (involved in mtDNA replication) results in premature aging in mice [71,130]. Studies show that introducing an error-prone version of mtDNA polymerase causes increased mtDNA mutation load [131] and a deficiency in mitochondrial respiratory complexes [131], and accelerates a premature aging phenotype in different mouse organs [71,131]. Other studies have reported that mutations in mtDNA are associated with aging and several chronic diseases [132–135]. For instance, much higher mtDNA mutation rates have been shown in Parkinson's, Alzheimer's, and cardiovascular diseases [128,136–138]. Furthermore, alteration in mtDNA reduces resistance to oxidative stress and increases risk of COPD, asthma, and other lung diseases [104,139,140]. More homoplasmic variants that lead to constant changes in electronic transport chain proteins have been observed in asthmatic patients [106]. Additionally, other studies have shown an imbalance between mtDNA and nDNA in asthma, COPD, and asthma–COPD overlap [141,142].

10. Cellular Senescence as a Therapeutic Target

Demonstrating that suppressing the accumulation of p16^{Ink4a} positive cells extends lifespan by decreasing growth hormone signals has helped to excite the field regarding the therapeutic potential of targeting senescent cells [126]. Subsequently, efforts to develop drugs that eliminate senescent cells (senolytics) without affecting normal cells have become a major focus in the field [68]. The idea is that senescent cells depend on their anti-apoptotic pathways to survive [68,132]. Senolytic cocktails of small molecules target the anti-apoptotic network [31,68,132]. Much work has focused on the use of dasatinib (D; a tyrosine kinase inhibitor) and quercetin (Q; a plant based flavonoid) [68,132]. D + Q were initially shown to effectively induce apoptosis in senescent cells of primary adipocyte progenitor cells and human umbilical vein endothelial cells but not in quiescent, proliferating, or differentiated cells [68,132]. In a mouse model, D + Q promotes physical function and reduces mortality in aged mice [143]. Human trials for D + Q in IPF suggest improvement in respiratory and physical function [62]. Recently, mice infected with SARS-CoV-2-related virus and treated with a senolytic showed reduced senescent cell burden and mortality while increasing antiviral antibodies [144].

Another approach to eliminating senescent cells is targeting the higher mitochondrial potential in senescent cells [145]. Mitochondria-targeted tamoxifen (MitoTam) is an anticancer agent that has been proven to inhibit oxidative phosphorylation and induce cell death in senescent cells [145]. MitoTam can selectively eliminate senescent cells in aging adults and premature or acute senescent cells at young ages [145].

Besides senolytics, strategies to develop drugs that target signaling pathways critical to senescent cells have investigated using senostatic (senomorphic) drugs [69,127]. Unlike senolytics, senostatics (senomorphics) can block paracrine signaling that activates nearby naïve cells (Figure 3) [69,127]. While senolytics induce apoptosis and eliminate senescent cells, senostatics (senomorphics) are geared towards inhibiting SASP release and signaling and/or cell-specific SASP factor expression (Figure 3). Studies show polyphenols, flavonoids, and phytochemicals are effective senostatic drugs inhibiting signals and suppressing SASP factors [69,127]. Diminishing PI3K-Akt signaling via a prodrug pan-PI3K inhibitor CL27c in aging lungs decreases inflammation and improves life expectancy in murine animal models of acute or glucocorticoid-resistant neutrophilic asthma [133,146]. Additionally, activation of AMPK by reducing cellular metabolic activity and increasing ATP synthesis blocks mTOR activation. Consequently, the signaling cascade that enhances proinflammatory cytokine secretions and p53 activation is terminated. NF-kB antioxidants and inhibitors are effective senostatics, suppressing SASP factors [134,147]. Using rapamycin or torin to inhibit the mTORC1 signaling pathway has been shown to rescue mitochondrial dysfunction [135,148]. Similarly, the antioxidant drug MitoQ is effective at targeting TNF-α-induced CXCL8 [72,119].

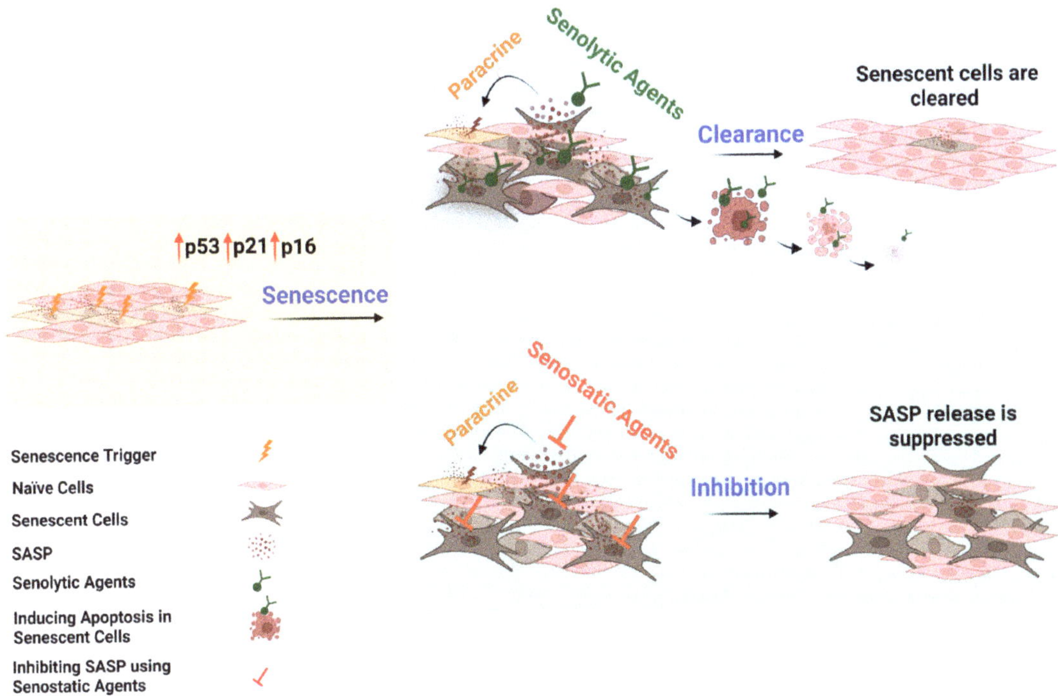

Figure 3. Shows different mechanisms between senolytic and senostatic agents targeting senescent cells. Left schematic figure shows normal cells exposed to DNA damaging agents resulting in upregulations of senescence signaling pathways. Top schematic figure shows senolytic agents selectively kill senescent cells in a living organism, inhibiting paracrine signaling with normal cell proliferation. Bottom schematic figure shows senostatic agents inhibit senescent cells releasing SASP, reducing paracrine signaling with normal cell proliferation. Figure 3 was created with BioRender.com accessed on 24 April 2022.

Furthermore, the widely prescribed and FDA-approved anti-diabetic drug metformin shows promise in the context of senescence [149–152]. Metformin can activate AMPK by blocking complex I, which drives ATP synthesis in the mitochondrial respiratory chain, and thus improving the AMP: ATP balance [78]. These emerging data provide substantial promise to the idea of modulating senescent cell burden and/or SASP portfolios or downstream signaling towards alleviating lung diseases associated with aging.

11. Conclusions and Future Insights

Cellular senescence is a hallmark of aging lungs and aging-associated lung diseases. While senescent cells have beneficial roles, with aging, an enhanced senescent cell burden and a pro-inflammatory and pro-fibrotic SASP can contribute to pathophysiology of diseases such as COPD, IPF, and even asthma. Senescence can be activated by multiple upstream mechanisms, and conversely can involve multiple, interactive downstream pathways. SASP effects on naïve cells can be cell and context dependent with multiple effects on remodeling relevant to lung disease. Thus, there is substantial enthusiasm in exploring the use of senolytics and senostatics in eliminating senescent cells or modulating SASP effects towards therapy for lung diseases. Here, what remains to be understood is the differences in senescent cell and SASP phenotypes in different diseases, further complicated by likely cell-type differences in senescence in the lung. Appreciating that aging may differentially influence different cell types in the lung, the contribution of senescence remains to be understood in a cell-specific fashion. Thus, understanding the relative roles of resident cells such as bronchial and alveolar epithelium, smooth muscle and fibroblasts in senescence and its downstream impact is critical. In this regard, the relative roles of different senescence pathways may also show cell dependent variability that remains to be understood. Thus, modulation of senescence as therapy may be a reality for multiple aging-associated lung diseases; the potential for future research is high.

Author Contributions: Conceive review, first draft preparation, and writing—A.A.; designing Figure 1, Figure 2, Figure 3 and figure legends—A.A.; shortening the first draft, editing, writing, and modifying sections—A.A. and M.L.K.N.; organizing references, re-editing, and re-writing sections—A.A.; editing and writing to final draft—Y.S.P.; final editing, reviewing, and approval, A.A., M.L.K.N., Y.S.P. and C.M.P. All authors have read and agreed to the published version of the manuscript.

Funding: This review is supported by the NIH grants T32 HL105355 (Aghali), R01 HL088029 (Prakash), R01 HL14602 (Pabelick), and Abel Tasman Talent Program Fellowship, in association with the Healthy Aging Alliance, provided by the University Medical Center Groningen (Netherlands) and the Mayo Clinic (Rochester, MN, USA) (Koloko Ngassie).

Institutional Review Board Statement: Not applicable.

Informed Consent Statement: Not applicable.

Data availability Statement: Links for data-generating Figure 2: Cellular Senescence Chronic Obstructive Pulmonary Disease https://pubmed.ncbi.nlm.nih.gov/?term=Cellular+Senescence+chronic+obstructive+pulmonary+disease&filter=years.2011-2021 (accessed on 22 March 2022). Cellular Senescence Idiopathic Pulmonary Fibrosis https://pubmed.ncbi.nlm.nih.gov/?term=Cellular+Senescence+Idiopathic+Pulmonary+Fibrosis&filter=years.2011-2021 (accessed on 22 March 2022).

Acknowledgments: Schematic figures were created using BioRender (BioRender, www.BioRender.com, 24 April 2022). Figure 2 was created using Public/Publisher Medline (National Library of Medicine, pubmed.ncbi.nlm.nih.gov, 22 March 2022).

Conflicts of Interest: The authors declare no conflict of interest.

References

1. López-Otín, C.; Blasco, M.A.; Partridge, L.; Serrano, M.; Kroemer, G. The Hallmarks of Aging. *Cell* **2013**, *153*, 1194–1217. [CrossRef] [PubMed]
2. Kuilman, T.; Peeper, D.S. Senescence-messaging secretome: SMS-ing cellular stress. *Nat. Cancer* **2009**, *9*, 81–94. [CrossRef] [PubMed]
3. Coppé, J.-P.; Desprez, P.-Y.; Krtolica, A.; Campisi, J. The Senescence-Associated Secretory Phenotype: The Dark Side of Tumor Suppression. *Annu. Rev. Pathol. Mech. Dis.* **2010**, *5*, 99–118. [CrossRef]
4. Parikh, P.; Wicher, S.; Khandalavala, K.; Pabelick, C.M.; Britt, R.D., Jr.; Prakash, Y.S. Cellular senescence in the lung across the age spectrum. *Am. J. Physiol.-Lung Cell. Mol. Physiol.* **2019**, *316*, L826–L842. [CrossRef] [PubMed]
5. Kirkwood, T.B.L. Understanding the Odd Science of Aging. *Cell* **2005**, *120*, 437–447. [CrossRef]
6. Ito, K.; Barnes, P.J. COPD as a Disease of Accelerated Lung Aging. *Chest* **2009**, *135*, 173–180. [CrossRef] [PubMed]
7. Mora, A.L.; Bueno, M.; Rojas, M. Mitochondria in the spotlight of aging and idiopathic pulmonary fibrosis. *J. Clin. Investig.* **2017**, *127*, 405–414. [CrossRef] [PubMed]
8. Yue, L.; Yao, H. Mitochondrial dysfunction in inflammatory responses and cellular senescence: Pathogenesis and pharmacological targets for chronic lung diseases. *J. Cereb. Blood Flow Metab.* **2016**, *173*, 2305–2318. [CrossRef]
9. Prakash, Y.; Pabelick, C.M.; Sieck, G. Mitochondrial Dysfunction in Airway Disease. *Chest* **2017**, *152*, 618–626. [CrossRef]
10. Zhou, W.-C.; Qu, J.; Xie, S.-Y.; Sun, Y.; Yao, H.-W. Mitochondrial Dysfunction in Chronic Respiratory Diseases: Implications for the Pathogenesis and Potential Therapeutics. *Oxidative Med. Cell. Longev.* **2021**, *2021*, 5188306. [CrossRef]
11. Hayflick, L.; Moorhead, P.S. The serial cultivation of human diploid cell strains. *Exp. Cell Res.* **1961**, *25*, 585–621. [CrossRef]
12. Shelton, D.N.; Chang, E.; Whittier, P.S.; Choi, D.; Funk, W.D. Microarray analysis of replicative senescence. *Curr. Biol.* **1999**, *9*, 939–945. [CrossRef]
13. Denchi, E.L.; Attwooll, C.; Pasini, D.; Helin, K. Deregulated E2F Activity Induces Hyperplasia and Senescence-Like Features in the Mouse Pituitary Gland. *Mol. Cell. Biol.* **2005**, *25*, 2660–2672. [CrossRef] [PubMed]
14. Campisi, J. The biology of replicative senescence. *Eur. J. Cancer* **1997**, *33*, 703–709. [CrossRef]

15. Storer, M.; Mas, A.; Robert-Moreno, À.; Pecoraro, M.; Ortells, M.C.; Di Giacomo, V.; Yosef, R.; Pilpel, N.; Krizhanovsky, V.; Sharpe, J.; et al. Senescence Is a Developmental Mechanism that Contributes to Embryonic Growth and Patterning. *Cell* **2013**, *155*, 1119–1130. [CrossRef] [PubMed]
16. Muñoz-Espín, D.; Cañamero, M.; Maraver, A.; Gómez-López, G.; Contreras, J.; Murillo-Cuesta, S.; Rodríguez-Baeza, A.; Varela-Nieto, I.; Ruberte, J.; Collado, M.; et al. Programmed Cell Senescence during Mammalian Embryonic Development. *Cell* **2013**, *155*, 1104–1118. [CrossRef] [PubMed]
17. Kuilman, T.; Michaloglou, C.; Mooi, W.J.; Peeper, D.S. The essence of senescence: Figure 1. *Genes Dev.* **2010**, *24*, 2463–2479. [CrossRef]
18. Parikh, P.; Britt, R.D., Jr.; Manlove, L.J.; Wicher, S.; Roesler, A.; Ravix, J.; Teske, J.; Thompson, M.A.; Sieck, G.C.; Kirkland, J.L.; et al. Hyperoxia-induced Cellular Senescence in Fetal Airway Smooth Muscle Cells. *Am. J. Respir. Cell Mol. Biol.* **2019**, *61*, 51–60. [CrossRef]
19. Kawamura, T.; Suzuki, J.; Wang, Y.V.; Menendez, S.; Morera, L.B.; Raya, A.; Wahl, G.M.; Belmonte, J.C.I. Linking the p53 tumour suppressor pathway to somatic cell reprogramming. *Nature* **2009**, *460*, 1140–1144. [CrossRef]
20. Barnes, P.J. Senescence in COPD and Its Comorbidities. *Annu. Rev. Physiol.* **2017**, *79*, 517–539. [CrossRef]
21. Greider, C.W. Telomeres Do D-Loop–T-Loop. *Cell* **1999**, *97*, 419–422. [CrossRef]
22. Birch, J.; Victorelli, S.; Rahmatika, D.; Anderson, R.K.; Jiwa, K.; Moisey, E.; Ward, C.; Fisher, A.J.; De Soyza, A.; Passos, J.F. Telomere Dysfunction and Senescence-associated Pathways in Bronchiectasis. *Am. J. Respir. Crit. Care Med.* **2016**, *193*, 929–932. [CrossRef] [PubMed]
23. d'Adda di Fagagna, F.; Reaper, P.M.; Clay-Farrace, L.; Fiegler, H.; Carr, P.; von Zglinicki, T.; Saretzki, G.; Carter, N.P.; Jackson, S.P. A DNA Damage Checkpoint Response in Telomere-Initiated Senescence. *Nature* **2003**, *426*, 194–198. [CrossRef] [PubMed]
24. Birch, J.; Anderson, R.K.; Correia-Melo, C.; Jurk, D.; Hewitt, G.; Marques, F.M.; Green, N.J.; Moisey, E.; Birrell, M.A.; Belvisi, M.G.; et al. DNA damage response at telomeres contributes to lung aging and chronic obstructive pulmonary disease. *Am. J. Physiol. Cell. Mol. Physiol.* **2015**, *309*, L1124–L1137. [CrossRef]
25. Córdoba-Lanús, E.; Cazorla-Rivero, S.; Espinoza-Jiménez, A.; De-Torres, J.P.; Pajares, M.J.; Aguirre-Jaime, A.; Celli, B.; Casanova, C. Telomere shortening and accelerated aging in COPD: Findings from the BODE cohort. *Respir. Res.* **2017**, *18*, 59. [CrossRef]
26. Passos, J.F.; Nelson, G.; Wang, C.; Richter, T.; Simillion, C.; Proctor, C.J.; Miwa, S.; Olijslagers, S.; Hallinan, J.; Wipat, A.; et al. Feedback between p21 and reactive oxygen production is necessary for cell senescence. *Mol. Syst. Biol.* **2010**, *6*, 347. [CrossRef]
27. Korolchuk, V.I.; Miwa, S.; Carroll, B.; von Zglinicki, T. Mitochondria in Cell Senescence: Is Mitophagy the Weakest Link? *EBioMedicine* **2017**, *21*, 7–13. [CrossRef]
28. Passos, J.F.; Saretzki, G.; Ahmed, S.; Nelson, G.; Richter, T.; Peters, H.; Wappler, I.; Birket, M.J.; Harold, G.; Schaeuble, K.; et al. Mitochondrial Dysfunction Accounts for the Stochastic Heterogeneity in Telomere-Dependent Senescence. *PLoS Biol.* **2007**, *5*, e110. [CrossRef]
29. van Deursen, J.M. The role of senescent cells in ageing. *Nature* **2014**, *509*, 439–446. [CrossRef]
30. Ovadya, Y.; Landsberger, T.; Leins, H.; Vadai, E.; Gal, H.; Biran, A.; Yosef, R.; Sagiv, A.; Agrawal, A.; Shapira, A.; et al. Impaired immune surveillance accelerates accumulation of senescent cells and aging. *Nat. Commun.* **2018**, *9*, 5435. [CrossRef]
31. Childs, B.G.; Durik, M.; Baker, D.J.; Van Deursen, J.M. Cellular senescence in aging and age-related disease: From mechanisms to therapy. *Nat. Med.* **2015**, *21*, 1424–1435. [CrossRef] [PubMed]
32. Herranz, N.; Gil, J. Mechanisms and functions of cellular senescence. *J. Clin. Investig.* **2018**, *128*, 1238–1246. [CrossRef] [PubMed]
33. Lasry, A.; Ben-Neriah, Y. Senescence-associated inflammatory responses: Aging and cancer perspectives. *Trends Immunol.* **2015**, *36*, 217–228. [CrossRef] [PubMed]
34. Martínez-Zamudio, R.I.; Robinson, L.; Roux, P.-F.; Bischof, O. SnapShot: Cellular Senescence Pathways. *Cell* **2017**, *170*, 816. [CrossRef] [PubMed]
35. Birch, J.; Barnes, P.J.; Passos, J.F. Mitochondria, telomeres and cell senescence: Implications for lung ageing and disease. *Pharmacol. Ther.* **2017**, *183*, 34–49. [CrossRef]
36. Herbig, U.; Jobling, W.A.; Chen, B.P.; Chen, D.J.; Sedivy, J.M. Telomere Shortening Triggers Senescence of Human Cells through a Pathway Involving ATM, p53, and p21CIP1, but Not p16INK4a. *Mol. Cell* **2004**, *14*, 501–513. [CrossRef]
37. Di Fagagna, F.D. Living on a break: Cellular senescence as a DNA-damage response. *Nat. Cancer* **2008**, *8*, 512–522. [CrossRef]
38. Panier, S.; Boulton, S.J. Double-strand break repair: 53BP1 comes into focus. *Nat. Rev. Mol. Cell Biol.* **2013**, *15*, 7–18. [CrossRef]
39. DiTullio, R.A., Jr.; Mochan, T.A.; Venere, M.; Bartkova, J.; Sehested, M.; Bartek, J.; Halazonetis, T.D. 53BP1 functions in an ATM-dependent checkpoint pathway that is constitutively activated in human cancer. *Nat. Cell Biol.* **2002**, *4*, 998–1002. [CrossRef]
40. Mochan, T.A.; Venere, M.; DiTullio, R.A.; Halazonetis, T.D. 53BP1, an activator of ATM in response to DNA damage. *DNA Repair* **2004**, *3*, 945–952. [CrossRef]
41. Childs, B.G.; Baker, D.J.; Kirkland, J.L.; Campisi, J.; Van Deursen, J.M. Senescence and apoptosis: Dueling or complementary cell fates? *EMBO Rep.* **2014**, *15*, 1139–1153. [CrossRef] [PubMed]
42. Silverman, J.; Takai, H.; Buonomo, S.B.; Eisenhaber, F.; de Lange, T. Human Rif1, ortholog of a yeast telomeric protein, is regulated by ATM and 53BP1 and functions in the S-phase checkpoint. *Genes Dev.* **2004**, *18*, 2108–2119. [CrossRef] [PubMed]
43. Fernandez-Capetillo, O.; Chen, H.-T.; Celeste, A.; Ward, I.; Romanienko, P.J.; Morales, J.; Naka, K.; Xia, Z.; Camerini-Otero, R.D.; Motoyama, N.; et al. DNA damage-induced G2–M checkpoint activation by histone H2AX and 53BP1. *Nat. Curell Biol.* **2002**, *4*, 993–997. [CrossRef] [PubMed]

44. Wang, B.; Matsuoka, S.; Carpenter, P.B.; Elledge, S.J. 53BP1, a Mediator of the DNA Damage Checkpoint. *Science* **2002**, *298*, 1435–1438. [CrossRef]
45. Ward, I.M.; Minn, K.; van Deursen, J.; Chen, J. p53 Binding Protein 53BP1 Is Required for DNA Damage Responses and Tumor Suppression in Mice. *Mol. Cell. Biol.* **2003**, *23*, 2556–2563. [CrossRef]
46. Romagosa, C.; Simonetti, S.; López-Vicente, L.; Mazo, A.; Lleonart, M.E.; Castellvi, J.; Cajal, S.R.Y. p16Ink4a overexpression in cancer: A tumor suppressor gene associated with senescence and high-grade tumors. *Oncogene* **2011**, *30*, 2087–2097. [CrossRef]
47. Sturmlechner, I.; Zhang, C.; Sine, C.C.; van Deursen, E.-J.; Jeganathan, K.B.; Hamada, N.; Grasic, J.; Friedman, D.; Stutchman, J.T.; Can, I.; et al. p21 produces a bioactive secretome that places stressed cells under immunosurveillance. *Science* **2021**, *374*. [CrossRef]
48. Gil, J.; Peters, G. Regulation of the INK4b–ARF–INK4a tumour suppressor locus: All for one or one for all. *Nat. Rev. Mol. Cell Biol.* **2006**, *7*, 667–677. [CrossRef]
49. Sharpless, N.E. INK4a/ARF: A multifunctional tumor suppressor locus. *Mutat. Res. Mol. Mech. Mutagen.* **2005**, *576*, 22–38. [CrossRef]
50. Debacq-Chainiaux, F.; Erusalimsky, J.D.; Campisi, J.; Toussaint, O. Protocols to detect senescence-associated beta-galactosidase (SA-βgal) activity, a biomarker of senescent cells in culture and in vivo. *Nat. Protoc.* **2009**, *4*, 1798–1806. [CrossRef]
51. Tilstra, J.S.; Robinson, A.R.; Wang, J.; Gregg, S.Q.; Clauson, C.L.; Reay, D.P.; Nasto, L.A.; Croix, C.M.S.; Usas, A.; Vo, N.; et al. NF-κB inhibition delays DNA damage–induced senescence and aging in mice. *J. Clin. Investig.* **2012**, *122*, 2601–2612. [CrossRef] [PubMed]
52. Hernandez-Segura, A.; de Jong, T.V.; Melov, S.; Guryev, V.; Campisi, J.; DeMaria, M. Unmasking Transcriptional Heterogeneity in Senescent Cells. *Curr. Biol.* **2017**, *27*, 2652–2660.e4. [CrossRef] [PubMed]
53. Casella, G.; Munk, R.; Kim, K.M.; Piao, Y.; De, S.; Abdelmohsen, K.; Gorospe, M. Transcriptome signature of cellular senescence. *Nucleic Acids Res.* **2019**, *47*, 7294–7305. [CrossRef] [PubMed]
54. Kurz, D.; Decary, S.; Hong, Y.; Erusalimsky, J. Senescence-associated (beta)-galactosidase reflects an increase in lysosomal mass during replicative ageing of human endothelial cells. *J. Cell Sci.* **2000**, *113*, 3613–3622. [CrossRef] [PubMed]
55. Hewitt, G.; von Zglinicki, T.; Passos, J.F. Cell Sorting of Young and Senescent Cells. *Methods Mol. Biol.* **2013**, *1048*, 31–47. [CrossRef]
56. Nicaise, A.M.; Wagstaff, L.J.; Willis, C.M.; Paisie, C.; Chandok, H.; Robson, P.; Fossati, V.; Williams, A.; Crocker, S.J. Cellular senescence in progenitor cells contributes to diminished remyelination potential in progressive multiple sclerosis. *Proc. Natl. Acad. Sci. USA* **2019**, *116*, 9030–9039. [CrossRef] [PubMed]
57. Nagano, T.; Nakano, M.; Nakashima, A.; Onishi, K.; Yamao, S.; Enari, M.; Kikkawa, U.; Kamada, S. Identification of cellular senescence-specific genes by comparative transcriptomics. *Sci. Rep.* **2016**, *6*, 31758. [CrossRef]
58. Zhang, L.; Dong, X.; Lee, M.; Maslov, A.Y.; Wang, T.; Vijg, J. Single-cell whole-genome sequencing reveals the functional landscape of somatic mutations in B lymphocytes across the human lifespan. *Proc. Natl. Acad. Sci. USA* **2019**, *116*, 9014–9019. [CrossRef]
59. Frescas, D.; Roux, C.M.; Aygun-Sunar, S.; Gleiberman, A.S.; Krasnov, P.; Kurnasov, O.V.; Strom, E.; Virtuoso, L.P.; Wrobel, M.; Osterman, A.L.; et al. Senescent cells expose and secrete an oxidized form of membrane-bound vimentin as revealed by a natural polyreactive antibody. *Proc. Natl. Acad. Sci. USA* **2017**, *114*, E1668–E1677. [CrossRef]
60. Schafer, M.J.; White, T.A.; Iijima, K.; Haak, A.J.; Ligresti, G.; Atkinson, E.J.; Oberg, A.L.; Birch, J.; Salmonowicz, H.; Zhu, Y.; et al. Cellular senescence mediates fibrotic pulmonary disease. *Nat. Commun.* **2017**, *8*, 14532. [CrossRef]
61. Yao, C.; Guan, X.; Carraro, G.; Parimon, T.; Liu, X.; Huang, G.; Mulay, A.; Soukiasian, H.J.; David, G.; Weigt, S.S.; et al. Senescence of Alveolar Type 2 Cells Drives Progressive Pulmonary Fibrosis. *Am. J. Respir. Crit. Care Med.* **2021**, *203*, 707–717. [CrossRef] [PubMed]
62. Justice, J.N.; Nambiar, A.M.; Tchkonia, T.; Lebrasseur, N.K.; Pascual, R.; Hashmi, S.K.; Prata, L.L.; Masternak, M.M.; Kritchevsky, S.B.; Musi, N.; et al. Senolytics in idiopathic pulmonary fibrosis: Results from a first-in-human, open-label, pilot study. *eBioMedicine* **2019**, *40*, 554–563. [CrossRef] [PubMed]
63. Cottage, C.T.; Peterson, N.; Kearley, J.; Berlin, A.; Xiong, X.; Huntley, A.; Zhao, W.; Brown, C.; Migneault, A.; Zerrouki, K.; et al. Targeting p16-induced senescence prevents cigarette smoke-induced emphysema by promoting IGF1/Akt1 signaling in mice. *Commun. Biol.* **2019**, *2*, 307. [CrossRef] [PubMed]
64. Rashid, K.; Sundar, I.K.; Gerloff, J.; Li, D.; Rahman, I. Lung cellular senescence is independent of aging in a mouse model of COPD/emphysema. *Sci. Rep.* **2018**, *8*, 1–14. [CrossRef] [PubMed]
65. Tsuji, T.; Aoshiba, K.; Nagai, A. Alveolar Cell Senescence in Patients with Pulmonary Emphysema. *Am. J. Respir. Crit. Care Med.* **2006**, *174*, 886–893. [CrossRef]
66. Amsellem, V.; Gary-Bobo, G.; Marcos, E.; Maitre, B.; Chaar, V.; Validire, P.; Stern, J.-B.; Noureddine, H.; Sapin, E.; Rideau, D.; et al. Telomere Dysfunction Causes Sustained Inflammation in Chronic Obstructive Pulmonary Disease. *Am. J. Respir. Crit. Care Med.* **2011**, *184*, 1358–1366. [CrossRef]
67. Rutten, E.P.; Gopal, P.; Wouters, E.F.; Franssen, F.M.; Hageman, G.J.; Vanfleteren, L.E.; Spruit, M.A.; Reynaert, N. Various Mechanistic Pathways Representing the Aging Process Are Altered in COPD. *Chest* **2016**, *149*, 53–61. [CrossRef]
68. Kirkland, J.L.; Tchkonia, T. Senolytic drugs: From discovery to translation. *J. Intern. Med.* **2020**, *288*, 518–536. [CrossRef]
69. Pignolo, R.J.; Passos, J.F.; Khosla, S.; Tchkonia, T.; Kirkland, J.L. Reducing Senescent Cell Burden in Aging and Disease. *Trends Mol. Med.* **2020**, *26*, 630–638. [CrossRef]

70. John-Schuster, G.; Günter, S.; Hager, K.; Conlon, T.M.; Eickelberg, O.; Yildirim, A. Inflammaging increases susceptibility to cigarette smoke-induced COPD. *Oncotarget* **2015**, *7*, 30068–30083. [CrossRef]
71. Kujoth, G.C.; Hiona, A.; Pugh, T.D.; Someya, S.; Panzer, K.; Wohlgemuth, S.E.; Hofer, T.; Seo, A.Y.; Sullivan, R.; Jobling, W.A.; et al. Mitochondrial DNA Mutations, Oxidative Stress, and Apoptosis in Mammalian Aging. *Science* **2005**, *309*, 481–484. [CrossRef] [PubMed]
72. Mizumura, K.; Cloonan, S.; Nakahira, K.; Bhashyam, A.R.; Cervo, M.; Kitada, T.; Glass, K.; Owen, C.A.; Mahmood, A.; Washko, G.R.; et al. Mitophagy-dependent necroptosis contributes to the pathogenesis of COPD. *J. Clin. Investig.* **2014**, *124*, 3987–4003. [CrossRef] [PubMed]
73. Paschalaki, K.; Rossios, C.; Pericleous, C.; MacLeod, M.; Rothery, S.; Donaldson, G.C.; Wedzicha, J.A.; Gorgoulis, V.; Randi, A.M.; Barnes, P.J. Inhaled corticosteroids reduce senescence in endothelial progenitor cells from patients with COPD. *Thorax* **2022**, *77*, 616–620. [CrossRef] [PubMed]
74. Zhang, M.; Zhang, Y.; Roth, M.; Zhang, L.; Shi, R.; Yang, X.; Li, Y.; Zhang, J. Sirtuin 3 Inhibits Airway Epithelial Mitochondrial Oxidative Stress in Cigarette Smoke-Induced COPD. *Oxidative Med. Cell. Longev.* **2020**, *2020*, 7582980. [CrossRef]
75. Okuda, R.; Aoshiba, K.; Matsushima, H.; Ogura, T.; Okudela, K.; Ohashi, K. Cellular senescence and senescence-associated secretory phenotype: Comparison of idiopathic pulmonary fibrosis, connective tissue disease-associated interstitial lung disease, and chronic obstructive pulmonary disease. *J. Thorac. Dis.* **2019**, *11*, 857–864. [CrossRef]
76. Woldhuis, R.R.; Heijink, I.H.; van den Berge, M.; Timens, W.; Oliver, B.G.G.; de Vries, M.; Brandsma, C.-A. COPD-derived fibroblasts secrete higher levels of senescence-associated secretory phenotype proteins. *Thorax* **2021**, *76*, 508–511. [CrossRef]
77. Shammas, M.A. Telomeres, lifestyle, cancer, and aging. *Curr. Opin. Clin. Nutr. Metab. Care* **2011**, *14*, 28–34. [CrossRef]
78. Finkel, T.; Hwang, P.M. The Krebs cycle meets the cell cycle: Mitochondria and the G_1–S transition. *Proc. Natl. Acad. Sci. USA* **2009**, *106*, 11825–11826. [CrossRef]
79. Weir, H.J.; Yao, P.; Huynh, F.K.; Escoubas, C.C.; Goncalves, R.L.; Burkewitz, K.; Laboy, R.; Hirschey, M.D.; Mair, W.B. Dietary Restriction and AMPK Increase Lifespan via Mitochondrial Network and Peroxisome Remodeling. *Cell Metab.* **2017**, *26*, 884–896.e5. [CrossRef]
80. Conti, V.; Corbi, G.; Manzo, V.; Pelaia, G.; Filippelli, A.; Vatrella, A. Sirtuin 1 and Aging Theory for Chronic Obstructive Pulmonary Disease. *Anal. Cell. Pathol.* **2015**, *2015*, 897327. [CrossRef]
81. Johnson, S.; Rabinovitch, P.S.; Kaeberlein, M. mTOR is a key modulator of ageing and age-related disease. *Nature* **2013**, *493*, 338–345. [CrossRef] [PubMed]
82. Mitani, A.; Ito, K.; Vuppusetty, C.; Barnes, P.J.; Mercado, N. Restoration of Corticosteroid Sensitivity in Chronic Obstructive Pulmonary Disease by Inhibition of Mammalian Target of Rapamycin. *Am. J. Respir. Crit. Care Med.* **2016**, *193*, 143–153. [CrossRef] [PubMed]
83. Freund, A.; Patil, C.K.; Campisi, J. p38MAPK is a novel DNA damage response-independent regulator of the senescence-associated secretory phenotype. *EMBO J.* **2011**, *30*, 1536–1548. [CrossRef] [PubMed]
84. Garcia, D.; Shaw, R.J. AMPK: Mechanisms of Cellular Energy Sensing and Restoration of Metabolic Balance. *Mol. Cell* **2017**, *66*, 789–800. [CrossRef] [PubMed]
85. Herzig, S.; Shaw, R.J. AMPK: Guardian of metabolism and mitochondrial homeostasis. *Nat. Rev. Mol. Cell Biol.* **2018**, *19*, 121–135. [CrossRef] [PubMed]
86. O'Neill, H.M.; Maarbjerg, S.J.; Crane, J.D.; Jeppesen, J.; Jørgensen, S.B.; Schertzer, J.D.; Shyroka, O.; Kiens, B.; van Denderen, B.J.; Tarnopolsky, M.A.; et al. AMP-activated protein kinase (AMPK) β1β2 muscle null mice reveal an essential role for AMPK in maintaining mitochondrial content and glucose uptake during exercise. *Proc. Natl. Acad. Sci. USA* **2011**, *108*, 16092–16097. [CrossRef] [PubMed]
87. Toyama, E.Q.; Herzig, S.; Courchet, J.; Lewis, T.L., Jr.; Losón, O.C.; Hellberg, K.; Young, N.P.; Chen, H.; Polleux, F.; Chan, D.C.; et al. Metabolism. AMP-activated protein kinase mediates mitochondrial fission in response to energy stress. *Science* **2016**, *351*, 275–281. [CrossRef]
88. Laker, R.C.; Drake, J.C.; Wilson, R.J.; Lira, V.A.; Lewellen, B.M.; Ryall, K.A.; Fisher, C.C.; Zhang, M.; Saucerman, J.J.; Goodyear, L.J.; et al. Ampk phosphorylation of Ulk1 is required for targeting of mitochondria to lysosomes in exercise-induced mitophagy. *Nat. Commun.* **2017**, *8*, 548. [CrossRef]
89. Yamakuchi, M.; Ferlito, M.; Lowenstein, C.J. miR-34a repression of SIRT1 regulates apoptosis. *Proc. Natl. Acad. Sci. USA* **2008**, *105*, 13421–13426. [CrossRef]
90. Ito, Y.; Inoue, A.; Seers, T.; Hato, Y.; Igarashi, A.; Toyama, T.; Taganov, K.D.; Boldin, M.P.; Asahara, H. Identification of targets of tumor suppressor microRNA-34a using a reporter library system. *Proc. Natl. Acad. Sci. USA* **2017**, *114*, 3927–3932. [CrossRef]
91. Kauppinen, A.; Suuronen, T.; Ojala, J.; Kaarniranta, K.; Salminen, A. Antagonistic crosstalk between NF-κB and SIRT1 in the regulation of inflammation and metabolic disorders. *Cell. Signal.* **2013**, *25*, 1939–1948. [CrossRef] [PubMed]
92. Renda, T.; Baraldo, S.; Pelaia, G.; Bazzan, E.; Turato, G.; Papi, A.; Maestrelli, P.; Maselli, R.; Vatrella, A.; Fabbri, L.M.; et al. Increased activation of p38 MAPK in COPD. *Eur. Respir. J.* **2008**, *31*, 62–69. [CrossRef] [PubMed]
93. Gaffey, K.; Reynolds, S.; Plumb, J.; Kaur, M.; Singh, D. Increased phosphorylated p38 mitogen-activated protein kinase in COPD lungs. *Eur. Respir. J.* **2012**, *42*, 28–41. [CrossRef]

94. Vallese, D.; Ricciardolo, F.L.; Gnemmi, I.; Casolari, P.; Brun, P.; Sorbello, V.; Capelli, A.; Cappello, F.; Cavallesco, G.N.; Papi, A.; et al. Phospho-p38 MAPK Expression in COPD Patients and Asthmatics and in Challenged Bronchial Epithelium. *Respiration* **2015**, *89*, 329–342. [CrossRef] [PubMed]
95. Lea, S.; Li, J.; Plumb, J.; Gaffey, K.; Mason, S.; Gaskell, R.; Harbron, C.; Singh, D. P38 MAPK and glucocorticoid receptor crosstalk in bronchial epithelial cells. *Klin. Wochenschr.* **2020**, *98*, 361–374. [CrossRef]
96. Sgalla, G.; Biffi, A.; Richeldi, L. Idiopathic pulmonary fibrosis: Diagnosis, epidemiology and natural history. *Respirology* **2015**, *21*, 427–437. [CrossRef]
97. Martinez, F.J.; Collard, H.R.; Pardo, A.; Raghu, G.; Richeldi, L.; Selman, M.; Swigris, J.J.; Taniguchi, H.; Wells, A.U. Idiopathic pulmonary fibrosis. *Nat. Rev. Dis. Prim.* **2017**, *3*, 17074. [CrossRef]
98. Sosulski, M.L.; Gongora, R.; Feghali-Bostwick, C.; Lasky, J.A.; Sanchez, C.G. Sirtuin 3 Deregulation Promotes Pulmonary Fibrosis. *Journals Gerontol. Ser. A* **2016**, *72*, 595–602. [CrossRef]
99. Pearce, S.F.; Rebelo-Guiomar, P.; D'Souza, A.R.; Powell, C.A.; Van Haute, L.; Minczuk, M. Regulation of Mammalian Mitochondrial Gene Expression: Recent Advances. *Trends Biochem. Sci.* **2017**, *42*, 625–639. [CrossRef]
100. Fernie, A.R.; Carrari, F.; Sweetlove, L.J. Respiratory metabolism: Glycolysis, the TCA cycle and mitochondrial electron transport. *Curr. Opin. Plant Biol.* **2004**, *7*, 254–261. [CrossRef]
101. Martínez-Reyes, I.; Chandel, N.S. Mitochondrial TCA cycle metabolites control physiology and disease. *Nat. Commun.* **2020**, *11*, 102. [CrossRef] [PubMed]
102. Bueno, M.; Lai, Y.-C.; Romero, Y.; Brands, J.; Croix, C.M.S.; Kamga, C.; Corey, C.; Herazo-Maya, J.D.; Sembrat, J.; Lee, J.; et al. PINK1 deficiency impairs mitochondrial homeostasis and promotes lung fibrosis. *J. Clin. Investig.* **2014**, *125*, 521–538. [CrossRef] [PubMed]
103. Kwong, F.N.K.; Nicholson, A.G.; Harrison, C.L.; Hansbro, P.M.; Adcock, I.M.; Chung, K.F. Is mitochondrial dysfunction a driving mechanism linking COPD to nonsmall cell lung carcinoma? *Eur. Respir. Rev.* **2017**, *26*, 170040. [CrossRef]
104. Zheng, S.; Wang, C.; Qian, G.; Wu, G.; Guo, R.; Li, Q.; Chen, Y.; Li, J.; Li, H.; He, B.; et al. Role of mtDNA haplogroups in COPD susceptibility in a southwestern Han Chinese population. *Free Radic. Biol. Med.* **2012**, *53*, 473–481. [CrossRef]
105. Calvo, S.E.; Clauser, K.; Mootha, V.K. MitoCarta2.0: An updated inventory of mammalian mitochondrial proteins. *Nucleic Acids Res.* **2015**, *44*, D1251–D1257. [CrossRef]
106. Xu, W.; Chen, R.; Hu, B.; Zein, J.G.; Liu, C.; Comhair, S.A.A.; Aldred, M.A.; Hawkins, G.A.; Meyers, D.A.; Bleecker, E.R.; et al. Mitochondrial DNA Variation and Severe Asthma. American Thoracic Society International Conference Abstracts B33. ASTHMA: MECHANISMS OF DISEASE II 2019. *Am. J. Respir. Crit. Care Med.* **2019**, *199*, 1–2.
107. Wiley, C.D.; Velarde, M.C.; Lecot, P.; Liu, S.; Sarnoski, E.A.; Freund, A.; Shirakawa, K.; Lim, H.W.; Davis, S.S.; Ramanathan, A.; et al. Mitochondrial Dysfunction Induces Senescence with a Distinct Secretory Phenotype. *Cell Metab.* **2015**, *23*, 303–314. [CrossRef]
108. Birch, J.; Passos, J.F. Targeting the SASP to combat ageing: Mitochondria as possible intracellular allies? *BioEssays* **2017**, *39*, 1600235. [CrossRef]
109. Miyamoto, S. Nuclear initiated NF-κB signaling: NEMO and ATM take center stage. *Cell Res.* **2010**, *21*, 116–130. [CrossRef]
110. Holt, J.I.; Harding, A.E.; Morgan-Hughes, J.A. Deletions of muscle mitochondrial DNA in patients with mitochondrial myopathies. *Nature* **1988**, *331*, 717–719. [CrossRef]
111. E Giles, R.; Blanc, H.; Cann, H.M.; Wallace, D.C. Maternal inheritance of human mitochondrial DNA. *Proc. Natl. Acad. Sci. USA* **1980**, *77*, 6715–6719. [CrossRef] [PubMed]
112. Imai, S.-I.; Guarente, L. It takes two to tango: NAD+ and sirtuins in aging/longevity control. *NPJ Aging Mech. Dis.* **2016**, *2*, 16017. [CrossRef] [PubMed]
113. Hoffmann, R.F.; Zarrintan, S.; Brandenburg, S.M.; Kol, A.; De Bruin, H.G.; Jafari, S.; Dijk, F.; Kalicharan, D.; Kelders, M.; Gosker, H.R.; et al. Prolonged cigarette smoke exposure alters mitochondrial structure and function in airway epithelial cells. *Respir. Res.* **2013**, *14*, 97. [CrossRef] [PubMed]
114. Schriner, S.E.; Linford, N.J.; Martin, G.M.; Treuting, P.; Ogburn, C.E.; Emond, M.; Coskun, P.E.; Ladiges, W.; Wolf, N.; Van Remmen, H.; et al. Extension of Murine Life Span by Overexpression of Catalase Targeted to Mitochondria. *Science* **2005**, *308*, 1909–1911. [CrossRef]
115. Soulitzis, N.; Neofytou, E.; Psarrou, M.; Anagnostis, A.; Tavernarakis, N.; Siafakas, N.; Tzortzaki, E.G. Downregulation of lung mitochondrial prohibitin in COPD. *Respir. Med.* **2012**, *106*, 954–961. [CrossRef]
116. Sureshbabu, A.; Ebhandari, V. Targeting mitochondrial dysfunction in lung diseases: Emphasis on mitophagy. *Front. Physiol.* **2013**, *4*, 384. [CrossRef]
117. Stewart, R.; Weir, E.; Montgomery, M.; Niewoehner, D. Hydrogen peroxide contracts airway smooth muscle: A possible endogenous mechanism. *Respir. Physiol.* **1981**, *45*, 333–342. [CrossRef]
118. Ballinger, S.W.; Patterson, C.; Yan, C.-N.; Doan, R.; Burow, D.L.; Young, C.G.; Yakes, F.M.; Van Houten, B.; Ballinger, C.A.; Freeman, B.A.; et al. Hydrogen Peroxide– and Peroxynitrite-Induced Mitochondrial DNA Damage and Dysfunction in Vascular Endothelial and Smooth Muscle Cells. *Circ. Res.* **2000**, *86*, 960–966. [CrossRef]
119. Harman, D. Aging: A Theory Based on Free Radical and Radiation Chemistry. *J. Gerontol.* **1956**, *11*, 298–300. [CrossRef]

120. Aravamudan, B.; Kiel, A.; Freeman, M.; Delmotte, P.; Thompson, M.; Vassallo, R.; Sieck, G.C.; Pabelick, C.M.; Prakash, Y.S. Cigarette smoke-induced mitochondrial fragmentation and dysfunction in human airway smooth muscle. *Am. J. Physiol.-Lung Cell. Mol. Physiol.* **2014**, *306*, L840–L854. [CrossRef]
121. Mishra, P.; Chan, D.C. Mitochondrial dynamics and inheritance during cell division, development and disease. *Nat. Rev. Mol. Cell Biol.* **2014**, *15*, 634–646. [CrossRef] [PubMed]
122. Russell, O.M.; Gorman, G.S.; Lightowlers, R.N.; Turnbull, D.M. Mitochondrial Diseases: Hope for the Future. *Cell* **2020**, *181*, 168–188. [CrossRef] [PubMed]
123. Bonora, M.; Pinton, P. Mitochondrial DNA keeps you young. *Cell Death Dis.* **2018**, *9*, 992. [CrossRef]
124. Zhang, X.; Sun, Y.; Dong, X.; Zhou, J.; Sun, F.; Han, T.; Lei, P.; Mao, R.; Guo, X.; Wang, Q.; et al. Mitochondrial DNA and genomic DNA ratio in embryo culture medium is not a reliable predictor for in vitro fertilization outcome. *Sci. Rep.* **2019**, *9*, 5378. [CrossRef]
125. Loeb, L.A.; Wallace, D.C.; Martin, G.M. The mitochondrial theory of aging and its relationship to reactive oxygen species damage and somatic mtDNA mutations. *Proc. Natl. Acad. Sci. USA* **2005**, *102*, 18769–18770. [CrossRef]
126. Krishnamurthy, J.; Torrice, C.; Ramsey, M.R.; Kovalev, G.I.; Al-Regaiey, K.; Su, L.; Sharpless, N.E. Ink4a/Arf expression is a biomarker of aging. *J. Clin. Investig.* **2004**, *114*, 1299–1307. [CrossRef]
127. Short, S.; Fielder, E.; Miwa, S.; von Zglinicki, T. Senolytics and senostatics as adjuvant tumour therapy. *eBioMedicine* **2019**, *41*, 683–692. [CrossRef]
128. Wallace, D.C. Mitochondrial genetic medicine. *Nat. Genet.* **2018**, *50*, 1642–1649. [CrossRef]
129. Singh, B.; Schoeb, T.R.; Bajpai, P.; Slominski, A.; Singh, K.K. Reversing wrinkled skin and hair loss in mice by restoring mitochondrial function. *Cell Death Dis.* **2018**, *9*, 735. [CrossRef]
130. Trifunovic, A.; Wredenberg, A.; Falkenberg, M.; Spelbrink, J.; Rovio, A.T.; Bruder, C.E.; Bohlooly-Y, M.; Gidlöf, S.; Oldfors, A.; Wibom, R.; et al. Premature ageing in mice expressing defective mitochondrial DNA polymerase. *Nature* **2004**, *429*, 417–423. [CrossRef]
131. Trifunovic, A.; Hansson, A.; Wredenberg, A.; Rovio, A.T.; Dufour, E.; Khvorostov, I.; Spelbrink, J.N.; Wibom, R.; Jacobs, H.T.; Larsson, N.-G. Somatic mtDNA mutations cause aging phenotypes without affecting reactive oxygen species production. *Proc. Natl. Acad. Sci. USA* **2005**, *102*, 17993–17998. [CrossRef] [PubMed]
132. Zhu, Y.I.; Tchkonia, T.; Pirtskhalava, T.; Gower, A.C.; Ding, H.; Giorgadze, N.; Palmer, A.K.; Ikeno, Y.; Hubbard, G.B.; Lenburg, M.; et al. The Achilles' heel of senescent cells: From transcriptome to senolytic drugs. *Aging Cell* **2015**, *14*, 644–658. [CrossRef] [PubMed]
133. Campa, C.C.; Silva, R.; Margaria, J.P.; Pirali, T.; Mattos, M.; Kraemer, L.R.; Reis, D.C.; Grosa, G.; Copperi, F.; Dalmarco, E.M.; et al. Inhalation of the prodrug PI3K inhibitor CL27c improves lung function in asthma and fibrosis. *Nat. Commun.* **2018**, *9*, 1–16. [CrossRef]
134. Nelson, G.; Wordsworth, J.; Wang, C.; Jurk, D.; Lawless, C.; Martin-Ruiz, C.; von Zglinicki, T. A senescent cell bystander effect: Senescence-induced senescence. *Aging Cell* **2012**, *11*, 345–349. [CrossRef] [PubMed]
135. Correia-Melo, C.; Marques, F.D.M.; Anderson, R.; Hewitt, G.; Hewitt, R.; Cole, J.; Carroll, B.M.; Miwa, S.; Birch, J.; Merz, A.; et al. Mitochondria are required for pro-ageing features of the senescent phenotype. *EMBO J.* **2016**, *35*, 724–742. [CrossRef]
136. Coskun, P.E.; Beal, M.F.; Wallace, D.C. Alzheimer's brains harbor somatic mtDNA control-region mutations that suppress mitochondrial transcription and replication. *Proc. Natl. Acad. Sci. USA* **2004**, *101*, 10726–10731. [CrossRef]
137. Coskun, P.; Wyrembak, J.; Schriner, S.E.; Chen, H.-W.; Marciniack, C.; LaFerla, F.; Wallace, D.C. A mitochondrial etiology of Alzheimer and Parkinson disease. *Biochim. Biophys. Acta (BBA) Gen. Subj.* **2012**, *1820*, 553–564. [CrossRef]
138. Corral-Debrinski, M.; Stepien, G.; Shoffner, J.M.; Lott, M.T.; Kanter, K.; Wallace, D.C. Hypoxemia Is Associated with Mitochondrial DNA Damage and Gene Induction. *JAMA* **1991**, *266*, 1812–1816. [CrossRef]
139. Zifa, E.; Daniil, Z.; Skoumi, E.; Stavrou, M.; Papadimitriou, K.; Terzenidou, M.; Kostikas, K.; Bagiatis, V.; Gourgoulianis, K.I.; Mamuris, Z. Mitochondrial genetic background plays a role in increasing risk to asthma. *Mol. Biol. Rep.* **2011**, *39*, 4697–4708. [CrossRef]
140. Velarde, M.C.; Flynn, J.; Day, N.U.; Melov, S.; Campisi, J. Mitochondrial oxidative stress caused by Sod2 deficiency promotes cellular senescence and aging phenotypes in the skin. *Aging* **2012**, *4*, 3–12. [CrossRef]
141. Carpagnano, G.E.; Lacedonia, D.; Malerba, M.; Palmiotti, G.A.; Cotugno, G.; Carone, M.; Foschino-Barbaro, M.P. Analysis of mitochondrial DNA alteration in new phenotype ACOS. *BMC Pulm. Med.* **2016**, *16*, 31. [CrossRef] [PubMed]
142. Carpagnano, G.E.; Lacedonia, D.; Carone, M.; Soccio, P.; Cotugno, G.; Palmiotti, G.A.; Scioscia, G.; Barbaro, M.P.F. Study of mitochondrial DNA alteration in the exhaled breath condensate of patients affected by obstructive lung diseases. *J. Breath Res.* **2016**, *10*, 26005. [CrossRef] [PubMed]
143. Xu, M.; Pirtskhalava, T.; Farr, J.N.; Weigand, B.M.; Palmer, A.K.; Weivoda, M.M.; Inman, C.L.; Ogrodnik, M.B.; Hachfeld, C.M.; Fraser, D.G.; et al. Senolytics improve physical function and increase lifespan in old age. *Nat. Med.* **2018**, *24*, 1246–1256. [CrossRef] [PubMed]
144. Camell, C.D.; Yousefzadeh, M.J.; Zhu, Y.; Prata, L.G.P.L.; Huggins, M.A.; Pierson, M.; Zhang, L.; O'Kelly, R.D.; Pirtskhalava, T.; Xun, P.; et al. Senolytics reduce coronavirus-related mortality in old mice. *Science* **2021**, *373*, eabe4832. [CrossRef] [PubMed]

145. Hubackova, S.; Davidova, E.; Rohlenova, K.; Stursa, J.; Werner, L.; Andera, L.; Dong, L.; Terp, M.; Hodny, Z.; Ditzel, H.J.; et al Selective elimination of senescent cells by mitochondrial targeting is regulated by ANT2. *Cell Death Differ.* **2018**, *26*, 276–290 [CrossRef] [PubMed]
146. Wang, Z.; Li, R.; Zhong, R. Extracellular matrix promotes proliferation, migration and adhesion of airway smooth muscle cells in a rat model of chronic obstructive pulmonary disease via upregulation of the PI3K/AKT signaling pathway. *Mol. Med. Rep.* **2018**, *18*, 3143–3152. [CrossRef] [PubMed]
147. Nelson, G.; Kucheryavenko, O.; Wordsworth, J.; Von Zglinicki, T. The senescent bystander effect is caused by ROS-activated NF-κB signalling. *Mech. Ageing Dev.* **2018**, *170*, 30–36. [CrossRef]
148. Pezze, P.D.; Nelson, G.; Otten, E.G.; Korolchuk, V.I.; Kirkwood, T.B.L.; Von Zglinicki, T.; Shanley, D.P. Dynamic Modelling of Pathways to Cellular Senescence Reveals Strategies for Targeted Interventions. *PLoS Comput. Biol.* **2014**, *10*, e1003728. [CrossRef]
149. Kim, J.; Yang, G.; Kim, Y.; Kim, J.; Ha, J. AMPK activators: Mechanisms of action and physiological activities. *Exp. Mol. Med.* **2016**, *48*, e224. [CrossRef]
150. Owen, M.R.; Doran, E.; Halestrap, A.P. Evidence that metformin exerts its anti-diabetic effects through inhibition of complex 1 of the mitochondrial respiratory chain. *Biochem. J.* **2000**, *348 Pt 3*, 607–614. [CrossRef]
151. Mimaki, M.; Wang, X.; McKenzie, M.; Thorburn, D.R.; Ryan, M.T. Understanding mitochondrial complex I assembly in health and disease. *Biochim. Biophys. Acta (BBA)-Bioenerg.* **2012**, *1817*, 851–862. [CrossRef] [PubMed]
152. Chenggui, W.; Xiaolei, Z.; Pan, Z.; Xu, D.; Zhou, Y.; Wu, Y.; Cai, N.; Tang, Q.; Wang, C.; Yan, M.; et al. Metformin protects against apoptosis and senescence in nucleus pulposus cells and ameliorates disc degeneration in vivo. *Cell Death Dis.* **2016**, *7*, e2441. [CrossRef]

cells

Review

Effects of Air Pollution on Cellular Senescence and Skin Aging

Ines Martic [1,2], Pidder Jansen-Dürr [1,2] and Maria Cavinato [1,2,*]

1 Institute for Biomedical Aging Research, Universität Innsbruck, 6020 Innsbruck, Austria; ines.martic@student.uibk.ac.at (I.M.); pidder.jansen-duerr@uibk.ac.at (P.J.-D.)
2 Center for Molecular Biosciences Innsbruck (CMBI), Innrain 58, 6020 Innsbruck, Austria
* Correspondence: maria.cavinato-nascimento@uibk.ac.at

Abstract: The human skin is exposed daily to different environmental factors such as air pollutants and ultraviolet (UV) light. Air pollution is considered a harmful environmental risk to human skin and is known to promote aging and inflammation of this tissue, leading to the onset of skin disorders and to the appearance of wrinkles and pigmentation issues. Besides this, components of air pollution can interact synergistically with ultraviolet light and increase the impact of damage to the skin. However, little is known about the modulation of air pollution on cellular senescence in skin cells and how this can contribute to skin aging. In this review, we are summarizing the current state of knowledge about air pollution components, their involvement in the processes of cellular senescence and skin aging, as well as the current therapeutic and cosmetic interventions proposed to prevent or mitigate the effects of air pollution in the skin.

Keywords: senescence; air pollution; skin aging; cosmetics

Citation: Martic, I.; Jansen-Dürr, P.; Cavinato, M. Effects of Air Pollution on Cellular Senescence and Skin Aging. *Cells* **2022**, *11*, 2220. https://doi.org/10.3390/cells11142220

Academic Editors: Nicole Wagner and Kay-Dietrich Wagner

Received: 1 May 2022
Accepted: 11 July 2022
Published: 17 July 2022

1. Introduction

The skin is the largest and outermost organ of the human body. As such, the skin represents the major protective barrier between the internal and external environment and protects the body from environmental damages. Additionally, the skin is important for the regulation of body temperature and water loss and participates in certain immune responses [1].

Skin aging is a result of two cumulative and overlaying mechanisms denominated as intrinsic and extrinsic aging. The process of intrinsic or chronological aging affects all tissues and organs of the body, is due to the passage of time, and is influenced by genetic background. The main signs of intrinsic aging in the skin are relatively mild, including the accumulation of fine wrinkles with moderate changes in skin pigmentation [1]. In addition, the skin is continuously exposed to environmental and lifestyle factors such as sunlight, pollution, cigarette smoke, and dietary habits. These factors, collectively denominated the skin exposome, are the major causes of the process of extrinsic skin aging. Major characteristics of extrinsic skin aging are coarse wrinkles, solar elastosis, and pigmentation irregularities [1,2]. Quality of life as well as emotional well-being are influenced by changes in the skin appearance [1,3].

From all factors of the exposome, sunlight is known to be the most harmful and for this purpose extrinsic skin aging is also referred as photoaging. Arguably, chronic exposition of the skin to sunlight leads to skin aging, which is characterized by increased oxidative stress, apoptosis, stimulation of melanogenesis and direct damage to DNA, membranes, and proteins [1,3,4].

A potentially serious, yet less recognized environmental factor is ambient (outdoor) air pollution. As a result of rapid industrialization and urbanization, environmental pollution is becoming a severe public health issue worldwide [5]. In 2019 the World Health Organization (WHO) determined that 99% of the world's population lives in places where air pollution levels exceed WHO limits and thus air pollution was designated as the world's largest single environmental health risk to humans [6]. Pollution has been strongly correlated to the degenerative processes of skin aging, particularly pigmentation issues, as well as to the onset of skin disorders [7,8].

Different compounds can be categorized as air pollutants, and as such each contribute differently to the impairment of the skin [9]. These compounds differ in chemical composition, reaction properties, ability to diffuse, and time of disintegration [10]. Given that the interest in the effects caused by exposure to air pollution has risen in the last few years, the number of publications about this topic is constantly increasing. Recent studies show that air pollutants such as particulate matter (PM) can cause skin tanning, skin aging [11,12], and inflammatory skin diseases such as atopic dermatitis (AD), acne, and allergic reactions [13]. Further investigations are needed to fully understand how skin is affected by the exposure to environmental stressors. Given that the skin is not exposed to one single source of environmental impact, it is especially important to examine how combined effects (e.g., UV light, air pollution, cigarette smoke) of different stressors damage the skin and contribute to aging of this tissue.

In this review we summarize the different components of air pollution and how they contribute to skin aging and cellular senescence. Furthermore, we will demonstrate how accumulation and persistence of senescent cells impact the process of skin aging. Additionally, we outline current interventions and cosmetic compounds used to prevent and mitigate effects of air pollution in the skin (Figure 1).

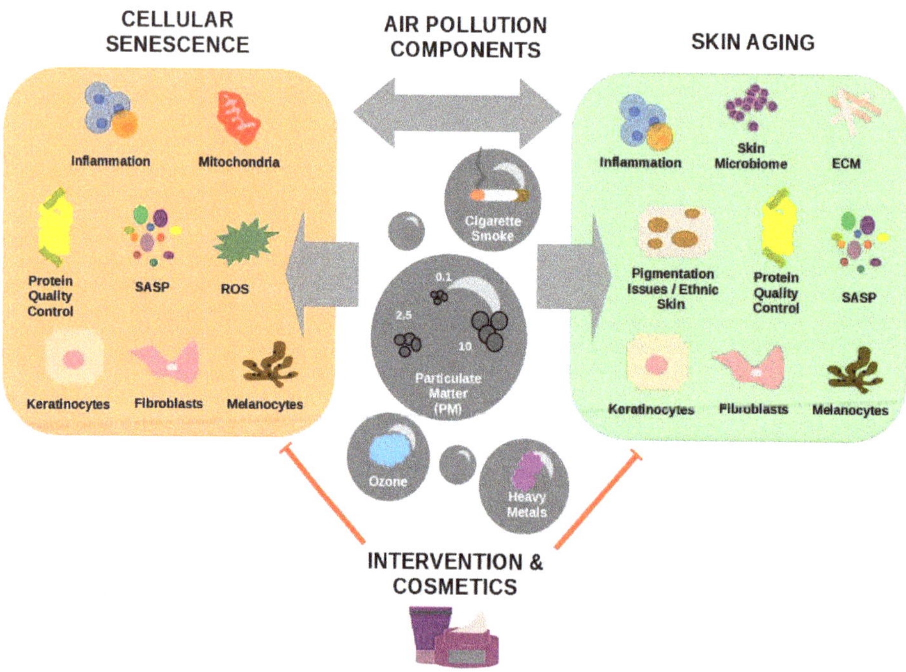

Figure 1. The interplay between air pollution, cellular senescence, and skin aging. The review focuses on the topics inflammation, protein quality control, mitochondrial dysfunction, reactive oxygen species (ROS), and senescence-associated secretory phenotype (SASP) in cutaneous skin cells during senescence induced by air pollution. Additionally, we discuss how air pollution components such as particulate matter (PM), ozone, heavy metals, and cigarette smoke impact skin aging. Finally, we summarize the latest information about interventions and cosmetics which were shown to prevent the impact of air pollution in the processes of cellular senescence and skin aging.

2. Cellular Senescence

Cellular senescence, which is considered one hallmark of aging, is defined by a state of proliferative arrest and is known to be involved in tumor suppression and progression, tissue remodeling, and embryonic development [14]. The accumulation and persistence of senescent cells is an important characteristic of aging of some tissues, including the skin [15]. Senescent cells contribute to the decline of tissue function and lead to age-related changes and pathologies [16]. This well-known interconnection between cellular senescence, skin aging, and skin diseases is often related to molecular processes such as inflammation (Figure 1).

Senescent cells are characterized by several characteristics such as a distinct morphology [17], DNA damage [18], cell cycle arrest [19], mitochondrial dysfunction [20], protein quality impairment [21], inflammation [22], and reactive oxygen species (ROS) generation [23]. Additionally, senescent cells are not able to proliferate as they reside in a cell cycle arrest, which is mostly caused by an accumulation of DNA damage [24,25]. If unrepaired, these damages lead to detrimental effects such as cellular dysfunction and cancer. Other features of senescent cells are the appearance of senescence-associated heterochromatin foci (SAHF) and decreased Lamin B1 expression [14].

Importantly, some types of senescent cells can also be recognized by other characteristics, such as increased senescence-associated β-galactosidase (SA-β-Gal) activity and the secretion of a set of pro-inflammatory factors, the so-called senescence-associated secretory phenotype (SASP) [26]. In general, SASP components can have beneficial effects such as recruitment of immune cells, promotion of anti-tumor response, and improved wound healing [26]. On the other hand, SASP components can contribute to the functional disruption of tissue structure in an autocrine and paracrine manner leading to the senescence of the neighboring cells via paracrine communication, immunosuppression, and inflammation [26,27]. The composition of the SASP is determined by the type of senescence and cell type but the most common cytokines upregulated in cutaneous skin cells during aging are interleukin (IL)-1 and IL-6, as well as the matrix metalloproteases (MMP)-1 and MMP-3 [26,28].

Another hallmark of senescent cells is the imbalance between their production of ROS and their ability to detoxify the reactive intermediates [29,30]. Elevated ROS levels cause damage to cellular molecules such as proteins, lipids, and nucleic acids as well as to organelles such as mitochondria and proteasomes [31]. Antioxidant enzymes such as glutathione and superoxide dismutase are able to stabilize or deactivate free radicals before they attack cellular components, avoiding the so-called oxidative stress. During senescence these defense mechanisms are decreased whereas the ROS is continuously produced and the cells become more prone to oxidative damage [32]. ROS can induce the mitogen-activated protein kinase (MAPK)-p38 pathway which leads to further activation of p53-p21 and cell cycle arrest and thus initiating or accelerating the senescence process [14]. Besides this, excessive ROS induce inflammation-related pathways such as the nuclear factor kappa-light-chain-enhancer of activated B cells (NF-κB) pathway, which regulates different intracellular responses such as apoptosis, cell proliferation, angiogenesis, metastasis, and tumor promotion [7].

Mitochondrial damage, another feature of senescent cells, results from accumulation of mitochondrial DNA mutations, dysfunction and structural alterations of important mitochondrial proteins and membranes, imbalance of fission and fusion, and impaired mitophagy which, in turn, results in increased ROS production and decreased energy generation by oxidative phosphorylation [20,33]. Mitochondria are the main intracellular sources of ROS and excessive ROS production can lead to fragmentation of mitochondria by modulation of mitochondrial fission and fusion proteins [34] and impairment of mitostasis is considered to be one of the causes of cellular senescence [20].

During senescence, the intracellular mechanisms of protein quality control, autophagy and the proteasome, are impaired. Autophagy is responsible for the elimination of damaged or excessive proteins and organelles via the lysosomes, whereas the proteasome especially degrades oxidized proteins [35]. Generally, the activity and function of both mechanisms are reduced with age and senescence [21,24]. For instance, decreased expression of genes that transcribe proteasome subunits as well as modifications of these subunits are considered as major drivers of the age-related impairment of proteasome activity [36]. Proteasome activity has been reported to be decreased in photoaging models of keratinocytes [37], melanocytes [17], and fibroblasts [31]. Furthermore, in senescent melanocytes and fibroblasts the impairment of the proteasome was compensated by an increase of autophagy as an alternative mechanism of protein quality control [17,31]. Impaired autophagy has been related to reduction in lysosomal proteolytic function, decreased rates of autophagolysosomal fusion, and impaired delivery of autophagy substrates to lysosomes, which leads to the intracellular accumulation of undigested material, exacerbates cellular impairment, and contributes to the development of senescence and in tissue to age-related diseases [21,38,39].

3. Cellular Senescence and Skin Aging

3.1. Main Characteristics of Skin Aging

The skin is constituted of many different cell types, including, among others, fibroblasts, keratinocytes, and melanocytes. Skin aging is a multifactorial process and most if not all skin cell types, when functionally impaired, can potentially contribute to the deterioration of the tissue [15]. Additionally, the skin is a useful system for the investigation of aging since this tissue undergoes morphological, biochemical, and functional modifications during the process of aging.

Skin aging is characterized by different features, one of which is the accumulation of damaged and dysfunctional macromolecules in the skin cells due to decreased function of autophagy and proteasome activity. Autophagy plays a role in extrinsic and intrinsic skin aging and regulates pigmentation, homeostasis, and the functions of fibroblasts, keratinocytes, and melanocytes [21,39]. In fibroblasts, it was shown that impaired autophagic flux induced by inhibition of lysosomal proteases lead to the decreased expression of hyaluron, elastin, and type 1 procollagen, and to the increased breakdown of collagen fibers which result in the impairment of dermal integrity and increased skin fragility [40]. In melanocytes, several groups showed that modulation of autophagy can induce cellular senescence and impairment of antioxidant defense mechanisms leading to changes in skin pigmentation [41,42]. Additionally, autophagy and oxidative stress are determinant factors in the fate of keratinocytes and control their progression to senescence, programmed cell death, or tumor formation [21,39].

During aging, skin cells also accumulate damaged mitochondria and mitochondrial DNA deletions leading to structural and functional alterations in the extracellular matrix (ECM) and the induction of inflammation which, in turn, accelerate the formation of skin wrinkles [18].

3.2. Inflammation and Skin Aging

Inflammation is one of the main pathways involved in the process of skin aging. Furthermore, chronic inflammation can lead to skin disorders such as AD and psoriasis [43]. AD is a skin inflammatory disease with symptoms such as pruritus, itching, pain, and sleep disturbance [44,45]. Several different pathways are involved in the activation of inflammation. Here, we will focus on three main activators of inflammation recognized to be involved in skin aging.

The first pathway involves the aryl hydrocarbon receptor (AhR), a transcription factor highly expressed in all cutaneous cell types which has been reported to play a key role in the maintenance of skin barrier, regulation of skin pigmentation, and skin immunity. In humans, the highly conserved AhR has a complex role. Different ligands such as polyaromatic hydrocarbons (PAH) can bind and activate AhR. This leads to AhR translocation to the nucleus where it regulates the expression of genes such as cytochrome P450 1A (*CYP1A*) which can, in turn, further induce oxidative damage by generating ROS [46]. Alterations in AhR signaling lead to dysregulated skin barrier function and can generate symptoms such as dryness, itchiness, and flakiness which are similar symptoms to AD [46–48]. In mice, depletion of AhR leads to transepidermal water loss, decreased expression of important barrier function proteins such as filaggrin and involucrin, and changes of the skin microbiome [49]. In chronic skin inflammation diseases such as psoriasis, activation of AhR can determine the severity of the symptoms [50]. In addition, the activation of AhR is crucial for melanocyte survival and melanogenesis which are events that can be linked to the appearance of senile lentigines [51].

The second important inflammatory pathway participating in skin aging is controlled by NF-κB, a transcription factor which resides in the cytoplasm and, if activated, translocates to the nucleus. Activation of the NF-κB signaling pathway is driven by the response to diverse stimuli, including ligands of various cytokine receptors, pattern-recognition receptors (PRRs), T-cell receptor (TCR) and B-cell receptors, as well as to a subset of TNF receptor (TNFR) superfamily members such as LTβR, BAFFR, CD40, and RANK [52]. Once translocated to the nucleus, NF-κB promotes the production and release of tumor necrosis factor-α (TNF-α), MMPs, and other SASP components such as IL-1α, and cyclooxygenases (COX)-1 and -2. These factors can reinforce the inflammation process and accelerate the skin aging process [53,54]. Importantly, the dysregulation of NF-κB plays an important role in the pathogenesis of chronic inflammatory diseases of the skin as well as in wound healing [55].

The third important inflammatory signaling pathway involved in skin aging is coordinated by nuclear factor erythroid 2-related factor 2 (Nrf2)/antioxidant response element (*ARE*). Nrf2 is an important transcription factor in the inflammation signaling cascade as well as in oxidative stress responses and is responsible for the expression of *ARE* genes such as heme oxygenase-1. Nrf2 can also negatively regulate NF-κB activation either by directly decreasing intracellular ROS levels or by inhibiting the translocation of NF-κB to the nucleus [56,57]. Recently, it was demonstrated that senescent melanocytes increased the expression of Nrf2 which was accompanied by decreased melanogenesis [58]. Nrf2 depletion in skin cells leads to the reduction of cell survival as well as induction of oxidative stress while mutations in the *Nrf2* gene, in turn, lead to the development of squamous cell carcinoma suggesting that pathways activated by Nrf2 are important for the maintenance of skin homeostasis [59].

3.3. Senescence of Skin Cells and Skin Aging

The accumulation of senescent cells in the epidermis and dermis is considered as a hallmark of aging. The occurrence of senescent cells can be accelerated by exposition of the skin to different sources of environmental and lifestyle factors [15,60]. Senescent fibroblasts display different senescence associated characteristics such as cell cycle arrest, decreased autophagy activity, increased SA-β-Gal activity, mitochondrial dysfunction, DNA damage, increased ROS generation, and increased expression of SASP factors [31,61]. The appearance and persistence of senescent fibroblasts in the dermis leads to the accumulation of elastotic material, which results from the incomplete degradation of elastic fibers, as well as to the decreased expression of extracellular matrix components simultaneously to the increased degradation of ECM. These events are manifested in dullness and reduced elasticity of the skin [1,62]. Keratinocytes are constantly renewed and more prone to apoptosis than senescence when confronted by stressors. Nevertheless, keratinocytes contribute greatly to the aging process by losing the ability to terminally differentiate or proliferate and responding differently to external stimuli [15]. Senescent keratinocytes express increased p15, IL-1α, and high mobility group A2 [27]. Other hallmarks of senescence such as SA-β-Gal, expression of p21, p53, and p16 are displayed in keratinocytes upon UVB exposure [16]. Senescence of melanocytes has been insufficiently explored but a recent study has reported that senescent melanocytes accumulate in human skin where they contribute to aging of this tissue by impairing proliferation of the neighboring keratinocytes [63]. Melanocytes' senescence can occur either by exhaustion of their replicative capacity, in a process called replicative senescence, or by stress-induced premature senescence in which senescence is triggered by chronic exposition of these cells to sublethal doses of a stressor agent such as UV or ROS-inducing agents [17,25]. Accumulated senescent melanocytes in skin is correlating with visible signs of skin aging such as facial wrinkling and deposition of elastin in the dermis [60]. Although melanocytes become age-dependently less active, darker pigmented spots, also described as age spots, solar lentigines, or lentigo senilis, are a common characteristic of aged skin, presumably occurring due to irregularities in melanocyte distribution, enhanced melanogenic signaling and decreased melanosome removal [15,64]. Besides the impairment of melanocytes, the dysregulated secretion of molecules produced by aged keratinocytes and fibroblasts cause the occurrence of senile lentigines [15,51]. For example, UV-induced senescent fibroblasts contribute to the formation of age spots by decreasing the expression and secretion of stromal cell-derived factor 1 [65], emphasizing the complexity of underlying factors driving skin aging and the development of senile lentigines.

3.4. Microbioma and Skin Aging

Another important but as yet less investigated change that occurs in the skin during aging is the alteration of the skin microbiome. The human skin microbiome consists of bacteria, fungi, viruses, archaea, and other microorganisms which are essential for body homeostasis and when dysregulated can contribute to diseases [45,66]. The skin of young individuals is rich in bacteria of the Firmicutes phylum but with age, these are replaced by Bacteroides and Proteobacteria. This imbalance of commensal skin microbes which produce immune factors affects the skin's immune system, explaining the increased risk of pathogenic invasions and age-related skin disorders [22,67].

3.5. Skin Aging in Different Ethnicities and Phototypes

In humans, melanin determines skin and hair color and is responsible for photoprotection against UV radiation. Melanin molecules surround the nuclei of the keratinocytes and melanocytes to protect their genetic material, by absorbing sunlight through the polymeric melanin molecule [68,69]. The differences in skin color are mainly determined by the type and amount of melanin produced by melanocytes which is a major factor of skin aging. Melanocytes can produce two different types of melanin. Eumelanin, a brown-black pigment, and pheomelanin, a yellow-red pigment [70]. Melanocytes transfer melanin to the neighboring keratinocytes, and melanin is degraded during keratinocytes differentiation. The ratio of melanin degradation determines the skin pigmentation and, consequently, the skin phototypes. For instance, in fair skin phototypes melanosomes are almost completely degraded, whereas in dark skins they barely undergo degradation and accumulate in the upper layers of the epidermis [1]. During the process of skin aging, the density and activity of melanocytes is continuously decreased, while the percentage of cells with impaired melanin synthesis is increased [15], leading to the appearance of skin pigmentation issues.

Given the differences in skin pigmentation and degradation of melanin, skin aging manifests differently among different skin phototypes and ethnicities. For instance in Asians and African Americans pigmentation changes are considered early manifestations of skin aging, whereas aging of Caucasian skin is characterized in its early stages by the appearance of wrinkles [71–73].

In general, highly pigmented skin structurally consists of a thicker dermis and stratum corneum which contribute to increased resistance and elasticity in comparison to light pigmented skin [71,74]. Recent studies demonstrated that highly pigmented skin exposed to UV displays a deposition of collagen and elastic network, a decrease in epidermal thickness [75], lightening of skin [76], as well as production of MMPs [77]. Upon skin aging, light pigmented skin exhibits thinner and less compact stratum corneum, deposition of elastic fibers, impairment of barrier function, reduction of sebum, production of MMPs, and loss of collagen [71,74]. These findings indicate that skin aging mechanisms seem to be similar between ethnicities. However, further investigations are necessary to understand the underlying molecular processes of different ethnic skin types.

4. Major Components of Air Pollution Affecting Skin Appearance

Pollution is defined as an environmental contamination by chemical, biological, or physical substances which can affect human health and ecosystems. Air pollutants are composed of organic and inorganic substances which are introduced into the atmosphere by residential wood heating, tobacco smoking, transportation, and industry, among other sources [78]. The composition of atmospheric pollution can also vary depending on the time of day, seasons, human activity, and geographic location [8,9]. Several reports suggest that air pollutant components are direct contributors to the process of aging [5,11,79–82]. Human exposition to air pollution contributes to increased mortality and hospital stays. The effects of air pollution can range from nausea, difficulty in breathing, skin irritation, birth defects, and reduced activity of immune system, to cancer. Results obtained from research with animal models and epidemiological studies suggest that the cardiovascular and respiratory system are the main affected systems by air pollution [10]. In the lungs, infiltration of air pollutant components, especially small particles which can reach bronchial tubes and deep lung, can induce a systemic immune response due to increased expression of IL-1, IL-6, IL-8, and monocyte chemoattractant protein-1 in macrophages and lung epithelial cells, leading to the development of respiratory diseases [5].

The skin can be affected by air pollution in two ways. Either directly, through the uptake of the air pollutants by the skin, especially by intrusions formed by hair follicles in the stratum corneum [5], or indirectly by the uptake of particles by the lungs which are further transported by the blood to the skin [9].

Exposure of the skin to urban air pollution activates mechanisms of cell detoxification, which, if active over a longer period of time, can lead to DNA and protein damage, elevated ROS levels and lipid peroxidation, resulting in skin alterations, such as impaired barrier function, pigment spots, wrinkles, and decreased skin hydration [83,84]. Additionally, air pollution induces inflammation, activates the AhR pathway, and leads to alterations of the skin microbiome [85,86]. Given the strong link between low-grade systemic inflammation and biological aging, it is possible that exposition of the skin to air pollution leads to premature cellular senescence and, consequently, to skin aging [85]. The literature regarding the effects of air pollution on different ethnic skin types is still elusive. The few broad ethnic studies show that exposition of the skin to air pollution mainly induces increased oxidative stress leading to skin pigmentation disorders and wrinkle formation [2,11,72]. The majority of the publications are exclusively related to Asian skin [87] and therefore very biased. Information on the effects of air pollution on different skin types is still lacking and therefore the awareness of different ethnicities in this field of research needs more attention.

In this review we will focus on the most investigated and relevant air pollutants for the process of skin aging: ozone, heavy metals, cigarette smoke, and PM.

4.1. Particulate Matter

Particulate matter (PM) is one of the main components of air pollution and is defined as a mixture of gas containing liquid and/or solid droplets varying in size and composition [86,88]. Particles contained in PM include substances such as metals, minerals, organic toxins, tobacco smoke, pollen, allergens, and smog [89]. These particles are classified according to their aerodynamic diameter (Dp) in three categories: particles with Dp bellow 10 μm are named "coarse particles" or PM 10 and include components of dust, soil, and dusty emission from industries; particles with Dp between 0.1 and 2.5 μm are called "particle matter" or PM 2.5 and are mainly derived from open fires, automobile exhausts, and power plants [90]; and particles with Dp bellow 0.1 μm are called "ultrafine particles" or PM 0.1 and mainly consist of emissions of diesel-powered engines [86]. The aerodynamic diameter of the particles is an important determinator of their ability to enter the body by alveolar–capillary barrier and travel across the blood [91]. The particle matter and ultrafine particles can penetrate the body either by systemic distribution through the blood circulation after entering the lungs' alveoli or by infiltrating the skin through hair follicles [92]. Recently, in contrast with what was believed before, it was demonstrated that coarse particles can also penetrate the stratum corneum or the respiratory system [53,93].

Exposure to particulate matter can cause cardiovascular and respiratory diseases, allergies, and cancer through different mechanism [83,94]. For instance, PM 2.5 exposure induces endoplasmatic reticulum stress response and apoptosis in the lung and liver of mice and this mechanism is believed to be one possible explanation for the development of metabolic, respiratory, and cardiovascular diseases in humans exposed to air pollution [95]. A study in mice showed that PM inhalation activates the TNF-α driven systemic inflammation and results in an impaired cardiac function which to a certain extent was prevented by a TNF-α inhibitor called inflimiximab [96]. In skin diseases such as psoriasis, the blockage of TNF-α leads to detrimental side-effects such as increased risk of infections and malignancies as well as to the formation of new and more psoriatic skin lesions [97].

A cross-sectional study which investigated the correlation of particulate matter exposure and aging demonstrated that leukocytes of peripheral blood from elderly humans exposed to daily high concentrations of PM 2.5 presented decreased telomere length, lower mitochondrial DNA content, and reduced sirtuin-1 expression [98]. In another study it was shown that exposure of human nasal epithelial cells to PM 2.5 leads to ROS production, degradation of tight junction proteins such as occludin, claudin-1, and E-cadherin, leading to sinonasal diseases through disrupted tissue integrity and permeability [45,99]. Similar downregulation of tight junctions as well as keratins and filaggrin were observed in pork skin in response to PM 2.5 exposure, leading to increased skin permeability [100,101]. Altogether, these studies demonstrate that short- and long-term exposure to PM can induce features of premature aging.

Particularly in the skin, PM 2.5 induces different detrimental processes such as DNA damage and lipid peroxidation. Exposure of the skin to PM results in the formation of senile lentigines, increased formation of ROS, and promotes the release of pro- inflammatory cytokines which all lead to accelerated skin aging and increased susceptibility to pathogen invasion [5]. ROS generation, induced by exposure of the skin to air pollution, can induce MMPs expression and increase their activity, especially MMP-1 and MMP-3 which are known to accelerate the skin aging process by degrading collagen and elastin. Additionally, the decreased expression of transforming growth factor (TGF) β, and reduced synthesis of collagen type 1 α chain (*COL1A1, COL1A2*) and elastin by fibroblasts are other contributors to the formation of wrinkles and skin aging induced by PM [80,82,102].

In human keratinocytes and mouse skin tissue, PM induced the expression of demethylases such as DNA demethylase 1 (TET1) and decreased the expression of DNA methylation-related proteins such as (DNMT)-1 and -3. These changes in the methylation pattern of DNA induced "skin senescence phenomenon" are characterized by features such as hyperkeratotic epidermis accompanied by the increased expression of keratin-10, an epidermal differentiation marker, and proliferating cell nuclear antigen, a proliferation marker [79]. These results demonstrate that PM-induced skin aging can be triggered by epigenetic modifications which could give rise to new strategies for therapeutics against skin aging.

Skin cells exposed to diesel particulate extract (DPE), which mainly contains PM and PAH [103], displayed dysregulation of proteins and lipids important for the maintenance of skin integrity, for the regulation of skin hydration and oxidative stress, such as NADPH oxidase (NOX), ceramide, plakins, transglutaminases, cystatins, and filaggrin [57,80]. Furthermore, DPE impaired mitochondrial oxidative phosphorylation and cell migration in these cells. These effects could be partially avoided or restored by the treatment of the cells with the antioxidant vitamin E [57].

PM causes inflammation in the skin through increased IL-8, MMP-1, ROS production, and neutrophil infiltration in the deep dermis [104]. Fibroblasts cultivated with conditioned medium obtained from immortalized human keratinocytes (HaCaT) treated with PM displayed increased nuclear translocation of p65 and p50 as well as increased ROS production, morphological changes, and secretion of SASP components including prostaglandin E2, COX-2, TNF-α, IL-1β, and IL-6 [93]. It is well established that the expression and release of TNF-α is increased upon UV exposure [105] as well in some skin diseases such as vitiligo [106].

Other studies have shown that the exposition of HaCaT cells or reconstructed human epidermis to PM induced upregulation of NF-κB, COX-1 as well as IL-1α leading to skin barrier dysfunction [81]. In addition, an increased autophagic activity shown by the turnover of the light chain 3 I (LC3) to LC3 II, expression of p62, and PM internalization in the autolysosomes was observed after exposure of fibroblasts to air pollutants such as PM [107,108].

The PM particles can transport organic chemicals and metals into cells which when localizing in the mitochondria can directly generate ROS [2]. Mostly, PM particles contain PAH which can penetrate the skin. PAH is an activator of AhR in melanocytes and keratinocytes and exposition of the skin to this chemical can influence epidermal turnover, melanogenesis, and differentiation [9,86]. PAH have synergistic effect with UVA and both contribute to skin carcinogenesis and to the appearance of skin pigmentation disorders such as senile lentigines [5,53,70,92,109,110]. Interestingly, exposure of melanocytes to PM induces apoptosis through cytochrome C release and activation of caspase-3 [111]. These events are linked to the disappearance of melanocytes which is considered one of the main pathogenic mechanisms of vitiligo.

The effects of PM 2.5 and 10 are linked to different skin diseases such as AD and allergies [45,112]. Several studies have demonstrated that AD skin, as well as healthy skin exposed to PM, display skin barrier disruption due to decreased expression of epidermal structural proteins such as filaggrin, E-cadherin, and cytokeratins [43,44,113]. Additionally, it is suggested that AD can be influenced by the exposure to air pollution resulting in imbalance of immune cell response and IgE production, activation of AhR/NF-κB, and the generation of ROS, and these effects can be prevented by improved air quality [5,44,85].

4.2. Cigarette Smoke

Cigarette smoke is composed of different chemical substances including reactive oxygen species, carbon monoxide, and reactive nitrogen species. Smoking is associated with lung cancer as well as with cutaneous squamous cell carcinoma. PAH, one important component of cigarette smoke, is a major contributor to cancer. PAH induces AhR and subsequently *CYP1A1* which leads to the formation of DNA adducts and can result in cancer development [114,115].

The effects of cigarette smoke are cumulative and associated with the appearance of signs of premature aging, especially in the facial skin. Among them the most common are deep wrinkles, dryness, leathery texture, sagging, premature graying, and orange to purple discoloration of the skin [8,53,88]. Exposition of the skin to cigarette smoke induces oxidative stress and the impairment of the antioxidant system. The consequences are transepidermal water loss, lipid peroxidation, cell death, and degeneration of connective tissue by MMP-1 and -3 [80,114,116]. In keratinocytes, cellular redox homeostasis and colony-forming potential was affected in response to exposure to tobacco smoke components such as PAH and PM [92]. In humans, a study comparing smokers and nonsmokers revealed the occurrence of shorter telomeres, a sign of cellular senescence, in peripheral white blood cells of smokers [98]. Synergistic effects may occur upon the combined exposure to UV and cigarette smoke leading to epidermal barrier disruption, increased erythema, and decreased elasticity [8,53,88]. Furthermore, a three months' whole body exposure of mice to tobacco smoke caused premature aging evidenced by the loss of collagen as well as hair loss and premature graying due to increased apoptosis and decreased melanogenesis, respectively [117]. Additionally, the combined exposure of tobacco smoke with ultraviolet light, exacerbated the effects of aging by the upregulation of p16 expression. These effects could be prevented by *N*-acetylcysteine (NAC) suggesting that premature aging triggered by cigarette smoke exposition is driven by ROS [5]. Other studies showed that exposure of skin cells, ex vivo skin biopsies, or reconstructed skin to cigarette smoke leads to production of pro-inflammatory cytokines and MMPs, lipid peroxidation, and downregulation of differentiation proteins such as loricrin and this, in turn, resulting in the impairment of skin barrier structure and function [114,115,118].

Besides the consequences of topical exposition of the skin to pollution, the inhalation of pollution has systemically effects to the skin. In a recent in vivo study, it was demonstrated that chronic inhalation of tobacco smoke leads to changes in the composition and deposition of elastin and fibrillin-rich microfibrils in the dermis which is accompanied by a higher stiffness of the skin [119].

In a study using a model of skin aging induced by tert-butyl hydroperoxide (tBHP), Wedel and colleagues showed that skin exposed to this chemical displayed downregulation of collagen synthesis and the increased secretion of MMPs. tBHP is categorized as an oxidative stress inducer which leads to the accumulation of intracellular ROS and deplete the antioxidant mechanisms and can be used as a proxy to study mechanisms that recapitulate the exposition of the skin to environmental stressors such as cigarette smoke [61].

4.3. Ozone

Stratospheric ozone has a protective role for earth-living organisms by filtering UV radiation. Ozone (O_3) in the troposphere reaches the skin surface and reacts with molecules such as lipids and proteins in the stratum corneum [9,120]. In general, ozone causes oxidative stress, increased lipid peroxidation, AhR induction, and depletion of Vitamins C and E in human skin [121,122]. A report has shown that in human keratinocytes, exposure to ozone leads to lactate dehydrogenase release, reduced cell proliferation, lipid peroxidation, and NF-κB activation and these effects can be prevented by pretreatment with antioxidant mixtures [123]. Another study has shown that the combination of UVA and ozone has a synergistic effect and causes increased oxidative stress of the skin and Nrf2 expression in keratinocytes leading to skin inflammation, formation of wrinkles, and pigment spots [8,9,104]. In human skin O_3 leads to activation of NF-κB, MMP-9, COX-2, and lipid peroxidation in the epidermis, whereas in the dermis expression of collagen-1 and -3 are downregulated. These consequences are accelerating the appearance of skin aging signs including the formation of wrinkles and senile lentigines as well as affecting wound healing [124,125].

4.4. Heavy Metals

The earth crest is formed by different components, among them heavy metals such as chromium, lead, cadmium, silver, nickel, mercury, manganese, and vanadium [8]. These metals can contaminate the water and food supply and cannot be degraded or destroyed. These substances are important trace elements for the maintenance of metabolic reactions but in higher concentrations they become toxic to the human body [10]. Heavy metals from industrial fumes reach plants through acidic rain. For instance, tobacco leaves are rich in cadmium which, when inhaled, cannot be excreted from the human body and leads to long term effects and damage to the lungs, kidneys, and bones [78]. A study showed that keratinocytes and skin explants exposed to dust particles containing heavy metals or heavy metals alone displayed increased expression of the pro-inflammatory cytokines IL-6, IL-8, caspase-14, and granulocyte macrophage colony-stimulating factor which are known to alter epidermal differentiation, ECM, apoptosis, DNA damage, lipid peroxidation, and skin immunity resulting in cutaneous inflammation and inflammation skin disorders [86,126].

Figure 2 shows a graphic representation of the complexity of air pollution damages to the skin and summarizes the most important signaling pathways regulated by different air pollutants on a cellular and on tissue level (Figure 2).

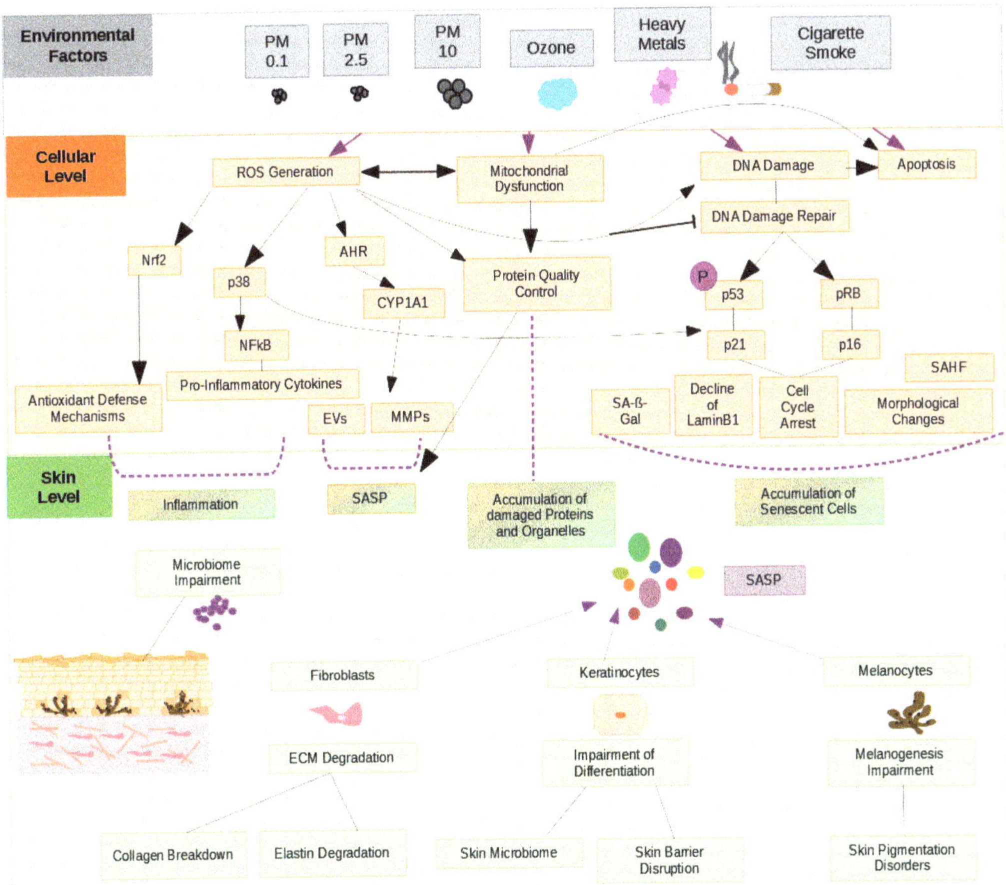

Figure 2. Schematic diagram showing the main signaling pathways involved in the process of cellular senescence which are contributing to skin aging. Pollutants act over many signaling pathways that control cell cycle progression and transcriptional regulation of genes involved in inflammation.

5. Therapeutics and Cosmetics

In recent years, due to increased industrialization and use of transport, the effects of air pollution on human health have increased logarithmically. As a consequence, a trend of development and use of anti-pollutant cosmetics and therapeutics originated in Asia, which is one of the most polluted places in the world. This trend spread to the western world and the demand in cosmetics and the personal care industry rose worldwide and led to increased research on the identification of new approaches and ingredients for anti-pollution cosmetics [127] (Table 1).

Table 1. A summary of the current interventions, an example of active ingredients, and which mechanisms are helping against air pollution-induced damage with corresponding references for detailed information.

Intervention	Active Ingredients	Mechanism	References
Sunscreen	Oxybenzone Zinc oxide	• Prevent synergistic effects of air pollution and UV	[1]
Washing and air filters		• Prevent deposition and penetration of pollutants on skin	[8,85]
Dietary habits	Phenolic compounds in plants	• Reduce cellular oxidative stress	[13]
Rinse-off, film-forming cosmetics, and emollients	BDDI *Aleurites fordii* oil copolymer Kaolin	• Reduced transepidermal water loss • Increased skin barrier function	[85]
Antioxidants	Vitamin C Vitamin E Ferulic acid	• Reduced ROS production and SASP • Prevent collagen degradation and hyperpigmentation	[1,85]
Botanicals	Algae	• Antibacterial and anti-inflammatory activity • Skin whitening agent • ROS scavenger	[127]

Several studies suggest different approaches to protect the skin from air pollution. Firstly, and most important, avoid exposure to air pollutants. Additionally, as the skin is exposed to air pollution and UV simultaneously, another measure to increase the protection of the skin is to use sunscreen, decreasing the possibility of occurrence of photoreaction [9,51]. Overwashing the skin can be harmful and break the skin barrier and homeostasis of the skin microbiome [9]. Indoor air ventilators or filters can also be helpful since they reduce the pollutant load in the air and on the skin [8]. One additional approach to protect the skin from external factors is the use of cosmetic preparations containing pro-, and prebiotics [128]. Components from our diet are also beneficial. UV-exposed skin revealed changes in the elastic and collagen network of the dermis including the composition and deposition of fibulin-5 which contributes to the structural formation of elastic fibers such as elastin. Additionally, the study showed that oral supplementation with green tea catechins in combination with vitamin C protected specifically fibulin-5 fibers against UV-induced changes [129]. Recently, reports have suggested that components from plants and other natural resources such as phenolic compounds, phytosterols, and saponins help against induced cellular oxidative stress by PM [10,13,85].

Recommendation of the topical use of cosmetics that protect or improve the skin barrier, enhance the antioxidant mechanism, and reduce inflammation can help avoid or mitigate air pollution skin damages [86]. Additionally, to remove chemicals and decrease the particle load deposited on the skin surface upon exposure to air pollution, the use of rinse-off products and emollients which reduce transepidermal water loss and increase skin barrier function is recommended [9,51,127]. One group selected E/Z-2-benzylidene-5,6-dimethoxy-3,3-dimethyl-indan-1-one (BDDI), an AhR antagonist, and could show that this component can transiently inhibit AhR activation in human skin in response to air pollution and prevent activation of genes which are relevant for wrinkle formation and skin carcinogenesis such as *CYP1*, *COX-2*, and *MMP-1* [9,130]. This molecule is also hypothesized to be a promising ingredient to prevent senile lentigines formation since it was demonstrated that it can decrease the production of SASP [51]. In general, the lack of testing compounds in advanced models for skin color, photoprotection, and anti-pollution properties is enormous [131]. Especially, the effects of anti-pollution cosmetics on different ethnic skin regarding pigmentation is still missing.

The deposition of environmental stressors on the skin can be prevented by film-forming cosmetics [127]. Lastly, protection against oxidative stress-inducing effects such as ROS generation and downregulation of antioxidant defense mechanisms in the epidermis can be reduced by the application of formulations with antioxidants. The well-known antioxidant Vitamin E has beneficial effects on the skin including decreased expression of Nrf2, restoration of mitochondrial complex I and complex IV, recovery of desmosomal protein integrity, and promotion of cellular migration [57]. Another commonly used antioxidant in cosmetics is L-ascorbic, acid also known as Vitamin C, which can neutralize free radicals, and prevent collagen degradation and hyperpigmentation, which are all signs of skin aging [85,102]. This antioxidant is mostly combined with other antioxidants such as Vitamin E, ferulic acid, L-selenomethionine, and phlorectin and displayed in vivo and in vitro anti-photoaging effects [53,102]. This combination of antioxidants decreases the nuclear translocation of NF-κB and thus promotes the inhibition of inflammation and the synthesis of pro-inflammatory cytokines, respectively [53].

The latest trend in the anti-pollution cosmetic field is the use of ingredients of natural or botanical origin such as algae and green tea leaves. There are several advantages of using naturally occurring ingredients which include increase of cosmetic efficiency and reduction of the risk for allergies and irritation. Furthermore, society demands more ethical, natural, and green formulations in cosmetic research [127]. In conclusion, different components and applications are established which help the skin to prevent or mitigate damages through environmental stress.

6. Conclusions

Research on the effects of air pollution on cellular senescence and skin aging has become a popular topic. Therefore, the purpose of this review is to highlight the current understanding of the connections between these topics, and to discuss recent research findings, approaches, and cosmetical interventions used to prevent or mitigate skin damages caused by exposure of the skin to air pollution. Our review demonstrates that cellular senescence and skin aging are closely regulated and connected and that they can, in fact, be induced by air pollution. For this reason, mechanisms of air pollution-induced skin aging have been the focus of many recent publications.

Author Contributions: Conceptualization, investigation, writing—review & editing, I.M. and M.C.; supervision, P.J.-D. All authors have read and agreed to the published version of the manuscript.

Funding: This work was supported by a grant from the Austrian Science Fund (FWF), P315820. Maria Cavinato and Ines Martic received funding by Tiroler Wissenschaftsförderung (ZAP746010 & F.33827/10-2021). Additionally, Ines Martic is financed by a scholarship from the University of Innsbruck (Doktoratsstipendium der Nachwuchsförderung).

Data Availability Statement: Not applicable.

Conflicts of Interest: The authors declare no conflict of interest.

Abbreviations

AD	atopic dermatitis
AhR	aryl hydrocarbon receptor
ARE	antioxidant response element
BAFFR	B-cell activating factor receptor
BDDI	E/Z-2-benzylidene-5,6-dimethoxy-3,3-dimethyl-indan-1-one
CD	cluster of differentiation
COL1A1	collagen type 1 alpha chain
COX-1 and -2	cyclooxygenases-1 and −2
CYP1A	cytochrome P450 1A
DNA	deoxyribonucleic acid
DNMT	DNA methyltransferase
Dp	aerodynamic diameter
DPE	diesel particulate extract
ECM	extracellular matrix
EVs	extracellular vesicles
HaCaT	immortalized human keratinocytes
IL	interleukin
LC3	light chain 3
LTβR	lymphotoxin β receptor
MAPK	mitogen-activated protein kinase
MMP	matrix metalloproteases
NAC	*N*-acetylcysteine
NF-κB	nuclear factor kappa-light-chain-enhancer of activated B cells
NOX	nicotinamide adenine dinucleotide phosphate oxidase
Nrf2	nuclear factor erythroid 2-related factor 2
O_3	ozone
PAH	polyaromatic hydrocarbons
PM	particulate matter
pRB	phosphorylated retinoblastoma protein
PRRs	pattern-recognition receptors
RANK	receptor activator of NF-κB
ROS	reactive oxygen species
SAHF	senescence-associated heterochromatin foci
SA-β-Gal	senescence-associated β-Galactosidase
SASP	senescence-associated secretory phenotype
tBHP	tert-butyl hydroperoxide
TCR	T-cell receptor
TET1	DNA demethylase 1
TGF	transforming growth factor
TNF-α	tumor necrosis factor alpha
TNFR	tumor necrosis factor receptor
UV	ultraviolet
WHO	world health organization

References

1. Cavinato, M. Cosmetics and Cosmeceuticals. In *Reference Module in Biomedical Sciences*; Ratan, S.I., Ed.; Elsevier: Amsterdam, The Netherlands, 2020; pp. 446–461, ISBN 9780128160756.
2. Vierkötter, A.; Krutmann, J. Environmental Influences on Skin Aging and Ethnic-Specific Manifestations. *Dermato-Endocrinology* **2012**, *4*, 227–231. [CrossRef] [PubMed]
3. Cavinato, M.; Jansen-Dürr, P. Molecular Mechanisms of UVB-Induced Senescence of Dermal Fibroblasts and Its Relevance for Photoaging of the Human Skin. *Exp. Gerontol.* **2017**, *94*, 78–82. [CrossRef] [PubMed]
4. Cavinato, M.; Wedel, S.; Jansen-Dürr, P. Aging of Cells In Vitro. In *Reference Module in Biomedical Sciences*; Elsevier: Amsterdam, The Netherlands, 2019; pp. 138–148, ISBN 9780128160763.
5. Kim, K.E.; Cho, D.; Park, H.J. Air Pollution and Skin Diseases: Adverse Effects of Airborne Particulate Matter on Various Skin Diseases. *Life Sci.* **2016**, *152*, 126–134. [CrossRef] [PubMed]

6. WHO. WHO Releases Country Estimates on Air Pollution Exposure and Health Impact (outdoor air pollution) in, Asia and Western Pacific regions. Available online: https://www.who.int/news-room/fact-sheets/detail/ambient- (accessed on 30 April 2022).
7. McDaniel, D.; Farris, P.; Valacchi, G. Atmospheric Skin Aging-Contributors and Inhibitors. *J. Cosmet. Dermatol.* **2018**, *17*, 124–137. [CrossRef]
8. Puri, P.; Nandar, S.; Kathuria, S.; Ramesh, V. Effects of Air Pollution on the Skin: A Review. *Indian J. Dermatol. Venereol. Leprol.* **2017**, *83*, 415. [CrossRef]
9. Krutmann, J.; Liu, W.; Li, L.; Pan, X.; Crawford, M.; Sore, G.; Seite, S. Pollution and Skin: From Epidemiological and Mechanistic Studies to Clinical Implications. *J. Dermatol. Sci.* **2014**, *76*, 163–168. [CrossRef]
10. Kampa, M.; Castanas, E. Human Health Effects of Air Pollution. *Environ. Pollut.* **2008**, *151*, 362–367. [CrossRef]
11. Grether-Beck, S.; Felsner, I.; Brenden, H.; Marini, A.; Jaenicke, T.; Aue, N.; Welss, T.; Uthe, I.; Krutmann, J. Air Pollution-induced Tanning of Human Skin*. *Br. J. Dermatol.* **2021**, *185*, 1026–1034. [CrossRef]
12. Shi, Y.; Zeng, Z.; Liu, J.; Pi, Z.; Zou, P.; Deng, Q.; Ma, X.; Qiao, F.; Xiong, W.; Zhou, C.; et al. Particulate Matter Promotes Hyperpigmentation via AhR/MAPK Signaling Activation and by Increasing α-MSH Paracrine Levels in Keratinocytes. *Environ. Pollut.* **2021**, *278*, 116850. [CrossRef]
13. Diao, P.; He, H.; Tang, J.; Xiong, L.; Li, L. Natural Compounds Protect the Skin from Airborne Particulate Matter by Attenuating Oxidative Stress. *Biomed. Pharmacother.* **2021**, *138*, 111534. [CrossRef]
14. Mohamad Kamal, N.S.; Safuan, S.; Shamsuddin, S.; Foroozandeh, P. Aging of the Cells: Insight into Cellular Senescence and Detection Methods. *Eur. J. Cell Biol.* **2020**, *99*, 151108. [CrossRef] [PubMed]
15. Bellei, B.; Picardo, M. Premature Cell Senescence in Human Skin: Dual Face in Chronic Acquired Pigmentary Disorders. *Ageing Res. Rev.* **2020**, *57*, 100981. [CrossRef] [PubMed]
16. Wang, A.S.; Dreesen, O. Biomarkers of Cellular Senescence and Skin Aging. *Front. Genet.* **2018**, *9*, 247. [CrossRef] [PubMed]
17. Martic, I.; Wedel, S.; Jansen-Dürr, P.; Cavinato, M. A New Model to Investigate UVB-Induced Cellular Senescence and Pigmentation in Melanocytes. *Mech. Ageing Dev.* **2020**, *190*, 111322. [CrossRef]
18. Ho, C.Y.; Dreesen, O. Faces of Cellular Senescence in Skin Aging. *Mech. Ageing Dev.* **2021**, *198*, 111525. [CrossRef]
19. Van Deursen, J.M. The Role of Senescent Cells in Ageing. *Nature* **2014**, *509*, 439–446. [CrossRef]
20. Vasileiou, P.; Evangelou, K.; Vlasis, K.; Fildisis, G.; Panayiotidis, M.; Chronopoulos, E.; Passias, P.-G.; Kouloukoussa, M.; Gorgoulis, V.; Havaki, S. Mitochondrial Homeostasis and Cellular Senescence. *Cells* **2019**, *8*, 686. [CrossRef]
21. Eckhart, L.; Tschachler, E.; Gruber, F. Autophagic Control of Skin Aging. *Front. Cell Dev. Biol.* **2019**, *7*, 143. [CrossRef]
22. Lee, Y.I.; Choi, S.; Roh, W.S.; Lee, J.H.; Kim, T.-G. Cellular Senescence and Inflammaging in the Skin Microenvironment. *Int. J. Mol. Sci.* **2021**, *22*, 3849. [CrossRef]
23. Davalli, P.; Mitic, T.; Caporali, A.; Lauriola, A.; D'Arca, D. ROS, Cell Senescence, and Novel Molecular Mechanisms in Aging and Age-Related Diseases. *Oxidative Med. Cell. Longev.* **2016**, *2016*, 3565127. [CrossRef]
24. Da Silva, P.F.L.; Schumacher, B. Principles of the Molecular and Cellular Mechanisms of Aging. *J. Investig. Dermatol.* **2021**, *141*, 951–960. [CrossRef] [PubMed]
25. Toussaint, O.; Medrano, E.; von Zglinicki, T. Cellular and Molecular Mechanisms of Stress-Induced Premature Senescence (SIPS) of Human Diploid Fibroblasts and Melanocytes. *Exp. Gerontol.* **2000**, *35*, 927–945. [CrossRef]
26. Birch, J.; Gil, J. Senescence and the SASP: Many Therapeutic Avenues. *Genes Dev.* **2020**, *34*, 1565–1576. [CrossRef] [PubMed]
27. Ritschka, B.; Storer, M.; Mas, A.; Heinzmann, F.; Ortells, M.C.; Morton, J.P.; Sansom, O.J.; Zender, L.; Keyes, W.M. The Senescence-Associated Secretory Phenotype Induces Cellular Plasticity and Tissue Regeneration. *Genes Dev.* **2017**, *31*, 172–183. [CrossRef] [PubMed]
28. Cuollo, L.; Antonangeli, F.; Santoni, A.; Soriani, A. The Senescence-Associated Secretory Phenotype (SASP) in the Challenging Future of Cancer Therapy and Age-Related Diseases. *Biology* **2020**, *9*, 485. [CrossRef] [PubMed]
29. Dumont, P.; Balbeur, L.; Remacle, J.; Toussaint, O. Appearance of Biomarkers of in Vitro Ageing after Successive Stimulation of WI-38 Fibroblasts with IL-1alpha and TNF-Alpha: Senescence Associated Beta-Galactosidase Activity and Morphotype Transition. *J. Anat.* **2000**, *197*, 529–537. [CrossRef]
30. Roy, A.; Sil, P.C. Tertiary Butyl Hydroperoxide Induced Oxidative Damage in Mice Erythrocytes: Protection by Taurine. *Pathophysiology* **2012**, *19*, 137–148. [CrossRef]
31. Cavinato, M.; Koziel, R.; Romani, N.; Weinmüllner, R.; Jenewein, B.; Hermann, M.; Dubrac, S.; Ratzinger, G.; Grillari, J.; Schmuth, M.; et al. UVB-Induced Senescence of Human Dermal Fibroblasts Involves Impairment of Proteasome and Enhanced Autophagic Activity. *J. Gerontol. Ser. A Biol. Sci. Med. Sci.* **2016**, *72*, glw150. [CrossRef]
32. Tang, L.; Wu, W.; Fu, W.; Hu, Y. The Effects of Phototherapy and Melanocytes on Keratinocytes. *Exp. Ther. Med.* **2018**, *15*, 3459–3466. [CrossRef]
33. Von Zglinicki, T.; Wan, T.; Miwa, S. Senescence in Post-Mitotic Cells: A Driver of Aging? *Antioxid. Redox Signal.* **2021**, *34*, 308–323. [CrossRef]
34. Wu, S.; Zhou, F.; Zhang, Z.; Xing, D. Mitochondrial Oxidative Stress Causes Mitochondrial Fragmentation via Differential Modulation of Mitochondrial Fission-Fusion Proteins. *FEBS J.* **2011**, *278*, 941–954. [CrossRef] [PubMed]
35. Jung, T.; Höhn, A.; Grune, T. The Proteasome and the Degradation of Oxidized Proteins: Part III—Redox Regulation of the Proteasomal System. *Redox Biol.* **2014**, *2*, 388–394. [CrossRef] [PubMed]

36. Löw, P. The Role of Ubiquitin–Proteasome System in Ageing. *Gen. Comp. Endocrinol.* **2011**, *172*, 39–43. [CrossRef] [PubMed]
37. Bulteau, A.-L.; Moreau, M.; Nizard, C.; Friguet, B. Impairment of Proteasome Function upon UVA- and UVB-Irradiation of Human Keratinocytes. *Free Radic. Biol. Med.* **2002**, *32*, 1157–1170. [CrossRef]
38. Aman, Y.; Schmauck-Medina, T.; Hansen, M.; Morimoto, R.I.; Simon, A.K.; Bjedov, I.; Palikaras, K.; Simonsen, A.; Johansen, T.; Tavernarakis, N.; et al. Autophagy in Healthy Aging and Disease. *Nat. Aging* **2021**, *1*, 634–650. [CrossRef]
39. Jeong, D.; Qomaladewi, N.P.; Lee, J.; Park, S.H.; Cho, J.Y. The Role of Autophagy in Skin Fibroblasts, Keratinocytes, Melanocytes, and Epidermal Stem Cells. *J. Investig. Dermatol.* **2020**, *140*, 1691–1697. [CrossRef]
40. Tashiro, K.; Shishido, M.; Fujimoto, K.; Hirota, Y.; Yo, K.; Gomi, T.; Tanaka, Y. Age-Related Disruption of Autophagy in Dermal Fibroblasts Modulates Extracellular Matrix Components. *Biochem. Biophys. Res. Commun.* **2014**, *443*, 167–172. [CrossRef]
41. Zhang, C.-F.; Gruber, F.; Ni, C.; Mildner, M.; Koenig, U.; Karner, S.; Barresi, C.; Rossiter, H.; Narzt, M.-S.; Nagelreiter, I.M.; et al. Suppression of Autophagy Dysregulates the Antioxidant Response and Causes Premature Senescence of Melanocytes. *J. Investig. Dermatol.* **2015**, *135*, 1348–1357. [CrossRef]
42. Kim, J.Y.; Kim, J.; Ahn, Y.; Lee, E.J.; Hwang, S.; Almurayshid, A.; Park, K.; Chung, H.; Kim, H.J.; Lee, S.; et al. Autophagy Induction Can Regulate Skin Pigmentation by Causing Melanosome Degradation in Keratinocytes and Melanocytes. *Pigment Cell Melanoma Res.* **2020**, *33*, 403–415. [CrossRef]
43. Pfisterer, K.; Shaw, L.E.; Symmank, D.; Weninger, W. The Extracellular Matrix in Skin Inflammation and Infection. *Front. Cell Dev. Biol.* **2021**, *9*, 1578. [CrossRef]
44. Stefanovic, N.; Irvine, A.D.; Flohr, C. The Role of the Environment and Exposome in Atopic Dermatitis. *Curr. Treat. Options Allergy* **2021**, *8*, 222–241. [CrossRef] [PubMed]
45. Celebi Sozener, Z.; Ozdel Ozturk, B.; Cerci, P.; Turk, M.; Gorgulu Akin, B.; Akdis, M.; Altiner, S.; Ozbey, U.; Ogulur, I.; Mitamura, Y.; et al. Epithelial Barrier Hypothesis: Effect of the External Exposome on the Microbiome and Epithelial Barriers in Allergic Disease. *Allergy* **2022**, *77*, 1418–1449. [CrossRef]
46. Fernández-Gallego, N.; Sánchez-Madrid, F.; Cibrian, D. Role of AHR Ligands in Skin Homeostasis and Cutaneous Inflammation. *Cells* **2021**, *10*, 3176. [CrossRef] [PubMed]
47. Kyoreva, M.; Li, Y.; Hoosenally, M.; Hardman-Smart, J.; Morrison, K.; Tosi, I.; Tolaini, M.; Barinaga, G.; Stockinger, B.; Mrowietz, U.; et al. CYP1A1 Enzymatic Activity Influences Skin Inflammation Via Regulation of the AHR Pathway. *J. Investig. Dermatol.* **2021**, *141*, 1553–1563.e3. [CrossRef] [PubMed]
48. Esser, C.; Bargen, I.; Weighardt, H.; Haarmann-Stemmann, T.; Krutmann, J. Functions of the Aryl Hydrocarbon Receptor in the Skin. *Semin. Immunopathol.* **2013**, *35*, 677–691. [CrossRef]
49. Haas, K.; Weighardt, H.; Deenen, R.; Köhrer, K.; Clausen, B.; Zahner, S.; Boukamp, P.; Bloch, W.; Krutmann, J.; Esser, C. Aryl Hydrocarbon Receptor in Keratinocytes Is Essential for Murine Skin Barrier Integrity. *J. Investig. Dermatol.* **2016**, *136*, 2260–2269. [CrossRef]
50. Di Meglio, P.; Duarte, J.H.; Ahlfors, H.; Owens, N.D.L.; Li, Y.; Villanova, F.; Tosi, I.; Hirota, K.; Nestle, F.O.; Mrowietz, U.; et al. Activation of the Aryl Hydrocarbon Receptor Dampens the Severity of Inflammatory Skin Conditions. *Immunity* **2014**, *40*, 989–1001. [CrossRef]
51. Nakamura, M.; Morita, A.; Seité, S.; Haarmann-Stemmann, T.; Grether-Beck, S.; Krutmann, J. Environment-Induced Lentigines: Formation of Solar Lentigines beyond Ultraviolet Radiation. *Exp. Dermatol.* **2015**, *24*, 407–411. [CrossRef]
52. Liu, T.; Zhang, L.; Joo, D.; Sun, S.-C. NF-KB Signaling in Inflammation. *Signal Transduct. Target. Ther.* **2017**, *2*, 17023. [CrossRef]
53. Burke, K.E. Mechanisms of Aging and Development—A New Understanding of Environmental Damage to the Skin and Prevention with Topical Antioxidants. *Mech. Ageing Dev.* **2018**, *172*, 123–130. [CrossRef]
54. Wang, Y.; Wang, L.; Wen, X.; Hao, D.; Zhang, N.; He, G.; Jiang, X. NF-KB Signaling in Skin Aging. *Mech. Ageing Dev.* **2019**, *184*, 111160. [CrossRef] [PubMed]
55. Kobielak, A.; Boddupally, K. Junctions and Inflammation in the Skin. *Cell Commun. Adhes.* **2014**, *21*, 141–147. [CrossRef] [PubMed]
56. Saha, S.; Buttari, B.; Panieri, E.; Profumo, E.; Saso, L. An Overview of Nrf2 Signaling Pathway and Its Role in Inflammation. *Molecules* **2020**, *25*, 5474. [CrossRef]
57. Rajagopalan, P.; Jain, A.P.; Nanjappa, V.; Patel, K.; Mangalaparthi, K.K.; Babu, N.; Cavusoglu, N.; Roy, N.; Soeur, J.; Breton, L.; et al. Proteome-Wide Changes in Primary Skin Keratinocytes Exposed to Diesel Particulate Extract—A Role for Antioxidants in Skin Health. *J. Dermatol. Sci.* **2018**, *91*, 239–249. [CrossRef] [PubMed]
58. Ko, H.; Kim, M.-M. H2O2 Promotes the Aging Process of Melanogenesis through Modulation of MITF and Nrf2. *Mol. Biol. Rep.* **2019**, *46*, 2461–2471. [CrossRef]
59. Gęgotek, A.; Skrzydlewska, E. The Role of Transcription Factor Nrf2 in Skin Cells Metabolism. *Arch. Dermatol. Res.* **2015**, *307*, 385–396. [CrossRef] [PubMed]
60. Waaijer, M.E.C.; Gunn, D.A.; Adams, P.D.; Pawlikowski, J.S.; Griffiths, C.E.M.; van Heemst, D.; Slagboom, P.E.; Westendorp, R.G.J.; Maier, A.B. P16INK4a Positive Cells in Human Skin Are Indicative of Local Elastic Fiber Morphology, Facial Wrinkling, and Perceived Age. *J. Gerontol. Ser. A Biol. Sci. Med. Sci.* **2016**, *71*, 1022–1028. [CrossRef] [PubMed]
61. Wedel, S.; Martic, I.; Hrapovic, N.; Fabre, S.; Madreiter-Sokolowski, C.T.; Haller, T.; Pierer, G.; Ploner, C.; Jansen-Dürr, P.; Cavinato, M. TBHP Treatment as a Model for Cellular Senescence and Pollution-Induced Skin Aging. *Mech. Ageing Dev.* **2020**, *190*, 111318. [CrossRef]

62. Rorteau, J.; Chevalier, F.P.; Bonnet, S.; Barthélemy, T.; Lopez-Gaydon, A.; Martin, L.S.; Bechetoille, N.; Lamartine, J. Maintenance of Chronological Aging Features in Culture of Normal Human Dermal Fibroblasts from Old Donors. *Cells* **2022**, *11*, 858. [CrossRef]
63. Victorelli, S.; Lagnado, A.; Halim, J.; Moore, W.; Talbot, D.; Barrett, K.; Chapman, J.; Birch, J.; Ogrodnik, M.; Meves, A.; et al. Senescent Human Melanocytes Drive Skin Ageing via Paracrine Telomere Dysfunction. *EMBO J.* **2019**, *38*, e101982. [CrossRef]
64. Barysch, M.J.; Braun, R.P.; Kolm, I.; Ahlgrimm-Siesz, V.; Hofmann-Wellenhof, R.; Duval, C.; Warrick, E.; Bernerd, F.; Nouveau, S.; Dummer, R. Keratinocytic Malfunction as a Trigger for the Development of Solar Lentigines. *Dermatopathology* **2019**, *6*, 1–11. [CrossRef] [PubMed]
65. Yoon, J.E.; Kim, Y.; Kwon, S.; Kim, M.; Kim, Y.H.; Kim, J.-H.; Park, T.J.; Kang, H.Y. Senescent Fibroblasts Drive Ageing Pigmentation: A Potential Therapeutic Target for Senile Lentigo. *Theranostics* **2018**, *8*, 4620–4632. [CrossRef] [PubMed]
66. Carmona-Cruz, S.; Orozco-Covarrubias, L.; Sáez-de-Ocariz, M. The Human Skin Microbiome in Selected Cutaneous Diseases. *Front. Cell. Infect. Microbiol.* **2022**, *12*, 834135. [CrossRef]
67. Kim, M.; Park, T.; Yun, J.I.; Lim, H.W.; Han, N.R.; Lee, S.T. Investigation of Age-Related Changes in the Skin Microbiota of Korean Women. *Microorganisms* **2020**, *8*, 1581. [CrossRef]
68. D'Mello, S.; Finlay, G.; Baguley, B.; Askarian-Amiri, M. Signaling Pathways in Melanogenesis. *Int. J. Mol. Sci.* **2016**, *17*, 1144. [CrossRef]
69. Yamaguchi, Y.; Brenner, M.; Hearing, V.J. The Regulation of Skin Pigmentation. *J. Biol. Chem.* **2007**, *282*, 27557–27561. [CrossRef] [PubMed]
70. Serre, C.; Busuttil, V.; Botto, J.-M. Intrinsic and Extrinsic Regulation of Human Skin Melanogenesis and Pigmentation. *Int. J. Cosmet. Sci.* **2018**, *40*, 328–347. [CrossRef] [PubMed]
71. Venkatesh, S.; Maymone, M.B.C.; Vashi, N.A. Aging in Skin of Color. *Clin. Dermatol.* **2019**, *37*, 351–357. [CrossRef]
72. Vierkötter, A.; Hüls, A.; Yamamoto, A.; Stolz, S.; Krämer, U.; Matsui, M.S.; Morita, A.; Wang, S.; Li, Z.; Jin, L.; et al. Extrinsic Skin Ageing in German, Chinese and Japanese Women Manifests Differently in All Three Groups Depending on Ethnic Background, Age and Anatomical Site. *J. Dermatol. Sci.* **2016**, *83*, 219–225. [CrossRef]
73. Wong, Q.Y.A.; Chew, F.T. Defining Skin Aging and Its Risk Factors: A Systematic Review and Meta-Analysis. *Sci. Rep.* **2021**, *11*, 22075. [CrossRef]
74. Vashi, N.A.; de Castro Maymone, M.B.; Kundu, R.V. Aging Differences in Ethnic Skin. *J. Clin. Aesthet. Dermatol.* **2016**, *9*, 31–38. [PubMed]
75. Langton, A.K.; Alessi, S.; Hann, M.; Chien, A.L.-L.; Kang, S.; Griffiths, C.E.M.; Watson, R.E.B. Aging in Skin of Color: Disruption to Elastic Fiber Organization Is Detrimental to Skin's Biomechanical Function. *J. Investig. Dermatol.* **2019**, *139*, 779–788. [CrossRef] [PubMed]
76. Chien, A.L.; Suh, J.; Cesar, S.S.A.; Fischer, A.H.; Cheng, N.; Poon, F.; Rainer, B.; Leung, S.; Martin, J.; Okoye, G.A.; et al. Pigmentation in African American Skin Decreases with Skin Aging. *J. Am. Acad. Dermatol.* **2016**, *75*, 782–787. [CrossRef]
77. Del Bino, S.; Duval, C.; Bernerd, F. Clinical and Biological Characterization of Skin Pigmentation Diversity and Its Consequences on UV Impact. *Int. J. Mol. Sci.* **2018**, *19*, 2668. [CrossRef]
78. Numan, M.; Brown, J.; Michou, L. Impact of Air Pollutants on Oxidative Stress in Common Autophagy-Mediated Aging Diseases. *Int. J. Environ. Res. Public Health* **2015**, *12*, 2289–2305. [CrossRef] [PubMed]
79. Ryu, Y.S.; Kang, K.A.; Piao, M.J.; Ahn, M.J.; Yi, J.M.; Bossis, G.; Hyun, Y.-M.; Park, C.O.; Hyun, J.W. Particulate Matter-Induced Senescence of Skin Keratinocytes Involves Oxidative Stress-Dependent Epigenetic Modifications. *Exp. Mol. Med.* **2019**, *51*, 1–14. [CrossRef]
80. Shin, K.-O.; Uchida, Y.; Park, K. Diesel Particulate Extract Accelerates Premature Skin Aging in Human Fibroblasts via Ceramide-1-Phosphate-Mediated Signaling Pathway. *Int. J. Mol. Sci.* **2022**, *23*, 2691. [CrossRef]
81. Lee, C.-W.; Lin, Z.-C.; Hu, S.C.-S.; Chiang, Y.-C.; Hsu, L.-F.; Lin, Y.-C.; Lee, I.-T.; Tsai, M.-H.; Fang, J.-Y. Urban Particulate Matter Down-Regulates Filaggrin via COX2 Expression/PGE2 Production Leading to Skin Barrier Dysfunction. *Sci. Rep.* **2016**, *6*, 27995. [CrossRef]
82. Park, S.-Y.; Byun, E.; Lee, J.; Kim, S.; Kim, H. Air Pollution, Autophagy, and Skin Aging: Impact of Particulate Matter (PM10) on Human Dermal Fibroblasts. *Int. J. Mol. Sci.* **2018**, *19*, 2727. [CrossRef]
83. Patatian, A.; Delestre-Delacour, C.; Percoco, G.; Ramdani, Y.; Di Giovanni, M.; Peno-Mazzarino, L.; Bader, T.; Bénard, M.; Driouich, A.; Lati, E.; et al. Skin Biological Responses to Urban Pollution in an Ex Vivo Model. *Toxicol. Lett.* **2021**, *348*, 85–96. [CrossRef]
84. Vierkötter, A.; Schikowski, T.; Ranft, U.; Sugiri, D.; Matsui, M.; Krämer, U.; Krutmann, J. Airborne Particle Exposure and Extrinsic Skin Aging. *J. Investig. Dermatol.* **2010**, *130*, 2719–2726. [CrossRef] [PubMed]
85. Damevska, K.; Simeonovski, V.; Darlenski, R.; Damevska, S. How to Prevent Skin Damage from Air Pollution Part 2: Current Treatment Options. *Dermatol. Ther.* **2021**, *34*, e15132. [CrossRef] [PubMed]
86. Mancebo, S.E.; Wang, S.Q. Recognizing the Impact of Ambient Air Pollution on Skin Health. *J. Eur. Acad. Dermatol. Venereol.* **2015**, *29*, 2326–2332. [CrossRef] [PubMed]
87. Peng, F.; Xue, C.-H.; Hwang, S.K.; Li, W.-H.; Chen, Z.; Zhang, J.-Z. Exposure to Fine Particulate Matter Associated with Senile Lentigo in Chinese Women: A Cross-Sectional Study. *J. Eur. Acad. Dermatol. Venereol.* **2017**, *31*, 355–360. [CrossRef]
88. Drakaki, E.; Dessinioti, C.; Antoniou, C.V. Air Pollution and the Skin. *Front. Environ. Sci.* **2014**, *2*, 11. [CrossRef]
89. Bae, Y.J.; Park, K.Y.; Han, H.S.; Kim, Y.S.; Hong, J.Y.; Han, T.Y.; Seo, S.J. Effects of Particulate Matter in a Mouse Model of Oxazolone-Induced Atopic Dermatitis. *Ann. Dermatol.* **2020**, *32*, 496. [CrossRef]

90. Pan, S.; Qiu, Y.; Li, M.; Yang, Z.; Liang, D. Recent Developments in the Determination of PM2.5 Chemical Composition. *Bull. Environ. Contam. Toxicol.* **2022**, *108*, 819–823. [CrossRef]
91. Arias-Pérez, R.D.; Taborda, N.A.; Gómez, D.M.; Narvaez, J.F.; Porras, J.; Hernandez, J.C. Inflammatory Effects of Particulate Matter Air Pollution. *Environ. Sci. Pollut. Res.* **2020**, *27*, 42390–42404. [CrossRef]
92. Soeur, J.; Belaïdi, J.-P.; Chollet, C.; Denat, L.; Dimitrov, A.; Jones, C.; Perez, P.; Zanini, M.; Zobiri, O.; Mezzache, S.; et al. Photo-Pollution Stress in Skin: Traces of Pollutants (PAH and Particulate Matter) Impair Redox Homeostasis in Keratinocytes Exposed to UVA1. *J. Dermatol. Sci.* **2017**, *86*, 162–169. [CrossRef]
93. Fernando, I.P.S.; Jayawardena, T.U.; Kim, H.-S.; Vaas, A.P.J.P.; De Silva, H.I.C.; Nanayakkara, C.M.; Abeytunga, D.T.U.; Lee, W.; Ahn, G.; Lee, D.-S.; et al. A Keratinocyte and Integrated Fibroblast Culture Model for Studying Particulate Matter-Induced Skin Lesions and Therapeutic Intervention of Fucosterol. *Life Sci.* **2019**, *233*, 116714. [CrossRef]
94. Estrella, B.; Naumova, E.N.; Cepeda, M.; Voortman, T.; Katsikis, P.D.; Drexhage, H.A. Effects of Air Pollution on Lung Innate Lymphoid Cells: Review of In Vitro and In Vivo Experimental Studies. *Int. J. Environ. Res. Public Health* **2019**, *16*, 2347. [CrossRef] [PubMed]
95. Laing, S.; Wang, G.; Briazova, T.; Zhang, C.; Wang, A.; Zheng, Z.; Gow, A.; Chen, A.F.; Rajagopalan, S.; Chen, L.C.; et al. Airborne Particulate Matter Selectively Activates Endoplasmic Reticulum Stress Response in the Lung and Liver Tissues. *Am. J. Physiol. Physiol.* **2010**, *299*, C736–C749. [CrossRef] [PubMed]
96. Marchini, T.; D'Annunzio, V.; Paz, M.L.; Cáceres, L.; Garcés, M.; Perez, V.; Tasat, D.; Vanasco, V.; Magnani, N.; Gonzalez Maglio, D.; et al. Selective TNF-α Targeting with Infliximab Attenuates Impaired Oxygen Metabolism and Contractile Function Induced by an Acute Exposure to Air Particulate Matter. *Am. J. Physiol. Circ. Physiol.* **2015**, *309*, H1621–H1628. [CrossRef] [PubMed]
97. Mylonas, A.; Conrad, C. Psoriasis: Classical vs. Paradoxical. The Yin-Yang of TNF and Type I Interferon. *Front. Immunol.* **2018**, *9*, 2746. [CrossRef] [PubMed]
98. Pieters, N.; Janssen, B.G.; Dewitte, H.; Cox, B.; Cuypers, A.; Lefebvre, W.; Smeets, K.; Vanpoucke, C.; Plusquin, M.; Nawrot, T.S. Biomolecular Markers within the Core Axis of Aging and Particulate Air Pollution Exposure in the Elderly: A Cross-Sectional Study. *Environ. Health Perspect.* **2016**, *124*, 943–950. [CrossRef] [PubMed]
99. Zhao, R.; Guo, Z.; Zhang, R.; Deng, C.; Xu, J.; Dong, W.; Hong, Z.; Yu, H.; Situ, H.; Liu, C.; et al. Nasal Epithelial Barrier Disruption by Particulate Matter $\leq$2.5 Mm via Tight Junction Protein Degradation. *J. Appl. Toxicol.* **2018**, *38*, 678–687. [CrossRef] [PubMed]
100. Dijkhoff, I.M.; Drasler, B.; Karakocak, B.B.; Petri-Fink, A.; Valacchi, G.; Eeman, M.; Rothen-Rutishauser, B. Impact of Airborne Particulate Matter on Skin: A Systematic Review from Epidemiology to in Vitro Studies. *Part. Fibre Toxicol.* **2020**, *17*, 35. [CrossRef]
101. Pan, T.-L.; Wang, P.-W.; Aljuffali, I.A.; Huang, C.-T.; Lee, C.-W.; Fang, J.-Y. The Impact of Urban Particulate Pollution on Skin Barrier Function and the Subsequent Drug Absorption. *J. Dermatol. Sci.* **2015**, *78*, 51–60. [CrossRef]
102. Kim, S.; Kim, J.; Lee, Y.I.; Jang, S.; Song, S.Y.; Lee, W.J.; Lee, J.H. Particulate Matter-induced Atmospheric Skin Aging Is Aggravated by UVA and Inhibited by a Topical L-ascorbic Acid Compound. *Photodermatol. Photoimmunol. Photomed.* **2022**, *38*, 123–131. [CrossRef]
103. Long, E.; Schwartz, C.; Carlsten, C. Controlled Human Exposure to Diesel Exhaust: A Method for Understanding Health Effects of Traffic-Related Air Pollution. *Part. Fibre Toxicol.* **2022**, *19*, 15. [CrossRef]
104. Molina-García, M.; Malvehy, J.; Granger, C.; Garre, A.; Trullàs, C.; Puig, S. Exposome and Skin. Part 2. The Influential Role of the Exposome, Beyond UVR, in Actinic Keratosis, Bowen's Disease and Squamous Cell Carcinoma: A Proposal. *Dermatol. Ther.* **2022**, *12*, 361–380. [CrossRef] [PubMed]
105. Bashir, M.M.; Sharma, M.R.; Werth, V.P. TNF-α Production in the Skin. *Arch. Dermatol. Res.* **2009**, *301*, 87–91. [CrossRef] [PubMed]
106. Singh, M.; Mansuri, M.S.; Kadam, A.; Palit, S.P.; Dwivedi, M.; Laddha, N.C.; Begum, R. Tumor Necrosis Factor-Alpha Affects Melanocyte Survival and Melanin Synthesis via Multiple Pathways in Vitiligo. *Cytokine* **2021**, *140*, 155432. [CrossRef]
107. Kim, H.; Park, S.-Y.; Moon, S.; Lee, J.; Kim, S. Autophagy in Human Skin Fibroblasts: Impact of Age. *Int. J. Mol. Sci.* **2018**, *19*, 2254. [CrossRef] [PubMed]
108. Yoon, S.; Lim, C.; Chung, H.-J.; Kim, J.-H.; Huh, Y.; Park, K.; Jeong, S. Autophagy Activation by Crepidiastrum Denticulatum Extract Attenuates Environmental Pollutant-Induced Damage in Dermal Fibroblasts. *Int. J. Mol. Sci.* **2019**, *20*, 517. [CrossRef] [PubMed]
109. Mokrzyński, K.; Krzysztyńska-Kuleta, O.; Zawrotniak, M.; Sarna, M.; Sarna, T. Fine Particulate Matter-Induced Oxidative Stress Mediated by UVA-Visible Light Leads to Keratinocyte Damage. *Int. J. Mol. Sci.* **2021**, *22*, 10645. [CrossRef] [PubMed]
110. Dimitrov, A.; Zanini, M.; Zucchi, H.; Boudah, S.; Lima, J.; Soeur, J.; Marrot, L. Vitamin C Prevents Epidermal Damage Induced by PM-associated Pollutants and UVA1 Combined Exposure. *Exp. Dermatol.* **2021**, *30*, 1693–1698. [CrossRef]
111. Suo, D.; Zeng, S.; Zhang, J.; Meng, L.; Weng, L. PM2.5 Induces Apoptosis, Oxidative Stress Injury and Melanin Metabolic Disorder in Human Melanocytes. *Exp. Ther. Med.* **2020**, *19*, 3227. [CrossRef]
112. Hieda, D.S.; Anastacio da Costa Carvalho, L.; Vaz de Mello, B.; de Oliveira, E.A.; Romano de Assis, S.; Wu, J.; Du-Thumm, L.; Viana da Silva, C.L.; Roubicek, D.A.; Maria-Engler, S.S.; et al. Air Particulate Matter Induces Skin Barrier Dysfunction and Water Transport Alteration on a Reconstructed Human Epidermis Model. *J. Investig. Dermatol.* **2020**, *140*, 2343–2352.e3. [CrossRef]
113. Sarama, R.; Matharu, P.K.; Abduldaiem, Y.; Corrêa, M.P.; Gil, C.D.; Greco, K.V. In Vitro Disease Models for Understanding Psoriasis and Atopic Dermatitis. *Front. Bioeng. Biotechnol.* **2022**, *10*, 803218. [CrossRef]

114. Percoco, G.; Patatian, A.; Eudier, F.; Grisel, M.; Bader, T.; Lati, E.; Savary, G.; Picard, C.; Benech, P. Impact of Cigarette Smoke on Physical-chemical and Molecular Proprieties of Human Skin in an Ex Vivo Model. *Exp. Dermatol.* **2021**, *30*, 1610–1618. [CrossRef] [PubMed]
115. Ono, Y.; Torii, K.; Fritsche, E.; Shintani, Y.; Nishida, E.; Nakamura, M.; Shirakata, Y.; Haarmann-Stemmann, T.; Abel, J.; Krutmann, J.; et al. Role of the Aryl Hydrocarbon Receptor in Tobacco Smoke Extract-Induced Matrix Metalloproteinase-1 Expression. *Exp. Dermatol.* **2013**, *22*, 349–353. [CrossRef] [PubMed]
116. Hoskin, R.; Pambianchi, E.; Pecorelli, A.; Grace, M.; Therrien, J.-P.; Valacchi, G.; Lila, M.A. Novel Spray Dried Algae-Rosemary Particles Attenuate Pollution-Induced Skin Damage. *Molecules* **2021**, *26*, 3781. [CrossRef] [PubMed]
117. D'Agostini, F.; Balansky, R.; Pesce, C.; Fiallo, P.; Lubet, R.A.; Kelloff, G.J.; De Flora, S. Induction of Alopecia in Mice Exposed to Cigarette Smoke. *Toxicol. Lett.* **2000**, *114*, 117–123. [CrossRef]
118. Lecas, S.; Boursier, E.; Fitoussi, R.; Vié, K.; Momas, I.; Seta, N.; Achard, S. In Vitro Model Adapted to the Study of Skin Ageing Induced by Air Pollution. *Toxicol. Lett.* **2016**, *259*, 60–68. [CrossRef]
119. Langton, A.K.; Tsoureli-Nikita, E.; Merrick, H.; Zhao, X.; Antoniou, C.; Stratigos, A.; Akhtar, R.; Derby, B.; Sherratt, M.J.; Watson, R.E.B.; et al. The Systemic Influence of Chronic Smoking on Skin Structure and Mechanical Function. *J. Pathol.* **2020**, *251*, 420–428. [CrossRef]
120. Petracca, B.; Nădăban, A.; Eeman, M.; Gooris, G.S.; Bouwstra, J.A. Effects of Ozone on Stratum Corneum Lipid Integrity and Assembly. *Chem. Phys. Lipids* **2021**, *240*, 105121. [CrossRef]
121. Valacchi, G.; Pagnin, E.; Corbacho, A.M.; Olano, E.; Davis, P.A.; Packer, L.; Cross, C.E. In Vivo Ozone Exposure Induces Antioxidant/Stress-Related Responses in Murine Lung and Skin. *Free Radic. Biol. Med.* **2004**, *36*, 673–681. [CrossRef]
122. Valacchi, G.; van der Vliet, A.; Schock, B.C.; Okamoto, T.; Obermuller-Jevic, U.; Cross, C.E.; Packer, L. Ozone Exposure Activates Oxidative Stress Responses in Murine Skin. *Toxicology* **2002**, *179*, 163–170. [CrossRef]
123. Valacchi, G.; Sticozzi, C.; Belmonte, G.; Cervellati, F.; Demaude, J.; Chen, N.; Krol, Y.; Oresajo, C. Vitamin C Compound Mixtures Prevent Ozone-Induced Oxidative Damage in Human Keratinocytes as Initial Assessment of Pollution Protection. *PLoS ONE* **2015**, *10*, e0131097. [CrossRef]
124. Krutmann, J.; Bouloc, A.; Sore, G.; Bernard, B.A.; Passeron, T. The Skin Aging Exposome. *J. Dermatol. Sci.* **2017**, *85*, 152–161. [CrossRef]
125. Valacchi, G.; Pecorelli, A.; Belmonte, G.; Pambianchi, E.; Cervellati, F.; Lynch, S.; Krol, Y.; Oresajo, C. Protective Effects of Topical Vitamin C Compound Mixtures against Ozone-Induced Damage in Human Skin. *J. Investig. Dermatol.* **2017**, *137*, 1373–1375. [CrossRef] [PubMed]
126. Chavatte, L.; Juan, M.; Mounicou, S.; Leblanc Noblesse, E.; Pays, K.; Nizard, C.; Bulteau, A.-L. Elemental and Molecular Imaging of Human Full Thickness Skin after Exposure to Heavy Metals. *Metallomics* **2020**, *12*, 1555–1562. [CrossRef] [PubMed]
127. Juliano, C.; Magrini, G. Cosmetic Functional Ingredients from Botanical Sources for Anti-Pollution Skincare Products. *Cosmetics* **2018**, *5*, 19. [CrossRef]
128. Mousavi, S.E.; Delgado-Saborit, J.M.; Adivi, A.; Pauwels, S.; Godderis, L. Air Pollution and Endocrine Disruptors Induce Human Microbiome Imbalances: A Systematic Review of Recent Evidence and Possible Biological Mechanisms. *Sci. Total Environ.* **2022**, *816*, 151654. [CrossRef]
129. Charoenchon, N.; Rhodes, L.E.; Nicolaou, A.; Williamson, G.; Watson, R.E.B.; Farrar, M.D. Ultraviolet Radiation-induced Degradation of Dermal Extracellular Matrix and Protection by Green Tea Catechins: A Randomized Controlled Trial. *Clin. Exp. Dermatol.* **2022**, *47*, 1314–1323. [CrossRef]
130. Tigges, J.; Haarmann-Stemmann, T.; Vogel, C.F.A.; Grindel, A.; Hübenthal, U.; Brenden, H.; Grether-Beck, S.; Vielhaber, G.; Johncock, W.; Krutmann, J.; et al. The New Aryl Hydrocarbon Receptor Antagonist E/Z-2-Benzylindene-5,6-Dimethoxy-3,3-Dimethylindan-1-One Protects against UVB-Induced Signal Transduction. *J. Investig. Dermatol.* **2014**, *134*, 556–559. [CrossRef]
131. Khmaladze, I.; Österlund, C.; Smiljanic, S.; Hrapovic, N.; Lafon-Kolb, V.; Amini, N.; Xi, L.; Fabre, S. A Novel Multifunctional Skin Care Formulation with a Unique Blend of Antipollution, Brightening and Antiaging Active Complexes. *J. Cosmet. Dermatol.* **2020**, *19*, 1415–1425. [CrossRef]

cells

Review

Age-Related Lysosomal Dysfunctions

Lena Guerrero-Navarro [1,2], Pidder Jansen-Dürr [1,2] and Maria Cavinato [1,2,*]

1 Institute for Biomedical Aging Research, Universität Innsbruck, 6020 Innsbruck, Austria; lena.guerrero-navarro@uibk.ac.at (L.G.-N.); pidder.jansen-duerr@uibk.ac.at (P.J.-D.)
2 Center for Molecular Biosciences Innsbruck, Innrain 58, 6020 Innsbruck, Austria
* Correspondence: maria.cavinato-nascimento@uibk.ac.at

Abstract: Organismal aging is normally accompanied by an increase in the number of senescent cells, growth-arrested metabolic active cells that affect normal tissue function. These cells present a series of characteristics that have been studied over the last few decades. The damage in cellular organelles disbalances the cellular homeostatic processes, altering the behavior of these cells. Lysosomal dysfunction is emerging as an important factor that could regulate the production of inflammatory molecules, metabolic cellular state, or mitochondrial function.

Keywords: lysosome; aging; senescence

Citation: Guerrero-Navarro, L.; Jansen-Dürr, P.; Cavinato, M. Age-Related Lysosomal Dysfunctions. *Cells* **2022**, *11*, 1977. https://doi.org/10.3390/cells11121977

Academic Editors: Kay-Dietrich Wagner and Nicole Wagner

Received: 1 May 2022
Accepted: 16 June 2022
Published: 20 June 2022

1. Aging and Senescence

Aging is a process associated with the detriment of normal physiological functions, which leads to the manifestation of diverse diseases such as cardiovascular and neurodegenerative diseases, joint degenerative diseases, and metabolic diseases such as diabetes, among others [1].

The aging process has been associated in many tissues to an increase in the number of senescent cells, which is considered as a hallmark of organismal aging. This rise in the ratio of senescent cells contributes to the pathogenesis of age-related diseases [1].

Senescence is a physiological process involved in the suppression of tumors, which also participates in the development of organisms and in wound healing, where these cells transiently appear and are removed [2]. However, the permanence of these metabolic active cells in aged tissues promotes the dysfunctional remodeling of the tissue and foments inflammation, contributing to tissue decline [2].

Precisely, due to their active metabolism, senescent cells affect their environment and neighboring cells, through the secretion of matrix metalloproteinases, growth factors, chemokines, cytokines, and other immunomodulatory molecules. The composition of this set of factors, commonly denominated as senescence-associated secretory phenotype (SASP), changes depending on the type of cell and the senescence-inducing stimuli [3].

Cells can undergo senescence in response to various intrinsic and extrinsic stimuli, giving rise to different subtypes of senescent cells. Replicative senescence occurs when the intrinsic stimuli that induces senescence is telomere shortening. When senescence is induced by an external harmful stimulus such as exposition to oxidant molecules or ionizing irradiation, it is referred to as stress-induced premature senescence (SIPS) [4]. Oncogene induced senescence occurs when a prooncogenic mutation or a loss-of function mutation in a tumor-suppressor gene induces a strong antiproliferative response. Recently, other types of senescence have been identified, for example, mitochondrial dysfunction-associated senescence (MiDAS), in which mitochondrial dysfunction induces a type of senescence with a characteristic SASP that lacks the IL-1 inflammatory arm [5].

As evidenced, senescence is heterogeneous, and some markers present in senescent cells also exist in other physiological conditions. For this reason, the identification of senescence requires the detection of several factors. These factors include SASP molecules, cell cycle arrest markers, chromatin remodeling, and SA-β-gal activity, among others [3].

One of the main characteristics of senescent cells is their inability to proliferate since their cell cycle can be arrested by several pathways. The principal pathways that are studied in this context are p53/21 and pRb/p16. p53, considered the guardian of the genome, is stabilized by post-translational modifications in response to cell damage, particularly damage to the genome, favoring the expression of CDK inhibitors such as p21 and p16. This leads to the dephosphorylation and consequent activation of pRb, which represses the activity of the E2F transcription factor family, stopping the cell cycle progression. The accumulation of p21, and particularly, of p16, are essential for the maintenance of senescence [2,6].

Age-related dysregulations in the organization of chromatin occur in senescent cells, with different structural changes occurring in different genes. Heterochromatin areas are present in target regions of E2F, reinforcing the cell cycle arrest, while other parts of the genome such as pericentromeric regions display a loss of heterochromatin, associated with genome reorganization and aberrant transcription [3,7]. Additionally, an increase in histone γ-H2AX foci associated with unrepaired DNA double-strand damage is a common characteristic of several types of senescent cells [2].

However, not only genomic damage has been reported as a feature of senescent cells. For instance, the nuclear membrane of different strains of human and murine cells is also impaired, with Laminb1 downregulation a common feature in senescent cells [8].

Additionally, senescent cells commonly present increased activity of the enzyme senescence-associated β-galactosidase detected at a pH of 6.0 (SA-β-gal). In fact, SA-β-gal is present in lysosomes, and in young cells, is detected at pH 4 and in senescent cells at a pH of 6 [9]. The fact that this enzyme works at a higher pH in senescent cells has not aroused much interest until recently, when damage to lysosomes associated with senescence and alterations in lysosomal acidification were evidenced [10–12].

Senescence is usually accompanied by other characteristics that can be linked to lysosomal function such as the loss of proteostasis, deregulated nutrient-sensing, mitochondrial dysfunction, and altered intercellular communication. Below, these hallmarks will be explained in the context of lysosomal dysfunction during aging.

2. Lysosomes

Lysosomes are heterogeneous organelles enclosed by a lipid bilayer and filled with hydrolytic enzymes. The lysosomes are traditionally described as the subcellular structures where the degradation of other organelles and macromolecules takes place, a fundamental process for maintaining cellular proteostasis [13].

There are several degradation processes in which the lysosomes are involved. If the substrate reaching the lysosomes comes from the extracellular environment, the degradation process is called endocytosis. If the material to be digested comes from the cell itself, the process is classified as autophagy. The lysosomes are also involved in plasma membrane repair through a mechanism called lysosomal exocytosis. Furthermore, degradation products from lysosomal catabolism can be sensed by metabolic complexes at the external side of its membrane, serving as a scaffold for metabolic regulation [13].

2.1. Lysosomal Structure and Components

To understand the lysosomal participation in these processes, the composition and structure of the organelle must be understood. The lysosome is delimited by a bilipidic membrane that encloses an acid lumen, where the degradative reactions take place. This lumen contains approximately 60 hydrolases that are active at acidic pH and participate in multistep catabolic processes to orchestrate the degradation of organelles and macromolecules to monomers [13]. Lysosomes have a role in protein degradation, since they contain proteases such as cathepsins. However, they are also involved in the degradation of other macromolecules such as lipids, which are catabolized by lipases such as the lysomal acid lipase LIPA [14]. Nucleases such as RNase T2 or DNase II participate in nucleic acid degradation [15], and enzymes such as the acid alpha-glucosidase are important for polysaccharide degradation [16]. Macromolecules catabolized into monomers inside the lysosomes can be reused in anaplerotic reactions, connecting lysosomal function with cellular metabolism.

The lysosomal membrane is covered in its internal part with a large glycocalyx, a thick layer of polysaccharides that prevent lysosomal membrane self-digestion, avoiding acid content leakage into the cytosol [13]. The lysosomal membrane contains key proteins for the fusion of the lysosome with autophagosomes, endosomes, or with the plasma membrane as well as pumps and channels that are fundamental for the transport of ions and molecules and to maintain lysosomal acidification [17].

Lysosomal acidification is tightly controlled. These organelles are acidified mainly by the action of the v-ATPase, which hydrolyses ATP to pump protons into the lysosomal lumen. This process generates a transmembrane voltage that is dissipated by the action of ion channels [17]. The correct functioning of these ion channels is important not just to dissipate the electrogenic gradient, but also for v-ATPase to continue introducing protons against the gradient. Indeed, an increase in lysosomal pH can have detrimental consequences on the digestion carried out by lysosomal hydrolases, since these enzymes are active at low pH (pH 4–5).

The compensation of the electrogenic gradient is achieved either through the action of a cation leaving the lysosome or by an anion entering the lysosome. In fact, the entry of chlorine (Cl^-) through channels such as the CLC-7 channel compensates for the entry of H+ in the acidification process, while cation channels such as TPRMLs and TPCs facilitate the efflux of K^+, Na^+, and Ca^{2+} [17].

Lysosomes are important calcium stores and the release of these ions is important to regulate endosome–lysosome fusion, autophagy, or lysosomal biogenesis [18–20]. Typically, lysosomal calcium efflux is regulated by Nicotinic acid adenine dinucleotide phosphate (NAADP), and the local calcium content allows the amplification of the signal increasing ER calcium release [21]. Moreover, contacts between the endolysosome and the ER have been described, being relevant for endosome maturation [18,22].

2.2. *Lysosomal Biogenetic Pathways and Metabolic Integration*

Lysosomal biogenesis is important for lysosomal adaptation to different situations and to replace disrupted or dysfunctional lysosomes. Similarly, under nutrient deprivation conditions, when autophagy is induced, lysosomal biogenesis is activated since the number of lysosomes must increase. Additionally, when a cell replicates, lysosomes have to be produced to be dispersed between the daughter cells. Diverse mechanisms participating in lysosome biogenesis have been described [23].

The mammalian target of rapamycin (mTORC) is a protein kinase involved in the regulation of cell growth and metabolism. mTORC is the main component of mTORC1 and mTORC2, two protein complexes that are distinguished by their accessory proteins. Aside from mTORC, mTORC1 contains the regulatory-associated protein of mTOR (RAPTOR), and mTORC2 is characterized by the presence of the rapamycin-insensitive subunit companion of mTOR (RICTOR) [24].

One mechanism that regulates lysosomal biogenesis depends on mTORC1 (mammalian target of rapamycin complex 1) activity, which converges in the activation or inhibition of TFEB (Figure 1). mTORC1 can sense the metabolic state of the cell. In nutrient-rich conditions, active mTORC1 binds with TFEB, the master transcriptional factor in the regulation of lysosome biogenesis and function, at the lysosomal surface [25]. mTORC1 phosphorylates TFEB, causing its binding, sequestration, and consequent inactivation by the regulatory 14-3-3 proteins in the cytosol. When the cell requires nutrients, mTORC1 is inactivated and TFEB translocates to the nucleus where it promotes the transcription of its target genes, thus enhancing lysosome biogenesis and autophagy [13,23]. This set of genes, involved in lysosome biogenesis, is known as the CLEAR (Coordinated Lysosomal

Expression and Regulation) gene network. TFEB is the main regulator of this network, but other transcription factors from the MiT/TFE family to which TFEB belongs such as TFE3, MITF, and TFEC can also regulate the transcription of these genes [26]. TFEB also regulates the expression of v-ATPase subunits, so mTORC1 indirectly regulates v-ATPase activity and the acidification process [27]. The v-ATPase subunit association is also a tightly regulated process. Although this mechanism has been better studied in yeast, it is known that in mammals, the assembly of v-ATPase is promoted by amino acid starvation and by glucose starvation [28]. Furthermore, TFEB enhances the expression of genes related to lipid catabolism as a response to cell starvation [29]. TFEB also enhances the transcription of PGC1-α, which is the master regulator of mitochondrial biogenesis and participates in mitochondrial clearance [30].

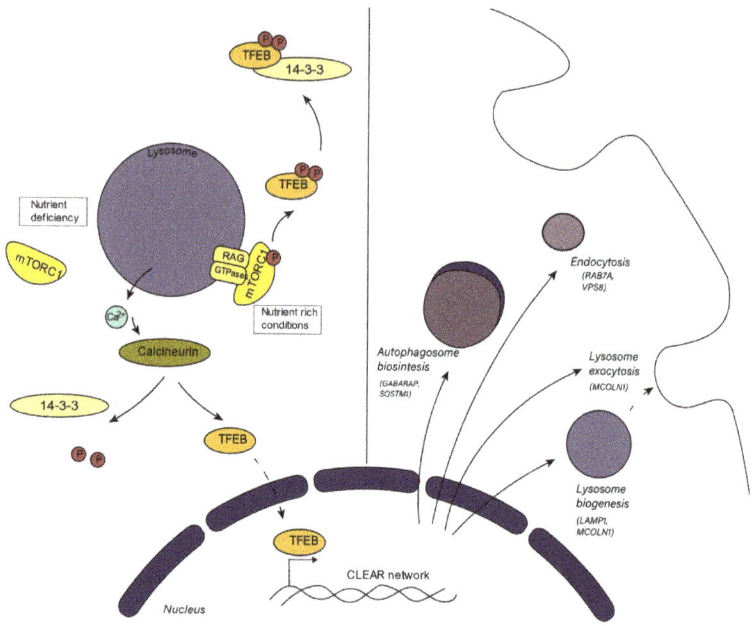

Figure 1. An overview of TFEB regulation through mTORC1 and calcineurin activity. Under nutrient rich conditions, mTORC1 phosphorylates TFEB, which is sequestered in the cytosol by 14-3-3 proteins. Under nutrient deficiency, mTORC1 is inactive, and calcineurin eliminates TFEB phosphates, allowing nuclear translocation, where TFEB induces the expression of genes related to autophagy, endocytosis, lysosome exocytosis, and lysosome biogenesis.

mTORC1 can regulate autophagy through TFEB activity, which induces the expression of genes involved in autophagy. In contrast, when nutrients are available, active mTORC1 phosphorylates and inhibits the activity of the complex ULK-ATG13-EIP200, required for autophagosome biogenesis, inhibiting autophagy [31].

Another described mechanism that regulates lysosomal biogenesis depends on protein kinase C (PKC). Similar to what has been described for mTORC1, the activation of PKC leads to phosphorylation and the consequent inactivation of the GSK3β kinase, which, in turn, fails to phosphorylate TFEB [32]. Unphosphorylated active TFEB translocates to the nucleus, where it performs its function. Additionally, the activation of PKC promotes lysosomal biogenesis through a second mechanism, in which the activation of JNK2/p38 leads to the phosphorylation of ZKSCAN3, a DNA-binding protein that upon phosphorylation leaves the nucleus, where it was repressing the expression of lysosomal genes [32].

Precisely, PKC can be activated by mTORC2, which emphasizes the role of mTOR in lysosomal biogenesis. Although mTORC1 function is better understood than the one of mTORC2, some mTORC2 effectors have been identified. Thus, mTORC2 can phosphorylate AGC kinases [33] including protein kinase B (AKT) and PKC. mTORC2 responds to growth factors, with the PI3K pathway important for mTORC2 activation [33], which can phosphorylate PKC, influencing lysosomal biogenesis. More research is needed to elucidate how mTORC2 regulates lysosomal biogenesis in response to growth factors, however, it appears that lysosomal positioning regulates the activation of mTORC1 and mTORC2 [34].

Lysosomal biogenesis can also be actively repressed. MYC, a transcriptional factor involved in cell proliferation, can, under certain conditions, occupy the promoter regions recognized by TFEB, disturbing the transcription of lysosomal biogenesis-related genes. Additionally, MYC interacts with histone deacetylases in these same promoter regions, interfering with the epigenetic regulation of lysosomal biogenesis [23,35].

Lysosomal biogenesis, which depends on TFEB, is closely related to the regulation of autophagy, and for this reason, to the physiological state of the lysosomes. Therefore, dysregulation of lysosomal biogenesis influences the degradative mechanisms that occur inside the lysosomes. The main publications that study the lysosomes in the context of aging are focused on autophagy, since the lysosomal state is tightly interconnected with the autophagic flux [36–38].

However, lysosomes also participate in the regulation of other molecular pathways, themselves being a subject of study in pathophysiological conditions. For instance, incorrect function of lysosomes can lead to metabolic alterations given that lysosomes act as a scaffold for metabolic integration, especially affecting mTORC1 function, which, in addition to autophagy, influences downstream processes such as translation, lipid synthesis, or energy metabolism [24]. Similarly, lysosomal disruption has important consequences in other processes with lysosomal participation such as endocytosis, mitophagy, or lysosomal exocytosis [13].

3. Processes in which the Lysosome Participates

3.1. Endocytosis

Endocytosis is a cellular process in which cellular membrane engulfing allows for the internalization of external material. There are several types of endocytosis: phagocytosis, macropinocytosis, clathrin-mediated endocytosis, caveolin-mediated endocytosis, and clathrin-and caveolin-independent endocytosis [39]. In each type of endocytosis, different membrane receptors participate in the engulfing process. After plasma membrane engulfment, an endosome is formed. The newly formed endosome matures to an early endosome that mediates receptor recycling to the membrane. The transition to late endosome is coordinated by small GTPases following a maturation pathway. This maturation allows for the progressive acidification of the endosome, which finally fuses with the lysosome, where it reaches the most acidic pH [40].

A general downregulation of endocytosis during aging or senescence has been observed, and some components important for endocytosis regulation such as βPIX or GIT also seem to be downregulated in senescent cells. βPIX is a p21-activated kinase that seems to control membrane ruffling. In fact, the knockdown of βPIX induces senescence in human dermal fibroblasts. In this case, the senescence phenotype is accompanied by a suppression of clathrin-mediated endocytosis, linked to the cleavage of amphiphysin 1, an endocytic adaptor important for actin polymerization. This dysfunctional endocytosis seems to be linked with persistent activated integrin signaling, which can be important for the senescent phenotype, as integrin signaling inhibition prevents senescence [41]. There is no information on the lysosomal dysfunction repercussions in endocytosis during senescence [42].

3.2. Autophagy

Autophagy is a cellular catabolic process that is lysosomal-dependent, in which the cell degrades its own material. It can be classified into three types: microautophagy, chaperone-mediated autophagy, and macroautophagy [43]. Macroautophagy is characterized by the formation of double-membrane vesicles, called autophagosomes, which sequester the substrates to be degraded. It is the most studied process, and it is the only way of eliminating whole organelles, so it is crucial for the removal of damaged organelles.

Macroautophagy requires the formation of a phagophore as a starting point, which leads to autophagosome development. The phagophore does not form from the budding of a pre-existing membrane, but the nucleation point starts at the cytosol and the ER seems to participate in lipid donation, establishing a structure called omegasome [44]. Autophagosome biogenesis occurs in three steps: nucleation, expansion, and closure [43]. LC3 is a protein involved in substrate selection and its lipidation is an important process during autophagy initiation. LC3 associates with the lipid phosphatidylethanol-amine (PE) in the phagosome, where it is processed by Atg3 and Atg7. The phagophore, which is a double membrane structure, is elongated in a process that involves these ATG proteins in a procedure regulated by kinases such as ULK and PI3KC3-CI. Finally, the fusion with the lysosome depends on RabGTPases and SNARE proteins [43,45,46].

Several studies have found that autophagy declines with aging in different organisms such as Drosophila and mice [47,48]. Particularly in humans, the expression of several autophagic proteins such as Atg5, Atg7, or BECN1 show a decline with aging [49]. In fact, in many animal models, a premature aging phenotype is observed when the activity of genes related to autophagy is silenced either by knockout or knockdown [50]. Additionally, in mice, the knockout of some autophagy-related genes such as Atg5, Atg9, or Atg13 results in a non-viable phenotype [51]. In contrast, the promotion or restoration of autophagy is often accompanied with a lifespan extension [52]. In fact, defects in autophagy are related to the development of age-related diseases such as atherosclerosis [53].

When it comes to cellular senescence, the role of autophagy becomes controversial. Autophagy can modulate the homeostasis of pro- and anti-senescence factors, either facilitating or impeding the development of the senescent phenotype depending on the stimuli, its duration, or the cell type [37]. In oncogene-induced senescence, for example, the inhibition of autophagy seems to ameliorate senescence [54]. However, the inhibition of autophagy in normal proliferating cells can facilitate the production of ROS promoting senescence. The selective induction of autophagy and the elimination of GATA4, a transcription factor that regulates senescence, may be beneficial for senescence reduction [37].

3.3. *Mitophagy and Mitochondrial Dysfunction*

Lysosomes and mitochondria are tightly interdependent [55]. Several mechanisms have been proposed to explain how lysosomal function interferes with mitochondrial function during aging. Mitophagy has emerged as the main link between these organelles.

Mitochondria are double-membrane-bound organelles that play a major role in cellular metabolism since aerobic respiration takes place in these organelles. The TCA cycle and fatty acid oxidation generate reducing agents that donate electrons to the mitochondrial electron transport chain for ATP production. Aside from their function in bioenergetics, mitochondria are organelles that influence cell fate since many enzymes that regulate apoptosis such as BCL-2 proteins or cytochrome c are found in these cellular compartments [56].

When the mitochondria are damaged and the mitochondrial membrane potential decreases, the respiratory chain does not work efficiently and ROS are produced. ROS generate oxidized proteins and lipids that further exacerbate mitochondrial dysfunction. These impaired mitochondria have to be removed, and the main mechanism to eliminate whole organelles is macroautophagy. The selective degradation of deficient mitochondria through macroautophagy is called mitophagy [56]. Thus, lysosomes are directly involved in the maintenance of mechanisms of mitochondria quality control.

Many authors have studied the effects of mitochondrial impairment on the lysosomes [57–59]. In T cells, the impairment of mitochondrial respiration enhances lysosomal biogenesis through TFEB. However, these lysosomes seem disrupted, as evidenced by lysosomal alkalinization, cathepsin B activity reduction, and lysosomal sphingolipid accumulation. Moreover, this same study demonstrated that mitochondrial impairment is linked with pro-inflammatory phenotypes [57]. This shows the important role played by lysosomal function in the induction of SASP on senescent cells, which are known for their partially dysfunctional mitochondria.

Moreover, the deletion of mitochondrial proteins such as AIF, OPA1, or PINK1, or the chemical inhibition of the electron transport chain result in lysosomal impairment. Antioxidant treatment can partially rescue the impaired lysosomes, suggesting that mitochondrial defects that favor ROS production lead to lysosomal defects [58]. Other authors suggest that, depending on the mitochondrial damage duration, lysosome biogenesis can be enhanced or downregulated. In particular, mitochondrial respiration dysfunctions lead to lysosomal biogenesis induction in short-term ETC inhibition. However, long-term respiration inhibition by rotenone, a complex I inhibitor, results in lysosomal biogenesis repression [59].

Similarly, in yeast, it has been found that lysosome-like vacuoles increase their pH during replicative aging, and this change affects normal mitochondrial function. A screening of relevant differentially expressed genes during this aging process identified VMA1 (a v-ATPase subunit) and VPH2 (an enzyme that participates in the v-ATPase assembly) as genes whose overexpression delayed mitochondrial impairment during aging. The overexpression of these genes has a pro-longevity effect, which is surprisingly not related to an enhanced autophagic flux but to the storage of neutral amino acids in the vacuole [60]. This shows the role of v-ATPase in the correct acidification and repercussion in the lysosomal function, which at the same time can resonate in the mitochondrial function.

Other authors have explored the repercussion of lysosomal inactivation on the mitochondria. Pyruvate seems to protect cells against senescence, as was already shown in the MiDAS [5]. However, in this study, the authors did not address how pyruvate could affect lysosomal function. Pyruvate deprivation promotes senescence and seems to enhance lysosomal inactivation through the acetylation of the v-ATPase, which leads to an accumulation of abnormal mitochondria. In this case, mitochondrial impairment is linked to mitophagy defects [61].

In yeast, it has been observed that the kinase Sch9 controls vacuolar v-ATPase. The lack of Sch9 is correlated with a more acidic cytosolic pH, which seems to be dependent on target of rapamycin 1 (TORC1), the mTORC1 homologue in yeast [62]. Genetic disruption of vma, a v-ATPase subunit, leads to mitochondrial impairment accompanied by iron–sulfur cluster deficiency. Increased iron uptake suppresses the mitochondrial dysfunction [56]. Iron is essential for mitochondrial function as it is used for the synthesis of cofactors that participate in oxidation–reduction reactions [63]. Thus, lysosomal function seems to be important for iron homeostasis.

Moreover, some lysosomal storage diseases that are the products of mutations in genes that code for specific lysosomal proteins are characterized by mitochondrial dysfunction phenotypes. For example, cells with mutations in lysosomal enzymes such as NPC1 display aberrant mitochondrial lipid content that influences metabolic processes such as glycolysis [64]. This shows that the correct catabolism in the lysosomes affects the function of other organelles such as the mitochondria, although these events do not always correlate with mitophagy, but mitochondrial metabolism may be affected.

3.4. Lysosomal Exocytosis

Lysosomes can fuse with the plasma membrane and secrete lysosomal content through exocytosis. This process was first believed to be reserved for specialized cells such as macrophages or melanocytes [65,66], but it has been observed that it occurs in all cells as a repair mechanism of the plasma membrane [67].

Most lysosomes are localized in the perinuclear area, but when the plasma membrane is damaged, lysosomes migrate to the cell periphery to fuse with the plasma membrane promoting the recovery of the membrane. The lysosomal fusion with the membrane is regulated by lysosomal calcium efflux through TRPML1. Precisely, lysosomal exocytosis is frequently studied by the presence of lysosomal proteins such as LAMP1 or TRPML1 in the plasma membrane [68].

Lysosomal exocytosis is a process that also allows to get rid of unprocessed materials and it has been described in the case of lysosomal enzymes. Cathepsin D has been found in extracellular media upon lysosomal alkalinization [69]. Other groups have reported that lysosomes that fuse with the plasma membrane can get rid of autophagosomes [70]. LC3 lipidation is important for lysosomal exocytosis, so its regulation must be closely linked to autophagy [70].

4. Lysosomal Age-Related Dysfunctions

As evidenced, lysosomal function is very important for cellular homeostasis, not only because of the recycling functions it exerts on damaged molecules and dysfunctional organelles, but also because of the impact it has on metabolic pathways or on organelles such as the mitochondria. Most studies are not focused on lysosomal function, but instead cover fields in which the lysosome is involved such as autophagy or mTORC1 status. However, some studies have reported dysfunctions associated with the lysosome in the context of aging (Figure 2).

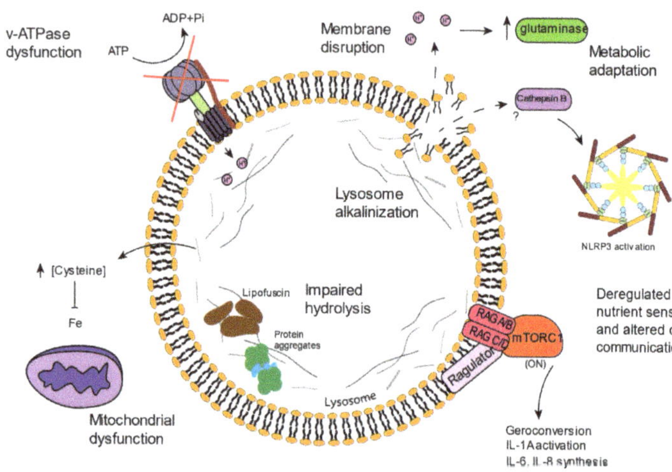

Figure 2. An overview of the lysosomal dysfunctions associated with aging or senescence. v-ATPase dysfunctions and membrane disruption impair the lysosomal acidification process, which shows increased alkalinization. Protons escaping the lysosome provoke cytosolic acidification, which alters cellular metabolism, and the presence of cathepsin B has been linked to NLRP3 inflammasome activation. Impaired hydrolysis inside the lysosome results in protein aggregation and lipofuscin storage. Incorrect amino acid storage is linked to mitochondrial dysfunction, and mTORC1 activation in the lysosome surface correlates with the synthesis of SASP molecules and geroconversion.

Precisely, mutations in enzymes related to lysosomal function such as in hydrolases or in lysosome membrane channels lead to lysosomal storage diseases, which are usually characterized by the increased accumulation of protein aggregates, leading to symptoms that resemble neurodegenerative diseases [71,72].

4.1. *v-ATPAse Dysfunction and Lysosomal Alkalinization*

Although some alterations have been described in luminal enzymes, the lysosomal membrane also has its relevance when it comes to aging. The correct v-ATPase function and lumen acidification has been correlated with delayed aging in yeast [60,62]. Many mutations in the v-ATPase subunits have been associated with neurodegenerative disorders, and animal models carrying mutations in v-ATPase subunits show lysosomal acidification problems accompanied by accelerated aging [73]. Cells lacking v-ATPase subunits also display mitochondrial dysfunctions, highlighting the important relation between lysosomes and mitochondria [74,75].

For instance, in *C. elegans*, it was recently demonstrated that several lysosomal genes such as v-ATPase subunits or lysosomal hydrolases are downregulated during aging. Moreover, these genes were found to be upregulated in long-lived mutants [76]. Furthermore, in *C. elegans*, the increase in lysosomal pH in older worms correlates with diminished proteostasis and the ceasing of reproduction. DAF-16, a gene previously known to regulate longevity, induces the expression of v-ATPase subunits and lysosomal acidification [77], evidencing the importance of correct lysosomal acidification for the promotion of longevity. Additionally, *C. elegans* lysosomes can release molecules that modulate longevity. In particular, the overexpression of the lysosomal acid lipase LIPA-4 promotes longevity, inducing the nuclear translocation of the fatty acid binding protein LBP-8 [78], indicating that lysosomal metabolism is crucial for lysosomes to act as signaling organelles, influencing the lifespan.

Precisely, the lysosomal state can modulate the metabolism, not only in *C. elegans*, but also on human cells. Recently, it has been described that the disruption of the lysosomal membrane during cellular senescence causes lysosomal alkalinization and, importantly, cytosolic acidification, which ultimately modifies cellular metabolism to counteract this acidification. In particular, acidic cytosolic pH is counteracted by increasing glutaminolysis [79]. Thus, lysosomal alkalinization results in metabolic cell adaptation, which foments the survival of senescent cells, denoting the important role lysosomes play in cellular metabolism and cell fate.

In macrophages, lysosomal exocytosis, which is an important functional mechanism, is propitiated by lysosomal alkalinization [68]. It would be interesting to know whether lysosomal exocytosis is also affected when lysosomal disruption and alkalinization occurs on senescent cells.

Importantly, the inhibition of lysosomal acidification by bafilomycin promotes iron deficiency, which results in impaired mitochondrial function and increased inflammation, linking correct lysosomal acidification to the functional mitochondrial state. Iron supplementation rescues these effects in cultured neurons and in a mouse model of impaired lysosomal acidification induced by the knockout of acid α-glucosidase, an essential enzyme for glycogen catabolism [75]. This evidences that lysosomal dysfunction led to imbalances in the compartmentalization of metabolites and ions, which have repercussions on other cellular functions.

4.2. Lysosomal Amino Acid Storage and Ion Homeostasis

Lysosomal dysfunctions lead to the accumulation of metabolites, and lysosomal storage diseases are an example of how the incorrect function of one lysosomal enzyme leads to storage problems, which conclude in signs and symptoms at the systemic level [80].

In yeast, aging has been linked to the incorrect storage of neutral amino acids in the vacuole [60]. Moreover, it has been proposed that the disruption of amino acid compartmentalization into the lysosome-like vacuole during yeast aging, especially impairment of vacuolar cysteine storage, causes cysteine accumulation in the cytosol. Cytosolic cysteine limits the iron bioavailability and leads to mitochondrial respiration impairment while cysteine depletion restores mitochondrial function [81].

Lysosomes receive iron through the endocytic pathway. The lysosomal enzyme STEAP3, whose activity depends on correct lysosomal acidification, is essential for iron reduction into Fe^{2+}, the form that is incorporated in iron-containing proteins. Moreover, lysosomes participate in the turnover of ferritin, the protein that stores iron inside the cell, and of mitochondria, organelles that contain great quantities of iron. Therefore, iron homeostasis is tightly regulated by the lysosome, and iron disturbances promote the accumulation of ROS as Fe^{2+} can react with hydrogen peroxide, inducing the formation of highly reactive species through the Fenton reaction [82]. Moreover, iron is indispensable for processes such as oxygen transport or collagen biosynthesis, and it is part of many mitochondrial complexes, as iron–sulfur clusters are essential for oxidation–reduction reactions [63,83]. For this reason, the dysregulation of iron homeostasis contributes to cardiovascular [84] and neurodegenerative diseases [85–87], among others.

Since disturbances in iron homeostasis lead to ROS formation, it is not surprising that iron imbalance contributes to the aging process. In fact, old people suffer from problems of absorbing iron at the systemic level, while an increase in iron at the cellular level has been observed in this same population. Indeed, senescent cells have a 10-fold increase in iron compared to young cells [88]. This iron accumulation during senescence is associated with impaired ferritin degradation in the lysosome and increased expression of cell cycle inhibitors [89]. Furthermore, lysosome involvement in iron homeostasis is linked to a type of cell death called ferroptosis. Ferroptosis, which has been linked to aging [90], is promoted by ROS generation, associated with iron disturbances in the lysosome.

Other ions have been associated with lysosomal function during aging. For example, decreased potassium levels in yeast are consistent with increased acidity in the vacuole and lifespan extension [91]. Ion storage depends on the correct lysosomal function and acidification, but further research is needed to elucidate how lysosomal storage affects the aging process.

4.3. Lipofuscin

Lipofuscin, hydrophobic yellow-brown granules composed of oxidized lipids and proteins that accumulate in the lysosomes, is a common feature of senescent cells [3]. ROS generation has been linked to lipofuscin accumulation, as this granulated material tends to incorporate metals such as iron, contributing to oxidative reactions [92]. ROS damages proteins that are prone to unfold, leading to protein aggregation, thus contributing to the generation of adducts of oxidized cellular molecules. The insolubilization and crosslinked structure of lipofuscin impedes its correct degradation, and lipofuscin structures tend to accumulate in lysosomes. However, upon the inhibition of macroautophagy, lipofuscin also accumulates in the cytosol [93]. It has been proposed that lipofuscin remains attached to the proteasome, which is unable to degrade these cross-linked aggregates [94].

During aging, autophagy and subsequent lysosomal degradation of lipofuscin is impaired, which promotes further aggregation of these granules. Additionally, not only do impaired lysosomes contribute to lipofuscin accumulation during senescence, but the lack of cell division impedes the distribution of these aggregates between the daughter cells, as would occur in proliferative cells. Moreover, the increase in ROS production and lipofuscin accumulation promotes mitochondrial dysfunction, which further exacerbates lysosomal impairment in a positive feedback mechanism [95]. Additionally, lipofuscin accumulation enhances the activity of caspase-3 and promotes lysosomal membrane disruption, having been linked to NLRP3 inflammasome activation and necroptosis induction [92,96].

4.4. Inflammation and Cell Death

Senescent cells show increased mitochondrial mass, and these mitochondria are often dysfunctional. Typically, the mitochondria of senescent cells display low membrane potential, lower mitochondrial ATP production, and increased ROS production [56]. The decreased levels of mitophagy observed in many senescent cells supports a possible mechanism that explains the increase in dysfunctional mitochondria. Lysosomal dysfunction impedes the correct degradation of mitochondria, exacerbating mitochondrial ROS production which, in turn, increases lysosomal damage in a feedback fashion. Both mitochondria and lysosomes have been correlated to SASP production. In fact, mtDNA depletion seems to induce MiDAS with a characteristic SASP pattern that lacks the IL-1 inflammatory arm [5], while lysosomal disruption seems to induce IL-1 activation, enhancing NLRP3 inflammasome function [97].

NLRP3 is an important protein complex for the innate immune system. The two-signal model is proposed to explain NLRP3 activation [98]. The first or priming signal normally consists of PAMPs (pathogen-associated molecular patterns) or DAMPs (damage-associated molecular patterns), which lead to NF-κB pathway activation. The second or activating signal can be supplied by diverse inputs such as extracellular ATP, changes in ionic flux, mitochondrial dysfunction, or lysosomal disruption, among others [98].

Lysosomal disruption with Leu-Leu-Omc triggers NLRP3 activation [99] and it has been hypothesized that the disruption of these organelles and the subsequent cathepsin B release into the cytosol can activate NLRP3 inflammasome, and this process is important for pro-IL-1β processing [97], linking the loss of lysosomal integrity to inflammation. Indeed, some authors have shown that treating senescent cells with lysosomal function inhibitors such as leupeptin induces inflammation [57].

Lysosomal membrane permeabilization is not only involved in the activation of pro-inflammatory pathways, but it has been suggested to be involved in the mechanisms of cell death, and recently, it has also been associated with pathophysiological conditions [12]. Indeed, some authors have proposed that lysosomal membrane permeabilization in association with lysosomal quality control mechanisms can determine cell fate, since lysosomal components that are released into the cytosol can trigger the activation of diverse cellular pathways [100]. Several studies have suggested that lysosomal damage is related to apoptosis, necroptosis, and ferroptosis. Apoptosis, the main type of programmed cell death, which is characterized by cytochrome c release from mitochondria, can be fostered by the release of lysosomal cathepsins, which can degrade Bcl-2 [101,102], one of the proteins involved in mitochondrial membrane stability during apoptosis. Similarly, cathepsin D release has been associated to the activation of RIPK1 during necroptosis [103,104], a type of cell death though to be non-programmed, but now known to be regulated by receptor-interacting protein kinases (RIPK). Ferroptosis is a regulated cell death type in which intracellular iron incites the formation of ROS. Lysosomes are closely related to ferroptosis, since these organelles constitute one of the main iron storage places. Lysosomal membrane disruption seems to foster the activation of this cell death pathway [105].

Lysosomal damage seems to be critical for the development of different cell death pathways, and cell death is an important mechanism for the development and maintenance of age-related diseases such as neurodegenerative or cardiovascular disorders [106,107], making lysosomal damage an interesting field to research in this context.

5. mTORC and Senescence

mTORC1 seems to be activated during aging and in some senescent cells [108,109]. This activation is evidenced by the phosphorylation of mTORC1 downstream targets. It has been reported that in replicative senescent fibroblasts, p70S6K undergoes phosphorylation by the activity of mTORC1 [110]. Moreover, it has been found that constitutive mTORC1 activation induces premature senescence in fibroblasts carrying tuberous sclerosis complex (TCS) mutations [111]. S6 kinases, phosphorylated and activated by mTORC1, are observed in aged muscle [112] and in the brains of Alzheimer's disease patients [113].

mTORC1 seems to be fundamental for the processing and activation of some SASP factors such as IL-6, IL-8, or IL-1A, which are, in turn, fundamental players in the inflammation induction that accompanies aging (Figure 2). Rapamycin treatment, which inhibits mTORC1, diminishes the pro-inflammatory phenotype of senescent cells through a reduction in the IL-1A and IL-6 levels [114]. Senescence induction seems to increase mTORC1 activity, which could be related to SASP induction through the TOR-Autophagy Spatial Coupling Compartment (TASCC). It has been suggested that mTORC1 localizes at this TASCC compartment in the Trans-Golgi and favors the synthesis of IL-6 and IL-8 cytokines [115]. Precisely, lysosomes originate from the Trans-Golgi network and mTORC1 activation occurs on the lysosomal membrane. Further studies are necessary to elucidate the relevance of the role of the mTORC1-lysosome on inflammation induction, but several authors have pointed out the role of lysosomes on inflammasome activation [80,96].

The mTOR pathway can positively or negatively regulate p53, being cell type and stress dependent [116]. In fact, mTORC1 and mTORC2 were reported to participate in the stabilization of cell cycle inhibitors (Figure 3). In MEFS, mTORC1 activation promotes the association of p53 mRNA with ribosomes, leading to p53 translation [117]. The overexpression of the microRNA miR-107 increases MTORC1 activity through PTEN inhibition, resulting in the activation of p16 [118]. Precisely, the loss of PTEN enhances p53 translation, triggered by mTORC1 [119]. Moreover, it has been reported that in PTEN-depleted cells, mTORC1 and mTORC2 bind and phosphorylate p53 at Ser15 [120]. This exhibits the participation of mTORC1 in senescent state regulation.

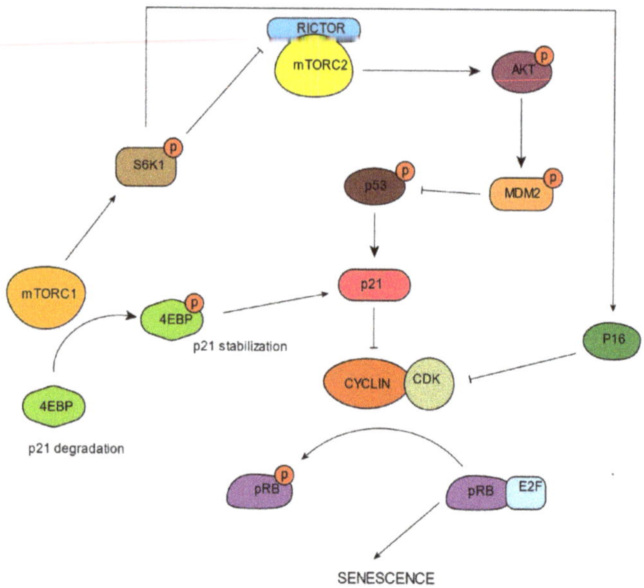

Figure 3. An overview of mTORC1 and mTORC2 on senescence control by the regulation of the p53/p21 and p16/pRb pathways. p21 is stabilized through mTORC1 dependent 4EBP phosphorylation, and through mTORC1 via S6K1 dependent mTORC2 inhibition, which has repercussions on p53 stabilization. p16 is activated downstream of mTORC1 by S6K1 phosphorylation, inhibiting the CDK–Cyclin complexes and impeding pRb phosphorylation. This results in the sequestration of E2F and the prevention of cell cycle progression, eventually leading to senescence.

Precisely, S6K1, an mTORC1 downstream effector, has been related to MDM2 inhibition, avoiding p53 degradation, and subsequently promoting p53 stabilization, an important process for senescence induction [121]. S6K1 has also been reported to upregulate p16 [111] and to phosphorylate RICTOR, inhibiting mTORC2 and AKT [122]. Inhibited AKT is unable to phosphorylate and activate MDM2, promoting p53 stabilization [123].

In some cancer cells, mTORC1 activation through 4E-BPI phosphorylation stabilizes p21 [124]. In U2OS cells supplemented with branched-chain amino acids, p21 is also increased via mTORC1 activation, promoting cellular senescence in the presence of DNA damage-inducing compounds [125]. Similarly, it has also been shown that treatment with the mTORC1 inhibitor rapamycin decreases p53 translation, downregulating p21 and preventing AKT-induced senescence in human fibroblasts [126].

6. Lysosomal Opportunities for Intervention in Aging

Strategies that increase autophagic activity have been proposed as a mechanism to eliminate senescent cells. In particular, the suppression of mTORC1 extends the lifespan of many animal models, and inhibitors of this pathway such as rapamycin or torin1 have been developed [127]. Other molecules such as metformin have been shown to increase autophagy and ameliorate inflammation [128].

It is suggested that the treatment of senescent cells with mTORC1 inhibitors ameliorates the senescence phenotype in many cases [129,130]. It has been pointed out that although senescent cells undergo a permanent loss of proliferative potential because their cell cycle is arrested, their cellular growth pathways remain active. For instance, among the several pathways deregulated during senescence, the presence of ROS has been linked to cell cycle block, along with an active mTORC1 [109]. Rapamycin, a mTORC1 inhibitor, could prevent the permanent loss of proliferation when cells are arrested by p21 or p16 function. The proliferation of cells arrested by p21 or p16 upregulation in the presence of rapamycin resumed the proliferation capacity once rapamycin was removed, suggesting that mTORC1 activation is important for the achievement of senescence [129].

Importantly, mTORC1 inhibition prevents geroconversion, defined as the transition from quiescence to senescence, supporting the importance of mTORC1 activity to achieve senescence [130]. During cellular starvation, contact inhibition or hypoxia, cells do not undergo senescence, and it is hypothesized that mTORC1 inhibition prevents the conversion into a senescent state. Precisely, the stimulation of mTORC1 during contact inhibition favors the geroconversion [130].

Therefore, mTORC1 has become one of the principal targets for the development of senolytic compounds. Rapamycin blocks the activation of mTORC1, having been shown not only to have effects on senescent cells, but also in several organisms, where it is able to extend the lifespan. Mice fed with a diet containing rapamycin exhibits lifespan extension [131]. Additionally, genetic interventions targeting TOR influence the animal lifespan. For instance, in *C. elegans*, TOR RNAi silencing promotes a lifespan extension [132]. Similarly, p70S6K deletion in mice increases their lifespan [133].

Part of the beneficial effects of mTORC1 inhibition is thought to be due to an increase in autophagic flux. However, although rapamycin can directly influence lysosomal status, it has been reported that the lysosomal calcium channel MCOLN1 is also directly activated by rapamycin [134]. This channel is important for lysosomal biogenesis since Ca^{2+} efflux through this channel creates a Ca^{2+} microdomain, which activates the phosphatase calcineurin, which, in turn, dephosphorylates TFEB and allows for its transport to the nucleus [19]. MCOLN1 status during senescence is unknown, but loss-of-function mutations of this gene lead to a lysosomal storage disease called Mucolipidosis IV [135]. Further research is needed in the context of aging and senescence, but it seems that TFEB activation ameliorates age-related diseases [136,137].

Since TFEB was reported as the master regulator of lysosome biogenesis, many efforts have been made to identify activators that promote lysosomal function and autophagy, since the correct function of lysosomes is the key to promoting longevity. Many compounds able to activate and upregulate TFEB affect TFEB downstream of the inhibition of mTORC1, but one compound called C1 has been identified as a direct activator of TFEB. Accordingly, C1 binds TFEB and promotes its nuclear translocation [138].

In some neurodegenerative disease models, increased TFEB function through genetic intervention induces the activation of autophagy and a re-establishment of proteostasis, ameliorating the protein accumulation characteristic of these diseases. In Huntington's disease, studies with in vitro models have shown that TFEB activation reduces HTT protein aggregation and decreases disease-related symptoms in mice [136]. In Parkinson's disease, TFEB activation also ameliorates α-synuclein toxicity through the stimulation of autophagy [137]. Alzheimer's disease is also related to impairment of proteostasis, and an accumulation of autophagosomes in the brains of Alzheimer's patients has been observed. TFEB activation could also have beneficial effects in this disease, since a reduction in tau pathology and neurodegeneration in a mouse model has been observed upon increased TFEB activity [11,139].

In contrast, TFEB overexpression seems to promote the development of some tumors [140–142]. Considering this, it is possible that the therapeutic use of TFEB activation may be limited by its oncogenic potential.

Other strategies do not focus on remodeling senescent metabolism, but on killing senescent cells. This is the case of senolytics, drugs that aspire to specifically remove senescent cells by targeting a mechanism that is normally upregulated in a specific senescent phenotype. However, as in the case of cancer treatments, it is difficult to address a treatment to a specific type of cell and avoid off-target effects. In recent years, an elegant proposal has been made, where the increased SA-β gal activity from the senescent lysosome was used to activate a senolytic compound [143]. The senolytic prodrug enters the cell and is sequestered in the lysosome, where SA-β gal can catalyze the cleavage of the prodrug, therefore selectively eliminating senescent cells.

7. Conclusions

The correct degradation of macromolecules is important to integrate distinct metabolic routes. The correct function of hydrolases is key to degrading polymeric macromolecules (e.g., proteins) to their monomeric building blocks (e.g., amino acids), which can be sensed by specialized proteins associated with mTORC1. In addition, these monomers act as substrates for anabolism, and their compartmentalization is important to regulate cellular metabolism, and thus, other cellular functions. Specific studies to investigate the role of lysosomal disruption in intracellular storage problems are needed. In some senescence models, lysosomal dysfunction and/or disruption has been reported, however, it is not known to which extent this event directly affects the phenotype of these cells. Similarly, lysosomal membrane disruption causes lysosome alkalinization and cytosolic acidification, which could also alter normal cellular functions besides glutaminolysis.

The emerging function of these organelles on the metabolic regulation or molecule storage must be studied in more depth, especially the relation between lysosomes and mitochondria. As has been observed, mitochondrial dysfunction can lead to lysosomal problems, and simultaneously, lysosomal dysfunction can disembogue in problems associated with mitochondria. Because of this, the understanding of signaling mechanisms existing between these organelles, especially the ones associated with monomer storage, is of special interest, since it can be fundamental to understand how senescence develops. However, future studies must discriminate between distinct mechanisms that take place in the lysosome, and in this way, it will be determined whether the relationship between lysosomes and mitochondria goes beyond mitophagy and can be associated with storage changes, lysosomal permeability, or other mechanisms.

Author Contributions: L.G.-N. and M.C. conceptualized the article and selected the topics to be covered. L.G.-N. and M.C. wrote and prepared the draft. P.J.-D. and M.C. provided critical commentary and corrected the text. M.C. supervised the work. P.J.-D. acquired funding. All authors have read and agreed to the published version of the manuscript.

Funding: This research was performed under the ARDRE COFUND doctoral training program, funded by the European Commission Horizon 2020 Marie Sklodowska-Curie research grant number 847681. Work in the Jansen-Dürr/Cavinato lab was supported by the Austrian Science Funds (FWF), P315820.

Data Availability Statement: Not applicable.

Conflicts of Interest: The authors declare no conflict of interest.

References

1. López-Otín, C.; Blasco, M.A.; Partridge, L.; Serrano, M.; Kroemer, G. The Hallmarks of Aging. *Cell* **2013**, *153*, 1194. [CrossRef] [PubMed]
2. Dodig, S.; Čepelak, I.; Pavić, I. Hallmarks of Senescence and Aging. *Biochem. Med.* **2019**, *29*, 483–497. [CrossRef] [PubMed]
3. Gorgoulis, V.; Adams, P.D.; Alimonti, A.; Bennett, D.C.; Bischof, O.; Bishop, C.; Campisi, J.; Collado, M.; Evangelou, K.; Ferbeyre, G.; et al. Cellular Senescence: Defining a Path Forward. *Cell* **2019**, *179*, 813–827. [CrossRef]
4. Toussaint, O.; Medrano, E.E.; von Zglinicki, T. Cellular and Molecular Mechanisms of Stress-Induced Premature Senescence (SIPS) of Human Diploid Fibroblasts and Melanocytes. *Exp. Gerontol.* **2000**, *35*, 927–945. [CrossRef]

5. Wiley, C.D.; Velarde, M.C.; Lecot, P.; Liu, S.; Sarnoski, E.A.; Freund, A.; Shirakawa, K.; Lim, H.W.; Davis, S.S.; Ramanathan, A.; et al. Mitochondrial Dysfunction Induces Senescence with a Distinct Secretory Phenotype. *Cell Metab.* **2016**, *23*, 303–314. [CrossRef] [PubMed]
6. Wang, A.S.; Dreesen, O. Biomarkers of Cellular Senescence and Skin Aging. *Front. Genet.* **2018**, *9*, 247. [CrossRef] [PubMed]
7. Swanson, E.C.; Manning, B.; Zhang, H.; Lawrence, J.B. Higher-Order Unfolding of Satellite Heterochromatin Is a Consistent and Early Event in Cell Senescence. *J. Cell Biol.* **2013**, *203*, 929–942. [CrossRef]
8. Freund, A.; Laberge, R.M.; Demaria, M.; Campisi, J. Lamin B1 Loss Is a Senescence-Associated Biomarker. *Mol. Biol. Cell* **2012**, *23*, 2066–2075. [CrossRef]
9. Lee, B.Y.; Han, J.A.; Im, J.S.; Morrone, A.; Johung, K.; Goodwin, E.C.; Kleijer, W.J.; DiMaio, D.; Hwang, E.S. Senescence-Associated Beta-Galactosidase Is Lysosomal Beta-Galactosidase. *Aging Cell* **2006**, *5*, 187–195. [CrossRef]
10. Carmona-Gutierrez, D.; Hughes, A.L.; Madeo, F.; Ruckenstuhl, C. The Crucial Impact of Lysosomes in Aging and Longevity. *Ageing Res. Rev.* **2016**, *32*, 2–12. [CrossRef]
11. Dehay, B.; Bové, J.; Rodríguez-Muela, N.; Perier, C.; Recasens, A.; Boya, P.; Vila, M. Pathogenic Lysosomal Depletion in Parkinson's Disease. *J. Neurosci.* **2010**, *30*, 12535–12544. [CrossRef] [PubMed]
12. Gómez-Sintes, R.; Ledesma, M.D.; Boya, P. Lysosomal Cell Death Mechanisms in Aging. *Ageing Res. Rev.* **2016**, *32*, 150–168. [CrossRef] [PubMed]
13. Settembre, C.; Fraldi, A.; Medina, D.L.; Ballabio, A. Signals from the Lysosome: A Control Centre for Cellular Clearance and Energy Metabolism. *Nat. Rev. Mol. Cell Biol.* **2013**, *14*, 283–296. [CrossRef]
14. Thelen, A.M.; Zoncu, R. Emerging Roles for the Lysosome in Lipid Metabolism. *Trends Cell Biol.* **2017**, *27*, 833. [CrossRef] [PubMed]
15. Fujiwara, Y.; Wada, K.; Kabuta, T. JB Special Review-Recent Topics in Ubiquitin-Proteasome System and Autophagy Lysosomal Degradation of Intracellular Nucleic Acids-Multiple Autophagic Pathways Overview of the Multiple Autophagic Pathways. *J. Biochem.* **2017**, *161*, 145–154. [CrossRef] [PubMed]
16. Stütz, A.E.; Wrodnigg, T.M. Carbohydrate-Processing Enzymes of the Lysosome: Diseases Caused by Misfolded Mutants and Sugar Mimetics as Correcting Pharmacological Chaperones. *Adv. Carbohydr. Chem. Biochem.* **2016**, *73*, 225–302. [CrossRef]
17. Mindell, J.A. Lysosomal Acidification Mechanisms. *Annu. Rev. Physiol.* **2012**, *74*, 69–86. [CrossRef]
18. Burgoyne, T.; Patel, S.; Eden, E.R. Calcium Signaling at ER Membrane Contact Sites. *Biochim. Biophys. Acta (BBA)—Mol. Cell Res.* **2015**, *1853*, 2012–2017. [CrossRef]
19. Medina, D.L.; di Paola, S.; Peluso, I.; Armani, A.; de Stefani, D.; Venditti, R.; Montefusco, S.; Scotto-Rosato, A.; Prezioso, C.; Forrester, A.; et al. Lysosomal Calcium Signalling Regulates Autophagy through Calcineurin and TFEB. *Nat. Cell Biol.* **2015**, *17*, 288–299. [CrossRef]
20. Borodkina, A.V.; Shatrova, A.N.; Deryabin, P.I.; Griukova, A.A.; Abushik, P.A.; Antonov, S.M.; Nikolsky, N.N.; Burova, E.B. Calcium Alterations Signal Either to Senescence or to Autophagy Induction in Stem Cells upon Oxidative Stress. *Aging* **2016**, *8*, 3400. [CrossRef]
21. Morgan, A.J.; Davis, L.C.; Wagner, S.K.T.Y.; Lewis, A.M.; Parrington, J.; Churchill, G.C.; Galione, A. Bidirectional Ca^{2+} Signaling Occurs between the Endoplasmic Reticulum and Acidic Organelles. *J. Cell Biol.* **2013**, *200*, 789. [CrossRef] [PubMed]
22. Friedman, J.R.; DiBenedetto, J.R.; West, M.; Rowland, A.A.; Voeltz, G.K. Endoplasmic Reticulum–Endosome Contact Increases as Endosomes Traffic and Mature. *Mol. Biol. Cell* **2013**, *24*, 1030. [CrossRef] [PubMed]
23. Yang, C.; Wang, X. Lysosome Biogenesis: Regulation and Functions. *J. Cell Biol.* **2021**, *220*, e202102001. [CrossRef] [PubMed]
24. Laplante, M.; Sabatini, D.M. MTOR Signaling in Growth Control and Disease. *Cell* **2012**, *149*, 274. [CrossRef]
25. Settembre, C.; Zoncu, R.; Medina, D.L.; Vetrini, F.; Erdin, S.; Erdin, S.; Huynh, T.; Ferron, M.; Karsenty, G.; Vellard, M.C.; et al. A Lysosome-to-Nucleus Signalling Mechanism Senses and Regulates the Lysosome via MTOR and TFEB. *EMBO J.* **2012**, *31*, 1095–1108. [CrossRef]
26. Sardiello, M.; Palmieri, M.; di Ronza, A.; Medina, D.L.; Valenza, M.; Gennarino, V.A.; di Malta, C.; Donaudy, F.; Embrione, V.; Polishchuk, R.S.; et al. A Gene Network Regulating Lysosomal Biogenesis and Function. *Science* **2009**, *325*, 473–477. [CrossRef]
27. Peña-Llopis, S.; Vega-Rubin-De-Celis, S.; Schwartz, J.C.; Wolff, N.C.; Tran, T.A.T.; Zou, L.; Xie, X.J.; Corey, D.R.; Brugarolas, J. Regulation of TFEB and V-ATPases by MTORC1. *EMBO J.* **2011**, *30*, 3242–3258. [CrossRef]
28. Zhang, C.S.; Jiang, B.; Li, M.; Zhu, M.; Peng, Y.; Zhang, Y.L.; Wu, Y.Q.; Li, T.Y.; Liang, Y.; Lu, Z.; et al. The Lysosomal V-ATPase-Ragulator Complex Is a Common Activator for AMPK and MTORC1, Acting as a Switch between Catabolism and Anabolism. *Cell Metab.* **2014**, *20*, 526–540. [CrossRef]
29. Settembre, C.; de Cegli, R.; Mansueto, G.; Saha, P.K.; Vetrini, F.; Visvikis, O.; Huynh, T.; Carissimo, A.; Palmer, D.; Jürgen Klisch, T.; et al. TFEB Controls Cellular Lipid Metabolism through a Starvation-Induced Autoregulatory Loop. *Nat. Cell Biol.* **2013**, *15*, 647–658. [CrossRef]
30. Martini-Stoica, H.; Xu, Y.; Ballabio, A.; Zheng, H. The Autophagy–Lysosomal Pathway in Neurodegeneration: A TFEB Perspective. *Trends Neurosci.* **2016**, *39*, 221. [CrossRef]
31. Kim, J.; Kundu, M.; Viollet, B.; Guan, K.-L. AMPK and MTOR Regulate Autophagy through Direct Phosphorylation of Ulk1. *Nat. Cell Biol.* **2011**, *13*, 132–141. [CrossRef] [PubMed]
32. Li, Y.; Xu, M.; Ding, X.; Yan, C.; Song, Z.; Chen, L.; Huang, X.; Wang, X.; Jian, Y.; Tang, G.; et al. Protein Kinase C Controls Lysosome Biogenesis Independently of MTORC1. *Nat. Cell Biol.* **2016**, *18*, 1065–1077. [CrossRef]

33. Fu, W.; Hall, M.N. Regulation of MTORC2 Signaling. *Genes* **2020**, *11*, 1045. [CrossRef] [PubMed]
34. Jia, R.; Bonifacino, J.S. Lysosome Positioning Influences MTORC2 and AKT Signaling. *Mol. Cell* **2019**, *75*, 26–38.e3. [CrossRef] [PubMed]
35. Annunziata, I.; van de Vlekkert, D.; Wolf, E.; Finkelstein, D.; Neale, G.; Machado, E.; Mosca, R.; Campos, Y.; Tillman, H.; Roussel, M.F.; et al. MYC Competes with MiT/TFE in Regulating Lysosomal Biogenesis and Autophagy through an Epigenetic Rheostat. *Nat. Commun.* **2019**, *10*, 3623. [CrossRef] [PubMed]
36. Aman, Y.; Schmauck-Medina, T.; Hansen, M.; Morimoto, R.I.; Simon, A.K.; Bjedov, I.; Palikaras, K.; Simonsen, A.; Johansen, T.; Tavernarakis, N.; et al. Autophagy in Healthy Aging and Disease. *Nat. Aging* **2021**, *1*, 634–650. [CrossRef] [PubMed]
37. Kang, C.; Elledge, S.J. How Autophagy Both Activates and Inhibits Cellular Senescence. *Autophagy* **2016**, *12*, 898–899. [CrossRef] [PubMed]
38. Rajendran, P.; Alzahrani, A.M.; Hanieh, H.N.; Kumar, S.A.; ben Ammar, R.; Rengarajan, T.; Alhoot, M.A. Autophagy and Senescence: A New Insight in Selected Human Diseases. *J. Cell Physiol.* **2019**, *234*, 21485–21492. [CrossRef] [PubMed]
39. Doherty, G.J.; Mcmahon, H.T. Mechanisms of Endocytosis. *Annu. Rev. Biochem.* **2009**, *78*, 857–902. [CrossRef]
40. Poteryaev, D.; Datta, S.; Ackema, K.; Zerial, M.; Spang, A. Identification of the Switch in Early-to-Late Endosome Transition. *Cell* **2010**, *141*, 497–508. [CrossRef]
41. Shin, E.Y.; Park, J.H.; You, S.T.; Lee, C.S.; Won, S.Y.; Park, J.J.; Kim, H.B.; Shim, J.; Soung, N.K.; Lee, O.J.; et al. Integrin-Mediated Adhesions in Regulation of Cellular Senescence. *Sci. Adv.* **2020**, *6*, eaay3909. [CrossRef] [PubMed]
42. Shin, E.Y.; Soung, N.K.; Schwartz, M.A.; Kim, E.G. Altered Endocytosis in Cellular Senescence. *Ageing Res. Rev.* **2021**, *68*, 101332. [CrossRef] [PubMed]
43. Parzych, K.R.; Klionsky, D.J. An Overview of Autophagy: Morphology, Mechanism, and Regulation. *Antioxid. Redox Signal.* **2014**, *20*, 460–473.e3. [CrossRef] [PubMed]
44. Axe, E.L.; Walker, S.A.; Manifava, M.; Chandra, P.; Roderick, H.L.; Habermann, A.; Griffiths, G.; Ktistakis, N.T. Autophagosome Formation from Membrane Compartments Enriched in Phosphatidylinositol 3-Phosphate and Dynamically Connected to the Endoplasmic Reticulum. *J. Cell Biol.* **2008**, *182*, 685. [CrossRef]
45. Hilverling, A.; Szegö, E.M.; Dinter, E.; Cozma, D.; Saridaki, T.; Falkenburger, B.H. Maturing Autophagosomes Are Transported Towards the Cell Periphery. *Cell Mol. Neurobiol.* **2022**, *42*, 155–171. [CrossRef]
46. Melia, T.J.; Lystad, A.H.; Simonsen, A. Autophagosome Biogenesis: From Membrane Growth to Closure. *J. Cell Biol.* **2020**, *219*, e202002085. [CrossRef]
47. Ott, C.; König, J.; Höhn, A.; Jung, T.; Grune, T. Macroautophagy Is Impaired in Old Murine Brain Tissue as Well as in Senescent Human Fibroblasts. *Redox Biol.* **2016**, *10*, 266–273. [CrossRef]
48. Simonsen, A.; Cumming, R.C.; Brech, A.; Isakson, P.; Schubert, D.R.; Finley, K.D. Promoting Basal Levels of Autophagy in the Nervous System Enhances Longevity and Oxidant Resistance in Adult Drosophila. *Autophagy* **2008**, *4*, 176–184. [CrossRef]
49. Lipinski, M.M.; Zheng, B.; Lu, T.; Yan, Z.; Py, B.F.; Ng, A.; Xavier, R.J.; Li, C.; Yankner, B.A.; Scherzer, C.R.; et al. Genome-Wide Analysis Reveals Mechanisms Modulating Autophagy in Normal Brain Aging and in Alzheimer's Disease. *Proc. Natl. Acad. Sci. USA* **2010**, *107*, 14164–14169. [CrossRef]
50. Hansen, M.; Rubinsztein, D.C.; Walker, D.W. Autophagy as a Promoter of Longevity: Insights from Model. *Nat. Rev. Mol. Cell Biol.* **2018**, *19*, 579. [CrossRef]
51. Kuma, A.; Komatsu, M.; Mizushima, N. Autophagy-Monitoring and Autophagy-Deficient Mice. *Autophagy* **2017**, *13*, 1619–1628. [CrossRef] [PubMed]
52. Pyo, J.O.; Yoo, S.M.; Ahn, H.H.; Nah, J.; Hong, S.H.; Kam, T.I.; Jung, S.; Jung, Y.K. Overexpression of Atg5 in Mice Activates Autophagy and Extends Lifespan. *Nat. Commun.* **2013**, *4*, 2300. [CrossRef] [PubMed]
53. Grootaert, M.O.J.; Moulis, M.; Roth, L.; Martinet, W.; Vindis, C.; Bennett, M.R.; de Meyer, G.R.Y. Vascular Smooth Muscle Cell Death, Autophagy and Senescence in Atherosclerosis. *Cardiovasc. Res.* **2018**, *114*, 622–634. [CrossRef] [PubMed]
54. Narita, M.; Young, A.R.J.; Narita, M. Autophagy Facilitates Oncogene-Induced Senescence. *Autophagy* **2009**, *5*, 1046–1047. [CrossRef]
55. Soto-Heredero, G.; Baixauli, F.; Mittelbrunn, M. Interorganelle Communication between Mitochondria and the Endolysosomal System. *Front. Cell Dev. Biol.* **2017**, *5*, 95. [CrossRef]
56. Chen, G.; Kroemer, G.; Kepp, O. Mitophagy: An Emerging Role in Aging and Age-Associated Diseases. *Front. Cell Dev. Biol.* **2020**, *8*, 200. [CrossRef]
57. Baixauli, F.; Acín-Pérez, R.; Villarroya-Beltrí, C.; Mazzeo, C.; Nuñez-Andrade, N.; Gabandé-Rodriguez, E.; Ledesma, M.D.; Blázquez, A.; Martin, M.A.; Falcón-Pérez, J.M.; et al. Mitochondrial Respiration Controls Lysosomal Function during Inflammatory T Cell Responses. *Cell Metab.* **2015**, *22*, 485–498. [CrossRef]
58. Demers-Lamarche, J.; Guillebaud, G.; Tlili, M.; Todkar, K.; Bélanger, N.; Grondin, M.; P'Nguyen, A.; Michel, J.; Germain, M. Loss of Mitochondrial Function Impairs Lysosomes. *J. Biol. Chem.* **2016**, *291*, 10263–10276. [CrossRef]
59. Fernández-Mosquera, L.; DIogo, C.V.; Yambire, K.F.; Santos, G.L.; Luna Sánchez, M.; Bénit, P.; Rustin, P.; Lopez, L.C.; Milosevic, I.; Raimundo, N. Acute and Chronic Mitochondrial Respiratory Chain Deficiency Differentially Regulate Lysosomal Biogenesis. *Sci. Rep.* **2017**, *7*, 45076. [CrossRef]
60. Hughes, A.L.; Gottschling, D.E. An Early Age Increase in Vacuolar PH Limits Mitochondrial Function and Lifespan in Yeast. *Nature* **2012**, *492*, 261–265. [CrossRef]

61. Kim, J.Y.; Lee, S.H.; Bae, I.H.; Shin, D.W.; Min, D.; Ham, M.; Kim, K.H.; Lee, T.R.; Kim, H.J.; Son, E.D.; et al. Pyruvate Protects against Cellular Senescence through the Control of Mitochondrial and Lysosomal Function in Dermal Fibroblasts. *J. Investig. Dermatol.* **2018**, *138*, 2522–2530. [CrossRef] [PubMed]
62. Wilms, T.; Swinnen, E.; Eskes, E.; Dolz-Edo, L., Uwineza, A.; van Essche, R.; Rosseels, J.; Zabrocki, P.; Cameroni, E.; Franssens, V.; et al. The Yeast Protein Kinase Sch9 Adjusts V-ATPase Assembly/Disassembly to Control PH Homeostasis and Longevity in Response to Glucose Availability. *PLoS Genet.* **2017**, *13*, e1006835. [CrossRef] [PubMed]
63. Paul, B.T.; Manz, D.H.; Torti, F.M.; Torti, S.v. Mitochondria and Iron: Current Questions. *Expert Rev. Hematol.* **2017**, *10*, 65. [CrossRef] [PubMed]
64. Torres, S.; Balboa, E.; Zanlungo, S.; Enrich, C.; Garcia-Ruiz, C.; Fernandez-Checa, J.C. Lysosomal and Mitochondrial Liaisons in Niemann-Pick Disease. *Front. Physiol.* **2017**, *8*, 982. [CrossRef]
65. DeSelm, C.J.; Miller, B.C.; Zou, W.; Beatty, W.L.; van Meel, H.; Takahata, Y.; Klumperman, J.; Tooze, S.A.; Teitelbaum, S.L.; Virgin, H.W. Autophagy Proteins Regulate the Secretory Component of Osteoclastic Bone Resorption. *Dev. Cell* **2011**, *21*, 966–974. [CrossRef] [PubMed]
66. Ganesan, A.K.; Ho, H.; Bodemann, B.; Petersen, S.; Aruri, J.; Koshy, S.; Richardson, Z.; Le, L.Q.; Krasieva, T.; Roth, M.G.; et al. Genome-Wide SiRNA-Based Functional Genomics of Pigmentation Identifies Novel Genes and Pathways That Impact Melanogenesis in Human Cells. *PLoS Genet.* **2008**, *4*, e1000298. [CrossRef]
67. Corrotte, M.; Castro-Gomes, T. Lysosomes and Plasma Membrane Repair. *Curr. Top. Membr.* **2019**, *84*, 1–16. [CrossRef]
68. Tancini, B.; Buratta, S.; Delo, F.; Sagini, K.; Chiaradia, E.; Pellegrino, R.M.; Emiliani, C.; Urbanelli, L. Lysosomal Exocytosis: The Extracellular Role of an Intracellular Organelle. *Membranes* **2020**, *10*, 406. [CrossRef]
69. Zhitomirsky, B.; Assaraf, Y.G. Lysosomal Accumulation of Anticancer Drugs Triggers Lysosomal Exocytosis. *Oncotarget* **2017**, *8*, 45117. [CrossRef]
70. Buratta, S.; Tancini, B.; Sagini, K.; Delo, F.; Chiaradia, E.; Urbanelli, L.; Emiliani, C. Lysosomal Exocytosis, Exosome Release and Secretory Autophagy: The Autophagic- and Endo-Lysosomal Systems Go Extracellular. *Int. J. Mol. Sci.* **2020**, *21*, 2576. [CrossRef]
71. Beck, M. The Link between Lysosomal Storage Disorders and More Common Diseases. *J. Inborn Errors Metab. Screen.* **2016**, *4*, 232640981668276. [CrossRef]
72. Toledano-Zaragoza, A.; Ledesma, M.D. Addressing Neurodegeneration in Lysosomal Storage Disorders: Advances in Niemann Pick Diseases. *Neuropharmacology* **2020**, *171*, 107851. [CrossRef]
73. Colacurcio, D.J.; Nixon, R.A. Disorders of Lysosomal Acidification-the Emerging Role of v-ATPase in Aging and Neurodegenerative Disease. *Ageing Res. Rev.* **2016**, *32*, 75–88. [CrossRef] [PubMed]
74. Baker, N.; Hamilton, G.; Wilkes, J.M.; Hutchinson, S.; Barrett, M.P.; Horn, D. Vacuolar ATPase Depletion Affects Mitochondrial ATPase Function, Kinetoplast Dependency, and Drug Sensitivity in Trypanosomes. *Proc. Natl. Acad. Sci. USA* **2015**, *112*, 9112–9117. [CrossRef] [PubMed]
75. Yambire, K.F.; Rostosky, C.; Watanabe, T.; Pacheu-Grau, D.; Torres-Odio, S.; Sanchez-Guerrero, A.; Senderovich, O.; Meyron-Holtz, E.G.; Milosevic, I.; Frahm, J.; et al. Impaired Lysosomal Acidification Triggers Iron Deficiency and Inflammation in Vivo. *eLife* **2019**, *8*, e51031. [CrossRef]
76. Sun, Y.; Li, M.; Zhao, D.; Li, X.; Yang, C.; Wang, X. Lysosome Activity Is Modulated by Multiple Longevity Pathways and Is Important for Lifespan Extension in *C. elegans*. *eLife* **2020**, *9*, e55745. [CrossRef]
77. Baxi, K.; Ghavidel, A.; Waddell, B.; Harkness, T.A.; de Carvalho, C.E. Regulation of Lysosomal Function by the DAF-16 Forkhead Transcription Factor Couples Reproduction to Aging in Caenorhabditis Elegans. *Genetics* **2017**, *207*, 83–101. [CrossRef]
78. Folick, A.; Oakley, H.D.; Yu, Y.; Armstrong, E.H.; Kumari, M.; Sanor, L.; Moore, D.D.; Ortlund, E.A.; Zechner, R.; Wang, M.C. Lysosomal Signaling Molecules Regulate Longevity in Caenorhabditis Elegans. *Science* **2015**, *347*, 83. [CrossRef]
79. Johmura, Y.; Yamanaka, T.; Omori, S.; Wang, T.W.; Sugiura, Y.; Matsumoto, M.; Suzuki, N.; Kumamoto, S.; Yamaguchi, K.; Hatakeyama, S.; et al. Senolysis by Glutaminolysis Inhibition Ameliorates Various Age-Associated Disorders. *Science* **2021**, *371*, 265–270. [CrossRef]
80. Simonaro, C.M. Lysosomes, Lysosomal Storage Diseases, and Inflammation. *J. Inborn Errors Metab. Screen.* **2016**, *4*, 1–8. [CrossRef]
81. Hughes, C.E.; Coody, T.K.; Jeong, M.Y.; Berg, J.A.; Winge, D.R.; Hughes, A.L. Cysteine Toxicity Drives Age-Related Mitochondrial Decline by Altering Iron Homeostasis. *Cell* **2020**, *180*, 296–310.e18. [CrossRef]
82. Rizzollo, F.; More, S.; Vangheluwe, P.; Agostinis, P. The Lysosome as a Master Regulator of Iron Metabolism. *Trends Biochem. Sci.* **2021**, *46*, 960–975. [CrossRef] [PubMed]
83. Zeidan, R.S.; Han, S.M.; Leeuwenburgh, C.; Xiao, R. Iron Homeostasis and Organismal Aging. *Ageing Res. Rev.* **2021**, *72*, 101510. [CrossRef] [PubMed]
84. Nordestgaard, B.G.; Adourian, A.S.; Freiberg, J.J.; Guo, Y.; Muntendam, P.; Falk, E. Risk Factors for Near-Term Myocardial Infarction in Apparently Healthy Men and Women. *Clin. Chem.* **2010**, *56*, 559–567. [CrossRef] [PubMed]
85. Rhodes, S.L.; Ritz, B. Genetics of Iron Regulation and the Possible Role of Iron in Parkinson's Disease. *Neurobiol. Dis.* **2008**, *32*, 183–195. [CrossRef]
86. Ayton, S.; Lei, P. Nigral Iron Elevation Is an Invariable Feature of Parkinson's Disease and Is a Sufficient Cause of Neurodegeneration. *BioMed Res. Int.* **2014**, *2014*, 581256. [CrossRef]
87. Liu, J.L.; Fan, Y.G.; Yang, Z.S.; Wang, Z.Y.; Guo, C. Iron and Alzheimer's Disease: From Pathogenesis to Therapeutic Implications. *Front. Neurosci.* **2018**, *12*, 632. [CrossRef]

88. Killilea, D.W.; Atamna, H.; Liao, C.; Ames, B.N. Iron Accumulation during Cellular Senescence in Human Fibroblasts in Vitro. *Antioxid. Redox Signal.* **2003**, *5*, 507–516. [CrossRef]
89. Masaldan, S.; Clatworthy, S.A.S.; Gamell, C.; Meggyesy, P.M.; Rigopoulos, A.T.; Haupt, S.; Haupt, Y.; Denoyer, D.; Adlard, P.A.; Bush, A.I.; et al. Iron Accumulation in Senescent Cells Is Coupled with Impaired Ferritinophagy and Inhibition of Ferroptosis. *Redox Biol.* **2018**, *14*, 100. [CrossRef]
90. Mazhar, M.; Din, A.U.; Ali, H.; Yang, G.; Ren, W.; Wang, L.; Fan, X.; Yang, S. Implication of Ferroptosis in Aging. *Cell Death Discov.* **2021**, *7*, 149. [CrossRef]
91. Sasikumar, A.N.; Killilea, D.W.; Kennedy, B.K.; Brem, R.B. Potassium Restriction Boosts Vacuolar Acidity and Extends Lifespan in Yeast. *Exp. Gerontol.* **2019**, *120*, 101–106. [CrossRef] [PubMed]
92. Höhn, A.; Jung, T.; Grimm, S.; Grune, T. Lipofuscin-Bound Iron Is a Major Intracellular Source of Oxidants: Role in Senescent Cells. *Free Radic. Biol. Med.* **2010**, *48*, 1100–1108. [CrossRef] [PubMed]
93. Höhn, A.; Grune, T. Lipofuscin: Formation, Effects and Role of Macroautophagy. *Redox Biol.* **2013**, *1*, 140. [CrossRef] [PubMed]
94. Reeg, S.; Grune, T. Protein Oxidation in Aging: Does It Play a Role in Aging Progression? *Antioxid. Redox Signal.* **2015**, *23*, 239. [CrossRef]
95. Moreno-García, A.; Kun, A.; Calero, O.; Medina, M.; Calero, M. An Overview of the Role of Lipofuscin in Age-Related Neurodegeneration. *Front. Neurosci.* **2018**, *12*, 464. [CrossRef]
96. Pan, C.; Banerjee, K.; Lehmann, G.L.; Almeida, D.; Hajjar, K.A.; Benedicto, I.; Jiang, Z.; Radu, R.A.; Thompson, D.H.; Rodriguez-Boulan, E.; et al. Lipofuscin Causes Atypical Necroptosis through Lysosomal Membrane Permeabilization. *Proc. Natl. Acad. Sci. USA* **2021**, *118*, e2100122118. [CrossRef]
97. Chevriaux, A.; Pilot, T.; Derangère, V.; Simonin, H.; Martine, P.; Chalmin, F.; Ghiringhelli, F.; Rébé, C. Cathepsin B Is Required for NLRP3 Inflammasome Activation in Macrophages, Through NLRP3 Interaction. *Front. Cell Dev. Biol.* **2020**, *8*, 167. [CrossRef]
98. Kelley, N.; Jeltema, D.; Duan, Y.; He, Y. The NLRP3 Inflammasome: An Overview of Mechanisms of Activation and Regulation. *Int. J. Mol. Sci.* **2019**, *20*, 3328. [CrossRef]
99. Hornung, V.; Bauernfeind, F.; Halle, A.; Samstad, E.O.; Kono, H.; Rock, K.L.; Fitzgerald, K.A.; Latz, E. Silica Crystals and Aluminum Salts Activate the NALP3 Inflammasome through Phagosomal Destabilization. *Nat. Immunol.* **2008**, *9*, 847–856. [CrossRef]
100. Zhu, S.Y.; Yao, R.Q.; Li, Y.; Zhao, P.; Ren, C.; Du, X.; Yao, Y.-M. Lysosomal Quality Control of Cell Fate: A Novel Therapeutic Target for Human Diseases. *Cell Death Dis.* **2020**, *11*, 817. [CrossRef]
101. Zhu, X.; Sun, Y.; Chen, D.; Li, J.; Dong, X.; Wang, J.; Chen, H.; Wang, Y.; Zhang, F.; Dai, J.; et al. Mastocarcinoma Therapy Synergistically Promoted by Lysosome Dependent Apoptosis Specifically Evoked by 5-Fu@nanogel System with Passive Targeting and PH Activatable Dual Function. *J. Control. Release* **2017**, *254*, 107–118. [CrossRef] [PubMed]
102. Gao, C.; Ding, Y.; Zhong, L.; Jiang, L.; Geng, C.; Yao, X.; Cao, J. Tacrine Induces Apoptosis through Lysosome- and Mitochondria-Dependent Pathway in HepG2 Cells. *Toxicol. Vitr.* **2014**, *28*, 667–674. [CrossRef] [PubMed]
103. Liu, S.; Li, Y.; Choi, H.M.C.; Sarkar, C.; Koh, E.Y.; Wu, J.; Lipinski, M.M. Lysosomal Damage after Spinal Cord Injury Causes Accumulation of RIPK1 and RIPK3 Proteins and Potentiation of Necroptosis. *Cell Death Dis.* **2018**, *9*, 476. [CrossRef] [PubMed]
104. Zou, J.; Kawai, T.; Tsuchida, T.; Kozaki, T.; Tanaka, H.; Shin, K.S.; Kumar, H.; Akira, S. Poly IC Triggers a Cathepsin D- and IPS-1-Dependent Pathway to Enhance Cytokine Production and Mediate Dendritic Cell Necroptosis. *Immunity* **2013**, *38*, 717–728. [CrossRef]
105. Gao, H.; Bai, Y.; Jia, Y.; Zhao, Y.; Kang, R.; Tang, D.; Dai, E. Ferroptosis Is a Lysosomal Cell Death Process. *Biochem. Biophys. Res. Commun.* **2018**, *503*, 1550–1556. [CrossRef]
106. Chi, H.; Chang, H.-Y.; Sang, T.-K. Molecular Sciences Neuronal Cell Death Mechanisms in Major Neurodegenerative Diseases. *Int. J. Mol. Sci.* **2018**, *19*, 3082. [CrossRef]
107. Patel, P.; Karch, J. Regulation of Cell Death in the Cardiovascular System. *Int. Rev. Cell Mol. Biol.* **2020**, *353*, 153–209. [CrossRef]
108. Nacarelli, T.; Azar, A.; Sell, C. Aberrant MTOR Activation in Senescence and Aging: A Mitochondrial Stress Response? *Exp. Gerontol.* **2015**, *68*, 66. [CrossRef]
109. Stallone, G.; Infante, B.; Prisciandaro, C.; Grandaliano, G. MTOR and Aging: An Old Fashioned Dress. *Int. J. Mol. Sci.* **2019**, *20*, 2774. [CrossRef]
110. Zhang, H.; Hoff, H.; Marinucci, T.; Cristofalo, V.J.; Sell, C. Mitogen-Independent Phosphorylation of S6K1 and Decreased Ribosomal S6 Phosphorylation in Senescent Human Fibroblasts. *Exp. Cell Res.* **2000**, *259*, 284–292. [CrossRef]
111. Barilari, M.; Bonfils, G.; Treins, C.; Koka, V.; de Villeneuve, D.; Fabrega, S.; Pende, M. ZRF1 Is a Novel S6 Kinase Substrate That Drives the Senescence Programme. *EMBO J.* **2017**, *36*, 736–750. [CrossRef] [PubMed]
112. Sandri, M.; Barberi, L.; Bijlsma, A.Y.; Blaauw, B.; Dyar, K.A.; Milan, G.; Mammucari, C.; Meskers, C.G.M.; Pallafacchina, G.; Paoli, A.; et al. Signalling Pathways Regulating Muscle Mass in Ageing Skeletal Muscle. The Role of the IGF1-Akt-MTOR-FoxO Pathway. *Biogerontology* **2013**, *14*, 303–323. [CrossRef] [PubMed]
113. Sun, Y.X.; Ji, X.; Mao, X.; Xie, L.; Jia, J.; Galvan, V.; Greenberg, D.A.; Jin, K. Differential Activation of MTOR Complex 1 Signaling in Human Brain with Mild to Severe Alzheimer's Disease. *J. Alzheimer's Dis.* **2014**, *38*, 437–444. [CrossRef] [PubMed]
114. Laberge, R.M.; Sun, Y.; Orjalo, A.V.; Patil, C.K.; Freund, A.; Zhou, L.; Curran, S.C.; Davalos, A.R.; Wilson-Edell, K.A.; Liu, S.; et al. MTOR Regulates the Pro-Tumorigenic Senescence-Associated Secretory Phenotype by Promoting IL1A Translation. *Nature Cell Biol.* **2015**, *17*, 1049–1061. [CrossRef]

115. Narita, M.; Young, A.R.J.; Arakawa, S.; Samarajiwa, S.A.; Nakashima, T.; Yoshida, S.; Hong, S.; Berry, L.S.; Reichelt, S.; Ferreira, M.; et al. Spatial Coupling of MTOR and Autophagy Augments Secretory Phenotypes. *Science* **2011**, *332*, 966–970. [CrossRef]
116. Khor, E.S.; Wong, P.F. The Roles of MTOR and MiRNAs in Endothelial Cell Senescence. *Biogerontology* **2020**, *21*, 517–530. [CrossRef]
117. Lee, C.H.; Inoki, K.; Karbowniczek, M.; Petroulakis, E.; Sonenberg, N.; Henske, E.P.; Guan, K.L. Constitutive MTOR Activation in TSC Mutants Sensitizes Cells to Energy Starvation and Genomic Damage via P53. *EMBO J.* **2007**, *26*, 4812–4823. [CrossRef]
118. Khor, E.S.; Wong, P.F. Endothelial Replicative Senescence Delayed by the Inhibition of MTORC1 Signaling Involves MicroRNA-107. *Int. J. Biochem. Cell Biol.* **2018**, *101*, 64–73. [CrossRef]
119. Alimonti, A.; Nardella, C.; Chen, Z.; Clohessy, J.G.; Carracedo, A.; Trotman, L.C.; Cheng, K.; Varmeh, S.; Kozma, S.C.; Thomas, G.; et al. A Novel Type of Cellular Senescence That Can Be Enhanced in Mouse Models and Human Tumor Xenografts to Suppress Prostate Tumorigenesis. *J. Clin. Investig.* **2010**, *120*, 681–693. [CrossRef]
120. Jung, S.H.; Hwang, H.J.; Kang, D.; Park, H.A.; Lee, H.C.; Jeong, D.; Lee, K.; Park, H.J.; Ko, Y.G.; Lee, J.S. mTOR Kinase Leads to PTEN-Loss-Induced Cellular Senescence by Phosphorylating P53. *Oncogene* **2019**, *38*, 1639–1650. [CrossRef]
121. Lai, K.P.; Leong, W.F.; Chau, J.F.L.; Jia, D.; Zeng, L.; Liu, H.; He, L.; Hao, A.; Zhang, H.; Meek, D.; et al. S6K1 Is a Multifaceted Regulator of Mdm2 That Connects Nutrient Status and DNA Damage Response. *EMBO J.* **2010**, *29*, 2994–3006. [CrossRef] [PubMed]
122. Julien, L.-A.; Carriere, A.; Moreau, J.; Roux, P.P. MTORC1-Activated S6K1 Phosphorylates Rictor on Threonine 1135 and Regulates MTORC2 Signaling. *Mol. Cell Biol.* **2010**, *30*, 908–921. [CrossRef] [PubMed]
123. Ogawara, Y.; Kishishita, S.; Obata, T.; Isazawa, Y.; Suzuki, T.; Tanaka, K.; Masuyama, N.; Gotoh, Y. Akt Enhances Mdm2-Mediated Ubiquitination and Degradation of P53. *J. Biol. Chem.* **2002**, *277*, 21843–21850. [CrossRef] [PubMed]
124. Llanos, S.; García-Pedrero, J.M.; Morgado-Palacin, L.; Rodrigo, J.P.; Serrano, M. ARTICLE Stabilization of P21 by MTORC1/4E-BP1 Predicts Clinical Outcome of Head and Neck Cancers. *Nat. Commun.* **2016**, *7*, 10438. [CrossRef]
125. Nakano, M.; Nakashima, A.; Nagano, T.; Ishikawa, S.; Kikkawa, U. Branched-Chain Amino Acids Enhance Premature Senescence through Mammalian Target of Rapamycin Complex I-Mediated Upregulation of P21 Protein. *PLoS ONE* **2013**, *8*, e80411. [CrossRef]
126. Astle, M.V.; Hannan, K.M.; Ng, P.Y.; Lee, R.S.; George, A.J.; Hsu, A.K.; Haupt, Y.; Hannan, R.D.; Pearson, R.B. AKT Induces Senescence in Human Cells via MTORC1 and P53 in the Absence of DNA Damage: Implications for Targeting MTOR during Malignancy. *Oncogene* **2012**, *31*, 1949–1962. [CrossRef]
127. Kucheryavenko, O.; Nelson, G.; von Zglinicki, T.; Korolchuk, V.I.; Carroll, B. The MTORC1-Autophagy Pathway Is a Target for Senescent Cell Elimination. *Biogerontology* **2019**, *20*, 331–335. [CrossRef]
128. Bharath, L.P.; Agrawal, M.; McCambridge, G.; Nicholas, D.A.; Hasturk, H.; Liu, J.; Jiang, K.; Liu, R.; Guo, Z.; Deeney, J.; et al. Metformin Enhances Autophagy and Normalizes Mitochondrial Function to Alleviate Aging-Associated Inflammation. *Cell Metab.* **2020**, *32*, 44–55.e6. [CrossRef]
129. Demidenko, Z.N.; Zubova, S.G.; Bukreeva, E.I.; Pospelov, V.A.; Pospelova, T.V.; Blagosklonny, M.v. Rapamycin Decelerates Cellular Senescence. *Cell Cycle* **2009**, *8*, 1888–1895. [CrossRef]
130. Leontieva, O.V.; Demidenko, Z.N.; Blagosklonny, M.v. Contact Inhibition and High Cell Density Deactivate the Mammalian Target of Rapamycin Pathway, Thus Suppressing the Senescence Program. *Proc. Natl. Acad. Sci. USA* **2014**, *111*, 8832–8837. [CrossRef]
131. Harrison, D.E.; Strong, R.; Dave Sharp, Z.; Nelson, J.F.; Astle, C.M.; Flurkey, K.; Nadon, N.L.; Erby Wilkinson, J.; Frenkel, K.; Carter, C.S.; et al. Rapamycin Fed Late in Life Extends Lifespan in Genetically Heterogeneous Mice. *Nature* **2009**, *460*, 392–395. [CrossRef]
132. Vellai, T.; Takacs-Vellai, K.; Zhang, Y.; Kovacs, A.L.; Orosz, L.; Müller, F. Genetics: Influence of TOR Kinase on Lifespan in C. Elegans. *Nature* **2003**, *426*, 620. [CrossRef] [PubMed]
133. Selman, C.; Withers, D.J. Mammalian Models of Extended Healthy Lifespan. *Philos. Trans. R. Soc. B Biol. Sci.* **2011**, *366*, 99. [CrossRef] [PubMed]
134. Zhang, X.I.; Chen, W.; Gao, Q.I.; Yang, J.I.; Yan, X.; Zhao, H.; Su, L.; Yang, M.; GaoID, C.; Yao, Y.; et al. Rapamycin Directly Activates Lysosomal Mucolipin TRP Channels Independent of MTOR. *PLoS Biol.* **2019**, *17*, e3000252. [CrossRef] [PubMed]
135. Boudewyn, L.C.; Walkley, S.U. Current Concepts in the Neuropathogenesis of Mucolipidosis Type IV. *J. Neurochem.* **2019**, *148*, 669–689. [CrossRef] [PubMed]
136. Tsunemi, T.; Ashe, T.D.; Morrison, B.E.; Soriano, K.R.; Au, J.; Roque, R.A.V.; Lazarowski, E.R.; Damian, V.A.; Masliah, E.; la Spada, A.R. PGC-1α Rescues Huntington's Disease Proteotoxicity by Preventing Oxidative Stress and Promoting TFEB Function. *Sci. Transl. Med.* **2012**, *4*, 142ra97. [CrossRef]
137. Decressac, M.; Mattsson, B.; Weikop, P.; Lundblad, M.; Jakobsson, J.; Björklund, A. TFEB-Mediated Autophagy Rescues Midbrain Dopamine Neurons from α-Synuclein Toxicity. *Proc. Natl. Acad. Sci. USA* **2013**, *110*, E1817–E1826. [CrossRef]
138. Song, J.X.; Sun, Y.R.; Peluso, I.; Zeng, Y.; Yu, X.; Lu, J.H.; Xu, Z.; Wang, M.Z.; Liu, L.F.; Huang, Y.Y.; et al. A Novel Curcumin Analog Binds to and Activates TFEB in Vitro and in Vivo Independent of MTOR Inhibition. *Autophagy* **2016**, *12*, 1372–1389. [CrossRef]
139. Kilpatrick, K.; Zeng, Y.; Hancock, T.; Segatori, L. Genetic and Chemical Activation of TFEB Mediates Clearance of Aggregated α-Synuclein. *PLoS ONE* **2015**, *10*, e0120819. [CrossRef]
140. Li, Y.; Hodge, J.; Liu, Q.; Wang, J.; Wang, Y.; Evans, T.D.; Altomare, D.; Yao, Y.; Murphy, E.A.; Razani, B.; et al. TFEB Is a Master Regulator of Tumor-Associated Macrophages in Breast Cancer. *J. Immunother. Cancer* **2020**, *8*, 543. [CrossRef]

141. Kim, J.H.; Lee, J.; Cho, Y.R.; Lee, S.Y.; Sung, G.J.; Shin, D.M.; Choi, K.C.; Son, J. TFEB Supports Pancreatic Cancer Growth through the Transcriptional Regulation of Glutaminase. *Cancers* **2021**, *13*, 483. [CrossRef] [PubMed]
142. Zhu, X.; Zhuo, Y.; Wu, S.; Chen, Y.; Ye, J.; Deng, Y.; Feng, Y.; Liu, R.; Cai, S.; Zou, Z.; et al. TFEB Promotes Prostate Cancer Progression via Regulating ABCA2-Dependent Lysosomal Biogenesis. *Front. Oncol.* **2021**, *11*, 236. [CrossRef]
143. Cai, Y.; Zhou, H.; Zhu, Y.; Sun, Q.; Ji, Y.; Xue, A.; Wang, Y.; Chen, W.; Yu, X.; Wang, L.; et al. Elimination of Senescent Cells by β-Galactosidase-Targeted Prodrug Attenuates Inflammation and Restores Physical Function in Aged Mice. *Cell Res.* **2020**, *30*, 574–589. [CrossRef] [PubMed]

cells

Review

Pyroptosis and Sarcopenia: Frontier Perspective of Disease Mechanism

Hongfu Jin [1,2], Wenqing Xie [1,2], Miao He [1,2], Hengzhen Li [1,2], Wenfeng Xiao [1,2,*] and Yusheng Li [1,2,*]

1 Department of Orthopedics, Xiangya Hospital, Central South University, Changsha 410008, China; 218112327@csu.edu.cn (H.J.); xiewenqing@csu.edu.cn (W.X.); hemiao@csu.edu.cn (M.H.); hengzhen@csu.edu.cn (H.L.)
2 National Clinical Research Center for Geriatric Disorders, Central South University, Xiangya Hospital, Changsha 410008, China
* Correspondence: xiaowenfeng@csu.edu.cn (W.X.); liyusheng@csu.edu.cn (Y.L.)

Abstract: With global ageing, sarcopenia, as an age-related disease, has brought a heavy burden to individuals and society. Increasing attention has been given to further exploring the morbidity mechanism and intervention measures for sarcopenia. Pyroptosis, also known as cellular inflammatory necrosis, is a kind of regulated cell death that plays a role in the ageing progress at the cellular level. It is closely related to age-related diseases such as cardiovascular diseases, Alzheimer's disease, osteoarthritis, and sarcopenia. In the process of ageing, aggravated oxidative stress and poor skeletal muscle perfusion in ageing muscle tissues can activate the nod-like receptor (NLRP) family to trigger pyroptosis. Chronic inflammation is a representative characteristic of ageing. The levels of inflammatory factors such as TNF-α may activate the signaling pathways of pyroptosis by the NF-κB-GSDMD axis, which remains to be further studied. Autophagy is a protective mechanism in maintaining the integrity of intracellular organelles and the survival of cells in adverse conditions. The autophagy of skeletal muscle cells can inhibit the activation of the pyroptosis pathway to some extent. A profound understanding of the mechanism of pyroptosis in sarcopenia may help to identify new therapeutic targets in the future. This review article focuses on the role of pyroptosis in the development and progression of sarcopenia.

Keywords: sarcopenia; aging; NLRP; pyroptosis; gasdermin

Citation: Jin, H.; Xie, W.; He, M.; Li, H.; Xiao, W.; Li, Y. Pyroptosis and Sarcopenia: Frontier Perspective of Disease Mechanism. *Cells* **2022**, *11*, 1078. https://doi.org/10.3390/cells11071078

Academic Editors: Nicole Wagner, Kay-Dietrich Wagner and Wan Lee

Received: 9 January 2022
Accepted: 18 March 2022
Published: 23 March 2022

1. Introduction

With the global proportion of elders over the age of 65 increasing, it is estimated that this situation will worsen by 2050 with the proportion of elders exceeding 1.3 billion and reaching up to 38% of the population [1]. The ageing population poses a huge challenge to public health, placing a great burden on the country and society. Ageing is an important factor leading to increased susceptibility to disease and disability [2]. The occurrence of many diseases is related to ageing [3–5]. Studies have shown that 23% of the total burden of global diseases can be attributed to diseases from elders over the age of 60. The main factors of age-related disease burden are cardiovascular diseases (30.3%), malignant tumors (15.1%), chronic respiratory disease (9.5%), musculoskeletal disease (7.5%), and others (6.6%) [6]. Sarcopenia, an age-related disease, causes disability and declining quality of life in the elderly. Among the quality-of-life assessment tools, such as the EuroQol-5D instrument (EQ-5D) and Short-form General Health Survey (SF-36), the outcomes show that a higher proportion of quality-of-life problems are reported by patients with sarcopenia [7]. In addition to being affected by age, sarcopenia is also affected by genetic and lifestyle factors. Disease progression involves a decline in muscle mass and increased consequences, such as falling, disability, weakness, and increased mortality [8]. At present, an increasing number of researchers are focusing on the morbidity mechanism and intervention measures to prevent and treat sarcopenia.

In the process of aging, changes occur at the cellular level, such as autophagy, mitochondrial dysfunction, cell senescence, and DNA methylation [1–3]. All of these changes can occur in various cell types, causing damage to cell structure and function [4]. Pyroptosis, a new regulated cell death, was discovered recently, and is different from other cell death such as apoptosis, necrosis, and others in morphological characteristics, occurrence, and mechanism [5,6]. Pyroptosis mainly relies on inflammation to activate the inflammasome of the caspase family, leading to the activation of gasdermin proteins, then the activated gasdermin proteins transposition to the membrane, where the membrane is breached, causing cell swelling, cytoplasmic alterations, and ultimately leading to membrane rupture [7]. Studies have shown that pyroptosis is comprehensively involved in infectious diseases, atherosclerotic diseases, and age-related diseases [8–10]. Its role in sarcopenia is attracting more and more attention from researchers. Deeply understanding the role of pyroptosis in the development and progression of sarcopenia is helpful in providing new ideas for its clinical prevention and treatment. This review article focuses on the role of pyroptosis in the disease mechanism of sarcopenia.

2. Sarcopenia

With the ageing of the human body, there is inevitably a gradual decline in muscle mass, quality, and strength. Studies have shown that skeletal muscle mass and strength from the age of 40 decreases linearly, and the loss of skeletal muscle mass in individuals who are 80 years old can reach up to 50%, greatly impacting the elderly [11]. Since sarcopenia was first defined by Rosenberg, it has been used to describe the age-related loss of skeletal muscle mass and strength [12]. Research on sarcopenia continues to be a major focal point of research worldwide. However, there is a lack of a single diagnostic criterion for sarcopenia. The diagnosis of sarcopenia consists of the following three key characteristics: (a) muscle strength, (b) muscle quantity or mass, and (c) physical performance. Musculoskeletal degeneration not only impacts the daily mobility of patients but also leads to an increase in the incidence of complications, mortality, and morbidity in major surgical procedures [13].

The aetiology of sarcopenia is multifactorial, and both environmental factors and internal factors play a role in the onset of sarcopenia [14]. Decline in the activity of the elderly is one of the important reasons for sarcopenia. Studies have shown that exercise (especially resistance training) is the most promising method for increasing muscle mass and strength in elderly individuals [15]. Elderly individuals are often at risk of malnutrition, and protein-energy malnutrition, in particular, is often observed. Nutritional interventions may prevent and reverse the progression of sarcopenia [16]. The increased burden of chronic diseases in the elderly, such as chronic kidney disease and cirrhosis, leads to pain and disability, which causes sarcopenia known as secondary sarcopenia [16,17]. Abnormal muscle development and low birth weight are also implicated in the decline of muscle mass and strength in adulthood [18]. In vivo, inflammatory pathway activation, mitochondrial dysfunction, denervation, satellite cell reduction, and endocrine disorders are considered to be internal factors that cause sarcopenia [14]. Inflammatory cytokines have been shown to promote the loss of muscle mass by stimulating protein metabolism and inhibiting muscle anabolism. The results of a meta-analysis and systematic review showed that serum IL-6 and TNF-α levels in patients with sarcopenia were not significantly different from those in the control group, while the level of C-reactive protein (CRP) in serum increased [19]. Mitochondrial population changes and dysfunction are considered to be the key contributors to sarcopenia, leading to the production of reactive oxygen species (ROS) and promoting inflammation [20]. Age-related degeneration of neuromuscular junctions is associated with loss of skeletal muscle mass, which is a prominent aspect of sarcopenia. Endocrine disorders, such as low levels of 2,5-(OH) vitamin D and decreased levels of growth hormone (GH) and insulin-like growth factor-1 (IGF-1), have been observed in patients with sarcopenia. Studies have shown that testosterone can improve muscle mass, strength, and function, offering new ideas to develop possible interventions for sarcopenia. Satellite cells, also called muscle stem cells, are crucial for the muscle regeneration process.

Satellite cells are reduced in the muscle tissues of patients with sarcopenia, suggesting that loss of satellite cells may be one of the pathogenic mechanisms of sarcopenia.

There is a link between apoptosis and sarcopenia. It is necessary to further clarify the apoptosis pathway related to sarcopenia. The increase in muscle protein decomposition is considered to be the result of the common effects of inflammation and sarcopenia [21]. A long-term low-grade inflammatory state is a significant feature of the ageing process, and involves concomitant chronic degenerative diseases, including sarcopenia. Several cytokines have been observed at high circulating levels, such as interleukin-6 (IL-6), C-reactive protein (CRP), and TNF-α [22]. TNF-α can induce apoptosis through a death receptor-mediated signaling pathway. Caspase-8 initiates caspase-mediated downstream cascade reactions, mediating crosstalk between the extrinsic and intrinsic apoptosis pathways [23]. The integrity of mitochondrial structure and function plays an important role in the clearance of mitochondrial oxidants, energy supply, protein repair, and degradation [23]. One possible consequence of mitochondrial dysfunction is the activation of apoptosis. Evidence has shown that mitochondrial-mediated apoptosis may be involved in the ageing process of skeletal muscle [24]. As a special form of apoptosis, pyroptosis occurs in aseptic inflammation, such as diabetes, atherosclerosis, acute liver injury, benign prostatic hyperplasia, Alzheimer's disease and other diseases [25–28]. This review article focuses on the role of pyroptosis in sarcopenia.

3. Pyroptosis

3.1. Characteristics of Pyroptosis

Pyroptosis, a newly regulated form of cell death, was discovered and has been confirmed in recent years, and it is characterized by the release of a large amount of proinflammatory cytokines [29]. It is different from other kinds of cell death. Pyroptosis and necroptosis are both inflammatory death pathways that allow the release of immunogenic cellular contents, which may act as activators of pattern-recognition receptors (PRRs). However, they differ in morphological characteristics, occurrence, and regulatory mechanisms [30]. Necroptosis occurs when there are obstacles in normal apoptotic pathways. Pyroptosis always occurs after sensing potential destructive injury [7]. For example, the inflammasome involved in the pyroptosis pathway can be activated in injured tissue, metabolic macrophages, monocytes, and so on [31]. Pyroptosis can be induced by gasdermin, but necroptosis is induced by membrane-associated mixed lineage kinase domain-like (MLKL), resulting in distinct cell morphologies. Pyroptosis is also distinct from apoptosis, although they have similar characteristics in some ways. As an inflammatory mode of regulated cell death, pyroptosis has unique characteristics differing from apoptosis, such as intact nuclei, DNA laddering, pore formation, cell swelling, and osmotic lysis. There are also differences in the activation of the caspase enzyme; apoptosis mainly activates caspase-2, 6, 7, and 8, whereas pyroptosis mainly activates caspase-1, 4, 5, and 11. Poly (ADP-ribose) polymerase (PARP) cleavage and caspase-activated DNase inhibitor (ICAD) cleavage both exist in apoptotic cells; however, pyroptosis is unique [5].

3.2. The Canonical Inflammasome Pathway of Pyroptosis

Many extracellular stimuli (such as bacteria, viruses, toxins, among others) can induce pyroptosis [32,33]. In the canonical inflammasome pathway of pyroptosis, the inflammasome sensing protein nod-like receptor (NLRP), which has an N-terminal caspase recruitment domain (CARD)/pyrin domain (PYD), can be stimulated by foreign substances [34]. Members of the NLRP family consist of nod-like receptor 1 (NLRP1), nod-like receptor 2 (NLRP2), nod-like receptor 3 (NLRP3), and nod-like receptor C4 (NLRC4) [7]. NLRP3 can identify most extrinsic stimulations. When NLRP3 is stimulated, it is indirectly activated through K+ flow. Activated NLRP3 recruits an apoptosis-associated speck-like protein containing a caspase recruitment domain (ASC) adaptor, further forming an ASC focus by connecting CARD/PYD with ASC. Then, the ASC focus recruits pro-caspase-1, leading to the activation of caspase-1 [32]. Activated caspase-1 promotes the formation and activation of IL-18 and IL-1β [35]. Gasdermin D (GSDMD) plays an important role in the downstream signaling pathways in pyroptosis, and can form pores in the plasma membrane [36]. The destruction of the plasma membrane leads to the release of cytosolic proteins, causing cell swelling and cytolysis [37]. Caspase-1 can cleave GSDMD to release the membrane pore-forming GSDMD-N domain, promoting not only the release of proinflammatory cytokines, such as IL-18 and IL-1β, leading to inflammatory reactions, but also the process of pyroptosis [38]. NLRP1, unlike NLRP3, can only identify microbial muramyl dipeptide (MDP) [39,40]. Evidence has shown that NLRP2 is related to pro-caspase-1, which is involved in the production of IL-1β [41]. NLRC4 can mediate the activation of caspase-1 and pyroptosis events caused by Gram-negative bacteria, and its functions can be seen as a host defense strategy against pathogens. The activation of caspase-1 and pyroptosis events mediated by NLRC4 can promote the fusion of pathogen-containing phagosomes and lysosomes, leading to bacterial degradation [42]. Absent in melanoma (AIM)-like receptors can activate caspase-1 to induce pyroptosis in the same way as NLRP. For example, AIM2 can be activated, especially by double-stranded DNA (dsDNA), triggering pyroptosis signaling pathways [43].

3.3. Non-Canonical Inflammasome Pathway of Pyroptosis

The noncanonical inflammasome pathway is mediated by caspase-4/5/11. All of these proteins can cleave GSDMD, leading to the activation of pyroptosis signaling pathways [44]. First, caspase-4/5/11 can directly recognize cytosolic lipopolysaccharide (LPS) [45]. Then, the combination of caspase-4/5/11 and cytosolic LPS results in GSDMD cleavage, causing K+ efflux, which is sufficient to induce the formation of NLRP3 and activate the canonical pathway of pyroptosis [46]. Caspase-4/5/11-induced GSDMD pore formation can cause the release of proinflammatory cytokines and the occurrence of pyroptosis.

4. Pyroptosis and Sarcopenia

4.1. Activation of NLRP Family Triggers Pyroptosis

Local changes in muscle cell bioenergetics, mitochondrial metabolism, and oxidative damage play a role in the aggravation of sarcopenia [47]. Oxidative damage occurring in aged myofibers can trigger metabolic dysfunction, which may limit substrate availability for contractile performance [48,49]. The age-related loss of motor unit innervation can promote mitochondrial dysfunction, which is a vital cause of increased oxidative metabolism in muscle. Mitochondrial dysfunction is considered to be a major marker of the ageing process [50]. An efficient skeletal muscle energy supplement is derived from mitochondria. A large number of studies have shown that mitochondrial dysfunction and population changes play a role in the reduction of muscle mass [51]. Age-dependent decline in skeletal muscle mass is associated with muscle tissue metabolism [52]. Mitochondrial function is primarily involved in the production of ATP through oxidative phosphorylation (OXPHOS) but is also involved in apoptosis, calcium homeostasis, and the production of reactive oxygen species (ROS), reactive nitrogen species (RNS), and so on [51]. Moreover, oxygen radicals, such as ROS, can activate the NLRP family [53], which can propagate cellular metabolic dysfunction into immune responses. NLRP3-mediated processing of caspase-1 can cleave and activate IL-1β and IL-18, leading to activation of the canonical pathway of pyroptosis [36]. NLRP3-dependent caspase-1 activity increases with ageing, as has been found in mouse skeletal muscle. NLRP3 is also required for proteolysis and inactivation of reduced glyceraldehyde-phosphate dehydrogenase (GAPDH) and decreased glycolytic myofiber size during ageing in skeletal muscle. Animal studies showed that NLRP3 was a contributor to the age-related decline in myofiber size. However, there was no evidence that systemic inflammation was associated with the NLRP3-dependent decrease in glycolytic potential and myofiber size during ageing [54]. IL-1β is an important inflammatory factor that is positively correlated with age [55]. Studies have shown that age-related decline in muscle strength in the general population is closely related to the existence and duration of inflammatory conditions [56]. The ageing process of muscle is accompanied by chronic, low-grade inflammation, which has been found in skeletal muscle cells [57]. Released IL-1β has a strong proinflammatory effect, which further leads to the production of inflammatory molecules, such as IL-6, IL-8, and TNF-α [58]. These cytokines have been found at high levels in the blood serum of patients with sarcopenia, which may be related to the occurrence of sarcopenia [59]. TNF-α can promote protein degradation, reduce protein synthesis, and inhibit muscle regeneration by preventing proliferation and differentiation of satellite cells [60]. Studies on samples of skeletal muscle collected from cattle showed that the priming and activation of the NLRP3 inflammasome during ageing may trigger and sustain a proinflammatory environment leading to sarcopenia [61].

4.2. *Lysosome-Induced Autophagy and Pyroptosis*

Lysosomes, important organelles in skeletal muscle cells, are essential for cell homeostasis, and are the command and control center of cell signaling, metabolism, and other processes [62]. It is generally recognized that lysosomes play a key role in regulating cell death under both physiological and pathological conditions [63]. Mitochondria-derived products, such as ROS and NOS, can cause lipid peroxidation (LPO) in the lysosomal membrane, which can promote lysosomal membrane permeabilization (LMP) [64]. LMP is a key step in regulated cell death [65]. Lysosomes participate in regulated cell death in different ways, such as necroptosis, ferroptosis, and pyroptosis [66]. Under the triggering of various inducers, lysosomes become unstable and release their contents. Cathepsin B released by lysosomes is a central modulator of inflammasome-dependent pyroptosis [67]. Cathepsin B can not only bind to NLRP3 directly but also activate NLRP3 through mitogen-activated protein kinase (MAPK) and the transforming growth factor β-activated kinase 1 (TAK1)/c-Jun N-terminal kinase (JNK) pathway [68]. In addition to the release of the cathepsin family, lysosomal LMP results in ROS production and K^+ efflux, leading to the activation of inflammasomes [69]. Therefore, rupture of the lysosomal membrane can be considered the upstream activator of NLRP, which can activate caspase-1 to trigger the canonical pathway of pyroptosis. LPS escapes the degradation of lysosomes, which allows LPS to activate the noncanonical pathway of pyroptosis.

As an important organelle, the lysosome also participates in autophagy, which is a protective mechanism in maintaining the integrity of intracellular organelles and the survival of cells in adverse conditions [70]. Studies have shown that autophagy can maintain muscle mass and enhance muscle strength to a certain extent [71]. Satellite cells, located between the basal lamina and the myofibrillar membrane, can regenerate and differentiate. Autophagy can activate satellite cells in response to muscle injury, leading to early compensatory regeneration of atrophic muscle [72]. When satellite cells are stimulated, they change from a resting state to an activated state with the ability to proliferate and differentiate [73]. With the ageing of skeletal muscles, stress from the imbalance of protein metabolism, increased oxidative stress and mitochondrial damage can lead to dysfunction of autophagy [74]. Autophagy can inhibit the activation of the NLRP3 inflammasome by removing damaged mitochondria to inhibit the production of oxidizing agents. Autophagy can also promote the degradation of inflammatory components, such as pro-IL-1β, NLRP3, caspase-1, and ASC [75] (Figure 1). All of these findings suggest that the autophagy of muscle cells can inhibit the activation of the pyrogenic pathway to some extent. Recently, studies have found that autophagy can also enhance the NLRP3 inflammasome, which in turn may cause pyroptosis [76]. Excessive activation of autophagy accelerates the decline in skeletal muscle mass [77].

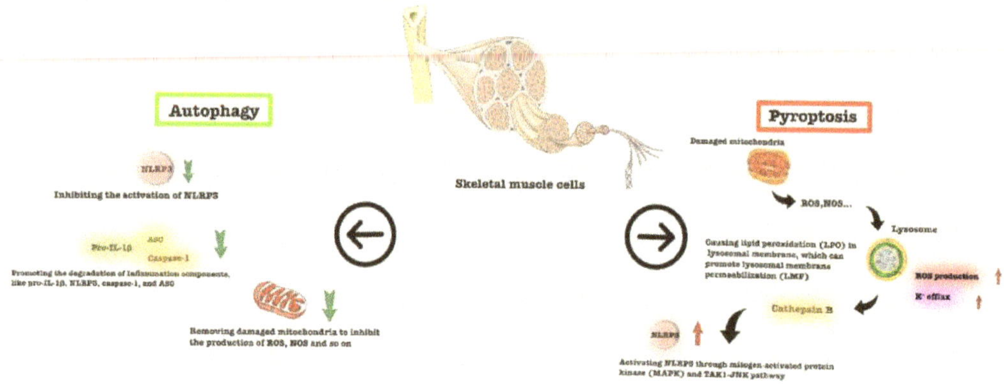

Figure 1. The autophagy of skeletal muscle cells in normal conditions can inhibit the activation of the pyroptosis to some extent. Autophagy can inhibit the activation of the nod-like receptor 3 (NLRP3) inflammasome by removing the damaged mitochondria, inhibiting the production of oxidizing agents, and promoting the degradation of inflammation components, such as pro-Interleukin-1β (pro-IL-1β), NLRP3, caspase-1, and apoptosis-associated speck-like protein containing a caspase recruitment domain (ASC).

4.3. *Poor Skeletal Muscle Perfusion and Pyroptosis*

Vascular dysfunction may become aggravated with increased age, and affects skeletal muscle perfusion, nutrient, and oxygen delivery to skeletal muscle, ultimately accelerating muscle mass loss and disability [78]. A positive correlation between the volume density of mitochondria and the number of capillaries supplying muscle fibers has been reported [79]. The number of capillaries supplying muscle fibers was also positively correlated with the size of muscle fibers [80]. The age-related decreases in fiber size are associated with capillary sparsity and oxygen-carrying capacity [81]. The heterogeneity of capillary spacing increases with age, and this seems to be related to the increased heterogeneity of muscle fiber size [82]. The distance between satellite cells and capillaries associated with type II fibers is longer in older people than in younger people, leading to the dysregulation of satellite cells and ultimately impairing muscle remodeling and regeneration [83]. Age-related biological processes such as oxidative stress, inflammation, and hormonal dysregulation can activate the NF-κB signaling pathway, increase P53/P21/P16 transcription, and inhibit autophagy. All of these factors may lead to dysfunction of vascular endothelial cells and vascular calcification. Vascular dysfunction may play a role in the development of sarcopenia [78]. ROS are mainly derived from endoplasmic reticulum (ER) stress, damaged mitochondria, and NADPH oxidase, and can trigger the activation of the NLRP3 inflammasome in endothelial cells, bridging the interaction between the NLRP3 inflammasome and endothelial dysfunction [84]. When the NLRP3 inflammasome is activated, it can further trigger endothelial cell pyroptosis, causing cell membrane rupture, cellular lysis, and the release of proinflammatory contents, including IL-1β, IL-18, and high-mobility group box 1 (HMGB1) [85]. IL-1β and IL-18, as the products of the NLRP3 inflammasome, can enhance the expression of inflammation-related genes of endothelial cells, such as adhesion molecules and chemokines, by activating the NF-κB pathway [86,87]. The inflammation-related pathological changes in the vascular endothelium may be involved in the early causative mechanisms in physical frailty and sarcopenia [88]. NLRP3 and ASC are phagocytosed by macrophages, leading to amplification of inflammation, suggesting that pyroptosis also plays a key role in transmitting inflammatory signals and amplifying the inflammatory reaction [89]. The loss of proper endothelial function may cause muscle tissue swelling, chronic inflammation, and the formation of thrombi [84]. Studies have shown that the dysfunction of vascular endothelial cells precedes a decline in muscle mass [90]. Vascular dysfunction can restrict blood flow and perfusion of skeletal muscle, blocking substrate transport to skeletal muscle and causing atrophy and dysfunction of skeletal muscle.

4.4. The NF-κB-GSDMD Axis Can Trigger Pyroptosis

The development and maintenance of skeletal muscles require a variety of signaling pathways. In the relevant signaling pathways, nuclear factor kappa B (NF-κB) is a key factor in the maintenance of skeletal muscle homeostasis [91]. NF-κB exists in the form of a dimer and is involved in the occurrence and progression of various diseases related to inflammation, apoptosis, and proliferation [91]. In an unstimulated state, the nuclear localization sequence of NF-κB is bound to inhibitory IκB proteins. IκB kinase (IKK) can be stimulated by various stimuli, resulting in the phosphorylation of IκB and causing the degeneration of IκB. Then, the NF-κB complexes liberated from IκB can translocate into the nucleus to exert their actions [92]. Increased NF-κB signaling has been observed during ageing, and the concentration of NF-κB protein in the muscles of elderly people was found to be four times that of young people [93]. NF-κB is a key molecular switch in the cellular response to oxidative stress that can enhance the expression of the NLRP3 inflammasome and cytokines [94]. The levels of inflammatory factors in the blood circulation of elderly individuals are higher than those in young individuals [95]. As a key inflammatory cytokine produced by various cells, such as macrophages, lymphocytes, and fibroblasts, TNF-α acts as a central regulator to promote the expression of other inflammatory cytokines, which can significantly reduce the expression levels of myosin heavy chain (MHC) and the surface area of myotube cells and activate creatine kinase (CK), greatly impairing skeletal muscle function [96,97]. When TNF-α binds to TNF receptors R1/R2, NF-κB, signaling can be activated through the RIP1/TRAF2/IKK and TRAF2/NIK pathways [98,99]. Studies have shown that triptolide can prevent LPS-induced inflammation and skeletal muscle atrophy in mice by inhibiting NF-κB/TNF-α and regulating the protein synthesis/degradation pathway [100]. NF-κB is an important transcription factor of GSDMD. When extracellular pyroptosis-related signals activate NLRP3 inflammasomes, GSDMD is subsequently cleaved, causing the release of the N-terminus of GSDMD, which forms nanoscale pores in the cell membrane, leading to the release of proinflammatory substances and cell swelling [101]. Studies have shown that NLRP3 activation is associated with several age-related diseases [102,103]. Oxidative stress triggers the NF-κB-GSDMD signaling axis, which is the key pathway that regulates pyroptosis in inflammatory diseases and endothelial dysfunction [91]. The exact mechanism of the NF-κB-GSDMD axis in sarcopenia remains to be further studied (Figure 2).

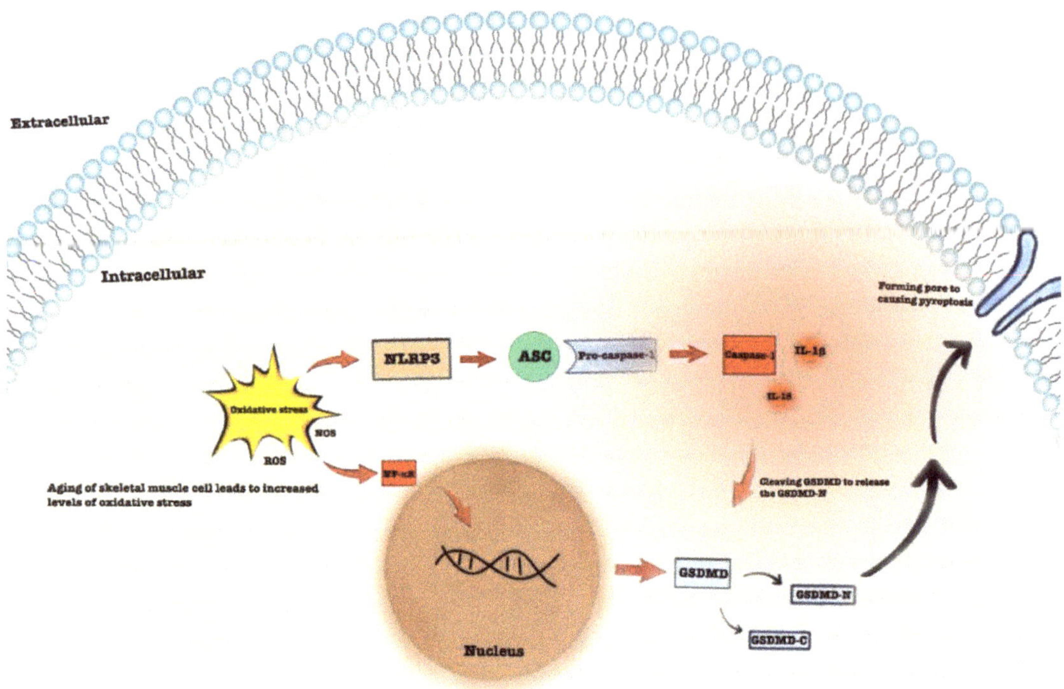

Figure 2. An increase in the level of oxidative stress caused by aging may trigger pyroptosis in skeletal muscle cells by the nuclear factor—kappaB (NF-κB)—gasdermin D (GSDMD) Axis. NF-κB is a transcription factor of GSDMD. When NLRP3 inflammasomes are activated by multiple activation mechanisms, the GSDMD will be subsequently cleaved, causing the release of the N-terminus of the GSDMD, which can form nanoscale pores in the cell membrane leading to the pyroptosis of skeletal muscle cells.

5. Perspectives in Pyroptosis and Sarcopenia

Sarcopenia, a form of age-related disease, causes a gradual inability to maintain skeletal muscle function and quality [93]. With the gradual increase in the incidence of sarcopenia in the elderly population and health care expenditures, sarcopenia has now become the focus of research and public policy. Although the importance of sarcopenia is concerning, sarcopenia remains to be further understood. Pyroptosis, a special type of cell death, is mediated by activation of inflammasome sensors, including the NLR family, the DNA receptor absent in melanoma 2 (AIM2), and the pyrin receptor. Under normal conditions, inflammasome sensors detect a variety of pathogen-associated molecular patterns (PAMPs) and danger-associated molecular patterns (DAMPs) and respond to intracellular and extracellular danger signals. However, the unbalanced activation of this essential physiological sentinel function leads to aggravated pathological inflammation. The NLRP3 inflammasome is the inflammasome most closely related to uncontrolled inflammation [104]. The potential role of pyroptosis in skeletal muscle pathology has received special attention in recent years given that serious pathological changes in skeletal muscles can lead to weakness and disability. The NLRP family may be normally expressed in skeletal muscle, but its activation in the ageing process may help to enhance the adverse environment, changing muscle synthesis metabolism. Age-related increases in cytokines, such as IL-1β and IL-18, have shown activation of the NLRP family [61]. At present, the treatment of sarcopenia is mainly concentrated on physiotherapy, such as muscle strengthening and gait training and nutritional intervention [105]. No pharmacological intervention for sarcopenia currently exists [18]. The study of the pathogenesis mechanism of sarcopenia will help to identify new therapeutic targets in the future.

New interventions may focus on interfering with the signaling pathway of pyroptosis (Figure 3). Melatonin has significant anti-inflammatory characteristics, and an increasing number of people have begun to focus on the role of melatonin in sarcopenia [106]. Studies have found that melatonin also reduces the expression of pyroptosis-related genes, including NLRP3, ASC, lysed Caspase1, NF-κB/GSDMD, GSDMD N-terminal, IL-1β and IL-18, in endothelial cells. The potential therapeutic effects of melatonin need to be further verified [103]. Bone morphogenetic protein 7 (BMP-7), a drug commonly used to treat patients with osteoporosis, was recently found to alleviate diabetes-induced inflammation-mediated pyroptosis, sarcopenia, and adverse muscle remodeling. Treatment with BMP-7 reduces caspase-1, IL-1β, and IL-18, markers of the pyroptosis cascade [107]. Carbenoxolone (CBX) attenuates intracellular lipid accumulation and aggravation of inflammation in the liver and skeletal muscle in obese mice induced by a high-fat diet by regulating the IκB-α/NF-κB pathway and NLRP3 inflammasome, which can significantly reduce the expression of p-IκB-α, p-NF-κB, p-IRS-1, NLRP3, and inflammatory factors and increase the expression of p-PI3K and p-AKT [108]. Trimetazidine has been used as an anti-anginal agent for decades; however, recent evidence suggests that trimetazidine may also improve skeletal muscle performance in humans and mice [109,110]. Trimetazidine can protect skeletal muscle cells from starvation or inflammation-induced loss of mass by inhibiting protein degradation and inducing autophagy [111]. Studies have shown that treatment with trimetazidine can reverse the pyroptosis of muscle in C2C12 mice induced by dexamethasone, showing a protective role in skeletal muscle. Therefore, trimetazidine may be a potential therapeutic agent for the treatment of glucocorticoid-induced skeletal muscle disability. However, there is no evidence for its treatment of primary sarcopenia [112].

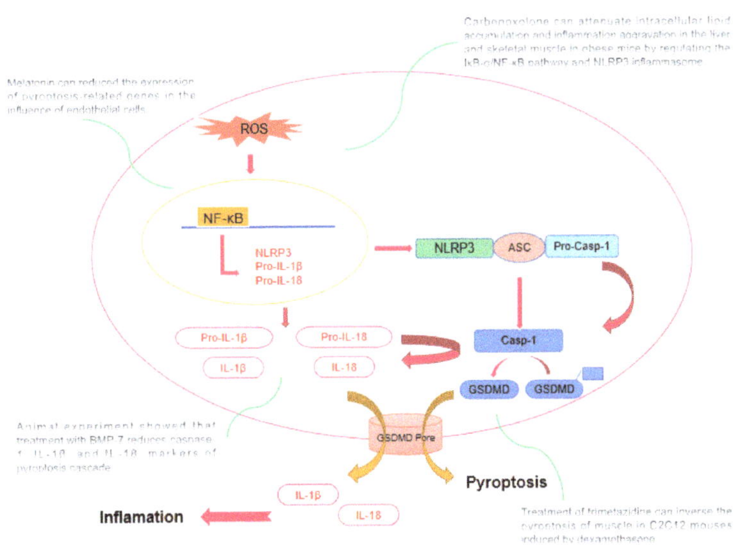

Figure 3. Possible interventions may focus on interfering with the signaling pathway of pyroptosis. a. Melatonin can reduce the expression of pyroptosis-related genes in the influence of endothelial cells. b. Treatment with bone morphogenetic protein 7 (BMP-7) reduces caspase-1, Interleukin-1β (IL-1β), and Interleukin-18 (IL-18), markers of pyroptosis cascade. c. Carbenoxolone can attenuate intracellular lipid accumulation and inflammation aggravation in the liver and skeletal muscle in obese mice induced by the high-fat diet by regulating the IkappaB-alpha (IκB-α)/ nuclear factor—kappaB (NF-κB) pathway and the nod-like receptor 3 (NLRP3) inflammasome. d. Treatment of trimetazidine can reverse the pyroptosis of muscle in C2C12 mouses induced by dexamethasone.

6. Conclusions

Sarcopenia, an age-related disease, can lead to a decline in skeletal muscle mass cause disability. With the gradual increase in the incidence of sarcopenia in the elderly population and health care expenditures, an increasing number of researchers are focusing on morbidity mechanisms and intervention measures to prevent and treat sarcopenia. Pyroptosis, also called inflammatory necrosis, may play a role in the mechanism of sarcopenia. In the process of ageing, aggravated oxidative stress and poor skeletal muscle perfusion in ageing muscle tissues can activate the NLRP family to trigger pyroptosis. Chronic, low-grade inflammation is a representative characteristic of the ageing process. The levels of inflammatory factors such as TNF-α may activate the signaling pathways of pyroptosis by the NF-κB-GSDMD axis, and this remains to be further studied. Autophagy is a protective mechanism in maintaining the integrity of intracellular organelles and the survival of cells in adverse conditions. The autophagy of skeletal muscle cells can inhibit the activation of the pyroptosis pathway to some extent. Study of the mechanism of pyroptosis in sarcopenia may be helpful for finding new therapeutic targets in the future.

Author Contributions: H.J. conceptualized this research and wrote the manuscript, M.H., W.X. (Wenqing Xie), and H.L. prepared the figures. W.X. (Wenfeng Xiao) and Y.L. decided on the content. All authors have read and agreed to the published version of the manuscript.

Funding: This work was supported by the National Natural Science Foundation of China (No. 81874030, 82072506), Hunan Young Talents of Science and Technology (No. 2021RC3025), Provincial Natural Science Foundation of Hunan (No. 2020JJ3060), Provincial Clinical Medical Technology Innovation Project of Hunan (No. 2020SK53709), Innovation-Driven Project of Central South University (No. 2020CX045), Wu Jieping Medical Foundation (320.6750.2020-03-14), National Clinical Research Center for Geriatric Disorders (Xiangya Hospital, No. 2021KFJJ02), National Clinical Research Center for Orthopedics, Sports Medicine and Rehabilitation (2021-NCRC-CXJJ-PY-40) and Exploration and Innovation Project for Undergraduate Students of Central South University (XCX2021046).

Institutional Review Board Statement: Not applicable.

Data Availability Statement: Not applicable.

Acknowledgments: We thank both Yilan Ding (E-mail: 8303180502@csu.edu.cn) and Jinglue Zhang (E-mail: zhangjinglue@csu.edu.cn) for their contributions during the revision process.

Conflicts of Interest: The authors declare that the research was conducted in the absence of any commercial or financial relationships that could be construed as a potential conflict of interest.

References

1. Liu, Z.; Leung, D.; Thrush, K.; Zhao, W.; Ratliff, S.; Tanaka, T.; Schmitz, L.L.; Smith, J.A.; Ferrucci, L.; Levine, M.E. Underlying features of epigenetic aging clocks in vivo and in vitro. *Aging Cell* **2020**, *19*, e13229. [CrossRef] [PubMed]
2. Kornicka, K.; Szłapka-Kosarzewska, J.; Śmieszek, A.; Marycz, K. 5-Azacytydine and resveratrol reverse senescence and ageing of adipose stem cells via modulation of mitochondrial dynamics and autophagy. *J. Cell. Mol. Med.* **2019**, *23*, 237–259. [CrossRef]
3. Schnekenburger, M.; Grandjenette, C.; Ghelfi, J.; Karius, T.; Foliguet, B.; Dicato, M.; Diederich, M. Sustained exposure to the DNA demethylating agent, 2′-deoxy-5-azacytidine, leads to apoptotic cell death in chronic myeloid leukemia by promoting differentiation, senescence, and autophagy. *Biochem. Pharmacol.* **2011**, *81*, 364–378. [CrossRef] [PubMed]
4. He, S.; Sharpless, N.E. Senescence in Health and Disease. *Cell* **2017**, *169*, 1000–1011. [CrossRef]
5. Fang, Y.; Tian, S.; Pan, Y.; Li, W.; Wang, Q.; Tang, Y.; Yu, T.; Wu, X.; Shi, Y.; Ma, P.; et al. Pyroptosis: A new frontier in cancer. *Biomed. Pharmacother.* **2020**, *121*, 109595. [CrossRef]
6. Shi, J.; Gao, W.; Shao, F. Pyroptosis: Gasdermin-Mediated Programmed Necrotic Cell Death. *Trends Biochem. Sci.* **2017**, *42*, 245–254. [CrossRef] [PubMed]
7. Frank, D.; Vince, J.E. Pyroptosis versus necroptosis: Similarities, differences, and crosstalk. *Cell Death Differ.* **2019**, *26*, 99–114. [CrossRef] [PubMed]
8. Zhaolin, Z.; Guohua, L.; Shiyuan, W.; Zuo, W. Role of pyroptosis in cardiovascular disease. *Cell Prolif.* **2019**, *52*, e12563. [CrossRef] [PubMed]
9. Xia, X.; Wang, X.; Cheng, Z.; Qin, W.; Lei, L.; Jiang, J.; Hu, J. The role of pyroptosis in cancer: Pro-cancer or pro-"host"? *Cell Death Dis.* **2019**, *10*, 650. [CrossRef] [PubMed]
10. Wang, S.; Yuan, Y.H.; Chen, N.H.; Wang, H.B. The mechanisms of NLRP3 inflammasome/pyroptosis activation and their role in Parkinson's disease. *Int. Immunopharmacol.* **2019**, *67*, 458–464. [CrossRef] [PubMed]

11. Metter, E.J.; Conwit, R.; Tobin, J.; Fozard, J.L. Age-associated loss of power and strength in the upper extremities in women and men. *J. Gerontol. A Biol. Sci. Med. Sci.* **1997**, *52*, B267–B276. [CrossRef] [PubMed]
12. Rosenberg, I.H. Epidemiologic and methodologic problems in determining nutritional status of older persons. In Proceedings of a Conference, Albuquerque, NM, USA, 19–21 October 1988. *Am. J. Clin. Nutr.* **1989**, *50*, 1121–1235.
13. Ardeljan, A.D.; Hurezeanu, R. Sarcopenia. In *StatPearls*; StatPearls Publishing Copyright© 2022; StatPearls Publishing LLC.: Treasure Island, FL, USA, 2022.
14. Walston, J.D. Sarcopenia in older adults. *Curr. Opin. Rheumatol.* **2012**, *24*, 623–627. [CrossRef] [PubMed]
15. Giallauria, F.; Cittadini, A.; Smart, N.A.; Vigorito, C. Resistance training and sarcopenia. *Monaldi Arch. Chest Dis.* **2016**, *84*, 738. [CrossRef]
16. Sieber, C.C. Malnutrition and sarcopenia. *Aging Clin. Exp. Res.* **2019**, *31*, 793–798. [CrossRef] [PubMed]
17. Aby, E.S.; Saab, S. Frailty, Sarcopenia, and Malnutrition in Cirrhotic Patients. *Clin. Liver Dis.* **2019**, *23*, 589–605. [CrossRef]
18. Dhillon, R.J.; Hasni, S. Pathogenesis and Management of Sarcopenia. *Clin. Geriatr. Med.* **2017**, *33*, 17–26. [CrossRef]
19. Bano, G.; Trevisan, C.; Carraro, S.; Solmi, M.; Luchini, C.; Stubbs, B.; Manzato, E.; Sergi, G.; Veronese, N. Inflammation and sarcopenia: A systematic review and meta-analysis. *Maturitas* **2017**, *96*, 10–15. [CrossRef] [PubMed]
20. Daussin, F.N.; Boulanger, E.; Lancel, S. From mitochondria to sarcopenia: Role of inflammaging and RAGE-ligand axis implication. *Exp. Gerontol.* **2021**, *146*, 111247. [CrossRef]
21. Marzetti, E.; Calvani, R.; Cesari, M.; Buford, T.W.; Lorenzi, M.; Behnke, B.J.; Leeuwenburgh, C. Mitochondrial dysfunction and sarcopenia of aging: From signaling pathways to clinical trials. *Int. J. Biochem. Cell Biol.* **2013**, *45*, 2288–2301. [CrossRef] [PubMed]
22. Chumlea, W.C.; Cesari, M.; Evans, W.J.; Ferrucci, L.; Fielding, R.A.; Pahor, M.; Studenski, S.; Vellas, B. Sarcopenia: Designing phase IIB trials. *J. Nutr. Health Aging* **2011**, *15*, 450–455. [CrossRef] [PubMed]
23. Li, H.; Zhu, H.; Xu, C.J.; Yuan, J. Cleavage of BID by caspase 8 mediates the mitochondrial damage in the Fas pathway of apoptosis. *Cell* **1998**, *94*, 491–501. [CrossRef]
24. Marzetti, E.; Leeuwenburgh, C. Skeletal muscle apoptosis, sarcopenia and frailty at old age. *Exp. Gerontol.* **2006**, *41*, 1234–1238. [CrossRef]
25. Toldo, S.; Abbate, A. The NLRP3 inflammasome in acute myocardial infarction. *Nat. Rev. Cardiol.* **2018**, *15*, 203–214. [CrossRef]
26. Jiang, M.Y.; Han, Z.D.; Li, W.; Yue, F.; Ye, J.; Li, B.; Cai, Z.; Lu, J.M.; Dong, W.; Jiang, X.; et al. Mitochondrion-associated protein peroxiredoxin 3 promotes benign prostatic hyperplasia through autophagy suppression and pyroptosis activation. *Oncotarget* **2017**, *8*, 80295–80302. [CrossRef] [PubMed]
27. Tan, M.S.; Tan, L.; Jiang, T.; Zhu, X.C.; Wang, H.F.; Jia, C.D.; Yu, J.T. Amyloid-β induces NLRP1-dependent neuronal pyroptosis in models of Alzheimer's disease. *Cell Death Dis.* **2014**, *5*, e1382. [CrossRef] [PubMed]
28. Han, Y.; Qiu, H.; Pei, X.; Fan, Y.; Tian, H.; Geng, J. Low-dose Sinapic Acid Abates the Pyroptosis of Macrophages by Down-regulation of lncRNA-MALAT1 in Rats With Diabetic Atherosclerosis. *J. Cardiovasc. Pharmacol.* **2018**, *71*, 104–112. [CrossRef] [PubMed]
29. Brault, M.; Oberst, A. Controlled detonation: Evolution of necroptosis in pathogen defense. *Immunol. Cell Biol.* **2017**, *95*, 131–136. [CrossRef] [PubMed]
30. Naderer, T.; Fulcher, M.C. Targeting apoptosis pathways in infections. *J. Leukoc. Biol.* **2018**, *103*, 275–285. [PubMed]
31. Broz, P.; Dixit, V.M. Inflammasomes: Mechanism of assembly, regulation and signalling. *Nat. Rev. Immunol.* **2016**, *16*, 407–420. [CrossRef] [PubMed]
32. Sun, Q.; Scott, M.J. Caspase-1 as a multifunctional inflammatory mediator: Noncytokine maturation roles. *J. Leukoc. Biol.* **2016**, *100*, 961–967. [CrossRef]
33. Latz, E.; Xiao, T.S.; Stutz, A. Activation and regulation of the inflammasomes. *Nat. Rev. Immunol.* **2013**, *13*, 397–411. [CrossRef] [PubMed]
34. Aachoui, Y.; Sagulenko, V.; Miao, E.A.; Stacey, K.J. Inflammasome-mediated pyroptotic and apoptotic cell death, and defense against infection. *Curr. Opin. Microbiol.* **2013**, *16*, 319–326. [CrossRef] [PubMed]
35. Davis, B.K.; Wen, H.; Ting, J.P. The inflammasome NLRs in immunity, inflammation, and associated diseases. *Annu. Rev. Immunol.* **2011**, *29*, 707–735. [CrossRef]
36. Shi, J.; Zhao, Y.; Wang, K.; Shi, X.; Wang, Y.; Huang, H.; Zhuang, Y.; Cai, T.; Wang, F.; Shao, F. Cleavage of GSDMD by inflammatory caspases determines pyroptotic cell death. *Nature* **2015**, *526*, 660–665. [CrossRef]
37. Franchi, L.; Park, J.H.; Shaw, M.H.; Marina-Garcia, N.; Chen, G.; Kim, Y.G.; Núñez, G. Intracellular NOD-like receptors in innate immunity, infection and disease. *Cell Microbiol.* **2008**, *10*, 1–8. [CrossRef]
38. Aglietti, R.A.; Estevez, A.; Gupta, A.; Ramirez, M.G.; Liu, P.S.; Kayagaki, N.; Ciferri, C.; Dixit, V.M.; Dueber, E.C. GsdmD p30 elicited by caspase-11 during pyroptosis forms pores in membranes. *Proc. Natl. Acad. Sci. USA* **2016**, *113*, 7858–7863. [CrossRef] [PubMed]
39. Gurcel, L.; Abrami, L.; Girardin, S.; Tschopp, J.; van der Goot, F.G. Caspase-1 activation of lipid metabolic pathways in response to bacterial pore-forming toxins promotes cell survival. *Cell* **2006**, *126*, 1135–1145. [CrossRef]
40. Bauernfried, S.; Scherr, M.J.; Pichlmair, A.; Duderstadt, K.E.; Hornung, V. Human NLRP1 is a sensor for double-stranded RNA. *Science* **2021**, *371*, eabd0811. [CrossRef]

41. Arian, S.; Rubin, J.; Chakchouk, I.; Sharif, M.; Mahadevan, S.K.; Erfani, H.; Shelly, K.; Liao, L.; Lorenzo, I.; Ramakrishnan, R.; et al. Reproductive Outcomes from Maternal Loss of Nlrp2 Are Not Improved by IVF or Embryo Transfer Consistent with Oocyte-Specific Defect. *Reprod. Sci.* **2021**, *28*, 1850–1865. [CrossRef] [PubMed]
42. Andrade, W.A.; Zamboni, D.S. NLRC4 biology in immunity and inflammation. *J. Leukoc. Biol.* **2020**, *108*, 1117–1127. [CrossRef] [PubMed]
43. Sharma, M.; de Alba, E. Structure, Activation and Regulation of NLRP3 and AIM2 Inflammasomes. *Int. J. Mol. Sci.* **2021**, *22*, 872. [CrossRef] [PubMed]
44. Al Mamun, A.; Ara Mimi, A.; Wu, Y.; Zaeem, M.; Abdul Aziz, M.; Aktar Suchi, S.; Alyafeai, E.; Munir, F.; Xiao, J. Pyroptosis in diabetic nephropathy. *Clin. Chim. Acta* **2021**, *523*, 131–143. [CrossRef] [PubMed]
45. Shi, J.; Zhao, Y.; Wang, Y.; Gao, W.; Ding, J.; Li, P.; Hu, L.; Shao, F. Inflammatory caspases are innate immune receptors for intracellular LPS. *Nature* **2014**, *514*, 187–192. [CrossRef]
46. Kearney, C.J.; Cullen, S.P.; Tynan, G.A.; Henry, C.M.; Clancy, D.; Lavelle, E.C.; Martin, S.J. Necroptosis suppresses inflammation via termination of TNF- or LPS-induced cytokine and chemokine production. *Cell Death Differ.* **2015**, *22*, 1313–1327. [CrossRef] [PubMed]
47. Hepple, R.T. Mitochondrial involvement and impact in aging skeletal muscle. *Front. Aging Neurosci.* **2014**, *6*, 211. [CrossRef]
48. Koopman, R.; Ly, C.H.; Ryall, J.G. A metabolic link to skeletal muscle wasting and regeneration. *Front. Physiol.* **2014**, *5*, 32. [CrossRef]
49. Narciso, L.; Fortini, P.; Pajalunga, D.; Franchitto, A.; Liu, P.; Degan, P.; Frechet, M.; Demple, B.; Crescenzi, M.; Dogliotti, E. Terminally differentiated muscle cells are defective in base excision DNA repair and hypersensitive to oxygen injury. *Proc. Natl. Acad. Sci. USA* **2007**, *104*, 17010–17015. [CrossRef]
50. López-Otín, C.; Blasco, M.A.; Partridge, L.; Serrano, M.; Kroemer, G. The hallmarks of aging. *Cell* **2013**, *153*, 1194–1217. [CrossRef]
51. Bellanti, F.; Lo Buglio, A.; Vendemiale, G. Mitochondrial Impairment in Sarcopenia. *Biology* **2021**, *10*, 31. [CrossRef]
52. Biolo, G.; Cederholm, T.; Muscaritoli, M. Muscle contractile and metabolic dysfunction is a common feature of sarcopenia of aging and chronic diseases: From sarcopenic obesity to cachexia. *Clin. Nutr.* **2014**, *33*, 737–748. [CrossRef] [PubMed]
53. Sharma, D.; Kanneganti, T.D. The cell biology of inflammasomes: Mechanisms of inflammasome activation and regulation. *J. Cell Biol.* **2016**, *213*, 617–629. [CrossRef] [PubMed]
54. McBride, M.J.; Foley, K.P.; D'Souza, D.M.; Li, Y.E.; Lau, T.C.; Hawke, T.J.; Schertzer, J.D. The NLRP3 inflammasome contributes to sarcopenia and lower muscle glycolytic potential in old mice. *Am. J. Physiol. Endocrinol. Metab.* **2017**, *313*, E222–E232. [CrossRef] [PubMed]
55. Chen, Z.H.; Jin, S.H.; Wang, M.Y.; Jin, X.L.; Lv, C.; Deng, Y.F.; Wang, J.L. Enhanced NLRP3, caspase-1, and IL- 1β levels in degenerate human intervertebral disc and their association with the grades of disc degeneration. *Anat. Rec. (Hoboken)* **2015**, *298*, 720–726. [CrossRef]
56. Beenakker, K.G.; Ling, C.H.; Meskers, C.G.; de Craen, A.J.; Stijnen, T.; Westendorp, R.G.; Maier, A.B. Patterns of muscle strength loss with age in the general population and patients with a chronic inflammatory state. *Ageing Res. Rev.* **2010**, *9*, 431–436. [CrossRef] [PubMed]
57. Franceschi, C.; Garagnani, P.; Parini, P.; Giuliani, C.; Santoro, A. Inflammaging: A new immune-metabolic viewpoint for age-related diseases. *Nat. Rev. Endocrinol.* **2018**, *14*, 576–590. [CrossRef]
58. Jimbo, K.; Park, J.S.; Yokosuka, K.; Sato, K.; Nagata, K. Positive feedback loop of interleukin-1beta upregulating production of inflammatory mediators in human intervertebral disc cells in vitro. *J. Neurosurg. Spine* **2005**, *2*, 589–595. [CrossRef]
59. Lang, C.H.; Frost, R.A.; Nairn, A.C.; MacLean, D.A.; Vary, T.C. TNF-alpha impairs heart and skeletal muscle protein synthesis by altering translation initiation. *Am. J. Physiol. Endocrinol. Metab.* **2002**, *282*, E336–E347. [CrossRef]
60. Langen, R.C.; Van Der Velden, J.L.; Schols, A.M.; Kelders, M.C.; Wouters, E.F.; Janssen-Heininger, Y.M. Tumor necrosis factor-alpha inhibits myogenic differentiation through MyoD protein destabilization. *FASEB J.* **2004**, *18*, 227–237. [CrossRef]
61. De Biase, D.; Piegari, G.; Prisco, F.; Cimmino, I.; d'Aquino, I.; Baldassarre, V.; Oriente, F.; Papparella, S.; Paciello, O. Implication of the NLRP3 Inflammasome in Bovine Age-Related Sarcopenia. *Int. J. Mol. Sci.* **2021**, *22*, 3609. [CrossRef] [PubMed]
62. Lawrence, R.E.; Zoncu, R. The lysosome as a cellular centre for signalling, metabolism and quality control. *Nat. Cell Biol.* **2019**, *21*, 133–142. [CrossRef] [PubMed]
63. Sharapova, T.N.; Romanova, E.A.; Sashchenko, L.P.; Yashin, D.V. FasL on the surface of Tag7 (PGRP-S)-activated lymphocytes induces necroptosis in HLA-negative tumor cells with the involvement of lysosomes and mitochondria. *Biochimie* **2018**, *152*, 174–180. [CrossRef]
64. Skoupilová, H.; Michalová, E.; Hrstka, R. Ferroptosis as a New Type of Cell Death and its Role in Cancer Treatment. *Klin. Onkol.* **2018**, *31*, 21–26. [CrossRef] [PubMed]
65. Vanden Berghe, T.; Linkermann, A.; Jouan-Lanhouet, S.; Walczak, H.; Vandenabeele, P. Regulated necrosis: The expanding network of non-apoptotic cell death pathways. *Nat. Rev. Mol. Cell Biol.* **2014**, *15*, 135–147. [CrossRef] [PubMed]
66. Alu, A.; Han, X.; Ma, X.; Wu, M.; Wei, Y.; Wei, X. The role of lysosome in regulated necrosis. *Acta Pharm. Sin. B* **2020**, *10*, 1880–1903. [CrossRef] [PubMed]
67. Orlowski, G.M.; Sharma, S.; Colbert, J.D.; Bogyo, M.; Robertson, S.A.; Kataoka, H.; Chan, F.K.; Rock, K.L. Frontline Science: Multiple cathepsins promote inflammasome-independent, particle-induced cell death during NLRP3-dependent IL-1β activation. *J. Leukoc. Biol.* **2017**, *102*, 7–17. [CrossRef] [PubMed]

68. Okada, M.; Matsuzawa, A.; Yoshimura, A.; Ichijo, H. The lysosome rupture-activated TAK1-JNK pathway regulates NLRP3 inflammasome activation. *J. Biol. Chem.* **2014**, *289*, 32926–32936. [CrossRef] [PubMed]
69. Orlowski, G.M.; Colbert, J.D.; Sharma, S.; Bogyo, M.; Robertson, S.A.; Rock, K.L. Multiple Cathepsins Promote Pro-IL-1β Synthesis and NLRP3-Mediated IL-1β Activation. *J. Immunol.* **2015**, *195*, 1685–1697. [CrossRef]
70. Mizushima, N.; Yoshimori, T.; Levine, B. Methods in mammalian autophagy research. *Cell* **2010**, *140*, 313–326. [CrossRef]
71. Le Grand, F.; Rudnicki, M.A. Skeletal muscle satellite cells and adult myogenesis. *Curr. Opin. Cell Biol.* **2007**, *19*, 628–633. [CrossRef] [PubMed]
72. Fiacco, E.; Castagnetti, F.; Bianconi, V.; Madaro, L.; De Bardi, M.; Nazio, F.; D'Amico, A.; Bertini, E.; Cecconi, F.; Puri, P.L.; et al. Autophagy regulates satellite cell ability to regenerate normal and dystrophic muscles. *Cell Death Differ.* **2016**, *23*, 1839–1849. [CrossRef]
73. Tang, A.H.; Rando, T.A. Induction of autophagy supports the bioenergetic demands of quiescent muscle stem cell activation. *Embo J.* **2014**, *33*, 2782–2797. [CrossRef] [PubMed]
74. García-Prat, L.; Martínez-Vicente, M.; Perdiguero, E.; Ortet, L.; Rodríguez-Ubreva, J.; Rebollo, E.; Ruiz-Bonilla, V.; Gutarra, S.; Ballestar, E.; Serrano, A.L.; et al. Autophagy maintains stemness by preventing senescence. *Nature* **2016**, *529*, 37–42. [CrossRef]
75. Lv, S.; Wang, H.; Li, X. The Role of the Interplay Between Autophagy and NLRP3 Inflammasome in Metabolic Disorders. *Front. Cell Dev. Biol.* **2021**, *9*, 634118. [CrossRef]
76. Cao, Z.; Wang, Y.; Long, Z.; He, G. Interaction between autophagy and the NLRP3 inflammasome. *Acta Biochim. Biophys. Sin.* **2019**, *51*, 1087–1095. [CrossRef]
77. Dalle Pezze, P.; Ruf, S.; Sonntag, A.G.; Langelaar-Makkinje, M.; Hall, P.; Heberle, A.M.; Razquin Navas, P.; van Eunen, K.; Tölle, R.C.; Schwarz, J.J.; et al. A systems study reveals concurrent activation of AMPK and mTOR by amino acids. *Nat. Commun.* **2016**, *7*, 13254. [CrossRef]
78. Jeon, Y.K.; Shin, M.J.; Saini, S.K.; Custodero, C.; Aggarwal, M.; Anton, S.D.; Leeuwenburgh, C.; Mankowski, R.T. Vascular dysfunction as a potential culprit of sarcopenia. *Exp. Gerontol.* **2021**, *145*, 111220. [CrossRef]
79. Mathieu-Costello, O.; Agey, P.J.; Quintana, E.S.; Rousey, K.; Wu, L.; Bernstein, M.H. Fiber capillarization and ultrastructure of pigeon pectoralis muscle after cold acclimation. *J. Exp. Biol.* **1998**, *201*, 3211–3220. [CrossRef]
80. Degens, H.; Turek, Z.; Hoofd, L.J.; Binkhorst, R.A. Capillary proliferation related to fibre types in hypertrophied aging rat M. plantaris. *Adv. Exp. Med. Biol.* **1994**, *345*, 669–676.
81. Barnouin, Y.; McPhee, J.S.; Butler-Browne, G.; Bosutti, A.; De Vito, G.; Jones, D.A.; Narici, M.; Behin, A.; Hogrel, J.Y.; Degens, H. Coupling between skeletal muscle fiber size and capillarization is maintained during healthy aging. *J. Cachexia Sarcopenia Muscle* **2017**, *8*, 647–659. [CrossRef] [PubMed]
82. Degens, H.; Morse, C.I.; Hopman, M.T. Heterogeneity of capillary spacing in the hypertrophied plantaris muscle from young-adult and old rats. *Adv. Exp. Med. Biol.* **2009**, *645*, 61–66. [PubMed]
83. Joanisse, S.; Nederveen, J.P.; Snijders, T.; McKay, B.R.; Parise, G. Skeletal Muscle Regeneration, Repair and Remodelling in Aging: The Importance of Muscle Stem Cells and Vascularization. *Gerontology* **2017**, *63*, 91–100. [CrossRef] [PubMed]
84. Bai, B.; Yang, Y.; Wang, Q.; Li, M.; Tian, C.; Liu, Y.; Aung, L.H.H.; Li, P.F.; Yu, T.; Chu, X.M. NLRP3 inflammasome in endothelial dysfunction. *Cell Death Dis.* **2020**, *11*, 776. [CrossRef]
85. Kono, H.; Kimura, Y.; Latz, E. Inflammasome activation in response to dead cells and their metabolites. *Curr. Opin. Immunol.* **2014**, *30*, 91–98. [CrossRef] [PubMed]
86. Mussbacher, M.; Salzmann, M.; Brostjan, C.; Hoesel, B.; Schoergenhofer, C.; Datler, H.; Hohensinner, P.; Basílio, J.; Petzelbauer, P.; Assinger, A.; et al. Cell Type-Specific Roles of NF-κB Linking Inflammation and Thrombosis. *Front. Immunol.* **2019**, *10*, 85. [CrossRef]
87. Barker, B.R.; Taxman, D.J.; Ting, J.P. Cross-regulation between the IL-1β/IL-18 processing inflammasome and other inflammatory cytokines. *Curr. Opin. Immunol.* **2011**, *23*, 591–597. [CrossRef]
88. Amarasekera, A.T.; Chang, D.; Schwarz, P.; Tan, T.C. Vascular endothelial dysfunction may be an early predictor of physical frailty and sarcopenia: A meta-analysis of available data from observational studies. *Exp. Gerontol.* **2021**, *148*, 111260. [CrossRef]
89. Franklin, B.S.; Bossaller, L.; De Nardo, D.; Ratter, J.M.; Stutz, A.; Engels, G.; Brenker, C.; Nordhoff, M.; Mirandola, S.R.; Al-Amoudi, A.; et al. The adaptor ASC has extracellular and 'prionoid' activities that propagate inflammation. *Nat. Immunol.* **2014**, *15*, 727–737. [CrossRef]
90. Hendrickse, P.; Degens, H. The role of the microcirculation in muscle function and plasticity. *J. Muscle Res. Cell Motil.* **2019**, *40*, 127–140. [CrossRef]
91. Lei, Q.; Yi, T.; Chen, C. NF-κB-Gasdermin D (GSDMD) Axis Couples Oxidative Stress and NACHT, LRR and PYD Domains-Containing Protein 3 (NLRP3) Inflammasome-Mediated Cardiomyocyte Pyroptosis Following Myocardial Infarction. *Med. Sci. Monit.* **2018**, *24*, 6044–6052. [CrossRef]
92. Gilmore, T.D. Introduction to NF-kappaB: Players, pathways, perspectives. *Oncogene* **2006**, *25*, 6680–6684. [CrossRef]
93. McKinnell, I.W.; Rudnicki, M.A. Molecular mechanisms of muscle atrophy. *Cell* **2004**, *119*, 907–910. [CrossRef]
94. Sies, H.; Berndt, C.; Jones, D.P. Oxidative Stress. *Annu. Rev. Biochem.* **2017**, *86*, 715–748. [CrossRef] [PubMed]
95. Schaap, L.A.; Pluijm, S.M.; Deeg, D.J.; Visser, M. Inflammatory markers and loss of muscle mass (sarcopenia) and strength. *Am. J. Med.* **2006**, *119*, 526.e9–526.e17. [CrossRef] [PubMed]

96. De Larichaudy, J.; Zufferli, A.; Serra, F.; Isidori, A.M.; Naro, F.; Dessalle, K.; Desgeorges, M.; Piraud, M.; Cheillan, D.; Vidal, H.; et al. TNF-α- and tumor-induced skeletal muscle atrophy involves sphingolipid metabolism. *Skelet. Muscle* **2012**, *2*, 2. [CrossRef] [PubMed]
97. Libert, C. Cytokine anniversary: TNF trailblazers five centuries apart. *Nature* **2015**, *523*, 158. [CrossRef]
98. Wajant, H.; Scheurich, P. TNFR1-induced activation of the classical NF-κB pathway. *FEBS J.* **2011**, *278*, 862–876. [CrossRef]
99. Zhou, J.; Liu, B.; Liang, C.; Li, Y.; Song, Y.H. Cytokine Signaling in Skeletal Muscle Wasting. *Trends Endocrinol. Metab.* **2016**, *27*, 335–347. [CrossRef] [PubMed]
100. Fang, W.Y.; Tseng, Y.T.; Lee, T.Y.; Fu, Y.C.; Chang, W.H.; Lo, W.W.; Lin, C.L.; Lo, Y.C. Triptolide prevents LPS-induced skeletal muscle atrophy via inhibiting NF-κB/TNF-α and regulating protein synthesis/degradation pathway. *Br. J. Pharmacol.* **2021**, *178*, 2998–3016. [CrossRef]
101. Ding, J.; Wang, K.; Liu, W.; She, Y.; Sun, Q.; Shi, J.; Sun, H.; Wang, D.C.; Shao, F. Pore-forming activity and structural autoinhibition of the gasdermin family. *Nature* **2016**, *535*, 111–116. [CrossRef]
102. Salminen, A.; Ojala, J.; Kaarniranta, K.; Kauppinen, A. Mitochondrial dysfunction and oxidative stress activate inflammasomes: Impact on the aging process and age-related diseases. *Cell Mol. Life Sci.* **2012**, *69*, 2999–3013. [CrossRef] [PubMed]
103. Zhang, Y.; Liu, X.; Bai, X.; Lin, Y.; Li, Z.; Fu, J.; Li, M.; Zhao, T.; Yang, H.; Xu, R.; et al. Melatonin prevents endothelial cell pyroptosis via regulation of long noncoding RNA MEG3/miR-223/NLRP3 axis. *J. Pineal Res.* **2018**, *64*, e12449. [CrossRef]
104. Bertheloot, D.; Latz, E.; Franklin, B.S. Necroptosis, pyroptosis and apoptosis: An intricate game of cell death. *Cell Mol. Immunol.* **2021**, *18*, 1106–1121. [CrossRef] [PubMed]
105. Marzetti, E.; Calvani, R.; Tosato, M.; Cesari, M.; Di Bari, M.; Cherubini, A.; Collamati, A.; D'Angelo, E.; Pahor, M.; Bernabei, R.; et al. Sarcopenia: An overview. *Aging Clin. Exp. Res.* **2017**, *29*, 11–17. [CrossRef]
106. Coto-Montes, A.; Boga, J.A.; Tan, D.X.; Reiter, R.J. Melatonin as a Potential Agent in the Treatment of Sarcopenia. *Int. J. Mol. Sci.* **2016**, *17*, 1771. [CrossRef]
107. Aluganti Narasimhulu, C.; Singla, D.K. Amelioration of diabetes-induced inflammation mediated pyroptosis, sarcopenia, and adverse muscle remodelling by bone morphogenetic protein-7. *J. Cachexia Sarcopenia Muscle* **2021**, *12*, 403–420. [CrossRef]
108. Chen, Y.; Qian, Q.; Yu, J. Carbenoxolone ameliorates insulin sensitivity in obese mice induced by high fat diet via regulating the IκB-α/NF-κB pathway and NLRP3 inflammasome. *Biomed. Pharmacother.* **2019**, *115*, 108868. [CrossRef]
109. Heggermont, W.A.; Papageorgiou, A.P.; Heymans, S.; van Bilsen, M. Metabolic support for the heart: Complementary therapy for heart failure? *Eur. J. Heart Fail.* **2016**, *18*, 1420–1429. [CrossRef]
110. Belardinelli, R.; Lacalaprice, F.; Faccenda, E.; Volpe, L. Trimetazidine potentiates the effects of exercise training in patients with ischemic cardiomyopathy referred for cardiac rehabilitation. *Eur. J. Cardiovasc. Prev. Rehabil.* **2008**, *15*, 533–540. [CrossRef]
111. Ferraro, E.; Giammarioli, A.M.; Caldarola, S.; Lista, P.; Feraco, A.; Tinari, A.; Salvatore, A.M.; Malorni, W.; Berghella, L.; Rosano, G. The metabolic modulator trimetazidine triggers autophagy and counteracts stress-induced atrophy in skeletal muscle myotubes. *FEBS J.* **2013**, *280*, 5094–5108. [CrossRef]
112. Wang, L.; Jiao, X.F.; Wu, C.; Li, X.Q.; Sun, H.X.; Shen, X.Y.; Zhang, K.Z.; Zhao, C.; Liu, L.; Wang, M.; et al. Trimetazidine attenuates dexamethasone-induced muscle atrophy via inhibiting NLRP3/GSDMD pathway-mediated pyroptosis. *Cell Death Discov.* **2021**, *7*, 251. [CrossRef] [PubMed]

cells

Review

Aging of the Immune System: Focus on Natural Killer Cells Phenotype and Functions

Ashley Brauning [1], Michael Rae [1], Gina Zhu [1], Elena Fulton [1], Tesfahun Dessale Admasu [1], Alexandra Stolzing [1,2,*] and Amit Sharma [1,*]

1 SENS Research Foundation, Mountain View, CA 94041, USA; ashley.brauning@sens.org (A.B.); michael.rae@sens.org (M.R.); ginazhu2@gmail.com (G.Z.); elenafulton@gmail.com (E.F.); tesfahun.admasu@sens.org (T.D.A.)

2 Centre for Biological Engineering, Wolfson School of Electrical, Material and Manufacturing Engineering, Loughborough University, Loughborough LE11 3TU, UK

* Correspondence: alexandra.stolzing@sens.org (A.S.); amit.sharma@sens.org (A.S.)

Abstract: Aging is the greatest risk factor for nearly all major chronic diseases, including cardiovascular diseases, cancer, Alzheimer's and other neurodegenerative diseases of aging. Age-related impairment of immune function (immunosenescence) is one important cause of age-related morbidity and mortality, which may extend beyond its role in infectious disease. One aspect of immunosenescence that has received less attention is age-related natural killer (NK) cell dysfunction, characterized by reduced cytokine secretion and decreased target cell cytotoxicity, accompanied by and despite an increase in NK cell numbers with age. Moreover, recent studies have revealed that NK cells are the central actors in the immunosurveillance of senescent cells, whose age-related accumulation is itself a probable contributor to the chronic sterile low-grade inflammation developed with aging ("inflammaging"). NK cell dysfunction is therefore implicated in the increasing burden of infection, malignancy, inflammatory disorders, and senescent cells with age. This review will focus on recent advances and open questions in understanding the interplay between systemic inflammation, senescence burden, and NK cell dysfunction in the context of aging. Understanding the factors driving and enforcing NK cell aging may potentially lead to therapies countering age-related diseases and underlying drivers of the biological aging process itself.

Keywords: natural killer cells (NK cells); aging; immunosenescence; senescence; elderly; frailty; cytokines; inflammation; immune system

Citation: Brauning, A.; Rae, M.; Zhu, G.; Fulton, E.; Admasu, T.D.; Stolzing, A.; Sharma, A. Aging of the Immune System: Focus on Natural Killer Cells Phenotype and Functions. *Cells* **2022**, *11*, 1017. https://doi.org/10.3390/cells11061017

Academic Editors: Kay-Dietrich Wagner and Nicole Wagner

Received: 17 February 2022
Accepted: 14 March 2022
Published: 17 March 2022

1. Introduction

Aging is the most significant cause of morbidity and mortality in the developed world, and increasingly in developing countries as well [1,2]. While the decline in smoking prevalence and advancements in medical science have resulted in an increased lifespan, there is an increasing population prevalence of the diseases and debilities of aging associated with the not-as-rapid increase in health span [2,3]. Immunosenescence, the age-related decline in immune function, is a critical cause of age-related ill-health, driving dramatic increases in vulnerability to infectious disease with age (including the dramatic age-related rise in morbidity and mortality from COVID-19) [4,5]. In addition to immunosenescence, chronic sterile low-grade inflammation associated with aging ("inflammaging") is likely contributing to the pathogenesis of multiple chronic diseases of aging [6,7].

NK cells are large granular innate immune cells that belong to the family of group 1 innate lymphocytes (ILC1) [8]. They play a vital role as immunological first responders, rapidly eliminating cells infected with viruses or other intracellular pathogens as well as pre-malignant cells [9], and have recently been identified as the principal actors in immune surveillance of senescent cells [10–12], which accumulate with age and are implicated in inflammaging, several age-related diseases of aging, and the underlying biological aging process itself [13,14]. Yet, comparatively little is known about the effects of aging on NK cells as compared to the cells in the adaptive immune system [15]. In this review, we discuss what is known about the phenotype and drivers of age-related NK cell dysfunction, areas of uncertainty, potential sources of new insight, and approaches to NK cell immune rejuvenation and fortification. Understanding and remediating the age-related decline in NK cell function could be an important means of ameliorating critical proximate causes of age-related ill-health and death and opposing one of the underlying drivers of aging.

2. NK Cells in Immunity: An Overview

NK cells eliminate target cells through direct cell-to-cell contact-based NK cell cytotoxicity (NKCC) and by secreting cytokines, such as interferon gamma (INF-γ) or tumor necrosis factor alpha (TNF-α) [16,17], which recruit additional immune cells to the site of infection or inflammation. NK cells represent $\approx$5–10% of peripheral blood lymphocytes (PBMC), with approximately 2 billion NK cells circulating in adult humans [18]. NK cells reside in different tissues, including bone marrow, lymph nodes, liver, skin, lung, and, to a lesser extent, secondary lymphoid organs [19]; however, this has not been extensively studied.

The human NK cell population can be subdivided based on the expression of the surface markers CD56 and CD16, with CD56^{dim} NK cells being largely cytotoxic and CD56^{bright} NK cells primarily responsible for the secretion of cytokines, such as IFN-γ and TNF-α [20,21]. However, in absolute numbers, CD56^{bright} NK cells, located predominantly in peripheral and lymphoid tissues [22,23], outnumber CD56^{dim} NK cells. The CD56^{dim} NK cell subpopulation represent at least 90% of peripheral blood NK cells, whereas the CD56^{bright} NK cells constitute the remaining [24]. NK cell phenotypes and their changes with age are described in greater depth in the subsequent sections.

The process of NK cell differentiation and maturation has been extensively reviewed elsewhere and is beyond the scope of this review [25]. In a brief summary, human NK cells develop from common lymphoid progenitor (CLP) cells derived from CD34$^+$ hematopoietic stem cells (HSCs), the common precursors of NK, T, B, and other lymphoid cells [26,27]. CLP cells express different markers, including CD38, CD7, CD10, CD127, and CD122. The expression of CD122 (IL-2Rβ/IL-15R) marks the irreversible fate of CLPs starting to differentiate into NK lineage [28–30]. NK cells can act as a bridge between innate and adaptive immunity, as they can interact with T and B cells via their costimulatory ligands, such as CD40L and OX40L, which further promote NK cell differentiation [31,32].

NK cells interact with target cells via activating and inhibitory receptors, which bind to the corresponding ligands on the target cell's surface. Depending on the balance of these ligand–receptor interactions, the NK cell determines whether or not to eliminate the target cell [33]. This review focuses on the alteration of the function of human NK cells with age; the NK cell receptors in humans and rodents have been extensively reviewed elsewhere [9,34]; we will therefore only provide a brief summary of this topic, as a background for what follows.

NKG2D, Ly49, and natural cytotoxicity receptors (NCRs), such as NKp46, are some of the main activating receptors for NK cells [16,35]. When the NKG2D or Ly49 interact with target cell ligands, such as MICA/MICB or ULBPs [36], their signal is mediated intracellularly through the YINM and immunoreceptor tyrosine-based activation motif (ITAM), respectively [36–38]. The phosphorylation of YINM or ITAM by tyrosine kinases leads to a cascade that results in the release of perforins and granzymes from NK cells to induce apoptosis in target cells [37,38]. In addition, NK cells also secrete cytokines and chemokines, which mediate their effector function directly on the target cells (cytotoxic signal) or through the recruitment of other immune cells, such as macrophages and neutrophils, to the site of inflammation [39,40].

The killer cell immunoglobulin-like receptors (KIR) and NKG2A are the major families of inhibitory receptors expressed on NK cells [41]. KIR bind to the HLA-A/B/C ligand, whereas NKG2A recognizes the inhibitory HLA-E [42,43]. These ligands bind to their respective receptors via immunoreceptor tyrosine-based inhibitory motifs (ITIM), leading to the inactivation of NK cells [44]. In addition, KIR and NK2GA are also known to coordinate NK cell tolerance [45]. Major histocompatibility complex class I (MHC-I) present non-self antigens for target engagement by CD8+ T-cells, but, conversely, are essential ligands for inhibitory receptors on NK cells for both mice and humans [46]. NK cells can recognize MHC- I directly (class Ia) or indirectly (class Ib). Reduced expression of MHC-I in cancer cells or cells infected with certain viruses lowers the inhibitory bar for attack by NK cells, enabling these cells' elimination [47].

Mice and humans differ in the mechanism by which MHC class Ia is recognized: human NK cells use KIR to recognize MHC Ia, while mice use lectin-like receptors of the Ly49 family [48–50]. However, mice and humans have similar mechanisms for recognizing MHC class Ib, which involve NKG2D in both species [48–50]. Despite the differences in the ligand recognition mechanisms, the intracellular signaling mechanism involved in NK cell activation is conserved in these two species [48].

3. Changes in NK Cell Numbers and Subpopulations with Age

Several studies have characterized changes in the absolute number of circulating NK cells as well as the distribution of NK cell subsets with increasing age. Age does not appear to influence NK progenitor numbers in the peripheral blood or in the bone marrow [51]. Nonetheless, although there is some disagreement in the field, most studies report an increase in NK cell number with age (11 of 13 reported studies) (Table 1).

Table 1. Changes in natural killer cell phenotype, subset, and cytotoxicity throughout the aging process. Summary of age-related changes in natural killer cell phenotype, subset distribution, and cytotoxicity. Age-related changes in natural killer cell phenotype, subset distribution, and cytotoxicity against cancer cell lines.

Young Donor Age Range	Old Donor Age Range	NK Cell Definition	NK Numbers	Aged Phenotype	Target Cell	Cytotoxic Potential	Assay	IL-2 Activation	Ref
25–34	75–84	*$CD16^+/Leu7^{\pm}$	Increased with age	$CD16^+$ and $Leu7^+$ increased	-	-	-	-	[52]
21–38	71–97	*$Leu7^+/Leu11a^{\pm}$	Increased with age	-	K562	Decreased cytotoxic potential with age in both subsets of NK cells	Cr Release Assay	-	[53]
20–60	80+	$CD16^+$ $CD56^+$ $CD5^{\pm}$	-	$CD16^+$ $CD57^+$ increased; $CD56^+/CD57^-$ decreased	K562	Increase in cytotoxic potential with age	Cr Release Assay	-	[54]
27 ± 6	81 ± 7	$CD16^+$	Increased with age	-	K562	Similar binding capacity, Decreased cytotoxicity with age	Flow Cytometry	-	[55]
19–36; 50–68	100–106	$CD16^+$ $CD57^{\pm}$	Increased with age	$CD16^+$ $CD57^-$ increased; no change in $CD16^+$ $CD57^+$	K562	No significant difference between age groups	Cr Release Assay	-	[56]
25–35	75–94	$CD16^+$ $CD56^+$ $KIRDL^+$ or $CD16^+$ $CD56^+$ $KIR2DL3^-$	Increased with age	$CD16^+$, $CD56^+$ and $GL138^+$ increased	K562	Decrease in lytic activity of $CD16^+$ cells	Cr Release Assay	Yes	[57]

Table 1. *Cont.*

Young Donor Age Range	Old Donor Age Range	NK Cell Definition	NK Numbers	Aged Phenotype	Target Cell	Cytotoxic Potential	Assay	IL-2 Activation	Ref
23–35	65–100	CD3$^-$ CD56$^+$	No change	-	K562	No significant difference between age groups	Cr Release Assay	Yes	[58]
23–35	67–95	CD3$^-$ CD56$^+$	No change		K562	Decrease in cytotoxic with age	Cr Release Assay	Yes	[59]
30 ± 2	85 ± 2	CD16$^+$	Increased with age	-	K562	Decrease in cytotoxic potential with age	Cr Release Assay	-	[60]
30 ± 2	85 ± 2	CD3$^-$ CD16$^+$ CD56$^+$	Increased with age	-	K562	Decrease in cytotoxic potential with age	Cr Release Assay	-	[61]
19–39	77–89	CD3$^-$ CD56^{dim} or CD3$^-$ CD56^{bright}	Increased with age	CD56^{dim} Increased; CD56^{bright} decreased	-	-	-	-	[62]
21–30	65+	CD3$^-$ CD56$^+$	-	CD94 and NKG2A decreased; KIR increased	P815	CD16 mediated cytotoxicity did not vary with age	Cr Release Assay	Yes	[63]
Cord blood; Young <60	60–75; 75–80	CD3$^-$ CD56^{dim} or CD3$^-$ CD56^{bright}	Increased in very old	CD56^{bright} decreased; NKp30 decreased; NKp46 decreased; KIRs increased	K562	No significant difference between age groups	Cr Release Assay	Yes	[64]
<60	60+	CD3$^-$ CD56^{dim} or CD3$^-$ CD56^{bright}	-	CD56^{dim} increased; CD56^{bright} decreased; NKG2A decreased; KLRG1 increased	-	-	-	-	[65]
Children <18; adults 19–59	60+	CD3$^-$ CD56^{dim} or CD3$^-$ CD56^{bright}	Increased in elderly, no difference between children and adults	CD56^{dim} increased, CD56^{bright} decreased, NKp30 and NKp46 decreased with age	K562	No significant difference between age groups	Flow Cytometry	Yes	[66]
20–34	70–86	CD16$^+$ CD56^{dim} or CD16$^+$ CD56^{bright}	-	CD56^{dim} increased, CD56^{bright} decreased, KLRG1 increased with age	MCF-10A	-	-	Yes	[67]
41–50, 51–60	61–70, 71–80	CD3$^-$ CD56^{dim} or CD3$^-$ CD56^{bright}	Increased with age	CD56^{dim} increased, CD56^{bright} decreased	-	-	-	Yes	[68]

* Leu7 = CD57 & Leu11a = CD16.

The markers used to label NK cells, or the gating strategy used to count them, may at least partially explain the discrepancy in the studies. The earliest studies relied on CD16 to distinguish NK cells from other lymphocytes, which may have resulted in an incomplete picture of total NK cell number. For example, Zang et al. used CD3 and CD16 expression to distinguish NK cells and reported no differences in peripheral NK cell numbers with age [69]. However, CD56 is now regarded as a more reliable and general marker of human NK cells than CD16. While CD16 is often expressed on the more mature subset, CD56^{dim} NK chambers, the CD56^{bright} NK cell subset sometimes lacks or expresses very low levels of CD16. It is possible that these earlier studies reported changes in CD56^{dim} NK cell populations. Later studies, which included CD56 to label NK cells, reported that absolute cell numbers increased with age (Table 1). With regards to changes in NK cell subpopulations, the studies listed in Table 1 almost unanimously report that there is a significant decrease in the percentage of CD56^{bright} cells (Table 1: six of six studies) [62,64–68] and a significant increase in CD56^{dim} cells (five of five studies) [62,65–68].

Unbiased approaches, such as single-cell RNA expression analysis, offer a possible approach to resolving some of the conflicting data on the effect of aging on NK cells. Transcriptional clustering has revealed two distinct signatures of splenic NK cells, which have a more active gene expression signature than NK cells present in the blood [70]. RNA-sequencing has confirmed the prevalence of several distinct subpopulations of NK cells in the human bone marrow and blood based on the expression of several conserved markers [71]. These findings expand upon the classical NK cell developmental model (CD56^{bright} $\rightarrow$ CD56^{dim}CD57^{-} $\rightarrow$ CD56^{dim}CD57^{+}) and identify two subpopulations in the CD57^{+} population, distinguished by the expression of *CX3CR1* and *HAVCR2* to differentiate the terminally differentiated NK cell population from the matured NK population [71]. In addition, single-cell RNA expression analysis of NK cells isolated from peripheral blood from healthy subjects reveals even more heterogeneity among NK cells, with up to ten subsets of NK cells identified by unsupervised clustering, with conserved makers per cluster independent of individual variation or stimulation [72]. Further, Smith et al. identified three novel NK cell populations; CD56^{neg} type I interferon-responding NK cells have increased granzyme B mRNA in response to IL2 and a small population with low ribosomal expression [72]. Similar methods were employed to analyze specific gene expression variances of NK cells in the context of aging, which demonstrated an increase in expanded late low-cytotoxic subsets with enriched apoptotic signaling pathways and decreased virus defense responses, as compared to the younger group [73].

An important limitation to our insight into the effect of aging on NK cell phenotype is that the great majority of studies, especially in humans, have only investigated age-related changes in circulating NK. Dogra et al. have substantially advanced our understanding by examining NK cells from multiple anatomic sites and compartments of 60 autopsies of subjects across a wide range of ages (5–92 years) and ethnic backgrounds [74]. Notably, across donors, the dominance of CD56^{bright}CD16^{-} vs. CD56^{dim}CD16^{+} was tissue-specific, whether pooled or at the individual donor level. CD56^{dim}CD16^{+} NK cells dominated in blood and in tissues rich in it, such as the bone marrow, spleen, and lungs. Moreover, CD56^{dim}CD16^{+} NK cells were substantially more likely to express CD57 (indicative of terminal differentiation) in tissues where this subset is dominant [74]

In contrast with previous reports (Table 1: [62,65–68]), Dogra et al. found no relationship between the ratio between CD56^{bright}CD16^{-} and CD56^{dim}CD16^{+} NK cell subsets and age in blood or any other tissue. It will be important to understand why this one impressive study came to such a different result on this front.

The frequency of terminally differentiated CD56^{dim}CD16^{+}CD57^{+} NK cells nominally rose with age in blood, bone marrow, spleen, and lung, but the trend was only statistically significant in the bone marrow [74]. The lack of a clear age-related change in this subset in blood is broadly consistent with limited and inconsistent prior reports [64,66]. Age was also unrelated to the level of granzyme B expression by CD56^{dim}CD16^{+} cells, nor the tissue-specific NK cell expression of the signaling adaptor FcεRIγ, an indicator of antibody-dependent cellular cytotoxicity (ADCC) potential [74]. Unlike its effects on T-cell differentiation and polyclonality (which clearly interfaces with aging, despite its uncertain effects on functional immune senescence [15]), cytomegalovirus serostatus did not impact NK cell frequency or tissue distribution [74].

Further investigation of age-related changes in the heterogeneity amongst NK cell populations may be vital in understanding the mechanism behind the loss of their functions with age.

4. Effect of Age on NK Cell Cytokine Production and Secretion

Decreased NK cell cytotoxic subtypes and function are associated with an increased risk of viral infections and cancers [75–77]. Dysfunctional NK cells are typically identified by decreased NK effector functions, such as impaired NKCC, as well as reduced IFN-γ secretion and expression of perforin and granzyme [78]. Release of cytokines, such as IFN-γ and TNF-α, activated the NK cells help coordinate a broader immune response, including the recruitment of macrophages [79] and neutrophils [80] at the site of cancer [81], infection [82], or senescent cells [83].

One potential cause of NK cell dysfunction with age may be changes in the systemic milieu [84,85]. Particularly notable is the well-established age-related decline in interleukin-2 (IL-2) production in humans [86]; IL-2 is a potent enhancer of both NKCC and NK cytokine secretion, such as TNF-α [87]. One study reported that TNF-α secretion is sustained in IL-2-stimulated NK cells from otherwise-healthy aging donors, despite their reduced IL-2 production [62]. Other studies have found that NK cells from older donors increase production of IFN-γ and/or TNF-α following IL-2 treatment [64,88], but to a lesser degree than that which occurs with younger donors [59,89]. Similar to IL-2, interleukin-12 (IL-12) enhances the cytotoxicity of NK cells and promotes IFN-γ secretion by NK cells [90,91]. Studies have reported a reduction in cytokines released by NK cells from older donors, despite activation with IL-2 [92–94], and chemokine production in older subjects, despite stimulation with IL-2 or IL-12 [95,96]. One caveat to these findings is that a reduction in cytokine or chemokine production per cell with age may possibly be attributable to the age-related shift from CD56^{bright} to CD56^{dim} NK cells, as the former are responsible for producing these mediators. Further investigation of age-related changes in the cytokine and chemokine production and secretion amongst NK cell populations may be vital in understanding the mechanism behind the decrease in NK cell activation and recruitment.

5. Changes in NK Cell Cytotoxic Activity with Age

A decline in NKCC with age has been well documented and has been linked to increased incidence of infectious diseases [73,97,98]. Age-related NKCC impairment has also been implicated in enabling the emergence of multiple noncommunicable diseases of aging [99]. For example, adult cancer incidence and mortality for all-site invasive cancers and for many individual cancer types rise progressively with age [100,101] and the decline in NK cell function with age is implicated in this increased risk [102]. NK cell surveillance provides a critical early defense against precancerous and cancerous cells [103,104]. Depressed NK cell activity in animals results in elevated tumor incidence [105] or increased metastasis [106,107]. Relatives of patients with familial melanoma [108] and individuals with a high family incidence of cancer [109] exhibit low NK cell activity compared to healthy controls; similarly, liver cirrhosis patients with reduced NKCC are prospectively at an elevated risk of progressing to liver cancer [110], and one study reported low NKCC to be longitudinally associated with a risk of cancer at all sites in the general Japanese population aged 40 and over [111]. A longitudinal study demonstrated a declining efficacy of NK cell cytotoxicity, coinciding with elevated risk of cancer in both males and females with age [84].

These epidemiological findings are supported by laboratory data. More than half of the studies (6 of 11 studies—Table 1) report a decrease with age in NK cell cytotoxicity against K562 cancer cells. The non-unanimous findings can be attributed to the lack in consistent NK cell markers, leading to potential contamination; in addition, some of the reports used IL-2, which augments the innate NKCC ability. One study [67] reported an age-related rise in expression of the inhibitory receptor KLRG1, particularly in CD56^{dim} NK cells; CD56^{dim} NK cells expressing high levels of KLRG1 had impaired cytotoxicity against the MCF-10A cancer line, which expresses high levels of KLRG1's ligand E-cadherin, and silencing RNA against KLRG1 enhanced these cells' NKCC against this target cell type. Notably, however, one study reported an increase in cytotoxic activity with age, associated with a shift from the $CD56^{+}57^{-}$ to the $CD56^{+}57^{+}$ phenotype [54]. The impact of aging on NK cell toxicity against senescent cells remains to be investigated. Identifying the cause(s) for the age-related decline in NKCC is a necessary step toward identifying interventions to retard or reverse this decline, and thus potentially mitigate its contribution to age-related diseases.

NK cell killing of target cells is dependent on the NK cell receptors necessary for target cell recognition. Age-related declines in the percentage of NK cells expressing the activating receptors, NKp30 or NKp46 [66,112,113], have been reported. One study reported an age-related increase in NK cells expressing the inhibitory receptor KLRG1, particularly but not only in the $CD56^{dim}$ subset [67]; contrarily, another group reported a decline in both NKG2A and KLRG1 [64]. The percentage of NK cells expressing inhibitory KIRs has been reported to increase with age [54]. Taken together, these results point towards an alteration in the ability to receive pro-activation signaling from the target cells, possibly indicating an overall shift towards a more inhibitory phenotype.

As a possible indication of the functional significance of such changes, researchers have investigated diversity and frequency of KIR genotypes in otherwise-healthy aging persons [114] and centenarians [115] as compared to younger controls. The first study did not reveal a clear enrichment of either KIR diversity or particular KIR genes in older compared to younger Irish subjects, suggesting that neither confers a late-life survivorship advantage [114]. The analysis of the HLA-DRB1 and KIR receptors/HLA ligand frequencies in centenarians and controls from Sicily also demonstrated no significant difference in KIR receptors [115].

A reduction in NK cell perforin expression has been reported to be the mechanism behind age-dependent decline in NK cell cytotoxic function [116]. However, one study found no significant differences in the perforin content, distribution, and utilization in the lytic assays in NK cells from young (≤35-year-old) and old (≥61-year-old) groups, but evidence of impaired perforin release through degranulation upon coculture with target cells [113]. Certain compounds, such as ionomycin and phorbol myristate acetate (PMA), are used in vitro to activate NK cells, bypassing the need for target cell surface receptors to induce NK cell degranulation [99]. NK cells from older donors maintain their sensitivity to this assay, suggesting that the age-related decline in NKCC may be due to loss of these discussed receptors and/or to cell signaling, and not the ability to produce cytotoxic granules when activated [88,89,99].

These results collectively indicate that NK cells from older subjects have undergone biological changes, which impact their ability to interact with other immune cells and their target cells. One hypothesis to explain the increase in the numbers of cytotoxic NK cells with age may be that it is an attempt to compensate for the age-related loss in effector function at the individual NK cell level.

6. Effect of Age on NK Cell Signal Transduction

While it has become clear that NK cell function declines with age, few studies provide clues about the underlying causes. The process of NK cell cytotoxicity is a complex and tightly coordinated series of events, which ultimately result in the release of lytic granules into the immunological synapse (IS) between the NK cell and a bound target cell (Figure 1) [117]. Several key checkpoints in this pathway involve the turnover of phosphoinositide molecules by phospholipase C (PLC), calcium mobilization and influx, polarization of secretory lysosomes to the IS, and fusion of these vesicles with the NK cell membrane [117–120]. Ca^{2+} release is also required for critical degranulation checkpoints, such as migration of the mitochondria to the IS (which controls subsequent Ca^{2+} influx) and fusion of secretory lysosomes with the NK cell membrane [121,122].

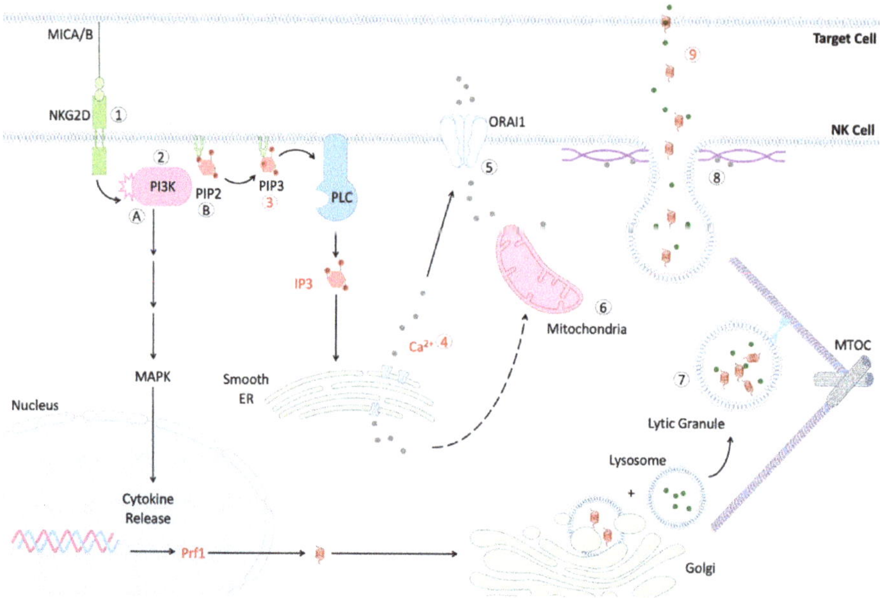

Figure 1. Scheme of NK cell signaling after formation of immunological synapse (IS). Interaction of activating receptor with cognate ligand leads to phosphorylation of cytoplasmic tail and recruitment of PI3K [61,120]. Pathway (**A**) activation of PI3K and PLC lead to downstream signaling via MAPK and NFκB pathways, respectively, thus facilitating the production and secretion of cytokines, such as IFN-γ. Activation of MAPK signaling also contributes to MTOC migration and the polarization of lytic granules. Pathway (**B**). Conversion of phosphatidylinositol-4, 5-bisphosphate (PIP2) to phosphatidylinositol-3, 4, 5-triphosphate (PIP3) allows for subsequent generation of the secondary messenger inositol trisphosphate (IP3) via phospholipase C (PLC) [117]. IP3 contributes to the release of internal calcium stores from the smooth endoplasmic reticulum [123], which in turn facilitates mitochondrial migration to IS and influx of extracellular calcium via the ORAI1 transporter [118,119]. Additionally, the microtubule organization center (MTOC) migrates towards the IS and allows for efficient transport of lytic granules (secretory lysosomes) to travel to the IS [124,125]. Lastly, lytic granules fuse with the NK cell membrane in a calcium-dependent manner and release perforin and granzymes into the IS [126]. These proteins form holes in the target cell membrane and induce apoptosis.

The second messenger, phosphoinositol phosphate 3 (IP3), is critical to this process, as it stimulates the release of internal Ca^{2+} stores, which leads to downstream activation events. The IP3 production, following exposure to K562 cancer cells, was significantly reduced in NK cells from older subjects [61]. As noted previously, one study reported that the Ca^{2+} mobilization in NK cells from older donors is reduced, relative to the level observed in NK cells from younger donors in response to IL-2 [62].The decline in the mobilization of Ca^{2+} resulting from the impaired production of IP3 might therefore contribute to the impaired cytotoxicity of the NK observed in older subjects [118].

In addition to the decreased Ca^{2+} mobilization, NK cells from older individuals exhibit diminished perforin expression and release, which likely contributes to the age-dependent decline in NK cell toxicity [113,116]. One study measured the abundance of intracellular perforin in unstimulated CD56^{+}CD16^{+} NK cells from various age groups and demonstrated an age-dependent decline in the percentage of perforin-positive NK cells in individuals that are 70 years and older, which was only partially remedied by treatment with IL-2 [116]. In another study with a much larger sample size, no decline in the expression of perforin with age was observed nor were changes in the fusion of secretory lysosomes with the NK cell membrane [113]. However, NK cells from older donors in this study did exhibit diminished perforin binding to the surface of K562 cancer cells [113]. As noted previously, the decline in IL-2 production with aging is well-known in humans [86]; thus, it is possible that in vitro experiments using activated NK cells could hide any difference in resting perforin expression between young and old NK cells [127].

7. NK Cell Exhaustion

The phenomenon of cell exhaustion is well known and characterized in T cells; however, it is not known whether NK cells undergo a similar state of exhaustion [128,129]. The impairment of the cytotoxic activity of NK cells in cancer patients is well known. For instance, a significant reduction in the numbers of NK cells expressing NKG2D, NKp30, NKp46, and perforin was observed in patients suffering from pancreatic, gastric, and colorectal cancers [130]. A similar loss in cytolytic function of NK cells is also observed in patients suffering from chronic viral infection [131]. An investigation of the molecular signaling in these NK cells may shed new light on any potential exhaustion phenotype in NK cells and help to distinguish them from senescent NK cells.

While NK cells themselves are short lived (around two weeks), human NK cells exhibit telomere shortening and a decrease in telomerase activity with age [132,133]. Cellular differentiation has a role in telomere shortening, leading to the more mature CD56^{bright} NK cells having shorter relative telomere length than the immature CD56^{dim} subset [132,134], but all subsets of NK cells have decreased telomere length with age, with CD56^{bright} cells demonstrating the greatest decrease [132]. Interestingly, NK cells expressing the activation markers, NKG2D and LFA-1, have shorter telomeres as compared to those without, while those expressing inhibitory markers, such as KIR or NKG2A, had an extremely heterogenous telomere length with no real pattern [134]. Taken together, these studies point to telomere attrition as a potential factor for diminished NK cell function with age, but more research would need to be conducted in order to determine any actual functional effects on cytotoxicity.

8. Impact of the Aging Systemic Milieu on NK Cells

To this point we have largely focused on predominantly cell-autonomous changes in NK cells with age, but changes in the local microenvironment during NK cell maturation or locally at the site of abnormal cells during aging may also skew NK cell behavior. Pro-inflammatory and anti-inflammatory cytokines work together to mediate proper immune responses and are vital to the healthy aging of an individual [135]. Chronic sterile systemic inflammation is often considered emblematic of dysfunction associated with aging, and is a powerful risk factor for morbidity and mortality in age-related diseases, such as sarcopenia [136], obesity [137], and Alzheimer's [138]. Following transplantation into old mice, NK cells isolated from young mice exhibit significant impairment in NK cell function in ex vivo assays, whereas the transfer of NK cells from old mice to young mice seems to fully restore their cytotoxic potential [139]. These findings suggest a strong influence of host tissue microenvironment on NK cell function in aging and are consistent with in vitro stimulation experiments of human NK cells [64,88]. However, the interpretation of studies in aging mice must be taken with caution, as mouse NK cells are functionally quite different from those of humans at the protein level (mouse have much lower perforin and granzyme levels) and in terms of localization patterns [140].

The aged environment might lack the stimulating factors necessary to achieve maximal NK function. With age, the levels of IL-2 [86] and IL-15 [141], vital cytokines for NK cell development and survival, decline, which may contribute to the decline in NK cell surveillance with age. In parallel, the levels of inflammatory cytokines, such as IL-6 [142] and GDF15 [143,144], increase with age, contributing to the aged inflammatory microenvironment. Treatment of NK cells in culture with interleukins 2, 12, or 15 (IL-2, IL-12,

IL-15) and interferon-α (IFN-α) can increase their cytotoxicity towards cancer cells and even toward cancer lines that are generally resistant to NK cell killing [145,146].

Studies are mixed on whether stimulation of aged NK cells with IL-2 ex vivo is sufficient to restore NKCC and related activities. One study reported full recovery of NKCC following IL-2 stimulation of in NK cells from old (60–80 years old) and even very old (80–100) vs. middle-aged (18–60) subjects [89]. However, others report that although NK cell activation by cytokine treatment does improve cytotoxicity in old-derived NK cells, it never reaches the level observed in NK cells derived from young donors, indicating a disruption in cell-level activation signaling with age [64,147]. One study demonstrated a decreased sensitivity of NK cells to IL-2 when derived from older donors [62], which may at least in part be explained by the use of the SENIEUR patient selection criteria in this study, which attempted to avoid confounding factors that elicit strong immunological responses, such as current infection, cancer, recent myocardial infarction or stroke, lymphoproliferative disorders, or laboratory findings outside the reference values for the subjects' age [148].

The evidence is similarly mixed on whether IL-2 stimulation of aged NK cells is sufficient to normalize production and secretion of NKCC effectors. One study reported an age-dependent decline in the percentage of perforin-positive NK cells derived from individuals that are 70 years and older, which was only partially remedied by treatment with IL-2 [116]. However, in another study with a much larger sample size, no decline in the expression of perforin with age was observed, nor were changes in the fusion of secretory lysosomes with the NK cell membrane; however, they did demonstrate diminished perforin binding to the surface of K562 cancer cells [113]. As the decline in IL-2 production with aging is well known in humans [86], it is possible that in vitro experiments using activated NK cells could hide any difference in resting perforin expression between young and old NK cells [127].

Epigenetics makers are known to be heavily influenced as organisms age. While there not many studies on the role of epigenetic modifications on the differentiation and function of NK cells, especially not in relation to age, it is very clear that the activation state of NK cells is epigenetically regulated [149]. NK cell activity is modulated by the complex interaction of activating and inhibiting receptors, and how much they can be activated seem to be modulated by epigenetics [150]. The exposure to cytokines, such as IL-6 or INF-γ, can influence NK cell activation; however, the strength of the response is modified by prior exposure to other stimuli, which alter the epigenetic state of the promotor for their receptors [151]. Exposure to factors that are known to influence systemic aging—such as CMV infection [152], exercise [153], and emotional stress [154], have also been demonstrated to change the epigenetic status of NK cells. More broadly speaking, long-term exposures to a variety of stresses lead to epigenetic modifications [149,152]

Interestingly, it seems that the differentiation of $CD56^{dim}$ NK cells affects epigenetic clocks, which use the level and pattern of DNA methylation in cells either in blood or other tissues to predict biological age and mortality [155]. Further, Jonkman et al. reported that cytotoxic NK cells as well as T-cells seem to be the important drivers of these clocks, and that blood $CD56^{dim}$ and $CD57^{bright}$ NK cells from the same donor are assessed by several such clocks to be biologically much older than $CD56^{bright}$ cells [155]. Certainly, our understanding of the role of the interaction of aged NK cells with the systemic milieu is incomplete, suggesting that additional cell-autonomous or microenvironmental factors remain to be identified.

9. Neuroendocrine Signaling and NK Cell Aging

Neuroendocrine factors affect NK cells, and many studies have demonstrated the influence of stress [156] and the nervous system on NK cell activity [157–159]. Some of these factors are known to change with age: for instance, glucocorticoid (GC) levels increase with age in human serum [160], even in relatively healthy aging people, as assessed by the SENIEUR criteria [161], and GCs were demonstrated to have an inhibitory effect on NK cell functions [162]. This effect is due to decreased gene expression for essential genes, such as perforin and granzyme B, leading to an overall reduction in cytotoxicity [116] and IFN-γ production [163].

The serotonin receptor agonist, quipazine, enhanced NK cell function, whereas various dopamine/serotonin antagonists inhibited CD16-mediated NK cell function [112]. In addition, NK cells express very high levels of dopamine receptors, and their activation seems to have an inhibitory function on human NK cells [164]. Similarly, epinephrine and norepinephrine seem to primarily inhibit NK cell cytotoxicity and cytokine production [165,166].

Neuropeptides present in the peripheral blood bind to and can modulate NK cell activity [167]. Galanin modulates IFN-γ secretion and sensitizes NK cells towards specific cytokines [168], while neuropeptide substance P increases the cytotoxicity of NK cells and helps with NK cell migration [168,169]. Both galanin and neuropeptide substance P decrease expression with age [170,171], suggesting that the observed decline with age may contribute to the age-related impairment in NK function.

Insulin-like growth factor 1 (IGF-1) is an important growth factor that promotes NK cell development from CD34+ cells and increases NK cell cytotoxicity [172]. IGF-1 levels decline with age [173,174]. Additionally, the secretome of senescent cells contains several pro-inflammatory factors, including TNF-α [175], which has been demonstrated to induce a state of IGF resistance [176]. Combined, the reduction in circulating IGF-1 and tissue resistance to IGF-1 signaling may contribute to the age-related changes to NK cells and the ensuing decrease in senescent cell clearing.

Other growth factors, such as platelet-derived growth factors D and DD (PDGF-D and PDGF-DD), are able to interact with the surface of NK cells to induce NK cell survival [177] and cytokine secretion [178]. There are conflicting results on the age-induced changes to levels of PDGF, with reports of decreased [179], unchanged [180], and increased [181] PDGF levels with age. These studies focus on PDGF in general rather than the PDGF-D and PDGF-DD specifically, and thus these levels would need to be examined in the context of age to further understand their role in NK cell aging.

In contrast to IGF-1 and PDGF, transforming growth factor beta (TGF-β) has been demonstrated to inhibit NK cell development and function [182,183]. TGF-β has been demonstrated to block IL-15-induced metabolic activity and proliferation of NK cells by inhibiting the mechanistic target of rapamycin (mTOR) activity [184] and the expression of NKp30 [185] and NKG2D receptors [186]. Senescent fibroblasts secrete TGF- β [187], and plasma TGF- β levels have been demonstrated to increase with age [188].

10. Reducing the Impacts of Secondary Aging on NK Cell Function

Ewald W. Busse is credited with introducing the conceptual distinction between primary aging, denoting age changes driven by intrinsic metabolic processes, and secondary aging, driven by supernumerary cellular and molecular damage inflicted through lifestyle and environmental exposures [189,190]. Recent years have observed rapid advances in the biology of aging, leading to the discovery, development, and translation from animal models of candidate interventions targeting primary aging processes, but drivers of secondary aging are already amenable to intervention today [191,192].

10.1. Micronutrient Deficiency

One recognized source of secondary aging in older persons is micronutrient deficiencies, resulting from some combination of malabsorption, low intake, and an increase in the aging body's requirements [193–197]. A dysregulation of essential nutrients, such as vitamin B_6, vitamin B_{12}, folic acid, and zinc, which are essential for cell-mediated immunity [198–200], are linked to various age-related pathologies and diseases, including inflammation [201,202], bone aging [203] and osteoporosis [204], and Alzheimer's disease [205,206], in addition to a decrease in NK cell count and/or cytotoxicity [197,207–209].

Nutrient supplementation studies demonstrate significant promise in mitigating these secondary aging effects. Studies have demonstrated that the age-related decrease in zinc levels correlate with a reduction in NK cell cytotoxicity [210]. Zinc supplementation in

NK cell cultures improves the age-related decline in proliferation/differentiation of $CD34^+$ cells, in part by restoring the transcription factor Gata-3 [51]. NK cell cytotoxicity in healthy elderly subjects is reported to be partially restored through supplementation with zinc [210,211].

Similarly, vitamin B_{12} and folate (vitamin B9) dysregulation have been correlated with lower NK cell counts and activity, and can be improved through supplementation, which increases NK cell counts and activity in aging rats [200,209]. Again, a nutritional formula containing 120 IU vitamin E, 3.8 μg vitamin B_{12}, 400 μg folic acid, and other nutrients increased the NK activity relative to the placebo in healthy subjects aged 70 years and older [212].

To our knowledge, the effect of vitamin D supplementation on NK cell function has not yet been tested in studies in aging humans, but the mechanisms of action of vitamin D in immunity are thought to principally involve the innate immune system [213]. A meta-analysis of individual participant data (IPD) from randomized controlled trials of vitamin D supplementation in subjects from birth up to 95 years of age reported a reduced risk of acute respiratory tract infections, particularly in trials involving daily or weekly vitamin D and in subjects with baseline deficient 25-hydroxyvitamin D levels [213]. Inconclusive data suggest that vitamin D deficiency may increase risk of infection with SARS-CoV-2 and/or adverse outcomes in COVID-19 [214], and evidence supports a potential role for vitamin D repletion in the prevention and amelioration of acute respiratory distress syndrome [215,216]. NK cell numbers and/or cytolytic activity were positively associated with serum vitamin D levels in a cross-sectional study in otherwise-healthy nonagenarians and centenarians [217].

In addition, the effect of supplementation with other nutrients whose deficiency is implicated in age-related immune dysfunction, such as vitamin E [218,219] on NK cell activity in aging subjects, remains to be investigated.

10.2. Stress Reduction

As noted above, GC levels rise with age [160] and are further elevated in association with elevated psychological stress [161] and impaired NK cell functions [162]. Consistent with this, extensive evidence has demonstrated that subjective stress levels and stressful life events predict impaired immunity [220], including increased susceptibility to induced rhinovirus infection [221] and less durable antibody titers after vaccination [222]. The increase in influenza-specific IFN-γ production following influenza vaccination and following four weeks of aerobic exercise appeared to be partially mediated by psychosocial factors [223]. Again, we are unaware of human challenge studies testing the effects of induced or self-reported stress on NK activity, but a study in mice subjected to selective REM sleep deprivation found that the ensuing impairment in NK cell numbers and activity was mediated by an upregulation in NK cell β–adrenergic receptors, possibly caused by an increase in GC levels, which was itself caused by the sleep restriction protocol [224].

A potential approach to reducing secondary aging of the immune system (including possibly NK function) in aging is therefore intervention against subjective stress and the associated elevation in GC levels. Modest evidence suggests that such modalities, as cognitive, psychological, or exercise interventions to lower chronic stress, may improve immune function [225]. Magnesium supplementation in older subjects has also been reported to lower nighttime GC levels [226,227].

10.3. Sleep Health and Circadian Alignment

The quality and quantity of sleep and the amplitude of circadian rhythms decline with age [228], and low objectively measured sleep quantity and/or quality are linked to neurodegenerative [229] and other diseases of aging [230,231], as well as total mortality [232–234]. Insufficient or circadian-misaligned sleep are known to cause impaired immune function [235,236]. A single night of imposed total [237] or even partial [238,239] sleep restriction, or nights of shorter vs. longer self-reported sleep [240,241] are followed by lower NK cell number and/or functions in humans the following day. Conversely, a night of recovery sleep allows NK activity to normalize [238,239], although IL-2 levels remained suppressed [238].

Consistent with this, animal studies demonstrate that chronic partial sleep restriction accelerates the growth and metastasis of experimental cancer associated with a decrease in tumor dendritic, NK, and T-cells [242,243], while better sleep is associated with longer survival in human cancer patients [244,245].

Therefore, the means to improve sleep quality or quantity, or to entrain circadian rhythms, may be a potential means to ameliorate secondary immunological aging and age impairments in NK function. In addition to cognitive-behavioral therapy and therapeutic sleep restriction for insomnia and sleep hygiene for more general poor sleep [246], some specific interventions demonstrated to improve sleep in older persons include magnesium supplementation [226,227] and creating a peripheral temperature gradient via wearing socks in bed [247,248] or through pre-bed hot baths [249].

Tobacco smoking and obesity also suppress NK cell activity, which reverses upon smoking cessation or weight loss, respectively, and should be pursued if the patient can be engaged [250].

11. Harnessing NK Cells as Immunosenolytics

Senescent cells are cells that have undergone an irreversible cell-cycle arrest in response to a range of stressors or as part of physiological processes, and they are known to secrete a pleiotropy of inflammatory, growth, and matrix-degrading factors, collectively termed the senescence-associated phenotype (SASP) [13,251]. By restricting tissue renewal with age and through promotion of inflammation, senescent cells have been implicated as drivers of the degenerative aging processes and contribute to specific diseases of aging [14].

Senolytics (drugs that selectively remove senescent cells from the body) have been demonstrated to increase the health span in aging laboratory animals and are advancing into clinical trials for human translation [10,252]. However, they bear a significant potential limitation. Such drugs rely on the differential activity of essential cell metabolic pathways, including anti-apoptotic, pro-survival, mitochondrial metabolic, kinase [83,253], and glutaminolytic pathways [254]. Such mechanisms of action necessarily perturb the healthy function of non-senescent cells, raising the specter of important side-effects, despite the overall positive effects on the health span and lifespan observed in animal models of aging and its diseases. A more physiological strategy that exploits the intrinsic immunosurveillance of senescent cells might offer greater potential to favorably impact aging and diseases of aging.

Recently, the discovery of NK cell-mediated immune surveillance of senescent cells has led to an interest in harnessing this effector function as an approach to combating aging and age-related diseases [10–12]. As noted previously, NK cells are the chief lymphocytes responsible for the clearance of senescent cells during wound resolution, development, and other physiological processes, and also from aging or diseased tissues. Thus, understanding the causes of the NK cell functional decline with age may allow us to develop therapeutic strategies to rejuvenate or fortify NK-mediated immunosurveillance of senescent cells.

One potential approach would be to reverse age-related deficits in the intrinsic capacities of this process. As we have discussed, NK function declines with age, and although the impact of aging on NK cell toxicity against senescent cells has not yet been elucidated, the known mechanisms driving and enforcing the decline in other NK cell functions may present targets for restoring the immunosurveillance of cancerous, virally infected, and senescent cells.

A second potential immunosurveillance-enhancing approach would be to target mechanisms whereby senescent cells evade immunological clearance. Such mechanisms include the shedding of NK binding ligands from the senescent cell surface [255] by matrix metalloproteinases (which are part of the SASP) [256] and upregulation of the inhibitory receptor HLA-E [257]. Although chronic inhibition of these mechanisms may be expected to lead to excessive or off-target removal of senescent or other cells, a more transient approach, similar to that which has been employed in most preclinical studies of senolytic drugs, may open a favorable therapeutic window.

A third approach would be to fortify the endogenous armamentarium of the immune system via transfer therapy of expanded and potentially engineered NK cells. Due to their significant involvement in the immune surveillance of cancer cells, NK cells have recently become a promising tool in cancer immunotherapy [258]. Several advancements have been made to enhance the efficacy of NK cells in response to the tumor microenvironment inhibition of NK cell activity [259], including the chimeric antigen receptor (CAR)-NK cell production and ex vivo activation of NK cells by cytokines [259,260].

CAR-T cells targeting a candidate senescent cell surface marker were reported to extend survival in a murine model of post-chemotherapy senescent cell accumulation, and to partially normalize liver fibrosis resulting from either treatment with CCl4 or from diet-induced nonalcoholic steatohepatitis (NASH) [261]. Similar strategies could be exploited to engineer NK cells to target senescent cells for elimination from tissues. NK cells, rather than T-cells, are the principal actors in the immune surveillance of senescent cells specifically, but even in the cancer context, CAR-NK cells offer a number of potential advantages over CAR-T therapy [262,263]. First, the profile of cytokines released by NK cells is thought to be substantially less likely to lead to a systemic inflammatory response; such a "cytokine release syndrome" is a critical side-effect of CAR-T therapy. The risk is further reduced by less expansion of CAR-NK cells after the infusion into patients. Additionally, CAR-NK cells do not persist as long in the body as CAR-T cells after mounting an attack on tumor cells in vivo, which lowers the risk of graft vs. host disease and malignant transformation. While CAR-T cells lose their cytotoxic capacity if expression of the CAR system is lost or its target antigen is not present on subpopulations of target cells, CAR-NK, regardless, would retain their native receptors as a backup system for targeting aberrant cells.

12. Perspective and Conclusions

NK cells play critical roles in combating cancer, disease, senescence, and infection, yet their function appears to decline with age. As part of immunosenescence, decline in NK cell cytotoxicity is likely a major cause of the increased susceptibility to viral infection, increased overall disease burden, and accumulation of senescent and cancer cells observed in the elderly [264]. Considering this, a growing number of studies have sought to probe specific changes that occur in NK cell biology and functionality with aging. Uncovering changes with age in gene expression, marker expression/distribution, degranulation, and internal or systemic signaling can lead to a better understanding of how NK cells age and the impact of their local environment. These insights can aid in the development of tailored rejuvenation and intervention strategies, which may restore the functionality of native NK cells or enable the augmentation of their forces with allogenic or autologus cells.

A recent study using mass cytometry (also known as or cytometry by time-of-flight, or CyTOF) has uncovered between 6000 and 30,000 distinct NK cell phenotypes within a given individual based on unique combinations of 35 cell surface antigens, whose expression is determined by both by genetics and the environment [265]. A further characterization of surface antigens and phenotypic expression changes with age will be important to developing better therapeutic interventions for improving NK cell functions for age-related diseases.

The transplantation of allogenic NK cells has demonstrated great promise in cancer treatment and as such, demonstrate promise for the transplantation of young allogenic or cord-blood-derived NK cells for treatment of various age-related diseases and aging in general [266,267]. Furthermore, CAR-T cells have proven to be powerful tools for immunotherapy for the treatment of cancer [268,269] and, recently, CAR-NK cell have emerged as an attractive alternative [263,268]. In recent clinical trials, CAR-NK cell immunotherapy has not only been found to be effective, it has also significantly improved affordability (by using off-the-shelf NK cells), reduced sensitivity to immune checkpoints, and lowered the probability of nonspecific neurotoxicity, as compared to CAR-T therapy [270,271].

In sum, much is known about the changes with age in NK cell function, but much less about the factors driving it. New tools and technologies will increasingly enable the understanding and therapeutic intervention of NK cell aging, allowing people of increasingly advanced ages to live lives more free of infection, cancer, senescence, and ultimately advance the widely emerging revolution in longevity therapeutics.

Author Contributions: Conceptualization, A.S. (Amit Sharma) and A.S. (Alexandra Stolzing); writing—original draft preparation, A.B. and E.F.; writing—review and editing, A.B., G.Z., M.R. and T.D.A.; analysis, A.S. (Amit Sharma), A.S. (Alexandra Stolzing) and M.R. All authors have read and agreed to the published version of the manuscript.

Funding: This work was funded by the SENS Research Foundation (SRF).

Institutional Review Board Statement: Not applicable.

Informed Consent Statement: Not applicable.

Data Availability Statement: Data are available on request.

Acknowledgments: Graphical abstract was created using BioRender.

Conflicts of Interest: The authors declare no conflict of interest.

References

1. Sierra, F. Geroscience and the challenges of aging societies. *Aging Med.* **2019**, *2*, 132–134. [CrossRef] [PubMed]
2. Rae, M.J.; Butler, R.N.; Campisi, J.; de Grey, A.D.; Finch, C.E.; Gough, M.; Martin, G.M.; Vijg, J.; Perrott, K.M.; Logan, B.J. The demographic and biomedical case for late-life interventions in aging. *Sci. Transl. Med.* **2010**, *2*, 40cm21. [CrossRef] [PubMed]
3. Austad, S.N. The Geroscience Hypothesis: Is It Possible to Change the Rate of Aging? In *Advances in Geroscience*; Sierra, F., Kohanski, R., Eds.; Springer International Publishing: Cham, Switzerland, 2016; pp. 1–36.
4. Akbar, A.N.; Gilroy, D.W. Aging immunity may exacerbate COVID-19. *Science* **2020**, *369*, 256–257. [CrossRef] [PubMed]
5. O'Driscoll, M.; Ribeiro Dos Santos, G.; Wang, L.; Cummings, D.A.T.; Azman, A.S.; Paireau, J.; Fontanet, A.; Cauchemez, S.; Salje, H. Age-specific mortality and immunity patterns of SARS-CoV-2. *Nature* **2021**, *590*, 140–145. [CrossRef] [PubMed]
6. Fülöp, T.; Dupuis, G.; Witkowski, J.M.; Larbi, A. The Role of Immunosenescence in the Development of Age-Related Diseases. *Rev. Investig. Clin.* **2016**, *68*, 84–91.
7. Barbé-Tuana, F.; Funchal, G.; Schmitz, C.R.R.; Maurmann, R.M.; Bauer, M.E. The interplay between immunosenescence and age-related diseases. *Semin. Immunopathol.* **2020**, *42*, 545–557. [CrossRef] [PubMed]
8. Zhang, Y.; Huang, B. The Development and Diversity of ILCs, NK Cells and Their Relevance in Health and Diseases. *Adv. Exp. Med. Biol.* **2017**, *1024*, 225–244. [CrossRef] [PubMed]
9. Vivier, E.; Tomasello, E.; Baratin, M.; Walzer, T.; Ugolini, S. Functions of natural killer cells. *Nat. Immunol.* **2008**, *9*, 503–510. [CrossRef] [PubMed]
10. Song, P.; An, J.; Zou, M.H. Immune Clearance of Senescent Cells to Combat Ageing and Chronic Diseases. *Cells* **2020**, *9*, 671. [CrossRef] [PubMed]
11. Antonangeli, F.; Zingoni, A.; Soriani, A.; Santoni, A. Senescent cells: Living or dying is a matter of NK cells. *J. Leukoc. Biol.* **2019**, *105*, 1275–1283. [CrossRef] [PubMed]
12. Kale, A.; Sharma, A.; Stolzing, A.; Desprez, P.Y.; Campisi, J. Role of immune cells in the removal of deleterious senescent cells. *Immun. Ageing* **2020**, *17*, 16. [CrossRef] [PubMed]
13. Van Deursen, J.M. The role of senescent cells in ageing. *Nature* **2014**, *509*, 439–446. [CrossRef] [PubMed]
14. Pignolo, R.J.; Passos, J.F.; Khosla, S.; Tchkonia, T.; Kirkland, J.L. Reducing Senescent Cell Burden in Aging and Disease. *Trends Mol. Med.* **2020**, *26*, 630–638. [CrossRef] [PubMed]
15. Nikolich-Žugich, J. The twilight of immunity: Emerging concepts in aging of the immune system. *Nat. Immunol.* **2018**, *19*, 10–19. [CrossRef] [PubMed]
16. Paul, S.; Lal, G. The Molecular Mechanism of Natural Killer Cells Function and Its Importance in Cancer Immunotherapy. *Front. Immunol.* **2017**, *8*, 1124. [CrossRef] [PubMed]
17. Martin-Fontecha, A.; Thomsen, L.L.; Brett, S.; Gerard, C.; Lipp, M.; Lanzavecchia, A.; Sallusto, F. Induced recruitment of NK cells to lymph nodes provides IFN-gamma for T(H)1 priming. *Nat. Immunol.* **2004**, *5*, 1260–1265. [CrossRef] [PubMed]
18. Blum, K.S.; Pabst, R. Lymphocyte numbers and subsets in the human blood. Do they mirror the situation in all organs? *Immunol. Lett.* **2007**, *108*, 45–51. [CrossRef] [PubMed]
19. Carrega, P.; Ferlazzo, G. Natural killer cell distribution and trafficking in human tissues. *Front. Immunol.* **2012**, *3*, 347. [CrossRef]

20. Michel, T.; Poli, A.; Cuapio, A.; Briquemont, B.; Iserentant, G.; Ollert, M.; Zimmer, J. Human CD56bright NK Cells: An Update. *J. Immunol.* **2016**, *196*, 2923–2931. [CrossRef]
21. Penack, O.; Gentilini, C.; Fischer, L.; Asemissen, A.M.; Scheibenbogen, C.; Thiel, E.; Uharek, L. CD56dimCD16neg cells are responsible for natural cytotoxicity against tumor targets. *Leukemia* **2005**, *19*, 835840. [CrossRef]
22. Carrega, P.; Bonaccorsi, I.; Di Carlo, E.; Morandi, B.; Paul, P.; Rizzello, V.; Cipollone, G.; Navarra, G.; Mingari, M.C.; Moretta, L.; et al. CD56(bright)perforin(low) noncytotoxic human NK cells are abundant in both healthy and neoplastic solid tissues and recirculate to secondary lymphoid organs via afferent lymph. *J. Immunol.* **2014**, *192*, 3805–3815. [CrossRef] [PubMed]
23. Lugthart, G.; Melsen, J.E.; Vervat, C.; van Ostaijen-Ten Dam, M.M.; Corver, W.E.; Roelen, D.L.; van Bergen, J.; van Tol, M.J.; Lankester, A.C.; Schilham, M.W. Human Lymphoid Tissues Harbor a Distinct $CD69^+CXCR6^+$ NK Cell Population. *J. Immunol.* **2016**, *197*, 78–84. [CrossRef] [PubMed]
24. Caligiuri, M.A. Human natural killer cells. *Blood* **2008**, *112*, 461–469. [CrossRef] [PubMed]
25. Yu, J.; Freud, A.G.; Caligiuri, M.A. Location and cellular stages of natural killer cell development. *Trends Immunol.* **2013**, *34*, 573–582. [CrossRef] [PubMed]
26. Klein Wolterink, R.G.; Garcia-Ojeda, M.E.; Vosshenrich, C.A.; Hendriks, R.W.; Di Santo, J.P. The intrathymic crossroads of T and NK cell differentiation. *Immunol. Rev.* **2010**, *238*, 126137. [CrossRef] [PubMed]
27. Cichocki, F.; Grzywacz, B.; Miller, J.S. Human NK Cell Development: One Road or Many? *Front. Immunol.* **2019**, *10*, 2078. [CrossRef] [PubMed]
28. Luetke-Eversloh, M.; Killig, M.; Romagnani, C. Signatures of human NK cell development and terminal differentiation. *Front. Immunol.* **2013**, *4*, 499. [CrossRef] [PubMed]
29. Rabinowich, H.; Pricop, L.; Herberman, R.B.; Whiteside, T.L. Expression and function of CD7 molecule on human natural killer cells. *J. Immunol.* **1994**, *152*, 517–526. [PubMed]
30. Higuchi, Y.; Zeng, H.; Ogawa, M. CD38 expression by hematopoietic stem cells of newborn and juvenile mice. *Leukemia* **2003**, *17*, 171–174. [CrossRef]
31. Zingoni, A.; Sornasse, T.; Cocks, B.G.; Tanaka, Y.; Santoni, A.; Lanier, L.L. Cross-talk between activated human NK cells and $CD4^+$ T cells via OX40-OX40 ligand interactions. *J. Immunol.* **2004**, *173*, 3716–3724. [CrossRef]
32. Blanca, I.R.; Bere, E.W.; Young, H.A.; Ortaldo, J.R. Human B cell activation by autologous NK cells is regulated by CD40-CD40 ligand interaction: Role of memory B cells and $CD5^+$ B cells. *J. Immunol.* **2001**, *167*, 6132–6139. [CrossRef] [PubMed]
33. Moretta, L.; Montaldo, E.; Vacca, P.; Del Zotto, G.; Moretta, F.; Merli, P.; Locatelli, F.; Mingari, M.C. Human natural killer cells: Origin, receptors, function, and clinical applications. *Int. Arch. Allergy Immunol.* **2014**, *164*, 253–264. [CrossRef] [PubMed]
34. Bryceson, Y.T.; March, M.E.; Ljunggren, H.G.; Long, E.O. Activation, coactivation, and costimulation of resting human natural killer cells. *Immunol. Rev.* **2006**, *214*, 73–91. [CrossRef] [PubMed]
35. Smith, K.M.; Wu, J.; Bakker, A.B.; Phillips, J.H.; Lanier, L.L. Ly-49D and Ly-49H associate with mouse DAP12 and form activating receptors. *J. Immunol.* **1998**, *161*, 7–10. [PubMed]
36. Cosman, D.; Mullberg, J.; Sutherland, C.L.; Chin, W.; Armitage, R.; Fanslow, W.; Kubin, M.; Chalupny, N.J. ULBPs, novel MHC class I-related molecules, bind to CMV glycoprotein UL16 and stimulate NK cytotoxicity through the NKG2D receptor. *Immunity* **2001**, *14*, 123–133. [CrossRef]
37. Jamieson, A.M.; Diefenbach, A.; McMahon, C.W.; Xiong, N.; Carlyle, J.R.; Raulet, D.H. The role of the NKG2D immunoreceptor in immune cell activation and natural killing. *Immunity* **2002**, *17*, 19–29. [CrossRef]
38. Martinez-Lostao, L.; Anel, A.; Pardo, J. How Do Cytotoxic Lymphocytes Kill Cancer Cells? *Clin. Cancer Res.* **2015**, *21*, 5047–5056. [CrossRef]
39. Gayoso, I.; Sanchez-Correa, B.; Campos, C.; Alonso, C.; Pera, A.; Casado, J.G.; Morgado, S.; Tarazona, R.; Solana, R. Immunosenescence of human natural killer cells. *J. Innate Immun.* **2011**, *3*, 337–343. [CrossRef]
40. Narni-Mancinelli, E.; Ugolini, S.; Vivier, E. Tuning the threshold of natural killer cell responses. *Curr. Opin. Immunol.* **2013**, *25*, 53–58. [CrossRef]
41. Canossi, A.; Aureli, A.; Del Beato, T.; Rossi, P.; Franceschilli, L.; De Sanctis, F.; Sileri, P.; di Lorenzo, N.; Buonomo, O.; Lauro, D.; et al. Role of KIR and CD16A genotypes in colorectal carcinoma genetic risk and clinical stage. *J. Transl. Med.* **2016**, *14*, 239. [CrossRef]
42. Lee, N.; Llano, M.; Carretero, M.; Ishitani, A.; Navarro, F.; Lopez-Botet, M.; Geraghty, D.E. HLA-E is a major ligand for the natural killer inhibitory receptor CD94/NKG2A. *Proc. Natl. Acad. Sci. USA* **1998**, *95*, 5199–5204. [CrossRef] [PubMed]
43. Braud, V.M.; Allan, D.S.; O'Callaghan, C.A.; Soderstrom, K.; D'Andrea, A.; Ogg, G.S.; Lazetic, S.; Young, N.T.; Bell, J.I.; Phillips, J.H.; et al. HLA-E binds to natural killer cell receptors CD94/NKG2A, B and C. *Nature* **1998**, *391*, 795–799. [CrossRef] [PubMed]
44. Veillette, A.; Latour, S.; Davidson, D. Negative regulation of immunoreceptor signaling. *Annu. Rev. Immunol.* **2002**, *20*, 669–707. [CrossRef] [PubMed]
45. Yokoyama, W.M.; Kim, S. Licensing of natural killer cells by self-major histocompatibility complex class I. *Immunol. Rev.* **2006**, *214*, 143–154. [CrossRef] [PubMed]
46. Kadri, N.; Thanh, T.L.; Höglund, P. Selection, tuning, and adaptation in mouse NK cell education. *Immunol. Rev.* **2015**, *267*, 167–177. [CrossRef]

47. Pahl, J.; Cerwenka, A. Tricking the balance: NK cells in anti-cancer immunity. *Immunobiology* **2017**, *222*, 11–20. [CrossRef]
48. Moesta, A.K.; Parham, P. Diverse functionality among human NK cell receptors for the C1 epitope of HLA-C: KIR2DS2, KIR2DL2, and KIR2DL3. *Front. Immunol.* **2012**, *3*, 336. [CrossRef]
49. Makrigiannis, A.P.; Anderson, S.K. Ly49 gene expression in different inbred mouse strains. *Immunol. Res.* **2000**, *21*, 39–47. [CrossRef]
50. Sternberg-Simon, M.; Brodin, P.; Pickman, Y.; Onfelt, B.; Karre, K.; Malmberg, K.J.; Hoglund, P.; Mehr, R. Natural killer cell inhibitory receptor expression in humans and mice: A closer look. *Front. Immunol.* **2013**, *4*, 65. [CrossRef]
51. Muzzioli, M.; Stecconi, R.; Moresi, R.; Provinciali, M. Zinc improves the development of human $CD34^+$ cell progenitors towards NK cells and increases the expression of GATA-3 transcription factor in young and old ages. *Biogerontology* **2009**, *10*, 593–604. [CrossRef]
52. Ligthart, G.J.; van Vlokhoven, P.C.; Schuit, H.R.; Hijmans, W. The expanded null cell compartment in ageing: Increase in the number of natural killer cells and changes in T-cell and NK-cell subsets in human blood. *Immunology* **1986**, *59*, 353–357. [PubMed]
53. Facchini, A.; Mariani, E.; Mariani, A.R.; Papa, S.; Vitale, M.; Manzoli, F.A. Increased number of circulating Leu 11+ (CD 16) large granular lymphocytes and decreased NK activity during human ageing. *Clin. Exp. Immunol.* **1987**, *68*, 340–347. [PubMed]
54. Krishnaraj, R.; Svanborg, A. Preferential accumulation of mature NK cells during human immunosenescence. *J. Cell. Biochem.* **1992**, *50*, 386–391. [CrossRef] [PubMed]
55. Vitale, M.; Zamai, L.; Neri, L.M.; Galanzi, A.; Facchini, A.; Rana, R.; Cataldi, A.; Papa, S. The impairment of natural killer function in the healthy aged is due to a postbinding deficient mechanism. *Cell. Immunol.* **1992**, *145*, 1–10. [CrossRef]
56. Sansoni, P.; Cossarizza, A.; Brianti, V.; Fagnoni, F.; Snelli, G.; Monti, D.; Marcato, A.; Passeri, G.; Ortolani, C.; Forti, E.; et al. Lymphocyte subsets and natural killer cell activity in healthy old people and centenarians. *Blood* **1993**, *82*, 2767–2773. [CrossRef] [PubMed]
57. Mariani, E.; Monaco, M.C.; Cattini, L.; Sinoppi, M.; Facchini, A. Distribution and lytic activity of NK cell subsets in the elderly. *Mech. Ageing Dev.* **1994**, *76*, 177–187. [CrossRef]
58. Kutza, J.; Murasko, D.M. Effects of aging on natural killer cell activity and activation by interleukin-2 and IFN-α. *Cell. Immunol.* **1994**, *155*, 195–204. [CrossRef] [PubMed]
59. Kutza, J.; Murasko, D.M. Age-associated decline in IL-2 and IL-12 induction of LAK cell activity of human PBMC samples. *Mech. Ageing Dev.* **1996**, *90*, 209–222. [CrossRef]
60. Mariani, E.; Sgobbi, S.; Meneghetti, A.; Tadolini, M.; Tarozzi, A.; Sinoppi, M.; Cattini, L.; Facchini, A. Perforins in human cytolytic cells: The effect of age. *Mech. Ageing Dev.* **1996**, *92*, 195–209. [CrossRef]
61. Mariani, E.; Mariani, A.R.; Meneghetti, A.; Tarozzi, A.; Cocco, L.; Facchini, A. Age-dependent decreases of NK cell phosphoinositide turnover during spontaneous but not Fc-mediated cytolytic activity. *Int. Immunol.* **1998**, *10*, 981–989. [CrossRef]
62. Borrego, F.; Alonso, M.C.; Galiani, M.D.; Carracedo, J.; Ramirez, R.; Ostos, B.; Pena, J.; Solana, R. NK phenotypic markers and IL2 response in NK cells from elderly people. *Exp. Gerontol.* **1999**, *34*, 253–265. [CrossRef]
63. Lutz, C.T.; Moore, M.B.; Bradley, S.; Shelton, B.J.; Lutgendorf, S.K. Reciprocal age related change in natural killer cell receptors for MHC class I. *Mech. Ageing Dev.* **2005**, *126*, 722–731. [CrossRef] [PubMed]
64. Le Garff-Tavernier, M.; Beziat, V.; Decocq, J.; Siguret, V.; Gandjbakhch, F.; Pautas, E.; Debre, P.; Merle-Beral, H.; Vieillard, V. Human NK cells display major phenotypic and functional changes over the life span. *Aging Cell* **2010**, *9*, 527–535. [CrossRef] [PubMed]
65. Hayhoe, R.P.; Henson, S.M.; Akbar, A.N.; Palmer, D.B. Variation of human natural killer cell phenotypes with age: Identification of a unique KLRG1-negative subset. *Hum. Immunol.* **2010**, *71*, 676–681. [CrossRef] [PubMed]
66. Almeida-Oliveira, A.; Smith-Carvalho, M.; Porto, L.C.; Cardoso-Oliveira, J.; Ribeiro Ados, S.; Falcao, R.R.; Abdelhay, E.; Bouzas, L.F.; Thuler, L.C.; Ornellas, M.H.; et al. Age-related changes in natural killer cell receptors from childhood through old age. *Hum. Immunol.* **2011**, *72*, 319–329. [CrossRef] [PubMed]
67. Muller-Durovic, B.; Lanna, A.; Covre, L.P.; Mills, R.S.; Henson, S.M.; Akbar, A.N. Killer Cell Lectin-like Receptor G1 Inhibits NK Cell Function through Activation of Adenosine 5′-Monophosphate-Activated Protein Kinase. *J. Immunol.* **2016**, *197*, 2891–2899. [CrossRef] [PubMed]
68. Gounder, S.S.; Abdullah, B.J.J.; Radzuanb, N.; Zain, F.; Sait, N.B.M.; Chua, C.; Subramani, B. Effect of Aging on NK Cell Population and Their Proliferation at Ex Vivo Culture Condition. *Anal. Cell. Pathol.* **2018**, *2018*, 7871814. [CrossRef] [PubMed]
69. Zhang, Y.; Wallace, D.L.; de Lara, C.M.; Ghattas, H.; Asquith, B.; Worth, A.; Griffin, G.E.; Taylor, G.P.; Tough, D.F.; Beverley, P.C.; et al. In vivo kinetics of human natural killer cells: The effects of ageing and acute and chronic viral infection. *Immunology* **2007**, *121*, 258–265. [CrossRef] [PubMed]
70. Crinier, A.; Milpied, P.; Escaliere, B.; Piperoglou, C.; Galluso, J.; Balsamo, A.; Spinelli, L.; Cervera-Marzal, I.; Ebbo, M.; Girard-Madoux, M.; et al. High-Dimensional Single-Cell Analysis Identifies Organ-Specific Signatures and Conserved NK Cell Subsets in Humans and Mice. *Immunity* **2018**, *49*, 971–986.e5. [CrossRef] [PubMed]
71. Yang, C.; Siebert, J.R.; Burns, R.; Gerbec, Z.J.; Bonacci, B.; Rymaszewski, A.; Rau, M.; Riese, M.J.; Rao, S.; Carlson, K.S.; et al. Heterogeneity of human bone marrow and blood natural killer cells defined by single-cell transcriptome. *Nat. Commun.* **2019**, *10*, 3931. [CrossRef]

72. Smith, S.L.; Kennedy, P.R.; Stacey, K.B.; Worboys, J.D.; Yarwood, A.; Seo, S.; Solloa, E.H.; Mistretta, B.; Chatterjee, S.S.; Gunaratne, P.; et al. Diversity of peripheral blood human NK cells identified by single-cell RNA sequencing. *Blood Adv.* **2020**, *4*, 1388–1406. [CrossRef] [PubMed]
73. Zheng, Y.; Liu, X.; Le, W.; Xie, L.; Li, H.; Wen, W.; Wang, S.; Ma, S.; Huang, Z.; Ye, J.; et al. A human circulating immune cell landscape in aging and COVID-19. *Protein Cell* **2020**, *11*, 740–770. [CrossRef] [PubMed]
74. Dogra, P.; Rancan, C.; Ma, W.; Toth, M.; Senda, T.; Carpenter, D.J.; Kubota, M.; Matsumoto, R.; Thapa, P.; Szabo, P.A.; et al. Tissue Determinants of Human NK Cell Development, Function, and Residence. *Cell* **2020**, *180*, 749–763.e13. [CrossRef] [PubMed]
75. Ogata, K.; An, E.; Shioi, Y.; Nakamura, K.; Luo, S.; Yokose, N.; Minami, S.; Dan, K. Association between natural killer cell activity and infection in immunologically normal elderly people. *Clin. Exp. Immunol.* **2001**, *124*, 392–397. [CrossRef] [PubMed]
76. Ogata, K.; Yokose, N.; Tamura, H.; An, E.; Nakamura, K.; Dan, K.; Nomura, T. Natural killer cells in the late decades of human life. *Clin. Immunol. Immunopathol.* **1997**, *84*, 269–275. [CrossRef]
77. Pierson, B.; Miller, J. CD56+bright and CD56+dim natural killer cells in patients with chronic myelogenous leukemia progressively decrease in number, respond less to stimuli that recruit clonogenic natural killer cells, and exhibit decreased proliferation on a per cell basis. *Blood* **1996**, *88*, 2279–2287. [CrossRef]
78. Judge, S.J.; Murphy, W.J.; Canter, R.J. Characterizing the Dysfunctional NK Cell: Assessing the Clinical Relevance of Exhaustion, Anergy, and Senescence. *Front. Cell. Infect. Microbiol.* **2020**, *10*, 49. [CrossRef]
79. Parameswaran, N.; Patial, S. Tumor necrosis factor-α signaling in macrophages. *Crit. Rev. Eukaryot. Gene Expr.* **2010**, *20*, 87–103. [CrossRef]
80. Vieira, S.M.; Lemos, H.P.; Grespan, R.; Napimoga, M.H.; Dal-Secco, D.; Freitas, A.; Cunha, T.M.; Verri, W.A., Jr.; Souza-Junior, D.A.; Jamur, M.C.; et al. A crucial role for TNF-alpha in mediating neutrophil influx induced by endogenously generated or exogenous chemokines, KC/CXCL1 and LIX/CXCL5. *Br. J. Pharm.* **2009**, *158*, 779–789. [CrossRef]
81. Billiau, A. Interferon-gamma: Biology and role in pathogenesis. *Adv. Immunol.* **1996**, *62*, 61–130. [CrossRef]
82. Boehm, U.; Klamp, T.; Groot, M.; Howard, J.C. Cellular responses to interferon-γ. *Annu. Rev. Immunol.* **1997**, *15*, 749–795. [CrossRef] [PubMed]
83. Soto-Gamez, A.; Quax, W.J.; Demaria, M. Regulation of Survival Networks in Senescent Cells: From Mechanisms to Interventions. *J. Mol. Biol.* **2019**, *431*, 2629–2643. [CrossRef] [PubMed]
84. Carson, W.E.; Fehniger, T.A.; Haldar, S.; Eckhert, K.; Lindemann, M.J.; Lai, C.F.; Croce, C.M.; Baumann, H.; Caligiuri, M.A. A potential role for interleukin-15 in the regulation of human natural killer cell survival. *J. Clin. Investig.* **1997**, *99*, 937–943. [CrossRef] [PubMed]
85. Cooper, M.A.; Fehniger, T.A.; Caligiuri, M.A. The biology of human natural killer-cell subsets. *Trends Immunol.* **2001**, *22*, 633–640. [CrossRef]
86. Pahlavani, M.A.; Richardson, A. The effect of age on the expression of interleukin-2. *Mech. Ageing Dev.* **1996**, *89*, 125–154. [CrossRef]
87. Wu, Y.; Tian, Z.; Wei, H. Developmental and Functional Control of Natural Killer Cells by Cytokines. *Front. Immunol.* **2017**, *8*, 930. [CrossRef] [PubMed]
88. Kaszubowska, L.; Foerster, J.; Schetz, D.; Kmiec, Z. CD56bright cells respond to stimulation until very advanced age revealing increased expression of cellular protective proteins SIRT1, HSP70 and SOD2. *Immun. Ageing* **2018**, *15*, 31. [CrossRef] [PubMed]
89. Kaszubowska, L.; Foerster, J.; Kaczor, J.J.; Schetz, D.; Slebioda, T.J.; Kmiec, Z. NK cells of the oldest seniors represent constant and resistant to stimulation high expression of cellular protective proteins SIRT1 and HSP70. *Immun. Ageing* **2018**, *15*, 12. [CrossRef]
90. Zhang, C.; Zhang, J.; Niu, J.; Zhou, Z.; Zhang, J.; Tian, Z. Interleukin-12 improves cytotoxicity of natural killer cells via upregulated expression of NKG2D. *Hum. Immunol.* **2008**, *69*, 490–500. [CrossRef]
91. Gately, M.K.; Renzetti, L.M.; Magram, J.; Stern, A.S.; Adorini, L.; Gubler, U.; Presky, D.H. The interleukin-12/interleukin-12-receptor system: Role in normal and pathologic immune responses. *Annu. Rev. Immunol.* **1998**, *16*, 495–521. [CrossRef]
92. Krishnaraj, R.; Bhooma, T. Cytokine sensitivity of human NK cells during immunosenescence. 2. IL2-induced interferon gamma secretion. *Immunol. Lett.* **1996**, *50*, 59–63. [CrossRef]
93. Krishnaraj, R. Senescence and cytokines modulate the NK cell expression. *Mech. Ageing Dev.* **1997**, *96*, 89–101. [CrossRef]
94. Mariani, E.; Pulsatelli, L.; Neri, S.; Dolzani, P.; Meneghetti, A.; Silvestri, T.; Ravaglia, G.; Forti, P.; Cattini, L.; Facchini, A. RANTES and MIP-1alpha production by T lymphocytes, monocytes and NK cells from nonagenarian subjects. *Exp. Gerontol.* **2002**, *37*, 219–226. [CrossRef]
95. Mariani, E.; Meneghetti, A.; Neri, S.; Ravaglia, G.; Forti, P.; Cattini, L.; Facchini, A. Chemokine production by natural killer cells from nonagenarians. *Eur. J. Immunol.* **2002**, *32*, 1524–1529. [CrossRef]
96. Mariani, E.; Pulsatelli, L.; Meneghetti, A.; Dolzani, P.; Mazzetti, I.; Neri, S.; Ravaglia, G.; Forti, P.; Facchini, A. Different IL-8 production by T and NK lymphocytes in elderly subjects. *Mech. Ageing Dev.* **2001**, *122*, 1383–1395. [CrossRef]
97. Orange, J.S.; Ballas, Z.K. Natural killer cells in human health and disease. *Clin. Immunol.* **2006**, *118*, 1–10. [CrossRef] [PubMed]
98. Jost, S.; Altfeld, M. Control of human viral infections by natural killer cells. *Annu. Rev. Immunol.* **2013**, *31*, 163–194. [CrossRef] [PubMed]
99. Hazeldine, J.; Lord, J.M. The impact of ageing on natural killer cell function and potential consequences for health in older adults. *Ageing Res. Rev.* **2013**, *12*, 1069–1078. [CrossRef] [PubMed]

100. Henley, S.J.; Singh, S.D.; King, J.; Wilson, R.J.; O'Neil, M.E.; Ryerson, A.B. Invasive Cancer Incidence and Survival-United States, 2013. *MMWR Morb. Mortal. Wkly. Rep.* **2017**, *66*, 69–75. [CrossRef] [PubMed]
101. Balducci, L.; Ershler, W.B. Cancer and ageing: A nexus at several levels. *Nat. Rev. Cancer* **2005**, *5*, 655–662. [CrossRef] [PubMed]
102. Moon, W.Y.; Powis, S.J. Does Natural Killer Cell Deficiency (NKD) Increase the Risk of Cancer? NKD May Increase the Risk of Some Virus Induced Cancer. *Front. Immunol.* **2019**, *10*, 1703. [CrossRef] [PubMed]
103. Kang, T.W.; Yevsa, T.; Woller, N.; Hoenicke, L.; Wuestefeld, T.; Dauch, D.; Hohmeyer, A.; Gereke, M.; Rudalska, R.; Potapova, A.; et al. Senescence surveillance of pre-malignant hepatocytes limits liver cancer development. *Nature* **2011**, *479*, 547–551. [CrossRef] [PubMed]
104. Waldhauer, I.; Steinle, A. NK cells and cancer immunosurveillance. *Oncogene* **2008**, *27*, 5932–5943. [CrossRef] [PubMed]
105. Gross, E.; Sunwoo, J.B.; Bui, J.D. Cancer Immunosurveillance and Immunoediting by Natural Killer Cells. *Cancer J.* **2013**, *19*, 483–489. [CrossRef] [PubMed]
106. López-Soto, A.; Gonzalez, S.; Smyth, M.J.; Galluzzi, L. Control of Metastasis by NK Cells. *Cancer Cell* **2017**, *32*, 135–154. [CrossRef] [PubMed]
107. Talmadge, J.E.; Meyers, K.M.; Prieur, D.J.; Starkey, J.R. Role of NK cells in tumour growth and metastasis in beige mice. *Nature* **1980**, *284*, 622–624. [CrossRef] [PubMed]
108. Hersey, P.; Edwards, A.; Honeyman, M.; McCarthy, W.H. Low natural-killer-cell activity in familial melanoma patients and their relatives. *Br. J. Cancer* **1979**, *40*, 113–122. [CrossRef]
109. Strayer, D.R.; Carter, W.A.; Mayberry, S.D.; Pequignot, E.; Brodsky, I. Low natural cytotoxicity of peripheral blood mononuclear cells in individuals with high familial incidences of cancer. *Cancer Res.* **1984**, *44*, 370–374.
110. Nakajima, T.; Mizushima, N.; Kanai, K. Relationship between natural killer activity and development of hepatocellular carcinoma in patients with cirrhosis of the liver. *Jpn. J. Clin. Oncol.* **1987**, *17*, 327–332.
111. Imai, K.; Matsuyama, S.; Miyake, S.; Suga, K.; Nakachi, K. Natural cytotoxic activity of peripheral-blood lymphocytes and cancer incidence: An 11-year follow-up study of a general population. *Lancet* **2000**, *356*, 1795–1799. [CrossRef]
112. Tarazona, R.; DelaRosa, O.; Alonso, C.; Ostos, B.; Espejo, J.; Pena, J.; Solana, R. Increased expression of NK cell markers on T lymphocytes in aging and chronic activation of the immune system reflects the accumulation of effector/senescent T cells. *Mech. Ageing Dev.* **2000**, *121*, 77–88. [CrossRef]
113. Hazeldine, J.; Hampson, P.; Lord, J.M. Reduced release and binding of perforin at the immunological synapse underlies the age-related decline in natural killer cell cytotoxicity. *Aging Cell* **2012**, *11*, 751–759. [CrossRef] [PubMed]
114. Maxwell, L.D.; Ross, O.A.; Curran, M.D.; Rea, I.M.; Middleton, D. Investigation of KIR diversity in immunosenescence and longevity within the Irish population. *Exp. Gerontol.* **2004**, *39*, 1223–1232. [CrossRef] [PubMed]
115. Listi, F.; Caruso, C.; Colonna-Romano, G.; Lio, D.; Nuzzo, D.; Candore, G. HLA and KIR frequencies in Sicilian Centenarians. *Rejuvenation Res.* **2010**, *13*, 314–318. [CrossRef] [PubMed]
116. Rukavina, D.; Laskarin, G.; Rubesa, G.; Strbo, N.; Bedenicki, I.; Manestar, D.; Glavas, M.; Christmas, S.E.; Podack, E.R. Age-related decline of perforin expression in human cytotoxic T lymphocytes and natural killer cells. *Blood* **1998**, *92*, 2410–2420. [CrossRef] [PubMed]
117. Dustin, M.L.; Long, E.O. Cytotoxic immunological synapses. *Immunol. Rev.* **2010**, *235*, 24–34. [CrossRef] [PubMed]
118. Maul-Pavicic, A.; Chiang, S.C.; Rensing-Ehl, A.; Jessen, B.; Fauriat, C.; Wood, S.M.; Sjoqvist, S.; Hufnagel, M.; Schulze, I.; Bass, T.; et al. ORAI1-mediated calcium influx is required for human cytotoxic lymphocyte degranulation and target cell lysis. *Proc. Natl. Acad. Sci. USA* **2011**, *108*, 3324–3329. [CrossRef] [PubMed]
119. McCarl, C.A.; Khalil, S.; Ma, J.; Oh-hora, M.; Yamashita, M.; Roether, J.; Kawasaki, T.; Jairaman, A.; Sasaki, Y.; Prakriya, M.; et al. Store-operated Ca^{2+} entry through ORAI1 is critical for T cell-mediated autoimmunity and allograft rejection. *J. Immunol.* **2010**, *185*, 5845–5858. [CrossRef] [PubMed]
120. Zweifach, A. Target-cell contact activates a highly selective capacitative calcium entry pathway in cytotoxic T lymphocytes. *J. Cell Biol.* **2000**, *148*, 603–614. [CrossRef]
121. Schwindling, C.; Quintana, A.; Krause, E.; Hoth, M. Mitochondria positioning controls local calcium influx in T cells. *J. Immunol.* **2010**, *184*, 184–190. [CrossRef]
122. Pores-Fernando, A.T.; Zweifach, A. Calcium influx and signaling in cytotoxic T-lymphocyte lytic granule exocytosis. *Immunol. Rev.* **2009**, *231*, 160–173. [CrossRef] [PubMed]
123. Segal, M.; Korkotian, E. Endoplasmic reticulum calcium stores in dendritic spines. *Front. Neuroanat.* **2014**, *8*, 64. [CrossRef] [PubMed]
124. Kopf, A.; Kiermaier, E. Dynamic Microtubule Arrays in Leukocytes and Their Role in Cell Migration and Immune Synapse Formation. *Front. Cell Dev. Biol.* **2021**, *9*, 635511. [CrossRef] [PubMed]
125. Banerjee, P.P.; Pandey, R.; Zheng, R.; Suhoski, M.M.; Monaco-Shawver, L.; Orange, J.S. Cdc42-interacting protein-4 functionally links actin and microtubule networks at the cytolytic NK cell immunological synapse. *J. Exp. Med.* **2007**, *204*, 2305–2320. [CrossRef] [PubMed]
126. Krzewski, K.; Coligan, J.E. Human NK cell lytic granules and regulation of their exocytosis. *Front. Immunol.* **2012**, *3*, 335. [CrossRef] [PubMed]
127. Rabinowich, H.; Goses, Y.; Reshef, T.; Klajman, A. Interleukin-2 production and activity in aged humans. *Mech. Ageing Dev.* **1985**, *32*, 213–226. [CrossRef]

128. Blank, C.U.; Haining, W.N.; Held, W.; Hogan, P.G.; Kallies, A.; Lugli, E.; Lynn, R.C.; Philip, M.; Rao, A.; Restifo, N.P.; et al. Defining 'T cell exhaustion'. *Nat. Rev. Immunol.* **2019**, *19*, 665–674. [CrossRef] [PubMed]
129. Wherry, E.J.; Kurachi, M. Molecular and cellular insights into T cell exhaustion. *Nat. Rev. Immunol.* **2015**, *15*, 486–499. [CrossRef]
130. Peng, Y.P.; Zhu, Y.; Zhang, J.J.; Xu, Z.K.; Qian, Z.Y.; Dai, C.C.; Jiang, K.R.; Wu, J.L.; Gao, W.T.; Li, Q.; et al. Comprehensive analysis of the percentage of surface receptors and cytotoxic granules positive natural killer cells in patients with pancreatic cancer, gastric cancer, and colorectal cancer. *J. Transl. Med.* **2013**, *11*, 262. [CrossRef]
131. Sun, C.; Sun, H.Y.; Xiao, W.H.; Zhang, C.; Tian, Z.G. Natural killer cell dysfunction in hepatocellular carcinoma and NK cell-based immunotherapy. *Acta Pharm. Sin.* **2015**, *36*, 1191–1199. [CrossRef]
132. Fali, T.; Papagno, L.; Bayard, C.; Mouloud, Y.; Boddaert, J.; Sauce, D.; Appay, V. New Insights into Lymphocyte Differentiation and Aging from Telomere Length and Telomerase Activity Measurements. *J. Immunol.* **2019**, *202*, 1962–1969. [CrossRef] [PubMed]
133. Mariani, E.; Meneghetti, A.; Formentini, I.; Neri, S.; Cattini, L.; Ravaglia, G.; Forti, P.; Facchini, A. Telomere length and telomerase activity: Effect of ageing on human NK cells. *Mech. Ageing Dev.* **2003**, *124*, 403–408. [CrossRef]
134. Ouyang, Q.; Baerlocher, G.; Vulto, I.; Lansdorp, P.M. Telomere length in human natural killer cell subsets. *Ann. N. Y. Acad. Sci.* **2007**, *1106*, 240–252. [CrossRef] [PubMed]
135. Rea, I.M.; Gibson, D.S.; McGilligan, V.; McNerlan, S.E.; Alexander, H.D.; Ross, O.A. Age and Age-Related Diseases: Role of Inflammation Triggers and Cytokines. *Front. Immunol.* **2018**, *9*, 586. [CrossRef] [PubMed]
136. Alemán, H.; Esparza, J.; Ramirez, F.A.; Astiazaran, H.; Payette, H. Longitudinal evidence on the association between interleukin-6 and C-reactive protein with the loss of total appendicular skeletal muscle in free-living older men and women. *Age Aging* **2011**, *40*, 469–475. [CrossRef]
137. Wang, T.; He, C. Pro-inflammatory cytokines: The link between obesity and osteoarthritis. *Cytokine Growth Factor Rev.* **2018**, *44*, 38–50. [CrossRef]
138. Swardfager, W.; Lanctôt, K.; Rothenburg, L.; Wong, A.; Cappell, J.; Herrmann, N. A Meta-Analysis of Cytokines in Alzheimer's Disease. *Biol. Psychiatry* **2010**, *68*, 930–941. [CrossRef]
139. Shehata, H.M.; Hoebe, K.; Chougnet, C.A. The aged nonhematopoietic environment impairs natural killer cell maturation and function. *Aging Cell* **2015**, *14*, 191–199. [CrossRef]
140. Sungur, C.M.; Murphy, W.J. Utilization of mouse models to decipher natural killer cell biology and potential clinical applications. *Hematology* **2013**, *2013*, 227–233. [CrossRef]
141. Quinn, L.S.; Anderson, B.G.; Strait-Bodey, L.; Wolden-Hanson, T. Serum and muscle interleukin-15 levels decrease in aging mice: Correlation with declines in soluble interleukin-15 receptor alpha expression. *Exp. Gerontol.* **2010**, *45*, 106–112. [CrossRef]
142. Jergovic, M.; Thompson-Ryder, H.L.; Asghar, A.; Nikolich-Zugich, J. The role of Interleukin-6 in age-related frailty syndrome. *J. Immunol.* **2020**, *204*, 59.17.
143. Liu, H.; Huang, Y.; Lyu, Y.; Dai, W.; Tong, Y.; Li, Y. GDF15 as a biomarker of ageing. *Exp. Gerontol.* **2021**, *146*, 111228. [CrossRef] [PubMed]
144. Tavenier, J.; Rasmussen, L.J.H.; Andersen, A.L.; Houlind, M.B.; Langkilde, A.; Andersen, O.; Petersen, J.; Nehlin, J.O. Association of GDF15 with Inflammation and Physical Function during Aging and Recovery after Acute Hospitalization: A Longitudinal Study of Older Patients and Age-Matched Controls. *J. Gerontol. Ser. A* **2021**, *76*, 964–974. [CrossRef]
145. Tomescu, C.; Chehimi, J.; Maino, V.C.; Montaner, L.J. Retention of viability, cytotoxicity, and response to IL-2, IL-15, or IFN-alpha by human NK cells after CD107a degranulation. *J. Leukoc. Biol.* **2009**, *85*, 871–876. [CrossRef] [PubMed]
146. Widowati, W.; Jasaputra, D.K.; Sumitro, S.B.; Widodo, M.A.; Mozef, T.; Rizal, R.; Kusuma, H.S.; Laksmitawati, D.R.; Murti, H.; Bachtiar, I.; et al. Effect of interleukins (IL-2, IL-15, IL-18) on receptors activation and cytotoxic activity of natural killer cells in breast cancer cell. *Afr. Health Sci.* **2020**, *20*, 822–832. [CrossRef] [PubMed]
147. Hammarsten, J.; Holm, J.; Schersten, T. Incidence of bacteria after skin wash of surgical patients in hospitals. *Lakartidningen* **1979**, *76*, 969–970.
148. Ligthart, G.J.; Corberand, J.X.; Fournier, C.; Galanaud, P.; Hijmans, W.; Kennes, B.; Müller-Hermelink, H.K.; Steinmann, G.G. Admission criteria for immunogerontological studies in man: The SENIEUR protocol. *Mech. Ageing Dev.* **1984**, *28*, 47–55. [CrossRef]
149. Wiencke, J.K.; Butler, R.; Hsuang, G.; Eliot, M.; Kim, S.; Sepulveda, M.A.; Siegel, D.; Houseman, E.A.; Kelsey, K.T. The DNA methylation profile of activated human natural killer cells. *Epigenetics* **2016**, *11*, 363–380. [CrossRef]
150. Schenk, A.; Bloch, W.; Zimmer, P. Natural Killer Cells—An Epigenetic Perspective of Development and Regulation. *Int. J. Mol. Sci.* **2016**, *17*, 326. [CrossRef] [PubMed]
151. Cichocki, F.; Miller, J.; Anderson, S.; Bryceson, Y. Epigenetic regulation of NK cell differentiation and effector functions. *Front. Immunol.* **2013**, *4*, 55. [CrossRef] [PubMed]
152. Merino, A.; Zhang, B.; Dougherty, P.; Luo, X.; Wang, J.; Blazar, B.R.; Miller, J.S.; Cichocki, F. Chronic stimulation drives human NK cell dysfunction and epigenetic reprograming. *J. Clin. Investig.* **2019**, *129*, 3770–3785. [CrossRef] [PubMed]
153. Zimmer, P.; Bloch, W.; Schenk, A.; Zopf, E.M.; Hildebrandt, U.; Streckmann, F.; Beulertz, J.; Koliamitra, C.; Schollmayer, F.; Baumann, F. Exercise-induced Natural Killer Cell Activation is Driven by Epigenetic Modifications. *Int. J. Sports Med.* **2015**, *36*, 510–515. [CrossRef] [PubMed]
154. Misale, M.S.; Witek Janusek, L.; Tell, D.; Mathews, H.L. Chromatin organization as an indicator of glucocorticoid induced natural killer cell dysfunction. *Brain Behav. Immun.* **2018**, *67*, 279–289. [CrossRef] [PubMed]

155. Jonkman, T.H.; Dekkers, K.F.; Slieker, R.C.; Grant, C.D.; Ikram, M.A.; van Greevenbroek, M.M.J.; Franke, L.; Veldink, J.H.; Boomsma, D.I.; Slagboom, P.E.; et al. Functional genomics analysis identifies T and NK cell activation as a driver of epigenetic clock progression. *Genome Biol.* **2022**, *23*, 24. [CrossRef]
156. Wyman, P.A.; Moynihan, J.; Eberly, S.; Cox, C.; Cross, W.; Jin, X.; Caserta, M.T. Association of Family Stress with Natural Killer Cell Activity and the Frequency of Illnesses in Children. *Arch. Pediatrics Adolesc. Med.* **2007**, *161*, 228–234. [CrossRef] [PubMed]
157. Holbrook, N.J.; Cox, W.I.; Horner, H.C. Direct suppression of natural killer activity in human peripheral blood leukocyte cultures by glucocorticoids and its modulation by interferon. *Cancer Res.* **1983**, *43*, 4019–4025. [PubMed]
158. Nair, M.P.; Schwartz, S.A. Immunomodulatory effects of corticosteroids on natural killer and antibody-dependent cellular cytotoxic activities of human lymphocytes. *J. Immunol.* **1984**, *132*, 2876–2882.
159. Parrillo, J.E.; Fauci, A.S. Comparison of the Effector Cells in Human Spontaneous Cellular Cytotoxicity and Antibody-Dependent Cellular Cytotoxicity: Differential Sensitivity of Effector Cells to In Vivo and In Vitro Corticosteroids. *Scand. J. Immunol.* **1978**, *8*, 99–107. [CrossRef]
160. Yiallouris, A.; Tsioutis, C.; Agapidaki, E.; Zafeiri, M.; Agouridis, A.P.; Ntourakis, D.; Johnson, E.O. Adrenal Aging and Its Implications on Stress Responsiveness in Humans. *Front. Endocrinol.* **2019**, *10*, 54. [CrossRef]
161. Luz, C.; Dornelles, F.; Preissler, T.; Collaziol, D.; da Cruz, I.M.; Bauer, M.E. Impact of psychological and endocrine factors on cytokine production of healthy elderly people. *Mech. Ageing Dev.* **2003**, *124*, 887–895. [CrossRef]
162. Gatti, G.; Cavallo, R.; Sartori, M.L.; del Ponte, D.; Masera, R.; Salvadori, A.; Carignola, R.; Angeli, A. Inhibition by cortisol of human natural killer (NK) cell activity. *J. Steroid Biochem.* **1987**, *26*, 49–58. [CrossRef]
163. Eddy, J.L.; Krukowski, K.; Janusek, L.; Mathews, H.L. Glucocorticoids regulate natural killer cell function epigenetically. *Cell. Immunol.* **2014**, *290*, 120–130. [CrossRef]
164. Duan, X.; Lu, J.; Wang, H.; Liu, X.; Wang, J.; Zhou, K.; Jiang, W.; Wang, Y.; Fang, M. Bidirectional factors impact the migration of NK cells to draining lymph node in aged mice during influenza virus infection. *Exp. Gerontol.* **2017**, *96*, 127–137. [CrossRef] [PubMed]
165. Hellstrand, K.; Hermodsson, S.; Strannegård, O. Evidence for a beta-adrenoceptor-mediated regulation of human natural killer cells. *J. Immunol.* **1985**, *134*, 4095–4099. [PubMed]
166. Sun, Z.; Hou, D.; Liu, S.; Fu, W.; Wang, J.; Liang, Z. Norepinephrine inhibits the cytotoxicity of NK92-MI cells via the β2-adrenoceptor/cAMP/PKA/p-CREB signaling pathway. *Mol. Med. Rep.* **2018**, *17*, 8530–8535. [CrossRef] [PubMed]
167. Stepien, H.; Zelazowski, P.; Döhler, K.D.; Pawlikowski, M. Neuropeptides and Natural Killer Cell Activity. In *Progress in Neuropeptide Research*; Birkhäuser: Basel, Switzerland, 1988; pp. 25–31. [CrossRef]
168. Koller, A.; Bianchini, R.; Schlager, S.; Münz, C.; Kofler, B.; Wiesmayr, S. The neuropeptide galanin modulates natural killer cell function. *Neuropeptides* **2017**, *64*, 109–115. [CrossRef] [PubMed]
169. Feistritzer, C.; Clausen, J.; Sturn, D.H.; Djanani, A.; Gunsilius, E.; Wiedermann, C.J.; Kähler, C.M. Natural killer cell functions mediated by the neuropeptide substance P. *Regul. Pept.* **2003**, *116*, 119–126. [CrossRef]
170. Mazzatenta, A.; Marconi, G.D.; Zara, S.; Cataldi, A.; Porzionato, A.; Di Giulio, C. In the carotid body, galanin is a signal for neurogenesis in young, and for neurodegeneration in the old and in drug-addicted subjects. *Front. Physiol.* **2014**, *5*, 427. [CrossRef]
171. Barbariga, M.; Rabiolo, A.; Fonteyne, P.; Bignami, F.; Rama, P.; Ferrari, G. The Effect of Aging on Nerve Morphology and Substance P Expression in Mouse and Human Corneas. *Investig. Ophthalmol. Vis. Sci.* **2018**, *59*, 5329–5335. [CrossRef]
172. Ni, F.; Sun, R.; Fu, B.; Wang, F.; Guo, C.; Tian, Z.; Wei, H. IGF-1 promotes the development and cytotoxic activity of human NK cells. *Nat. Commun.* **2013**, *4*, 1479. [CrossRef]
173. Bidlingmaier, M.; Friedrich, N.; Emeny, R.T.; Spranger, J.; Wolthers, O.D.; Roswall, J.; Körner, A.; Obermayer-Pietsch, B.; Hübener, C.; Dahlgren, J.; et al. Reference intervals for insulin-like growth factor-1 (igf-i) from birth to senescence: Results from a multicenter study using a new automated chemiluminescence IGF-I immunoassay conforming to recent international recommendations. *J. Clin. Endocrinol. Metab.* **2014**, *99*, 1712–1721. [CrossRef] [PubMed]
174. Sonntag, W.E.; Ramsey, M.; Carter, C.S. Growth hormone and insulin-like growth factor-1 (IGF-1) and their influence on cognitive aging. *Ageing Res. Rev.* **2005**, *4*, 195–212. [CrossRef]
175. Salminen, A. Feed-forward regulation between cellular senescence and immunosuppression promotes the aging process and age-related diseases. *Ageing Res. Rev.* **2021**, *67*, 101280. [CrossRef]
176. O'Connor, J.C.; McCusker, R.H.; Strle, K.; Johnson, R.W.; Dantzer, R.; Kelley, K.W. Regulation of IGF-I function by proinflammatory cytokines: At the interface of immunology and endocrinology. *Cell. Immunol.* **2008**, *252*, 91–110. [CrossRef] [PubMed]
177. Ma, S.; Tang, T.; Wu, X.; Mansour, A.G.; Lu, T.; Zhang, J.; Wang, L.S.; Caligiuri, M.A.; Yu, J. PDGF-D-PDGFRβ signaling enhances IL-15-mediated human natural killer cell survival. *Proc. Natl. Acad. Sci. USA* **2022**, *119*, e2114134119. [CrossRef] [PubMed]
178. Barrow, A.D.; Edeling, M.A.; Trifonov, V.; Luo, J.; Goyal, P.; Bohl, B.; Bando, J.K.; Kim, A.H.; Walker, J.; Andahazy, M.; et al. Natural Killer Cells Control Tumor Growth by Sensing a Growth Factor. *Cell* **2018**, *172*, 534–548.e19. [CrossRef] [PubMed]
179. Drubaix, I.; Giakoumakis, A.; Robert, L.; Robert, A.M. Preliminary data on the age-dependent decrease in basic fibroblast growth factor and platelet-derived growth factor in the human vein wall and in their influence on cell proliferation. *Gerontology* **1998**, *44*, 9–14. [CrossRef] [PubMed]
180. Benatti, B.B.; Silvério, K.G.; Casati, M.Z.; Sallum, E.A.; Nociti, F.H., Jr. Influence of aging on biological properties of periodontal ligament cells. *Connect. Tissue Res.* **2008**, *49*, 401–408. [CrossRef]

181. Zhou, L.; Dong, J.; Yu, M.; Yin, H.; She, M. Age-dependent increase of NF-κB translocation and PDGF-B expression in aortic endothelial cells of hypercholesterolemic rats. *Exp. Gerontol.* **2003**, *38*, 1161–1168. [CrossRef]
182. Slattery, K.; Gardiner, C.M. NK Cell Metabolism and TGF–Implications for Immunotherapy. *Front. Immunol.* **2019**, *10*, 2915. [CrossRef]
183. Wilson, E.B.; El-Jawhari, J.J.; Neilson, A.L.; Hall, G.D.; Melcher, A.A.; Meade, J.L.; Cook, G.P. Human tumour immune evasion via TGF-β blocks NK cell activation but not survival allowing therapeutic restoration of anti-tumour activity. *PLoS ONE* **2011**, *6*, e22842. [CrossRef] [PubMed]
184. Viel, S.; Marçais, A.; Guimaraes, F.S.-F.; Loftus, R.; Rabilloud, J.; Grau, M.; Degouve, S.; Djebali, S.; Sanlaville, A.; Charrier, E. TGF-β inhibits the activation and functions of NK cells by repressing the mTOR pathway. *Sci. Signal.* **2016**, *9*, ra19. [CrossRef] [PubMed]
185. Castriconi, R.; Cantoni, C.; Della Chiesa, M.; Vitale, M.; Marcenaro, E.; Conte, R.; Biassoni, R.; Bottino, C.; Moretta, L.; Moretta, A. Transforming growth factor β1 inhibits expression of NKp30 and NKG2D receptors: Consequences for the NK-mediated killing of dendritic cells. *Proc. Natl. Acad. Sci. USA* **2003**, *100*, 4120–4125. [CrossRef] [PubMed]
186. Lazarova, M.; Steinle, A. Impairment of NKG2D-Mediated Tumor Immunity by TGF-β. *Front. Immunol.* **2019**, *10*, 2689. [CrossRef] [PubMed]
187. Frippiat, C.; Chen, Q.M.; Zdanov, S.; Magalhaes, J.P.; Remacle, J.; Toussaint, O. Subcytotoxic H_2O_2 stress triggers a release of transforming growth factor-beta 1, which induces biomarkers of cellular senescence of human diploid fibroblasts. *J. Biol. Chem.* **2001**, *276*, 2531–2537. [CrossRef] [PubMed]
188. Forsey, R.J.; Thompson, J.M.; Ernerudh, J.; Hurst, T.L.; Strindhall, J.; Johansson, B.; Nilsson, B.O.; Wikby, A. Plasma cytokine profiles in elderly humans. *Mech. Ageing Dev.* **2003**, *124*, 487–493. [CrossRef]
189. Anstey, K.; Stankov, L.; Lord, S. Primary aging, secondary aging, and intelligence. *Psychol. Aging* **1993**, *8*, 562–570. [CrossRef] [PubMed]
190. Busse, E.W.; Pfeiffer, E. *Behavior and Adaptation in Late Life*; Little, Brown: Boston, MA, USA, 1977.
191. Zhimulev, I.F.; Belyaeva, E.S.; Bgatov, A.V.; Baricheva, E.M.; Vlassova, I.E. Cytogenetic and molecular aspects of position effect variegation in Drosophila melanogaster. II. Peculiarities of morphology and genetic activity of the 2B region in the T(1;2)dorvar7 chromosome in males. *Chromosoma* **1988**, *96*, 255–261. [CrossRef] [PubMed]
192. Zealley, B.; de Grey, A.D. Strategies for engineered negligible senescence. *Gerontology* **2013**, *59*, 183–189. [CrossRef] [PubMed]
193. Cabrera, Á.J.R. Zinc, aging, and immunosenescence: An overview. *Pathobiol. Aging Age-Relat. Dis.* **2015**, *5*, 25592. [CrossRef]
194. Prasad, A.S.; Fitzgerald, J.T.; Hess, J.W.; Kaplan, J.; Pelen, F.; Dardenne, M. Zinc deficiency in elderly patients. *Nutrition* **1993**, *9*, 218–224. [PubMed]
195. Stover, P.J. Vitamin B12 and older adults. *Curr. Opin. Clin. Nutr. Metab. Care* **2010**, *13*, 24–27. [CrossRef] [PubMed]
196. Wolters, M.; Ströhle, A.; Hahn, A. Age-associated changes in the metabolism of vitamin B(12) and folic acid: Prevalence, aetiopathogenesis and pathophysiological consequences. *Z. Gerontol. Geriatr.* **2004**, *37*, 109–135. [CrossRef] [PubMed]
197. Wintergerst, E.S.; Maggini, S.; Hornig, D.H. Contribution of selected vitamins and trace elements to immune function. *Ann. Nutr. Metab.* **2007**, *51*, 301–323. [CrossRef] [PubMed]
198. Haase, H.; Mocchegiani, E.; Rink, L. Correlation between zinc status and immune function in the elderly. *Biogerontology* **2006**, *7*, 421–428. [CrossRef] [PubMed]
199. Elmadfa, I.; Meyer, A.L. The Role of the Status of Selected Micronutrients in Shaping the Immune Function. *Endocr. Metab. Immune Disord. Drug Targets* **2019**, *19*, 1100–1115. [CrossRef] [PubMed]
200. Mikkelsen, K.; Apostolopoulos, V. Vitamin B12, Folic Acid, and the Immune System. In *Nutrition and Immunity*; Springer: Cham, Switzerland, 2019; pp. 103–114. [CrossRef]
201. Nour Zahi Gammoh, L.R. Zinc in Infection and Inflammation. *Nutrients* **2017**, *9*, 624. [CrossRef]
202. Mertens, K.; Lowes, D.A.; Webster, N.R.; Talib, J.; Hall, L.; Davies, M.J.; Beattie, J.H.; Galley, H.F. Low zinc and selenium concentrations in sepsis are associated with oxidative damage and inflammation. *Br. J. Anasthesia* **2015**, *114*, 990–999. [CrossRef]
203. Busse, B.; Bale, H.A.; Zimmermann, E.A.; Panganiban, B.; Barth, H.D.; Carriero, A.; Vettorazzi, E.; Zustin, J.; Hahn, M.; Ager, J.W., 3rd; et al. Vitamin D deficiency induces early signs of aging in human bone, increasing the risk of fracture. *Sci. Transl. Med.* **2013**, *5*, 193ra188. [CrossRef] [PubMed]
204. Chung, M.; Lee, J.; Terasawa, T.; Lau, J.; Trikalinos, T.A. Vitamin D with or without calcium supplementation for prevention of cancer and fractures: An updated meta-analysis for the U.S. Preventive Services Task Force. *Ann. Intern. Med.* **2011**, *155*, 827–838. [CrossRef]
205. Kruman, I.I.; Kumaravel, T.S.; Lohani, A.; Pedersen, W.A.; Cutler, R.G.; Kruman, Y.; Haughey, N.; Lee, J.; Evans, M.; Mattson, M.P. Folic Acid Deficiency and Homocysteine Impair DNA Repair in Hippocampal Neurons and Sensitize Them to Amyloid Toxicity in Experimental Models of Alzheimer's Disease. *J. Neurosci.* **2002**, *22*, 1752–1762. [CrossRef] [PubMed]
206. Clarke, R.; Smith, A.D.; Jobst, K.A.; Refsum, H.; Sutton, L.; Ueland, P.M. Folate, vitamin B12, and serum total homocysteine levels in confirmed Alzheimer disease. *Arch. Neurol.* **1998**, *55*, 1449–1455. [CrossRef] [PubMed]
207. Partearroyo, T.; Úbeda, N.; Montero, A.; Achón, M.; Varela-Moreiras, G. Vitamin B12 and Folic Acid Imbalance Modifies NK Cytotoxicity, Lymphocytes B and Lymphoprolipheration in Aged Rats. *Nutrients* **2013**, *5*, 4836–4848. [CrossRef] [PubMed]

208. Vassiliou, A.G.; Jahaj, E.; Pratikaki, M.; Keskinidou, C.; Detsika, M.; Grigoriou, E.; Psarra, K.; Orfanos, S.E.; Tsirogianni, A.; Dimopoulou, I.; et al. Vitamin D deficiency correlates with a reduced number of natural killer cells in intensive care unit (ICU) and non-ICU patients with COVID-19 pneumonia. *Hell. J. Cardiol.* **2021**, *62*, 381. [CrossRef] [PubMed]
209. Tamura, J.; Kubota, K.; Murakami, H.; Sawamura, M.; Matsushima, T.; Tamura, T.; Saitoh, T.; Kurabayshi, H.; Naruse, T. Immunomodulation by vitamin B12: Augmentation of $CD8^+$ T lymphocytes and natural killer (NK) cell activity in vitamin B12-deficient patients by methyl-B12 treatment. *Clin. Exp. Immunol.* **2001**, *116*, 28–32. [CrossRef] [PubMed]
210. Mariani, E.; Neri, S.; Cattini, L.; Mocchegiani, E.; Malavolta, M.; Dedoussis, G.V.; Kanoni, S.; Rink, L.; Jajte, J.; Facchini, A. Effect of zinc supplementation on plasma IL-6 and MCP-1 production and NK cell function in healthy elderly: Interactive influence of +647 MT1a and −174 IL-6 polymorphic alleles. *Exp. Gerontol.* **2008**, *43*, 462–471. [CrossRef] [PubMed]
211. Gill, H.S.; Rutherfurd, K.J.; Cross, M.L. Dietary probiotic supplementation enhances natural killer cell activity in the elderly: An investigation of age-related immunological changes. *J. Clin. Immunol.* **2001**, *21*, 264–271. [CrossRef] [PubMed]
212. Bunout, D.; Barrera, G.; Hirsch, S.; Gattas, V.; de la Maza, M.P.; Haschke, F.; Steenhout, P.; Klassen, P.; Hager, C.; Avendaño, M.; et al. Effects of a nutritional supplement on the immune response and cytokine production in free-living Chilean elderly. *JPEN J. Parenter. Enter. Nutr.* **2004**, *28*, 348–354. [CrossRef]
213. Martineau, A.R.; Jolliffe, D.A.; Hooper, R.L.; Greenberg, L.; Aloia, J.F.; Bergman, P.; Dubnov-Raz, G.; Esposito, S.; Ganmaa, D.; Ginde, A.A.; et al. Vitamin D supplementation to prevent acute respiratory tract infections: Systematic review and meta-analysis of individual participant data. *BMJ* **2017**, *356*, i6583. [CrossRef] [PubMed]
214. Chen, J.; Mei, K.; Xie, L.; Yuan, P.; Ma, J.; Yu, P.; Zhu, W.; Zheng, C.; Liu, X. Low vitamin D levels do not aggravate COVID-19 risk or death, and vitamin D supplementation does not improve outcomes in hospitalized patients with COVID-19: A meta-analysis and GRADE assessment of cohort studies and RCTs. *Nutr. J.* **2021**, *20*, 89. [CrossRef]
215. Dancer, R.C.; Parekh, D.; Lax, S.; D'Souza, V.; Zheng, S.; Bassford, C.R.; Park, D.; Bartis, D.G.; Mahida, R.; Turner, A.M.; et al. Vitamin D deficiency contributes directly to the acute respiratory distress syndrome (ARDS). *Thorax* **2015**, *70*, 617–624. [CrossRef] [PubMed]
216. Al-Beltagi, M.; Rowiesha, M.; Elmashad, A.; Elrifaey, S.M.; Elhorany, H.; Koura, H.G. Vitamin D status in preterm neonates and the effects of its supplementation on respirat.tory distress syndrome. *Pediatr. Pulmonol.* **2020**, *55*, 108–115. [CrossRef] [PubMed]
217. Mariani, E.; Ravaglia, G.; Forti, P.; Meneghetti, A.; Tarozzi, A.; Maioli, F.; Boschi, F.; Pratelli, L.; Pizzoferrato, A.; Piras, F.; et al. Vitamin D, thyroid hormones and muscle mass influence natural killer (NK) innate immunity in healthy nonagenarians and centenarians. *Clin. Exp. Immunol.* **1999**, *116*, 19–27. [CrossRef] [PubMed]
218. Lewis, E.D.; Meydani, S.N.; Wu, D. Regulatory role of vitamin E in the immune system and inflammation. *IUBMB Life* **2019**, *71*, 487–494. [CrossRef] [PubMed]
219. Wu, D.; Meydani, S.N. Age-associated changes in immune function: Impact of vitamin E intervention and the underlying mechanisms. *Endocr. Metab. Immune Disord. Drug Targets* **2014**, *14*, 283–289. [CrossRef]
220. Madison, A.A.; Shrout, M.R.; Renna, M.E.; Kiecolt-Glaser, J.K. Psychological and Behavioral Predictors of Vaccine Efficacy: Considerations for COVID-19. *Perspect. Psychol. Sci.* **2021**, *16*, 191–203. [CrossRef] [PubMed]
221. Cohen, S.; Tyrrell, D.A.; Smith, A.P. Psychological stress and susceptibility to the common cold. *N. Engl. J. Med.* **1991**, *325*, 606–612. [CrossRef]
222. Glaser, R.; Sheridan, J.; Malarkey, W.B.; MacCallum, R.C.; Kiecolt-Glaser, J.K. Chronic stress modulates the immune response to a pneumococcal pneumonia vaccine. *Psychosom. Med.* **2000**, *62*, 804–807. [CrossRef]
223. Kohut, M.L.; Lee, W.; Martin, A.; Arnston, B.; Russell, D.W.; Ekkekakis, P.; Yoon, K.J.; Bishop, A.; Cunnick, J.E. The exercise-induced enhancement of influenza immunity is mediated in part by improvements in psychosocial factors in older adults. *Brain Behav. Immun.* **2005**, *19*, 357–366. [CrossRef]
224. De Lorenzo, B.H.; de Oliveira Marchioro, L.; Greco, C.R.; Suchecki, D. Sleep-deprivation reduces NK cell number and function mediated by β-adrenergic signalling. *Psychoneuroendocrinology* **2015**, *57*, 134–143. [CrossRef]
225. Vedhara, K.; Ayling, K.; Sunger, K.; Caldwell, D.M.; Halliday, V.; Fairclough, L.; Avery, A.; Robles, L.; Garibaldi, J.; Welton, N.J.; et al. Psychological interventions as vaccine adjuvants: A systematic review. *Vaccine* **2019**, *37*, 3255–3266. [CrossRef] [PubMed]
226. Held, K.; Antonijevic, I.A.; Künzel, H.; Uhr, M.; Wetter, T.C.; Golly, I.C.; Steiger, A.; Murck, H. Oral Mg^{2+} supplementation reverses age-related neuroendocrine and sleep EEG changes in humans. *Pharmacopsychiatry* **2002**, *35*, 135–143. [CrossRef] [PubMed]
227. Abbasi, B.; Kimiagar, M.; Sadeghniiat, K.; Shirazi, M.M.; Hedayati, M.; Rashidkhani, B. The effect of magnesium supplementation on primary insomnia in elderly: A double-blind placebo-controlled clinical trial. *J. Res. Med. Sci.* **2012**, *17*, 1161–1169. [PubMed]
228. Mander, B.A.; Winer, J.R.; Walker, M.P. Sleep and Human Aging. *Neuron* **2017**, *94*, 19–36. [CrossRef] [PubMed]
229. Abbott, S.M.; Videnovic, A. Chronic sleep disturbance and neural injury: Links to neurodegenerative disease. *Nat. Sci. Sleep* **2016**, *8*, 55–61. [CrossRef] [PubMed]
230. Huang, T.; Mariani, S.; Redline, S. Sleep Irregularity and Risk of Cardiovascular Events: The Multi-Ethnic Study of Atherosclerosis. *J. Am. Coll. Cardiol.* **2020**, *75*, 991–999. [CrossRef]
231. Garbarino, S.; Lanteri, P.; Bragazzi, N.L.; Magnavita, N.; Scoditti, E. Role of sleep deprivation in immune-related disease risk and outcomes. *Commun. Biol.* **2021**, *4*, 1304. [CrossRef] [PubMed]
232. Guida, J.L.; Alfini, A.J.; Gallicchio, L.; Spira, A.P.; Caporaso, N.E.; Green, P.A. Association of objectively measured sleep with frailty and 5-year mortality in community-dwelling older adults. *Sleep* **2021**, *44*, zsab003. [CrossRef]

233. Cappuccio, F.P.; D'Elia, L.; Strazzullo, P.; Miller, M.A. Sleep duration and all-cause mortality: A systematic review and meta-analysis of prospective studies. *Sleep* **2010**, *33*, 585–592. [CrossRef]
234. Kripke, D.F.; Langer, R.D.; Elliott, J.A.; Klauber, M.R.; Rex, K.M. Mortality related to actigraphic long and short sleep. *Sleep Med.* **2011**, *12*, 28–33. [CrossRef]
235. Besedovsky, L.; Lange, T.; Haack, M. The Sleep-Immune Crosstalk in Health and Disease. *Physiol. Rev.* **2019**, *99*, 1325–1380. [CrossRef] [PubMed]
236. Besedovsky, L.; Lange, T.; Born, J. Sleep and immune function. *Pflug. Arch.* **2012**, *463*, 121–137. [CrossRef] [PubMed]
237. Matsumoto, Y.; Mishima, K.; Satoh, K.; Tozawa, T.; Mishima, Y.; Shimizu, T.; Hishikawa, Y. Total sleep deprivation induces an acute and transient increase in NK cell activity in healthy young volunteers. *Sleep* **2001**, *24*, 804–809. [PubMed]
238. Irwin, M.; McClintick, J.; Costlow, C.; Fortner, M.; White, J.; Gillin, J.C. Partial night sleep deprivation reduces natural killer and cellular immune responses in humans. *FASEB J.* **1996**, *10*, 643–653. [CrossRef] [PubMed]
239. Irwin, M.; Mascovich, A.; Gillin, J.C.; Willoughby, R.; Pike, J.; Smith, T.L. Partial sleep deprivation reduces natural killer cell activity in humans. *Psychosom. Med.* **1994**, *56*, 493–498. [CrossRef] [PubMed]
240. Fondell, E.; Axelsson, J.; Franck, K.; Ploner, A.; Lekander, M.; Bälter, K.; Gaines, H. Short natural sleep is associated with higher T cell and lower NK cell activities. *Brain Behav. Immun.* **2011**, *25*, 1367–1375. [CrossRef] [PubMed]
241. Shakhar, K.; Valdimarsdottir, H.B.; Guevarra, J.S.; Bovbjerg, D.H. Sleep, fatigue, and NK cell activity in healthy volunteers: Significant relationships revealed by within subject analyses. *Brain Behav. Immun.* **2007**, *21*, 180–184. [CrossRef]
242. De Lorenzo, B.H.P.; Novaes, E.B.R.R.; Paslar Leal, T.; Piqueira Garcia, N.; Martins Dos Santos, R.M.; Alvares-Saraiva, A.M.; Perez Hurtado, E.C.; Braga Dos Reis, T.C.; Duarte Palma, B. Chronic Sleep Restriction Impairs the Antitumor Immune Response in Mice. *Neuroimmunomodulation* **2018**, *25*, 59–67. [CrossRef] [PubMed]
243. Huang, J.; Song, P.; Hang, K.; Chen, Z.; Zhu, Z.; Zhang, Y.; Xu, J.; Qin, J.; Wang, B.; Qu, W.; et al. Sleep Deprivation Disturbs Immune Surveillance and Promotes the Progression of Hepatocellular Carcinoma. *Front. Immunol.* **2021**, *12*, 727959. [CrossRef] [PubMed]
244. Collins, K.P.; Geller, D.A.; Antoni, M.; Donnell, D.M.; Tsung, A.; Marsh, J.W.; Burke, L.; Penedo, F.; Terhorst, L.; Kamarck, T.W.; et al. Sleep duration is associated with survival in advanced cancer patients. *Sleep Med.* **2017**, *32*, 208–212. [CrossRef]
245. Palesh, O.; Aldridge-Gerry, A.; Zeitzer, J.M.; Koopman, C.; Neri, E.; Giese-Davis, J.; Jo, B.; Kraemer, H.; Nouriani, B.; Spiegel, D. Actigraphy-measured sleep disruption as a predictor of survival among women with advanced breast cancer. *Sleep* **2014**, *37*, 837–842. [CrossRef] [PubMed]
246. Edinger, J.D.; Arnedt, J.T.; Bertisch, S.M.; Carney, C.E.; Harrington, J.J.; Lichstein, K.L.; Sateia, M.J.; Troxel, W.M.; Zhou, E.S.; Kazmi, U.; et al. Behavioral and psychological treatments for chronic insomnia disorder in adults: An American Academy of Sleep Medicine clinical practice guideline. *J. Clin. Sleep Med.* **2021**, *17*, 255–262. [CrossRef] [PubMed]
247. Raymann, R.J.; Swaab, D.F.; Van Someren, E.J. Skin temperature and sleep-onset latency: Changes with age and insomnia. *Physiol. Behav.* **2007**, *90*, 257–266. [CrossRef] [PubMed]
248. Ko, Y.; Lee, J.Y. Effects of feet warming using bed socks on sleep quality and thermoregulatory responses in a cool environment. *J. Physiol. Anthr.* **2018**, *37*, 13. [CrossRef] [PubMed]
249. Haghayegh, S.; Khoshnevis, S.; Smolensky, M.H.; Diller, K.R.; Castriotta, R.J. Before-bedtime passive body heating by warm shower or bath to improve sleep: A systematic review and meta-analysis. *Sleep Med. Rev.* **2019**, *46*, 124–135. [CrossRef] [PubMed]
250. Deng, X.; Terunuma, H.; Nieda, M. Immunosurveillance of Cancer and Viral Infections with Regard to Alterations of Human NK Cells Originating from Lifestyle and Aging. *Biomedicines* **2021**, *9*, 557. [CrossRef]
251. d'Adda di Fagagna, F. Living on a break: Cellular senescence as a DNA-damage response. *Nat. Rev. Cancer* **2008**, *8*, 512–522. [CrossRef] [PubMed]
252. Kirkland, J.L.; Tchkonia, T. Clinical strategies and animal models for developing senolytic agents. *Exp. Gerontol.* **2015**, *68*, 19–25. [CrossRef] [PubMed]
253. Kirkland, J.L.; Tchkonia, T. Senolytic drugs: From discovery to translation. *J. Intern. Med.* **2020**, *288*, 518–536. [CrossRef] [PubMed]
254. Johmura, Y.; Yamanaka, T.; Omori, S.; Wang, T.W.; Sugiura, Y.; Matsumoto, M.; Suzuki, N.; Kumamoto, S.; Yamaguchi, K.; Hatakeyama, S.; et al. Senolysis by glutaminolysis inhibition ameliorates various age-associated disorders. *Science* **2021**, *371*, 265–270. [CrossRef]
255. Muñoz, D.P.; Yannone, S.M.; Daemen, A.; Sun, Y.; Vakar-Lopez, F.; Kawahara, M.; Freund, A.M.; Rodier, F.; Wu, J.D.; Desprez, P.Y.; et al. Targetable mechanisms driving immunoevasion of persistent senescent cells link chemotherapy-resistant cancer to aging. *JCI Insight* **2019**, *5*, e124716. [CrossRef] [PubMed]
256. Zingoni, A.; Cecere, F.; Vulpis, E.; Fionda, C.; Molfetta, R.; Soriani, A.; Petrucci, M.T.; Ricciardi, M.R.; Fuerst, D.; Amendola, M.G.; et al. Genotoxic Stress Induces Senescence-Associated ADAM10-Dependent Release of NKG2D MIC Ligands in Multiple Myeloma Cells. *J. Immunol.* **2015**, *195*, 736–748. [CrossRef] [PubMed]
257. Pereira, B.I.; Devine, O.P.; Vukmanovic-Stejic, M.; Chambers, E.S.; Subramanian, P.; Patel, N.; Virasami, A.; Sebire, N.J.; Kinsler, V.; Valdovinos, A.; et al. Senescent cells evade immune clearance via HLA-E-mediated NK and CD8$^+$ T cell inhibition. *Nat. Commun.* **2019**, *10*, 2387. [CrossRef] [PubMed]
258. Valipour, B.; Velaei, K.; Abedelahi, A.; Karimipour, M.; Darabi, M.; Charoudeh, H.N. NK cells: An attractive candidate for cancer therapy. *J. Cell. Physiol.* **2019**, *234*, 19352–19365. [CrossRef] [PubMed]

259. Barrow, A.D.; Colonna, M. Tailoring Natural Killer cell immunotherapy to the tumour microenvironment. *Semin. Immunol.* **2017**, *31*, 30–36. [CrossRef] [PubMed]
260. Bachanova, V.; Miller, J.S. NK cells in therapy of cancer. *Crit. Rev. Oncog.* **2014**, *19*, 133–141. [CrossRef] [PubMed]
261. Amor, C.; Feucht, J.; Leibold, J.; Ho, Y.J.; Zhu, C.; Alonso-Curbelo, D.; Mansilla-Soto, J.; Boyer, J.A.; Li, X.; Giavridis, T.; et al. Senolytic CAR T cells reverse senescence-associated pathologies. *Nature* **2020**, *583*, 127–132. [CrossRef] [PubMed]
262. Islam, R.; Pupovac, A.; Evtimov, V.; Boyd, N.; Shu, R.; Boyd, R.; Trounson, A. Enhancing a Natural Killer: Modification of NK Cells for Cancer Immunotherapy. *Cells* **2021**, *10*, 1058. [CrossRef]
263. Wang, L.; Dou, M.; Ma, Q.; Yao, R.; Liu, J. Chimeric antigen receptor (CAR)-modified NK cells against cancer: Opportunities and challenges. *Int. Immunopharmacol.* **2019**, *74*, 105695. [CrossRef] [PubMed]
264. Lang, P.O.; Mitchell, W.A.; Lapenna, A.; Pitts, D.; Aspinall, R. Immunological pathogenesis of main age-related diseases and frailty: Role of immunosenescence. *Eur. Geriatr. Med.* **2010**, *1*, 112–121. [CrossRef]
265. Horowitz, A.; Strauss-Albee, D.M.; Leipold, M.; Kubo, J.; Nemat-Gorgani, N.; Dogan, O.C.; Dekker, C.L.; Mackey, S.; Maecker, H.; Swan, G.E.; et al. Genetic and Environmental Determinants of Human NK Cell Diversity Revealed by Mass Cytometry. *Sci. Transl. Med.* **2013**, *5*, 208ra145. [CrossRef] [PubMed]
266. Benjamin, J.E.; Gill, S.; Negrin, R.S. Biology and clinical effects of natural killer cells in allogeneic transplantation. *Curr. Opin. Oncol.* **2010**, *22*, 130–137. [CrossRef] [PubMed]
267. Lupo, K.B.; Matosevic, S. Natural Killer Cells as Allogeneic Effectors in Adoptive Cancer Immunotherapy. *Cancers* **2019**, *11*, 769. [CrossRef] [PubMed]
268. Rezvani, K.; Rouce, R.; Liu, E.; Shpall, E. Engineering Natural Killer Cells for Cancer Immunotherapy. *Mol. Ther.* **2017**, *25*, 1769–1781. [CrossRef] [PubMed]
269. Labanieh, L.; Majzner, R.G.; Mackall, C.L. Programming CAR-T cells to kill cancer. *Nat. Biomed. Eng.* **2018**, *2*, 377–391. [CrossRef]
270. Pfefferle, A.; Huntington, N.D. You Have Got a Fast CAR: Chimeric Antigen Receptor NK Cells in Cancer Therapy. *Cancers* **2020**, *12*, 706. [CrossRef] [PubMed]
271. Xie, G.; Dong, H.; Liang, Y.; Ham, J.D.; Rizwan, R.; Chen, J. CAR-NK cells: A promising cellular immunotherapy for cancer. *EBioMedicine* **2020**, *59*, 102975. [CrossRef] [PubMed]

cells MDPI

Review

Induction of Accelerated Aging in a Mouse Model

Nanshuo Cai †, Yifan Wu † and Yan Huang *

MOE Key Laboratory of Gene Function and Regulation, Guangzhou Key Laboratory of Healthy Aging Research and State Key Laboratory of Biocontrol, School of Life Sciences, Sun Yat-sen University, Guangzhou 510275, China; cainsh@mail2.sysu.edu.cn (N.C.); wuyf97@mail2.sysu.edu.cn (Y.W.)
* Correspondence: huangy336@mail.sysu.edu.cn
† These authors contributed equally to this work.

Abstract: With the global increase of the elderly population, the improvement of the treatment for various aging-related diseases and the extension of a healthy lifespan have become some of the most important current medical issues. In order to understand the developmental mechanisms of aging and aging-related disorders, animal models are essential to conduct relevant studies. Among them, mice have become one of the most prevalently used model animals for aging-related studies due to their high similarity to humans in terms of genetic background and physiological structure, as well as their short lifespan and ease of reproduction. This review will discuss some of the common and emerging mouse models of accelerated aging and related chronic diseases in recent years, with the aim of serving as a reference for future application in fundamental and translational research.

Keywords: aging; aging-related disease; mouse model

Citation: Cai, N.; Wu, Y.; Huang, Y. Induction of Accelerated Aging in a Mouse Model. *Cells* **2022**, *11*, 1418. https://doi.org/10.3390/cells11091418

Academic Editors: Nicole Wagner and Kay-Dietrich Wagner

Received: 26 March 2022
Accepted: 20 April 2022
Published: 22 April 2022

1. Introduction

The world's population continues to grow, with those over 65 being the fastest growing age group. In 2019, one out of every eleven people in the worldwide population was over the age of 65. By 2050, this proportion will increase to one in every six people over the age of 65. In some regions, such as in Europe and North America, one in four people will be over 65 years old. Aging is a key risk factor for multiple disorders. More than 75% of the elderly suffer from at least one chronic disease, and the problem of "unhealthy longevity" for the elderly is prominent [1,2].

As countries face the challenges of aging populations, there is a need to promote healthy aging and provide adequate social protection. To promote a healthy aging process and prevent aging-related health problems, correctly understanding aging mechanisms and developing effective and affordable intervention strategies for anti-aging have great social significance and huge economic benefits. During this process, an animal model of aging is a powerful tool for us to study the mechanism of aging.

Generally, aging models are divided into two categories: natural aging models and accelerated aging models. In the process of aging, naturally aging mice develop many phenotypes similar to normal human aging like cataracts [3] and muscle weakness [4]. For example, the most well-known strain of mice, C57BL/6, have a lifespan of two to three years, while naked mole-rats can live up to 30 years. One longevity mechanism of naked mole-rats is their effective DNA damage repair system. However, research on natural aging model is time-consuming, labor-intensive, and expensive, with large individual variations compared to the accelerated aging models [5,6]. As such, the induced accelerated aging model is favored by researchers because of its convenient source, short modeling time, and relatively controllable aging effect. For aging and aging-related chronic disorders, mice have become one of the most important animal models for studying various aging-related human diseases due to their similarities to humans in terms of genetic background and the structure and function of various organs or systems, as well as the advantages of short lifespan and ease of reproduction [6,7]. With the help of mouse models, researchers have revealed the mechanisms of aging as well as the pathogenesis of various chronic diseases, and many therapeutic approaches for such chronic diseases have been validated in mouse models preclinically.

In this review, we discuss some common and emerging mouse models of accelerated aging and its related chronic diseases in recent years, with the aim of serving as a reference for future applications.

2. Systemic-Induced Accelerated Aging Mouse Model

In this section, we will review some commonly used mouse models of accelerated systemic aging including drug treatment, genetic engineered models, irradiation induction, etc. They are characterized by an aging phenotype in multiple tissues or organs, reflecting systemic aging. The systemic-induced accelerated aging mouse models are summarized in Table 1.

Table 1. Summary of systemic-induced accelerated aging mouse models.

Type	Subdivision	Phenotypes
D-galactose-induced senescence model	Brain	Cognitive impairment Mitochondrial dysfunction Neuronal degeneration Apoptosis Depressive and anxious
	Heart	Cardiac fibrosis Collagen accumulation Fibroblasts disordered arrangement
	Kidney	Kidney index ↓ Uric acid & Cys-C ↑ Glomerular and tubular damage ↑
	Liver	Liver fibrosis Glycogen levels ↓ Lipid deposition ↑
	Reproductive system	Estrogen and progesterone ↓ Ovarian follicle regression Uterine wall endometrial gland atrophy Disrupt estrous cycles
	Intestinal flora	Disturbance
	Lung	Oxidative stress ↑ Fibrotic status Chronic inflammation
SAMP mice	SAMP 1	Aging amyloidosis Immune dysfunction Renal atrophy Hearing loss Senile pulmonary hyperinflation
	SAMP 6	Senile osteoporosis Myeloid progenitor cell senescence
	SAMP 8	Astrogliosis Microgliosis Neurodegeneration Amyloid accumulation MAPT hyperphosphorylation
	SAMP 10	Learning and memory impairment Cerebral cortex and limbic system atrophy

Table 1. *Cont.*

Rps9 D95N mouse		Altered fur Cataracts Hunched posture Body composition function & body weight ↓ Fat mass & muscle strength ↓ Shortened lifespan Mouse urinary syndrome Extramedullary hematopoiesis
Progeria syndrome mouse	*Lama*$^{-/-}$	Short lifespan Growth retardation Muscular dystrophy Altered lipid metabolism
	Wrn$^{\Delta hel/\Delta hel}$	Short lifespan Abnormal hyaluronic acid excretion Metabolic abnormalities Increased genomic instability and cancer incidence
	Csa$^{-/-}$, *Csb*$^{-/-}$	Short lifespan Reduced fat mass Photoreceptor cell loss Neural pathology
	Xpd$^{TTD/TTD}$	Short lifespan Trichothiodystrophy
	Bub1b$^{H/H}$, *Xpg*$^{-/-}$	Brain atrophy Neuronal loss Neurofibrillary deposition of Aβ or senile plaques
	Bub1b$^{H/H}$, *Bub1b*$^{+/GTTA}$	Mean muscle fiber diameter ↓ Muscle fiber size variation ↑ Intermuscular fibrosis Regenerative capacity of skeletal muscles ↓
Mitochondrial DNA polymerase mutant mouse		Lifespan ↓ Weight loss Subcutaneous fat ↓ Hair loss Kyphosis Osteoporosis Anemia Fertility ↓ Spermatogonia depletion Heart enlargement
Total body irradiation (TBI) model		Progressive premature frailty Cognitive decline Whole blood antioxidant capacity ↓ RBC glutathione ↓ Thymic involution Articular cartilage and bone degeneration Ovarian environment damage
Ozone-induced senescence model		Cognitive decline Memory impairment AD symptoms Lung tumor growth ↑
Chronic jet-lag mouse		Accelerated initial tumor growth Shortened mouse survival Induce spontaneous hepatocellular carcinoma Obesity Depression Addiction Abnormal cardiac structure Impaired cardiac function

2.1. The D-Galactose-Induced Senescence Model

D-galactose is a common aldohexose that exists naturally in the body and in daily foods [8]. After ingestion, a healthy adult can metabolize and eliminate a maximum daily dose of 50 g of galactose from the body within about eight hours [9]. However, when galactose accumulates to high levels, reactive oxygen species (ROS) are generated by mitochondrial respiratory chain enzymes, xanthine oxidase, lipoxygenase, cyclooxygenase, nitric oxide synthase, and peroxidase. Increased ROS can subsequently lead to elevated oxidative stress and inflammation, inducing mitochondrial dysfunction and apoptosis [10]. Meanwhile, the elevated mitochondrial ROS level can lead to the activation of many biochemical pathways, such as the polyol pathway, the formation of advanced glycation end products (AGEs), the activation of protein kinase C, and the hexosamine pathway [11,12]. Overall, D-galactose-induced methods can increase aging markers such as AGEs, receptors for advanced glycation end products (RAGEs), aldose reductase (AR), sorbitol dehydrogenase (SDH), decreased telomerase activity, shortened telomere, β-site amyloid precursor protein cleaving enzyme 1 (BACE-1), amyloid β (Aβ), aging-related pathways (p16, p21, p53, etc.), and positive senescence-associated β-galactosidase (SA-β-gal) staining [13]. Multiple tissues and organs, including the brain, heart, lung, liver, kidney, reproductive system, gastrointestinal system, and so on, manifest aging phenotypes after D-galactose treatment [14].

Previous works have shown that D-galactose induces brain aging by increasing mitochondrial dysfunction, oxidative stress, inflammation, apoptosis, and decreasing the expression of brain-derived neurotrophic factor. D-galactose injections may induce brain aging similar to human brain aging in many ways, including mitochondrial dysfunction, increased oxidative stress, decreased ATP production, neuronal degeneration and apoptosis, and cognitive deficits [13,15–17]. D-galactose increases the neuro-inflammation markers via activating NF-κB, leading to memory impairment [18,19]. Besides learning and memory inhibition, D-galactose-treated mice also exhibit depressive and anxious behaviors [20].

The leading cause of death in elderly people worldwide is cardiovascular disease [21]. D-galactose treatment increases the risk of cardiovascular disease, which is associated with excess ROS and oxidative stress. Persistent oxidative stress has been revealed to be related to decreased ferric reducing antioxidant power and lower activity of Cu-Zn superoxide dismutase, leading to myocardial damage [22]. Studies have shown that galactose reduces endogenous hydrogen sulphide producing enzyme cystathionine γ-lyase (CSE) [23] and antioxidant enzymes such as catalase (CAT), haem oxygenase-1 (HO-1), superoxide dismutase (SOD), glutathione peroxidase (GSH-Px), and nitric oxide synthase (NOS), leading to decreased total antioxidant capacity and inducing lipid peroxidation markers including malondialdehyde (MDA), lipid hydroperoxides (L-OOH), and conjugated dienes (CD) in cardiac tissue [24,25]. D-galactose increases whole heart weight and left ventricle weight, which is associated with hypertension and aging. At the same time, the heart tissue showed enlarged myocardial fibers, blurred structure, shortened distortion, widening of the interval, and obvious capillaries of myocardial interstitial congestion [26]. D-galactose treatment resulted in cardiac fibrosis, significant accumulated collagen, and disordered arrangement of fibroblasts compared with the control. D-galactose also increased cardiac apoptosis markers [27]. Excessive D-galactose can be transformed to advanced AGEs via the Maillard reaction [28]. AGEs bind to the receptors, RAGE, increasing ROS production via NADPH oxidase. NADPH oxidase further activates p38 MAP kinases, causing transcription factor NF-κB translocating into the nucleus, where transcription of inflammatory cascades like tumor necrosis factor alpha (TNF-α) are enhanced [29]. D-galactose treatment can also increase fibrotic markers such as connective tissue growth factor (CTGF), transforming growth factor β1 (TGF-β1), phosphorylated mitogen-activated protein kinase 1/2 (p-MEK1/2), phosphorylated extracellular signal-regulated kinase 1/2 (p-ERK1/2), matrix metalloproteinase (MMP), and pathological specific protein 1 (SP1) [30].

D-galactose treatment also increases oxidative markers (e.g., MDA) and decreases antioxidant enzymes (e.g., SOD, GSH-Px and NOS) in lung, liver, and renal tissues [26,31–35]. D-galactose treatment can affect lung elastic constitution. The primary effects of treatment on the lungs are increased alveolar size and reduced elastic recoil, which may facilitate airway closure [36]. D-galactose treatment is considered to successfully mimic the natural aging process by increasing oxidative stress, fibrotic status, and chronic inflammation in the lungs.

The liver is the main site where D-galactose is metabolized, thus excess D-galactose in the body may significantly affect the liver. As previously mentioned, high levels of D-galactose can react with free amines in the amino acids to form AGEs, which is found to be involved in the progression of various liver diseases [12]. High levels of D-galactose lead to the accumulation of its final metabolite, galactitol, which will eventually lead to ROS accumulation through the p38 MAPK/NRF2/HO-1 signaling pathway and cause cellular osmotic stress in the liver [37]. D-galactose treatment significantly increased apoptotic proteins including Bax, procaspase-3, and caspase-3, and raised the ratio of Bax/Bcl-2 in the liver tissue [34,38,39]. Various changes similar to natural aging were also observed in the D-galactose-treated livers. Compared with controls, the livers of D-galactose-treated animals had lower levels of glycogen and more lipid deposition. Masson's trichrome staining showed obvious collagen fibers around blood vessels and in the interphase of liver tissue, and irregular pseudo-lobules around liver tissue, indicating a state of liver fibrosis [40,41].

D-galactose treatment significantly reduced the renal index of animals, and markers of acute kidney injury such as uric acid and cystatin C (Cys-C) were also increased [42]. In the kidney, D-galactose administration resulted in an increase in the TUNEL-positive cells and DNA fragmentation. In addition, p21 expression and the staining intensity of SA-β-gal were also increased in kidney cells [43]. Extensive glomerular and tubular damages were detected in D-galactose-treated animals, as the number of tubules with cellular necrosis from the renal cortices and outer medulla were significantly increased [43,44].

In the male reproductive system, D-galactose induced oxidative stress, marked by an increase in MDA levels in the prostate, testis, epididymis, and decrease in SOD activity in the testis. Peroxidation in the testicular and epididymal mitochondrial fractions was also significantly increased after D-galactose treatment [45]. Female reproductive aging is characterized by decreased levels of estrogen, progesterone, inhibin B, anti-Müllerian hormone, and androgens, which include free testosterone, dehydroepiandrosterone (DHEAS), and androstenedione [46]. Compared to the control groups, D-galactose administration produced aging-associated changes like reduced estrogen and progesterone levels and increased FSH and luteinizing hormone (LH) levels. Ovarian follicle regression and atrophy on the uterine wall and endometrial gland were observed after D-galactose treatment, which indicates disrupted estrous cycles and damaged uterine and ovarian tissues [47].

D-galactose injection can lead to changes in the level of oxidative stress that affect the microbial environment in the intestine and lead to intestinal flora disorder [48]. The ecology of intestinal flora is closely related to aging, and intestinal ecological disturbance can lead to accelerated aging and a shortened lifespan [49]. Transferring gut microbiota from aged to young germ-free mice triggered innate immune and inflammatory responses. Effects of aging include increased differentiation of $CD4^+$ T cells in the spleen, upregulation of intestinal inflammatory cytokine such as TNF-α, and increased cycling of bacterial-derived inflammatory cytokines [50,51]. In addition, D-galactose administration significantly decreased the small intestine propulsion rates and prolonged gastrointestinal transit time [49]. The aging gut triggers chronic inflammation, leading to gut dysplasia and intestinal dysplasia in turn leads to defective epithelial function, predisposing the host to infection, neoplasia, and increased mortality [52].

2.2. *Senescence-Accelerated Mouse/Prone*

Researchers from Kyoto University found an aging phenotype from a subset of pups while maintaining an inbred line of AKR/J mice. Characteristics of these mice include hair loss, reduced activity, and a shortened lifespan. These aging traits are thought to develop as a result of elevated oxidative stress and are inherited by their offspring. Further, the accelerated aging mice were grouped into several distinct subtypes according to their phenotypes [53].

Senescence-accelerated mouse/prone (SAMP) is a group of inbred mouse strains that typically exhibit accelerated aging [54]. Meanwhile, since it shows various aging-related diseases similar to humans, such as aging amyloidosis, senile osteoporosis, learning/memory impairment, etc., in specific lines, it is widely used for aging research. Cellular senescence in various cell types, including astrocytes, endothelial cells, progenitor cells, retinal epithelial cells, and fibroblasts, was found in the aging SAMP mice [55–57]. SAMP mice exhibit an increase in ROS generated by mitochondria or other cellular sites, which not only causes damage to mitochondria, but also triggers degradation that leads to the aging outcome [58].

SAMP1 mice are characterized by aging amyloidosis, immune dysfunction, renal atrophy, hearing loss, and senile pulmonary hyperinflation [59,60]. The lifespan of the SAMP1 mice is about 40% shorter than senescence-accelerated resistant mice (SAMR1), and various signs of aging appear early in appearance.

The SAMP6 mice model is a senile osteoporosis model animal that tends to develop osteoporosis at an early stage with aging due to its low bone density in instar [61]. Bone marrow transplantation experiments have revealed that the cause of SAMP6 osteoporosis is abnormalities in bone marrow stem cells [62]. The incidence of spontaneous leg fractures due to osteoporosis is high in adult SAMP6 mice. Cellular senescence in myeloid progenitors disrupts their differentiation in favor of adipogenesis over osteogenesis. This mechanism is thought to contribute to low osteoblast activity and osteoporosis in SAMP6 mice and the elderly [61,63,64].

The SAMP8 mice develop age-associated deficits in learning and memory and also exhibit various age-related neuropathological changes similar to aging humans [65,66]. Neuropathological changes including astrogliosis, microgliosis, and neurodegeneration occur as early as five months of age [67]. SAMP8 mice also showed accumulated amyloid and age-related microtubule-associated protein tau (MAPT) hyperphosphorylation, as well as increased nitric oxide synthase activity, further demonstrating their feasibility as a brain aging model [68,69]. In addition, decreased activities in SOD, CAT, glutathione reductase, and GSH-Px, and increased activity in acyl-CoA oxidase were detected in SAMP8 mice at 1–12 months of age [70]. SAMP6 and SAMP8 mice also develop many other age-related diseases, including retinal degeneration, testosterone deficiency [71], myocardial fibrosis [72], and hepatic lipid deposition [73].

The SAMP10 mice exhibit learning and memory impairment with aging, and atrophy is observed in the cerebral cortex and limbic system, so it is considered to be a spontaneous model animal for aging and brain degeneration. The atrophy of the frontal cerebral cortex and olfactory bulb is marked in SAMP10 mice [74,75].

So far, more than ten strains of SAMP mice have been identified and widely used in aging studies, each of which can develop various age-related diseases such as renal fibrosis (shrinking of the kidneys), immune dysfunction, and degenerative joint disease like osteoarthritis (OA), etc.

2.3. *Rps9 D95N Mouse*

Rps9 D95N is a ribosome ambiguity mutation that causes error-prone protein synthesis in mammalian ribosomes, resulting in increased error-prone translation. Rps9 D95N mutant mice exhibit features of accelerated aging, including morphological (altered fur, cataracts, and hunched posture), physiological (body composition and function, body weight, fat mass, and muscle strength), and pathological (shortened lifespan, mouse urinary syndrome and extramedullary hematopoiesis), which also complements and explains the link between accumulation of erroneous proteins resulting from protein mistranslation and individual aging [76].

2.4. *Progeria Syndrome Mouse*

Mouse models of progeria syndrome have emerged as an attractive tool for evaluating intervention strategies for unhealthy aging due to their short lifespan, relatively simple strategies by single gene deletion or mutation, and their strong phenotypic similarities to normal aging [7]. To understand important mechanisms of the aging process, progeria mice ($Lmna^{-/-}$, $Wrn^{\Delta hel/\Delta hel}$, $Csa^{-/-}$ or $Csb^{-/-}$) from common progeria types such as Hutchinson-Gilford Progeria, Werner Syndrome, and Cockayne Syndrome are being widely used.

The most common phenotypes in mouse models of progeria can be observed in bones, joints, skin, nervous system, adipose tissue, skeletal muscle, cardiovascular system, liver, kidney, and the hematopoietic system. Less common lesions occur in the gonads, eyes, and occasionally in the gastrointestinal tract [77].

Some human progeria syndromes like Werner Syndrome exhibit osteoporosis. Current data from a mouse model of progeria indicate that senile osteoporosis is the result of reduced bone turnover and loss of bone mass due to defects in osteoblast progenitor cells, osteoblast differentiation, or osteoblast function [77]. Degenerative joint disease is another major symptom that affects the elderly. Mutation in the *Xpd* gene of nucleotide excision repair (NER) leads to a short lifespan, causing trichothiodystrophy (TTD) in humans. As expected, $Xpd^{TTD/TTD}$ mice exhibited a significant decrease in subchondral bone plate thickness compared to that observed in wild-type mice. Surprisingly, female $Xpd^{TTD/TTD}$ mice exhibited less cartilage damage and fewer lost articular cartilage, compared to WT females [78].

During the aging process, skeletal muscle was inevitably accompanied with a reduction in muscle mass, known as sarcopenia or age-related muscle atrophy, and macroscopic examination of sarcopenic mice will show weight loss and marked reduction in muscle mass. Sarcopenia is characterized by the loss of muscle fibers and smaller fiber cross-sectional area that is defined as fiber atrophy [79]. The progeria mouse model $Bub1b^{H/H}$ and $Bub1b^{+/GTTA}$ mice showed decreased mean muscle fiber diameter, increased myofiber size variation, increased intermuscular fibrosis, and impaired regenerative capacity in skeletal muscles [80].

Common brain aging diseases include Alzheimer's disease (AD) and cognitive dysfunction syndrome. Brain atrophy, neuronal loss, neurofibrillary deposition of Aβ or senile plaques, intraneuronal tauopathies (neurofibrillary tangles, NFTs), cerebrovascular amyloid angiopathy, neuronal lipofuscinosis, vascular and meningeal calcification, decreased white matter integrity, and astrogliosis are common age-related neurological pathologies [81]. The brains of several mouse models of progeria show neurodegenerative changes at an early stage. Similar to age-related gliosis, $Bub1b^{H/H}$ mice have increased numbers of astrocytes and microglia at one and five months of age, respectively. Additionally, $Xpg^{-/-}$ mice developed more astrocytes and increased activation of microglia in the brain and spinal cord [82].

2.5. Mitochondrial DNA Polymerase Mutant Mouse

PolgA is the catalytic subunit of mtDNA polymerase encoded in the nucleus. Mice that were knocked-in proofreading-deficient *PolgA* developed a mutator phenotype with an over threefold increase in the mtDNA levels with point mutations and deletions. The mtDNA polymerase mutant mouse manifest reduced lifespan and premature phenotypes including weight loss, hair loss, reduced subcutaneous fat, kyphosis, osteoporosis, anemia, infertility, depletion of spermatogonia, heart hypertrophy, etc.

Besides, the mtDNA polymerase deficient mouse showed age-related loss of skeletal muscle likely contributing to sarcopenia, and age-related loss of spiral ganglion neurons, which is recognized as a feature of presbycusis.

Mutations in mtDNA may contribute to premature aging of the organism through apoptotic loss of critical, irreplaceable cells and thus induce tissue dysfunction. The *PolgA* mutation mice showed increased TUNEL-positive cells in intestinal epithelial, thymic, and testicular tissues. Compared to wild type mice, cleaved caspase-3 levels were significantly higher in the cytoplasmic fractions of viscera from *PolgA* mutation mice. Thus, mtDNA mutation-induced senescence was associated with the activation of apoptotic pathways in these tissues [83,84].

2.6. Total Body Irradiation (TBI) Model

Radiotherapy and chemotherapy are commonly used in anti-tumor therapy. However, they result in long-term damage to a variety of organ systems, including the cardiovascular system, gastrointestinal tract, lung, liver, musculoskeletal system, and neurological network. These treatments cause tissue phenotypes resembling accelerated aging [85,86]. Studies showed that sub-lethal total body irradiation (TBI) in mice induces progressive premature frailty and damage various tissues and organs. Radiation-induced frailty can be used to predict increased mortality and is associated with cognitive decline. Compared to control, TBI mice gradually developed frailty at a time approximately two times earlier and the frailty phenotype of irradiated mice was similar to those without irradiation at a higher age [87].

TBI may induce cellular and tissue damage by direct ionization effects and indirect free radicals generated during water radiolysis [88]. After TBI, changes of whole blood antioxidant capacity and red blood cell (RBC) glutathione (GSH) levels happened within two months in mice. The GSH levels and GSH/oxidized glutathione (GSSG) ratio in RBC decreased chronically after TBI ≥ 1 Gy [89]. Following a single and relatively low dose of TBI, the number and ability of hematopoietic stem cells (HSCs) to generate T cells is markedly and permanently impaired, thereby accelerating aging-related thymic involution [90].

Acute stressors such as radiation can cause senescent cells to accumulate in the elderly and animals. These senescent cells continue to secrete proteins that induce chronic inflammation, break tissue homeostasis, and disrupt organ function, a phenomenon called senescence-associated secretory phenotype (SASP) [91]. Individuals that received acute total body irradiation of approximately 1.5 Gy or more will suffer from bone marrow dysfunction and pancytopenia, the magnitude of which increases gradually with increasing intensity of irradiation [92].

After TBI of adult rats exposed prior to skeletal maturation, late degenerative joint damages in articular cartilage and trabecular bone were detected with a non-invasive imaging techniques [93]. In vitro, radiation exposure has been shown to cause acute degenerative alterations in the cartilage matrix composition and metabolism via lower proteoglycan (PG) content and compressive stiffness [94,95]. Radiation can also cause damage to the ovarian environment as well [96]. In adult female mice, drastic primordial follicle loss was observed in serial sections of ovaries even at the lowest dose of irradiation [97].

2.7. Ozone-Induced Senescence Model

The aging model produced by ozone damage can cause senescence in many tissues such as the heart, kidney, lung, and skin. Recent findings suggest that the ozone may drive neurological disorders such as cognitive decline, memory impairment, and AD symptoms [98]. Acute ozone exposure has effects on the nervous system of mice, causing reactive microgliosis and increased expression of Aβ in cortical and limbic regions of the brain [99]. Ozone exposure stimulates lung tumor growth and exacerbates the production of lung inflammation [100,101]. Lung epithelial injury caused by ozone exposure can lead to inflammation response, which results in the generation of multiple cytokines and chemokines and leads to neutrophil influx [102]. Studies indicate that TNF-α is required for ozone-induced airway hyperresponsiveness and inflammation, and the requirement appears to depend on the strength and duration of the ozone exposure [103]. Moreover, the antioxidants metallothionein (MT) and HO-1 are strongly induced by acute ozone exposure [104].

2.8. Chronic Jet-Lag Mouse

Jet-lag has repeatedly been shown to hasten death in animals [105].

When the core component of the circadian clock genes *Bmal1* was knocked out, *Bmal1*$^{-/-}$ mice had reduced lifespan and displayed a number of symptoms of premature aging including sarcopenia, cataracts, less subcutaneous fat, organ shrinkage, and others. The early aging phenotype is associated with observed high levels of reactive oxygen species in certain tissues of the *Bmal1*$^{-/-}$ mice [106].

A variety of research studies used rodents treated with chronic jet-lag by alternating the timing of light/dark cycles. The circadian rhythm regulates crucial cellular activities including metabolism and hormone secretion through circadian clock genes, which can modulate orchestrated physiological rhythms in tissues and the whole body [107]. Studies showed that the jet-lag-related dysregulation of the innate immune system is associated with altered or abolished rhythms in the expression of clock genes in the suprachiasmatic nucleus (SCN), liver, thymus, and peritoneal macrophages in mice [108]. Body temperature rhythm, corticosterone level in serum, and expression of mPER1 protein were significantly altered in the SCN in jet-lag mice compared to controls.

Dysregulation of the central clock genes led to increased c-Myc expression and enhanced cell proliferation and metabolic dysfunction and promoted growth and progression of tumors [109–112]. Chronic jet-lag increased the level of β-galactosidase staining in the liver as well as induced spontaneous hepatocellular carcinoma (HCC) and early breast cancer in mice [113–115]. Besides, chronic jet-lag during pregnancy can result in abnormal cardiac structure and impaired cardiac function in offspring [116].

Jet-lag can augment the risk for metabolic syndromes like obesity [117]. Jet-lag treatment induced more fat accumulation and significantly larger adipocytes, and also increased body weight and altered the metabolic gene profile in the mouse liver [118]. Furthermore, decreased microbial abundance, richness, and diversity in both feces and jejunal contents were observed in mice subjected to chronic jet-lag [119].

Jet-lag treatment aggravated depression-like behaviors as well as corticosterone-induced depression-like behavior. Jet-lag treatment decreased the mRNA expressions of telomere repeat binding factor 2 (*Trf2*) and telomerase reverse transcriptase (*Tert*) in the liver, hippocampus, spleen, and muscle. Moreover, jet-lag treatment significantly decreased the number of mitochondria, the level of NAD^+, and the mRNA expressions of *COX1* and *ND1* [114]. Results from preclinical studies have suggested that the circadian system may play a critical role in drug addiction [120]. Interestingly, the expression levels of the *Cat*, *Gpx1*, and *Sod1* genes decreased more in the livers of mice subjected to both D-galactose and jet-lag, suggesting a synergistic effect of jet-lag and D-galactose on the aging process [121].

3. Tissue-, Organ-, or System-Specific Mouse Models of Aging-Related Diseases

Among the leading effects of ageing is the heightened incidence of various aging-related diseases, and mouse models continue to serve an essential part in the study of the pathogenesis and treatment of these illnesses. A variety of commonly used and emerging mouse models have been developed for different aging-related diseases, with the aim of reproducing as closely as possible the progression of the diseases in humans (Figure 1).

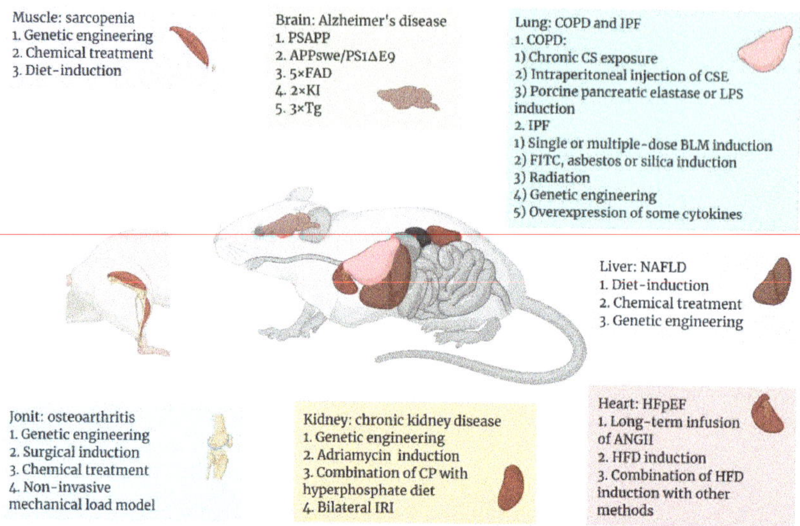

Figure 1. Aging-related diseases include Alzheimer's disease (AD), sarcopenia, heart failure with preserved ejection fraction (HFpEF), non-alcoholic fatty liver disease (NAFLD), chronic kidney disease (CKD), osteoarthritis, chronic obstructive pulmonary disease (COPD), and idiopathic pulmonary fibrosis (IPF), which occur in different tissues or organs. With the help of mice as model organisms, researchers have used different methods to establish disease models in different tissues or organs of mice. In recent years, some common or emerging methods of modeling aging-related diseases are shown above, and these methods are classified according to the organ to which the disease mainly affects. Figure 1 was created with BioRender software (https://biorender.com/ (accessed on 20 April 2022)).

3.1. Model of Aging Brain or Nerve System

Although there are some genetic similarities between mice and humans, the two still diverge in the expression levels of certain age-related genes, which leads to potentially different aging process in the central nervous system of humans and mice [122]. Therefore, changing the expression levels of those specific genes that contribute to the aging process of the brain is the main strategy for modeling the aging brain in mice. For now, to ensure that experimental results can be extrapolated to humans, research on such models has mainly concentrated on evolutionarily conserved mechanisms that modulate aging. These conserved mechanisms include genomic instability, epigenetic changes, telomere attrition, mitochondrial dysfunction, loss of proteostasis, and dysregulated nutrient sensing [122,123].

For instance, as a neurodegenerative disease, AD represents one of the most common neurological disorders. The number of people affected by the disease has been recorded as over tens of millions worldwide, and the number will continue to rise. Also, it is the most common cause of dementia [124]. Over the past decades of research, with the help of animal models, substantial advances have been made, expanding our understanding of this disease.

So far, the etiology of AD is not yet completely understood. It is thought that genetic elements play key roles in the onset of such disease [125]. The typical histopathological features of AD are Aβ deposits and NFTs in the brain. However, wild-type mice do not spontaneously exhibit symptoms of AD [126]. Because AD-related proteins differ in sequence, pathogenicity, and number of isoforms between rodents and humans [127], the main strategy for modeling AD in mice is to construct transgenic mice that cause amyloid deposition or NFTs in the brain.

Common AD-associated mutant proteins in humans include Aβ, presenilin 1 (PS1), apolipoprotein E (ApoE), and MAPT. In the 1990s, the first transgenic AD model mice stably expressing the mutant human Aβ precursor protein (APP) were constructed [128]. After this, transgenic mice carrying multiple human AD-related mutant proteins emerged. One of the most commonly used AD models today is hAPP/PS1 lines, which carry both mutated human APP and PSEN1, including transgenic strains PSAPP, APPswe/PS1ΔE9, 5XFAD, and 2xKI [129]. Compared to monogenic lines containing only mutated APP or PS1, these transgenic mice exhibit earlier and faster onset of amyloid accumulation and cognitive impairment [130]. However, such AD models do not exhibit signs of NFTs, which can be imitated by mouse models that express human MAPT. Based on this, Oddo et al. constructed a 3xTg model combining the human APP, PS1, and MAPT mutations [131]. Since this model can show both Aβ deposition and NFTs in the brain, it is considered to be the well-established transgenic model of AD. Consistent with patients with AD, some of these transgenic mouse models (e.g., PS19, APPswe/PS1ΔE9) showed increased levels of NF-κB pathway-related proteins (IKKβ, p65, and COX-2) in the brain, indicating upregulation of brain inflammation in these mice. Therefore, the use of such mouse models will also help us to further clarify the relationship between neuroinflammation and the pathogenesis of AD [132,133].

3.2. Model of Aging Muscle

The aging of the body is accompanied by the aging of the skeletal muscles. Among other things, sarcopenia, which is a widespread progressive skeletal muscle disorder, is associated with an increased probability of adverse consequences like falls, fractures, physical disability, and death, and its risk increases with age [134]. Modeling of sarcopenia in mice is divided into two main categories: models of genetic engineering and chemical or dietary-induced models.

For genetic engineering models, the majority of research has employed knockout (KO) mice. For example, mitofusion2 (Mfn2) is one of the important protein components mediating mitochondrial fusion, and Mfn2 KO mice exhibit mitochondrial dysfunction in skeletal muscle cells and specific atrophy of type IIb glycolytic fibers [135]. Collagen VI, an extracellular matrix (ECM) protein, has a critical role in skeletal muscle. Six-month-old *Col6α1*$^{-/-}$ mice exhibit alterations of the diaphragm consistent with aged wild-type mice, such as abnormal tricarboxylic acid (TCA) cycle and decreased autophagy in diaphragm cells, indicating the *Col6α1*$^{-/-}$ mouse could be considered as a premature model of skeletal muscle aging [136]. Additionally, interleukin 10 (IL-10) [137], SOD1 [138], and NOD-like receptor protein 3 (NLRP3) [139] deficient mice have also been employed in studying the pathogenesis of sarcopenia as well as the intervention of therapeutically targeting such a disease. Overexpression of certain proteins can also lead to sarcopenia, such as TNF-α transgenic mice that exhibit reduced muscle mass, muscle fiber diameter, and Pax7^{+} muscle stem-cell content [140].

For chemical or diet-induced sarcopenia models, dexamethasone is a common inducer, which is essentially a glucocorticoid that triggers muscle atrophy in mice. It is shown that dexamethasone induces upregulation of ubiquitin ligases in muscle, including muscle atrophy F-box (MAFbx) and muscle ring finger 1 (MuRF1), which may further mediate the degradation of muscle atrophy-associated proteins [141,142]. In addition, diet-induced sarcopenia mouse models allow us to investigate the co-occurrence of sarcopenia with other disorders, e.g., sarcopenia can also develop from the reduction in muscle mass and strength caused by certain chronic diseases. Fabián et al. [143] treated mice with hepatotoxin 5-diethoxycarbonyl-1,4-dihydrocollidine (DDC), thereby inducing sarcopenia secondary to chronic liver disease (CLD), as evidenced by reduced muscle strength and motility, as well as the reduction in muscle fiber size and its type of transformation in mice.

3.3. Model of Aging Heart

Heart failure is a complex disease that can eventually result from almost all cardiovascular disorders, like myocardial infarction, atherosclerosis, and hypertension. Based on the left ventricular ejection fraction, heart failure is clinically classified into two major categories: heart failure with reduced ejection fraction or preserved ejection fraction (HFpEF) [144]. Among them, due to the increasing morbidity and mortality of HFpEF in recent years and the lack of effective therapeutic options for this disease, research on HFpEF has received increasing attention and as a result, some mouse models of HFpEF have been developed.

Long-term infusion of angiotensin II (ANGII) into mice based on a mini-osmotic pump is one of the most common methods of modeling HFpEF. Elevated levels of ANGII in the mouse circulatory system can trigger vasoconstriction, hypertension, aldosterone secretion, TGF-β-mediated inflammation, and fibrosis, and ultimately cardiac hypertrophy [144]. These symptoms closely resemble those exhibited by HFpEF in humans. Furthermore, in addition to causing obesity, a high-fat diet (HFD) is also known to induce a host of cardiac-related symptoms, including left ventricular hypertrophy, HFpEF, and diastolic dysfunction [145–148]. In addition, Withaar et al. [149] constructed a model with HFD and ANGII treatment, which exhibited higher levels of cardiac fibrosis as well as more severe diastolic dysfunction and cardiac hypertrophy compared to the single-treatment group with HFD or ANGII. Also, Schiattarella et al. [150] developed a model in which both HFD and the constitutive NO synthase inhibitor N^{ω}-nitro-L-arginine methyl ester (L-NAME) were imposed. Although L-NAME caused an increase in diastolic and systolic blood pressure, the HFD-L-NAME group exhibited more significant cardiomyocyte hypertrophy and a reduction of myocardial capillary density.

3.4. Model of Aging Liver

The major liver disorders in relation to aging include alcoholic liver disease (ALD), non-alcoholic fatty liver disease (NAFLD), and liver fibrosis, with NAFLD becoming one of the most widespread CLDs and an indication of the need of a liver transplant in recent years [151]. NAFLD is marked by excessive fatty deposits within the liver, and some patients may develop non-alcoholic steatohepatitis (NASH), which is a more aggressive form of the disease with histological manifestations including steatosis, hepatocellular swelling, and lobular inflammation, eventually leading to liver fibrosis, cirrhosis, and even liver cancer [152].

Studies have shown that a HFD represents a major contributor to the onset and progression of obesity and its associated metabolic diseases [153]. Therefore, in addition to being used for HFpEF modeling, HFD can also be used for the modeling of mouse models of NAFLD. Mice fed with HFD for a prolonged period of time develop insulin resistance, hepatic steatosis, inflammation, and liver fibrosis [154]. In addition to HFD, diet-induced models of NAFLD include high fructose diet, high cholesterol and bile salt diet, methionine- and choline-deficient diet, as well as a choline-deficient, L-amino acid-defined (CDAA) diet [155–158]. Among them, Keisuke et al. [159] established a NASH model for the CDAA diet combined with intraperitoneal injection of lipopolysaccharide (LPS) in C57BL/6J mice. This model showed more severe NASH-associated pathologic phenotypes and significant NF-κB activation compared to the mice fed with the CDAA diet only. In addition, certain chemical drugs can be used to induce NAFLD in mouse models, including streptozotocin, carbon tetrachloride (CCl_4), and tetracycline. CCl_4, for example, can accelerate the process of steatosis and fibrosis by generating reactive oxygen radicals, resulting in the damage of hepatocyte structure and function, while metabolites of CCl_4 can facilitate the release of pro-inflammatory cytokines, further aggravating liver injury [160]. However, the treatment of chemical drugs is often combined with dietary induction, as Kubota et al. [161] administered CCl_4 subcutaneously eight times to C57BL/6N mice fed with HFD, and showed that mice in the HFD-CCl_4 group exhibited more significant steatohepatitis compared to mice fed with only HFD. There are also genetic models of NAFLD, such as *ob/ob* (leptin deficient), *db/db* (leptin receptor deficient), and melanocortin receptor 4 knockout ($Mc4r^{-/-}$) mice. As with chemical induction, these genetically engineered mice usually require a combination of dietary induction to better model NAFLD [154]. For example, Mc4r is a G protein-coupled receptor expressed in the hypothalamic nucleus and is associated with modulation of food intake and metabolism [162]. $Mc4r^{-/-}$ mice fed with HFD exhibit significant hepatic fibrosis with histological features of NASH similar to humans, i.e., inflammatory cell infiltration and hepatocyte ballooning [163].

3.5. Model of Chronic Kidney Disease

Chronic kidney disease (CKD) is clinically recognized as a premature aging disease that causes progressive systemic inflammation, vascular disease, muscle atrophy, and organismal weakness [164]. In recent years, the morbidity and mortality of CKD have been increasing year by year. Research in human and various animal models have shown that CKD exhibits features of cellular senescence, like significantly higher levels of p16 and p21 in CKD patients in comparison with age-matched healthy groups. Furthermore, high levels of p21 predict a poor prognosis in patients with chronic renal failure [165]. This implies a close relationship between CKD and cellular senescence.

Diabetic nephropathy (DN) serves as an important cause of CKD. Various mouse models associated with diabetic nephropathy are available, yet so far none of them can fully mimic human signs of the disease [166]. Many attempts have been made to better reproduce the signs of human DN. For example, *db/db* DBA/2J was used to establish a DN model, which exhibited glomerulosclerosis, loss of pedicles, and thickened glomerular basement membranes after growing to 12 weeks of age. Also, the *db/db* DBA/2J mice showed markedly higher albumin and albumin-to-creatinine ratio (ACR) in urine in comparison with *db/db* BLKS/J mice [167]. Focal segmental glomerulosclerosis (FSGS) is a widespread primary glomerular disease marked by podocyte impairment and loss, along with significant proteinuria [168]. Maimaitiyiming et al. [169] administered a single high dose of adriamycin to C57BL6/J mice, which induced signs such as proteinuria, glomerulosclerosis, and increased levels of inflammation. In addition, some data suggested that acute kidney injury (AKI) can increase the risk of CKD in patients [170]. Therefore, researchers have also developed mouse models for the transition from AKI to CKD to investigate the underlying mechanisms. For example, 129S1/SVlmj mice were modeled with AKI and CKD using a single low-dose injection of cisplatin (CP) in combination with a hyperphosphate diet, resulting in the development of CKD signs such as low creatinine clearance, kidney interstitial fibrosis, hyperphosphatemia, and vascular calcification [171]. In addition, Wei et al. [172] established a model of bilateral ischemia-reperfusion injury to CKD using C57BL/6 mice that exhibited incomplete recovery from AKI followed by a progressive decrease in glomerular filtration rate, elevated plasma creatinine, and CKD-related histopathological changes, including bilateral interstitial fibrosis and worsening of proteinuria.

3.6. Model of Osteoarthritis

Osteoarthritis (OA), the most prevalent type of arthritis, is a synovial joint disorder characterized with cartilage degeneration and osteophytes [173]. Osteoarthritis can have a variety of causes, including obesity, inflammation, trauma, or genetic factors, the most important of which is aging [174]. Aging of the body induces senescence in a variety of osteoarthritis-related cells, including osteoblasts, bone lining cells, cartilage cells, and bone marrow cells, etc. The development of osteoarthritis can be further promoted by various pro-inflammatory or pro-aging SASP factors secreted by certain senescent cells. These studies imply an inextricable relationship between osteoarthritis and aging of the body [175,176].

The spontaneous model is the simplest model for establishing OA in mice, i.e., no treatment is applied to the mice and they are allowed to develop OA spontaneously with age. The progression of OA in these mice is very similar to the progression of nontraumatic OA in humans, mimicking the natural wear and tear of the joint throughout its life course [177]. However, not all wild-type mice develop OA, and even when they do, they require a longer rearing period, so spontaneous OA models in mice are often combined with transgenic means. For example, the STR/ort mouse is one of the commonly used spontaneous OA models, which has a high incidence of OA because it often exhibits bone dislocation, elevated levels of oxidative stress, and elevated expression of various inflammatory factors including IL-1β, IL-12, and macrophage inflammatory protein-1β, and can develop OA spontaneously early in life [178,179]. Surgical induction is another commonly used method to induce OA in mice, which allows for the induction of OA at a specific joint. The most common surgical approach is anterior cruciate ligament transection (ACLT), which causes instability of the ACLT-treated joint and induces post-traumatic osteoarthritis (PTOA), mimicking the progression and pathogenesis of PTOA in humans. McCulloch et al. [180] used a combination of destabilized medial meniscotibial ligament (DMM) and cartilage scratch to model PTOA in C57BL/6 mice, and showed that mice that underwent both procedures exhibited more severe osteophytes and synovitis than mice in the single procedure treatment group. Furthermore, meniscectomy and oophorectomy are also commonly used to model OA in mice [179]. In addition, OA can be induced in mice by chemical treatment, of which collagenase is most commonly used, and the administration of collagenase to the kneecap of mice leads to a series of osteoarthritis-like lesions caused by patellar malalignment [181]. Another method of mouse OA modeling has been developed in recent years, the non-invasive mechanical load model, which has the advantage of no surgery requirement, thus avoiding the potential artefact of surgical intervention, infection, or variation due to surgery or healing. However, the age, sex (hormonal status), and strain of the mice may have a significant impact on the modeling results of such methods [182,183]. For example, using an indenter, researchers have applied steady pressure to the tibial plateau of mice, which caused damage to the synovium, meniscus, ligaments, and articular cartilage, thus triggering histopathological changes in PTOA such as articular cartilage loss, chondrocyte disintegration, meniscal hyperplasia, and mineralization [183].

3.7. Model of Aging Lung

Another serious problem associated with the increase in the aging population is the rise in morbidity and mortality from aging-related lung diseases, of which chronic obstructive pulmonary disease (COPD) and idiopathic pulmonary fibrosis (IPF) are the main types [184,185]. In humans, COPD includes chronic obstructive bronchitis with fibrosis and small airway obstruction, and emphysema with airspace enlargement, lung parenchymal destruction, loss of lung elasticity, and small airway closure [186]. In the case of IPF, increased deposition of ECM in the pulmonary interstitium is manifested, ultimately resulting in damage to the lung structure and function [185]. For COPD, smoking is the main cause, while the etiology of IPF is not yet clear. Taken together, we still know very little about the mechanisms in the development of both diseases. To address this issue, more established mouse models for the two aging-related lung diseases are available.

COPD mouse models have been established mainly by drug induction. Among them, cigarettes, as the primary contributor of COPD, are also most commonly used for the induction of mouse COPD models [187]. Chronic cigarette smoke (CS) exposure can be divided into two types, airway-only or systemic exposure, both of which can lead to COPD signs such as respiratory impairment, emphysema, small airway and vascular remodeling, or pulmonary hypertension in the lungs of mice [188]. However, this method requires a long modeling time, often lasting several months, leading to the development of some short-term mouse models of CS-induced COPD. For example, He et al. [189] used a 28-day CS exposure in combination with intraperitoneal injection of CS extract in C57BL/6 mice, which not only shortened the modeling time, but also showed comparable inflammatory levels and pathophysiological changes to long-term CS modeling. In addition, elastase is also widely used for COPD modeling. The most common model utilizes porcine pancreatic elastase, which mimics the release of neutrophil-derived elastase during COPD, leading to alveolar tissue rupture, emphysema, and further inflammatory responses. These models have low cost and short modeling time compared to CS exposure [190,191]. Furthermore, LPS is also commonly used in the induction of COPD models in mice [192,193]. Mebratu et al. [194] treated C57BL/6 mice with intranasal LPS and elastase, which resulted in COPD signs such as mucocyte hyperplasia and emphysema. Additionally, LPS and CSE co-treatment was also shown to significantly enhance the nuclear translocation of NF-κB, thereby promoting the inflammatory response in COPD model mice [195].

For mouse models of IPF, the most common method is the transtracheal administration of bleomycin (BLM), a chemotherapeutic agent derived from *Streptomyces verticillis*, which also has the side effect of causing acute lung damage and fibrosis in humans. BLM can chelate with metal ions such as iron ions and react with oxygen to produce superoxide, which can cause DNA breakage. On the other hand, BLM can also induce lipid peroxidation, leading to tissue damage. Such a process would ultimately result in severe inflammatory response and pulmonary fibrosis [196,197]. Moreover, BLM induces activation of the transcription factor NF-κB, which to some extent mediates the enhancement of pulmonary inflammation in BLM-treated mice [198,199]. A single transtracheal administration of BLM triggers acute lung injury and neutrophil-driven inflammatory response that lasts seven to ten days and changes to a fibrotic response by approximately day 14 after administration, with fibrosis levels peaking between 21–28 days [196,200]. However, there are some limitations to this modeling approach, most notably, the spontaneous regression of the fibrotic phenotype in mice modeled in this way after 28 days of modeling. To address this issue, many studies have used multiple-dose BLM administrations. For example, Redente et al. [201] developed a fortnightly three-dose BLM drip model, which showed a more significant fibrosis phenotype in comparison with mice treated with a single dose, and exhibited persistent and progressive characteristics. Similar to human IPF, this model produced a phenotype of alveolar epithelial cell remodeling. In addition to BLM, fluorescein isothiocyanate (FITC), asbestos, and silica can also be used for modeling pulmonary fibrosis, but the histopathological features of these pulmonary fibrosis models generated using these substances differ significantly from those of human IPF. For instance, FITC-induced pulmonary fibrosis does not produce fibroblastic lesions, while lung diseases induced by asbestos and silica are more similar to human asbestosis and silicosis, respectively [200]. One of the most commonly used methods for non-chemical-induced IPF is radiation induction [202]. A single exposure to radiation can induce the fibrotic process by triggering DNA breaks that lead to alveolar epithelial cell death and tissue damage, ultimately causing inflammation and fibrotic responses in the lung. Nevertheless, it should be noted that this modeling method takes longer and also lacks some of the histopathological features of human IPF [203]. In addition, some transgenic mice can be used for modeling pulmonary fibrosis. For example, the surfactant protein C (SP-C) gene (*Sftpc*) is expressed only in alveolar epithelial type II cells (AEC2s), and mutations in this gene are thought to be associated with familial interstitial lung disease [204,205]. Nureki et al. [206] selectively expressed human mutant SP-C within AEC2s in mice, which led to diffuse lung injury, and subsequently elevated levels of lung inflammatory response and spontaneously remodeled lung fibrosis. In addition, mice with defects in some telomere-related genes have been used for modeling pulmonary fibrosis, such as *Tert* [207] and *Trf1* [208] deficient mice. However, it has also been previously shown that although telomerase deficiency leads to telomere shortening in mice, there is no difference in the phenotype of pulmonary fibrosis produced by wild-type and telomere deficient mice treated with BLM at the same time [209]. Therefore, further studies are needed to determine whether telomere-associated protein defects can cause or promote the development of pulmonary fibrosis. Mucin 5B is encoded by the *Muc5b* gene and has an essential function in mucociliary clearance and host defense. Hancock et al. [210] demonstrated that the presence of excessive *Muc5b* within mouse AEC2s enhanced the level of BLM-induced pulmonary fibrosis. Further, Nedd4-2, a ubiquitin ligase, can be engaged in a variety of cellular processes related to epithelial homeostasis, and its dysregulation might correlate with the development of chronic lung disease and fibrosis. Studies have shown that conditional knockout of Nedd4-2 in epithelial cells can lead to pulmonary fibrosis [211,212]. Finally, overexpression of some cytokines in mice can also be applied to model pulmonary fibrosis. For example, adenovirus-based delivery of TGF-β1 in mice leads to alveolar collapse associated with surfactant dysfunction, which further results in the development of fibrosis [213].

4. Discussion

In this review, we summarize the accelerated aging mouse models, both in systematic whole-body and in specific organs. Although systemic-induced accelerated aging mouse models mimic natural aging, it is not fully equal to natural aging. Models of aging organs can mimic specific diseases to some extent while not representing the overall aging process in the body.

When assessing the effects of inducing an accelerated aging model with the markers of aging, not all markers show significant or consistent differences and some even exhibit contradictory results. Therefore, diverse mouse models have been employed to recapitulate different features of diseases, which implies the complexity of the pathogenesis of various illnesses. Accordingly, multiple and diverse mouse models are necessary to elucidate the pathogenesis of the various disorders that develop during human aging.

For instance, *Lmna* mutant mice (i.e., $Lmna^{L530P/L530P}$ and $Lmna^{G609G/G609G}$ mice) can serve as a model for accelerated aging, but the cardiovascular phenotype exhibited by such mice is far from the cardiovascular alterations of humans that occur during natural aging. While cardiomyocytes expand and arterial walls thicken with age in humans, *Lmna* mutant mice exhibit cardiomyocyte atrophy and depletion in vascular smooth muscle of the aortic arch [214]. Moreover, the D-galactose-induced premature aging model can be highly variable among rodents of different species or ages, e.g., intraperitoneal administration of 300 mg/kg D-galactose for two months to four-week-old Wistar rats did not influence their anxiety levels, spatial cognition, memory, or neurogenesis [215].

Due to the space limit, we may miss some aging mouse models. The development process has been proved to affect the aging process. For example, intrauterine growth retardation can cause the offspring to have an increased incidence of type 2 diabetes in adulthood, thus affecting their lifespan. These mouse models can also be a premature senescence model for energy metabolism studies. Low protein restriction, total calorie restriction, and maternal glucocorticoid exposure to the pregnant rodents are commonly used to induce intrauterine growth retardation. Also, induction of uteroplacental insufficiency by a uterine artery ligation surgery can limit the nutrient supply to the fetus. These treatments lead to epigenetic reprogramming to offspring in utero, causing their predisposition to diabetes and other metabolic disorders in adults [216].

It should be noted that accelerated aging is not the same as natural aging. The stimuli induced by the above induction methods can trigger adaptive responses in mice, and when these adaptive responses are exhausted, the failure of various tissues and organs in mice is induced. The various aging phenotypes produced by this process are quite different from those of natural aging and the accelerated aging models cannot fully reflect the pathogenesis of various aging-related diseases in humans. Taking the various AD mouse models mentioned above as examples, although some transgenic models can display the histopathological features of AD, the pathogenesis of AD not only involves mutated genes such as *APP* and *PSEN*, but also environmental factors (e.g., heavy metals) or chronic stress [125,217]. It is also worth noting that the organ background of the mice themselves has an important influence on the phenotypic outcome of the various models, which is particularly reflected in the modeling results in mice of different ages. For example, in the BLM-induced pulmonary fibrosis model, older mice were more susceptible to stimuli than younger mice and showed a more pronounced fibrotic phenotype and higher TGF-β expression in their lungs, exhibiting less spontaneous regression of fibrosis [218,219]. Therefore, when conducting aging-related disease studies, we must select mice of an appropriate age or genetic background for disease modeling according to the study objectives. Because of ethics-related issues, long lifespan, environmental effects, and various constraints, it will remain difficult to study humans as subjects for aging and its relevant disorders, at least for the foreseeable future. In this review, we summarize some of the commonly used and emerging aging-related mouse models in recent years. We found that although mice are the most commonly used experimental animals today, and the utilization of mouse models has contributed to our understanding of pathogenesis of aging and related diseases to a certain extent, there are few mouse models that can well replicate the aging process or the development of aging-related diseases in humans, which has seriously hindered the progress of research in this field. The methods of induced aging may require further research and improvement, for instance, exploring the possibility of combined effects. As the research progresses, we will hopefully get more positive results.

Author Contributions: N.C. and Y.W. conceived and wrote the manuscript; Y.H. conceived, supervised, and revised the manuscript. All authors contributed to the discussions and presentations. All authors have read and agreed to the published version of the manuscript.

Funding: This research was funded by the National Natural Science Foundation (81871109, 82071587, 31930058), and Guang Dong Basic and Applied Basic Research Foundation (2020A1515010462).

Acknowledgments: We thank the colleagues of Zhou Songyang's lab.

Conflicts of Interest: The authors declare no conflict of interest.

References

1. DESA | United Nations. Available online: https://www.un.org/en/desa (accessed on 20 March 2022).
2. World Population Prospects 2019. Available online: https://population.un.org/wpp/ (accessed on 20 March 2022).
3. Graw, J. Mouse models of cataract. *J. Genet.* **2009**, *88*, 469–486. [CrossRef] [PubMed]
4. Brooks, S.V.; Faulkner, J.A. Skeletal muscle weakness in old age: Underlying mechanisms. *Med. Sci. Sports Exerc.* **1994**, *26*, 432–439. [CrossRef] [PubMed]
5. Gurkar, A.U.; Niedernhofer, L.J. Comparison of mice with accelerated aging caused by distinct mechanisms. *Exp. Gerontol.* **2015**, *68*, 43–50. [CrossRef] [PubMed]
6. Vanhooren, V.; Libert, C. The mouse as a model organism in aging research: Usefulness, pitfalls and possibilities. *Ageing Res. Rev.* **2013**, *12*, 8–21. [CrossRef] [PubMed]
7. Harkema, L.; Youssef, S.A.; de Bruin, A. Pathology of Mouse Models of Accelerated Aging. *Vet. Pathol.* **2016**, *53*, 366–389. [CrossRef] [PubMed]
8. Acosta, P.B.; Gross, K.C. Hidden sources of galactose in the environment. *Eur. J. Pediatrics* **1995**, *154*, S87–S92. [CrossRef] [PubMed]
9. Morava, E. Galactose supplementation in phosphoglucomutase-1 deficiency; review and outlook for a novel treatable CDG. *Mol Genet. Metab.* **2014**, *112*, 275–279. [CrossRef]
10. Ullah, F.; Ali, T.; Ullah, N.; Kim, M.O. Caffeine prevents d-galactose-induced cognitive deficits, oxidative stress, neuroinflammation and neurodegeneration in the adult rat brain. *Neurochem. Int.* **2015**, *90*, 114–124. [CrossRef]
11. Wu, D.M.; Lu, J.; Zheng, Y.L.; Zhou, Z.; Shan, Q.; Ma, D.F. Purple sweet potato color repairs d-galactose-induced spatial learning and memory impairment by regulating the expression of synaptic proteins. *Neurobiol. Learn Mem.* **2008**, *90*, 19–27. [CrossRef]
12. Azman, K.F.; Safdar, A.; Zakaria, R. D-galactose-induced liver aging model: Its underlying mechanisms and potential therapeutic interventions. *Exp. Gerontol.* **2021**, *150*, 111372. [CrossRef] [PubMed]
13. Shwe, T.; Pratchayasakul, W.; Chattipakorn, N.; Chattipakorn, S.C. Role of D-galactose-induced brain aging and its potential used for therapeutic interventions. *Exp. Gerontol.* **2018**, *101*, 13–36. [CrossRef]
14. Azman, K.F.; Zakaria, R. D-Galactose-induced accelerated aging model: An overview. *Biogerontology* **2019**, *20*, 763–782. [CrossRef]
15. Li, W.; Wang, S.; Wang, H.; Wang, J.; Jin, F.; Fang, F.; Fang, C. Astragaloside IV Prevents Memory Impairment in D-galactose-induced Aging Rats Via the AGEs/RAGE/ NF-kappaB Axis. *Arch. Med. Res.* **2022**, *53*, 20–28. [CrossRef]
16. Banji, O.J.; Banji, D.; Ch, K. Curcumin and hesperidin improve cognition by suppressing mitochondrial dysfunction and apoptosis induced by D-galactose in rat brain. *Food Chem. Toxicol.* **2014**, *74*, 51–59. [CrossRef]
17. Khedr, N.F.; Werida, R.H.; Abo-Saif, M.A. Candesartan protects against d-galactose induced—Neurotoxicity and memory deficit via modulation of autophagy and oxidative stress. *Toxicol. Appl. Pharm.* **2022**, *435*, 115827. [CrossRef]
18. Long, J.; Wang, X.; Gao, H.; Liu, Z.; Liu, C.; Miao, M.; Cui, X.; Packer, L.; Liu, J. D-galactose toxicity in mice is associated with mitochondrial dysfunction: Protecting effects of mitochondrial nutrient R-alpha-lipoic acid. *Biogerontology* **2007**, *8*, 373–381. [CrossRef]

19. Kumar, A.; Dogra, S.; Prakash, A. Effect of carvedilol on behavioral, mitochondrial dysfunction, and oxidative damage against D-galactose induced senescence in mice. *Naunyn-Schmiedeberg's Arch. Pharmacol.* **2009**, *380*, 431–441. [CrossRef]
20. Haider, S.; Liaquat, L.; Shahzad, S.; Sadir, S.; Madiha, S.; Batool, Z.; Tabassum, S.; Saleem, S.; Naqvi, F.; Perveen, T. A high dose of short term exogenous D-galactose administration in young male rats produces symptoms simulating the natural aging process. *Life Sci.* **2015**, *124*, 110–119. [CrossRef]
21. Sacco, R.L.; Roth, G.A.; Reddy, K.S.; Arnett, D.K.; Bonita, R.; Gaziano, T.A.; Heidenreich, P.A.; Huffman, M.D.; Mayosi, B.M.; Mendis, S.; et al. The Heart of 25 by 25: Achieving the Goal of Reducing Global and Regional Premature Deaths from Cardiovascular Diseases and Stroke: A Modeling Study from the American Heart Association and World Heart Federation. *Circulation* **2016**, *133*, e674–e690. [CrossRef]
22. Cebe, T.; Yanar, K.; Atukeren, P.; Ozan, T.; Kuruç, A.I.; Kunbaz, A.; Sitar, M.E.; Mengi, M.; Aydın, M.S.; Eşrefoğlu, M.; et al. A comprehensive study of myocardial redox homeostasis in naturally and mimetically aged rats. *Age* **2014**, *36*, 9728. [CrossRef]
23. Chang, Y.M.; Chang, H.H.; Lin, H.J.; Tsai, C.C.; Tsai, C.T.; Chang, H.N.; Lin, S.L.; PadmaViswanadha, V.; Chen, R.J.; Huang, C.Y. Inhibition of Cardiac Hypertrophy Effects in D-Galactose-Induced Senescent Hearts by Alpinate Oxyphyllae Fructus Treatment. *Evid. Based Complement Altern. Med.* **2017**, *2017*, 2624384. [CrossRef]
24. Dehghani, A.; Hafizibarjin, Z.; Najjari, R.; Kaseb, F.; Safari, F. Resveratrol and 1,25-dihydroxyvitamin D co-administration protects the heart against D-galactose-induced aging in rats: Evaluation of serum and cardiac levels of klotho. *Aging Clin. Exp. Res.* **2019**, *31*, 1195–1205. [CrossRef]
25. Wang, L.F.; Cao, Q.; Wen, K.; Xiao, Y.F.; Chen, T.T.; Guan, X.H.; Liu, Y.; Zuo, L.; Qian, Y.S.; Deng, K.Y.; et al. CD38 Deficiency Alleviates D-Galactose-Induced Myocardial Cell Senescence Through NAD(+)/Sirt1 Signaling Pathway. *Front. Physiol.* **2019**, *10*, 1125. [CrossRef]
26. Lei, L.; Ou, L.; Yu, X. The antioxidant effect of *Asparagus cochinchinensis* (Lour.) Merr. shoot in D-galactose induced mice aging model and in vitro. *J. Chin. Med. Assoc.* **2016**, *79*, 205–211. [CrossRef]
27. Sun, S.L.; Guo, L.; Ren, Y.C.; Wang, B.; Li, R.H.; Qi, Y.S.; Yu, H.; Chang, N.D.; Li, M.H.; Peng, H.S. Anti-apoptosis effect of polysaccharide isolated from the seeds of *Cuscuta chinensis* Lam on cardiomyocytes in aging rats. *Mol. Biol. Rep.* **2014**, *41*, 6117–6124. [CrossRef]
28. Hegab, Z.; Gibbons, S.; Neyses, L.; Mamas, M.A. Role of advanced glycation end products in cardiovascular disease. *World J. Cardiol.* **2012**, *4*, 90–102. [CrossRef]
29. Bo-Htay, C.; Palee, S.; Apaijai, N.; Chattipakorn, S.C.; Chattipakorn, N. Effects of d-galactose-induced ageing on the heart and its potential interventions. *J. Cell. Mol. Med.* **2018**, *22*, 1392–1410. [CrossRef]
30. Chang, Y.M.; Tamilselvi, S.; Lin, H.J.; Tsai, C.C.; Lin, Y.M.; Day, C.H.; Viswanadha, V.P.; Chang, H.N.; Kuo, W.W.; Huang, C.Y. Alpinia oxyphylla Miq extract ameliorates cardiac fibrosis associated with D-galactose induced aging in rats. *Environ. Toxicol.* **2019**, *34*, 172–178. [CrossRef]
31. Yeh, S.L.; Wu, T.C.; Chan, S.T.; Hong, M.J.; Chen, H.L. Fructo-oligosaccharide attenuates the production of pro-inflammatory cytokines and the activation of JNK/Jun pathway in the lungs of D-galactose-treated Balb/cJ mice. *Eur. J. Nutr.* **2014**, *53*, 449–456. [CrossRef]
32. Feng, Y.; Yu, Y.H.; Wang, S.T.; Ren, J.; Camer, D.; Hua, Y.Z.; Zhang, Q.; Huang, J.; Xue, D.L.; Zhang, X.F.; et al. Chlorogenic acid protects D-galactose-induced liver and kidney injury via antioxidation and anti-inflammation effects in mice. *Pharm. Biol.* **2016**, *54*, 1027–1034. [CrossRef]
33. Chen, H.L.; Wang, C.H.; Kuo, Y.W.; Tsai, C.H. Antioxidative and hepatoprotective effects of fructo-oligosaccharide in d-galactose-treated Balb/cJ mice. *Br. J. Nutr.* **2011**, *105*, 805–809. [CrossRef]
34. Shahroudi, M.J.; Mehri, S.; Hosseinzadeh, H. Anti-Aging Effect of Nigella Sativa Fixed Oil on D-Galactose-Induced Aging in Mice. *J. Pharmacopunct.* **2017**, *20*, 29–35. [CrossRef]
35. Fan, Y.; Xia, J.; Jia, D.; Zhang, M.; Zhang, Y.; Huang, G.; Wang, Y. Mechanism of ginsenoside Rg1 renal protection in a mouse model of d-galactose-induced subacute damage. *Pharm. Biol.* **2016**, *54*, 1815–1821. [CrossRef]
36. Miller, M.R. Structural and physiological age-associated changes in aging lungs. *Semin. Respir. Crit. Care Med.* **2010**, *31*, 521–527. [CrossRef] [PubMed]
37. Gao, J.; Yu, Z.; Jing, S.; Jiang, W.; Liu, C.; Yu, C.; Sun, J.; Wang, C.; Chen, J.; Li, H. Protective effect of Anwulignan against D-galactose-induced hepatic injury through activating p38 MAPK-Nrf2-HO-1 pathway in mice. *Clin. Interv. Aging* **2018**, *13*, 1859–1869. [CrossRef]
38. Xu, L.Q.; Xie, Y.L.; Gui, S.H.; Zhang, X.; Mo, Z.Z.; Sun, C.Y.; Li, C.L.; Luo, D.D.; Zhang, Z.B.; Su, Z.R.; et al. Polydatin attenuates d-galactose-induced liver and brain damage through its anti-oxidative, anti-inflammatory and anti-apoptotic effects in mice. *Food Funct.* **2016**, *7*, 4545–4555. [CrossRef]
39. Zhang, X.; Wu, J.Z.; Lin, Z.X.; Yuan, Q.J.; Li, Y.C.; Liang, J.L.; Zhan, J.Y.; Xie, Y.L.; Su, Z.R.; Liu, Y.H. Ameliorative effect of supercritical fluid extract of Chrysanthemum indicum Linnén against D-galactose induced brain and liver injury in senescent mice via suppression of oxidative stress, inflammation and apoptosis. *J. Ethnopharmacol.* **2019**, *234*, 44–56. [CrossRef]
40. Huang, C.C.; Chiang, W.D.; Huang, W.C.; Huang, C.Y.; Hsu, M.C.; Lin, W.T. Hepatoprotective Effects of Swimming Exercise against D-Galactose-Induced Senescence Rat Model. *Evid. Based Complement Altern. Med* **2013**, *2013*, 275431. [CrossRef]
41. Wang, H.; Hu, L.; Li, L.; Wu, X.; Fan, Z.; Zhang, C.; Wang, J.; Jia, J.; Wang, S. Inorganic nitrate alleviates the senescence-related decline in liver function. *Sci. China Life Sci.* **2018**, *61*, 24–34. [CrossRef]

42. Mo, Z.Z.; Liu, Y.H.; Li, C.L.; Xu, L.Q.; Wen, L.L.; Xian, Y.F.; Lin, Z.X.; Zhan, J.Y.; Chen, J.N.; Xu, F.F.; et al. Protective Effect of SFE-CO_2 of Ligusticum chuanxiong Hort Against d-Galactose-Induced Injury in the Mouse Liver and Kidney. *Rejuvenation Res.* **2017**, *20*, 231–243. [CrossRef]
43. Liu, C.M.; Ma, J.Q.; Lou, Y. Chronic administration of troxerutin protects mouse kidney against D-galactose-induced oxidative DNA damage. *Food Chem. Toxicol.* **2010**, *48*, 2809–2817. [CrossRef] [PubMed]
44. Zhuang, Y.; Ma, Q.; Guo, Y.; Sun, L. Protective effects of rambutan (*Nephelium lappaceum*) peel phenolics on H_2O_2-induced oxidative damages in HepG2 cells and d-galactose-induced aging mice. *Food Chem. Toxicol.* **2017**, *108*, 554–562. [CrossRef]
45. Liao, C.H.; Chen, B.H.; Chiang, H.S.; Chen, C.W.; Chen, M.F.; Ke, C.C.; Wang, Y.Y.; Lin, W.N.; Wang, C.C.; Lin, Y.H. Optimizing a Male Reproductive Aging Mouse Model by D-Galactose Injection. *Int. J. Mol. Sci.* **2016**, *17*, 98. [CrossRef]
46. Djahanbakhch, O.; Ezzati, M.; Zosmer, A. Reproductive ageing in women. *J. Pathol.* **2007**, *211*, 219–231. [CrossRef]
47. Ahangarpour, A.; Najimi, S.A.; Farbood, Y. Effects of Vitex agnus-castus fruit on sex hormones and antioxidant indices in a d-galactose-induced aging female mouse model. *J. Chin. Med. Assoc.* **2016**, *79*, 589–596. [CrossRef]
48. Wang, F.; Zhou, H.; Deng, L.; Wang, L.; Chen, J.; Zhou, X. Serine Deficiency Exacerbates Inflammation and Oxidative Stress via Microbiota-Gut-Brain Axis in D-Galactose-Induced Aging Mice. *Mediat. Inflamm* **2020**, *2020*, 5821428. [CrossRef]
49. Liu, X.; Wu, C.; Han, D.; Liu, J.; Liu, H.; Jiang, Z. Partially Hydrolyzed Guar Gum Attenuates d-Galactose-Induced Oxidative Stress and Restores Gut Microbiota in Rats. *Int. J. Mol. Sci.* **2019**, *20*, 4861. [CrossRef]
50. Fransen, F.; van Beek, A.A.; Borghuis, T.; Aidy, S.E.; Hugenholtz, F.; van der Gaast-de Jongh, C.; Savelkoul, H.F.J.; De Jonge, M.I.; Boekschoten, M.V.; Smidt, H.; et al. Aged Gut Microbiota Contributes to Systemical Inflammaging after Transfer to Germ-Free Mice. *Front. Immunol.* **2017**, *8*, 1385. [CrossRef]
51. Yin, R.; Liu, S.; Jiang, X.; Zhang, X.; Wei, F.; Hu, J. The Qingchangligan Formula Alleviates Acute Liver Failure by Regulating Galactose Metabolism and Gut Microbiota. *Front. Cell Infect Microbiol.* **2021**, *11*, 771483. [CrossRef]
52. Kim, S.; Jazwinski, S.M. The Gut Microbiota and Healthy Aging: A Mini-Review. *Gerontology* **2018**, *64*, 513–520. [CrossRef]
53. Takeda, T.; Hosokawa, M.; Higuchi, K. Senescence-accelerated mouse (SAM): A novel murine model of senescence. *Exp. Gerontol.* **1997**, *32*, 105–109. [CrossRef]
54. Higuchi, K. Genetic characterization of senescence-accelerated mouse (SAM). *Exp. Gerontol.* **1997**, *32*, 129–138. [CrossRef]
55. Garcia-Matas, S.; Gutierrez-Cuesta, J.; Coto-Montes, A.; Rubio-Acero, R.; Diez-Vives, C.; Camins, A.; Pallas, M.; Sanfeliu, C.; Cristofol, R. Dysfunction of astrocytes in senescence-accelerated mice SAMP8 reduces their neuroprotective capacity. *Aging Cell* **2008**, *7*, 630–640. [CrossRef] [PubMed]
56. Beck, J.; Horikawa, I.; Harris, C. Cellular Senescence: Mechanisms, Morphology, and Mouse Models. *Vet. Pathol.* **2020**, *57*, 747–757. [CrossRef]
57. Lecka-Czernik, B.; Moerman, E.J.; Shmookler Reis, R.J.; Lipschitz, D.A. Cellular and molecular biomarkers indicate precocious in vitro senescence in fibroblasts from SAMP6 mice. Evidence supporting a murine model of premature senescence and osteopenia. *J. Gerontol. Biol. Sci. Med. Sci.* **1997**, *52*, B331–B336. [CrossRef] [PubMed]
58. Manczak, M.; Jung, Y.; Park, B.S.; Partovi, D.; Reddy, P.H. Time-course of mitochondrial gene expressions in mice brains: Implications for mitochondrial dysfunction, oxidative damage, and cytochrome c in aging. *J. Neurochem.* **2005**, *92*, 494–504. [CrossRef] [PubMed]
59. Takeda, T.; Hosokawa, M.; Takeshita, S.; Irino, M.; Higuchi, K.; Matsushita, T.; Tomita, Y.; Yasuhira, K.; Hamamoto, H.; Shimizu, K.; et al. A new murine model of accelerated senescence. *Mech. Ageing Dev.* **1981**, *17*, 183–194. [CrossRef]
60. Takeda, T.; Hosokawa, M.; Higuchi, K. Senescence-accelerated mouse (SAM): A novel murine model of accelerated senescence. *J. Am. Geriatr. Soc.* **1991**, *39*, 911–919. [CrossRef]
61. Matsushita, M.; Tsuboyama, T.; Kasai, R.; Okumura, H.; Yamamuro, T.; Higuchi, K.; Higuchi, K.; Kohno, A.; Yonezu, T.; Utani, A.; et al. Age-related changes in bone mass in the senescence-accelerated mouse (SAM). SAM-R/3 and SAM-P/6 as new murine models for senile osteoporosis. *Am. J. Pathol.* **1986**, *125*, 276–283.
62. Takada, K.; Inaba, M.; Ichioka, N.; Ueda, Y.; Taira, M.; Baba, S.; Mizokami, T.; Wang, X.; Hisha, H.; Iida, H.; et al. Treatment of senile osteoporosis in SAMP6 mice by intra-bone marrow injection of allogeneic bone marrow cells. *Stem Cells* **2006**, *24*, 399–405. [CrossRef]
63. Stein, G.H.; Drullinger, L.F.; Soulard, A.; Dulić, V. Differential roles for cyclin-dependent kinase inhibitors p21 and p16 in the mechanisms of senescence and differentiation in human fibroblasts. *Mol. Cell. Biol.* **1999**, *19*, 2109–2117. [CrossRef] [PubMed]
64. Al-Azab, M.; Wang, B.; Elkhider, A.; Walana, W.; Li, W.; Yuan, B.; Ye, Y.; Tang, Y.; Almoiliqy, M.; Adlat, S.; et al. Indian Hedgehog regulates senescence in bone marrow-derived mesenchymal stem cell through modulation of ROS/mTOR/4EBP1, p70S6K1/2 pathway. *Aging* **2020**, *12*, 5693–5715. [CrossRef] [PubMed]
65. Morley, J.E.; Armbrecht, H.J.; Farr, S.A.; Kumar, V.B. The senescence accelerated mouse (SAMP8) as a model for oxidative stress and Alzheimer's disease. *Biochim. Biophys. Acta* **2012**, *1822*, 650–656. [CrossRef] [PubMed]
66. Karasawa, N.; Nagatsu, I.; Sakai, K.; Nagatsu, T.; Watanabe, K.; Onozuka, M. Immunocytochemical study of catecholaminergic neurons in the senescence-accelerated mouse (SAM-P8) brain. *J. Neural Transm.* **1997**, *104*, 1267–1275. [CrossRef] [PubMed]
67. Sureda, F.X.; Gutierrez-Cuesta, J.; Romeu, M.; Mulero, M.; Canudas, A.M.; Camins, A.; Mallol, J.; Pallas, M. Changes in oxidative stress parameters and neurodegeneration markers in the brain of the senescence-accelerated mice SAMP-8. *Exp. Gerontol.* **2006**, *41*, 360–367. [CrossRef]

68. Manich, G.; Mercader, C.; del Valle, J.; Duran-Vilaregut, J.; Camins, A.; Pallàs, M.; Vilaplana, J.; Pelegrí, C. Characterization of amyloid-β granules in the hippocampus of SAMP8 mice. *J. Alzheimers Dis.* **2011**, *25*, 535–546. [CrossRef]
69. Anderson, R.; Lagnado, A.; Maggiorani, D.; Walaszczyk, A.; Dookun, E.; Chapman, J.; Birch, J.; Salmonowicz, H.; Ogrodnik, M.; Jurk, D.; et al. Length-independent telomere damage drives post-mitotic cardiomyocyte senescence. *EMBO J.* **2019**, *38*, e100492. [CrossRef]
70. Alvarez-Garcia, O.; Vega-Naredo, I.; Sierra, V.; Caballero, B.; Tomas-Zapico, C.; Camins, A.; Garcia, J.J.; Pallas, M.; Coto-Montes, A. Elevated oxidative stress in the brain of senescence-accelerated mice at 5 months of age. *Biogerontology* **2006**, *7*, 43–52. [CrossRef]
71. Ota, H.; Akishita, M.; Akiyoshi, T.; Kahyo, T.; Setou, M.; Ogawa, S.; Iijima, K.; Eto, M.; Ouchi, Y. Testosterone deficiency accelerates neuronal and vascular aging of SAMP8 mice: Protective role of eNOS and SIRT1. *PLoS ONE* **2012**, *7*, e29598. [CrossRef]
72. Karuppagounder, V.; Arumugam, S.; Babu, S.S.; Palaniyandi, S.S.; Watanabe, K.; Cooke, J.P.; Thandavarayan, R.A. The senescence accelerated mouse prone 8 (SAMP8): A novel murine model for cardiac aging. *Ageing Res. Rev.* **2017**, *35*, 291–296. [CrossRef]
73. Niimi, K.; Takahashi, E.; Itakura, C. Adiposity-related biochemical phenotype in senescence-accelerated mouse prone 6 (SAMP6). *Comp. Med.* **2009**, *59*, 431–436. [PubMed]
74. Shimada, A.; Ohta, A.; Akiguchi, I.; Takeda, T. Inbred SAM-P/10 as a mouse model of spontaneous, inherited brain atrophy. *J. Neuropathol. Exp. Neurol.* **1992**, *51*, 440–450. [CrossRef] [PubMed]
75. Shimada, A.; Hosokawa, M.; Ohta, A.; Akiguchi, I.; Takeda, T. Localization of atrophy-prone areas in the aging mouse brain: Comparison between the brain atrophy model SAM-P/10 and the normal control SAM-R/1. *Neuroscience* **1994**, *59*, 859–869. [CrossRef]
76. Shcherbakov, D.; Nigri, M.; Akbergenov, R.; Brilkova, M.; Mantovani, M.; Petit, P.I.; Grimm, A.; Karol, A.A.; Teo, Y.; Sanchón, A.C.; et al. Premature aging in mice with error-prone protein synthesis. *Sci. Adv.* **2022**, *8*, eabl9051. [CrossRef] [PubMed]
77. Brennan, T.A.; Egan, K.P.; Lindborg, C.M.; Chen, Q.; Sweetwyne, M.T.; Hankenson, K.D.; Xie, S.X.; Johnson, F.B.; Pignolo, R.J. Mouse models of telomere dysfunction phenocopy skeletal changes found in human age-related osteoporosis. *Dis. Model Mech.* **2014**, *7*, 583–592. [CrossRef]
78. Botter, S.M.; Zar, M.; van Osch, G.J.; van Steeg, H.; Dolle, M.E.; Hoeijmakers, J.H.; Weinans, H.; van Leeuwen, J.P. Analysis of osteoarthritis in a mouse model of the progeroid human DNA repair syndrome trichothiodystrophy. *Age* **2011**, *33*, 247–260. [CrossRef] [PubMed]
79. Edstrom, E.; Altun, M.; Bergman, E.; Johnson, H.; Kullberg, S.; Ramirez-Leon, V.; Ulfhake, B. Factors contributing to neuromuscular impairment and sarcopenia during aging. *Physiol. Behav.* **2007**, *92*, 129–135. [CrossRef]
80. Baker, D.J.; Jeganathan, K.B.; Cameron, J.D.; Thompson, M.; Juneja, S.; Kopecka, A.; Kumar, R.; Jenkins, R.B.; de Groen, P.C.; Roche, P.; et al. BubR1 insufficiency causes early onset of aging-associated phenotypes and infertility in mice. *Nat. Genet.* **2004**, *36*, 744–749. [CrossRef]
81. Serrano-Pozo, A.; Frosch, M.P.; Masliah, E.; Hyman, B.T. Neuropathological alterations in Alzheimer disease. *Cold Spring Harb. Perspect. Med.* **2011**, *1*, a006189. [CrossRef]
82. Hartman, T.K.; Wengenack, T.M.; Poduslo, J.F.; van Deursen, J.M. Mutant mice with small amounts of BubR1 display accelerated age-related gliosis. *Neurobiol. Aging* **2007**, *28*, 921–927. [CrossRef]
83. Trifunovic, A.; Wredenberg, A.; Falkenberg, M.; Spelbrink, J.N.; Rovio, A.T.; Bruder, C.E.; Bohlooly, Y.M.; Gidlöf, S.; Oldfors, A.; Wibom, R.; et al. Premature ageing in mice expressing defective mitochondrial DNA polymerase. *Nature* **2004**, *429*, 417–423. [CrossRef] [PubMed]
84. Kujoth, G.C.; Hiona, A.; Pugh, T.D.; Someya, S.; Panzer, K.; Wohlgemuth, S.E.; Hofer, T.; Seo, A.Y.; Sullivan, R.; Jobling, W.A.; et al. Mitochondrial DNA mutations, oxidative stress, and apoptosis in mammalian aging. *Science* **2005**, *309*, 481–484. [CrossRef] [PubMed]
85. Cupit-Link, M.C.; Kirkland, J.L.; Ness, K.K.; Armstrong, G.T.; Tchkonia, T.; LeBrasseur, N.K.; Armenian, S.H.; Ruddy, K.J.; Hashmi, S.K. Biology of premature ageing in survivors of cancer. *ESMO Open* **2017**, *2*, e000250. [CrossRef] [PubMed]
86. Robison, L.L.; Hudson, M.M. Survivors of childhood and adolescent cancer: Life-long risks and responsibilities. *Nat. Rev. Cancer* **2014**, *14*, 61–70. [CrossRef]
87. Fielder, E.; Weigand, M.; Agneessens, J.; Griffin, B.; Parker, C.; Miwa, S.; von Zglinicki, T. Sublethal whole-body irradiation causes progressive premature frailty in mice. *Mech. Ageing Dev.* **2019**, *180*, 63–69. [CrossRef]
88. Kasmann, L.; Dietrich, A.; Staab-Weijnitz, C.A.; Manapov, F.; Behr, J.; Rimner, A.; Jeremic, B.; Senan, S.; De Ruysscher, D.; Lauber, K.; et al. Radiation-induced lung toxicity—Cellular and molecular mechanisms of pathogenesis, management, and literature review. *Radiat. Oncol.* **2020**, *15*, 214. [CrossRef]
89. Sun, L.; Inaba, Y.; Sogo, Y.; Ito, A.; Bekal, M.; Chida, K.; Moritake, T. Total body irradiation causes a chronic decrease in antioxidant levels. *Sci. Rep.* **2021**, *11*, 6716. [CrossRef]
90. Xiao, S.; Shterev, I.D.; Zhang, W.; Young, L.; Shieh, J.H.; Moore, M.; van den Brink, M.; Sempowski, G.D.; Manley, N.R. Sublethal Total Body Irradiation Causes Long-Term Deficits in Thymus Function by Reducing Lymphoid Progenitors. *J. Immunol.* **2017**, *199*, 2701–2712. [CrossRef]
91. Citrin, D.E.; Shankavaram, U.; Horton, J.A.; Shield, W., 3rd; Zhao, S.; Asano, H.; White, A.; Sowers, A.; Thetford, A.; Chung, E.J. Role of type II pneumocyte senescence in radiation-induced lung fibrosis. *J. Natl. Cancer Inst.* **2013**, *105*, 1474–1484. [CrossRef]
92. Singh, V.K.; Beattie, L.A.; Seed, T.M. Vitamin E: Tocopherols and tocotrienols as potential radiation countermeasures. *J. Radiat. Res.* **2013**, *54*, 973–988. [CrossRef]

93. Hutchinson, I.D.; Olson, J.; Lindburg, C.A.; Payne, V.; Collins, B.; Smith, T.L.; Munley, M.T.; Wheeler, K.T.; Willey, J.S. Total-body irradiation produces late degenerative joint damage in rats. *Int. J. Radiat. Biol.* **2014**, *90*, 821–830. [CrossRef] [PubMed]
94. Cornelissen, M.; de Ridder, L. Dose- and time-dependent increase of lysosomal enzymes in embryonic cartilage in vitro after ionizing radiation. *Scanning Microsc.* **1990**, *4*, 769–773; discussion 764–773. [PubMed]
95. Cornelissen, M.; Thierens, H.; de Ridder, L. Effects of ionizing radiation on the size distribution of proteoglycan aggregates synthesized by chondrocytes in agarose. *Scanning Microsc.* **1993**, *7*, 1263–1267; discussion 1267–1268. [PubMed]
96. Onder, G.O.; Balcioglu, E.; Baran, M.; Ceyhan, A.; Cengiz, O.; Suna, P.A.; Yıldız, O.G.; Yay, A. The different doses of radiation therapy-induced damage to the ovarian environment in rats. *Int. J. Radiat. Biol.* **2021**, *97*, 367–375. [CrossRef]
97. Pesty, A.; Doussau, M.; Lahaye, J.B.; Lefèvre, B. Whole-body or isolated ovary (60)Co irradiation: Effects on in vivo and in vitro folliculogenesis and oocyte maturation. *Reprod. Toxicol.* **2010**, *29*, 93–98. [CrossRef]
98. Chang, H.H.; Hao, H.; Sarnat, S.E. A Statistical Modeling Framework for Projecting Future Ambient Ozone and its Health Impact due to Climate Change. *Atmos. Environ.* **2014**, *89*, 290–297. [CrossRef]
99. Tyler, C.R.; Noor, S.; Young, T.L.; Rivero, V.; Sanchez, B.; Lucas, S.; Caldwell, K.K.; Milligan, E.D.; Campen, M.J. Aging Exacerbates Neuroinflammatory Outcomes Induced by Acute Ozone Exposure. *Toxicol. Sci.* **2018**, *163*, 123–139. [CrossRef]
100. Stober, V.P.; Garantziotis, S. Assessment of Ozone-Induced Lung Injury in Mice. *Methods Mol. Biol.* **2018**, *1809*, 301–314. [CrossRef]
101. Tomita, K.; Okawara, T.; Ohira, C.; Morimoto, A.; Aihara, R.; Kurihara, T.; Fukuyama, T. An Acceptable Concentration (0.1 ppm) of Ozone Exposure Exacerbates Lung Injury in a Mouse Model. *Am. J. Respir. Cell Mol. Biol.* **2021**, *65*, 674–676. [CrossRef]
102. Hazucha, M.J.; Folinsbee, L.J.; Bromberg, P.A. Distribution and reproducibility of spirometric response to ozone by gender and age. *J. Appl. Physiol.* **2003**, *95*, 1917–1925. [CrossRef]
103. Matsubara, S.; Takeda, K.; Jin, N.; Okamoto, M.; Matsuda, H.; Shiraishi, Y.; Park, J.W.; McConville, G.; Joetham, A.; O'Brien, R.L.; et al. Vgamma1+ T cells and tumor necrosis factor-alpha in ozone-induced airway hyperresponsiveness. *Am. J. Respir. Cell Mol. Biol.* **2009**, *40*, 454–463. [CrossRef] [PubMed]
104. Vasu, V.T.; Oommen, S.; Lim, Y.; Valacchi, G.; Hobson, B.; Eiserich, J.P.; Leonard, S.W.; Traber, M.G.; Cross, C.E.; Gohil, K. Modulation of ozone-sensitive genes in alpha-tocopherol transfer protein null mice. *Inhal. Toxicol.* **2010**, *22*, 1–16. [CrossRef] [PubMed]
105. Davidson, A.J.; Sellix, M.T.; Daniel, J.; Yamazaki, S.; Menaker, M.; Block, G.D. Chronic jet-lag increases mortality in aged mice. *Curr. Biol.* **2006**, *16*, R914–R916. [CrossRef] [PubMed]
106. Kondratov, R.V.; Kondratova, A.A.; Gorbacheva, V.Y.; Vykhovanets, O.V.; Antoch, M.P. Early aging and age-related pathologies in mice deficient in BMAL1, the core componentof the circadian clock. *Genes Dev.* **2006**, *20*, 1868–1873. [CrossRef]
107. Ortiz-Tudela, E.; Bonmatí-Carrión Mde, L.; De la Fuente, M.; Mendiola, P. Chronodisruption and ageing. *Rev. Esp. Geriatr. Gerontol.* **2012**, *47*, 168–173. [CrossRef] [PubMed]
108. Castanon-Cervantes, O.; Wu, M.; Ehlen, J.C.; Paul, K.; Gamble, K.L.; Johnson, R.L.; Besing, R.C.; Menaker, M.; Gewirtz, A.T.; Davidson, A.J. Dysregulation of inflammatory responses by chronic circadian disruption. *J. Immunol.* **2010**, *185*, 5796–5805. [CrossRef]
109. Filipski, E.; Delaunay, F.; King, V.M.; Wu, M.W.; Claustrat, B.; Gréchez-Cassiau, A.; Guettier, C.; Hastings, M.H.; Francis, L. Effects of chronic jet lag on tumor progression in mice. *Cancer Res.* **2004**, *64*, 7879–7885. [CrossRef]
110. Filipski, E.; Li, X.M.; Lévi, F. Disruption of circadian coordination and malignant growth. *Cancer Causes Control* **2006**, *17*, 509–514. [CrossRef]
111. Filipski, E.; Lévi, F. Circadian disruption in experimental cancer processes. *Integr. Cancer Ther.* **2009**, *8*, 298–302. [CrossRef]
112. Papagiannakopoulos, T.; Bauer, M.R.; Davidson, S.M.; Heimann, M.; Subbaraj, L.; Bhutkar, A.; Bartlebaugh, J.; Vander Heiden, M.G.; Jacks, T. Circadian Rhythm Disruption Promotes Lung Tumorigenesis. *Cell Metab.* **2016**, *24*, 324–331. [CrossRef]
113. Kettner, N.M.; Voicu, H.; Finegold, M.J.; Coarfa, C.; Sreekumar, A.; Putluri, N.; Katchy, C.A.; Lee, C.; Moore, D.D.; Fu, L. Circadian Homeostasis of Liver Metabolism Suppresses Hepatocarcinogenesis. *Cancer Cell* **2016**, *30*, 909–924. [CrossRef] [PubMed]
114. Shen, Q.; Wu, J.; Ni, Y.; Xie, X.; Yu, C.; Xiao, Q.; Zhou, J.; Wang, X.; Fu, Z. Exposure to jet lag aggravates depression-like behaviors and age-related phenotypes in rats subject to chronic corticosterone. *Acta Biochim. Biophys. Sin.* **2019**, *51*, 834–844. [CrossRef] [PubMed]
115. Van Dycke, K.C.; Rodenburg, W.; van Oostrom, C.T.; van Kerkhof, L.W.; Pennings, J.L.; Roenneberg, T.; van Steeg, H.; van der Horst, G.T. Chronically Alternating Light Cycles Increase Breast Cancer Risk in Mice. *Curr. Biol.* **2015**, *25*, 1932–1937. [CrossRef] [PubMed]
116. Chaves, I.; van der Eerden, B.; Boers, R.; Boers, J.; Streng, A.A.; Ridwan, Y.; Schreuders-Koedam, M.; Vermeulen, M.; van der Pluijm, I.; Essers, J.; et al. Gestational jet lag predisposes to later-life skeletal and cardiac disease. *Chronobiol. Int.* **2019**, *36*, 657–671. [CrossRef] [PubMed]
117. Pfeffer, M.; Korf, H.W.; Wicht, H. Synchronizing effects of melatonin on diurnal and circadian rhythms. *Gen. Comp. Endocrinol.* **2018**, *258*, 215–221. [CrossRef] [PubMed]
118. Oike, H.; Sakurai, M.; Ippoushi, K.; Kobori, M. Time-fixed feeding prevents obesity induced by chronic advances of light/dark cycles in mouse models of jet-lag/shift work. *Biochem. Biophys. Res. Commun.* **2015**, *465*, 556–561. [CrossRef]
119. Matenchuk, B.A.; Mandhane, P.J.; Kozyrskyj, A.L. Sleep, circadian rhythm, and gut microbiota. *Sleep Med. Rev.* **2020**, *53*, 101340. [CrossRef]

120. Doyle, S.E.; Feng, H.; Garber, G.; Menaker, M.; Lynch, W.J. Effects of circadian disruption on methamphetamine consumption in methamphetamine-exposed rats. *Psychopharmacology* **2015**, *232*, 2169–2179. [CrossRef]
121. Xu, Y.; Wu, T.; Jin, Y.; Fu, Z. Effects of age and jet lag on D-galactose induced aging process. *Biogerontology* **2009**, *10*, 153–161. [CrossRef]
122. Loerch, P.M.; Lu, T.; Dakin, K.A.; Vann, J.M.; Isaacs, A.; Geula, C.; Wang, J.; Pan, Y.; Gabuzda, D.H.; Li, C.; et al. Evolution of the aging brain transcriptome and synaptic regulation. *PLoS ONE* **2008**, *3*, e3329. [CrossRef]
123. Bishop, N.A.; Lu, T.; Yankner, B.A. Neural mechanisms of ageing and cognitive decline. *Nature* **2010**, *464*, 529–535. [CrossRef] [PubMed]
124. Esquerda-Canals, G.; Montoliu-Gaya, L.; Güell-Bosch, J.; Villegas, S. Mouse Models of Alzheimer's Disease. *J. Alzheimer's Dis.* **2017**, *57*, 1171–1183. [CrossRef] [PubMed]
125. Dorszewska, J.; Prendecki, M.; Oczkowska, A.; Dezor, M.; Kozubski, W. Molecular Basis of Familial and Sporadic Alzheimer's Disease. *Curr. Alzheimer Res.* **2016**, *13*, 952–963. [CrossRef] [PubMed]
126. Teich, A.F.; Patel, M.; Arancio, O. A reliable way to detect endogenous murine β-amyloid. *PLoS ONE* **2013**, *8*, e55647. [CrossRef]
127. Myers, A.; McGonigle, P. Overview of Transgenic Mouse Models for Alzheimer's Disease. *Curr. Protoc. Neurosci.* **2019**, *89*, e81. [CrossRef]
128. Games, D.; Adams, D.; Alessandrini, R.; Barbour, R.; Berthelette, P.; Blackwell, C.; Carr, T.; Clemens, J.; Donaldson, T.; Gillespie, F.; et al. Alzheimer-type neuropathology in transgenic mice overexpressing V717F beta-amyloid precursor protein. *Nature* **1995**, *373*, 523–527. [CrossRef]
129. Hall, A.M.; Roberson, E.D. Mouse models of Alzheimer's disease. *Brain Res. Bull* **2012**, *88*, 3–12. [CrossRef]
130. Sasaguri, H.; Nilsson, P.; Hashimoto, S.; Nagata, K.; Saito, T.; De Strooper, B.; Hardy, J.; Vassar, R.; Winblad, B.; Saido, T.C. APP mouse models for Alzheimer's disease preclinical studies. *EMBO J.* **2017**, *36*, 2473–2487. [CrossRef]
131. Oddo, S.; Caccamo, A.; Shepherd, J.D.; Murphy, M.P.; Golde, T.E.; Kayed, R.; Metherate, R.; Mattson, M.P.; Akbari, Y.; LaFerla, F.M. Triple-transgenic model of Alzheimer's disease with plaques and tangles: Intracellular Abeta and synaptic dysfunction. *Neuron* **2003**, *39*, 409–421. [CrossRef]
132. Zhang, X.; Luhrs, K.J.; Ryff, K.A.; Malik, W.T.; Driscoll, M.J.; Culver, B. Suppression of nuclear factor kappa B ameliorates astrogliosis but not amyloid burden in APPswe/PS1dE9 mice. *Neuroscience* **2009**, *161*, 53–58. [CrossRef]
133. Sun, X.-Y.; Li, L.-J.; Dong, Q.-X.; Zhu, J.; Huang, Y.-R.; Hou, S.-J.; Yu, X.-L.; Liu, R.-T. Rutin prevents tau pathology and neuroinflammation in a mouse model of Alzheimer's disease. *J. Neuroinflamm.* **2021**, *18*, 131. [CrossRef] [PubMed]
134. Xie, W.-Q.; He, M.; Yu, D.-J.; Wu, Y.-X.; Wang, X.-H.; Lv, S.; Xiao, W.-F.; Li, Y.-S. Mouse models of sarcopenia: Classification and evaluation. *J. Cachexia Sarcopenia Muscle* **2021**, *12*, 538–554. [CrossRef] [PubMed]
135. Sebastián, D.; Sorianello, E.; Segalés, J.; Irazoki, A.; Ruiz-Bonilla, V.; Sala, D.; Planet, E.; Berenguer-Llergo, A.; Muñoz, J.P.; Sánchez-Feutrie, M.; et al. Mfn2 deficiency links age-related sarcopenia and impaired autophagy to activation of an adaptive mitophagy pathway. *EMBO J.* **2016**, *35*, 1677–1693. [CrossRef] [PubMed]
136. Capitanio, D.; Moriggi, M.; De Palma, S.; Bizzotto, D.; Molon, S.; Torretta, E.; Fania, C.; Bonaldo, P.; Gelfi, C.; Braghetta, P. Collagen VI Null Mice as a Model for Early Onset Muscle Decline in Aging. *Front. Mol. Neurosci.* **2017**, *10*, 337. [CrossRef] [PubMed]
137. Ko, F.; Abadir, P.; Marx, R.; Westbrook, R.; Cooke, C.; Yang, H.; Walston, J. Impaired mitochondrial degradation by autophagy in the skeletal muscle of the aged female interleukin 10 null mouse. *Exp. Gerontol.* **2016**, *73*, 23–27. [CrossRef] [PubMed]
138. Ahn, B.; Smith, N.; Saunders, D.; Ranjit, R.; Kneis, P.; Towner, R.A.; Van Remmen, H. Using MRI to measure in vivo free radical production and perfusion dynamics in a mouse model of elevated oxidative stress and neurogenic atrophy. *Redox. Biol.* **2019**, *26*, 101308. [CrossRef]
139. Sayed, R.K.A.; Fernández-Ortiz, M.; Diaz-Casado, M.E.; Aranda-Martínez, P.; Fernández-Martínez, J.; Guerra-Librero, A.; Escames, G.; López, L.C.; Alsaadawy, R.M.; Acuña-Castroviejo, D. Lack of NLRP3 Inflammasome Activation Reduces Age-Dependent Sarcopenia and Mitochondrial Dysfunction, Favoring the Prophylactic Effect of Melatonin. *J. Gerontol. Ser. A Biol. Sci. Med. Sci.* **2019**, *74*, 1699–1708. [CrossRef]
140. Li, J.; Yi, X.; Yao, Z.; Chakkalakal, J.V.; Xing, L.; Boyce, B.F. TNF Receptor-Associated Factor 6 Mediates TNFα-Induced Skeletal Muscle Atrophy in Mice During Aging. *J. Bone Min. Res.* **2020**, *35*, 1535–1548. [CrossRef]
141. Sun, H.; Gong, Y.; Qiu, J.; Chen, Y.; Ding, F.; Zhao, Q. TRAF6 inhibition rescues dexamethasone-induced muscle atrophy. *Int. J. Mol. Sci.* **2014**, *15*, 11126–11141. [CrossRef]
142. Wada, S.; Kato, Y.; Okutsu, M.; Miyaki, S.; Suzuki, K.; Yan, Z.; Schiaffino, S.; Asahara, H.; Ushida, T.; Akimoto, T. Translational suppression of atrophic regulators by microRNA-23a integrates resistance to skeletal muscle atrophy. *J. Biol. Chem.* **2011**, *286*, 38456–38465. [CrossRef]
143. Campos, F.; Abrigo, J.; Aguirre, F.; Garcés, B.; Arrese, M.; Karpen, S.; Cabrera, D.; Andía, M.E.; Simon, F.; Cabello-Verrugio, C. Sarcopenia in a mice model of chronic liver disease: Role of the ubiquitin-proteasome system and oxidative stress. *Pflug. Arch. Eur. J. Physiol.* **2018**, *470*, 1503–1519. [CrossRef] [PubMed]
144. Noll, N.A.; Lal, H.; Merryman, W.D. Mouse Models of Heart Failure with Preserved or Reduced Ejection Fraction. *Am. J. Pathol.* **2020**, *190*, 1596–1608. [CrossRef] [PubMed]
145. Meng, Q.; Lai, Y.C.; Kelly, N.J.; Bueno, M.; Baust, J.J.; Bachman, T.N.; Goncharov, D.; Vanderpool, R.R.; Radder, J.E.; Hu, J.; et al. Development of a Mouse Model of Metabolic Syndrome, Pulmonary Hypertension, and Heart Failure with Preserved Ejection Fraction. *Am. J. Respir. Cell Mol. Biol.* **2017**, *56*, 497–505. [CrossRef] [PubMed]

146. Manrique, C.; DeMarco, V.G.; Aroor, A.R.; Mugerfeld, I.; Garro, M.; Habibi, J.; Hayden, M.R.; Sowers, J.R. Obesity and insulin resistance induce early development of diastolic dysfunction in young female mice fed a Western diet. *Endocrinology* **2013**, *154*, 3632–3642. [CrossRef] [PubMed]
147. Aroor, A.R.; Habibi, J.; Kandikattu, H.K.; Garro-Kacher, M.; Barron, B.; Chen, D.; Hayden, M.R.; Whaley-Connell, A.; Bender, S.B.; Klein, T.; et al. Dipeptidyl peptidase-4 (DPP-4) inhibition with linagliptin reduces western diet-induced myocardial TRAF3IP2 expression, inflammation and fibrosis in female mice. *Cardiovasc. Diabetol.* **2017**, *16*, 61. [CrossRef]
148. Cannon, M.V.; Sillje, H.H.; Sijbesma, J.W.; Khan, M.A.; Steffensen, K.R.; van Gilst, W.H.; de Boer, R.A. LXRα improves myocardial glucose tolerance and reduces cardiac hypertrophy in a mouse model of obesity-induced type 2 diabetes. *Diabetologia* **2016**, *59*, 634–643. [CrossRef]
149. Withaar, C.; Meems, L.M.G.; Markousis-Mavrogenis, G.; Boogerd, C.J.; Silljé, H.H.W.; Schouten, E.M.; Dokter, M.M.; Voors, A.A.; Westenbrink, B.D.; Lam, C.S.P.; et al. The effects of liraglutide and dapagliflozin on cardiac function and structure in a multi-hit mouse model of heart failure with preserved ejection fraction. *Cardiovasc. Res.* **2021**, *117*, 2108–2124. [CrossRef]
150. Schiattarella, G.G.; Altamirano, F.; Tong, D.; French, K.M.; Villalobos, E.; Kim, S.Y.; Luo, X.; Jiang, N.; May, H.I.; Wang, Z.V.; et al. Nitrosative stress drives heart failure with preserved ejection fraction. *Nature* **2019**, *568*, 351–356. [CrossRef]
151. Mikolasevic, I.; Filipec-Kanizaj, T.; Mijic, M.; Jakopcic, I.; Milic, S.; Hrstic, I.; Sobocan, N.; Stimac, D.; Burra, P. Nonalcoholic fatty liver disease and liver transplantation—Where do we stand? *World J. Gastroenterol.* **2018**, *24*, 1491–1506. [CrossRef]
152. Younossi, Z.M.; Koenig, A.B.; Abdelatif, D.; Fazel, Y.; Henry, L.; Wymer, M. Global epidemiology of nonalcoholic fatty liver disease—Meta-analytic assessment of prevalence, incidence, and outcomes. *Hepatology* **2016**, *64*, 73–84. [CrossRef]
153. Barrera, F.; George, J. The role of diet and nutritional intervention for the management of patients with NAFLD. *Clin. Liver Dis.* **2014**, *18*, 91–112. [CrossRef] [PubMed]
154. Santhekadur, P.K.; Kumar, D.P.; Sanyal, A.J. Preclinical models of non-alcoholic fatty liver disease. *J. Hepatol.* **2018**, *68*, 230–237. [CrossRef] [PubMed]
155. Denda, A.; Kitayama, W.; Kishida, H.; Murata, N.; Tamura, K.; Kusuoka, O.; Tsutsumi, M.; Nishikawa, F.; Kita, E.; Nakae, D.; et al. Expression of inducible nitric oxide (NO) synthase but not prevention by its gene ablation of hepatocarcinogenesis with fibrosis caused by a choline-deficient, L-amino acid-defined diet in rats and mice. *Nitric Oxide Biol. Chem.* **2007**, *16*, 164–176. [CrossRef] [PubMed]
156. Takahashi, Y.; Soejima, Y.; Fukusato, T. Animal models of nonalcoholic fatty liver disease/nonalcoholic steatohepatitis. *World J. Gastroenterol.* **2012**, *18*, 2300–2308. [CrossRef] [PubMed]
157. Kohli, R.; Kirby, M.; Xanthakos, S.A.; Softic, S.; Feldstein, A.E.; Saxena, V.; Tang, P.H.; Miles, L.; Miles, M.V.; Balistreri, W.F.; et al. High-fructose, medium chain trans fat diet induces liver fibrosis and elevates plasma coenzyme Q9 in a novel murine model of obesity and nonalcoholic steatohepatitis. *Hepatology* **2010**, *52*, 934–944. [CrossRef] [PubMed]
158. Matsuzawa, N.; Takamura, T.; Kurita, S.; Misu, H.; Ota, T.; Ando, H.; Yokoyama, M.; Honda, M.; Zen, Y.; Nakanuma, Y.; et al. Lipid-induced oxidative stress causes steatohepatitis in mice fed an atherogenic diet. *Hepatology* **2007**, *46*, 1392–1403. [CrossRef] [PubMed]
159. Nakanishi, K.; Kaji, K.; Kitade, M.; Kubo, T.; Furukawa, M.; Saikawa, S.; Shimozato, N.; Sato, S.; Seki, K.; Kawaratani, H.; et al. Exogenous Administration of Low-Dose Lipopolysaccharide Potentiates Liver Fibrosis in a Choline-Deficient l-Amino-Acid-Defined Diet-Induced Murine Steatohepatitis Model. *Int. J. Mol. Sci.* **2019**, *20*, 2724. [CrossRef]
160. Zhong, F.; Zhou, X.; Xu, J.; Gao, L. Rodent Models of Nonalcoholic Fatty Liver Disease. *Digestion* **2020**, *101*, 522–535. [CrossRef]
161. Kubota, N.; Kado, S.; Kano, M.; Masuoka, N.; Nagata, Y.; Kobayashi, T.; Miyazaki, K.; Ishikawa, F. A high-fat diet and multiple administration of carbon tetrachloride induces liver injury and pathological features associated with non-alcoholic steatohepatitis in mice. *Clin. Exp. Pharmacol. Physiol.* **2013**, *40*, 422–430. [CrossRef]
162. Hainer, V.; Aldhoon Hainerová, I.; Kunešová, M.; Taxová Braunerová, R.; Zamrazilová, H.; Bendlová, B. Melanocortin pathways: Suppressed and stimulated melanocortin-4 receptor (MC4R). *Physiol. Res.* **2020**, *69*, S245–S254. [CrossRef]
163. Itoh, M.; Suganami, T.; Nakagawa, N.; Tanaka, M.; Yamamoto, Y.; Kamei, Y.; Terai, S.; Sakaida, I.; Ogawa, Y. Melanocortin 4 receptor-deficient mice as a novel mouse model of nonalcoholic steatohepatitis. *Am. J. Pathol.* **2011**, *179*, 2454–2463. [CrossRef] [PubMed]
164. Dai, L.; Qureshi, A.R.; Witasp, A.; Lindholm, B.; Stenvinkel, P. Early Vascular Ageing and Cellular Senescence in Chronic Kidney Disease. *Comput. Struct. Biotechnol. J.* **2019**, *17*, 721–729. [CrossRef] [PubMed]
165. Wang, Y.; Wang, Y.; Yang, M.; Ma, X. Implication of cellular senescence in the progression of chronic kidney disease and the treatment potencies. *Biomed. Pharmacother.* **2021**, *135*, 111191. [CrossRef] [PubMed]
166. Deb, D.K.; Sun, T.; Wong, K.E.; Zhang, Z.; Ning, G.; Zhang, Y.; Kong, J.; Shi, H.; Chang, A.; Li, Y.C. Combined vitamin D analog and AT1 receptor antagonist synergistically block the development of kidney disease in a model of type 2 diabetes. *Kidney Int.* **2010**, *77*, 1000–1009. [CrossRef] [PubMed]
167. Østergaard, M.V.; Pinto, V.; Stevenson, K.; Worm, J.; Fink, L.N.; Coward, R.J. DBA2J db/db mice are susceptible to early albuminuria and glomerulosclerosis that correlate with systemic insulin resistance. *Am. J. Physiol. Ren. Physiol.* **2017**, *312*, F312–F321. [CrossRef] [PubMed]
168. Fogo, A.B. Causes and pathogenesis of focal segmental glomerulosclerosis. *Nat. Rev. Nephrol.* **2015**, *11*, 76–87. [CrossRef] [PubMed]

169. Maimaitiyiming, H.; Zhou, Q.; Wang, S. Thrombospondin 1 Deficiency Ameliorates the Development of Adriamycin-Induced Proteinuric Kidney Disease. *PLoS ONE* **2016**, *11*, e0156144. [CrossRef]
170. Chawla, L.S.; Kimmel, P.L. Acute kidney injury and chronic kidney disease: An integrated clinical syndrome. *Kidney Int.* **2012**, *82*, 516–524. [CrossRef]
171. Shi, M.; McMillan, K.L.; Wu, J.; Gillings, N.; Flores, B.; Moe, O.W.; Hu, M.C. Cisplatin nephrotoxicity as a model of chronic kidney disease. *Lab. Investig.* **2018**, *98*, 1105–1121. [CrossRef]
172. Wei, J.; Zhang, J.; Wang, L.; Jiang, S.; Fu, L.; Buggs, J.; Liu, R. New mouse model of chronic kidney disease transitioned from ischemic acute kidney injury. *Am. J. Physiol. Ren. Physiol.* **2019**, *317*, F286–F295. [CrossRef]
173. Loeser, R.F.; Goldring, S.R.; Scanzello, C.R.; Goldring, M.B. Osteoarthritis: A disease of the joint as an organ. *Arthritis Rheum.* **2012**, *64*, 1697–1707. [CrossRef] [PubMed]
174. Johnson, V.L.; Hunter, D.J. The epidemiology of osteoarthritis. *Best Pract. Res. Clin. Rheumatol.* **2014**, *28*, 5–15. [CrossRef] [PubMed]
175. Coryell, P.R.; Diekman, B.O.; Loeser, R.F. Mechanisms and therapeutic implications of cellular senescence in osteoarthritis. *Nat. Rev. Rheumatol.* **2021**, *17*, 47–57. [CrossRef] [PubMed]
176. Jeon, O.H.; Kim, C.; Laberge, R.M.; Demaria, M.; Rathod, S.; Vasserot, A.P.; Chung, J.W.; Kim, D.H.; Poon, Y.; David, N.; et al. Local clearance of senescent cells attenuates the development of post-traumatic osteoarthritis and creates a pro-regenerative environment. *Nat. Med.* **2017**, *23*, 775–781. [CrossRef] [PubMed]
177. Little, C.B.; Hunter, D.J. Post-traumatic osteoarthritis: From mouse models to clinical trials. *Nat. Rev. Rheumatol.* **2013**, *9*, 485–497. [CrossRef] [PubMed]
178. Staines, K.A.; Poulet, B.; Wentworth, D.N.; Pitsillides, A.A. The STR/ort mouse model of spontaneous osteoarthritis—An update. *Osteoarthr. Cartil.* **2017**, *25*, 802–808. [CrossRef]
179. Bapat, S.; Hubbard, D.; Munjal, A.; Hunter, M.; Fulzele, S. Pros and cons of mouse models for studying osteoarthritis. *Clin. Transl. Med.* **2018**, *7*, 36. [CrossRef] [PubMed]
180. McCulloch, K.; Huesa, C.; Dunning, L.; Litherland, G.J.; Van 'T Hof, R.J.; Lockhart, J.C.; Goodyear, C.S. Accelerated post traumatic osteoarthritis in a dual injury murine model. *Osteoarthr. Cartil.* **2019**, *27*, 1800–1810. [CrossRef]
181. Alves-Simões, M. Rodent models of knee osteoarthritis for pain research. *Osteoarthr. Cartil.* **2022**. [CrossRef]
182. Poulet, B. Non-invasive Loading Model of Murine Osteoarthritis. *Curr. Rheumatol. Rep.* **2016**, *18*, 40. [CrossRef]
183. Stiffel, V.; Rundle, C.H.; Sheng, M.H.C.; Das, S.; Lau, K.-H.W. A Mouse Noninvasive Intraarticular Tibial Plateau Compression Loading-Induced Injury Model of Posttraumatic Osteoarthritis. *Calcif. Tissue Int.* **2020**, *106*, 158–171. [CrossRef] [PubMed]
184. Easter, M.; Bollenbecker, S.; Barnes, J.W.; Krick, S. Targeting Aging Pathways in Chronic Obstructive Pulmonary Disease. *Int. J. Mol. Sci.* **2020**, *21*, 6924. [CrossRef] [PubMed]
185. Liu, R.-M.; Liu, G. Cell senescence and fibrotic lung diseases. *Exp. Gerontol.* **2020**, *132*, 110836. [CrossRef]
186. Vlahos, R.; Bozinovski, S. Recent advances in pre-clinical mouse models of COPD. *Clin. Sci.* **2013**, *126*, 253–265. [CrossRef]
187. Ghorani, V.; Boskabady, M.H.; Khazdair, M.R.; Kianmeher, M. Experimental animal models for COPD: A methodological review. *Tob. Induc. Dis.* **2017**, *15*, 25. [CrossRef] [PubMed]
188. Vlahos, R.; Bozinovski, S. Preclinical murine models of Chronic Obstructive Pulmonary Disease. *Eur. J. Pharmacol.* **2015**, *759*, 265–271. [CrossRef]
189. He, S.; Li, L.; Sun, S.; Zeng, Z.; Lu, J.; Xie, L. A Novel Murine Chronic Obstructive Pulmonary Disease Model and the Pathogenic Role of MicroRNA-21. *Front. Physiol.* **2018**, *9*, 503. [CrossRef]
190. Serban, K.A.; Petrache, I. Mouse Models of COPD. *Methods Mol. Biol.* **2018**, *1809*, 379–394. [CrossRef]
191. Antunes, M.A.; Rocco, P.R. Elastase-induced pulmonary emphysema: Insights from experimental models. *An. Acad. Bras. Cienc.* **2011**, *83*, 1385–1396. [CrossRef]
192. Brass, D.M.; Hollingsworth, J.W.; Cinque, M.; Li, Z.; Potts, E.; Toloza, E.; Foster, W.M.; Schwartz, D.A. Chronic LPS inhalation causes emphysema-like changes in mouse lung that are associated with apoptosis. *Am. J. Respir. Cell Mol. Biol.* **2008**, *39*, 584–590. [CrossRef]
193. Cheng, Q.; Fang, L.; Feng, D.; Tang, S.; Yue, S.; Huang, Y.; Han, J.; Lan, J.; Liu, W.; Gao, L.; et al. Memantine ameliorates pulmonary inflammation in a mice model of COPD induced by cigarette smoke combined with LPS. *Biomed. Pharmacother.* **2019**, *109*, 2005–2013. [CrossRef] [PubMed]
194. Mebratu, Y.A.; Tesfaigzi, Y. IL-17 Plays a Role in Respiratory Syncytial Virus-induced Lung Inflammation and Emphysema in Elastase and LPS-injured Mice. *Am. J. Respir. Cell Mol. Biol.* **2018**, *58*, 717–726. [CrossRef] [PubMed]
195. Li, D.; Wang, J.; Sun, D.; Gong, X.; Jiang, H.; Shu, J.; Wang, Z.; Long, Z.; Chen, Y.; Zhang, Z.; et al. Tanshinone IIA sulfonate protects against cigarette smoke-induced COPD and down-regulation of CFTR in mice. *Sci. Rep.* **2018**, *8*, 376. [CrossRef] [PubMed]
196. Moore, B.B.; Hogaboam, C.M. Murine models of pulmonary fibrosis. *Am. J. Physiol. Lung Cell. Mol. Physiol.* **2008**, *294*, L152–L160. [CrossRef] [PubMed]
197. Claussen, C.A.; Long, E.C. Nucleic Acid Recognition by Metal Complexes of Bleomycin. *Chem. Rev.* **1999**, *99*, 2797–2816. [CrossRef] [PubMed]

198. Hou, J.; Ma, T.; Cao, H.; Chen, Y.; Wang, C.; Chen, X.; Xiang, Z.; Han, X. TNF-α-induced NF-κB activation promotes myofibroblast differentiation of LR-MSCs and exacerbates bleomycin-induced pulmonary fibrosis. *J. Cell. Physiol.* **2018**, *233*, 2409–2419. [CrossRef] [PubMed]
199. Chitra, P.; Saiprasad, G.; Manikandan, R.; Sudhandiran, G. Berberine attenuates bleomycin induced pulmonary toxicity and fibrosis via suppressing NF-κB dependant TGF-β activation: A biphasic experimental study. *Toxicol. Lett.* **2013**, *219*, 178–193. [CrossRef]
200. Carrington, R.; Jordan, S.; Pitchford, S.C.; Page, C.P. Use of animal models in IPF research. *Pulm. Pharmacol. Ther.* **2018**, *51*, 73–78. [CrossRef]
201. Redente, E.F.; Black, B.P.; Backos, D.S.; Bahadur, A.N.; Humphries, S.M.; Lynch, D.A.; Tuder, R.M.; Zemans, R.L.; Riches, D.W.H. Persistent, Progressive Pulmonary Fibrosis and Epithelial Remodeling in Mice. *Am. J. Respir. Cell Mol. Biol.* **2021**, *64*, 669–676. [CrossRef]
202. Morgan, G.W.; Breit, S.N. Radiation and the lung: A reevaluation of the mechanisms mediating pulmonary injury. *Int. J. Radiat. Oncol. Biol. Phys.* **1995**, *31*, 361–369. [CrossRef]
203. Degryse, A.L.; Lawson, W.E. Progress toward improving animal models for idiopathic pulmonary fibrosis. *Am. J. Med. Sci.* **2011**, *341*, 444–449. [CrossRef]
204. Nogee, L.M.; Dunbar, A.E., 3rd; Wert, S.E.; Askin, F.; Hamvas, A.; Whitsett, J.A. A mutation in the surfactant protein C gene associated with familial interstitial lung disease. *N. Engl. J. Med.* **2001**, *344*, 573–579. [CrossRef] [PubMed]
205. Thomas, A.Q.; Lane, K.; Phillips, J., 3rd; Prince, M.; Markin, C.; Speer, M.; Schwartz, D.A.; Gaddipati, R.; Marney, A.; Johnson, J.; et al. Heterozygosity for a surfactant protein C gene mutation associated with usual interstitial pneumonitis and cellular nonspecific interstitial pneumonitis in one kindred. *Am. J. Respir. Crit. Care Med.* **2002**, *165*, 1322–1328. [CrossRef] [PubMed]
206. Nureki, S.-I.; Tomer, Y.; Venosa, A.; Katzen, J.; Russo, S.J.; Jamil, S.; Barrett, M.; Nguyen, V.; Kopp, M.; Mulugeta, S.; et al. Expression of mutant Sftpc in murine alveolar epithelia drives spontaneous lung fibrosis. *J. Clin. Investig.* **2018**, *128*, 4008–4024. [CrossRef] [PubMed]
207. Povedano, J.M.; Martinez, P.; Flores, J.M.; Mulero, F.; Blasco, M.A. Mice with Pulmonary Fibrosis Driven by Telomere Dysfunction. *Cell Rep.* **2015**, *12*, 286–299. [CrossRef] [PubMed]
208. Naikawadi, R.P.; Disayabutr, S.; Mallavia, B.; Donne, M.L.; Green, G.; La, J.L.; Rock, J.R.; Looney, M.R.; Wolters, P.J. Telomere dysfunction in alveolar epithelial cells causes lung remodeling and fibrosis. *JCI Insight* **2016**, *1*, e86704. [CrossRef]
209. Degryse, A.L.; Xu, X.C.; Newman, J.L.; Mitchell, D.B.; Tanjore, H.; Polosukhin, V.V.; Jones, B.R.; McMahon, F.B.; Gleaves, L.A.; Phillips, J.A., 3rd; et al. Telomerase deficiency does not alter bleomycin-induced fibrosis in mice. *Exp. Lung Res.* **2012**, *38*, 124–134. [CrossRef] [PubMed]
210. Hancock, L.A.; Hennessy, C.E.; Solomon, G.M.; Dobrinskikh, E.; Estrella, A.; Hara, N.; Hill, D.B.; Kissner, W.J.; Markovetz, M.R.; Grove Villalon, D.E.; et al. Muc5b overexpression causes mucociliary dysfunction and enhances lung fibrosis in mice. *Nat. Commun.* **2018**, *9*, 5363. [CrossRef]
211. He, H.; Huang, C.; Chen, Z.; Huang, H.; Wang, X.; Chen, J. An outlined review for the role of Nedd4-1 and Nedd4-2 in lung disorders. *Biomed. Pharmacother.* **2020**, *125*, 109983. [CrossRef]
212. Duerr, J.; Leitz, D.H.W.; Szczygiel, M.; Dvornikov, D.; Fraumann, S.G.; Kreutz, C.; Zadora, P.K.; Seyhan Agircan, A.; Konietzke, P.; Engelmann, T.A.; et al. Conditional deletion of Nedd4-2 in lung epithelial cells causes progressive pulmonary fibrosis in adult mice. *Nat. Commun.* **2020**, *11*, 2012. [CrossRef] [PubMed]
213. Beike, L.; Wrede, C.; Hegermann, J.; Lopez-Rodriguez, E.; Kloth, C.; Gauldie, J.; Kolb, M.; Maus, U.A.; Ochs, M.; Knudsen, L. Surfactant dysfunction and alveolar collapse are linked with fibrotic septal wall remodeling in the TGF-β1-induced mouse model of pulmonary fibrosis. *Lab. Investig. J. Tech. Methods Pathol.* **2019**, *99*, 830–852. [CrossRef] [PubMed]
214. Mounkes, L.C.; Kozlov, S.; Hernandez, L.; Sullivan, T.; Stewart, C.L. A progeroid syndrome in mice is caused by defects in A-type lamins. *Nature* **2003**, *423*, 298–301. [CrossRef] [PubMed]
215. Cardoso, A.; Magano, S.; Marrana, F.; Andrade, J.P. D-Galactose High-Dose Administration Failed to Induce Accelerated Aging Changes in Neurogenesis, Anxiety, and Spatial Memory on Young Male Wistar Rats. *Rejuvenation Res.* **2015**, *18*, 497–507. [CrossRef] [PubMed]
216. Pinney, S.E. Intrauterine Growth Retardation—A Developmental Model of Type 2 Diabetes. *Drug Discov. Today. Dis. Models* **2013**, *10*, e71–e77. [CrossRef]
217. Bisht, K.; Sharma, K.; Tremblay, M.-È. Chronic stress as a risk factor for Alzheimer's disease: Roles of microglia-mediated synaptic remodeling, inflammation, and oxidative stress. *Neurobiol Stress* **2018**, *9*, 9–21. [CrossRef]
218. Sueblinvong, V.; Neujahr, D.C.; Mills, S.T.; Roser-Page, S.; Ritzenthaler, J.D.; Guidot, D.; Rojas, M.; Roman, J. Predisposition for disrepair in the aged lung. *Am. J. Med. Sci.* **2012**, *344*, 41–51. [CrossRef]
219. Hecker, L.; Logsdon, N.J.; Kurundkar, D.; Kurundkar, A.; Bernard, K.; Hock, T.; Meldrum, E.; Sanders, Y.Y.; Thannickal, V.J. Reversal of persistent fibrosis in aging by targeting Nox4-Nrf2 redox imbalance. *Sci. Transl. Med.* **2014**, *6*, 231ra47. [CrossRef]

MDPI
St. Alban-Anlage 66
4052 Basel
Switzerland
Tel. +41 61 683 77 34
Fax +41 61 302 89 18
www.mdpi.com

Cells Editorial Office
E-mail: cells@mdpi.com
www.mdpi.com/journal/cells

www.ingramcontent.com/pod-product-compliance
Lightning Source LLC
Chambersburg PA
CBHW061621220926
51850CB00005B/561
* 9 7 8 3 0 3 6 5 7 1 6 9 0 *